AF323860

MECHANICAL ENGINEERING
Emerging Vistas

MECHANICAL ENGINEERING
Emerging Vistas

Editors
B.D. Gupta
R.K. Upadhyay
R.P. Sinha
Jagvir Singh

Narosa Publishing House

New Delhi   Chennai   Mumbai   Kolkata

Editors
B.D. Gupta
R.K. Upadhyay
R.P. Sinha
Jagvir Singh
Department of Mechanical Engineering
Anand Engineering College
Agra, Uttar Pradesh

Preface

In the fast growing Global economy, Mechanical Engineering, an evergreen branch of Engineering, has been playing a vital role in the development of mankind from the ancient time. Growth of Mechanical Science has led the development of Mechanics, Automobiles, Aerospace Engineering, and Industrial productivity to new heights in due course of time. Researchers, scientists and technocrats are toiling day and night to improve the human life by incorporation of innovation and creativity; and developing machines, materials for the present world. New discoveries and researches in the field of mechanical engineering have a great value for mankind.

The present volume contains the work of researchers and scientists in the diversified field of Mechanical Engineering. In present scenario survival of mankind can not be possible without integration of Engineering and Technology This volume covers the ongoing research in various dimensions of Mechanical Engineering, covering Thermal Engineering, Manufacturing Engineering and manufacturing Processes, Fluid Dynamics and Computation Fluid Flow, Industrial Engineering, Human Factor Engineering, Soft Computing, Composite Materials and Smart Materials. Research papers on Virtual Prototyping, Fuzzy Model for predicting energy demand in India, Biodiesel Production for sustainable development in Indian perspective, Smart Materials for society, Nano Technology, Green Technology ,Wastage Utilization for Bio Gas Production, Hand Transmitted Vibrations, Whole Body Vibrations, Geothermal Energy, Finite Element Modeling, Suplly Chain Management, Electrochemical Machining, Flexible Manufacturing Systems, Lean Manufacturing and Solar Power Air Conditioner make this volume unique as it covers almost all the topics of vital importance in the growing field of Mechanical Engineering in integration with emerging Technologies..

We are thankful to all the researchers, scientitists, and technocrats who have contributed a lot through the outcome of their research findings to be published in this volume. We are also thankful to Sh. P.K.Gupta and Sh. Y.K.Gupta for their constant support for making this work published. Last but not the least Narosa Publishing House deserves a greatt appreciation for publishing the current volume of this book.

B.D. Gupta
R.K. Upadhyay
R.P. Sinha
Jagvir Singh

Contents

Experimental Study of Friction Stir Processing (FSP) and Its Effect on Microstructure and Hardness of Al6061 Plates

K. Hans Raj[1], Rahul Swarup Sharma[1], Pritam Singh[1], Rajat Setia[1], R.K. Sharma,[2] Amitabh Pal[2] and B. Choudhury[2]

[1]Dayalbagh Educational Institute, Dayalbagh, Agra-282110
[2]Aerial Delivery Research and Development Establishment(DRDO), Agra-282001

Abstract— Production of Bulk Nano-structured Materials (BNM) using Severe Plastic Deformation (SPD) is currently being pursued by researchers all around the world. Recently FSP is gaining prominence on account of its ability to create BNM in the processed specimen. Friction Stir Processing (FSP) is invented in 1991 along with Friction Stir Welding (FSW). In the present study an improvised Vertical Milling Machine is designed that can perform FSP on thin metallic sheets. Al6061 sheets of 2mm thick are subjected to FSP and its effect on microstructure is studied. It is observed that very fine grains are formed in these sheets. Further effect of FSP on hardness is studied with the help of micro hardness tester. It is found that the hardness is increased by 50% on the processed AL6061 sheets on account of FSP. The study clearly brings out the advantage of severe plastic deformation caused on account of FSP and is a step forward for developing Bulk Nanostructured Materials (BNM) materials in future.

Keywords : Friction Stir Processing (FSP), Bulk Nano Materials (BNM)

I. INTRODUCTION

Bulk Nanostructured Materials (BNM)

Bulk Nanostructured Materials (BNM) are defined as solids with nanoscale (typically 1-100 nm) substructures. The processing of metals has a long history dating back to 300 BC. The legendry steel from Ancient India, known as "Wootz Steel" is characterized by a pattern of bands or sheets of micro carbides within a tempered martensite or pearlite matrix. It is believed that Damascus blades were forged directly from small cakes of Wootz Steel. Reibold et al. [1] have used high-resolution transmission electron microscopy to examine a sample of Damascus sabre steel from the seventeenth century and found that it contains carbon nanotubes as well as cementite nanowires. The first work reported in modern times on SPD is credited to P. W. Bridgman who proposed High Pressure Torsion [2]. Since then many different SPD processes are discovered and analyzed.

In recent years, BNM have been the subjects of intensive research due to their superior physical and mechanical properties [3-5]. These include a combination of high strength and ductility at room temperature, high strain rate and low temperature superplasticity. These superior mechanical and physical properties make BNM attractive for numerous advanced applications in medical, aerospace, sporting goods, and transportation industries.

Various methods for producing BNM have been developed, which, based on their approaches, can be classified into two categories. The first is the "bottom up" approach, which builds materials atom by atom. Methods in this category include inert gas condensation [3], high-energy ball milling [6], spray conversion processing [7], sputtering [8], physical vapor deposition

[9], chemical vapor deposition [10], and electrodeposited nanocrystals [11]. Nano-powder production technologies are well developed, and various metallic, as well as ceramic, powders are commercially available. Powder consolidation, however, has been largely unsuccessful. The heat and pressure applied during the powder consolidation also promotes the grain growth, which makes materials lose their nano characteristics. So far, only penny-sized BNM have been produced, which are too small for any structural applications. Other problems with powder consolidation methods include high cost, contamination, and porosity. The second approach for producing BNM is the "top down" approach, which refines coarse-grained metals through severe plastic deformation (SPD).

Severe Plastic Deformation (SPD)

Severe plastic deformation (SPD) is a generic term describing a group of metal-working techniques involving very large strains which are imposed without introducing any significant changes in the overall dimensions of the specimen or work-piece. Processing by SPD provides the opportunity to achieve remarkable grain refinement in bulk crystalline solids. Typically, materials processed by SPD have grain size in sub micrometer and nano-meter range [12–15]. Number of SPD techniques are available for producing UFG [18] material like Equal Channel Angular Pressing (ECAP) [19], Accumulative Roll Bonding (ARB) [16], Repetitive Corrugation and Straightening (RCS) [17] and High Pressure Torsion (HPT) [20].

In addition, in recent studies Friction Stir Processing (FSP) started gaining prominence. The microstructure of a friction stir processed zone (FSP zone consists of fine recrystallized grains resulting from a severe plastic deformation of the magnitude experienced in equal-channel angular pressing (ECAP) and high-pressure torsion (HPT). It was also noted that recrystallized grain sizes in aluminum alloys ranged from 1 to 10 μm. These sizes were approximately 10 to 100 times smaller than in the original workpiece materials. This result suggests that materials with very fine grains, on the scale of microns may be produced through FSP.

FSP has been studied not as a welding technique but as a new grain refinement process. In previous research on influences of the microstructure of the starting material on the FSP zone, it was reported that commercially pure

aluminum (1050) with very fine grain sizes from 1 to 2 μm was produced through FSP. In the present study an improvised Vertical Milling Machine is developed that can perform FSP on thin metallic sheets. Al6061 sheets of 2mm thick are subjected to FSP and its effect on microstructure is studied.

In addition, effect of FSP on hardness is studied with the help of micro hardness tester. It is found that the hardness is increased by 50% on the processed AL6061 sheets on account of FSP.

The basic principle of FSP in the present research is schematically illustrated in Fig. 1. Monolithic cold-rolled plates of aluminum alloy (AL6061) are used as starting materials (work pieces).

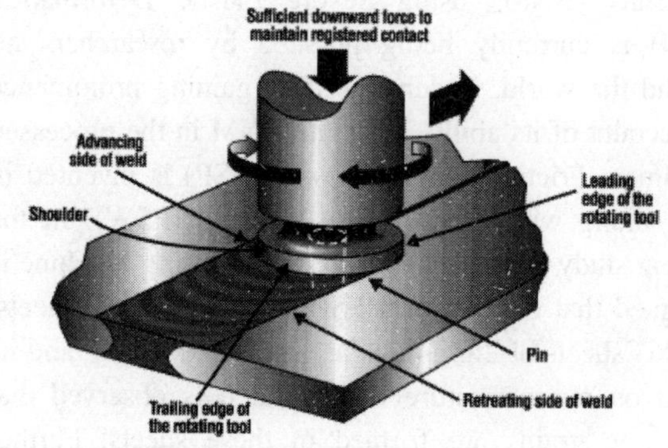

Fig. 1 Friction Stir Processing

In the process shown in Fig. 1, a cylindrical rod rotating at high speed (probe) is inserted into the plate to be processed. The friction of the tool on the part softens the material, which turns into a paste and is stirred. This is why the process is called "friction stir". The shoulder that is pressed with a large force on the plates prevents the stirred metal from being expelled and produces a forging effect at the back of the material that was just softened out of shape and stirred.

Since its discovery, FSP has evolved as a technique of choice in the routine processing of aluminium components. There have been widespread benefits resulting from the application of FSP for example in aerospace, shipbuilding, automotive and railway industries.

The benefits of FSP are summarized as below:

- This process can be used to improve microstructure to improve the surface properties of metals.
- Because no melting of materials is involved, friction stir processing avoids the weaknesses caused by distortion and metallurgical reactions observed in conventional surface treatment.
- The friction heating is generated locally, so there is no widespread softening of the assembly.
- The process is completed in a few seconds with very high reproducibility.

II. EXPERIMENTATION OF FSP WITH IMPROVISED VERTICAL MILLING MACHINE

An improvised vertical milling machine as shown in fig. 2 is manufactured as per design specified with a 3phase 1HP motor with rpm range of 550 to 950 and feed range of 10 to 80 mm/min for doing the FSP experiments. Initially one 77mm long, 34mm broad and 2mm thick aluminum alloy (Al6061) piece is processed using the ingenious setup shown in fig. 2.

Fig. 2 Friction Stir Processing Machine Setup

Fig. 3 shows the grain structure before the friction stir process of the AL6061 plates. It is explicit that the structure is coarse and unrefined. The grain size of this

unprocessed material is found to be in the range of 10 μm. The standard Hardness of AL6061 is 95 BHN.

Fig. 4 shows the grain structure after the FSP of the AL6061 plates. The grain size and the microstructure is in the range of 1 μm. This shows a considerable reduction of the grain size. The hardness after the FSP was found to be in the range of 140 BHN, which clearly indicates the increment of about 50%.

Fig. 3 Grain structure before FSP of rolled Al6061

Fig. 4 Grain structure of AL6061 plate after FSP

III. RESULT AND DISCUSSION

The microstructure of friction stir processed Al6061 plate shows the effect of stirring and formation of smaller grains. The observations by computerized microscope revealed the grain size of processed plate to be approximately 1 •m. The average Brinnel-hardness is observed with micro-hardness tester. An increase of 50% is observed in Brinnel Hardness Number (BHN) value. This clearly depicts the advantage of FSP on Al6061. The equivalent strain is also a measure of strength of the material and it can be inferred that the yield strength of AL6061 plates has increased on account of FSP.

IV. CONCLUSION

In the present study an improvised Vertical Milling Machine is indigenously designed and got developed that can perform FSP on thin metallic sheets. Al6061 sheets of 2mm thick were subjected to FSP and its effect on microstructure was studied. It was observed that coarse grains were transformed into very fine grains in these sheets. Further effect of FSP on hardness is studied with the help of micro hardness tester. It was found that the hardness is increased by 50% on the processed AL6061 sheets on account of FSP. The study clearly brings out the advantage of severe plastic deformation caused on account of FSP and is a step forward for developing Bulk Nanostructured Materials (BNM) materials in future.

ACKNOWLEDGMENTS

We deeply appreciate the inspiration and guidance provided by Most Revered Professor P. S. Satsangi, Chairman, Advisory Committee on Education, Dayalbagh.

Partial support for this research from the Department of Science & Technology (DST), Government of India under grant No. SR/S3/MERC/104/2008, All India Council of Technical Education (AICTE), under research promotion scheme (RPS), file number 8023/BOR/RID/RPS-153/2008-09, UGC major research project and funding from ADRDE (DRDO) under file no. ADRDE/QMS/PM/8/MTL/180/147, dated 02/03/2009, is duly acknowledged.

REFERENCES

[1] M. Reibold1,2, P. Paufler1, A. A. Levin1, W. Kochmann1, N. Pätzke1 & D. C. Meyer1, Materials: Carbon nanotubes in an ancient Damascus saber, Nature 444, p.286 (16 November 2006).

[2] P.W. Bridgman, "Effect of high shearing stress combined with high hydrostatic pressure", Physical Review, Vol. 48, pp. 825-847, (15 November 1935).

[3] Birringer, R., Gleiter, H., H. P. Klein, Marquardt, P., "Nanocrystalline materials: an approach to a novel solid structure with gas-like disorder", Physics letters, 102 A, 365- 369, 1984.

[4] Gleiter, H., "Nanocrystalline materials", Progress in Materials Science, 33(4), 223-315, 1989.

[5] Froes, F. H., Suryanarayana, C., "Nanocrystalline Metals for Structural Applications", JOM Journal of the Minerals, Metals and Materials Society, 41(6), 12-17, 1989.

[6] Koch, C.C., "Top-down synthesis of nanostructured materials: Mechanical and thermal processing methods", Review of Advance Material Science, 5, 91–99, 2003.

[7] Kear, B.H. and McCandish, L.E., "Chemical Processing and Properties of Nanostructured WC-Co Materials," Nanostructured Materials, 3, 19-30, 1993.

[8] Chang, H., Altstetter, C. J. and Averback, R. S., "Characteristics of nanophase TiAl produced by inert gas condensation", Journal of Materials Research, 7, 2962-2970, 1992.

[9] Bickerdike, R., Clark, D., Easterbrook, J. N., Hughes, G., Mair, W. N., Partridge,P. G. and Ranson, H. C., "The deposition of pyrolytic carbon in the pores of bonded and unbonded carbon powders", International Journal of Rapid Solidification, 1, 305, 1985.

[10] Chang, W. G., Skandan, S., Danforth, C., Rose, M., Balogh, A. G., Hahn, H. and Kear, B., "Nanostructured ceramics synthesized by chemical vapor condensation", Nanostructured Materials, 6, 321-324, 1995.

[11] Erb, U., "Electrodeposited nanocrystals: Synthesis, properties and industrial applications", Nanostructured Materials, 6(5-8), 533-538, 1995.

[12] A.P. Zhilyaev, T.G. Langdon,. "Plastic working of metals by simple shear", Russian Metallurgy (Metally), Sci. 53 (2008) 893–979.

[13] R.Z. Valiev, R.K. Islamgaliev, I.V. Alexandrov, "Materials science: nanomaterial advantage", Nature, 419 (6910), pp. (2002) 887-889.

[14] T. C. Lowe and R. Z. Valiev (Eds.), "Investigations and Applications of Severe Plastic Deformation", Kluwer Academia Publisher, Dordrecht, (2000) 509-520.

[15] M. Furukawa, Z. Horita and T. G. Langdon, "Application of High pressure torsion to Aluminum and Copper Single Crystals," Materials Science Forum, 539-543, (2007) 2853-2858.

[16] M. Zehetbauer and R. Z. Valiev (Eds.), "Nanomaterials by Severe Plastic Deformation: Fundamentals, Processing, Applications", Wiley-VCH, Weinheim, Germany, (2004) 272-320.

[17] Y. T. Zhu, P. B. Berbon, A. H. Chokshi, Z. Horita, S. V. Raj, and K. Xia (Eds.), "The Langdon Symposium: Flow and Forming of Crystalline Materials", Elsevier, (2005),123-130.

[18] M. J. Zehetbauer and Y. T. Zhu (Eds.), "Bulk Nanostructured Materials", Wiley- VCH Weinheim, Germany, (2007),140-150.

[19] K. Hans Raj, Rahul Swarup Sharma, R. Setia and S. S. Sharma, "Study and Influence of friction and channel angle in ECAP", In the Proceedings of 4th International Conference on Tribology of Manufacturing. (ICTMP 2010), Nice, France, June 13-15, vol. 2, pp (2010) 517-527.

[20] Rahul swarup sharma, Atul Dayal, K.Hans Raj , "finite element analysis of pure aluminium (AL99) processed by high pressure torsion (HPT)", NSC 2010.

Rapid Prototyping Technologies, Applications and Part Deposition Planning

Pulak M. Pandey

Associate Professor Department of Mechanical Engineering Indian Institute of Technology Delhi

I. INTRODUCTION

Prototyping or model making is one of the important steps to finalize a product design. It helps in conceptualization of a design. Before the started around mid-1970s, when a soft prototype modeled by 3D curves and surfaces could be stressed in virtual environment, simulated and tested with exact material and other properties. Third and the latest trend of prototyping, i.e., Rapid Prototyping (RP) by layer-by-layer material deposition, started during early 1980s with the enormous growth in Computer Aided Design and Manufacturing (CAD/CAM) technologies when almost unambiguous solid models with knitted information of edges and surfaces could define a product and also manufacture it by CNC machining. The historical development of RP and related technologies is presented in table 1.

Table 1: Historical development of Rapid Prototyping and related technologies (after Chua and Leong, 2000)

Year of inception	Technology
1770	Mechanization
1946	First computer
1952	First Numerical Control (NC) machine tool
1960	First commercial laser
1961	First commercial Robot
1963	First interactive graphics system (early version of Computer Aided Design)
1988	First commercial Rapid Prototyping system

II. BASIC PRINCIPLE OF RAPID PROTOTYPING PROCESSES

RP process belong to the generative (or additive) production processes unlike subtractive or forming processes such as lathing, milling, grinding or coining etc. in which form is shaped by material removal or plastic deformation. In all commercial RP processes, the part is fabricated by deposition of layers contoured in a (x-y) plane two dimensionally. The third dimension (z) results from single layers being stacked up on top of each other, but not as a continuous z-coordinate. Therefore, the prototypes are very exact on the x-y plane but have stair-stepping effect in z-direction. If model is deposited with very fine layers, i.e., smaller z-stepping, model looks like original. RP can be classified into two fundamental process steps namely generation of mathematical layer information and generation of physical layer model. Typical process chain of various RP systems is shown in figure 1.

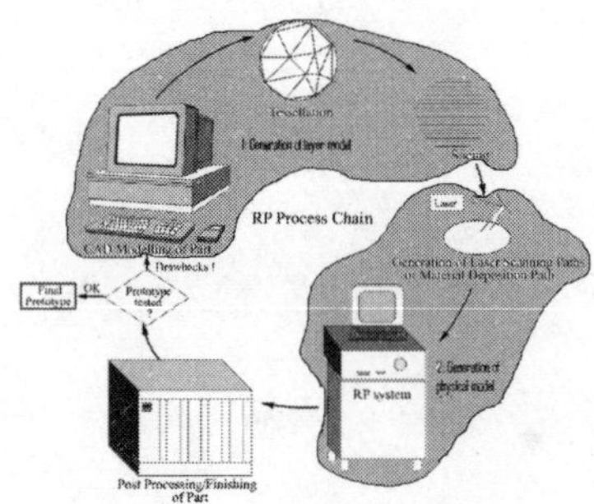

Figure 1: RP process chain showing fundamental process steps

It can be seen from figure 1 that process starts with 3D modeling of the product and then STL file is exported by tessellating the geometric 3D model. In tessellation various surfaces of a CAD model are piecewise approximated by a series of triangles (figure 2) and co-ordinate of vertices of triangles and their surface normals are listed. The number and size of triangles are decided by facet deviation or chordal error as shown in figure 2. These STL files are checked for defects like flip triangles, missing facets, overlapping facets, dangling edges or faces etc. and are repaired if found faulty. Defect free STL files are used as an input to various slicing softwares. At this stage choice of part deposition orientation is the most important factor as part building time, surface quality, amount of support structures, cost etc. are influenced. Once part deposition orientation is decided and slice thickness is selected, tessellated model is sliced and the generated data in standard data formats like SLC (stereolithography contour) or CLI (common layer interface) is stored. This information is used to move to step 2, i.e., generation of physical model. The software that operates RP systems generates laser-scanning paths (in processes like Stereolithography, Selective Laser Sintering etc.) or material deposition paths (in processes like Fused Deposition Modeling). This step is different for different processes and depends on the basic deposition principle used in RP machine. Information computed here is used to deposit the part layer-by-layer on RP system platform. The generalized data flow in RP is given in figure 3.

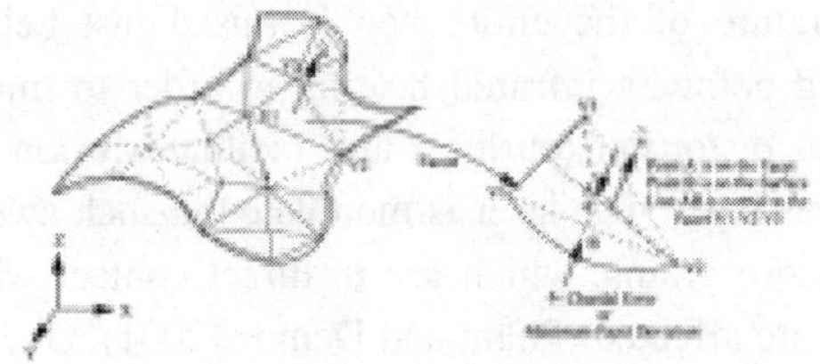

Figure 2: Tessellation of a typical surface of CAD model (after Pandey et al. 2003b)

The final step in the process chain is the post-processing task. At this stage, generally some manual operations are necessary therefore skilled operator is required. In cleaning, excess elements adhered with the part or support structures are removed. Sometimes the surface of the model is finished by sanding, polishing or painting for better surface finish or aesthetic appearance. Prototype is then tested or verified and suggested

engineering changes are once again incorporated during the solid modeling stage.

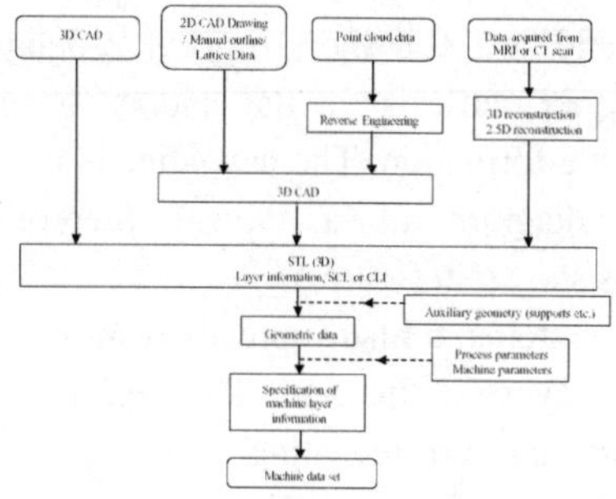

Figure 3: Generalized illustration of data flow in RP (after Gebhardt, 2003)

III. RAPID PROTOTYPING PROCESSES

The professional literature in RP contains different ways of classifying RP processes. However, one representation based on German standard of production processes classifies RP processes according to state of aggregation of their original material and is given in figure 4.

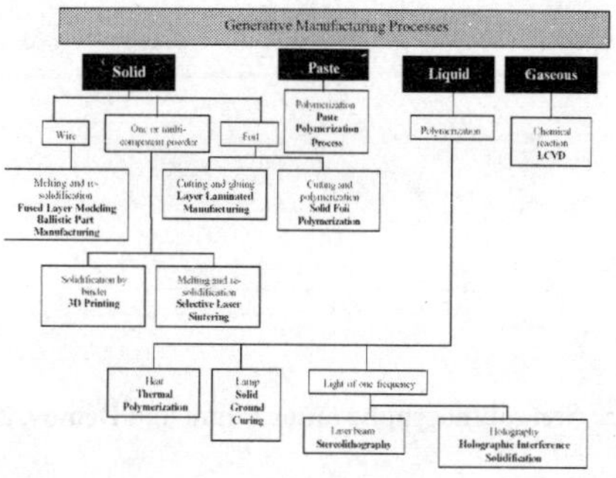

Figure 4: Classification of RP processes (after Gebhardt, 2003)

Here, few important RP processes namely Stereolithography (SL), Selective Laser Sintering (SLS), Fused Deposition Modeling (FDM) and Laminated Object Manufacturing (LOM) are described.

Stereolithography

In this process photosensitive liquid resin which forms a solid polymer when exposed to ultraviolet light is used as a fundamental concept. Due to the absorption and scattering of beam, the reaction only takes place near the surface and voxels of solid polymeric resin are formed. A SL machine consists of a build platform (substrate), which is mounted in a vat of resin and a UV Helium-Cadmium or Argon ion laser. The laser scans the first layer and platform is then lowered equal to one slice thickness and left for short time (dip-delay) so that liquid polymer settles to a flat and even surface and inhibit bubble formation. The new slice is then scanned. Schematic diagram of a typical Stereolithography apparatus is shown in figure 5.

In new SL systems, a blade spreads resin on the part as the blade traverses the vat. This ensures smoother surface and reduced recoating time. It also reduces trapped volumes which are sometimes formed due to excessive polymerization at the ends of the slices and an island of liquid resin having thickness more than slice thickness is formed (Pham and Demov, 2001). Once the complete part is deposited, it is removed from the vat and then excess resin is drained. It may take long time due to high viscosity of liquid resin. The 'green' part is then post-cured in an UV oven after removing support structures.

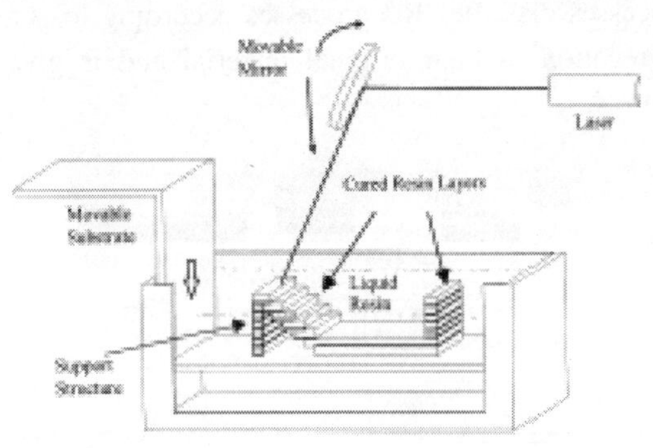

Figure 5: Stereolithography (after Pham and Demov, 2001)

Overhangs or cantilever walls need support structures as a green layer has relatively low stability and strength. These overhangs etc. are supported if they exceed a certain size or angle, i.e., build orientation. The main functions of these structures are to support projecting parts and also to pull other parts down which due to shrinkage tends to curl up (Gebhardt, 2003). These support structures are generated during data processing and due to these data grows heavily specially with STL files, as cuboid shaped support element need information about at least twelve triangles. A solid support is very difficult to remove later and may damage the model. Therefore a new support structure called "fine point" was developed by 3D Systems (figure 6) and is company's trademark.

Build strategies have been developed to increase build speed and to decrease amount of resin by depositing the parts with a higher proportion of hollow volume. These strategies are devised as these models are used for making cavities for precision castings. Here walls are designed hollow connected by rod-type bridging elements and skin is introduced that close the model at the top and the bottom. These models require openings to drain out uncured resin.

Figure 6: Fine point structure for Stereolithography (after Gebhardt, 2003)

Selective Laser Sintering

In Selective Laser Sintering (SLS) process, fine polymeric powder like polystyrene, polycarbonate or polyamide etc. (20 to 100 micrometer diameter) is spread on the substrate using a roller. Before starting CO_2 laser scanning for sintering of a slice the temperature of the entire bed is raised just below its melting point by infrared heating in order to minimize thermal distortion (curling) and facilitate fusion to the previous layer. The laser is modulated in such away that only those grains, which are in direct contact with the beam, are affected (Pham and Demov, 2001). Once laser scanning cures a slice, bed is lowered and powder feed chamber is raised so that a covering of powder can be spread evenly over the build area by counter rotating roller. In this process support structures are not required as the unsintered powder remains at the places of support structure. It is cleaned away and can be recycled once the model is complete. The schematic diagram of a typical SLS apparatus is given in figure 7.

Fused Deposition Modeling

In Fused Deposition Modeling (FDM) process a movable (x-y movement) nozzle on to a substrate deposits thread of molten polymeric material. The build material is heated slightly above (approximately 0.5°C) its melting temperature so that it solidifies within a very short time (approximately 0.1 s) after extrusion and cold-welds to the previous layer as shown in figure 8. Various important factors need to be considered and are steady nozzle and material extrusion rates, addition of support structures for overhanging features and speed of the nozzle head, which affects the slice thickness. More recent FDM systems include two nozzles, one for part material and other for support material. The support material is relatively of poor quality and can be broken easily once the complete part is deposited and is removed from substrate. In more recent FDM technology, water-soluble support structure material is used. Support structure can be deposited with lesser density as compared to part density by providing air gaps between two consecutive roads.

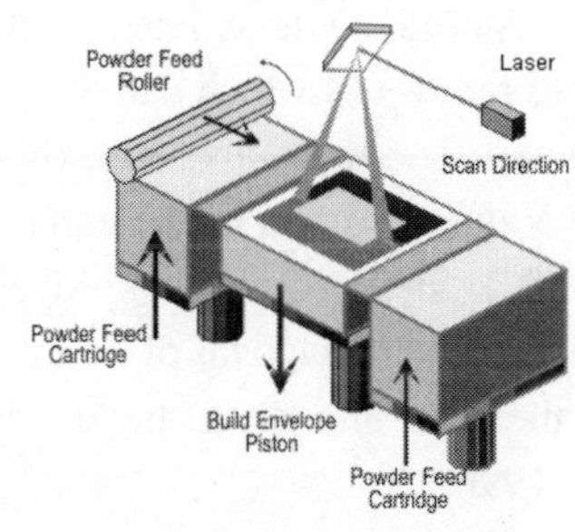

Figure 7: Selective Laser Sintering System

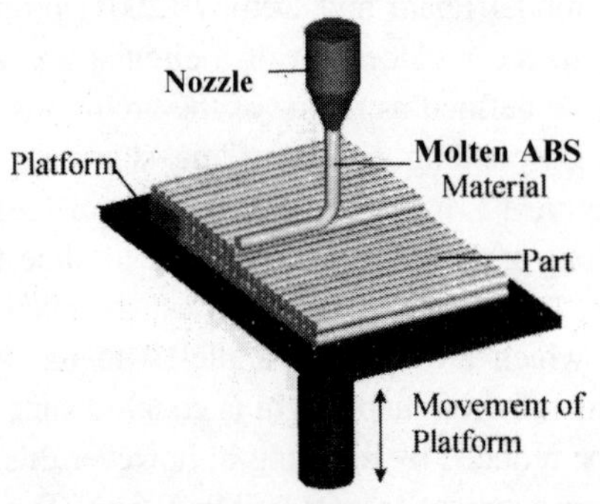

Figure 8: Fused Deposition Modeling Process (after Pham and Demov, 2001)

Laminated Object Manufacturing

Typical system of Laminated Object Manufacturing (LOM) has been shown in figure 9. It can be seen form the figure that the slices are cut in required contour from roll of material by using a 25-50 watt CO_2 laser beam. A new slice is bonded to previously deposited slice by using a hot roller, which activates a heat sensitive adhesive. Apart from the slice unwanted material is also hatched in rectangles to facilitate its later removal but remains in place during the build to act as supports. Once one slice is completed platform can be lowered and roll of material can be advanced by winding this excess onto a second roller until a fresh area of the sheet lies over the part. After completion of the part they are sealed with a urethane lacquer, silicone fluid or epoxy resin to prevent later distortion of the paper prototype through water absorption.

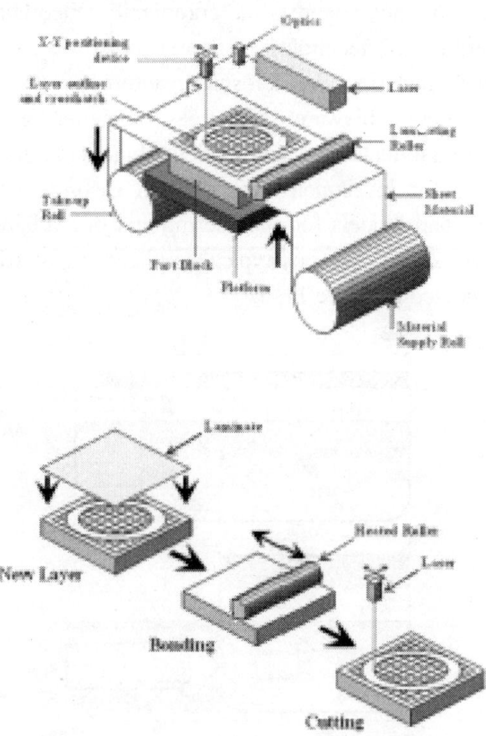

Figure 9: Laminated Object Manufacturing Process

In this process, materials that are relatively cheaper like paper, plastic roll etc. can be used. Parts of fiber-reinforced glass ceramics can be produced. Large models can be produced and the building speed is 5-10 times as compared to other RP processes. The limitation of the process included fabrication of hollow models with undercuts and reentrant features. Large amount of scrap is formed. There remains danger of fire hazards and drops of the molten materials formed during the cutting also need to be removed (Pham and Demov, 2001).

IV. APPLICATIONS OF RP TECHNOLOGIES

RP technology has potential to reduce time required from conception to market up to 10-50 percent (Chua and Leong, 2000) as shown in figure 10. It has abilities of enhancing and improving product development while at the same time reducing costs due to major breakthrough in manufacturing (Chua and Leong, 2000). Although poor surface finish, limited strength and accuracy are the limitations of RP models, it can deposit a part of any degree of complexity theoretically. Therefore, RP technologies are successfully used by various industries like aerospace, automotive, jewelry, coin making, tableware, saddletrees, biomedical etc. It is used to fabricate concept models, functional models, patterns for investment and vacuum casting, medical models and models for engineering analysis (Pham and Demov, 2001). Various typical applications of RP are summarized in figure 11.

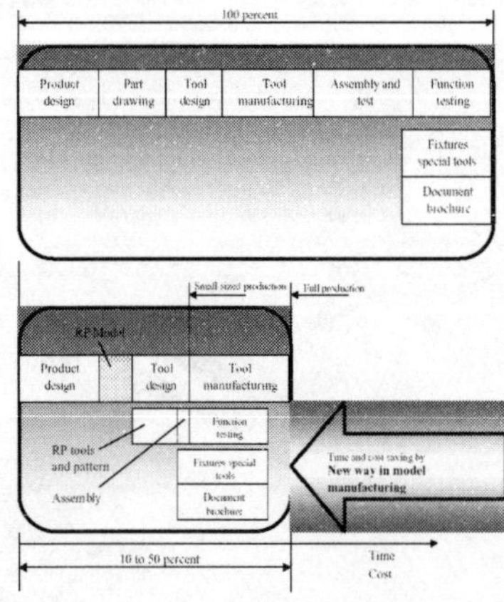

Figure 10: Result of introduction of RP in design cycle (after Chua and Leong, 2001)

V. PART DEPOSITION PLANNING

A defect less STL file is used as an input to RP software like QuickSilce or RPTools for further processing. At this stage, designer has to take an important decision about the part deposition orientation. The part deposition orientation is important because part accuracy, surface quality, building time, amount of support structures and hence cost of the part is highly influenced (Pandey et al., 2004b). In this section various factors influencing accuracy of RP parts and part deposition orientation are discussed.

Factors influencing accuracy

Accuracy of a model is influenced by the errors caused during tessellation and slicing at data preparation stage. Decision of the designer about part deposition orientation also affects accuracy of the model.

Errors due to tessellation: In tessellation surfaces of a CAD model are approximated piecewise by using triangles. It is true that by reducing the size of the triangles, the deviation between the actual surfaces and approximated triangles can be reduced. In practice, resolution of the STL file is controlled by a parameter namely chordal error or facet deviation as shown in figure 2. It has also been suggested that a curve with small radius (r) should be tessellated if its radius is below a threshold radius (r_o) which can be considered as one tenth of the part size, to achieve a maximum chordal error of $(r/r_o)^\alpha$. Value of α can be set equal to 0 for no improvement and 1 for maximum improvement. Here part size is defined as the diagonal of an imaginary box drawn around the part and α is angle control value (Williams et al., 1996).

Errors due to slicing: Real error on slice plane is much more than that is felt, as shown in figure 12(a). For a spherical model Pham and Demov (2001) proposed that error due to the replacement of a circular arc with stair-steps can be defined as radius of the arc minus length up to the corresponding corner of the staircase, i.e., cusp height (figure 12 (b)). Thus maximum error (cusp height) results along z direction and is equal to slice thickness. Therefore, cusp height approaches to maximum for surfaces, which are almost parallel with the x-y plane. Maximum value of cusp height is equal to slice thickness and can be reduced by reducing it; however this results in drastic improvement in part building time. Therefore, by using slices of variable thicknesses (popularly known as adaptive slicing, as shown in figure 13), cusp height can be controlled below a certain value.

Except this, mismatching of height and missing features are two other problems resulting from the slicing. Although most of the RP systems have facility of slicing with uniform thickness only, adaptive slicing scheme, which can slice a model with better accuracy and surface finish without loosing important features must be selected. Review of various slicing schemes for RP has been done by Pandey et al. (2003a).

Part building

During part deposition generally two types of errors are observed and are namely curing errors and control errors. Curing errors are due to over or under curing with respect to curing line and control errors are caused due to variation in layer thickness or scan position control. Figures 14 illustrate effect of over curing on part geometry and accuracy. Adjustment of chamber temperature and laser power is needed for proper curing. Calibration of the system becomes mandatory to minimize control errors Shrinkage also causes dimensional inaccuracy and is taken care by choosing proper scaling in x, y and z directions. Polymers are also designed to have almost negligible shrinkage factors. In SL and SLS processes problem arises with downward facing layers as these layers do not have a layer underneath and are slightly thicker, which generate dimensional error. If proper care is not taken in setting temperatures, curling is frequently observed.

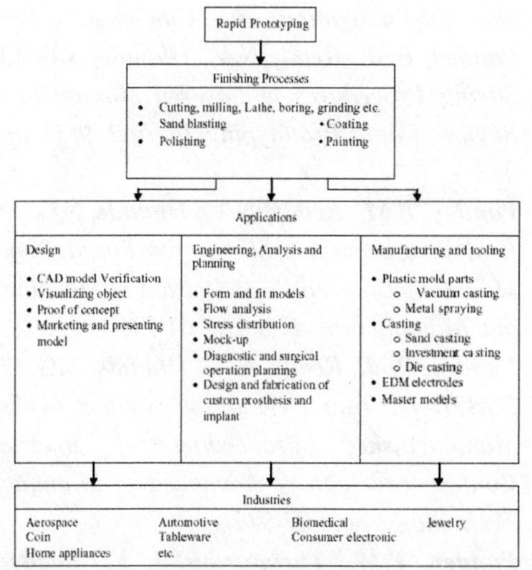

Figure 11(a): Typical application areas of RP parts (after Chual and Leong, 2000)

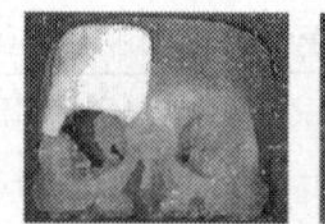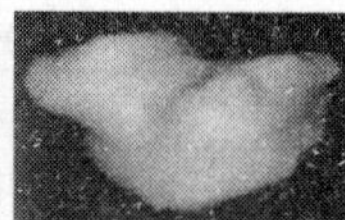

(b) SL model with the resection template Silicon implant molded from a tool (after Pham and Demov, 2001)

Figure 11: Applications of RP processes

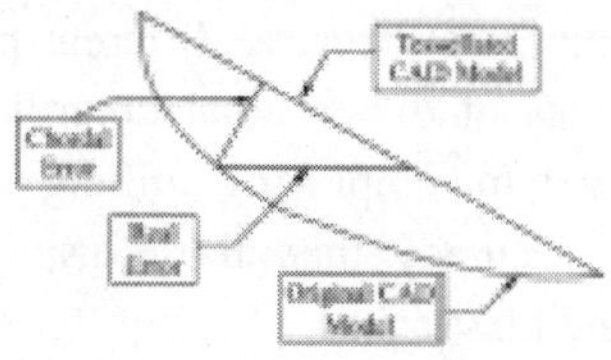

(a) Real error slice plane (after Pandey et al., 2003a)

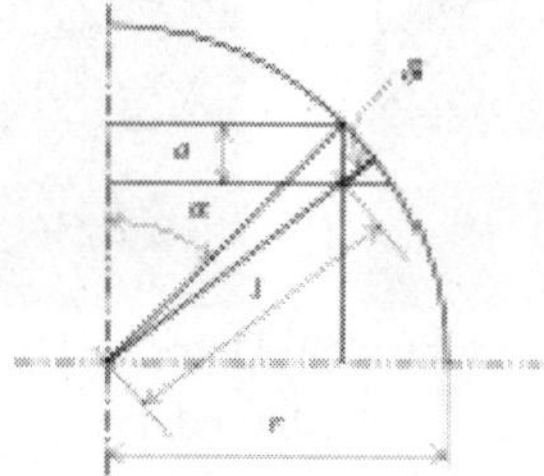

(b) Error due to replacement of arcs with stair-steps, cusp height δ (after Pham and Demov, 2001)

Figure 12: slicing error

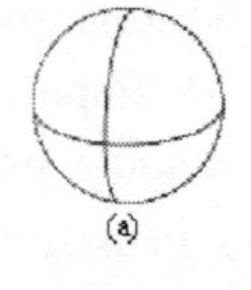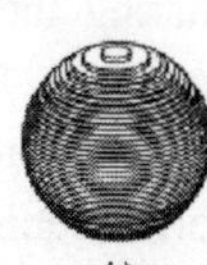

Figure 13: Slicing of a ball, (a) No slicing (b) Thick slicing (c) This slicing (d) Adaptive slicing (after Pham and Demov, 2001)

Part finishing

Poor surface quality of RP parts is a major limitation and is primarily due to staircase effect. Surface roughness can be controlled below a predefined threshold value by using an adaptive slicing (Pandey et al., 2003b). Further, the situation can be improved by finding out a part deposition orientation that gives minimum overall average part surface roughness (Singhal et al., 2005). However, some RP applications like exhibition models, tooling or master pattern for indirect tool production etc. require additional finishing to improve the surface appearance of the part. This is generally carried by sanding and polishing RP models which leads to change in the mathematical definitions of the various features of the model. The model accuracy is mainly influenced by two factors namely the varying amount of material removed by the finishing process and the finishing technique adopted. A skilled operator is required as the amount of material to be removed from different surfaces may be different and inaccuracies caused due to deposition can be brought down. A finishing technique selection is important because different processes have different degrees of dimensional control. For example models finished by employing milling will have less influence on accuracy than those using manual wet sanding or sand blasting.

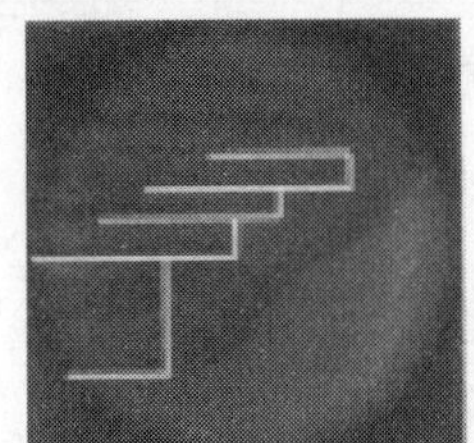 

(a) Thicker bottom layer (b) Deformed hole
boundary

Figure 14: Over-curing effects on accuracy in Stereolithography (after Pham and Demov, 2001)

Selection of part deposition orientation

This is one of the crucial decisions taken before slicing the part and initiating the process of deposition for a particular RP process. This decision is important because it has potential to reduce part building time, amount of supports required, part quality in terms of surface finish or accuracy and cost as well. Selection of part deposition orientation is process specific where in designer and RP machine operators should consider number of different process specific constraints. This may be a difficult and time consuming task as designer has to trade-off among various conflicting objectives or process outcomes. For example better part surface quality can be obtained but it will lead to increase in the building time. Pandey et al. (2004b) handled conflicting situation of the abovementioned two objectives and proposed use of multi-objective genetic algorithm for finding out optimum part deposition orientations (pareto optimal solutions) for FDM process. In their work, amount of support structures were also minimized implicitly. Thrimurthullu et al. (2004) converted multi-objective problem into single objective problem and then solved by using real coded genetic algorithm. Singhal et al. (2005) made an attempt to find out optimum part deposition orientation for SL process by using optimization tool box of MATLAB 6.5 for minimizing overall part surface roughness. Except these, researchers suggested to find out a suitable part deposition orientation for objectives like maximum accuracy, minimum building time, support structure or cost. A through review of the various part deposition orientation studies has been done by Pandey et al. (2004a). Pham and Demov (2001) discussed guidelines for selection of part deposition orientation for SL and SLS processes.

VI. SUMMARY

This paper provides an overview of RP technology in brief and emphasizes on their ability to shorten the product design and development process. Classification of RP processes and details of few important processes is given. The description of various stages of data preparation and model building has been presented. An attempt has been made to include some important factors to be considered before starting part deposition for proper utilization of potentials of RP processes.

REFERENCES

[1] **Chua, C.K., Leong, K.F.** *(2000) Rapid Prototyping: Principles and Applications in Manufacturing, World Scientific.*

[2] **Gebhardt, A.,** *(2003) Rapid Prototyping, Hanser Gardner Publications, Inc., Cincinnati.*

[3] **Pandey, P.M., Reddy N.V., Dhande, S.G.** *(2003a) Slicing Procedures in Layered Manufacturing: A Review, Rapid Prototyping Journal, 9(5), pp. 274-288.*

[4] **Pandey, P.M., Reddy, N.V., Dhande, S.G.** *(2003b) Real Time Adaptive Slicing for Fused Deposition Modelling, International Journal of Machine Tools and Manufacture, 43(1), pp 61-71.*

[5] **Pandey, P.M., Reddy, N.V., Dhande, S.G.** *(2004a) Part Deposition Orientation Studies in Layered Manufacturing, Proceeding of International Conference on Advanced Manufacturing Technology, pp. 907-912.*

[6] **Pandey, P.M., Thrimurthullu, K., Reddy, N.V.** *(2004b) Optimal Part Deposition Orientation in FDM using Multi-Criteria GA, International Journal of Production Research, 42(19), pp. 4069-4089.*

[7] **Pham, D.T., Dimov, S.S.** *(2001) Rapid Manufacturing, Springer-Verlag London Limited.*

[8] **Singhal, S.K., Pandey, A.P., Pandey, P.M., Nagpal, A.K.** *(2005) Optimum Part Deposition Orientation in Stereolithography, Computer Aided Design and Applications, 2 (1-4).*

[9] **Thrimurthullu, K., Pandey, P.M., Reddy, N.V.** *(2004) Part Deposition Orientation in Fused Deposition Modeling, International Journal of Machine Tools and Manufacture, 2004, 44, pp. 585-594.*

[10] **Williams, R.E., Komaragiri., S.N., Melton, V.L., Bishu, R.R.** *(1996) Investigation of the Effect of Various Build Methods on the Performance of Rapid Prototyping (Stereolithography), Journal of Materials Processing Technology, 61, (1-2), pp. 173-178.*

Performance Evaluation of Domestic Refrigerator Using Different Refrigerants

A.K.Ahluwalia[1] and A.K.Saluja[2]
[1]HOD, Pant Polytechnic, Okhla, New Delhi.
[2] ADVISOR, Sachdeva Institute of Technology, Mathura.

Abstract In the recent years, a Non Azeotropic Refrigerant Mixture (NARM) of Propane (HC290) and Isobutane (HC600a) is considered to be a prominent substitute of CFC12 because it gives zero Ozone Depletion Potential (ODP) due to no chlorine atom and also have very low Global Warming Potential (GWP). In this paper, theoretically, vapour compression cycle analysis has been carried out and a computer programme developed for pure CFC12, new eco-friendly alternative HFC134a and NARM of HC290/HC600a having compositions 45/55, 50/50, 55/45, 60/40, 65/35, 70/30 by mass and parameters, viz., pressure ratio, discharge temperature, volumetric efficiency, mass flow rate, refrigerating capacity, power, heat rejected and COP have been evaluated and presented for comparisons.

Key words: Vapour compression cycle analysis, CFC12, HFC134a, NARM of HC290/HC600a.

I. INTRODUCTION

The chlorofluoro carbons (CFCs) are commonly used refrigerants because of their favourable thermodynamic and transport properties, but are considered to have a damaging effect on ozone layer. Therefore, Montreal Protocol recommended the ban on the use of CFCs. Thus there is a search of alternative refrigerants which should perform satisfactorily, perhaps even better than the currently used CFC12, should be harmless to human health and friendly to the environment. Therefore, any alternative has to have a low or preferably zero ODP and relatively low GWP. The most promising alternatives to CFC12 has been HFC134a, though it gives good performance characteristics still it is not a long term environment friendly due to high GWP. Therefore, attention is diverted towards the Non Azeotropic Refrigerant Mixtures (NARM) which serves double purpose of vapour pressure requirement and increase the system performance due to gliding temperature effect. In this race Propane (HC290) and Isobutane (HC600a) mixture is a leading NARM because it has no Chlorine (Cl) atom resulting zero ODP and at the same time has very low GWP.

Before conducting experiments, a theoretical cycle analysis is necessary, because it gives preliminary idea about the suitability of the proposed refrigerant in place of traditional refrigerant and modifications, if required. Therefore in this paper, theoretical cycle analysis has been carried out for CFC12, HFC134a and NARM of HC290/HC600a. It is an attempt to screen out the best mixture ratio of HC290/HC600a to replace FCF12.

For the cycle analysis, the data of 165 litre, domestic refrigerator has been considered. The schematic diagram Fig. 1, consist of usual four components, namely the compressor, condenser, expansion device (capillary tube) and the evaporator (freezer). The design specifications which are used in this cycle are mentioned in Table 1.

The various parameters calculated are pressure ratio, discharge temperature, volumetric efficiency, mass flow rate, refrigerating capacity, power required, heat rejected, COP with CFC12, and its alternative HFC134a and NARM of HC290/HC600a having compositions 45/55, 50/50, 55/45, 60/40, 65/35, 70/30 by mass.

II. VAPOUR COMPRESSION CYCLE FOR PURE REFRIGERANTS

Fig. 2 shows the vapour compression cycle on P-h diagram for pure refrigerant (or azeotropic refrigerant) for which temperature and pressure remain constant during evaporation and condensation. The cycle consists of the following processes.

1-2 : Superheating in suction line at constant pressure .

2-3 : Pressure loss at compressor suction due to wire drawing.

3-4 : Superheating inside the compressor during suction at constant pressure.

4-5 : Isentropic compression.

5-6 : Pressure loss at compressor discharge due to wire drawing.

6-7 : Heat rejected in the condenser with subcooling at constant pressure.

7-8 : Throttling.

8-1 : Heat abstracted in evaporator with superheating at constant pressure.

III. VAPOUR COMPRESSION CYCLE FOR NON AZEOTROPIC MIXTURES

Unlike pure refrigerants (or azeotropic mixtures), the temperature and vapour liquid composition of non–azeotropic mixtures (HC290/HC600a) do not remain constant at constant pressure as refrigerant evaporates or condenses.

Fig. 3 presents a simplified vapour compression cycle on P-h diagram for a non–azeotropic refrigerant mixture. Constant temperature lines are included in the two phase region to illustrate the temperature change at constant pressure. The vapour and liquid compositions are always different in two phase region and change continuously at constant pressure between the bubble point and dew point. The pure refrigerant condenser and evaporator temperatures are compared with the mean condenser and evaporator temperatures of a non-azeotropic mixture.

IV. RESULTS AND DISCUSSION

Performance analysis of pure CFC12, HFC134a, and mixture of HC290/HC600a (45/55, 50/50, 55/45, 60/40, 65/35, 70/30) have been carried out at constant condenser temperature of 55^0C and evaporator temperature ranging from -25^{0}C to 0^{0}C. The parameters, viz., refrigerating capacity, power, COP, pressure ratio, discharge temperature, volumetric efficiency, mass flow rate, heat rejected have been evaluated. For this purpose required properties for pure refrigerants have been taken from ASHRAE [1] and have been correlated where as for NARM, pressure and temperature do not remain constant in the two phase region and therefore liquid mole fraction (x_1, x_2) and vapour mole fraction (y_1, y_2) have been calculated as given by Reid et al. [2]. Results have been presented in the Fig. 4 to Fig. 11. The following effects have been observed:

In general refrigerating capacity increases with increase in more volatile component HC290 in mixture (Fig. 4). However, 70/30 mixture have better effect, and 65/35 have 0.1% to 3.7% higher refrigerating capacity than CFC12 while 60/40 gives on lower side. So to retrofit 65/35 is better. The refrigerating capacity of HFC134a is 4.6% to 37.7% lower than CFC12.

As compared to CFC12, power required is slightly on higher side 0.2% to 3.5% with 65/35 mixture while 60/40 mixture require slightly lower (Fig. 5). The power required by HFC134a is 2% to 34% lower than CFC12.

For all considered mixtures, COP is almost equal to CFC12 (Fig. 6). For 70/30, 65/35, although power required is more, but refrigerating capacity is also more but in case of 60/40, 55/45, 50/50 and 45/55, power and refrigerating capacity both are on lower side, so all these leads to equal COP to CFC12. COP of HFC134a is 5.1% to 2.6% lower than CFC12.

Pressure ratio decreases with increase HC290 in mixture. 65/35 and 70/30 have low pressure ratio than CFC12 but 65/35 is near to CFC12 (0.7% to 1.8% lower, (Fig. 7). The ratio of more volatile refrigerant increases pressure in evaporator which result in low pressure ratio. HFC134a require 15.6% to 33.8% more pressure ratio than CFC12. Low pressure ratio is always suggested for light designing and avoid buckling, leakage past the piston.

Discharge temperature for all range of mixture is quite lower than CFC12 (Fig.8), resulting to reduce thermal stresses. Discharge temperature of HFC134a is lower than CFC12 but higher than 65/35.

Volumetric efficiency increases with HC290 in mixture (Fig. 9), which may be attributed due to reduced pressure ratio. 70/30 and 65/35 is comparable to CFC12. For 65/35, its value is lower by 0.8% to 3.9% than CFC12. HFC134a gives 3.5% to 26.3% lower value than CFC12.

Mass Flow rate for CFC12 is very high in comparison of all mixture ratios (Fig.10). Which is attributed due to high volumetric efficiency and lower specific volume. 65/35 mixture require 57.5% to 58.4% lower mass flow rate than CFC12 with better cooling capacity. HFC134a needs 23.5% to 49.4% lower mass flow rate than CFC12.

Heat rejected for 70/30 is high (Fig. 11) as expected because higher refrigerating capacity and power required. 45/55, 50/50, 55/45 and 60/40 have low heat rejection due to low power and low refrigerating capacity. Heat rejection for 65/35 is slightly higher 1.2% to 3.8% than CFC12. Heat rejection decides the condenser size and hence, same condenser can be used with 65/35 mixture. Heat rejection with HFC134a is 3% to 35.5% lower than CFC12.

Table 2 shows the effect of condenser temperature on refrigerating capacity. For all condenser temperature varying from 55°C to 35°C, it has been observed that 65/35 gives better cooling capacity (0.1% to 7.6%) than CFC12 and much better (60.8% to 35.5%) than HFC134a.

Table 3 shows the effect of condenser temperature on power requirements. For condenser temperature varying from 55°C to 35°C, it has been observed that power required for 65/35 is almost same as required by CFC12 (within 0.8% to 5%).

V. CONCLUSION

A comparative analysis of pure CFC12, HFC134a and mixture of HC290/HC600a having mass fraction 45/55, 50/50, 55/45, 60/40, 65/35 and 70/30 have been carried out at constant condenser temperature of 55^0C and evaporator temperature ranging from -25^0C to 0^0C. For retrofitting, it has been observed that 65/35 mixture gives better cooling capacity, same COP as CFC12 with the same arrangement, although power required is slightly higher. Same condensing unit may work satisfactorily with 65/35 because heat load on condenser is slightly higher. Other parameters as pressure ratio, discharge temperature, mass flow rate are on encouraging side for 65/35 than CFC12 but volumetric efficiency is slightly lower. The refrigerating capacity and COP of HFC134a is on quite lower side which is the main criteria for retrofitting, although power required is low. On the other hand pressure ratio are higher and volumetric efficiency are lower than CFC12 and all mixtures. Discharge temperature and mass flow rate are

lower than CFC12 but higher than all mixtures. Only heat rejection is lower than CFC12 and 65/35 mixture. Thus all above results support to 65/35 mixture of HC290/HC600a as a drop in substitute for CFC12 in vapour compression system. Although, this attempt is a theoretical approach, but may be very useful for further experimental investigations.

NOMENCLATURE:

COP	Coefficient of performance	v	Volume (m³/kg)
h	Enthalpy (kJ/kg)	V_c	Clearance volume (m³)
HREM	Heat removed in evaporator (Watt)	V_s	Stroke volume (m³)
HREJ	Heat rejected in condenser (Watt)	W	Mass fraction
m	Mass flow rate (kg/sec)	x	Liquid mole fraction
N	R.P.M.	y	Vapour mole fraction
P	Pressure (bar)	η_v	Volumetric efficiency
POW	Power (Watt)		
P.R	Pressure ratio	Subscript	
s	Entropy (kJ/kg)	CON	Condenser
T	Temperature (Kelvin)	EVA	Evaporator

REFERENCES

{1} ASHRAE HAND BOOK. 1997. Fundamentals. American Society of Heating, Refrigeration and Air conditioning Engineers, New York.

{2} Reid, C.R., Prausnitz, J.M. and Poling, B.E. 1988. The Properties of Gasses and Liquids, 4th ed. Mc Graw-Hill.

Table 1. Specifications of the Refrigerator

Compressor Speed	2880 r.p.m
Stroke Volume	4.49 c.c
Clearance Ratio	0.035
Suction Temperature	$32\ ^0$C
Degree of Superheat in Evaporator	$5\ ^0$C

Degree of Subcooling in Condenser	11 ^{0}C
Pressure loss at Compressor Suction	0.3 bar
Pressure loss at Compressor Discharge	0.3 bar
Compressor and Motor Efficiency	0.55

Table 2. Refrigerating capacity (Watt)

COND/EVA (IN °C)	R12	R134a	R290/R600a (45/55)	R290/R600a (50/50)	R290/R600a (55/45)	R290/R600a (60/40)	R290/R600a (65/35)	R290/R600a (70/30)
55/-25	60.42	37.62	46.24	44.65	49.74	55.01	60.48	66.18
55/-23.3	70.17	47.06	48.23	53.53	59.02	64.71	70.59	76.71
55/-20	91.21	67.74	66.64	72.79	79.14	85.69	92.45	99.43
50/-25	67.82	45.57	47.19	52.49	57.98	63.66	69.54	75.64
50/-23.3	78.88	56.52	56.96	62.71	68.66	74.79	81.13	87.69
50/-20	100.36	77.78	75.91	82.56	89.39	96.41	103.64	111.09
45/-25	75.26	53.61	54.73	60.45	66.34	72.42	78.69	85.23
45/-23.3	86.90	65.26	65.12	71.32	77.69	84.24	90.99	98.01
45/-20	109.52	87.87	85.29	92.41	99.71	107.19	114.89	122.81
40/-25	82.69	64.81	62.33	68.46	74.74	81.23	87.92	94.78
40/-23.3	94.92	73.98	73.34	79.97	86.75	93.73	100.92	108.29
40/-20	118.65	97.94	94.71	102.30	110.06	118.00	126.15	134.53
35/-25	90.09	69.66	69.96	76.48	83.16	90.01	97.07	104.35
35/-23.3	102.88	82.68	81.58	88.62	95.82	103.19	110.79	118.58
35/-20	127.72	107.94	104.14	112.18	120.39	128.77	137.41	146.20

TABLE 3. POWER (Watt)

COND/EVA (IN °C)	R12	R134a	R290/R600a (45/55)	R290/R600a (50/50)	R290/R600a (55/45)	R290/R600a (60/40)	R290/R600a (65/35)	R290/R600a (70/30)
55/-25	69.02	45.31	46.24	51.62	57.22	63.07	69.20	75.63
55/-23.3	76.28	53.81	53.26	58.79	64.58	70.59	76.88	83.49
55/-20	90.13	70.08	66.64	72.50	78.59	84.92	91.54	98.44
50/-25	70.39	49.53	49.65	54.88	60.31	65.97	71.86	78.03
50/-23.3	77.01	57.31	56.07	61.44	67.02	72.82	78.85	85.17
50/-20	89.85	72.42	68.52	74.14	80.03	86.11	92.43	99.03
45/-25	70.94	52.60	52.10	57.16	62.39	67.82	73.46	79.35
45/-23.3	76.99	59.74	57.99	63.18	68.53	74.08	79.85	85.87
45/-20	88.72	73.60	69.43	74.86	80.44	86.24	92.26	98.53
40/-25	70.71	54.59	53.66	58.51	63.51	68.69	74.06	79.66
40/-23.3	76.18	61.09	59.03	63.98	69.08	74.37	79.85	85.55
40/-20	86.80	73.71	69.44	74.59	79.89	85.39	91.08	96.99
35/-25	69.74	55.60	54.39	59.00	63.75	68.65	73.74	79.03

35/-23.3	74.62	61.46	59.22	63.91	68.75	73.73	78.91	84.28
35/-20	84.12	72.83	68.60	73.44	78.45	83.60	88.93	94.46

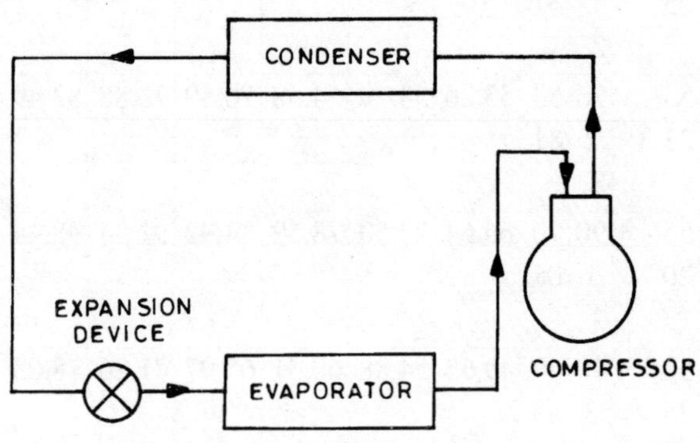

Fig. 1 SCHEMATIC DIAGRAM

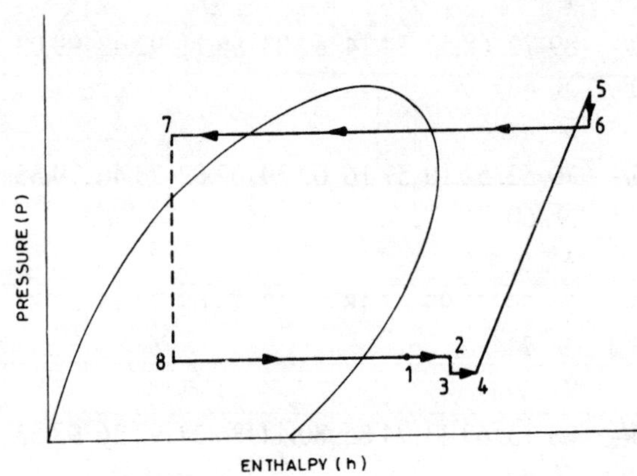

Fig. 2 P-h DIAGRAM FOR PURE (OR AZEOTROPIC MIXTURE) REFRIGERANT

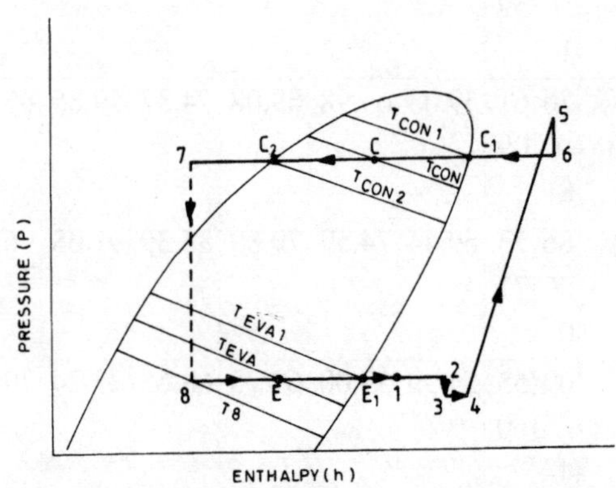

Fig. 3 P-h DIAGRAM FOR NON-AZEOTROPIC MIXTURE

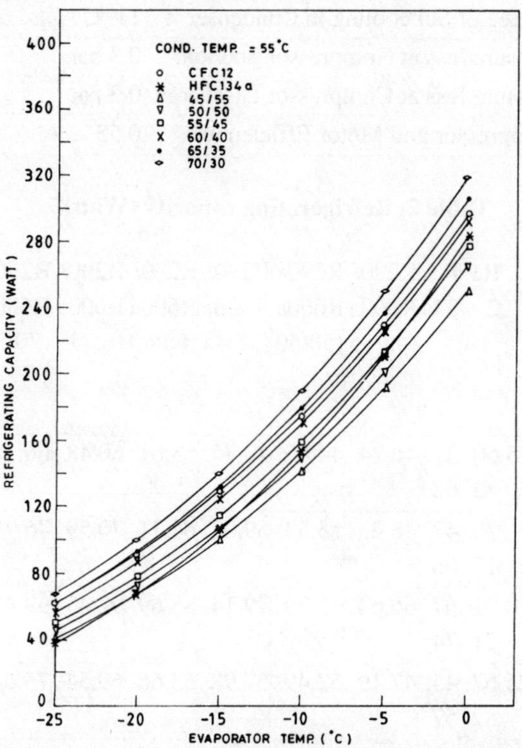

Fig. 4 EFFECT OF COMPOSITION OF MIXTURE ON REFRIGERATING CAPACITY WITH VARIATION IN EVAPORATOR TEMP.

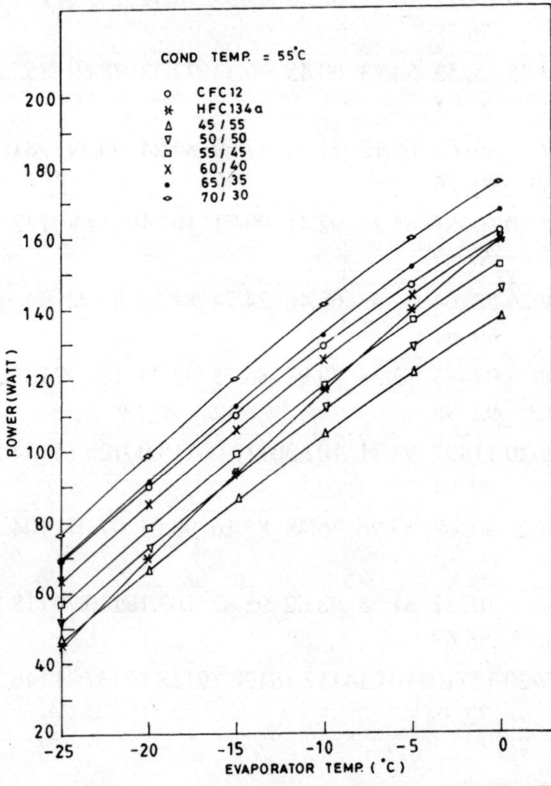

Fig. 5 EFFECT OF COMPOSITION OF MIXTURE ON POWER WITH VARIATION IN EVAPORATOR TEMP.

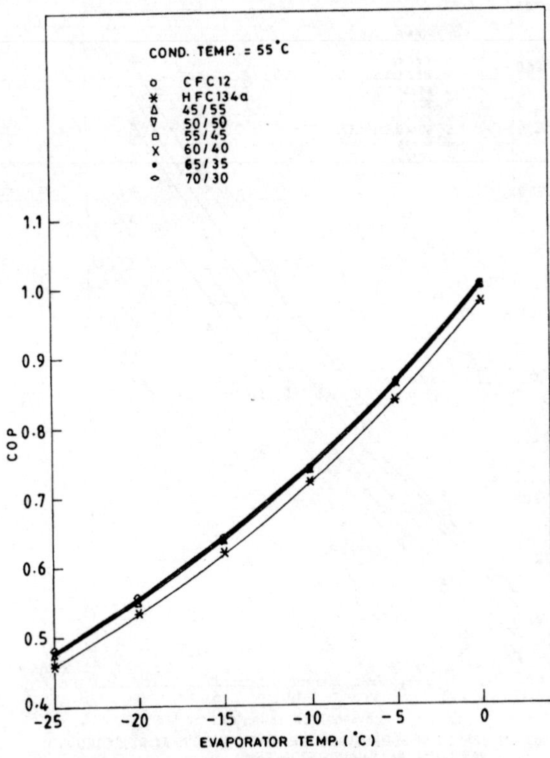

Fig. .6 EFFECT OF COMPOSITION OF MIXTURE ON COP WITH VARIATION IN EVAPORATOR TEMP.

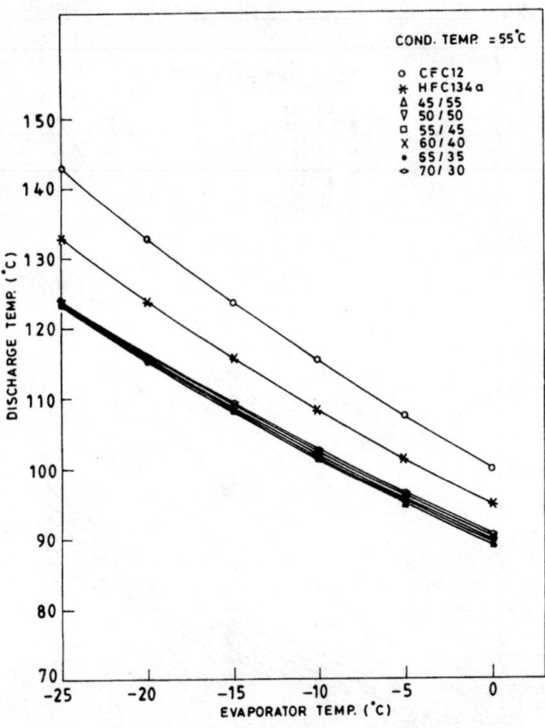

Fig. .8 EFFECT OF COMPOSITION OF MIXTURE ON DISCHARGE TEMPERATURE WITH VARIATION IN EVAPORATOR TEMP.

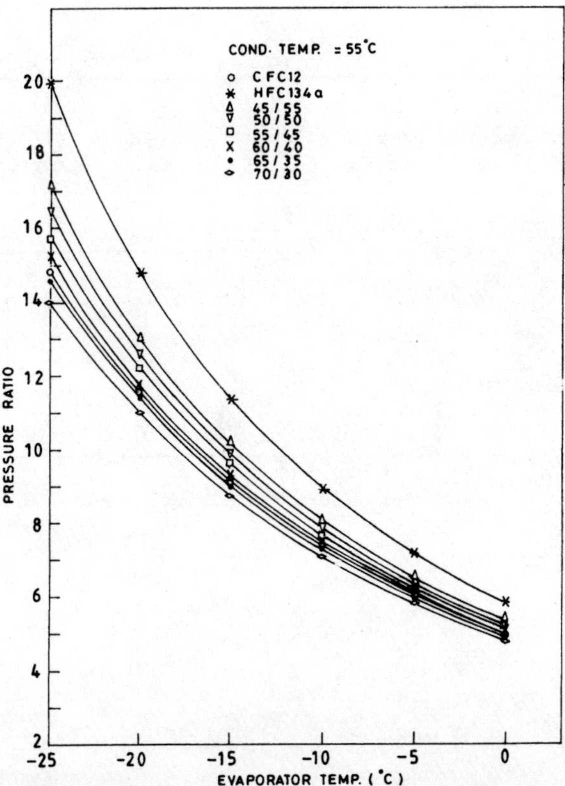

Fig. .7 EFFECT OF COMPOSITION OF MIXTURE ON PRESSURE RATIO WITH VARIATION IN EVAPORATOR TEMP.

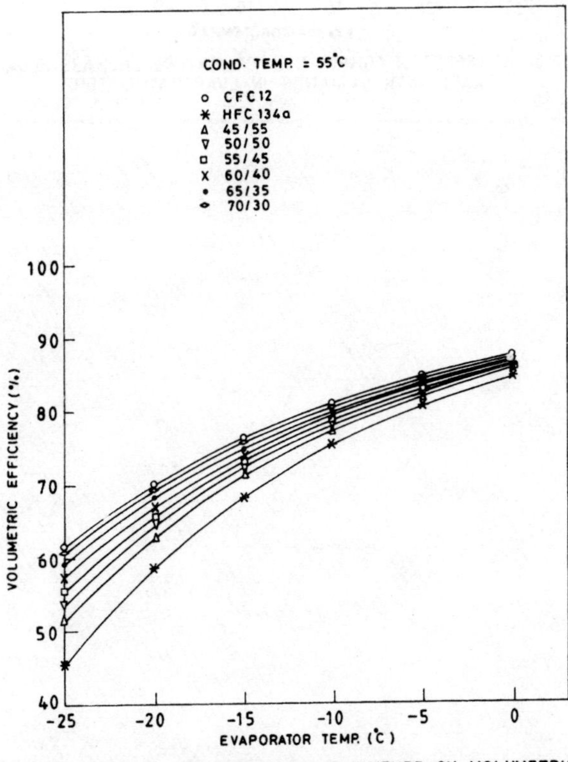

Fig. .9 EFFECT OF COMPOSITION OF MIXTURE ON VOLUMETRIC EFFICIENCY WITH VARIATION IN EVAPORATOR TEMP.

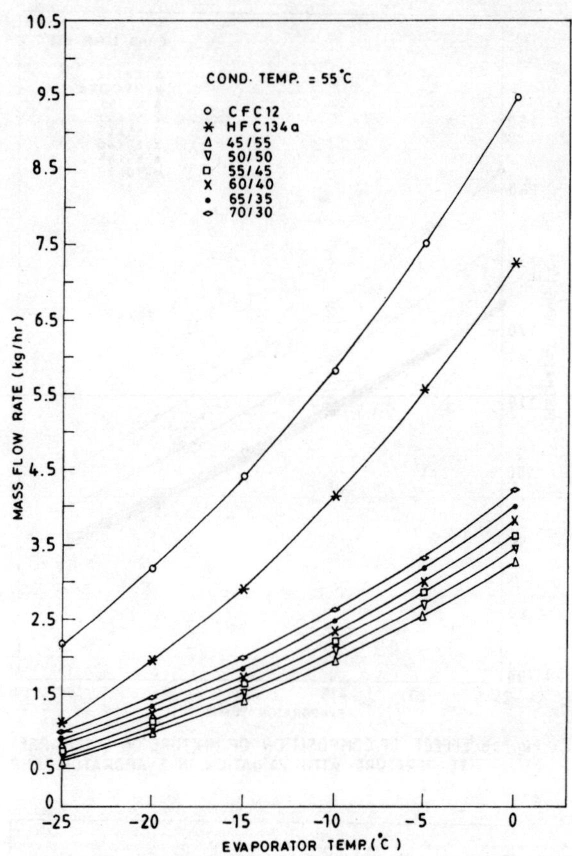

Fig. .10 EFFECT OF COMPOSITION OF MIXTURE ON MASS FLOW
RATE WITH VARIATION IN EVAPORATOR TEMP.

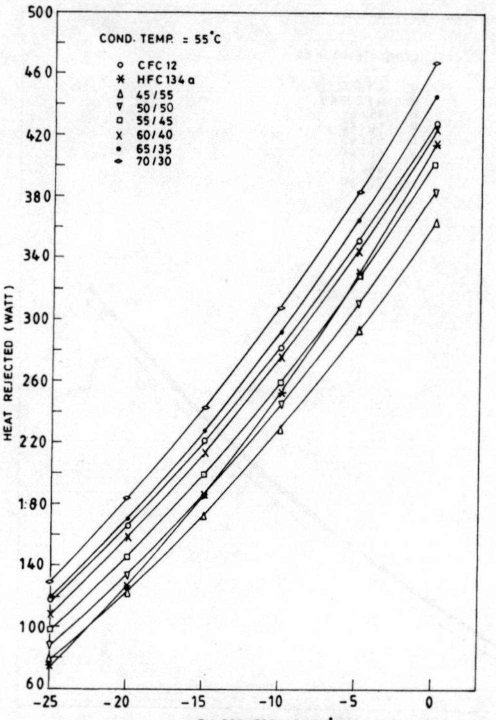

Fig. .11 EFFECT OF COMPOSITION OF MIXTURE ON HEAT REJECTED WITH
VARIATION IN EVAPORATOR TEMP.

Appreciation of Various Technologies for the Development of Parachute System

Balraj Gupta

Head (R&D and Projects) Sharda Group of Institutions Agra

I. INTRODUCTION

Parachutes are broadly used in variety of applications such as Aircraft Brake Parachute, Controlled Aerial Delivery system, Paratroopers Parachute, Cargo Parachute, Armament Parachute, Pilot Seat Ejection Parachute, Space capsule recovery Parachute. etc.

The parachute goes through various extensive testing to qualify the system. Aerodynamic parameters of a parachute are generated by wind tunnel testing. Parachute structural integrity tests are carried out in RTRS test. Simulation of load on the parachute, speed reductions and parachute deployment loads are measured using various instrumentation in airdrops trial from AN-32 & IL-76 a/c and helicopters. Based on the application and estimated parachute load various textile materials have been developed within the country for example textile material made from Nylon 6/66, Kevlar, Dacron and Vectra. In space recovery parachutes, the materials are to be evaluated for survival in the predicted and specified environmental levels.

II. TECHNOLOGY INVOLVEMENT

Parachute technology is a complete package of all kind of engineering. Involvement of these makes the parachute system more effective, reliable and high performance system. The brief appreciation of technologies involved is described in the following manner.

Textile Engineering

Textile material viz., fabrics, tape, cordages, sewing threads, coated and laminated fabrics etc have got wide area of application in parachutes. Since parachute systems area used for strategic aerospace applications for our armed forces and space applications, there is hardly any scope for failure. Further these materials are required to be engineered resulting into textile materials with light weight and high strength. It is also required to be highly protective to prolonged exposure of environment factors i.e., temperature, moisture, sunlight, humidity etc.

In the design, the aim is always at achieving the high ratio of strength to mass. Other characteristics, flexibility, wear resistance and effect of environment exposure are also considered.

Basic Material used

For development of textile materials, the following material are mainly used

Fibres	Nylon, Polyster, Cotton etc
High perfocemance fibres	Para-aramid, Vectran, Spectra, Dyneema, etc.
Coating Polysters	Polyurethan (PU), Poly Vinyl Chloride, Neroprene etc.
Films	Tedlar, Mylar, PU, etc.
Finishes	Flame Retardency, Water Repellency, UV resistance, Siliconising, Heat Settling, etc.

Electronics & Instrumentations

As in modern days parachute are pertaining complex task. Therefore electronics are widely used in a parachute system specially in controlled delivery of payload. The capability of controlled Aerial Delivery

System (CADS) is to deliver a payload to a predefined target. The inherent advantage of CADS is safe and precise delivery of system without endangering of the aircraft. The system has its own control and guidance unit, which works automatically. The CADS, with its Air Borne Unit (ABU), steers its flight path toward predetermined target by operating two of its control lanyards. The system uses Global Positioning system (GPS) to get the current heading for its entire control operation. The system control can also be taken in manual mode by ground operator during terminal phase of flight. CADS development need a suitable size parachute and a mathematical model of parafoil/payload system in terms of turn rate, glide ratio and descent rate with respect to different brake conditions and a Control Law (CLAW).

Proper instrumentation is required during the design validation and qualification testing to measure various parameters like load & velocity profile, oscillation, deployment time etc. Load cell, Accelerometer, Wind sensors etc are widely used during RTRS and Flight trials.

Aeronautical Engineering

A parachute, being a flexible body, the effect of air is predominant. All kind of parachutes are dependent on the only one fact, "How does the air flow over it?" Not only the parachutes but also the payload attached to it is also affected by the air. Therefore it is very clear that knowing the characteristics of air flow over the parachutes and the payload, the performance of the entire system can be easily predicted.

There is some different kind of parachutes such as Ram Air Parachute which is similar to the wing of an aircraft. The cross section of the parachute is an airfoil. Therefore the flight dynamics and the aerodynamics are the prime mover in designing such parachutes.

Mathematical Simulation

i. Trajectory Simulation

The three dimensional trajectory simulations are important for the parachute payload system to locate the position during the delivery. Applications of various mathematical models are used based on the requirements of the parachute. Such as 4 DOF model is sufficient to

predict the positional accuracy of round canopy parachute/payload system, whereas people go for 9 DOF model for gliding parachute/payload system. **Fig 1** shows the trajectory simulation of ejection seat which tells the position of pilot and the ejection seat after ejected safely from the aircraft.

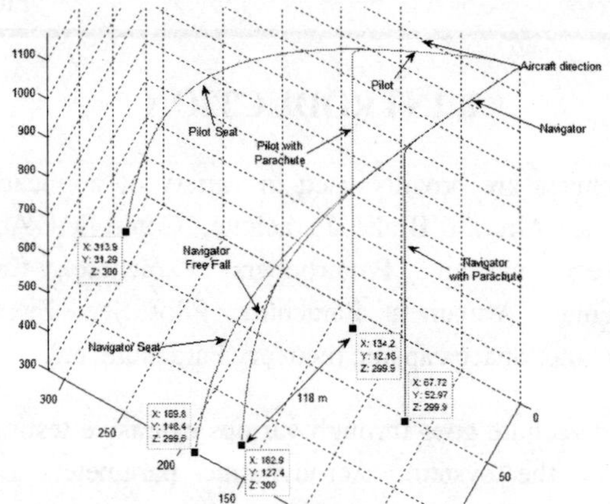

Fig 1. 3-D trajectory simulation of ejection seat

ii. Stability Simulation

The stability simulation of the parachute payload combination actually tells that the behavior of the payload alone, parachute alone and the effect of each on each other. It also determines the stable system configuration and zone of flight.

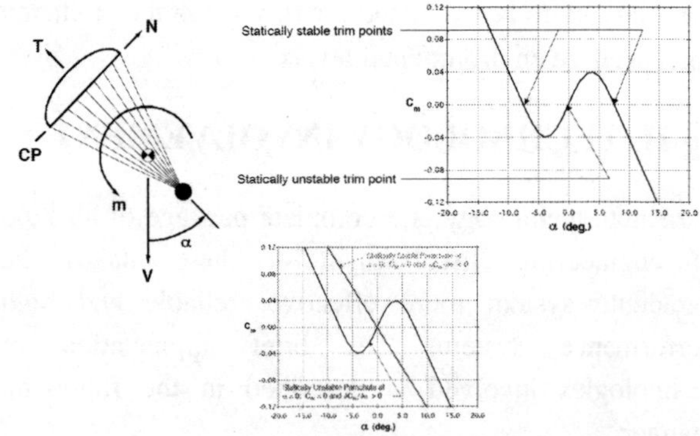

Fig 2. Static stability simulation of parachute payload system

iii. Stress Analysis

A simulation on load experiencing by a parachute can be done by considering the parachute type, its canopy filling time, velocity of deployment, inflation type and trajectory. The distribution of load on the parachute gives the stress on individual components.

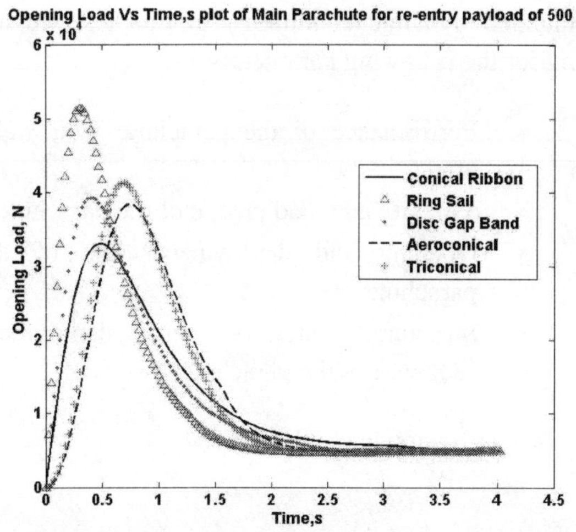

Fig 3. Peak load simulation on various parachute

Computational Fluid Dynamics

Many parachutes have various kinds of slots on it from where the air can flow across. Most of the characteristics of parachutes depend on the nature of airflow around and across the slots and therefore the prediction of behavior of such flow is very useful to design the parachutes.

In India, the accuracy in design and qualification is achieved based on Wind Tunnel Testing, RTRS Testing, and flight trials. The accuracy levels of these testing are fairly good. But the cost, effort, manpower and time to achieve such accuracy are very high. Also before going for any trial, a kind of strong prediction is needed. CFD applications are very useful for such prediction. It can reduce the number of trials thus cost and effort. By the same time it increases the accuracy of the system and help in redesign. The visualization of the flow parameters, such as pressure, temperature, density, velocity and the stream lines etc. are the best features of any CFD tool and it can give the actual feel of the system behavior.

The parachutes are functioning behind the payload, therefore the wake of the payload or primary body is also comes in the consideration and hence the CFD analysis of the primary body is also required to be done. **Fig 4** shows the pressure distribution behind the parent vehicle for the purpose of recovery of telemetry package from the vehicle moving at supersonic speed which determines the ejection force and ejection direction of the parachute by drogue gun.

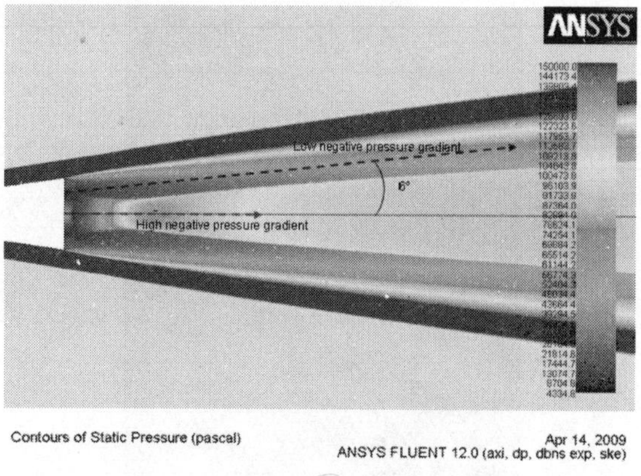

Fig 4. CFD simulation of flow behind the parent vehicle at supersonic speed

In heavy drop parachute system, there are one to five main parachutes used in cluster. The main parachutes are used to decelerate the payload to desired rate of descent. In the system, brake parachutes are also used with the main parachutes to stabilize the payload when coming out from the aircraft. **Fig 5** shows CFD study to know the effect of wake of the load carrying platform on the brake parachute.

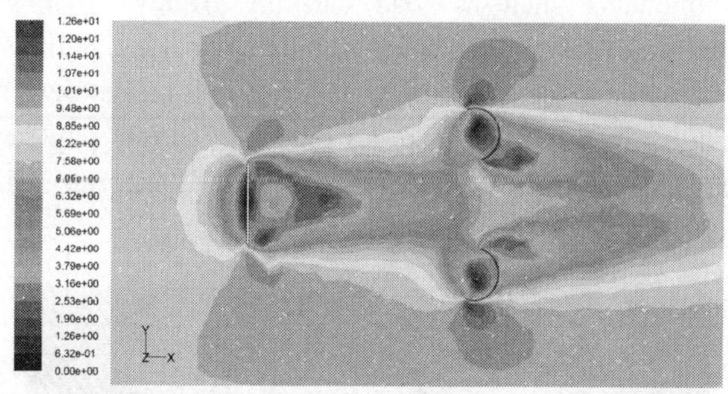

Fig 5. CFD simulation of flow over parachute and payload combination

Wind Tunnel Testing

Wind tunnel testing is one of the best methods to validate the parachute design. Before fabrication of full scale prototype parachutes, the model parachutes are tested in Wind tunnel, which depicts the characteristics of the parachutes i.e., coefficient of drag, oscillatory and opening behavior of the parachute along with wake effect of the payload. Tests are carried out for rigid, semi rigid or flexible models. **Fig 6** shows a flexible model of Ram air parachute of Control Aerial Delivery System in a Wind tunnel test section.

Fig 6. Ram air parachute model testing in wind tunnel

Mechanical Engineering

Mechanical components and structures are essential part of any parachute systems. From small mechanical links to heavy drop platforms are widely used with the system. Heavy Drop platforms are used for dropping heavy payload like jeep, tanks etc. Mechanical components such as load carrying trolley, high g-platforms, parachute disengaging unit, shackles, D-links, load cell connects etc. are widely used on proper design analysis, fabrication and testing. Fig shows payload carrying platform on trolley of AN-32 aircraft HD system.

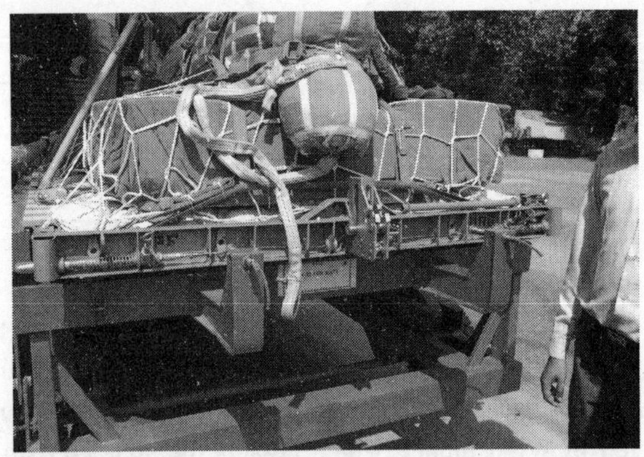

Fig 7. Load carrying trolley and platform of AN-32 aircraft system

Rail Track Rocket Sled (RTRS) Tests

As a part of development & qualification of the deceleration sub-system the RTRS tests with simulated dynamic conditions are carried out. Usually the parachutes test parameters are kept at 25% of higher side of simulated dynamic conditions. In this test we are looking for the following parameters

- Performance of the parachute with wake effect
- Velocity and load profile of the parachute
- Opening and deployment time of the parachute
- Structural integrity and deployment sequence of the parachute.

Fig 8 shows SRE module in RTRS test.

Fig 8. RTRS test of SRE module

After the design validation the parachute system goes through the qualification test which is done by dropping the system from transport aircraft or helicopters. The aircraft and helicopter are chosen for drop test based on the payload mass, flight speed and the altitude required.

Drop Test from Aircraft

Design validation of deceleration system in simulated conditions is a part of the qualification tests. The parachutes are tested for dummy drop tests by airdropping the parachutes with the required dummy load at the desired forward speeds. The dummy payload consists of rubber dummies placed on the skid-board, which are inter-laced with chains and straps. Primarily individual parachutes of the deceleration systems are tested in dummy drop tests to check the performance of individual parachutes. The instrumented trials give the load & velocity profiles, deployment time etc to validate the parachute design. For qualification of a parachute 10 such dummy drops are required.

The videography by high speed cameras gives the inflation characteristics of the parachutes. Parameters like inflation and deployment time can also be obtained from the video data.

Fig 9. Drop Test from aircraft

Once the individual parachutes are qualified, the complete deceleration system is tested in dummy drop trials in simulated conditions. This trial shows the sequence of operation of different subsystems. These kinds of trials are usually conducted with the available aircraft AN-32 & IL-76 of IAF.

Drop Test from Helicopter

When the deceleration system is required to be tested with actual payload, we usually go for Helicopter to test the system. The objectives of such trials are

- Verification of deployment sequence/ opening behavior of the parachutes.
- Verification of functioning of other subsystems
- Measurement of the peak load.
- Measurement of rate of descent of the system.
- Measurement of deployment time.
- To study the stability, functioning and structural integrity of the deceleration system.

One of the examples of such tests is drop test of SRE module from helicopter shown in the **Fig10**.

Fig 10. Drop Test of SRE parachute system from helicopter

III. PARACHUTE SYSTEM UNDER DEVELOPMENT IN INDIA

Parachutes have many applications which need to be explored. The followings are the future programmes undergoing in India to enhance the parachute technology.

- **Recovery system for Human Space Flight (HSP) programme:** A crew carrying space capsule can be recovered by parachute based earth landing system.
- **Controlled Aerial Delivery Heavy System:** The capability of the system is to deliver a heavy payload of range up to 3000 kg to a predefined target automatically.
- **Tandem Combat Free Fall Parachute (TCFF) System:** TCFF systems are capable of two personnel jumping with single parachute. The passenger jumper can be a doctor or any non jumper who may be required in the operational areas. Similarly trained Mine Detector Dogs or other search dogs can also be made to jump in this tandem system.
- **Smart Ammunition delivery:** Parachute can retard the ammunition speed to low subsonic speed for effective searching of targets.
- **Powered Parachutes:** A gliding parachute with a propulsion system can cover a very high range to deliver the payload and can also be used as UAVs.
- **Telemetry Recovery:** A telemetry package can be recovered when separated from the parent vehicle moving at high super sonic speed.

IV. CONCLUSION

A parachute system is a complete package of various technology, without support of them a reliable parachute system cannot be made. Hence the performance, capability, reliability and cost effectiveness of parachute system can replace any other complicated mechanical systems.

Modeling of Equal Channel Angular Pressing (ECAP) with Finite Element Method

Rahul Swarup Sharma, K. Hans Raj Ankit Sahai, Rajat Setia, Shanti Swaroop Sharma

Department of Mechanical Engineering, Dayalbagh Educational Institute,
Dayalbagh, Agra

Abstract—**The growing need for superior quality materials lead to the development of new processes for producing Ultra-Fine Grain (UFG) materials. Severe Plastic Deformation (SPD) is one such process for producing nano structured material from bulk material. A number of researchers are working on Equal Channel Angular Pressing (ECAP), which is an important technique for inducing SPD in materials. Equal-Channel Angular Pressing is an effective tool for attaining ultrafine grain sizes in bulk materials with improved mechanical properties. In this work an effort is made to perform Finite Element Modeling (FEM) to depict the change in equivalent strain in the Al6061 specimen up to one pass. Also the study is extended to analyze the effect of friction and channel angle on equivalent strain and forging force using the current FE Model in Forge-2007 Environment. The FE simulations provide insight into the ECAP process and are helpful in developing UFG materials economically in industry.**

Index Terms— **Equal-channel angular pressing, Severe Plastic Deformation, Finite Element Modeling, Ultra-Fined Grained materials.**

I. INTRODUCTION

THE grain size of a polycrystalline metal plays a critical role that dictates the mechanical properties of the material. At low temperatures, the strength is related to the grain size 'd' through the Hall-Petch relationship [1, 2], which is of the form

$$\sigma_y = \sigma_o + k_y\, d^{-\frac{1}{2}} \qquad [1]$$

where σ_y is the yield stress, σ_0 is a friction stress and k_y is a constant of yielding. It follows from Eq. [1] that a reduction in grain size is beneficial because it leads to an increase in the strength of the material. Grain refinement is achieved traditionally through the application of thermomechanical treatments that lead typically to grain sizes of the order of a few microns. However, these procedures cannot refine the grains to the submicrometer (0.1 to 1.0 µm) or nanometer (<100 nm) range. For these ultrafine grain sizes, it is necessary to use alternative techniques based on the application of severe plastic deformation (SPD) in which material is subjected to high strains under high imposed pressures without incurring any significant changes in the overall dimensions of the work piece.

Many SPD techniques like Equal Channel Angular Pressing (ECAP) [3-7], High Pressure Torsion (HPT) [3], Accumulative Roll Bonding (ARB) [8], Twist Extrusion (TE) [9-10], etc, have been developed and analyzed. Currently, one of the most attractive SPD processing technique is Equal-Channel Angular Pressing (ECAP). Starting from the mid-1990s, processing by ECAP has attracted the close attention of researchers from many different laboratories and the process has experienced active development. These developments include not only the application of ECAP to many different metals and alloys but also the establishment of the basic principles of ECAP process. ECAP is used to produce an ultrafine grained (UFG) structure in metals in order to improve their mechanical and physical properties by introducing very high shear deformations into the material under super imposed hydrostatic pressure.

In this paper, Finite Element Modeling of Equal Channel Angular Pressing (ECAP) process is attempted. The simulation results are analysed upto one pass to study the effect of channel angle and friction on the evolution of equivalent strain and forging force. The advantage of using finite element modeling is illustrated in the present work.

II. PRINCIPLE OF ECAP TECHNOLOGY

In ECAP (fig.1) a well lubricated round or square metallic billet is pressed through a sharply bent channel by a ram through a die constrained within two intersecting channels which are placed at an abrupt angle that is equal to, or very close to 90^0. Since the cross-section of the intersecting channel does not change, plastic deformation occurs only at the channel bend, the mode of deformation_close to simple shear. The cubic block changing its shape into a parallelepiped. In this way, the complete billet undergoes deformation of uniform strain, except the small_end regions. The ram retreats after the extrusion stroke, and the deformed billet is withdrawn from the second channel. The billet can be subjected to multiple pressing and entry, various passes are performed by rotating the billets around the longitudinal axis at _different angles _of 0^0, 90^0 and 180^0.

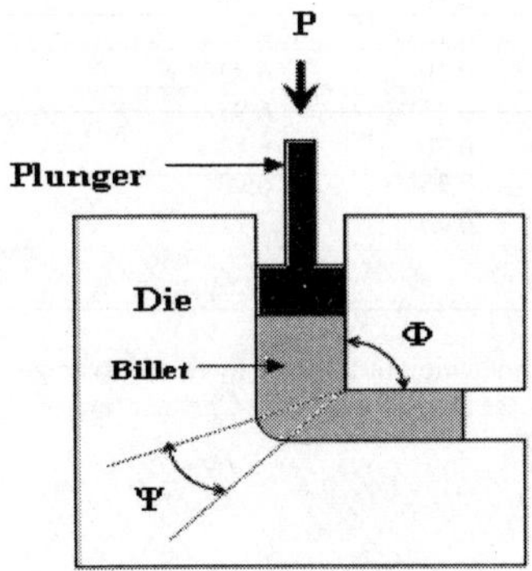

Fig. 1: Schematic illustration of a typical single turn ECAP process.

The sample or billet emerges from the die without any change in the cross-sectional dimensions. Thus, this process is distinct from the more conventional metal working processes such as rolling and extrusion where there is a concomitant reduction in the cross-sectional dimensions of the work piece. The strain imposed on the sample in a single passage through the die is dependent primarily upon the angle Ô between the two separate parts of the channel within the die. There is also a minor dependence upon the angle Ø at the outer arc of curvature where the two channels intersect. The strain increment that the material undergoes after each pass can be expressed in terms of punch pressure (P) and the flow stress of the material (σ_{fs}) and depends on the intersection angle (Ö) between two channels as shown in Eq. 2.

$$\Delta\varepsilon = \frac{P}{\sigma_{fs}} = \frac{2}{\sqrt{3}}\cot\varphi \qquad [2]$$

The shear strain value greatly depends on the number of passes (N) and the curvature angle at the channel intersection and is given by the below equation:

$$\varepsilon_N = N\sqrt{3}\left[2\cot(\varphi/2+\psi/2)+\psi\,\mathrm{cosec}(\varphi/2+\psi/2)\right] \qquad [3]$$

From the above equation it can be inferred that the deformed billet experiences a shear strain value of nearly equal to 1, considering the frequently practiced values of Ö and Ø. The degree of grain refinement in ECAP method depends on various factors like processing parameters, phase composition, and initial microstructure of a material. Although stress and reaction forces are routinely available from FE simulations, only few papers mentioned these results and used them to evaluate material behaviour, tool pressure, and the process force [11-12]. The majority of ECAP simulations used a 2-D model assuming plane strain. For round billets and rectangular billets with friction and heat transfer, 3-D simulation is more appropriate.

In this paper effort is made to study the influence of channel intersection angle and friction on ECAPed billet using FE analysis. Separate FE models are developed using channel angles of 90^0, 105^0 and 120^0.

III. FE MODELLING OF ECAP

The finite element method (FEM) is one of the most important numerical methods that can be used to explain the deformation process during the ECAP and related processes. The first finite element (FE) simulation of ECAP was performed in 1997 [13]. As many future simulations, it assumed a 2-D plane strain model and investigated the effect of channel geometry and friction on the plastic flow of the material. Another early FE analysis of ECAP also evaluated the effect of friction but, more importantly, it dealt with a hot process [14]. The attempts to use models with microstructure together with FE analysis are rare. The vast majority of FE simulations that followed these early publications were carried out with a view to assess the influence of the tool geometry, process conditions, and materials properties on strain homogeneity of the ECAPed billets and, less frequently to assess the process force. FE simulations help to understand and critically assess the existing ECAP process with a better insight into influence of different process parameters. Recently, Suo et al. [15] have done some 3D analyses to trace the homogeneity during the ECAP processes after the first pass. Zhang et al. [16] studied the distribution of strain in the cross-section of the sample of pure Al during the 3D FEM simulations for the multiple passes.

In this work, the modeling of ECAP process is done in FORGE environment. The effect of friction and channel angle on equivalent strain and punch force required for ECAP is studied. FORGE is capable of modeling three dimensional situations of metal forming with automatic mesh regeneration thermal and friction. The material is assumed to be homogeneous, isotropic and incompressible. The dies are assumed to be rigid and the material used is an H13 tool steel. The dimension of the plunger is 20mm (width) x 20mm (height). The three dimensional workpiece (billet) considered has the dimensions of 20 mm (width) x 20 mm (breadth) and 105 mm (height). The material of the billet was assumed to be Al6061. FE simulations are carried out for ECAP (channel intersection angle, $Ö = 90^0$) and the curvature angle, $\varnothing = 0^0$ with different values of friction (0, 0.1, 0.15, 0.2, 0.25, 0.30, 0.35 and 0.40). Recently, Hans Raj et al. [17] analyzed the influence of friction and channel angle in ECAP. The elastic strains in comparison to visco-plastic ones are considered to be negligible. A fully automated remeshing procedure is incorporated into the analysis. The material behavior is assumed to follow that of Norton-Hoff law written in following tensorial form:

$$s = 2K(T,\overline{\varepsilon},...)(\sqrt{3}\,\dot{\overline{\varepsilon}})^{m-1}\dot{\varepsilon} \qquad [4]$$

where s = shear stress, K = material consistency, T = temperature, $\overline{\varepsilon}$ = equivalent strain, $\dot{\overline{\varepsilon}}$ = equivalent strain rate and $\dot{\varepsilon}$ = strain rate.

The flow stress in case of Al6061 is directly interpolated from the material data file available in FORGE software for various temperatures, strains, strain-rates. The values of K and m are not explicitly keyed in by the user in the data file for defining the rheology law. Generalized coulomb friction law is used in the current analysis given by:

$$\tau = \mu\sigma_n \text{ if } \mu\sigma_n < \bar{m}\frac{\sigma_0}{\sqrt{3}} \text{ and } \tau = \bar{m}\frac{\sigma_0}{\sqrt{3}} \frac{\overline{\Delta V}}{\Delta V} \text{ if } \mu\sigma_n > \bar{m}\frac{\sigma_0}{\sqrt{3}} \quad [5]$$

where,

τ = friction stress tangential to the surface

μ = coefficient of friction

σ_n = compressive stress normal to the surface (contact pressure)

$\bar{m}$ = Tresca coefficient

The variation of forging force required and equivalent strain in the end product with channel intersection angle ($\ddot{O}$) of 90^0, 105^0 and 120^0 and different coefficients of friction are obtained, table-1, for ECAP process at constant temperature (20^0 C) under constant ram velocity of 10 mm/s.

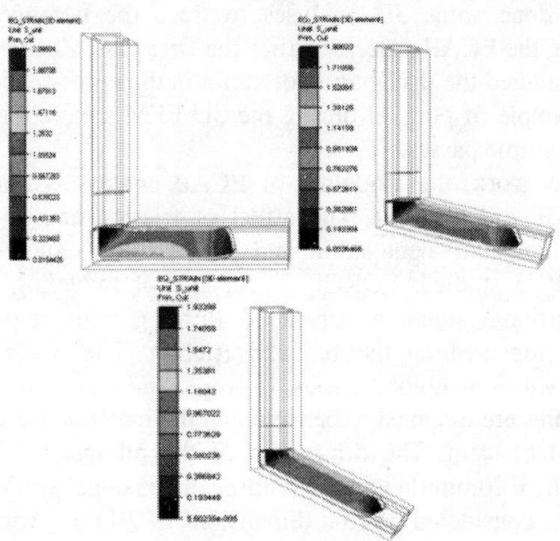

Fig. 2: Equivalent Strain contours for $\hat{O}$ = 90^0, 105^0 and 120^0

Channel Angle ($\ddot{O}$)	Friction	Forging Force	Equivalent Strain
	0.00	08.03	1.27
	0.10	10.81	1.30
	0.15	10.89	1.33
	0.20	10.83	1.35
90^0	0.25	10.89	1.36
	0.30	10.98	1.38
	0.35	10.98	1.43
	0.40	10.92	1.45
	0.00	6.35	0.92
	0.10	9.12	0.96
	0.15	9.14	1.02
	0.20	9.06	1.04
105^0	0.25	9.16	1.10
	0.30	9.15	1.05
	0.35	9.17	1.18
	0.40	9.18	1.23
120^0	0.00	4.76	0.75
	0.10	6.58	0.76
	0.15	6.71	0.80
	0.20	6.82	0.84
	0.25	6.93	0.87
	0.30	6.88	0.91
	0.35	6.90	1.02
	0.40	7.11	1.10

Table 1: The FE evaluation of forming force and average equivalent strain during ECAP process after 1st pass with Channel Intersection Angle ($\ddot{O}$) of 90^0, 105^0 and 120^0.

IV. RESULTS AND DISCUSSION

During ECAP, billet experiences severe plastic deformation by simple shear at the region where two channels intersect. The equivalent plastic strain distributions obtained during simulation for ECAP through route A during 1st pass is depicted in Fig._2. The equivalent strain distribution is inhomogeneous in the inner region of the billet. The equivalent strain in the front and the end parts of the billet is smaller than the one in central part of the billet. Almost little deformation is obtained in the end part of the billet, so the equivalent strain is the minimum whereas intensive shear deformation is obtained in the center part of the billet, so it exhibits large values of equivalent strain.

The effect of friction can be clearly predicted from Fig. 3 which indicates that its effect is negligible on the evolution of plastic strain, though the average equivalent strain increases slowly with increase of friction. The effect of channel angle can also be analyzed from Fig. 3 which clearly indicates that for $\ddot{O}$ = 90^0 and at μ = 0.4, average equivalent strain is maximum and therefore channel angle 90^0 is best suited among other angles for the evolution of plastic strain. Hence, with increase in channel angle (CA), plastic strain decreases.

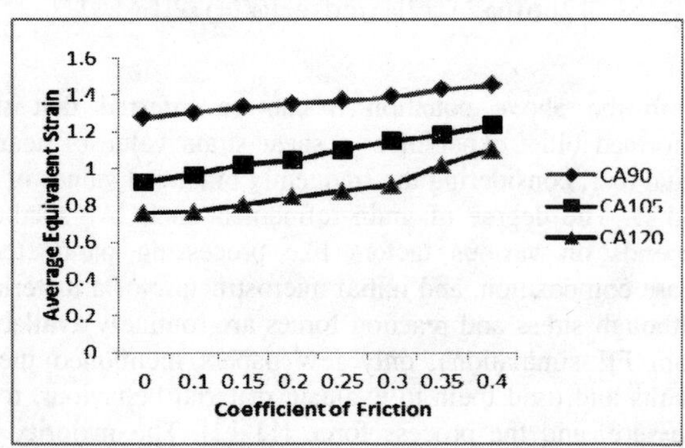

Fig. 3: Effect of Friction and Channel Angle (CA) on evolution of Average Equivalent Strain

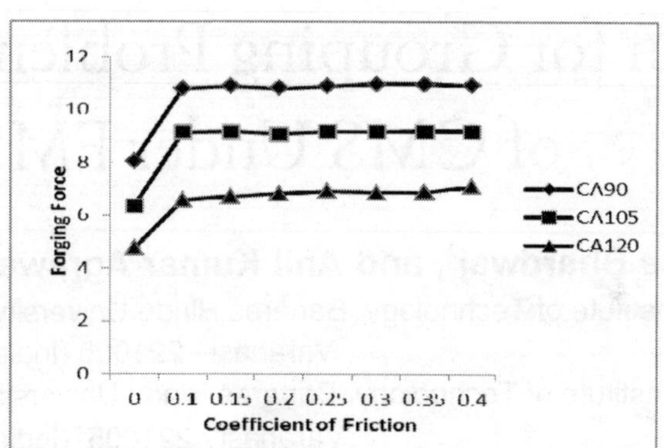

Fig. 4: Effect of Friction and Channel Angle (CA) on variation of Forging Force

The effect of channel angle and coefficient of friction on forging force can be studied from Fig. 4 which indicates that with increase of channel angle, forging force decreases. The value of forging force increases rapidly from 0 friction to 0.1 and after that it almost remains constant in all the observed channel angles.

V. CONCLUSIONS

The single pass finite element analysis was carried out in order to investigate the plastic deformation behavior of the work piece during ECAP. The simulation study illustrated the role of friction and channel angle on evolution of plastic strain and forging force (Table-1). The channel angle 90^0 is best among other angles for evolution of strain and high value of plastic strain is the indication of smaller grain size. Channel angle plays a significant role in the evolution of strain and the effect of friction is negligible. The average equivalent strain decreases with increase of channel angle. Front and back end of workpiece exhibit very less strain as these regions experience very less deformation while central part of the work piece exhibits high values of plastic strain. Also, the value of forging force decreases with increase in channel angle. ECAP opens up possibilities for forming UFG materials that can be of high usage in automobile, military and space industries.

ACKNOWLEDGMENTS

We deeply appreciate the inspiration and guidance provided by most revered Professor P. S. Satsangi, Chairman of Advisory Committee on Education, Dayalbagh.

The authors also gratefully acknowledge the help of Professor J. L. Chenot and Dr. L.Fourment of CEMEF Laboratory, France. Partial support for this research from the Department of Science & Technology (DST), Government of India under grant No. SR/S3/MERC/104/2008, All India Council of Technical Education (AICTE), under research promotion scheme (RPS), file number 8023/BOR/RID/RPS-153/2008-09, UGC major research project and funding from ADRDE (DRDO) is duly acknowledged.

REFERENCES

[1] E.O. Hall: Proc. Roy. Soc. B, 1951, vol. 64, pp. 747–53

[2] N.J. Petch: J. Iron Steel Inst., 1953, vol. 174, pp. 25–28.

[3] V. M. Segal, V. I. Reznikov, A. E. Drobyshevskiy and V. I. Kopylov, "Plastic working of metals by simple shear", *Russian Metallurgy (Metally)*, 1, PP. 99-105, 1981.

[4] R. Z. Valiev, "Materials science: nanomaterial advantage", *Nature*, 419, 6910, PP. 887-889 2002.

[5] T. C. Lowe and R. Z. Valiev (Eds.), "Investigations and Applications of Severe Plastic Deformation", Kluwer Academia Publisher, Dordrecht, 2000.

[6] M. Furukawa, Z. Horita and T. G. Langdon, "Application of Equal-Channel Angular Pressing to Aluminum and Copper Single Crystals," *Materials Science Forum*, 539-543, PP. 2853-2858, 2007.

[7] T. G. Langdon, "The Principles of Grain Refinement in Equal-Channel Angular Pressing," *Materials Science and Engineering*, A462, PP. 3-11, 2007.

[8] P.Sherstnev, I.Flitta, C. Sommitsch, M. Hacksteiner, T. Ebner "The effect of the initial rolling temperature on the microstructure evolution during and after hot rolling of AA6082" *International Journal of Material Forming 1*, Supplement 1, PP. 339-342, 2008.

[9] D. Orlov, Y. Beygelzimer, V. Varyukhin, S. Synkov, N. Tsuji and Z. Horita, "Microstructure evolution in pure Al processed with twist extrusion", *Materials Transactions*, vol. 50, 96-100, 2009.

[10] Y. Beygelzimer, V. Varyukhin, S. Synkov and D. Orlov, "Useful properties of twist   extrusion",*Materials Science and Engineering*, A 503, PP. 14-17, 2009.

[11] Z. A. Khan, Chakkingal, U. Chakkingal and P. Venugopal, "Analysis of Forming Loads,   Microstructure Development and Mechanical Property Evolution During Equal Channel Angular Extrusion of a Commercial Grade Aluminium Alloy" *Journal of Materials Processing Technology*, 135, 59, 2003.

[12] A. Rosochowski, L. Olejnik and M. Richert, "Metal forming technology for producing bulk nanostructured metals", *Journal of Steel and Related Materials – Steel GRIPS, 2, Suppl. Metal Forming* , PP. 35-44, 2004.

[13] P. B. Prangnell, C. Harris and S. M. Roberts, "Finite element modeling of equal channel angular Extrusion", *Scripta Materialia*, 37, PP. 983–989, 1997.

[14] D. P. DeLo and S. L. Semiatin, "Finite-element modeling of non-isothermal equal-channel angular extrusion", *Metallurgical and Materials Transactions*, 30A, PP. 1391–1402, 1999.

[15] T. Suo, Y. Li, Y. Guo and Y.Liu, "The simulation of deformation distribution during ECAP using 3d finite element method", *Material Science and Engineering A466*, PP. 166-171, 2007.

[16] W. Wei, W. Zhang, K. X. Wei, Y. Zhong, G. Zheng and J. Hu, "Finite Element analysis of deformation behavior in continuous ECAP process", *Material Science and Egineering, A 516*, PP. 111- 118, 2009.

[17] K. Hans Raj, R. S. Sharma, R. Setia and S. S. Sharma, "Study and influence of friction and channel angle in ECAP", In the Proceedings of 4th International Conference on Tribology of Manufacturing. (ICTMP 2010), Nice, France, June 13-15, vol. 2, PP. 517-527, 2010.

Ant Colony Optimization for Grouping Problem of CMS Under FMS

Prabhas Bhardwaj[1], and Anil Kumar Agrawal[2]
[1]Reader, Mechanical Engineering Department, Institute of Technology, Banaras Hindu University, Varanasi - 221005 (India).
[2]Professor, Mechanical Engineering Department, Institute of Technology, Banaras Hindu University, Varanasi - 221005 (India).

Abstract—Cellular Manufacturing System (CMS) is gaining a lot of attraction in its application to Flexible Manufacturing (FMS).It is helpful to meet the customer's fast changing demand and variety of products while enjoying the advantage of Mass Production System(MPS). This particular talk is related to the application of Ant Colony Optimization (ACO) to the grouping problem of CMS under FMS. These are the finding of the research work of the authors. ACO is a meta-heuristic approach can be applied to variety of the problems, especially those which are hard to solve through mathematical modeling. Cell Formation Problem (CFP) of CMS is one of this kind and so solved by the ACO. Authors have developed six ACO algorithms to solve simple and generalized grouping problem (SGP and GGP). This talk presents the findings of the author's research work and includes the ACO methodology for SGP, results which shows the superiority of this meta-heuristic.

I. INTRODUCTION

Cellular Manufacturing System (CMS) is having major challenges of formation of cell of machine groups and part families. In real life, it is very hard to get disjoint cells as solution of Cell Formation Problem (CFP). In this type of problem, Exceptional Elements (EE) and Voids (V) play an important role. Exceptional elements are those which belongs to a cell but require a machine type of some other cell. EE appears as 1 outside the cell formed *i.e.* 1 appeared in the off-diagonal space of Part Machine Incidence matrix (PMIM). Void is one which appears as 0 in the formed cells of PMIM. Void is related to the part which is not requiring a particular type of machine of the parent cell. Both EE and V have impact on Material Handling Cost (MHC). AS EE and V increases, MHC increases. Objective to solve CFP is to minimize the total number of EE and V in order to have least MHC. This can be achieved by either minimizing the total material handling cost by making a mathematical model or maximizing efficiency measure by directly minimizing EE and V. As CFP is very hard problem and the solution of a mathematical model of moderate size problem cannot be obtained in the real time. So, solution through branch and bound method of a mathematical modeling of a real life problem is not justifiable. Researchers are reporting to use heuristic or meta-heuristic to solve such type of problem. Much work can be found on meta-heuristics such as Genetic Algorithm (GA), Tabu Search (TS), Simulated Annealing (SA) I(Chen *etal.*,1995 and Liu and Wu, 1993), Swarm Particle Optimization (SPO), etc to solve CFP. Here, authors present his research work related to the Ant Colony Optimization (ACO) for CFP.

Prabhakaran et al. (2005) suggested Ant Colony System (ACS) for CFP that minimized the cell load variations and intercell movement. They demonstrated the superiority of ACS over GA. Islier (2005) used the ACO to model CFP. He used efficiency measure, due to Chandrasekharan and Rajagopalan (1986), to compare the results of ACO with GA, SA and TS. He showed that the ACO approach is better over the others. Inspired by the work of Islier (2005), Authors has developed five algorithms to solve Simple Grouping Problem(SGP) and one for Generalized Grouping Problem(GGP). Besides this, authors have developed two new efficiency measures to evaluate the goodness of a particular solution. Here, in this talk, only the work of ACO(Islier, 2005, Kao *et al.* 2008, Agrawal, 2010 and Agrawal *et al.,* 2011) for SGP is presented.

The remainder of the paper is consisting of section 2 for introducing SGP, Section 3 for introduction of ACO and modeling of ACO to CFP for simple grouping, section 4 devoted to algorithms and their solutions to different problems, section 5 the conclusion.

II. SIMPLE GROUPING PROBLEM

Researchers have used PMIM as basic data for the problem of grouping. PMIM can be used to show grouping problem with or without FMS. Considering PMIM without FMS deals with SGP where all parts have one only and one process plan. Figure 1 shows the PMIM for SGP. Figure 1 shows incidence matrix with 5 parts and 4 machines types. Figure 1 is developed for SGP and has only one process plans for every part. To explain incidence matrix of SGP clearly, consider the part 1of figure 2. This part is requiring machine types 3 and 4 and is shown by 1 while there is non-requirement of machine type 1 and 2 represented by blank entry. Likewise, the whole incidence matrix can be read as having 5 process plans for 5 parts. SGP is least flexible and easy to model in comparison to GGP. Disjoint groups can easily be made by the GGP in comparison to SGP. The objective of solving SGP is to form groups of machines and families of parts and thus different cells. Cells are formed in such a way that, ideally, all parts of a cell are processed on the machines of that cell only.

Part		1	2	3	4	5
	1		1		1	1
	2	1		1		
	3		1		1	
	4	1		1		1

Machine type

Fig 1. PMIM for SGP problems

	Parts	2	4	5	1	3
	1	1	1	1		
	3	1	1			
	2				1	1
	4			1	1	1

Machines

Fig 2. Formed cells of PMIM of Figure 1

Figure 2 shows the solution of the CFP of Figure 1. Figure 2 clearly shows the formed cells as two in number. First cell is having group of machines of type 1 and 3 that are supposed to process part family of part 2, 4 and 5. Part 2 and 4 can be processed completely in cell 1 but part 5 is requiring machine 4 and so this part visit machine type 4 in cell 2. Such type of the movement is known intercellular movement and associated entry, as 1, in PMIM of figure 2 is called exceptional element. Exceptional cost additional MHC. Figure 2 also shows that part 5 is not requiring machine type 3. This is what is called as void and amount to intracellular MHC cost. This particular example is for non-ideal case of CFP. Ideally, objective is to have Exceptional elements and voids zero; which is not possible in real life problem. Here in this talk, ACO approach is discussed for SGP only.

III. ACO AND ITS MODELING TO SGP(BHARDWAJ 2008)

Ant Algorithm or Ant System uses ants' behavior of real ants to reach a food source. Ants try to discover the shortest/best path and follow it in their journey. This concept can be easily understood with the help of the figure 3 (Islier, 2005).

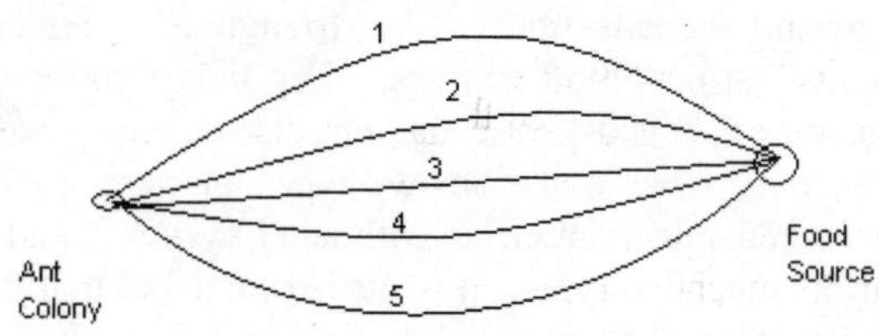

Fig 3. Routes from ant colony to food source Islier(2005).

In figure 3, it is clear that the ants those comes out from the colony can reach the food source using either of the routes labeled as 1, 2, …, 5. They, Ants, move randomly in all directions in the beginning in order to search of food,. An ant may follow route 1, other route

2 and so on. The ants, while moving between source and food, keep on laying pheromone so that they can trace back their path from the food source to the colony. When they find the food source, they return back to the colony following the same path and again laying pheromone on their path, thus intensifying the pheromone deposits. This helps other ants to follow this path by sensing the intensity of pheromone dropped by ants earlier.

Pheromone also evaporates by the time. Let the size of these arcs represents routes as well as the length of corresponding routes. So with time, shortest route *i.e.* path 3 shall have more intensity of pheromone as other paths shall loose comparatively more pheromone from evaporation because of long time taken on these lengthy routes. Ultimately, path 3 becomes the path most of the ants follow.

In ACO, food source is considered to be the objective and the routes followed by ants as alternative solutions. Hence, ant algorithm can be seen as an approach to develop the optimal solution in relation to the specified objective. Concepts used by ACO, as modeled for CFP are demonstrated in the next section. This section also explains the concept of memory and visibility.

ACO for GT problem

As ACO is communication oriented and easy to apply, it is better than many of the random search based approaches. It involve positive feedback, distributed computation and constructive greedy heuristic as main characteristics.

Islier (2005) has modeled ACO for the GT problems. He presented his model having different types of holes for machines and parts. This can be explained with the help of figure 4. In this figure, all circular and square grey-shaded elements represent the holes as exit or entry point to the nests, which are underground and have underground connections. A triangular element represents various food sources. For the purpose of demonstration, it is assumed that circular elements with label as *t, u, v, w,* and *x* shows types of parts to be produced and square elements with label as *a, b, c* and *d* for various machine types. It is further assumed that the triangular elements with label as *1, 2* and *3* symbolize the groups to be formed. Following paragraph shows the mechanism of ACO for CFP.

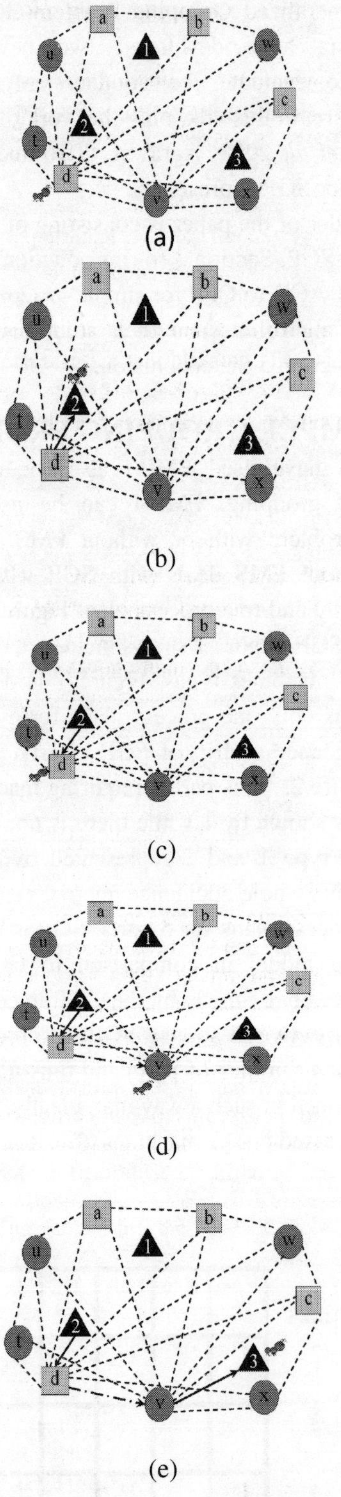

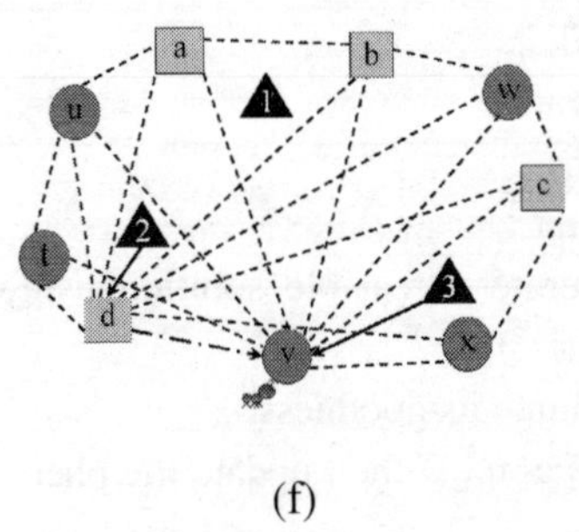

(f)

Figure 4. ACO Mechanism (Agrawal *et al.*, 2010).

First, an ant (suppose) at nest *d* (figure 4(a)) comes out comes out to the surface from the related hole *d* and goes to food source 2 (figure 4(b)). It, then, collect food from this food source, it goes back to nest *d* (figure 4(c)) to place the food there. Now, this ant goes underground to nest *v* (figure 4(d)) and from here it comes out to the surface. Form this point of exit, it reaches to another food source *3* (figure 4(e)).

Then, It is once again returns back to its nest of origination *v* (figure 4(f)) to place the picked up food there. In this way, the ant is assumed to move from one nest to the other in search of food up, picking it up from the various food sources and storing food to various nests. In this way, through its movement, ant links the nests to various food sources, and in turn can be understood to link various parts and machines to various groups. So, machine type *d* will get assigned to group *2* and part type *v* to group *3*. This can be read as *d2v3*.

Let us assume that the ant, at the end, completes a tour as *a1b1c3d2t2u2v3w1x3* is generated. Standard sequence for machines and parts helps us to understand it as *abcdtuvw with* the solution string corresponding to the generated as *113222313*. These numerical values represent the group number to which the various part types and machines are assigned when read in the order of the above mentioned sequence.

As explained in above paragraphs, an ant deposits pheromone in the path of its movement while it traverses and generates a feasible solution. The amount of pheromone laid decides the biasness for the next ant to follow the route of this ant. Higher pheromone value on a path will results higher probabilities for the corresponding route to be selected by the next ant. For all paths, amount of pheromone is shown by a data structure known as "pheromone matrix". This matrix is having a size of a *(m+n)*k* (where *m* is a machine type, *n* is a part type, *and k* being the number of cells). This matrix holds values of the pheromone accumulated by ants on their movement from a machine or a part to any cell.

The pheromone is supposed to be evaporated with time. Evaporation ratio shows the amount of evaporated pheromone to the original value of pheromone *i.e* evaporation ratio decides the amount of the scent value after an ant passed from that path. Lower values of the evaporation ratio lead to wrong accumulation of data. On the other hand, it represents the loss of useful information. Thus amount of pheromone, in some way, can be seen to relate and dominate memory power of ants (Agrawal *etal.* 2010).

Researchers believe that the real ants are semi-blind. Artificial ants are also assumed to be semi-blind. This has been modeled with visibility ratio. Their movement, in the algorithm, is dictated by visibility ratio. Visibility ratio shows the visibility power of the ants. Suppose, at any decision point, there are two choices for an ant. Either it can follow a random path (decided through roulette wheel method) or the ant may decide to choose any visible path (apparent solution) apparent from the pheromone value (Islier, 2005 and Bhardwaj, 2008). A random selection is made by generating a random number *r*. If *r* is less than visibility ratio, then apparent solution is chosen or otherwise the random one. So a higher value of visibility ratio gives higher chances for choosing apparent solution. So, a very high visibility ratio may lead myopic solution.

IV ANT COLONY OPTIMIZATION (ACO) (AGRAWAL *ET AL.*, 2010)

Author has developed five algorithms for CFP. These are as follows…

1. ACO-E: ACO with elitist ant, 2. ACO-MA: basic Multiple with ant; 3. ACO-R: ACO with ranking; 4. ACO-TS: ACO with Tabu Search; and 5. ACO-SE: ACO with Shared Experience.

Here, in this paper of the invited talk, only the ACO-E and ACO-MA are discussed in detail while the performance of all five have been detailed out.

ACO with elitist ant approach (ACO-E)

In ACO algorithm, artificial ants are supposed to posses memory and remember their path with the help of pheromone. As opposed to real ants, artificial ants, here, lay pheromone only while returning back from the food source to the colony. Ant which brings better results, are only allowed to deposit pheromones. This means the ants bringing inferior solution are not allowed and thus not treated as "Elite Ant". Such ants that are called Elitist Ants are supposed to have a better solution than other. Other ants which traverse inferior paths and "not so elitist" are not allowed to deposit pheromones. This is the extension of the approach that has originally been proposed by Islier (2005).

The heuristic for ACO-E approach with single ant at a time, is given below.

initialize;
i = 0;
repeat
generate a feasible solution;
evaluate its goodness η;
if η > n_{max}, update η_{max} and the pheromone matrix;
evaporate pheromone;
i = i+1;
until i = σ;

Here initialization involves declaration of the heuristic parameters and reading of the incidence matrix. The parameters are:

> k: anticipated number of cells,
>
> T: threshold value for initial tours,
>
> γ: evaporation ratio for the pheromone trail,
>
> v: visibility ratio, and
>
> σ: maximum number of iterations.

As the iterations proceed, better solutions are found and the pheromone matrix is changed according to updating rule. For updating of the pheromone matrix, the following relationship is used.

$$\zeta = \zeta + d\zeta$$

> $D\zeta = \eta$ x scaling factor
>
> where ζ is the pheromone value and $d\zeta$ is the change in value of pheromone.

Scaling factor is used to weigh the efficiency currently obtained and to highlight the pheromone value to a reasonably visible level.

As iteration proceeds, a particular value of pheromone becomes very high for a particular solution in comparison to others. This particular value is associated to the best path that has been discovered by the ants and gives the best solution of the heuristic.

Advanced ant colony optimization with multiple elitist ants

In this section, Only ACO-MA has been discussed though authors have developed five algorithms. These algorithms are basically the advanced version of ACO-E. In the heuristic of Section 4.1, only an ant was used for performing the iterations and bringing the best solution of the heuristic. Now a multiple ant system is shown in which *an* number of ants move independently and simultaneously, all deciding their paths from a single pheromone matrix generated in the previous iteration()Agrawal *et al.*, 2010). The ants which bring better results are considered. Their solution are compared to the best solution obtained in the previous iterations. If the solution is better, they are allowed to update the pheromone matrix. In the last, η_{max} is updated to the new highest value of efficiency. The heuristic is as follows.

initialize;

i = 0;
repeat
j = 0;
repeat
generate a feasible solution using pheromone matrix 1;
evaluate its goodness;
if(η > η_{max}) then update the pheromone matrix 2;
j = j+1;
until j = an;
update η_{max} as the best efficiency achieved so far;
copy pheromone matrix 2 in matrix 1;
evaporate pheromone matrix 1;
i = i+1;
until i = σ/an;

From the above elaboration, collective effort and team work is envisaged in the heuristic. Multiple ants are sharing their experiences to each other and are understanding to settle for only the positive experiences. In essence, multiple ant approach avoids biasness and looks for taking quantum improvement. This heuristic tries to avoid sub-optimal solution. In case of single ant, a small improvement results into modification of pheromone matrix. Further moves are now based on this pheromone matrix. Multiple ant approach simply tries to avoid it as all the ants, in a particular iteration, shall generate moves based on the pheromone matrix of the last iteration (Bhardwaj, 2008).

V PERFORMANCE EVALUATION OF PROPOSED ACO BASED HEURISTICS

This section is showing the comparative performance of heuristics described in the preceding section using some test problems. For this purpose, total of 7 problems have been used for analyzing the comparative performance of all five approaches. in the reported literature, the number of parts or machines in the grouping problems generally varies from 5 to 33. Therefore, the random problems have been envisaged to have the number of parts or machines in this range only. The packing of 1's in the generated PMIM was taken as average of 50% (Agrawal *et al*, 2010). Because of randomness of the process of generation of these problems, it is less than 50% for some ones; while for the others, it is more than that. This was done, with a purpose, to avoid problems representing extreme cases of grouping — most favorable or highly unfavorable (Bhardwaj, 2008).

For a better judgment about their comparative performance, the efficiency values (Chandrasekhar and

Rajgopalan, 1986) at different iterations of these heuristics are noted along with. The performance is being measured in terms of number of iterations (number of total tours made by all the ants put together) required to attain the best grouping efficiency (Bhardwaj, 2008).

The result of comparative performance of heuristics is summarized in table 1. In this table, maximum efficiency achieved with number of moves or iterations, at which this maximum efficiency was obtained, is reported along.

Table 1. Comparative performance of ACO heuristics (Bhardwaj, 2008).

	ACO-E		ACO-MA		ACO-SE		ACO-R		ACO-TS	
	η_{max}	no. of moves	η_{max}	no. of moves	η_{max}	no. of moves	η_{max}	no. of moves	η_{max}	no. of moves
Problem 1	0.700	3500	0.750	1050	0.750	300	0.760	2500	0.750	310
Problem 2	0.603	1700	0.644	2000	0.644	2500	0.650	1300	0.647	3250
Problem 3	0.660	2500	0.700	2500	0.733	3900	0.720	300	0.713	1200
Problem 4	0.650	3000	0.670	1000	0.650	750	0.673	1200	0.703	1500
Problem 5	0.620	5230	0.623	1600	0.620	2750	0.643	1400	0.680	2000
Problem 6	0.611	2300	0.651	3100	0.642	3900	0.655	3550	0.705	1330
Problem 7	0.623	3550	0.641	280	0.652	630	0.660	3600	0.683	6500
Average	0.638	3111	0.668	1647	0.670	2104	0.680	1979	0.697	2299

From table 1, it can be observed that ACO-MA produces better results as compared to that with ACO-E, both in terms of grouping efficiency and number of iterations. ACO-MA has further shown to generally converge fast. ACO-MA, sometimes, does not yield the best efficiency in comparison to some of the other proposed variations of the ACO approach. On an average, improved results have been found with ACO-SE over ACO-MA. But, ACO-SE requires more number of iterations. The point to be noted is that ACO-SE has resulted comparatively better just for Problems 3 and 7, but inferior solutions for Problems 4, 5 and 6. ACO-R is, on an average and also for most of the problems, better to both ACO-MA and ACO-SE in terms of grouping efficiency and convergence speed both. ACO-R has reported the best grouping efficiency for Problems 1 and 2; whereas ACO-SE for Problems 3. ACO-TS has brought the best results in four cases, namely for Problems 4 to 7. However, this heuristic generally requires more moves as compared to all the other heuristics. Nevertheless, ACO-TS heuristic should be preferred because of its characteristic of yielding better results on an average. At a later stage, the same experimentation with 20 problems has strengthened the belief that ACO-TS the best. Because this attribute of ACO-TS, the performance of this heuristic has been tested against a method already reported in the related literature in the next section,

VI ACO-TS HEURISTIC VERSUS THE ASSIGNMENT METHOD (AGRAWAL *ET AL, 2010* AND BHARDWAJ, 2008)

In the present section, performance of ACO-TS heuristic is compared with the ones obtained from the assignment model (SAM) due to Srinivasan *et al.* (1990). SAM is a powerful heuristic and has been reported by them to provide close solutions to that obtained from Kusiak's (1987) p-median formulation which is an optimization framework. For this purpose, 10 problems are taken from the literature. Table 2 shows the solutions from the two approaches. In this table, G stands for number of groups formed, E for the number of exceptional elements, V for number of voids, and S represents the sum of E and V. Sum is the total number of undesired elements.

It can be noted from table 2 that first 5 problems yields the same solutions from SAM and ACO-TS. For the sixth problem (Chandrasekharan and Rajagopalan 1989), ACO-TS has higher value of S, but E value. It is so because the efficiency measure by Kumar and Chandrasekhar (1990) has more concern for exceptional elements. In the seventh problem, S is the same but E is less again due to the very characteristic of the efficiency measure used. In the last 3 problems, ACO-TS is found to perform better. An important characteristic of ACO-TS can be observed from the last problem (Kusiak and Chow 1987) where more groups have been formed. Here to avoid excessive voids, exceptional elements have been tolerated by ACO-TS.

VII. CONCLUSION

In the present talk, two of five ACO based heuristics have been presented for simple grouping problems. Five algorithm are as (i) Single Elitist Ant Approach(ACO-E), (ii) Multiple Elitist Ant (ACO-MA) Approach, (iii) Multiple Ant Colony Optimization with Ranking (ACO-R), (iv) Ant Colony Optimization with Tabu Search (ACO-TS) and (v) Ant Colony Optimization with Shared Experience (ACO-SE). Comparative performance of all five algorithms has also been presented.

Basic single elitist ant approach is very much similar to the approach due to Islier (2005) except the elitist ant concept. In the approach of ACO-E, only an ant makes a move to search the food (i.e. a solution) at a time,. In realty ant system, ants move in groups simultaneously and independently. Based on this fact, ACO-MA was developed and it was found to be faster in convergence in comparison to the first one. In ACO-MA, every elitist ant was considered to be alike in transforming pheromone matrix. Rest three are not presented in this paper of the talk.

Table 2. Comparison of performance of ACO-TS against SAM.

Problem Reference	Matrix Size	SAM				ACO-TS			
		G	*E*	*V*	*S*	*G*	*E*	*V*	*S*
Pannerselvam and Balasubramanian, 1985	10 X 5	2	0	6	6	2	0	6	6
Seifoddini and Wolfe, 1986	8 X 12	4	10	1	11	4	10	1	11
McAuley, 1972	12X 10	3	3	7	10	3	3	7	10
Chandrasekharan and Rajagopalan,1986	8 X 20	3	9	0	9	3	9	0	9
Chandrasekharan and Rajagopalan,1989	24X 40	7	0	0	0	7	0	0	0
Chandrasekharan and Rajagopalan,1989	24X 40	7	10	9	19	7	9	11	20
Askin and Subramanian, 1987	14X 24	5	9	15	24	5	6	18	24
King and Nakornchai, 1982	5 X 7	2	2	4	6	2	2	3	5
Siefoddini, 1989	11X 22	3	15	31	46	3	10	15	25
Kusiak and Chow, 1987	7 X 11	2	3	18	21	3	6	6	12

Comparative study of all the five algorithms shows that ACO-Ts is best among all. Though, ACO-MA, ACO-R ACO-S are better than ACO-E.

The performance of ACO-TS was also compared with the assignment method of Srinivasan *et al.* (1990). Here also ACO-TS was found to result better grouping solutions in general

REFERENCES

Agrawal A., K., Bhardwaj P., & Srivastava V , 2010, Ant colony optimization for group technology applications, Int J Adv Manuf Technol, DOI 10.1007/s00170-010-3097-1

Agrawal A., K., Bhardwaj P., & Srivastava V , 2011, On some measures for grouping efficiency, Int J Adv Manuf Technol, DOI 10.1007/s00170-011-3201-1

Askin, R.G., and Subramanian, S.P., 1987, A cost based heuristic for group technology configuration, *International Journal of Production Research*, **25**(1), 101-113.

Bhardwaj, P., 2008, Group Technology Application for Flexible Manufacturing Systems: Some Models and Methodologies, Unpublished Ph.D. Thesis, Department of Mechanical Engineering, Institute of Technology, Banaras Hindu University, Varanasi, INDIA.

Chandrasekharan MP, Rajagopalan R (1986) An ideal seed nonhierarchical clustering algorithm for cellular manufacturing. Int J Prod Res 24(2):451–464

Chandrasekharan, M.P., and Rajagopalan, R., 1989, Groupability: an analysis of the properties of binary data matrices for group technology, *International Journal of Production Research*, **27**, 1035-1052.

Chen, C.L., Cotruvo, N.A., and Baek, W., 1995, A simulated annealing solution to cell formation problem, *International Journal of Production Research*, **33**, 2601-2614.

Islier AA (2005) Group technology by an ant system algorithm. Int J Prod Res 43(5):913–932

Kao, Y., and Li, Y. L., 2008, Ant colony recognition systems for part clustering problems, *International Journal of Production Research*, **46**(15),4237 — 4258

King, J.R., and Nakornchai, V., 1982, Machine-component group formation in group technology: review and extension, *International Journal of Production Research*, **20**(2), 117-123.

Kusiak, A., 1987, Artificial intelligence and operations research in flexible manufacturing systems, *INFOR*, **25**, 2-12.

Kusiak, A., and Chow, W.S., 1987, Efficient solving of the group technology problem, *Journal of Manufacturing Systems*, **6**, 117-124.

Liu, C., and WU, J., 1993, Machine cell formation: using simulated annealing algorithm, *International Journal of Computer Integrated Manufacturing*, **6**, 335-349.

McAuley, J., 1972, Machine grouping for efficient production, Production Engineer, **51**(2), 53-57.

Pannerselvam, R, and Balasubramanian, K.N., 1985, Algorithm grouping of operation sequences, *Engineering Cost of Production Economics*. **13**(6), 567-579.

Prabhakaran G., Muruganandam A., Asokan P., and Girish BS, 2005, Machine cell formation for cellular manufacturing system using ant colony system approach, Int J Adv Manuf Technol, 25, 1013–1019

Seifoddini, H., and Wolfe, P.M., 1986, Selection of a threshold value based on the material handling cost in machine-component grouping, *I.I.E. Transactions*, **19**(3),, 266-270.

Srinivasan, G., Narendran, T.T., and Mahadevan, B., 1990, An assignment model for the part-families problem in group technology, *International Journal of Production Research*, **28**(1), 145-152.

Deformation Mechanism of Thin Target on Normal Impact of the BluntNosed Cylindrical Projectile

Abdul Aziz[1] and R. Ansari[2]
[1]Mechanical Engineering Department, I.I.M.T. Engineering College, Meerut, U.P.
[2]Mechanical Engineering Department, Aligarh Muslim University Aligarh, U.P.

Abstract- The results of experimental investigation carried out to study the response of normal impact of rigid blunt ended projectile of diameter 12.8mm and length 25.6mm on thin aluminum plate of thickness 0.81mm when impacted at different velocities within sub-ordnance velocity range. Impact of blunt projectile on the thin target caused the failure of the target by dishing followed by shearing. The residual velocity of the projectile for a particular thickness of target increases with impact velocity. Energy absorbed by the target of particular thickness is almost constant at varying impact energy.
The study of deformation mechanism and determination of ballistic performance of a target is of important in amour design for defense applications.

I. INTRODUCTION

Plate impact is a highly complex phenomena which involves the effect of strain rate, hydrodynamics, elastic, viscous and plastic wave motions, thermal strain softening (adiabatic shear), fracture initiation and propagation, fracture surface sliding, crushing, shattering, and even erosion and impact explosion at very high velocities (Backman and Goldsmith, 1978). No single analytical model has thus far been constructed that is capable of predicting all features of the event and that incorporates all the mechanisms cited and perhaps others that might be significant, for all ranges of impact velocity, types of bullet motion and angles of incidence, and for the various materials that have been employed for both projectile and target. Our study is concerned with the normal impact on thin Aluminum target by rigid blunt-nosed cylindrical projectile striking at a velocity in the vicinity of the ballistic limit. This region is of special interest because, in addition to perforation, embedment or ricochet of the striker, extensive deformations of the target are observed.

II. LITERATURE REVIEW

Target Element Phenomena

An alternative method of defining a velocity range involves the phenomenology of the target. Elastic deformations of the elements can be achieved only at extremely low striker velocities that are employed in practice for a restricted range of laboratory experiments. The limiting velocity v_{EA} of this domain in the case of normal contact of a plane-ended striker and target is that required to produce the compressive yield stress σ_{YC} in the either object, given by

$$v_{EA} = \frac{\sigma_{YC}[\rho_t c_{Dt} + \rho_p c_{Op}]}{(\rho_t c_{Dt})(\rho_p c_{Op})} \quad (1)$$

where $c_D = \sqrt{[(\lambda + 2G)/\rho]}$ and $c_O = \sqrt{(E/\rho)}$ are the dilatational and rod wave velocities, λ and $G = E/2(1+v)$ are the Lame's constants, E is Young's modulus, v is Poisson's ratio, ρ is the mass density, and subscripts p and t refer to projectile and target, respectively [3,4].

III. EXPERIMENTAL SETUP

The present experimental program was intended to study the mechanism of   plate deformation due to impact of rigid projectile on thin target, to assist the experiment and study and to obtained suitable results in the range of sub-ordnance velocities, pneumatic gun barallel diameter **12.8mm**, hardened steel projectile (Shape : **Cylindrical blunt-nosed** , Diameter = **12.8mm** , L/d ratio = **2**,   Mass = 25.08 gm, and aluminium plate targets of thickness **0.81mm**). Measurements were taken of the initial and residual velocities. The experimental setup is shown in the Fig.[1]

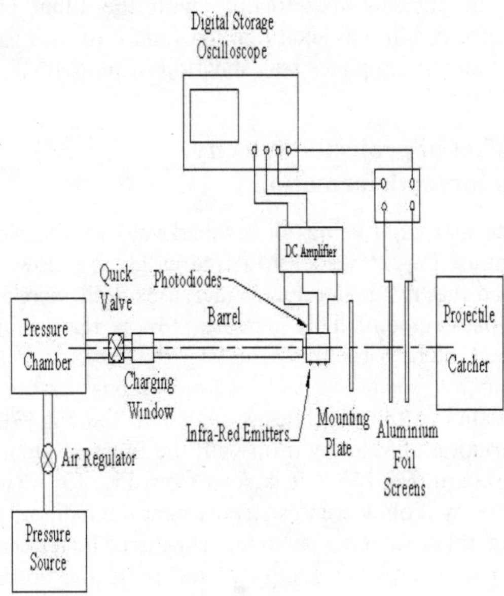

Fig. 1 Schematic diagram of experimental setup

The equipment setup essentially consists of:
- A pneumatic gun for propelling the projectile through the barrel.
- Projectile and its carrier.
- An arrangement for measuring the impact velocity of the projectile.
- An arrangement for measuring the residual velocity of projectile.
- Clamping fixture for the target plate.

I. PNEUMATIC GUN

It consists of a high pressure cylinder of 190mm internal diameter and 215mm outer diameter and 690mm in length covers with 30mm thick plates on both ends these cover plates are tied with 8 tie rods of 19mm diameter and 800mm length. This cylinder is used as a pressure regulator in such a way that compressed air is taken into cylinder up to the desired pressure from the high pressure air compressor through a check valve which is fitted in the compressor to cylinder line. A hand operated stop valve is positioned along the direction of barrel. The projectile is placed at the start of barrel through the charging window, which is fitted in between the stop valve and the barrel. Charging window is consists of

a sleeve and a cover with a rectangular ventilated slot of 70mmX19mm. The barrel is fixed down to a heavy bed by two suitable fixtures and its ends. It is also restricted the axial moment of the barrel.

II. PROJECTILE AND ITS CARRIER

To study the effect of projectile velocity on target plate, projectile of diameter **12.8mm** is used on **0.81mm** thick target plate and constant length to diameter ratio   l/d = 2, the diameter of projectile is taken according the barrel available in the market. At different velocities projectile fired from single gun having barrel of 12.8mm internal diameter.

Arrangement For Measuring Impact Velocity Of Projectile

It was difficult to measure the velocity of projectile at the moment of impact. Therefore the velocity has been measured near the exit of barrel and it is assume that the velocity at the moment of impact is same as this exit velocity.

The velocity calculated from the time taken for the projectile to interrupt three infrared lights, which maintained at a fixed distance. The three infrared light emitting diodes LED are fitted to a single block so that the infrared beams project just horizontally to the barrel. At the same time three photo transistors are fitted in the same block opposite to the infrared LED in such a manner so that the infra red beams are incident straightway on the photo transistors with maximum intensity so as to result in an efficient and accurate manner.

The time to interrupt, these light beam are recorded on a digital oscilloscope (TDS 224 Tektronicx). Suitable electronic circuit is employed which amplify and shape the photo-transistor pulses and trigger the input channels of the digital oscilloscope. The distance between the points from where the start and stop pulses are able to trigger the channels is obtained by slowly moving the projectile by hand and measuring its two positions corresponding to the starting and stopping of oscilloscope.

RESIDUAL VELOCITY MEASUREMENT

Residual velocity is required to calculate the energy absorbed during the perforation. The projectile after perforation of the target may not travel along the same line as before the impact. Normally it deviates a little from the path of its travel due to one or other reason. The method used to measure the impact velocity was thus not suitable for measuring the residual velocity of the projectile. For measuring the residual velocity, two sets of aluminum foils were placed at a fixed distance and time is taken for the projectile to pass through it, each set is comprises of two aluminum foil of 8micron thickness, which were pasted on either side of a 3.3mm thick mica sheet having an opening of 100mm diameter in the center that work as an open switch. The set of foil is then connected to the oscilloscope along with 5volt DC supply these set of foil act as switch and give signal in form of pulse when the foil were perforated by the projectile. The distance between the two set of foil is divided by the time taken to pass the projectile through it, gives the residual velocity.

TARGET PLATE CLAMPING FIXTURE

A rigid fixture is used to clamp the target plate in such a manner that it can rotate in different angles with the barrel without displacing the center of the target plate. The fixture consists of a 23mm thick 300mmX330mm plate with eight tapped hole on 230mm pcd and a 100mm through hole at the center. This plate is fixed at the horizontal axis and held at all four corners with suitable sliding mechanism to fix at different angles without displacement of the centre of the target plate.

IV. EXPERIMENTAL METHOD

Experiment on aluminium plate of thickness 0.81 mm, was carried out to study the response of the plates subjected to impact of the projectile. Circular plates of 255mm diameter were cut out of the aluminium sheets. Eight holes of 12.7 mm (1/2") diameter were made at the circumference of each plate. The plates along with mild steel ring of 8mm thickness, as shown in figure, were bolted on the specimen holding ring. It was observed after the test that there was no deformation in any of these holes in the plate. The edge of the plate was thus considered as fixed. After fixing the plate along with mild steel ring, the span of the plate becomes 205mm in diameter.Two aluminium foil screens were placed behind the target at a fixed distance from each other.

This distance was kept 118mm for normal (0° obliquity). Spacers of 50mm were placed between the screens to keep these distances fixesd. Each screen was made of two aluminum foils fixed in front and back of a mica sheet of thickness of 3.3mm to cover a circular hole of diameter 100mm cut in it. Two of these screens were placed behind the target for measuring the residual velocity in each experiment.

In the experiment, 0.81 mm plates were impacted by 12.8 mm diameter blunt ended projectile of 25.6 mm length at different velocities up to 106 m/s. The mass of the projectile was 25.08 gm. Results of the plate response and the measured residual velocities are presented and effect of varying impact velocity on these results is discussed. On the average 7 runs were made for each plate thickness. Impact and residual velocities were measured in each run with the help of 4-channel digital oscilloscope with a voltage amplifier circuit.The projectile after the impact was collected in wooden box stuffed with cotton rag to avoid any damage to the projectile and the equipment.

Experiments on Aluminum Plates

In the present experiments with the blunt ended projectiles, it was seen that the failure mechanism of the aluminum plates was shearing of plug along with dishing.

Effect of projectile velocity on target deformation

The measured values of residual velocity are plotted against impact velocity in figures given below. It is seen that residual velocity increases with increase in impact velocity. This increase is more rapid initially (near the ballistic limit) and later the curve of residual velocity versus impact velocity tends to become parallel to the 45° line as shown in the Fig.[3].The variation of velocity drop with the impact velocity is shown in Fig [4] . It is seen from Fig. [3], that the velocity drop decreases steeply near the ballistic limit and then on it tends to be constant.The effects of impact energy on energy of the projectile absorbed by the plates during perforation are shown in Fig. [4]. The absorbed energy is almost constant at different impact energy/velocities of the projectile for a particular plate thickness, in the velocity range employed. The absorbed energy increases with increase of plate thickness. The dishing and bulging are also sensitive to the velocity of impact. For same plate thickness the dishing varies with impact velocity; it reduces with increase in velocity of impact above ballistic limit.

Mechanism of Deformationin Aluminum Plates

It is observed during experiment that in perforation of the aluminum plates by blunt ended projectile, deformation is by the yielding of the plate around the tip of the projectile and dishing in rest of the plate, followed by the plug shearing, as shown in Fig.[2(a), 2(b)]. The plug forms same as the diameter of projectile (d_p) in case of normal impact. Initially the projectile pushes the plate outwards, and this causes the dishing. With further movement of the projectile the plug is formed and move with the projectile velocity

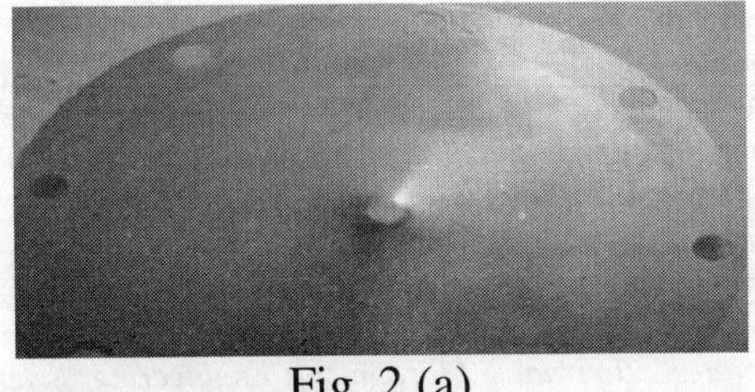

Fig. 2 (a)

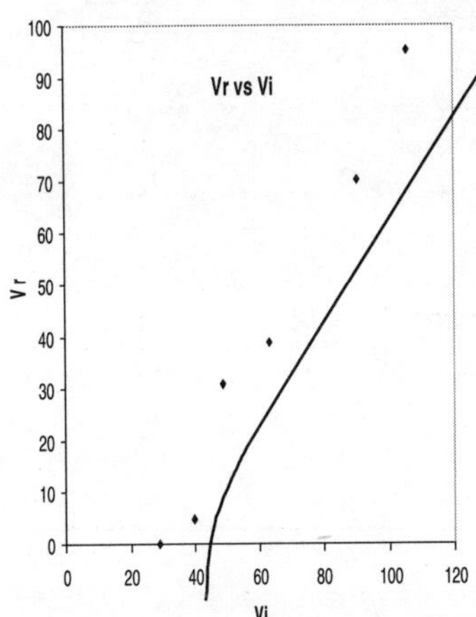

Fig. 2(b)

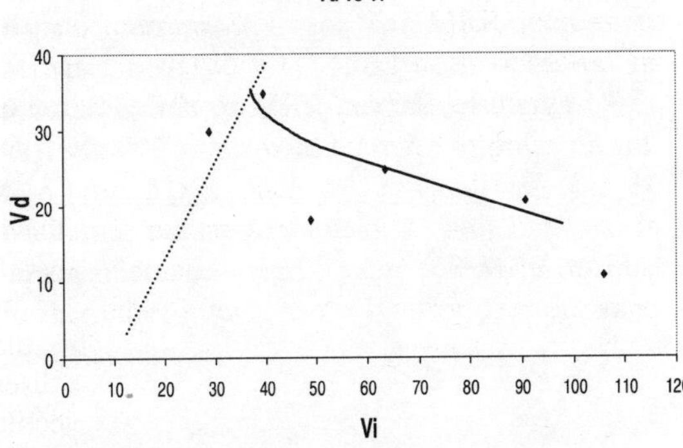

Fig 3. Variation of residual velocity(Vr)

with respect to input velocity(Vi)

Fig 4. Trend of velocity drop (Vd) with respect to input velocity(Vi)

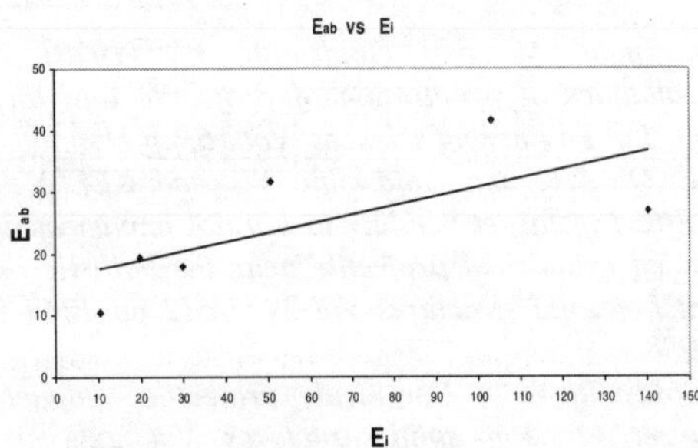

Fig 5. Variation of absorbed Energy (E_{ab}) with respect to input energy (E_i)

V. CONCLUSION

Experimental and analytical study has been conducted to study the response of normal impact of rigid blunt ended projectile of diameter 12.8mm and length 25.6mm on thin aluminum plate of thickness 0.81mm when impacted at different velocities within sub-ordnance velocity range. It was observed that aluminum plates when impacted with the blunt projectile, fail by dishing followed by shearing by plug equal to the diameter of the projectile.

The residual velocity of the projectile for a particular plate thickness increases with the impact velocity. This increase is rapid near the ballistic limit and then the curve tends to be linear. With increases in impact velocity the velocity drop decreases steeply near the ballistic limit and then it becomes constant. The absorbed energy for the particular plate thickness is almost constant at varying impact energy.

An analytical model is proposed which considers two types of deformation of the plate independently, i.e. the contact region of the plate and the projectile and in the rest of the plate. Deformation in the contact region is modeled as the formation of plug by shearing whose diameter is equal to the diameter of projectile in the normal impact. The deformation in the rest of the plate is considered as radial stretching (dishing) and modeled with the help of plate profile. The model predicts the ballistic limit of the plate and residual velocity of projectile for a given impact velocity which is an agreement with the experimental results.

VI. REFERENCES

[1] *Backman, M. and Goldsmith, W. (1978). "The mechanics of penetration of projectile into target." Int. J.of Enginering sciences Vol.16, pp.1-94.*

[2] *JENQ S.T. and Goldsmith, W and KELLY J.M. "Effect of target bending in normal impact of a flat-ended cylindrical projectile near the ballistic limit." Int.J.of solid structures Vol 24, No12 pp 1243-1266, 1988.*

[3] *Goldsmith,W." Non-ideal projectile impact on target" Int.J.of solid structures Vol 22, pp .95-395.1989.*

[4] *Wasifullah Khan "Oblique impact of projectile on thin plates" M.Tech. desertation , A.M.U. Aligarh 2002.*

[5] *W.U. Khan, Ansari. R, and Gupta. N.K"Oblique impact of projectile on thin Aluminium plates" .Defence Science Journal , Vol.53, April 2003,pp. 139-146.*

[6] *M Raguraman, ADeb, G Jagadeesh" A numerical study of blique impact of projectile on thin Aluminium plates" Journal Mechanical Engineering Science, Vol.233,C 2009*

Aerodynamic Analysis of Combat Search and Rescue (CSAR) Basket using CFD

**Gunjan Kumari, AK Kashya p, Ashutosh Kumar,
Rajeev Jain, RK Sharma , A K Saxena**
Scientist, Aerial Delivery Research and Development Establishment (ADRDE), Agra, U.P.

Abstract - The objective of the present work is to demonstrate the aerodynamic behavior of Combat Search and Rescue Basket that is used as a helicopter under slung store carrier using CFD technique. Accurate prediction of aerodynamic forces and moments is necessary for designing most efficient and light weight basket with required payload capacity. CFD simulations have been carried out at different payload conditions and different operation conditions. The flow patterns over this were studied using FLUENTCFD Package. Longitudinal static derivatives and aerodynamic forces are estimated. Stability criteria have also been studied.

NOMENCLATURE

C_D	=	**Coefficient of drag**
C_L	=	**Coefficient of lift**
C_l	=	**Coefficient of rolling moment**
C_m	=	**Coefficient of pitching moment**
C_n	=	**Coefficient of yawing moment**
C_X	=	**Force coefficient in X direction**
C_Y	=	**Force coefficient in Y direction**
C_Z	=	**Force coefficient in Z direction**
α	=	**Angle of attack**
F_X	=	**Force in X direction**
F_Y	=	**Force in Y direction**
F_Z	=	**Force in Z direction**
L	=	**Length of basket**
A	=	**Reference area**
EB	=	**CSAR Basket without payload**
BFL	=	**CSAR Basket fully loaded**
BM	=	**CSAR Basket with men on stretchers**

I.   INTRODUCTION

Combat Search and Rescue (CSAR) Basket is designed to be used where transport of food, equipments, and armament stores can't be delivered by any other means e.g. in confined spaces,on slopes and in wooded terrain etc. Typically, it is shaped to accommodate an adult in a face up position and it is used in search and rescue operations. After the supply of equipments or providing food, the basket may be wheeled, carried by hand, mounted on an ATV,   lifted or lowered on high angle ropes, or hoisted by helicopter. It is mainly used by armed forces. It plays a vital role during war time. Fig. 1 shows the front view of CSAR Basket. All safety measures are considered to design the basket.

II.   GEOMETRICAL DESCRIPTION & 3D MODELING OF CSAR BASKET

CSAR Basket comprises of a rectangular frame structure having net to provide full cover inside the basket. It is mainly divided into three parts as shown in Fig. 1:-

- Suspension assembly
- Cradle
- Basket body

It is a framework of tubes made of light weight material preferably Al alloy and welded together to form a cage structure. It can carry desired payload. The advantage of using pipe is to reduce weight of overall system and easy manufacturing. The tube elements have circular cross sections. The bottom portion of the basket is arrayed with small size tubes which act as a base. Cradle portion is designed to connect with suspension assembly to lift the system along with stores or accessories. There are mainly three operational conditions for basket e.g. CSAR

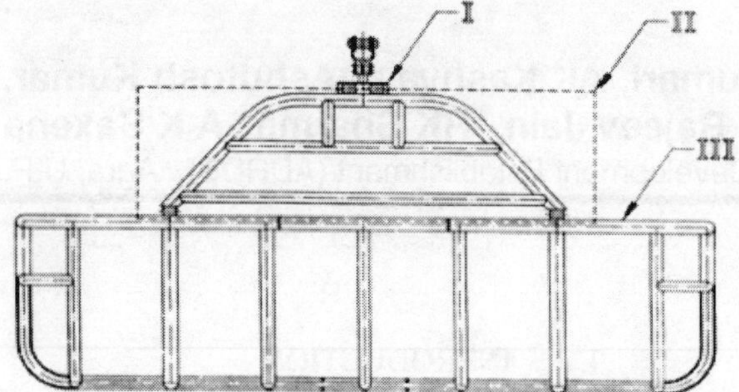

Fig. 1 Front assembly view of CSAR Basket.
I=Suspension assembly, II=Cradle, III=Basket body

Basket without payload, CSAR Basket fully loaded and CSAR Basket with men on stretchers. Fig. 2, 3 and 4 show the 3D model of CSAR basket with respective cases. CATIA software is used for 3D modeling.

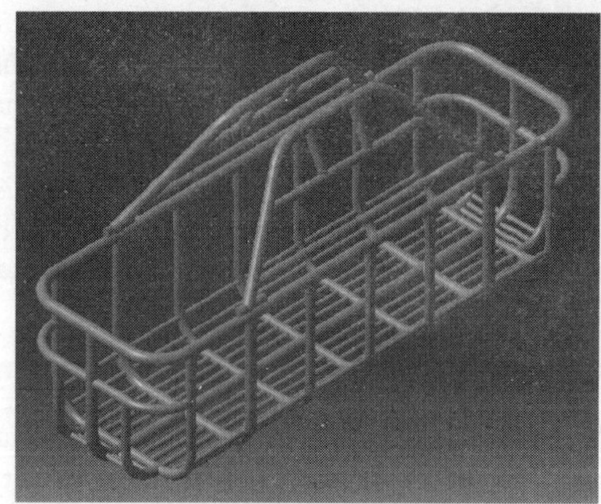

Fig. 2 3D modeling of CSAR basket without payload

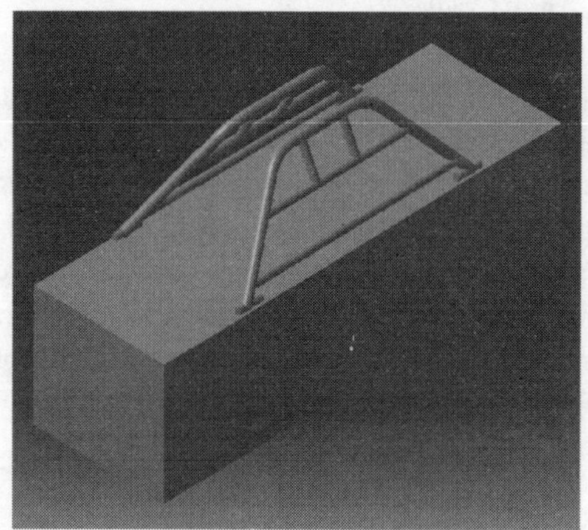

Fig. 3 3D modeling of CSAR basket fully loaded

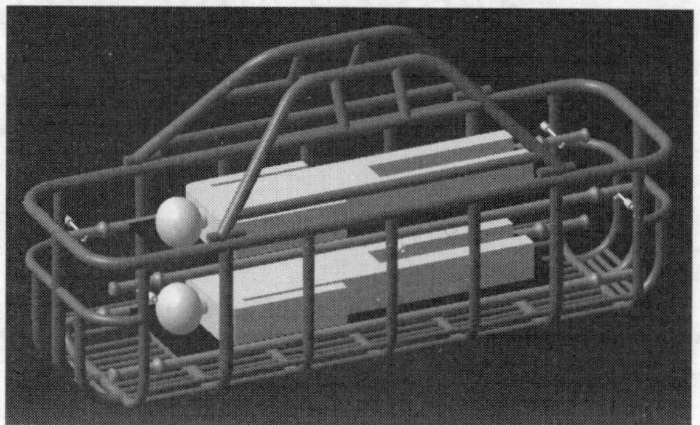

Fig. 4 3D modeling of CSAR basket with men on stretchers

III.   GOVERNING EQUATION AND NUMERICAL APPROACH

Flow solver Fluent is used for the CFD analysis, which solves flow equation in implicit manner. In implicit option of the density-based solver, each equation in the coupled set of governing equations is linearized implicitly with respect to all dependent variables in the set. This will result in a system of linear equations with N equations for each cell in the domain, where N is the number of coupled equations in the set. Because there are N equations per cell, this is sometimes called a "block" system of equations. For turbulent flows, Reynolds Averaged Navier-Stokes (RANS) equations are used. They are derived from the Navier - Stokes equations by introducing a time averaging procedure. The laws of motion are then expressed for the mean variable. Here standard k-ϵ model is used. The standard k-ϵ model [1] is a semi-empirical model based on model transport equations for the turbulence kinetic energy (k) and its dissipation rate (ϵ).

A.   *Transport Equations for the Standard* k-ϵ *model [1]*

The turbulence kinetic energy, k, and its rate of dissipation, ϵ, are obtained from the following transport equations:

$$\frac{\partial}{\partial t}(\rho k) + \frac{\partial}{\partial x_i}(\rho k u_i) = \frac{\partial}{\partial x_j}\left[\left(\mu + \frac{\mu_t}{\sigma_k}\right)\frac{\partial k}{\partial x_j}\right] + G_k + G_b - \rho\epsilon - Y_M + S_k$$

$$\frac{\partial}{\partial t}(\rho\epsilon)+\frac{\partial}{\partial x_i}(\rho\epsilon u_i)=\frac{\partial}{\partial x_j}\left[\left(\mu+\frac{\mu_t}{\sigma_\epsilon}\right)\frac{\partial\epsilon}{\partial x_j}\right]+C_{1\epsilon}\frac{\epsilon}{k}(G_k+C_{3\epsilon}G_b)-C_{2\epsilon}\rho\frac{\epsilon^2}{k}+S_\epsilon$$

Where,

- G_k represents the generation of turbulence kinetic energy due to the mean velocity gradients.
- G_b is the generation of turbulence kinetic energy due to buoyancy.
- Y_M represents the contribution of the fluctuating dilatation in compressible turbulence to the overall dissipation rate.
- $C_{1\epsilon}$, $C_{2\epsilon}$, and $C_{3\epsilon}$ are constants. σ_k and σ_ϵ are the turbulent Prandtl numbers for k and ϵ, respectively.
- S_k and S_ϵ are user-defined source terms.

IV.   CFD ANALYSIS

For CFD simulation, CAD geometries of respective cases are modified. The small surfaces, curves and holes have been neglected for the analysis purpose as their contributions will be very small and create problem during mesh generation. As the CSAR basket is symmetric about XY plane, only half model is considered. A rectangle computational domain with width 20 times of width of CSAR basket and height 20 times the height of CSAR basket is considered. The upstream and downstream domains are extended by a value equal to 20 and 30 times the length of CSAR basket. This half basket model is exported as IGES file format for meshing. Gambit software is used as a preprocessor for geometry cleanup and mesh generation. An unstructured 3D tetrahedral mesh is generated. Fig. 5 shows the grid of half symmetric model of CSAR basket. Mesh is clustered near the basket wall to capture the physics of flow near the wall of geometry. Boundary conditions are defined for CFD simulation. All boundaries that represent the internal and external walls of the basket are defined as 'Wall' boundary type. The 'Pressure far field' condition is applied to the outer boundary domain. Standard atmospheric pressure 0.1013 MPa and temperature 303.2 K is used. Symmetry plan (XY) is defined as 'Symmetry' boundary condition. The half symmetric model of CSAR basket is imported in 'FLUENT' solver. The grid check was performed and was found satisfactory. A density based, implicit, symmetric, steady and absolute velocity formulation solver is used.

Analysis has been carried out for velocities of 30, 40, 50 and 60 m/s at various angles of attack (0° to 40°) with increment of 10°. The convergence criteria of 1E-05 are maintained with second order accuracy. The aerodynamic forces F_X, F_Y and F_Z are estimated for all flight conditions. The momentum Cl, C_m, and C_n are calculated about CG of CSAR basket. The reference area A and length L has been taken as basket front cross section area and length of basket respectively.

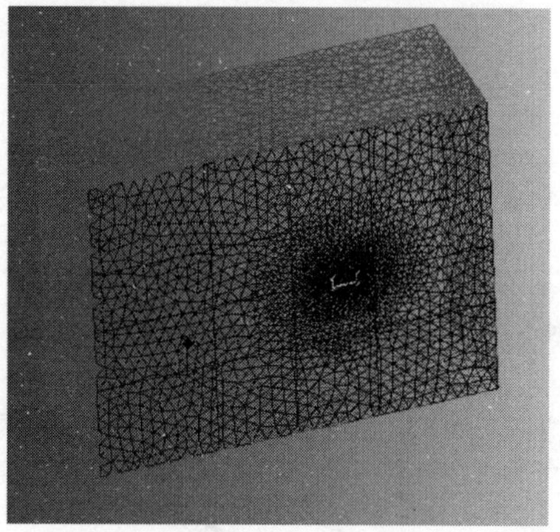

Fig. 5 Grid of half symmetric model of basket

V.   RESULTS AND DISCUSSION

Fig. 6 shows the comparison of force in X direction for all three cases. Force F_X reaches maximum of its value when angle of attack reaches 20° for different velocities. But, F_Y and F_Z increase with increase in velocity and angle of attack.

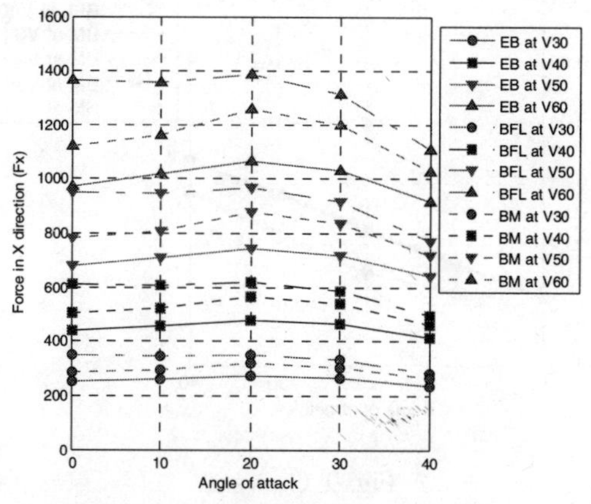

Fig. 6 Fx vs α

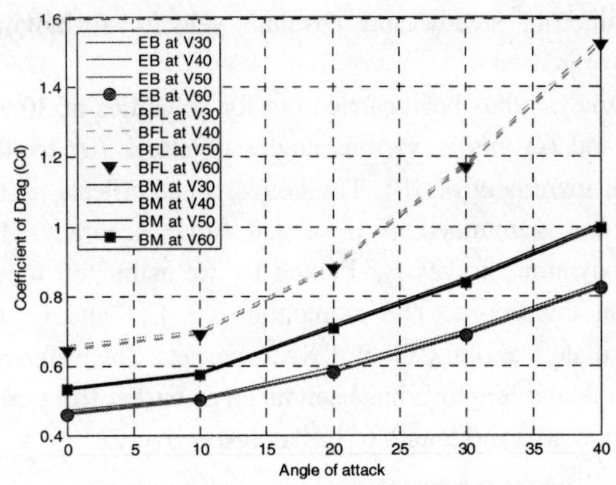

Fig. 7 C_D vs α

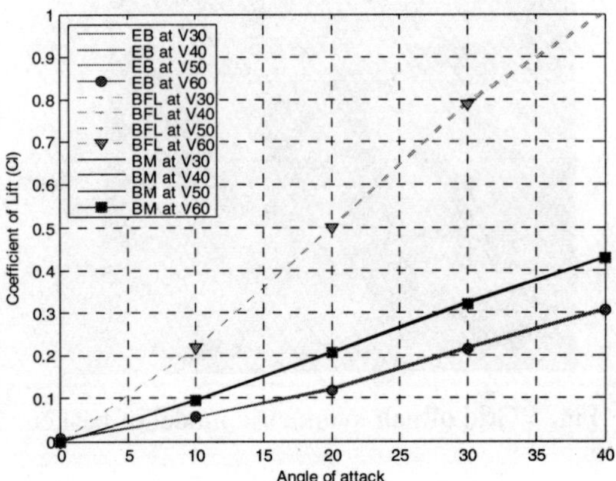

Fig. 8 C_L vs α

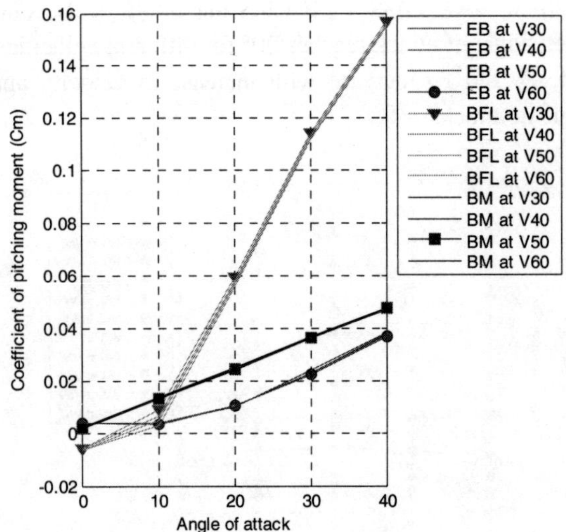

Fig. 9 C_m vs α

Fig. 7 and 8 follow the standard trend of C_D vs α and C_L vs α. With increase in angle of attack, C_L increases. But there is small change in C_L with increase in velocity for each case. As the CSAR basket is a framework of pipes, flow separates at each joint of pipes and doesn't contribute to lift. Due to flow separation at each joint of pipes, there is considerable amount of increase in drag with increase in angle of attack. With increase in angle of attack, a longitudinal force decreases but there is substantial increase in lateral force. That contributes to the overall increase in drag force. Fig. 9 shows the plot of C_m vs α. C_m is calculated with respect to CG of EB, BFL and BM. For CSAR basket without payload case, Pitching moment coefficient first decreases for small angle of attack upto $10°$ then increases with further increase in AOA. This implies that basket is longitudinally stable for small angle less than $10°$. It becomes longitudinally unstable for larger angle. From plot 9, it has been found that basket is longitudinally unstable for BFL and BM cases.

The computed flow fields obtained after CFD analysis for different velocities and various angles of attack have been plotted as contour plots. Fig. 10 to 15 confer the fluid (air) flow physics and behavior around CSAR basket. Fig. 10, 12 and 15 show the sectional velocity magnitude plot. These plots highlight the change of flow characteristics around basket on symmetric plane $Z = 0$. Small and big wake regions are created behind the tubes in CSAR Basket without payload case. Velocity increases at fillet portion of the basket as shown in velocity contour plots. For typical case of V = 30 m/s and AOA = $0°$, maximum velocity around basket is 39.17 m/s.

Fig. 11, 13 and 14 show the pressure contour plot at different velocities. A sectional plane at Y= -0.650 m and $Z = 0$ position is created for CSAR basket without payload and CSAR basket fully loaded case respectively for clear visualization of flow behavior. Plots show the high pressure region near front face of basket and low pressure region at back portion and behind the tubes. Maximum pressure 0.1019 MPa for velocity 30 m/s and AOA $0°$ has been observed. Consecutive high and low pressure regions at the base of CSAR basket can be seen as base is a framework of small and big tubes. As angle of attack increases the corresponding change in flow field can be clearly seen as it is visible in following contour plots.

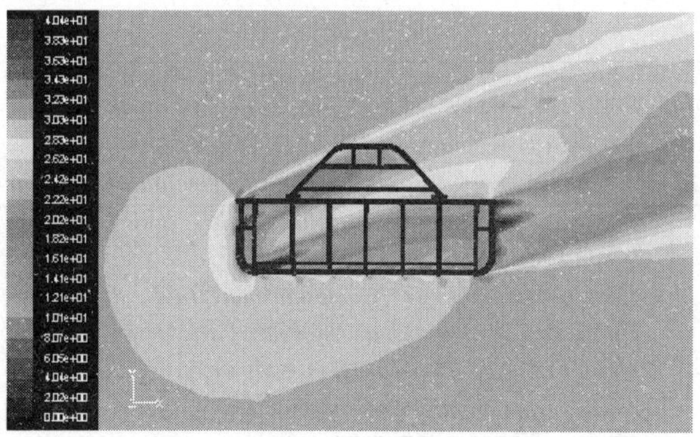

Fig. 10   Velocity contour plot of CSAR basket without payload at V = 30 m/s and A = 20º

Fig. 13   Pressure contour plot of basket fully loaded at V = 30 m/s and A = 20º

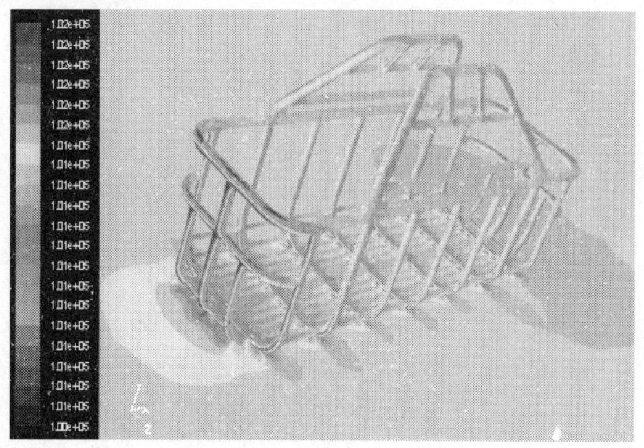

Fig. 11   Pressure contour plot of CSAR basket without payload at V = 30 m/s and A = 20º

Fig. 14   Pressure contour plot of CSAR basket with men on stretchers at V = 60 m/s and A = 30º

Fig. 12   Velocity contour plot of CSAR basket fully loaded at V = 30m/s and A = 20º

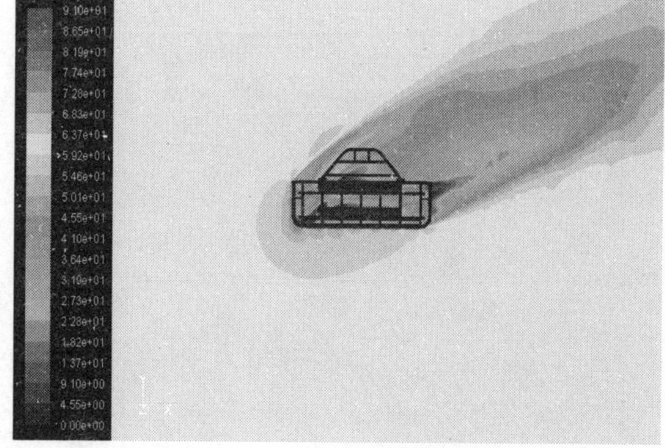

Fig. 15   Velocity contour plot of basket with men on stretchers at V = 60 m/s and A = 30º

VI.   CONCLUSIONS

A detailed CFD simulation of basket is done for clear understanding of flow pattern over the frame structure of CSAR basket. The aerodynamic data are estimated for given flight conditions. It is observed that flow separation occurs over all individual tubes that increases overall drag force. Coefficient of drag (C_D) increases with increase in angle of attack. It is also observed that drag force reaches to its maximum value at 20° and then decreases with further increase in velocity for CSAR Basket without payload, CSAR Basket fully loaded and CSAR Basket with men on stretchers cases. But F_Y and F_Z increase beyond 20° that implies increase in rolling moment with increase in AOA. Drag coefficient is lowest for EB case and highest for BFL case. Basket is longitudinally stable for small angles less than 10°.

VII.   REFERENCES

[1]   Frank M. White "Fluid Mechanics" Fourth edition, McGraw companies © 2001.

[2]   Michael v. cook "Flight Dynamics Principles (A Linear Systems Approach toAircraft Stability and Control)" Second edition, USA, Elsevier© 2007.

[3]   John D.Anderson, Jr "Computational fluid dynamics" University of Maryland , WCB/Mc Graw-Hill©1999.

[4]   Menter, F.R., "Two-equation eddy-viscosity turbulence models for engineering applications", AIAA-journal., 32(8), pp.1598-2605, 1994.

[5]   Dr.-Ing. S. F. Hoerner, "FLUID – DYNAMIC DRAG (Theoretical, Experimental and Statical information)" Washington,D.C.-August 1957.

[6]   David G. Hull " Fundamentals of airplane flight mechanics" © Springer - Verlag berlin Heidelberg 2007.

FE Modeling of Rolling Process

**K. Hans Raj, Rahul Swarup Sharma, Ankit Sahai, Atul Dayal,
Shanti Swaroop Sharma and Sanjeev Sharma**

Department of Mechanical Engineering, Faculty of Engineering,
Dayalbagh Educational Institute, Dayalbagh, Agra

Abstract: **Rolling is commonly used for making thin metallic sheets for various applications by industries. It is an important process to induce Severe Plastic Deformation (SPD) in materials. In this work, it is applied for development of Ultra-Fine Grained Materials (UFG) with improved mechanical properties. Finite Elements Method (FEM) is one of the popular methods used in industry for design and analysis of forged and rolled components. A 3D- FE model for rolling is developed in Forge-2007 environment to investigate equivalent stress, strain rate, and rolling energy for cold rolled strips of Al6061 material. The percentage reduction in thickness is varied from 10 to 50. The analysis brought out the effects of severe plastic deformation on equivalent strain which is an indicator for yield strength of material. The FE simulations thus proved to be of immense benefit in estimating the microstructure and strength of rolled specimen which are governed by equivalent strain. The current work which helps in development of Al6061 sheets with better mechanical properties is useful for aeronautical industry.**

Index Terms— Rolling Process, SPD, FEM, UFG Materials

I. INTRODUCTION

THE rolling process is one of the most popular. esses in manufacturing industries. Among all kinds of rolling processes, flat rolling is the most common process In industries as 40 % of rolling products are produced with this type of rolling. Therefore, many scientists have tried to enhance the quality and quantity of products by optimizing this process by identifying parameters affecting it. Throughout the century, rolling process has been analyzed by various analytical and numerical methods such as the slab method, the slip-line field method, the upper bound method, the boundary element method and the Finite element method.

In comparison with other methods for analyzing the rolling process, the finite element method [1 to 4] is the most widely used one

II. ROLLING PROCESS

A schematic illustration of the flat rolling process is shown in Figure1. A strip thickness enters the roll gap and is reduced to 't_r' by a pair of rotating rolls, each roll being powered through its own shaft by electric motors. The surface speed of the roll is Vr.

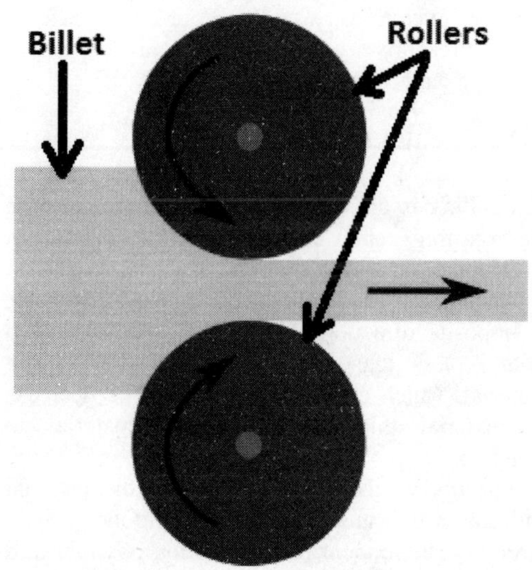

Fig-1 A simple Rolling process.

Cold rolling occurs with the metal below its recrystallization temperature (usually at room temperature), which increases the strength via strain hardening up to 20%. It also improves the surface finish and holds tighter tolerances. Common cold-rolled products include sheets, strips, bars, and rods. These products are usually smaller than the same products that are hot rolled because of the smaller size of the work pieces and their greater strength, as compared to hot rolled stock. Cold rolling cannot reduce the thickness of a work piece as much as hot rolling in a single pass. Cold-rolled sheets and strips come in various conditions: full-hard, half-hard, quarter-hard, and skin-rolled. Full-hard rolling reduces the thickness by 50%, while the others involve less of a reduction. Quarter-hard is defined by its ability to be bent back onto itself along the grain boundary without breaking. Half-hard can be bent 90°, while full-hard can only be bent 45°, with the bend radius approximately equal to the material thickness. Skin-rolling, also known as ask in-pass, involves the least amount of reduction: 0.5-1%. It is used to produce a smooth surface, a uniform thickness, and reduce the yield-point phenomenon . Skin-rolled stock is usually used in subsequent cold-working processes where good ductility is required. Other shapes can be cold-rolled if the cross-section is relatively uniform and the transverse dimension is relatively small, approximately less than 50 mm (2.0 in). This may be a cost-effective alternative to extruding or machining the profile, if the volume is in several tons. Cold rolling shapes requires a series of shaping operations, usually along the lines of sizing, breakdown, roughing, semi-roughing, semi-finishing, and finishing.

III ROLLING PRINCIPAL

Flat rolling is the most basic form of rolling with the starting and ending material having a rectangular cross-section. The material is fed in between two rollers, called working rolls that rotate in opposite directions. The gap between the two rolls is less than the thickness of the starting material, which causes it to deform. The decrease in material thickness causes the material to elongate.

The friction at the interface between the material and the rolls causes the material to be pushed through. The amount of deformation possible in a single pass is limited by the friction between the rolls; if the change in thickness is too great the rolls just slip over the material and do not draw it in.

It is well known that the Hall-Petch relationship was used to describe the dependence of flow stress on the grain size of the materials with high angle boundaries as described in Eq. 1. Two parameters have been often used to describe the relationship between flow stress and microstructure for understanding of the work hardening dislocation mechanism. These parameters are the average misorientation and the average sub (grain) size.

$$\sigma_f = \sigma_0 + k_H d^{-1/2} \qquad \text{............ [1]}$$

Where σ_f is the yield stress, σ_0 is the material constants for the starting stress for dislocation movement, K_H is the strengthening coefficient, and d is the average grain diameter.

In the 1960's, Li [6] reported that there is no effect of misorientation on flow stress based on the theoretical calculation, when the misorientation is over a certain value. Recently Hansen [7] has reported that the contribution of cell boundaries to the flow stress is dependent on the misorientation in cold-rolled materials. Finite element simulations for studying various process parameters are done by several research groups [8, 9, and 10].

IV. FE MODELING OF ROLLING PROCESS

In this work, the modeling of rolling processes is done in FORGE environment. This paper explains the effect of percentage reduction in thickness of billet on equivalent strain, strain rate, and energy for rolling processes. FORGE is capable of modeling 3-D situations of metal forming (including thermal and friction effects) with automatic mesh regeneration. The material is assumed to be homogeneous, isotropic. The dies (rollers) are assumed to be rigid. The dimension of the three dimensional work piece (billet) is 100 mm (Length) x 30mm (width) x 10mm. and the rollers 50mm in diameter and 100m in length. The material of the billet is Al6061. FE simulations are carried out for Rolling processes (Coefficient of friction, 0.4; rolling speed, 1.6 rpm), with different values of percentage reduction in thickness of billet ranging from 10 to 50. The material behavior is assumed to follow that of Norton-Hoff law as in equation 2.

$$s = 2k(T, \bar{\varepsilon}, ...)(\sqrt{3}\,\dot{\bar{\varepsilon}})^{m-1}\,\dot{\varepsilon} \,...... \,(2)$$

Where,

S = shear stress,
K = material consistency,
$\bar{\varepsilon}$ = equivalent strain. ,
$\dot{\varepsilon}$ = strain rate,
$\dot{\bar{\varepsilon}}$ = equivalent strain rate

The flow stress in case of Al6061 is directly interpolated from the material data file available in FORGE software for various temperatures, strains, and strain rates. Generalized coulomb friction law is used in the current analysis given by equation (4):

$$\tau = \mu\sigma_n \quad if \; \mu\sigma_n \langle \overline{m}\frac{\sigma_0}{\sqrt{3}}$$

and

$$\tau = \overline{m}\frac{\sigma_0}{\sqrt{3}}\frac{\Delta V}{\Delta V} if \quad \mu\sigma_n \rangle \overline{m}\frac{\sigma_0}{\sqrt{3}} \quad ......(4)$$

Where,

τ = friction stress tangential to the surface

μ = coefficient of friction

σ = compressive stress normal to the surface (Contact pressure)

$\overline{m}$ = Tresca coefficient

The variation of equivalent strain, strain rate and energy in the end product obtained for various percentage reductions in thickness in rolling processes are reported in table-1. The input parameters of rolling process are 1.6rpm, coefficient of friction is 0.4, and billet temperature is taken as 100°C for billet.

TABLE1: FE analysis results of average equivalent strain, strain rate and energy during rolling Process for various percentage reductions in thickness.

Reduction in Thickness (%)	Average Equivalent Strain	Strain Rate	Rolling Energy (KJ)
10	0.113	0.159	0.384
20	0.192	0.199	0.889
30	0.280	0.250	1.543
40	0.395	0.308	2.325
50	0.595	0.398	3.456

V. RESULTS AND DISCUSSION

A brief review of FE modeling of rolling process is attempted. Finite element model for this process is developed in FORGE environment. FE analysis results of rolling process are shown in table-1. For percentage reduction in thickness of 50%in one pass, maximum equivalent strain of 0.595 is observed. The equivalent strain contours are shown in figure-2. The process of rolling is more effective in terms of setup time, processing time and deformation. Large equivalent strain is achieved in single pass in rolling process as the percentage reduction in thickness increases. Also from the table-1, we see that as the percentage reduction in thickness increases the strain rate, and rolling energy increase.

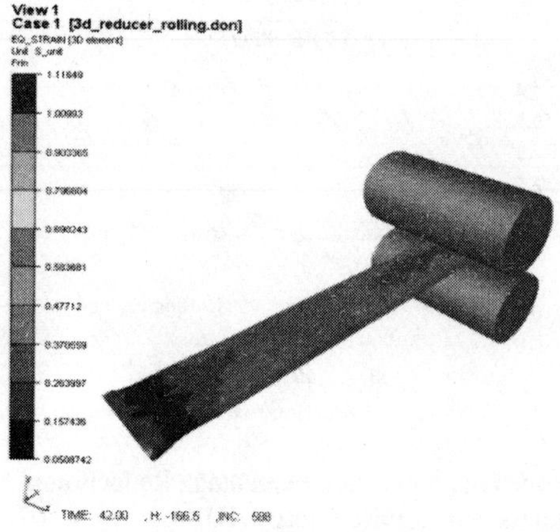

Figure. 2 Equivalent Strain Contours of rolling process for percentage reduction in thickness is 50.

The graphs are plotted for rolling process between various percentage reductions in thickness and for average equivalent strain, strain rate, and rolling energy as shown in figures 3, 4, and 5.

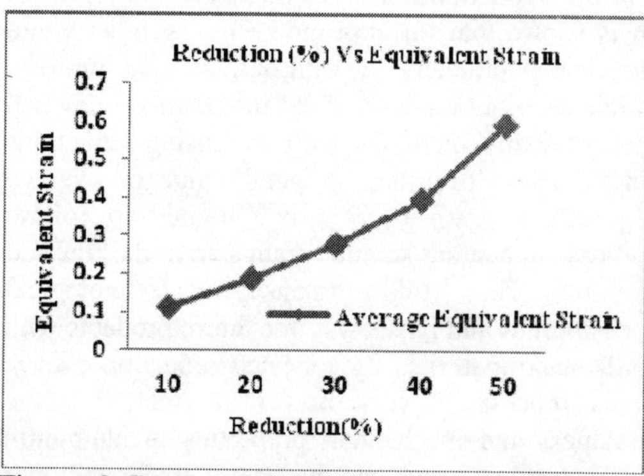

Figure. 3 Graphs between Percentage Reduction in Thickness and Average Equivalent Strain.

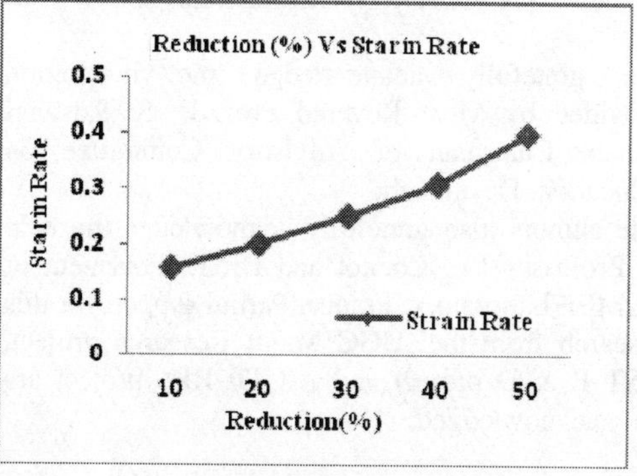

Figure. 4 Graph between Percentage Reduction in Thickness and Strain Rate

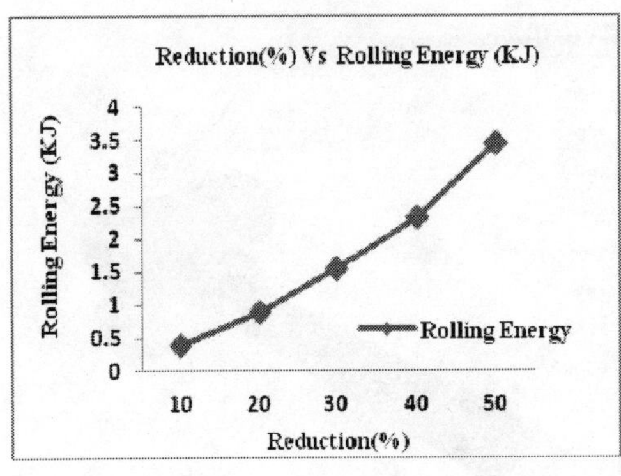

Figure-5. Graph between Percentage Reduction in Thickness and Rolling Energy (KJ)

VI. CONCLUSIONS

Rolling is a popular metal forming process that uses Severe Plastic Deformation (SPD).It is emerging as an important process for creating ultra-fine grained materials. Three dimensional FE modeling of rolling process is realized in this work. Finite element simulations of rolling process on various reductions in billet thickness are reported. It is shown that for appropriate tool geometry and process parameters, a well-defined and uniform strain distribution is obtained in this process which progressively increases with increasing reductions in thickness of billet. A good value of average equivalent strain (0.595) is obtained in rolling process indicating smaller grain size in the finished product. Thus rolling opens new technological possibilities and paves way for future products with bulk nano-materials by repeated reduction cycles. This process is very useful in improving the hardness and mechanical properties of aluminum alloy products and can be easily integrated into forging industry.

VII. ACKNOWLEDGEMENTS

We gratefully acknowledge the inspiration provided by Most Revered Prof. P. S. Satsangi Sahab, Chairman of Advisory Committee on Education, Dayalbagh.
The authors also gratefully acknowledge the help of Professor J.L. Chenot and Dr. L. Fourment of CEMEF Laboratory, France. Partial support for this research from the, UGC Major Research project, DST R & D project and AICTE RPS project are duly acknowledged.

VII. REFERENCES

[1]   Alejandro Rivera "Non-linear finite element method simulation and modeling of the cold and hot rolling processes" Thesis submitted to the Faculty of the Virginia Polytechnic Institute and State University,2007.

[2]   S. Abdelkhalek1, P. Montmitonnet, N. Legrand, P. Buessler "Manifested flatness predictions in thin strip cold rolling" International Journal of Material Forming 1, Supplement 1,PP.339-342, 2008.

[3]   P.Sherstnev, I.Flitta, C. Sommitsch, M. Hacksteiner, T. Ebner "The effect of the initial rolling temperature on the microstructure evolution during and after hot rolling of AA6082" International Journal of Material Forming 1, Supplement 1,PP. 339-342, 2008.

[4]   J. C. Lin "Prediction of Rolling Force and Deformation in Three-Dimensional Cold Rolling by Using the Finite-Element Method and a Neural Network" International Journal of Advanced Manufacturing Technology, Vol. 20, PP. 799–806, 2002

[5]   M.Ould, O.Aberkane "Micromechanical modeling of the rolling of a A1050P aluminium sheet" International Journal Mater Form, Vol.2, PP.25–36, 2009.

[6]   B.L. Li, W.Q. Cao, Q. Liu, W. Liu "Flow stress and microstructure of the cold-rolled IF-steel",Materials Science and Engineering A356,PP. 37- 42, 2003.

[7]   E.Lopez-Chipres, E.Garcia-Sanchez, E.Ortiz-Cuellar, M.A.L. Hernandez-Rodriguez, and R. Colas "Optimization of the Severe Plastic Deformation Processes for the Grain Refinement of Al6060 Alloy Using 3D FEM Analysis" Journal of Materials Engineering and Performance revised form October, Vol. 14, 2010.

[8]   A.Tajyar and K. Abrinia "FEM Simulation of Reshaping of Thick Tubes indifferent passes", International Journal of Recent Trends in Engineering, Vol. 1, No. 5, May 2009.

[9]   A.R. Shahani1,S.A. Nodamai and I.Salehinia1 "Parametric Study of Hot Rolling Process by the Finite Element Method" Transaction B: Mechanical Engineering Vol. 16, No. 2, pp. 130-139, Sharif University of Technology, April 2009.

[10] A.A.Akbari Mousavi, S. M. Ebrahimi "Optimum rolling parameters of asymmetric cold rolling", VIII International Conference on Computational Plasticity Complas viii o Cimne, Barcelona, 2005.

Three Variables Fuzzy Time Series Model for Predicting Energy Demand in India

[1]**Akash Tomar**, [1]**S. K. Gaur**, [1]**Man Mohan**, [2]**Sunil Pasbola**

[1]Dayalbagh Educational Institute, Dayalbagh, Agra
[2]R.B.S. College,Bichpuri,Agra

Abstract-**Forecasting plays an important role in planning. During last two decades various approaches have been developed for time series forecasting. The present paper focuses on three variable fuzzy time series model to forecasts energy demand in Indian context. The interested variable is main variable that is energy demand (in quadrillion btu) and remaining two are secondary variable. These two secondary variables are Gross Domestic Product (in million dollars) and Population (in million).History of past 24 years is used for making new forecasts. The model appears to have better accuracy as compared to simple linear regression models. The model can be used in situations where the historical data is subjective or fuzzy.**

I INTODUCTION

Predictions of future events and conditions are called forecasts, and the act of making such predictions is called forecasting[1],[2],[3]. Demand predictions are basically divided into three categories:[3]Long Term Predictions- Long Term Predictions are longer than a year. Medium Term Predictions- Medium term predictions are from week to a year.Short Term Predictions- Short term predictions are from one hour to one week. In our daily life, people often use forecasting techniques to model and predict economy, population growth, fuzzy time series for TAIFEX and daily temperature in Taipei, Taiwan.

In this paper, we present a new modified method to predict Electrical load demand based on the m-factors high-order fuzzy time series. The proposed
method constructs m-factor high-order fuzzy logical relationships based onthe historical data to increase the forecasting accuracy rate.

stocks, insurance/re-insurance, portfolio analysis and etc. However, in the real world, and event can be effected by many factors[16]. Therefore, if we consider more factors for predictions, with higher complexities then we can better forecasting results.

Most forecasting methods use statistical techniques or artificial intelligence algorithms such as regression, neural networks, fuzzy logic, and expert systems[5],[6],[13]. Two of the methods, so-called end-use and econometric approach are broadly used for medium- and long-term forecasting. A variety of methods, which include the so-called similar day approach, various regression models, time series, neural networks, statistical learning algorithms, fuzzy logic, and expert systems, have been developed for short-term forecasting.[6],[8],[13]

In recent years, many researchers used fuzzy time series to handle prediction problems. Song and Chissom[4] presented the concept of fuzzy time series based on the concepts of fuzzy set theory to forecast the historical enrollments of the University of Alabama. Huarng [9] presented the definition of two kinds of intervals in the universe of discourse to forecast the TAIFEX (Taiwan future exchange). Chen [10] presented a method for forecasting based on high-order fuzzy time series. Lee [15] presented a method for temperature prediction based on two-factor high order fuzzy time series. Melike [12] proposed forecasting method using first order fuzzy time series for forecasting enrollments in University of Alabama. Lee [15] Presented handling of forecasting problems using two-factor high order

II FUZZY TIME SERIES

Time series analysis plays vital role in most of the actuarial related problems. As most of the actuarial issues are born with uncertainty, therefore, each observation of a fuzzy time series is assumed to be a fuzzy variable along with associated membership function. Based on fuzzy relation, and fuzzy inference rules, efficient modeling and forecasting of fuzzy time series is possible, see [14] and [11]. This field of fuzzy time series analysis is not very mature concept for many antecedents and single consequent.For example, in designing two-factor kth order fuzzy time series model with X be the primary and Y be second fact. We assume that there are k antecedent.

$((X_1,Y_1)(X_2,Y_2),.........,(X_k,Y_k))$ and one consequent X_{k+1}.

If $(X_1=x_1,Y_1=y_1),(X_2=x_2,Y_2=y_2)$,
................,$(X_k=x_k,Y_k=y_k)->(X_{k+1}=x_{k+1})$

In the similar way, we can define m-factor i=1,2,...,m and kthorder fuzzy time series as

If $(X_{11}=x_{11},X_{12}=x_{12},.......,X_{1k}=x_{1k})$,
    $(X_{21}=x_{21},X_{22}=x_{22},.......,X_{2K}=x_{2k}),.......$,
    $(X_{m1}=x_{m1},X_{m2}=x_{m2},........,X_{mk}=x_{mk})$
then$(X_{m+1,k+1}=x_{m+1.k+1})$ for i=1,2,3,....., m , j=1,2,3,.......,k

III m-VARIABLE FUZZY TIME SERIES MODEL FOR PREDICTING THE LONG TERM ENERGY DEMAND

m-variable fuzzy time series model proposed by Chain[7] has been used to predict energy of demand of India[17].It has been observed that energy demand is dependent on G.D.P[18] and population[19] .Hence Energy demand has been considered as the main predictable variable while G.D.P. and population as the two secondary variables.A window size of three years is taken to predict the demand of fourth year.The energy demand in 10^{15} Btu(1Btu=0.0002928 kilowatt-hrs). G.D.P. in million dollors equivalent and population in million is given in Table 1.

Table 1. Indian Energy Demand
Vis-à-Vis G.D.P. and population

Year	energy Demand in 10^{15} Btu (X)	G.D.P in million dollars equivalent (Y1)	Population in million (Y2)
1980	04.041	181765.0	679.00
1981	04.627	187996.0	692.00
1982	04.739	194,767.2	708.00
1983	04.993	212278.6	723.00
1984	05.654	206512.9	739.00
1985	05.914	227,211.7	755.00
1986	06.368	243522.2	771.00
1987	06.449	273290.5	788.00
1988	07.047	291,103.4	805.00
1989	07.462	292,013.5	822.00
1990	07.879	316937.4	839.00
1991	08.374	266864.9	856.00
1992	08.850	244175.0	872.00
1993	09.289	273937.9	889.00
1994	09.995	322552.6	905.30

due to the time and space complexities in most of the actuarial related issue, thus we can extend this

1995	11.443	355162.7	922.60
1996	11.042	385411.8	938.10
1997	11.637	409674.5	954.40
1998	12.166	413824.6	970.60
1999	12.988	450476.2	986.80
2000	13.462	460195.4	1002.70
2001	13.937	478290.4	1023.295
2002	13.844	501917.9	1040.285
2003	14.285	601826.9	1057.251
2004	15.538	695858.4	1074.159
2005	16.341	805.732.0	1090.973
2006	17.677	906268.0	1107.624

Let $Y(t),(t=...,0,1,2,...)$ be the universe of discourse and $Y(t) \in R$. Assume that fi(t), i=1,2,... is defined in the universe of discourse Y(t) and F(t) is a collection of f(ti),(i=...,0,1,2,...) , then F(t) is called a fuzzy time series of Y(t), i=1,2,... ... Using fuzzy relation, we define F(t)=F(t−1)oR(t,t−1) , where R(t,t−1) is a fuzzy relation and " o " is the max–min composition operator, then F(t) is caused by F(t−1) where F(t) and F(t−1) are fuzzy sets.

For forecasting purpose, we can define relationship among present and future state of a time series with the help of fuzzy sets. Assume the fuzzified data of the ith and (i+1)th day are Aj and Ak , respectively, whereAj,Ak∈U, then Aj →Ak represented the fuzzy logical relationship between Aj and Ak . Let F(t) be a fuzzy time series. If F(t) is caused by F(t−1) , F(t−2) ,..., F(t−n) , then the fuzzy logical relationship is represented by F(t−n),...,F(t−2),F(t−1)→F(t) is called the one-factor nth order fuzzy time series forecasting model. Let F(t) be a fuzzy time series.

 If F(t) is caused by (F1(t−1),F2(t−1)) , (F1(t−2),F2(t−2)) ,...,(F1(t−n),F2(t−n)), then this fuzzy logical relationship is represented by

(F1(t-n),F2 (t-n))............ F1((t-2),F2 (t-2)),(F1(t-1) F2(t-1))-> F(t)

is called the two-factors nth order fuzzy time series forecasting model, where F1 (t) and F2(t) are called the main factor and the secondary factor fuzzy time

series' respectively. In the similar way, we can define m-factor nth order fuzzy logical relationship as

(F1(t-n), F2(t-n),........., Fm(t-n)),

(F1(t-2), F2(t-2),........., Fm(t-2)),

(F1(t-1), F2(t-1),........., Fm(t-1)) ->F(t)

Here F1(t) is called the main factor and F2(t),F3(t),...,Fm(t) are called secondary factor fuzzy time series'. Here we can implement any of the fuzzy membership function to define the fuzzy time series in above equations. Comparative study by using different membership functions is also possible. We have used triangular membership function due to low computational cost.

Using fuzzy composition rules, we establish a fuzzy inference system for fuzzy time series forecasting with higher accuracy. The accuracy of forecast can be improved by considering higher number of factors and higher dependence on history.

Following is an extended method for handling forecasting problems based on m-factors high-order fuzzy time series.

Step 1) Define the universe of discourse U of the main factor U=[Dmin−D1,Dmax−D2] , where Dmin and Dmax are the minimum and the maximum values of the main factor of the known historical data, respectively, and D1, D2 are two proper positive real numbers to divide the universe of discourse into n equal length intervals , ..., u1 ,u2 ,.......,ul. . Define the universes of discourse Vi

,i=1,2,...,m−1 of the secondary factors Vi=[(Ei)min−Ei1,(Ei)max −Ei2], where (Ei)min=[(E1)min,(E2)min,.......,(Em)min], and (Ei)max=[(E1)max,(E2)max,.......,(Em)max] are the minimum and maximum values of the secondary-factors of the known historical data, respectively, and Ei1 , Ei2 are vectors of proper positive numbers to divide each of the universe of discourse Vi, i=1,2,...,m−1 into equal length intervals termed as v1,l,v2,l,...,vm−1,l, l=1,2,...,p, where v1,l=[v1,1,v1,2,...v1,p] represents n intervals of equal length of universe of discourse V1 for first secondary-factor fuzzy time series. Thus we have (m−1)×l matrix of intervals for secondary-factors.

Step 2) Define the linguistic term Ai represented by fuzzy sets of the main factor shown as follows:

A1=1/u1+0.5/u2+0/u3+0/u4+.........+0/ul-2+0/ul-1+0/ul
A2=0.5/u1+1/u2+0.5/u3+0/u4+.........+0/ul-2+0/ul-1+0/ul
A3=0/u1+0.5/u2+1/u3+0.5/u4+.........+0/ul-2+0/ul-1+0/ul

.

An=0/u1+0/u2+0/u3+0/u4+.........+0/ul-2+0/ul-1+0/ul

Similarly, for ith secondary fuzzy time series,we define the linguistic term Bi,j,i=1,2,.....,m-1, j=1,2,....n represented by fuzzy sets of the secondary factors,

Bi,1=1/Vi,1+.5/Vi,2+0/Vi,3+0/Vi,4+.....+0/Vi,l-2+0/Vi,l-1+0/Vi,l
Bi,2=.5/Vi,1+1/Vi,2+.5/Vi,3+0/Vi,4+.....+0/Vi,l-2+0/Vi,l-1+0/Vi,l
Bi,3=0/Vi,1+.5/Vi,2+1/Vi,3+.5/Vi,4+.....+0/Vi,l-2+0/Vi,l-1+0/Vi,l

.

.

Bi,n=0/Vi,1+.0/Vi,2+0/Vi,3+0/Vi,4+.....+0/Vi,l-2+.5/Vi,l-1+1/Vi,l

Step 3) Fuzzify the historical data described as follows. Find out the interval ul ,l=1,2,...,p to which the value of the main factor belongs

Case 1) If the value of the main factor belongs to u1 , then the value of the main factor is fuzzified into 1/A1+0.5/A2+0/A3 denoted by X1.

Case 2) If the value of the main factor belongs to ul ,l=2,3,...,p−1 then the value of the main factor is fuzzified into

0.5/Ai-1+1/Ai+0.5/Ai+1 denoted by Xi

Case 3) If the value of the main factor belongs to Un , then the value of the main factor is fuzzified into

0/An-2+.5/An-1+1/An denoted by Xn

Now, for ith secondary-factor, find out the interval Vi,l to which the value of the secondary-factor belongs.

Case 1) If the value of the ith secondary-factor belongs to Vi,l , then the value of the secondary-factor is fuzzified into

1/Bi,1+0.5/Bi,2+0/Bi,3, denoted by Yi,1 = [Y1,1,Y2,1,...,Ym−1,1].

Case 2) If the value of the ith secondary-factor belongs to vi,l,l=2,3,..,p−1, then the value of the ith secondary-factor is fuzzified into 0.5/Bi,j-1+1/Bi,j+0.5/Bi,j+1,j=i=2,3,...,n-1 denoted by Yi,j where j=2,3,....,n-1.

Case 3) If the value of the ith secondary-factor belongs to Vi,p , then the value of the secondary-factor is fuzzified into

0/Bi,n-2+0.5/Bi,n-1+1/Bi,n,denoted by Yi,n.

Step 4) To forecast the data of time t,we must decide the window basis w.Then,we can get the criterion vectorC(t) and the operation matrix Ow(t)at time t which are expressed as follows:[20]

$$C(t)=f(t-1)=[C1,C2,\ldots\ldots,Cm]$$

$$Ow(t)=\begin{pmatrix} f(t-2) \\ f(t-3) \\ \vdots \\ F(t-?) \end{pmatrix}$$

$$=\begin{pmatrix} O11 & O12 & O1m \\ O21 & O22 & O2m \\ \vdots & \vdots & \vdots \\ O(w-1)1 & O(w-1)2 & O(w-1)3 \end{pmatrix}$$

Let S(t) is a second factor vector.Then fuzzy relationship matrix R(t) between C(t),Ow(t) and S(t) is equal to R(t)=Ow(t)oS(t)oC(t). Fuzzified forecasted demand f(t) between time t and time t-1 described by

$$f(t)=[Max.(R11,R21\ldots..R(w-1)1)Max.(R12R22R(w-1)2\ldots\ldots.Max.(R1mR2m\ldots\ldots R(w-1)m)]$$

Table 2. Actual verses forecasted energy demand

Year	Actual Energy Demand In 10^{15}Btu(Ai)	Forecasted Energy Demand In 10^{15} Btu(Fi)	$\left\| \dfrac{Fi\text{-}Ai}{Ai} \right\|$ relative error
1980	04.041		
1981	04.627		
1982	04.739		
1983	04.993	04.937	0.011
1984	05.654	04.937	0.126
1985	05.914	05.563	0.059
1986	06.358	06.187	0.026
1987	06.449	06.187	0.040
1988	07.047	06.187	0.115
1989	07.462	06.656	0.107
1990	07.879	07.229	0.082
1991	08.374	07.750	0.074
1992	08.850	08.590	0.030
1993	09.289	08.400	0.086
1994	09.995	09.310	0.068
1995	11.443	10.300	0.096
1996	11.042	10.300	0.063
1997	11.637	11.000	0.056
1998	12.166	11.700	0.033
1999	12.988	11.700	0.093
2000	13.462	11.900	0.069
2001	13.937	13.580	0.025
2002	13.844	13.180	0.049
2003	14.285	12.800	0.028
2004	15.538	14.500	0.064
2005	16.341	15.520	0.050
2006	17.677	15.520	0.012

Mean Relative Error=0.0609

Step 5)Defuzzify the fuzzified forecasted variations of the main factor fuzzy time series.Following are the rules

1) If the grades of membership of the fuzzified forecasted variation are all zero, the we set the forecasted variation term 0. 2)If the maximum membership of the forecasted variation f(t) occurred at Ui and the midpoints of Ui is mi,then the forecasted variation is mi.If the maximum membership of the fuzzified forecasted variation f(t) occurred at U1,U2,Uk and the mid points are m1,m2........mk respectively then the forcasted variation is (m1,m2.....mk)/k.

Following the above steps the Energy demand has been forecasted and mean relative error is obtained.This is shown in Table 2.

IV CONCLUSION

Proposed method can successfully be applied where the previous data are not in precise form of actual data are not maintained.

In this project a multi factor or multivariable fuzzy time series based long term prediction of power demand is proposed.

The proposed methodology for demand forecasting uses this fact that energy demand of an electrical system has time series behavior.

The forecasting ability is expected to be better than traditional forecasting method.

Future research involves applying the proposed measure to dealing with more complicated applications.

V. REFERENCES

[1]John P Mentzer,Mark A. Moon "sales forecasting management"Sage publications.

[2]Gerorge Atsalakis,Camelia usenic, and Christos H.Skiadas "Time series prediction of the greek manufacturing index for the non-metallic minerals sector using a neuro-fuzzy approach".

[3]Eugene A Feinberg (State University of New York, Stony Brook) and Dora Genethliou (State university of New York, stony brook) of Load forecasting chapter 12.

[4]Q. Song and B.S.chissom "Forecasting enrollment with fuzzy time series"part-2 FSS vol.62 no.1 pp.1-8 (1994).

[5]J.Sullivan,W.H.Woostall "A comparison of fuzzy forecasting and Markov model"; FSS vol.64 no.3 pp.279-293(1994).

[6]M.Y. Day , "Research of applying GA to fuzzy forecasting focus on sales forecasting";M.S.Thesis,Tankun University;Taipie Taiwan,R.C.CC(1995).

[7]S.M.Chan "Forecasting enrolment base on fuzzy time series";FSS vol.81 no.3 pp.311-319(1996)

[8]J.R.Hwang,S.M.Chan and C.H.Lin "Handling forecasting problem using fuzzy time series analysis";FSS vol.100 no.2 pp.217-226(1998)

[9]K. Huarng, "Heuristic models of fuzzy time series for forecasting," Fuzzy Sets Systems, vol. 123, no. 3, pp. 369–386, 2001a.

[10]S. M. Chen, "Forecasting Enrollments Based on High-Order Fuzzy Time Series," Cybernetic Systems, Vol. 33, No. 1, pp. 1–16, 2002.

[11]R. R. Yager and P. P. D. Filev, Essentials of FUZZY MODELING and Control, John Wiley and Sons, Inc. 2002.

[12]Melike Sah and Y. D. Konstsntin, "Forecasting Enrollment Model based on first-order fuzzy time series," Published in proc., International Conference on Computational Intelligence, Istanbul, Turkey, 2004

[13]Ruey-Chyn Tsaur "Further examination to fuzzy exponential smoothing method"Journal of the Chinese institute of industrial Engineers,vol.22,No. 6,pp.521-530(2005).

[14]G. J. Klir and B. Yuan, Fuzzy Sets and Fuzzy Logic: Theory and Applications, Prentice Hall, India, 2005, Ch. 4.

[15]L. W. Lee, L. W. Wang, S. M. Chen, "Handling Forecasting Problems Based on Two-Factors High-Order Time Series," IEEE Transactions on Fuzzy Systems, Vol. 14, No. 3, pp.468-477, Jun. 2006.

[16]Tahseen A. Jilani, S.M.Aqil Burney, and C.Ardil, Multivariate High Order Fuzzy Time Series Forecasting for Car Road Accidents. World Academy of Science, Engineering and Technology 25 2007

[17]Energy Information Administration International energy annual 2006 Table posted :December 19,2008

[18]Prediction of net energy consumption based on economicindicators,linkinghub.elsevier.com/retrieve/pii/ S0301421507001796

[19] http://en.wikipedia.org/wiki/Energy_crisis

[20]Krishan Pal "Two variable fuzzy time series analysis for predicting sales volume of an industrial set up" unpublished M.tech dissertation, Department of Mechanical Engineering, Dayalbagh Educational Institute,2006

Virtual Prototyping of 2000 m^3 Aerostat

Imtiaj Khan

Scientist, Aerial Delivery Research and Development Establishment (ADRDE), DRDO, Agra -282 001

Abstract— Today, manufacturers are under pressure to reduce time to market and optimize products to higher levels of performance and reliability. A much higher number of products are being developed in the form of virtual prototypes in which engineering simulation software are used to predict performance prior to constructing physical prototyping. Performance of thousands of design alternatives can be explored without investing the time and money required to build physical prototypes. The ability to explore a wide range of design alternatives leads to improvements in performance and design quality. Yet the time required to bring the product to market is usually reduced substantially because virtual prototypes can be produced much faster than physical prototypes. Various subsystems of aerostat are being modeled using CATIA V5 software. Geometrical properties of system are being calculated. This paper focus on the modeling procedure adapted for virtual prototyping of Aerostat.

I. INTRODUCTION

Virtual Prototyping is the creation of three dimensional model of a product in digital form. ADRDE has developed a 2000 m^3 Aerostat which has a ceiling of 1000 meters. For this task there was a need to develop various systems of aerostat using CAD (Computer Aided Design) software. CAD Modeling is the creation, manipulation and storage of geometric objects to represent virtual objects. The process begins with the use of specialized 3D software. For this purpose CATIA V5 software was used. This development would help us in future to carry out various studies such as modification of payload structures, repositioning of lightning arrangement and futuristic design of large Aerostat system.

II. APPROACH CARRIED OUT FOR MODELING

To make the designing process simple and quick, the software package has been classified into different modules. i.e. each step of designing is completed in a different module. Generally a typical design task can be accomplished using various modules of CATIA V5 as tabulated in Table 1.

Parametric design strategy was used while modeling each part and bottom up approach was followed in assembling each part. Later on the geometric properties, mass properties and the moment of inertia of the assembled components were evaluated.

Table 1: Modules of CATIA V5

Design Task	Mode in CATIAV5
Sketching using basic sketch entities	Sketcher
Converting sketches into Parts	Part Mode
Assembling different parts and analyzing	Assembly Design
Drawing views	Drafting
Converting sketches into surfaces	Generative shape design
Animation of assembled system	DMU Navigator
Mechanism/ Motion	DMU Kinematics

III. VIRTUAL PROTOTYPING OF AEROSTAT

3D Modeling of balloon (e.g. hull, ballonet, fines, truss, patches, payload, sensors, accessories etc) , Winching & Mooring system(WMS), Lighting Protection System(LPS), Aerostat Health Monitoring System (AHMS) and Fiber optic rotary Joint (FORJ) is carried out using 'Sketcher', 'Part mode' and 'Generative shape design' of CATIA V5 software [1]. In later stage these parts are assembled using Assembly design module. Positioning of various components on aerostat is carried out based on operational and center of gravity requirements. Finally interface drawing of various sub-systems are prepared using 'drafting module'.

The balloon is nearly perfectly gas tight aerodynamically shaped compartment inflated with helium [2]. It is an airborne platform to mount various payloads for surveillance and reconnaissance. The empennage consists of three aerodynamically shaped fins. The windscreen is an aerodynamically shaped, air-inflated protective housing for payloads [3].

A lightning protection system is installed along the top and each side of the aerostat to intercept lightning discharges from the atmosphere and conduct the associated currents harmlessly to ground through the copper shield on the tether. Balloon along with Lightening protection system is shown in Fig.1.

Winching and mooring system (Fig.2) has two major parts i.e. Winch System which is used to hoist, hold, de-hoist the aerostat in controlled manner and Mooring System which is used to hold the aerostat effectively at ground in no flight condition. Aerostat can be moored during integration of various payloads and maintenance purposes.

Fiber Optic rotary joint (Fig.3) is mounted at the confluence point of the balloon i.e. airborne mechanical swivel to decouple the rotary motion of balloon. Various subsystems of aerostat are shown in Fig.4.

Table 2: Aerostat parameters

Helium Volume	2000 m³
Length	34 meter
Maximum Diameter	10 meter
Overall Width (lower fin span)	28.47 meter
Overall Height	15.25 meter
Ballonet Volume	1200 m³
Endurance	7 days
Payload capacity	300 kg
Operating altitude	1000 meters

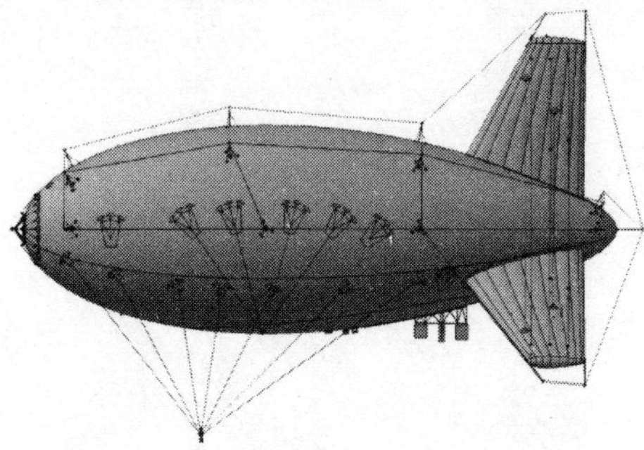

Fig.1. Balloon with LPS

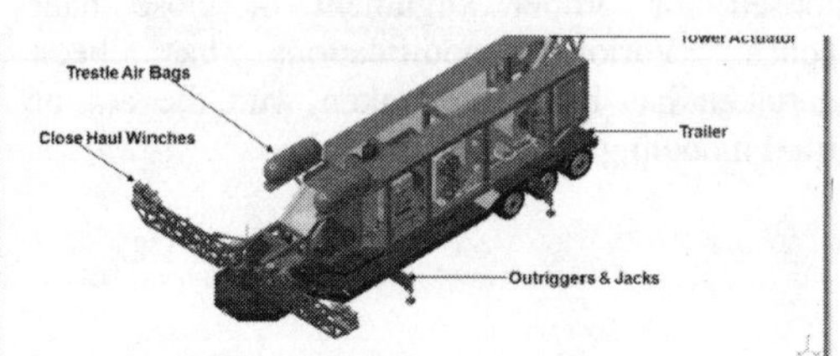

Fig.2 Winching and mooring system

Fig.3. FORJ

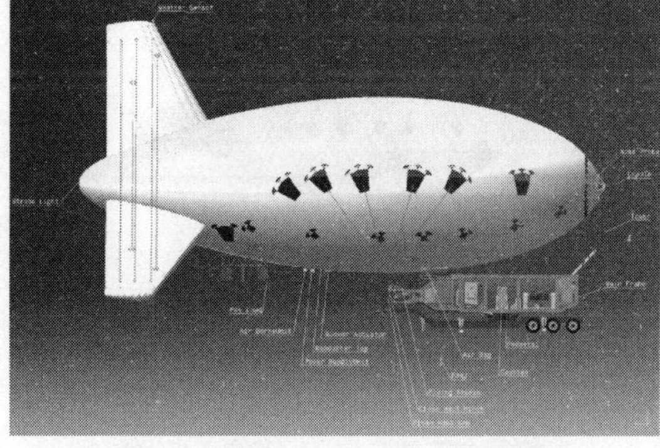

Fig.4. Aerostat system

IV. INTERFACE DRAWINGS

Interface drawings (Fig.5) have helped us to study the interface of Balloon with payloads, Winching and mooring system, AHMS components and lighting protection system. Interface drawing thus generated has helped us to integrate various systems on aerostat. Problems caused due to mismatch interference etc are detected during the early stage itself and remedial action can be taken to avoid delay in later stage e.g. while fitment, contour of Nose cradle have slight mismatch with balloon .This problem was solved by modifying the contour shape of nose cradle. Due to misalignment of cordages repositioning of patches is also being proposed for proper alignment of close haul winches. Various modifications has been undertaken has been undertaken with the aid of Virtual modeling.

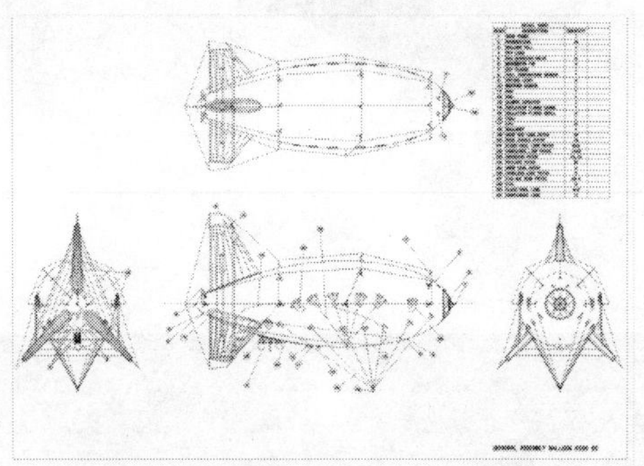

Fig.5. Interface drawing of balloon

V. RESULTS OF VIRTUAL PROTOTYPING

Full animation of complete aerostat system is being prepared which shows the aerostat launching operation. In this animation balloon is being launched from moored condition to deployed condition of balloon. Overall system behavior can be easily understood using this animation. Fig.6 and Fig.7 shows close resemblance between virtual and physical models.

Fig.6. Virtual model: Virtual 3D model

Fig.7. Physical model: Field Trials

VI. ESTIMATION OF GEOMETRIC PROPERTIES

Mass, c.g. and moment of Inertia of entire system is obtained and tabulated in Table 3.

Table 3: Geometrical properties

Parameter	Symbol	Value
Mass (kg)	M	898.081
Centre of Gravity (mm)	Gx	19332.087
	Gy	1.043
	Gz	-2398.752
Moment of Inertia (kg-mm^2)	Ixx	17989.845
	Iyy	72776.194
	Izz	65572.639

VII. CONCLUSIONS

Virtual modeling has helped us in developing the prototype thus facilitated superior design quality, reduced expenditures and material wastage. Any modification at any design stage can easily be incorporated in the CAD model. Various iterative concepts can be studied at the design stage itself. Model generated can be used for detailed FEM and CFD analysis. In the whole process virtual prototyping has integrated the "virtual" and "real time" methods in order to achieve dramatic improvements in productivity, lead time, and agility.

VIII. REFERENCES

[1]   *Catia v5 User manual*

[2]   *Khoury GA and Gillett JD, "Airship Technology" Book, Cambridge University Press, 2000*

[3]   *Jonathan I. Miller and Meyer Nahon, "The Design of Robust Helium Aeostats" AIAA 2005-7441*

Finite Element Analysis of Excavator Arm links

[1]Abdul Aziz, [2]Rupendra Singh Rajpurohit, [1]Saswat Kumar Das

[1]IIMT Engg. College Meerut, [2]BIT Meerut

Abstract—**The important characteristics of CAD system includes fully dimensional, associative, centralized and integrated data base which is always rich in information needed for both design and manufacturing process. The centralized concept implies that any change made in geometric modeling automatically reflects in the existing views or any other view that may be defined later. The integrated concept implies that a geometric model of an object can be utilized at various phases of a product cycle. The associativity concept implies that input information can be retrieved in various forms.**

At present CAD systems compose of 2D and 3D modes which are user friendly and are based on Interactive Computer Graphics (ICG). As the name suggests it displays information and data in form of graphics. The designer is in a position to enter data in form of commands which are converted by the software and hardware into graphical form. 3D mode helps to get a clear picture of the real situation from all dimensions. Thus we have created the arm links drawings, and modified them and analyzed the 3D model by using CATIA & ANSYS Software up to load of 2000N at various operations of the Excavator. 3D images were obtained showing maximum and minimum stress and strain zones which help in locating the weak zones and to modify the link for better performance and least exposure to stress and strain during working. In present time, the rate of changes produced in the design is very high. So it is imperative to switch over to this tool for cost effectiveness and efficiency.

I. INTRODUCTION

Excavators are the material handling machine generally used in Civil Engineering. applications. They are open chain mechanism and are subjected to varying degrees of stresses during various operations.

FEM techniques can be effectively applied to these links to achive better design and to provide safety during operations.

In our work we had taken Excavator and focused on its three links i.e. front arm, back arm & bucket. During our analysis we had designed and modeled these three links or parts of Excavator on CATIA Software.

Later they were analyzed using ANSYS software using Tresca and Von Misses theory as criterion for failure. Highly stressed zones were obtained which has to be reconsidered before finalizing the design.

II. ANALYSIS PROCEDURE

After drafting the three links of the mechanism namely front arm, back arm & bucket of Excavator using CATIA a 3D model was prepared.Once a 3D model was obtained it was discritized into 3D mesh model using ANSYS and the model was tested at 2000N load during digging and lifting of load.During analysis failure due to stresses was analysed using and Von Mises Theory

Terminology

Excavators are also called diggers and 360 degree excavators, sometimes abbreviated simply to a 360. Tracked excavators are sometimes called track hoes by analogy to the backhoe. Even though the 'back' in backhoe refers to the action of the bucket (which pulls "back" toward the machine) and not the location of the shovel, excavators are also occasionally referred to as front hoes or even just "hoes".

Introduction to CAD

Computer Aided Design (CAD) has been acknowledged as a key for improving manufacturing techniques and productivity, along with meeting recent critical design requirements.

The important characteristics of CAD system includes fully dimensional, associative, centralized and integrated data base which is always rich in information needed for both design and manufacturing process.

At present CAD systems compose of 2D and 3D modes which are user friendly and are based on Interactive Computer Graphics (ICG). As the name suggests it displays information and data in form of graphics. The designer is in a position to enter data in form of commands which are converted by the software and hardware into graphical form. 3D mode helps to get a clear picture of the real situation from all dimensions. Thus the user can create drawings, modify them and explore further possibilities and options. In present time, the rate of changes produced in the design is very high. So it is imperative to switch over to this tool for cost effectiveness and efficiency.

Drafting of different links and body of Excavator

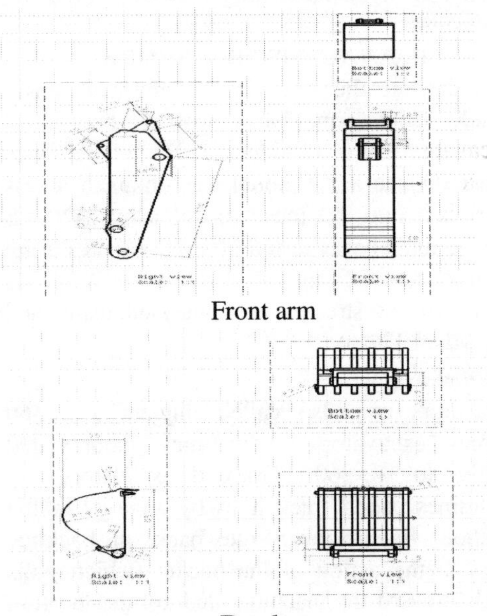

Front arm

Bucket

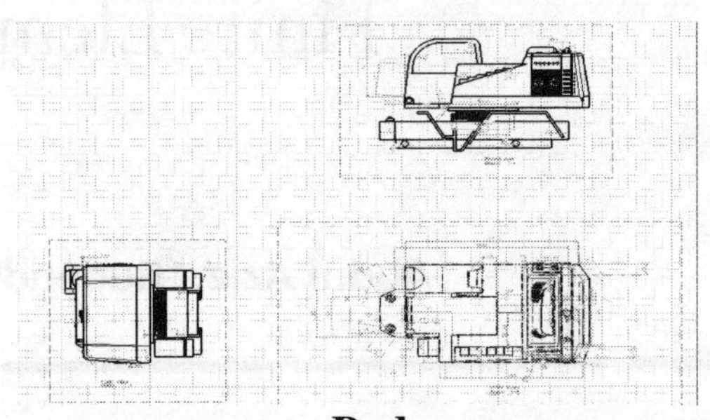

Body

Modeling of different parts of Excavator

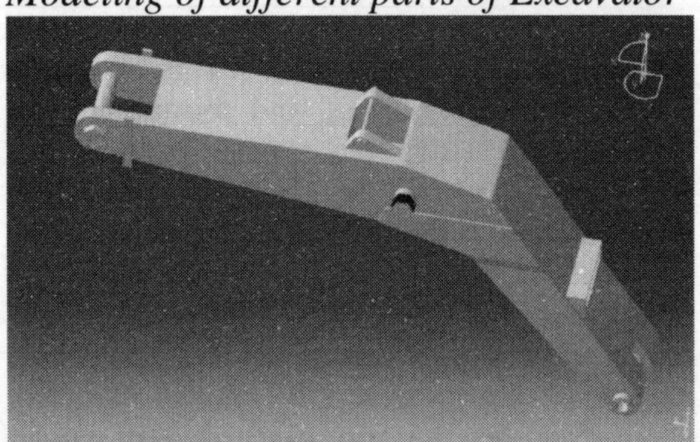

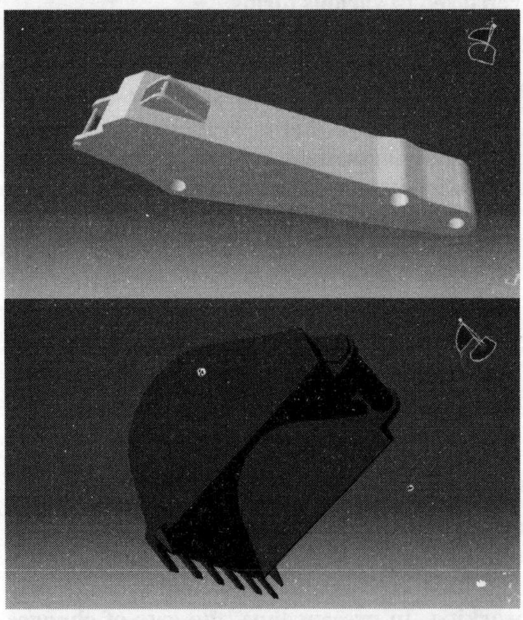

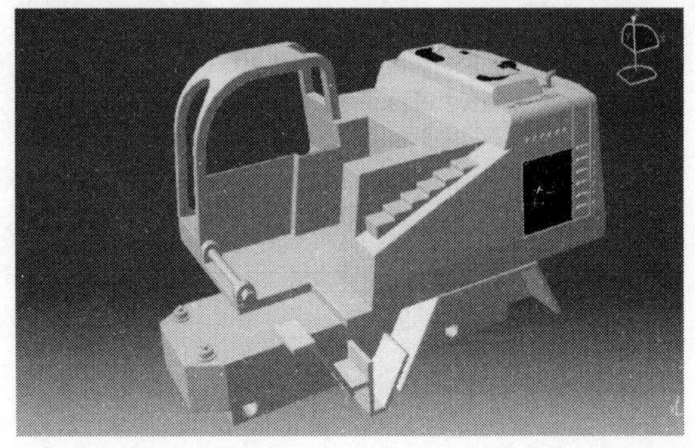

III. ANALYSIS

An analytical solution is a mathematical expression that gives the values of the desired unknown quantity at any location in a body, and as a consequence, it is valid for an infinite number of locations in the body. However, analytical solutions can be obtained only for simple engineering problems. It is extremely difficult, and many a times impossible, to obtain the exact analytical mathematical solution for many complex engineering problems. In such cases, the technique known as Finite Element Method is used.

In FEM, the continuum (or structure) is divided into finite number of smaller units known as elements. This process of dividing the structure into finite number of elements in known as discretization. These elements are considered interconnected at joints which are known as nodes or nodal points. Instead of solving the problem for each element and combined to obtain the solution the solution for the original continuum. This approach is known as going from part to whole. Though it will be theoretically correct to satisfy the continuity requirement all along the edges of elements, this will lead to more complicated analysis. Hence in FEM, in order to make analysis simpler, it is assumed that the elements are connected at a finite number of joints called nodes or nodal points. It is only at nodes the continuity equation are required to be satisfied.

In FEM, the amount of data to be handled is dependent upon the number of elements into which the original continuum is divided. For a large number of elements it is a formidable task to handle the volume of data manually and hence in such cases the use of commuters is inevitable. It has been found that the accuracy of solution, in general, increases with the number of elements taken. However more number of elements will result in increased computation

IV. ANSYS MESH MORPHER

By working with a mesh and not the solid model, the ANSYS Mesh Morpher allows parameterization of models created from CAD data, nonparametric geometry data such as IGES or STEP, or mesh files such as the ANSYS .cdb file. Read a mesh into FE Modeler and then create an initial configuration to synthesize geometry from the existing mesh.

Applications

Perform rapid modifications of a design (evolution of existing designs)

Perform concepts analysis at the CAE level

Explore many design alternatives

Von Mises yield criterion

The von Mises yield criterion suggests that the yielding of materials begins when the second deviatoric stress invariant J_2 reaches a critical value k. For this reason, it is sometimes called the J_2-*plasticity* or J_2 flow theory. Because the von Mises yield criterion is independent of the first stress invariant, I_1, it is applicable for the analysis of plastic deformation for ductile materials such as metals, as the onset of yield for these materials do not depend on the volumetric component of the stress tensor (mean stress or hydrostatic stress). Although formulated by Maxwell in 1865, it is generally attributed to von Mises (1913).[1] Huber (1904), in a paper in polish, anticipated to some extent this criterion.[2] This criterion is referred also as the Maxwell-Huber-Hencky-von Mises theory.

Mathematical formulation

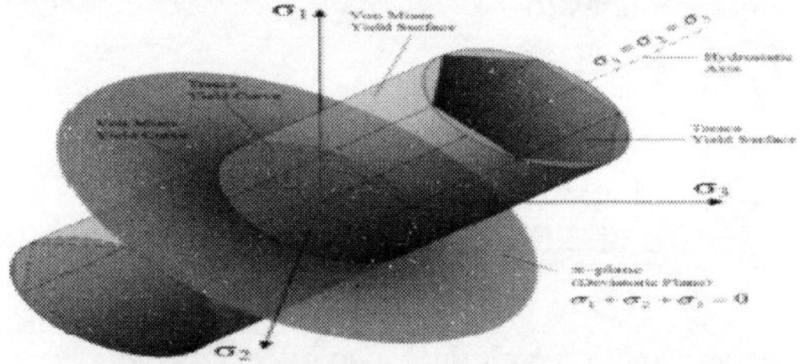

A Von Mises and Tresca Yield surface in principal stress coordinates.

Mathematically the yield function for the von Mises condition is expressed as:

$$f(J_2) = \sqrt{J_2} - k = 0$$

An alternative form is:

$$f(J_2) = J_2 - k^2 = 0$$

Substituting the value of J_2 into this equation we obtain the von Mises yield criterion as a function of the principal stresses

$$(\sigma_1 - \sigma_2)^2 + (\sigma_2 - \sigma_3)^2 + (\sigma_1 - \sigma_3)^2 = 6k^2$$

or

$$(\sigma_1^2 + \sigma_2^2 + \sigma_3^2) - \sigma_1\sigma_2 - \sigma_2\sigma_3 - \sigma_1\sigma_3 = 3k^2$$

or as a function of the stress tensor components

$$(\sigma_{11} - \sigma_{22})^2 + (\sigma_{22} - \sigma_{33})^2 + (\sigma_{11} - \sigma_{33})^2 + 6(\sigma_{23}^2 + \sigma_{31}^2 + \sigma_{12}^2) = 6k^2$$

This equation defines the yield surface as a circular cylinder (See Figure) whose yield curve, or intersection with the deviatoric plane, is a circle with radius $\sqrt{2}k$. This implies that the yield condition is independent of hydrostatic stresses.

3.10.4) Physical interpretation of the von Mises yield criterion

Hencky (1924) offered a physical interpretation of von Mises criterion suggesting that yielding begins when the elastic energy of distortion reaches a critical value. [3] For this, the von Mises criterion is also known as the *maximum distortion strain energy* criterion. This comes from the relation between J_2 and the elastic strain energy of distortion W_D:

$$W_D = \frac{J_2}{2G},$$

where G is the elastic shear modulus defined as:

$$G = \frac{E}{2(1 + \nu)}$$

Von Mises criterion is also known as the *maximum octahedral shear stress* in view of the relation between J_2 and the octahedral shear stress, τ_{oct}. In 1937 [4] Arpad L. Nadai suggested that yielding begins when the octahedral shear stress reaches a critical value k:

$$\tau_{oct} = \sqrt{\frac{2}{3} J_2} = \sqrt{\frac{2}{3}} k$$

which reduces to

$$J_2 - k^2 = 0$$

Bucket Analysis

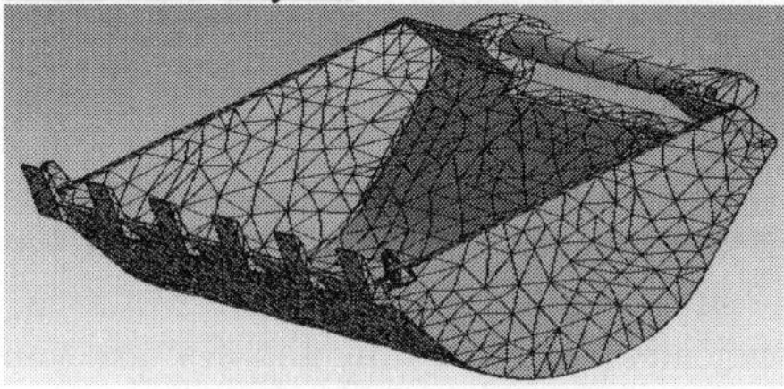

Conditions at 2000N load when starts to lift

Von mises Stress at 2000N load when starts to lift

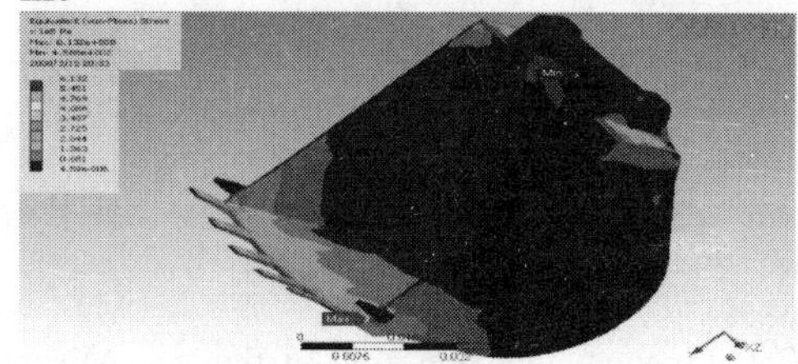

Principle Stress at 2000N load when starts to lift

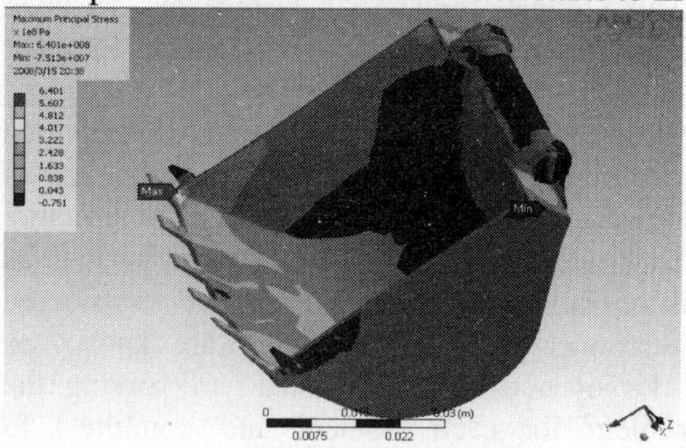

Strain at 2000N load when starts to lift

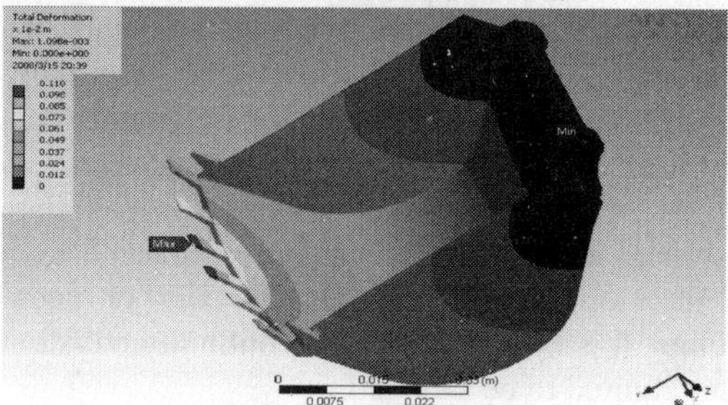

Boundary Conditions at 2000N load (Lifting Load)

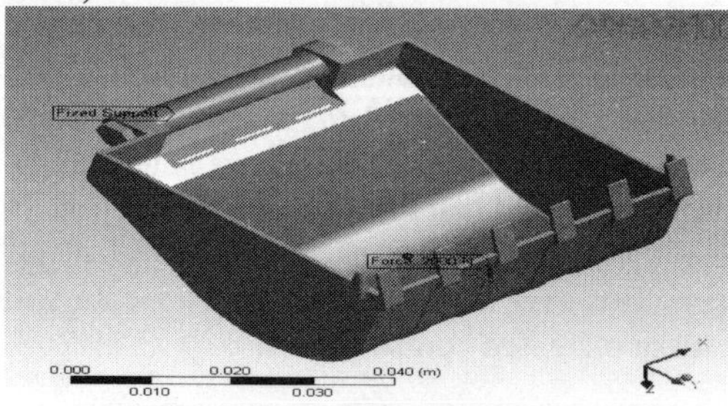

Vonmises Stress at 2000N load

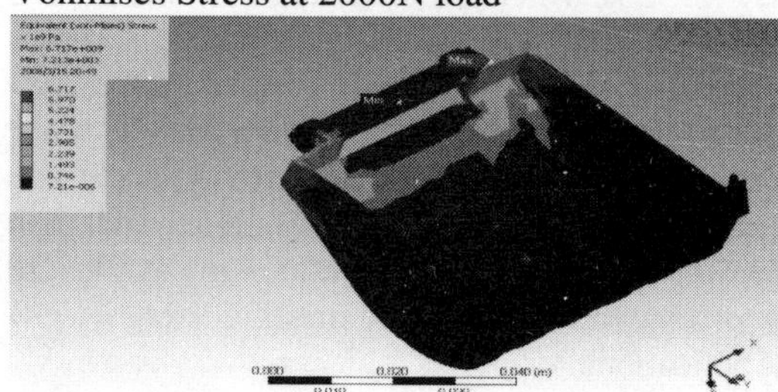

Strain at 2000N load

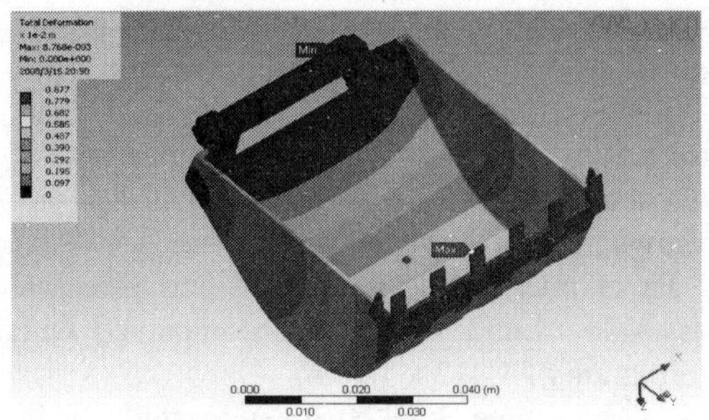

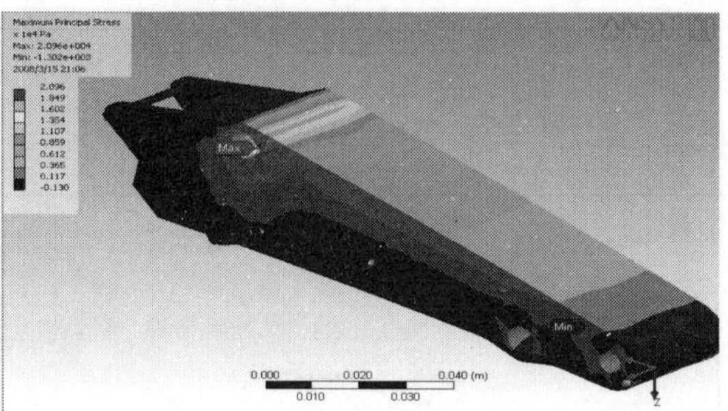

Connecting Arm Geometry (front arm)

Strain when a load of 2000N is lifted

Meshing of front arm

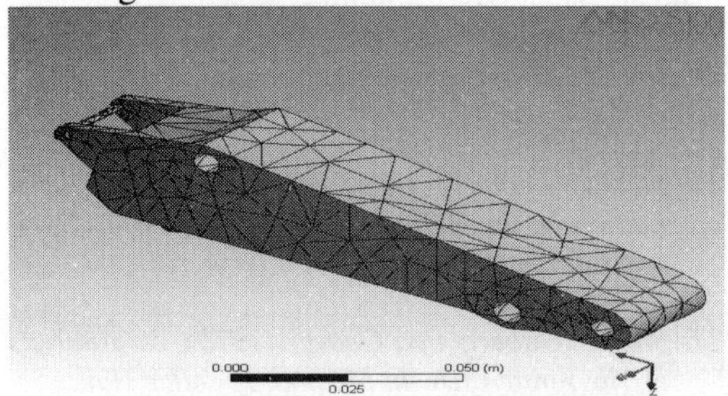

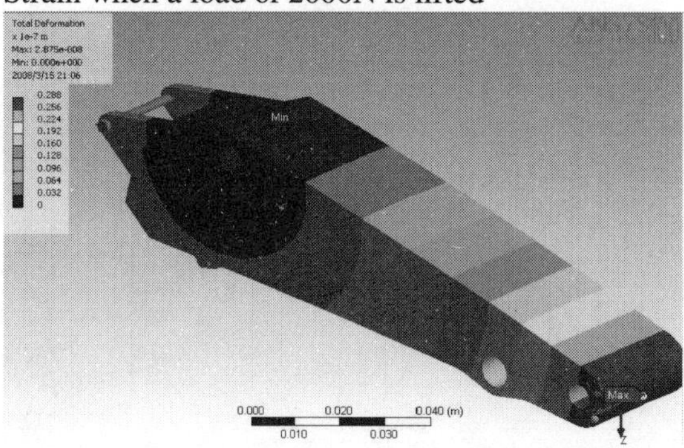

Vonmises Stress when load of 2000N is lifted

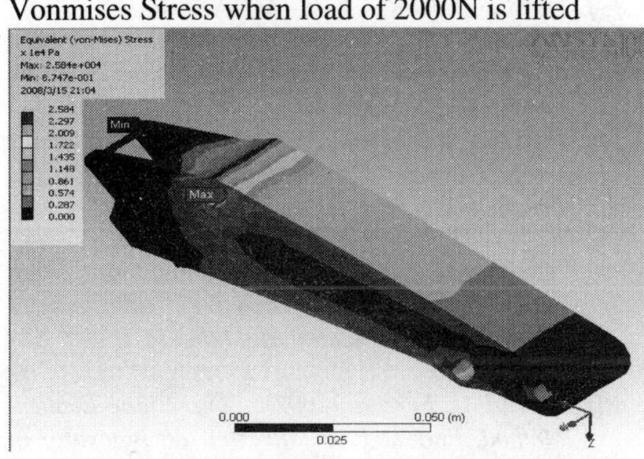

Moving Arm Geometry (back arm)
Meshing of back arm

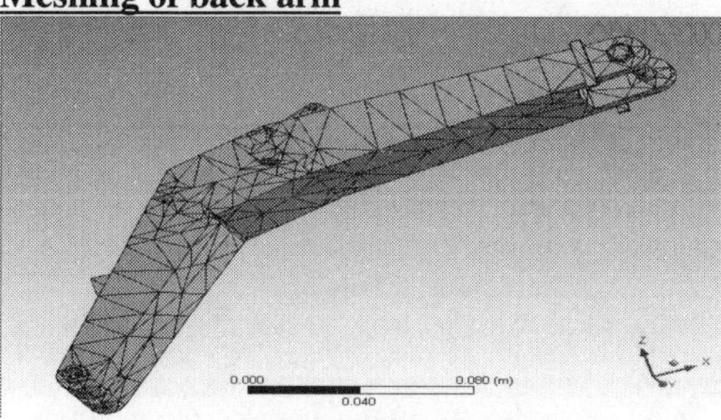

Vonmises Stress when a load of 2000N is lifted

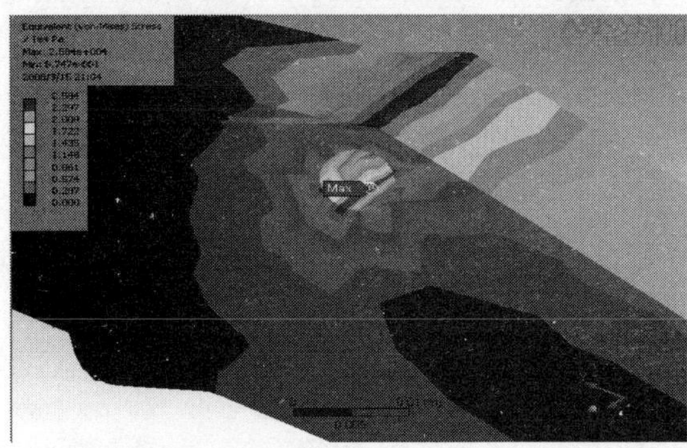

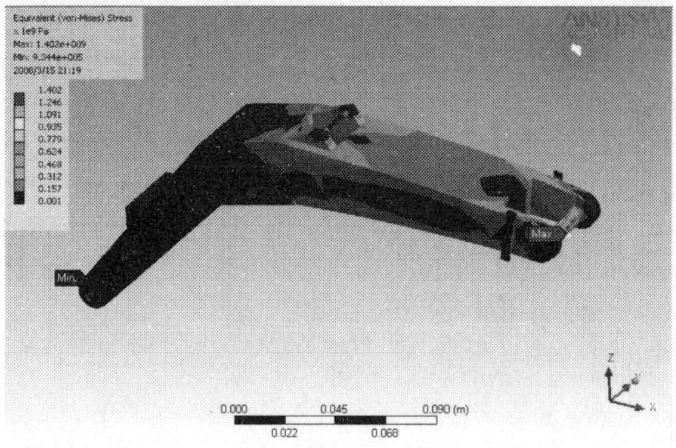

Principle Stress when a load of 2000N is lifted

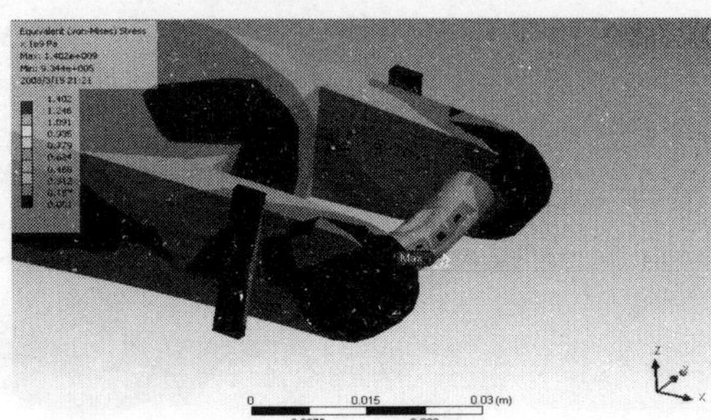

Principle Stress when a load of 2000N is lifted

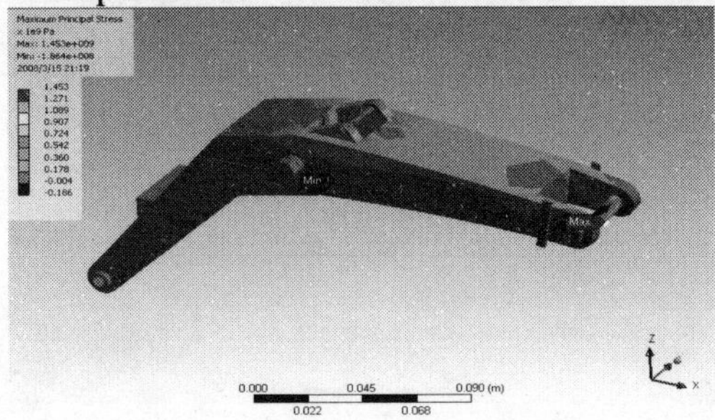

Strain when load of 2000N is lifted

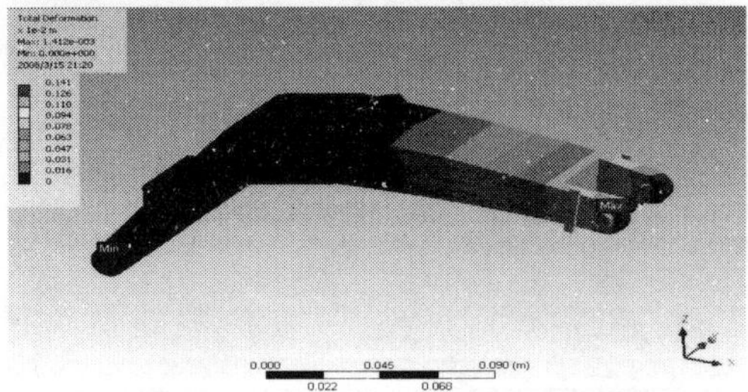

V. CONCLUSION

Von Mises Stress developed during the lifting operations were in the range 4.59×10^2 Pa to 6.132×10^8 Pa. Max stress was developed in the front extreme corners of the bucket below base of extreme right and left teeth. Max principal stress values obtained is 6.401×10^8 Pa at front end and $- 7.51 \times 10^7$ Pa at the rear end and maximum deformation of 1.098×10^{-3} m is observed as the middle teeth.

Connecting arm is subjected to Max von-misses stress value of 2.58×10^4 Pa. and principle stress of 2.096×10^4 at hinge hole and max deformation 2.875×10^{-8} m at front hinge hole of connecting arm.

In moving arm Von Mises stress value is 1.402×10^9 Pa at bucket side hinge rod and principal stress value 1.453×10^9 Pa accompanied with deformation of 1.412×10^{-3} m.

Peak stress zone bucket is suggested to be reinforced to reduce the stress value hinge hole end should be increased in thickness and also the hinge rod should be increased in diameter to avoid stress and deformation.

VI. REFERENCES

[1]. Matrix Analysis Of Framed Structures, 3rd Edition by Jr. William Weaver, James M. Gere, 3rd Edition, Springer-Verlag New York, LLC, 3, First edition 1966

[2]. Strang, Gilbert; Fix, George (1973). An Analysis of The Finite Element Method. Prentice Hall.

[3]. Hastings, J. K., Juds, M. A., Brauer, J. R., Accuracy and Economy of Finite Element Magnetic Analysis, 33rd Annual National Relay Conference, April 1985.

[4]. P. Solin, K. Segeth, I. Dolezel: Higher-Order Finite Element Methods, Chapman & Hall/CRC Press, 2003

[5]. B, Ivo; U Banerjee, J. E. Osborn (June 2004). "Generalized Finite Element Methods: Main Ideas, Results, and Perspective". International Journal of Computational Methods 1 (1): 67–103.

[6]. Ronald L. Ferrari (2007). "The Finite-Element Method, Part 2: P. P. Silvester, an Innovator in Electromagnetic Numerical Modeling".

[7]. Giuseppe Pelosi (2007). "The finite-element method, Part I: R. L. Courant: Historical Corner.

[8]. E. Stein (2009), Olgierd C. Zienkiewicz, a pioneer in the development of the finite element method in engineering science. P 264-272.

[9]. K. Al-Fadhalah, A. Elkholy, M Majeed,failure,(2010), "Aanalysis of Grade-80 alloy steel towing chain". 64-74

Analysis of Thin-Walled Structured Beams Using LS-DYNA and FE Modeling Techniques

Anurag Porwal[1], Pawan Sharma[2], Manish Chauhan[2]
[1]Analyst CAE, AT Technologies, Pune
[2]Anand Engineering College, Agra

Abstract- **This paper deals with investigation of the dynamic crushing behaviors of steel beams with box cross sections. In order to reveal the effect of material properties, including strain hardening ratio and strain rate effect, length of the beam, and initial impact velocity on the crushing behaviors of the steel beams, systematic parametric studies were conducted. A number of finite element models were constructed with various sets of parameters and used for crashworthiness analyses. Maximum crushing force, mean force, and specific energy absorption (SEA) were recorded after analyses and compared to reflect the influences of parameters. For FE modeling, Hyper mesh was used and an explicit finite element solver, LS-DYNA® was used in this study for analyses.**

Key-Words: Beams, Strain Hardening, Finite Element models, Hyper mesh, Impact velocity

I. INTRODUCTION

Now days, we find application of thin-walled box-sectional beams as energy absorbers in civil engineering, automotive engineering, shipbuilding, and other industries because of their excellent energy absorption capacity during impacts. Such beams are commonly made of steel, which is a strain rate sensitive material [1]. Jones [1] and Han et al. [2] discussed the effects of strain hardening and stain rate sensitivity on crash responses of circular steel results of peak crushing force, mean crushing force, and energy absorption capacity (SEA). Hyper mesh was used to generate the FE models and LS DYNA runs the numerical analyses.

tubes. In this paper, the strain hardening and strain rate effects on steel box-sectional beams are examined.

Crush behaviours of box-sectional beams under axial compression have been thoroughly illustrated in previous researches [3-5]. In those literatures, researchers derived equations indicated that mean crushing force can be evaluated based on beam's material properties and cross-sectional dimensions. Nevertheless, peak crushing force occurs in crashes receives more attention because in automotive engineering the peak crushing force directly affects occupant comfort and human safety. According to previous researches, the peak crushing force is also related to the beam's total length and initial impact velocity as well as the beam's material and cross-sectional dimensions. One objective of this study is to reveal the effects of beam's length and initial velocity on the peak crushing force.

In this study, a number of finite element (FE) models were created for the steel box-sectional beams with different lengths. These FE models then were used for crashworthiness analyses and during the analyses, different initial impact velocities were assigned. The effects of material's strain hardening ratio and strain sensitivity rate were firstly discussed. Next, for the steel beams with box sections, which have both strain hardening and rate sensitivity, the effects of beam length and impact velocity on the beam's crushing behaviors were demonstrated based on the analyses

II. FINITE ELEMENT MODELING AND NUMERICAL ANALYSIS

Finite element models were created for the steel box - sectional beams. These beams have fixed cross-sectional dimensions (45mm×45mm and wall thickness (t) is 2.0mm) and various lengths, which varied from 200mm to 400mm. The commercial meshing software, Hyper mesh was used for generating FE modeling and explicit code, LS-DYNA, was used for the dynamic analyses. As shown in figure 1, the FE models were meshed with 4-node quadrilateral shell element. To define the type of shell element, a Belytschko-Tsay shell element card with five integration points through its thickness was included in hyper mesh. For defining material to such type of beam, MATL 3 was included which represents the material used is MAT_PLASTIC_KINEMATIC according to the material models present in LS DYNA. The material properties of the mild steel are: density = 7830kg/m^3, Young's modulus E = 207GPa, Poisson's ratio = 0.3, yield stress σ_y = 200MPa. On finding the effects of strain hardening and strain rate, 4 cases were considered: case 1-steel with neither strain hardening nor strain ratio (O/O); case 2-steel with strain hardening but rate effects (H/O); case 3-steel with rate effects but strain hardening (O/R); case 4-steel with both strain hardening and rate effects (H/R). Different settings corresponding to above cases are listed in table 1.

Table1. Material models for strain hardening and rate effects

	Strain hardening effects		Strain rate effects	
Parameters	Et	B	C	P
Yes	1.5	1	40	5
No	0	0	0	0

In table 1, E_t is tangent modulus, which is ratio between material's ultimate stress and its initial yield stress (σ_u/σ_0), and it determines the strain hardening effect together with hardening parameter β (as shown in figure 2). C and P are two strain rate parameters and in this LS-DYN model the influence of material strain rate sensitivity is considered using Cowper and Symond model [1]:

$$\sigma_0' = \sigma_0 \left\{ 1 + \left(\frac{\varepsilon}{C} \right)^{1/P} \right\} \quad - - 1$$

Where ε is the strain rate and from that equation it is shown that during dynamic crushing, the original yield strength σ_0 is replaced by the dynamic flow stress σ_0' due to the strain rate effects.

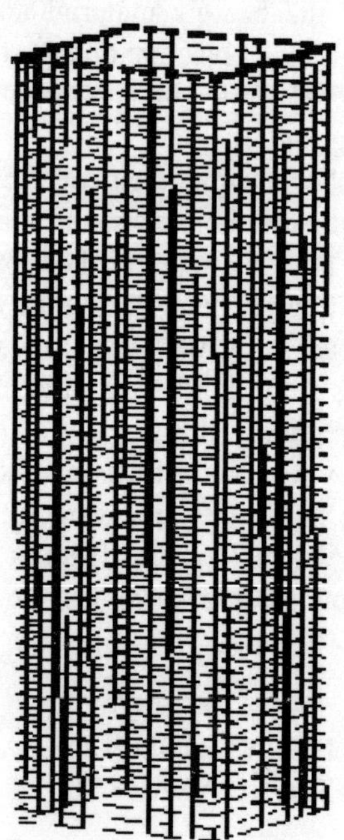

Figure1. FE model for a steel box -sectional beam

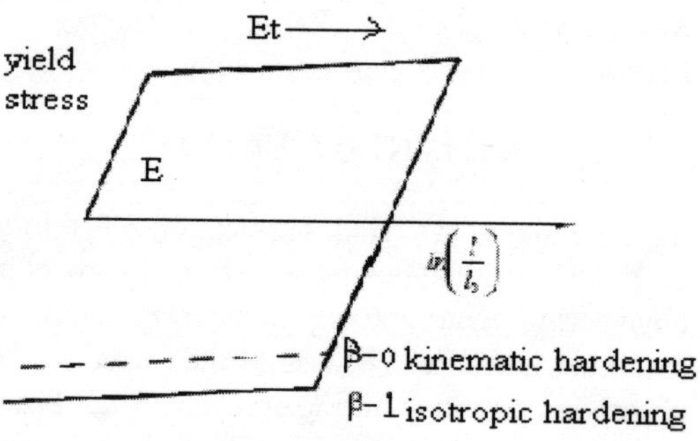

Figure2. Strain hardening effects modeled in LS-DYNA [6]

Dynamic crushing analyses were performed on the FE models and during the analysis, these beam models impact onto a rigid wall at different initial speeds (5m/s, 10m/s, and 15m/s). Analysis results were collected afterwards and the peak crushing force, mean crushing force, and SEA are considered as most important data in sketching the beams' crushing behaviors. Important results are listed throughout table 2 to 4.

Table-2 Summary of dynamic crushing analysis on steel beams (L = 210mm)

Initial Velocity	Material Case #	Peak Force	Mean Force	SEA
5m/s	1 (O/O)	68.6kN	30.8kN	3.6kJ/kg
	2 (H/O)	68.8kN	47.2kN	5.8kJ/kg
	3 (O/R)	149.4kN	54.3kN	7.2kJ/kg
	4 (H/R)	168.3kN	95.8kN	10.6kJ/kg
10m/s	1 (O/O)	81.7kN	33.9kN	8.5kJ/kg
	2 (H/O)	81.9kN	47.1kN	10.7kJ/kg
	3 (O/R)	155.2kN	64kN	17kJ/kg
	4 (H/R)	168kN	94.4kN	21kJ/kg
15m/s	1 (O/O)	82.1kN	35.6kN	12.2kJ/kg
	2 (H/O)	82.6kN	47.2kN	17.2kJ/kg
	3 (O/R)	205.5kN	64.6kN	25kJ/kg
	4 (H/R)	205.5kN	97.5kN	32.3kJ/kg

Table3. Summary of dynamic crushing analysis on steel beams (L = 300mm)

Initial Velocity	Material Case #	Peak Force	Mean Force	SEA
5m/s	1 (O/O)	72kN	32.1kN	3.5kJ/kg
	2 (H/O)	72.3kN	46.3kN	5kJ/kg
	3 (O/R)	145.9kN	55kN	6.9kJ/kg
	4 (H/R)	150.3kN	86.6kN	9.1kJ/kg
10m/s	1 (O/O)	81.7kN	36.9kN	8.7kJ/kg
	2 (H/O)	81.9kN	50.2kN	11kJ/kg
	3 (O/R)	153.3kN	66.2kN	16.9kJ/kg
	4 (H/R)	161.1kN	92.9kN	20kJ/kg
15m/s	1 (O/O)	82.1kN	36.2kN	12.kJ/kg
	2 (H/O)	82.6kN	48.9kN	14.3kJ/kg
	3 (O/R)	205.6kN	69.5kN	25.7kJ/kg
	4 (H/R)	205.6kN	94.7kN	30.9kJ/kg

Table4. Summary of dynamic crushing analysis on steel beams (L = 400mm)

Initial Velocity	Material Case #	Peak Force	Mean Force	SEA
5m/s	1 (O/O)	71.9kN	31.5kN	3.6kJ/kg
	2 (H/O)	72.1.4kN	46.6kN	5.2kJ/kg
	3 (O/R)	144.4kN	55.1kN	6.9kJ/kg
	4 (H/R)	153.2kN	87.5kN	9.5kJ/kg
10m/s	1 (O/O)	81.7kN	34.8kN	11.9kJ/kg
	2 (H/O)	81.8kN	49.2kN	17.5kJ/kg
	3 (O/R)	178.3kN	66.3kN	16.8kJ/kg
	4 (H/R)	159.3kN	95.5kN	20.8kJ/kg
15m/s	1 (O/O)	82.1kN	34.6kN	11.9kJ/kg
	2 (H/O)	82.4kN	50.3.2kN	17.5kJ/kg
	3 (O/R)	210.1kN	66.9kN	25kJ/kg
	4 (H/R)	210.1kN	97.7kN	32kJ/kg

Effects of Strain Rate and Strain Hardening

It is known that steel is a strain rate sensitive material with high strain hardening ratio, thus the effects of both strain rate and hardening have to be considered in analysis of steel beams. From table 2 to 4, following observations can be made:

1)   Strain hardening doesn't apparently affect the peak crushing force but cause increased mean crushing force and higher energy absorption capability. Obviously, the strain hardening effect is helpful in improving box-sectional beam's crushing performance.

2)   Compare to the models without strain rate effect, the peak crushing force, mean crushing force, and SEA are much higher for the steel beams with strain rate effect. The higher peak crushing force due to the strain rate effect may affect occupant's conformability and safety in vehicle design.

3)   For the steel beam models with strain rate effect, peak crushing force increases much faster at higher impact velocity (v > 10m/s) than lower velocity (v < 5m/s). This is because under the effect of strain rate, with the increase of impact velocity, the corner crushing load grows larger, which causes the highly increasing peak crushing force [2].

The conclusions drawn here perfectly accord with the observations made by Han et al. [2] from a series of quasi-static and dynamic impact experiments. Furthermore, according to Jones [1], the strain hardening and strain rate effects could be decoupled as

$$\sigma_0' = \sigma_0 \left\{ 1 + \left(\frac{\dot{\varepsilon}}{C} \right)^{\frac{1}{q}} \right\} g(\varepsilon) \qquad (2)$$

Where g(ε) stands for strain hardening and in our study it is taken as linear (through E_t).

Effects of beam length and impact velocity

As illustrated before, both strain rate and strain hardening effects have to be accounted when modeling steal beams. In this section, the analysis results from the models with strain rate and strain hardening were plotted in order to find the effects of beam length and impact velocity on the model crushing behaviors. Important results are plotted throughout figure 3 to 5.

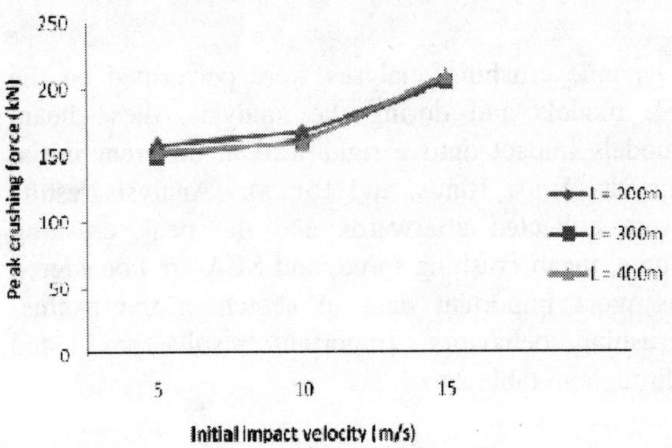

Figure3. Peak crushing forces under various beam lengths and impact velocities

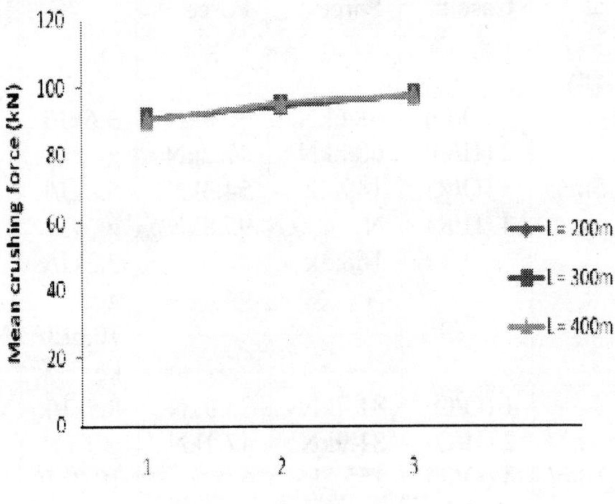

Figure.4 Mean crushing forces under various beam lengths and impact velocities

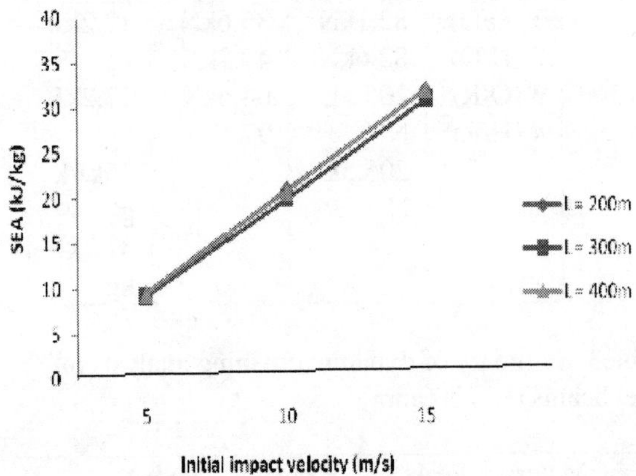

Figure5. SEA under various beam lengths and impact velocities

Effects of beam length

From figure 3 to 5 it is shown that the effect of beam length on its crushing response is very small. Peak crushing force, mean crushing forces, and SEA are almost invariant for beams with different lengths (from 200 – 400mm). Therefore, in predicting thin-walled box-sectional beams' crushing behaviors, the effect of beam length can be neglected and these beams will show a similar collapse mode under different lengths.

Effects of impact velocity

Unlike the beam length, the initial impact velocity apparently affects the peak crushing force and the SEA. As mentioned before, the increasing initial velocity leads to a higher peak crushing force. At higher initial velocity (>10m/s), the increment of peak crushing force becomes higher because the corner crushing load grows larger at high impact velocities. Figure 5 shows that when the initial velocity increases, the SEA increases linearly. However, the initial velocity doesn't arouse a noticeable change in the mean crushing force: the mean crushing force slightly increases with the increasing velocity.

Discussions

From above discussion, it can be seen that in predicting the thin-walled box-sectional beam's crushing behaviors, its mean crushing force can be assumed as independent of beam length and initial velocity. According to Wierzbicki and Abramowicz's theory [3], the mean crushing force can be evaluated based on the beam's material and its cross-sectional dimensions, as shown in Eq. (3).

$$P_m = 52.22 \left(\frac{\sigma_0 t^2}{4}\right)\left(\frac{a}{t}\right)^{0.37} \quad - - \ (3)$$

In our study, without regard to the effects of strain hardening and strain rate, the mean crushing forces obtained from different lengths and under different impact velocities vary from 30.8 to 36.2kN. Substitutes a = 45mm, t = 2mm and yield stress = 200 MPa into above equation, the evaluated mean crushing force value equals 33.1kN, which agrees to our analysis results very well. Similarly, when the strain hardening ratio takes 1.5, the mean crushing force is calculated as 49.6kN using Eq. (3), while the

results yielded from the crashworthiness analysis lie between 46.2 and 50.3kN (table 2 to 4).

Practically, as shown in figure 4, with the effect of strain rate, the mean crushing forces are slightly enhanced due as the impact velocity increased. The tiny effect of the impact velocity on the mean crushing force due to the strain rate sensitivity is described by Jones [1] as

$$P_m = 52.22 \left(\frac{\sigma_0 t^2}{4}\right)\left(\frac{a}{t}\right)^{0.37}\left\{1+(0.33V_0/aC)^{\frac{1}{P}}\right\} - - - - - - \ (4)$$

After substituting C = 40 and P = 5 in that equation, the mean crushing forces under different impact velocities 5m/s, 10m/s, and 15m/s are 57.9kN, 62.1kN, and 64.9kN, respectively. The mean crushing forces obtained from the dynamic numerical analyses vary from 54.3 to 55.1kN (V_0 = 5m/s), 64 to 66.3kN (V_0 = 10m/s), and 64. 6 to 69.5kN (V_0 = 15m/s). Both sets of results still correlate to each other. Similar observations can be obtained considering the effects of both strain hardening and strain rate. By taking strain hardening ratio as 1.5, the mean crushing forces evaluated from Eq. (4) with the different impact velocities increase 1.5 times, which are 86.8kN, 93.2kN, and 97.4kN,. And the numerical results lie between 87.5 to 90.8kN (V_0 = 5m/s), 92.9 to 95.5kN (V_0 = 10m/s), and 94.7 to 97.7kN (V_0 = 15m/s). The comparison results verify Jones' crash theories and again testify that during the axial crushing, the effects of strain hardening and strain rate can be decoupled. The relationship between the SEA and initial impact velocity is almost linearly, therefore, the energy absorption capability SEA under different impact velocities can be easily predicted employing the linear relationship. Nevertheless, the relationship between the peak crushing force and the initial velocity has to be derived based on the results listed in table 2-4 by performing a nonlinear regression analysis.

2. Conclusions

In this paper, the effects of strain hardening and strain rate on the crash response of the steel thin-walled beams with box sections are studied. It is concluded that the strain hardening improves the mean crushing force as well as the SEA while doesn't affect the peak crushing force. In another hand, the strain rate effect leads to much higher peak crushing force, mean crushing force, and SEA. A systemic crash analyses were performed to investigate the influences of the beam length and initial impact velocity on the crushing behaviors of the thin-walled box-sectional beam. The results show that the beam length does not significantly affect such beam's crushing behavior and as the initial velocity increased, both peak crushing force and SEA increased apparently. This study also verifies that LS-DYNA is an efficient tool in computational modeling and analysis.

References

[1] N. Jones, Structural impact, Cambridge University Press, 1997.

[2] H.P. Han, F. Taheri, N. Pegg, Quasi-static and dynamic crushing behaviors of aluminum and steel tubes with a Cut out, Thin-Walled Structures, 45 (2007) 283 – 300.

[3] T. Wierzbicki, W. Abramowicz, On the crushing mechanisms of thin-walled structures, Journal of Applied Mechanics, 50 (1983) 727 – 734.

[4] W. Abramowicz, T. Wierzbicki, Axial crushing of multi-corner sheet metal columns, Journal of Applied Mechanics, 53 (1989) 113 – 120.

[5] T. Wierzbicki, L. Recke, W. Abramowicz, T. Gholmai, Stress profiles in thin-walled prismatic columns subjected to crushing loading – I. Compression, Computers & Structures, 51(6) (1994) 611 – 623.

[6] J. Hallquist, LS-DYNA 3D: Theoretical Manual, Livermore Software Technology Corporation, (1993)

Factors Affecting Wave Propagation Speed in Hyperconcentrated Slurry Flows Carrying Solid Particles

Sanjeev Kumar Sharma

Assistant Professor Skyline Institute of Engineering & Technology Greater Noida-201306

Abstract : **Theory of water hammer in homogeneous and heterogeneous flows have been described and mathematical equations for the wave propagation speed & the extra pressure due to water hammer in both flows have been developed.**

Key words : Slurry, Homogeneous flow, heterogeneous flow, Wave propagation Speed, Water Hammer

I. INTRODUCTION

Pipeline transport has been a advanced technology for transporting a large quantity of bulk material. Principal applications include the transportation of phosphorus concentrate and tailings, coal, coal ash, dredging and filling, collection and disposal of solid wastes and materials waste which involves several industrial sectors such as metallurgy, chemical industry, energy, building materials & others.

Experts believe that transportation through pipe lines will become the fifth means of transport following highways, railways, water and air freight. Slurry transport through pipelines ensures a dust free environment, makes possible full automation, demands substantially less space and requires a minimum of operating staff. However frequently utilized efficient slurry transport still posses serious engineering problems through the properties of slurry .These Include, but are not limited to, particle settling, attrition, pipe/fitting/impeller wear, degradation of flocculated or friable solids and pump ability of the slurry. The concentrations of slurries being transported are increasingly higher and the distance of pipelines becomes longer and longer. For instances, the U.S. has the world largest and longest coal slurry pipeline-the Black Mesa pipeline from Arizona to Nevada, over a distance of 273 miles. It has an 18 inch diameter pipeline that transports 5 million tons of coal per year approximately. Numerous short-distance pipelines are being used to transport tails and coal ash. Slurries in these pipelines are pressurized by a number of pump stations at various sections along the length of pipe, and therefore designer must take conditions for both steady and unsteady flow due to the operation of valves, problems in power supplies or mechanical failure into consideration[1] .

Fluid flow principle indicates that a sudden change of velocity in a closed conduit will result in an instant variation of pressure, which is called "water hammer". In case of slurry flow with hyperconcentrated solid particles it is called slurry hammer. Slurry hammers behave differently from water hammers due to different densities and elastic moduli at different points.

Slurry is usually classified into types, homogeneous and heterogeneous slurry. Homogeneous slurry is one which does not exhibit a measurable concentration gradient of solids along the vertical axis of the pipe. However, in practical sense a better definition would be that it is slurry in which the inertia effect of suspended particles is relatively minor. Heterogeneous flow is characterized by a pronounced solid concentration gradient across the vertical axis of the pipe, and it is encountered in wide variety of commercial applications ranging from dredging to coal transportation.

II. WAVE PROPAGATION SPEED FOR SLURRY HAMMER

Wave propagation speed for slurry hammer is analyzed in a similar way by taking the typical case of rapid closure of a valve at the end of a pipe as an example. When the valve is closed the pressure in the pipeline will increase due to fluid motion caused by inertia. The principle of continuity requires that the net mass inflow due to inertia be equal to the expansion of the pipe walls plus the volume compression of the water and particle. Both the expansion and compression are very small because of very high elastic moduli of the pipe material, water & solids. In order to hold the surplus slurry the expansion front of the pipe will propagate rapidly upwards at a certain speed (Fig. 1).

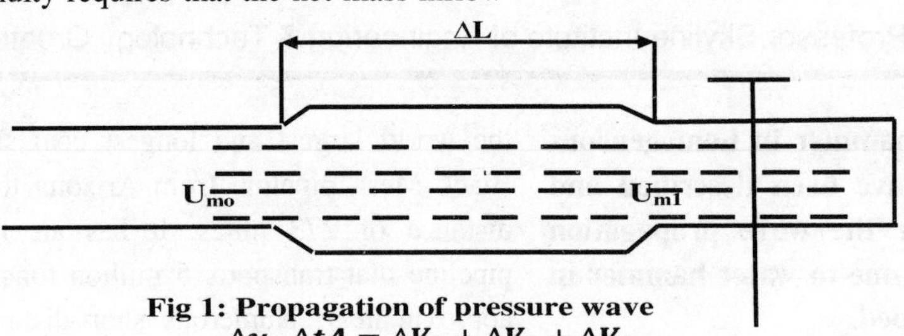

Fig 1: Propagation of pressure wave
$$\Delta u_m \Delta t A = \Delta V_S + \Delta V_L + \Delta V_P$$

This speed is called the wave propagation speed, which is denoted by a_m. An equation for calculating the wave propagation speed can be derived based on principle of continuity, i.e. the increment of the slurry for the expanded segment ΔL equals the sum of the volumetric increment of the pipe and the volumetric compression of water and that of solid particles in the expanded segment ΔL .

Equation for wave propagation speed in a homogeneous flow

Suppose the initial velocity of the homogeneous flow is u_{mo} and the velocity after a time interval Δt is u_{mi} . Let the increment of the liquid for the expanded segment ΔL is

$$V = A\Delta t(U_{mo} - U_{mi}) = A\Delta t\, \Delta u_m \tag{1}$$

in which A is the pipe cross-sectional area and Δt is the time interval. Suppose the increment of pressure due to the change in velocity is P , and the volumetric compression of water in the expanded segment ΔL can be derived from the definition of elastic modulus of liquid E_L .

$$E_L = P \Big/ \frac{\Delta V_L}{(1 - C_v)A\Delta L} \tag{2}$$

in which C_v is the solid concentration by volume and ΔV_L is the volumetric compression of the liquid. From eq.(2) it is found that

$$\Delta V_L = \frac{P}{E_L(1 - C_v)A\Delta L} \tag{3}$$

Similarly the volumetric compression of the solid particles in the expanded segment ΔL can be found out as

$$\Delta V_S = \frac{P}{E_S} C_V A\Delta L \tag{4}$$

In which E_S is the elastic modulus of solids. The volumetric increment ΔV_P can be found out according to the increment of tangential strain, diametrical strain and cross-sectional area due to increment of pressure in the conduit[1] .

The tangential strain
$$\varepsilon_T = \frac{\lambda_T}{E_P} = \frac{1}{E_P}\frac{PD}{2e} \tag{5}$$

in which E_P is the elastic modulus of the pipe material; λ_T is the tension stress in the pipe wall; D is the pipe diameter and e is the pipe wall thickness. The diametrical strain is

$$\Delta R = \frac{D}{2}\varepsilon_T = PD^2 \Big/ 4E_P e \tag{6}$$

The increment of pipe cross- sectional area is given by

$$\Delta A = \pi D\Delta R = \pi\frac{D^2}{4}\frac{PD}{eE_P} = A\frac{PD}{eE_P} \tag{7}$$

If the axial strain is neglected, the volumetric increment in the segment ΔL Of the pipe can be expressed as

$$\Delta V_P = \frac{PD}{eE_P} A\Delta L \tag{8}$$

According to law of continuity,

$$\Delta u_m A\Delta t = \frac{P}{E_L(1 - C_v)A\Delta L A} + \frac{P}{E_S}C_V A\Delta L + \frac{PD}{eE_P} \tag{9}$$

From definition of wave propagation speed we
$$\frac{\Delta L}{\Delta t} = a_m \tag{10}$$

According to the momentum law,

$$AP\Delta t = \rho_m A\Delta L \Delta u_m \tag{11}$$

$$\Delta u_m = \frac{P}{\rho_m a_m} \tag{12}$$

Where the density of homogeneous flow ρ_m is given by

$\rho_m = \rho_s C_v + (1 - C_v)\rho_L$, in which ρ_s and ρ_L are the densities of the solids and water respectively.

Simultaneous solution of eq.(9), (10), and (12) gives the wave propagation speed for homogeneous flow a_{m1} as follows:

$$a_{m1} = \sqrt{\frac{E_L/\rho_m}{1 - C_v + \frac{E_L}{E_S}C_v + \frac{E_L}{E_P}\frac{D}{e}}} \tag{13}$$

Equation for wave propagation speed in a heterogeneous flow

In heterogeneous flows the velocity of solid particles lags behind the surroundings no matter whether the flow is steady or not. Thus, in continuity equation of solid liquid flow the velocity of the solids and the liquids must be considered separately [2]. Suppose the initial velocities of the solids and liquids are u_{s0} and u_{L0} respectively, and velocities after time interval Δt are u_{s1} and u_{L1} respectively. The continuity equation of non homogenous is

$$[C_v(u_{s0} - u_{s1}) + (1 - C_v)(u_{L0} - u_{L1})]A\Delta t$$
$$= \frac{PC_v\Delta LA}{E_S} + \frac{P(1 - C_v)\Delta LA}{E_L} + \frac{PDA\Delta L}{E_P e} \tag{14}$$

The momentum equation of heterogeneous flow is

$$AP\Delta t = \rho_s C_v A\Delta L\Delta u_s + \rho_L(1 - C_v)A\Delta L\Delta u_L \tag{15}$$

Because the increment of pressure on the cross section is uniform, the impulse can also be assumed to be volumetrically uniform, i.e.

$$C_v AP\Delta t = \rho_s C_v A\Delta L\Delta u_s \tag{16}$$

$$(1 - C_v)AP\Delta t = \rho_L(1 - C_v)A\Delta L\Delta u_s \tag{17}$$

Flow equation can be obtained from the above two equations and the definition of wave propagation speed $a_m = \dfrac{\Delta L}{\Delta t}$

$$\Delta u_s = \frac{P}{\rho_s}\frac{1}{a_m} \tag{18}$$

$$\Delta u_L = \frac{P}{\rho_L}\frac{1}{a_m} \tag{19}$$

Substitute equations (18) and (19) into equation (13) and simplify the result. The formula of the wave propagation speed for heterogeneous flow a_{m2} is obtained.

$$a_{m2} = \sqrt{\frac{\left(\frac{C_v}{\rho_s} + \frac{1 - C_v}{\rho_L}\right)E_L}{1 - C_v + \frac{E_L}{E_S}C_v + \frac{E_L}{E_P}\frac{D}{e}}} \tag{20}$$

Comparison of wave propagation speed equations for homogeneous and heterogeneous flows

The above analysis of the equations for wave propagation speed for the two kind of flows are based on the assumptions that the motion of the solid particles exhibits completely different patterns in the two kinds of flows [3]. To make a detailed analysis of the difference between equations (13) and (20), we have

$$\frac{a_{m2}}{a_{m1}} = \sqrt{\frac{C_v\rho_m}{\rho_s} + \frac{(1 - C_v)\rho_m}{\rho_L}} \tag{21}$$

Substitute $\rho_m = \rho_s C_v + (1 - C_v)\rho_L$ into equation (21), the following is obtained:

$$\frac{a_{m2}}{a_{m1}} = \sqrt{\left[\left(2 - \frac{\rho_s}{\rho_L} - \frac{\rho_L}{\rho_s}\right)(C_v^2 - C_v) + 1\right]} \tag{22}$$

Equation (22) shows that, with other conditions being identical, the ratio of wave propagation speeds for two kind of flows is decided by the concentration of solid particles and the ratio of densities of two liquids, i.e.

$$\frac{a_{m2}}{a_{m1}} = f\left(\frac{\rho_s}{\rho_L}C_v\right).$$

Fig. 2A, 2B, 2C indicates the relationship between $\frac{a_{m2}}{a_{m1}}$ and C_V with $\frac{\rho_s}{\rho_L}$ being a parameter. Fig. (2A, 2B, 2C) indicates that when $\frac{\rho_s}{\rho_L}$ increases, for example, $\frac{\rho_s}{\rho_L} = 4.8$ for iron concentrate, the ratio of wave propagation speeds calculated by the two equations increases rapidly with the solid concentrations. The wave propagation speed calculated for heterogeneous flow is 30% greater than that for homogeneous flow. So when the solid density is large the calculating formula should be chosen with great care in order to avoid significant error.

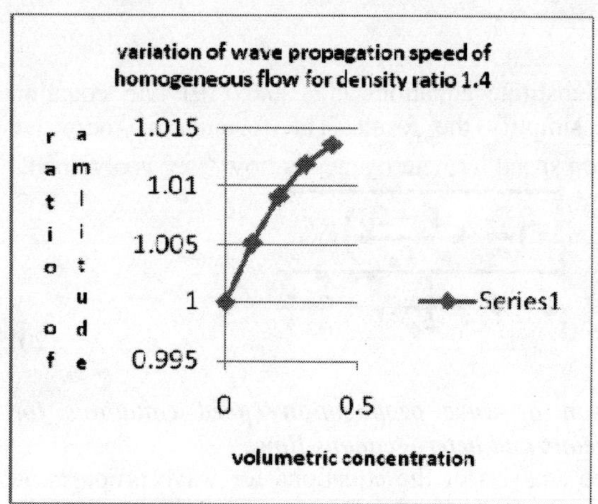

Fig 2A

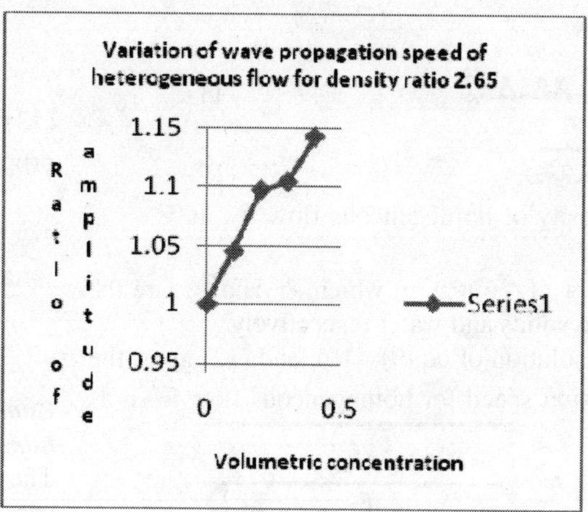

Fig 2B

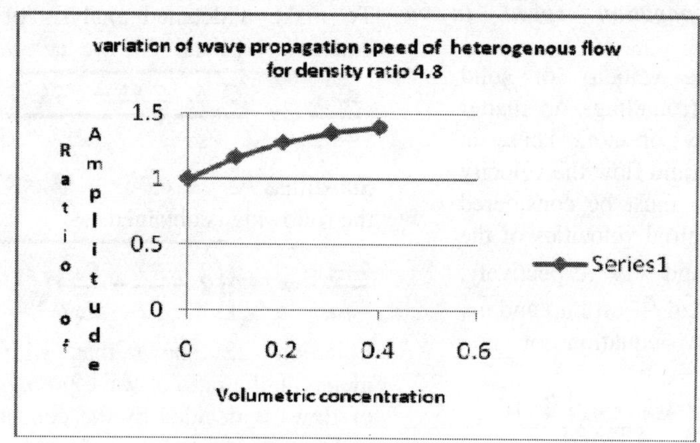

Fig 2C

Factors affecting wave propagation speed.
The impact of the physical properties of the solid particles on wave propagation speed can be found out by comparing the equations of wave propagation speed for homogeneous flow with that for clear water [4]:

$$\frac{a_{m1}}{a_0} = \sqrt{\left[1 + \frac{C_V\left(\frac{E_L}{E_S} - 1\right)}{1 + \frac{E_L D}{E_P e}}\right]} \cdot \sqrt{\left[C_V \frac{\rho_S}{\rho_L} + (1 - C_V)\right]} \quad (23)$$

Equation (23) shows that $\frac{a_{m1}}{a_0}$ is a function of relative density $\frac{\rho_S}{\rho_L}$, relative elastic modulus $\frac{E_S}{E_L}$, deflection of pipe $\frac{E_L D}{E_P e}$ and the solid concentration by volume C_V.

III. CONCLUSIONS

From this study the following conclusions can be drawn:

1. The slurry water hammers in pipelines with hyperconcentrated slurry flows carrying solid particles are divided into two types, i.e. homogeneous flow hammer and heterogeneous flow hammer. In engineering practice, most of the slurries can be regarded as homogeneous flow, as the transported particles are very fine due to considerations of slurry stability.

2. Factors influencing wave propagation speed include deflection of pipe, elastic moduli of water and solid, solid concentration by volume, and solid density.

IV REFERENCES

[1]   Wang Shuren, *Construction of Hydraulic Power Station, Beijing: Publishing House of Tsinghua University, 1984.*

[2]   *Wood, D. J., Kao Ten-Yu, Unsteady flow of solid liquid suspensions, ASC, 1966, 92(EM6):117*

[3]   *Wasp, E. J., Solid-Liquid Flow Slurry Pipe Line Transportation, Hydraulic Publishing House, 1980, 126-132.*

[4]   *Thomas, D. G. , Transport characteristics of suspensions part IV, AICHE journal , 1962,8:373*

Strategic Planning of Production System of a Small Industrial Unit Using Fuzzy Cognitive Map

Bhupesh Kumar[1], G. P.Mishra[2], S.K.Gaur[1], D.S Mishra[1]
[1]Department of Mechanical Engineering, Faculty of Engineering
[2]Technical College Dayalbagh Educational Institute, Agra, Uttar Pradesh

Abstract-**Small scale industries are essential links in the chain of Indian economy. These have helped in conserving Indian art, culture, craftsmanship and skill. These industries have assumed an important distinction on account of certain features namely employment potential, short gestation period, utilization of local resources, export earnings, etcetera. Importance of manufacturing and its effects on the general standard of living cannot be denied. Manufacturing industries generate a substantial proportion of wealth in most of the developed or developing countries. The manufacturing sectors of many countries also play a key role in international trade because manufactured exports provide a major source of international currencyFuzzy cognitive maps (FCMs) have become an important means for describing a particular domain showing the concepts and relationship between them.An attempt has been made in the present paper to focus the different strategic planning situations of a production unit of small scale industry of Agra region using FCM to investigate various decision situations.**

Keywords: Small scale industrial system, Fuzzy Cognitive Map, Strategic planning, Production system.

I.   INTRODUCTION

Small scale industries have been the back bone of Indian economy and the nurturing ground of Indian art, culture and skill. On one hand, it provides employment while on the other it caters to the need of consumers, medium and large scale industries. [1]

Small scale industries constitute integral part of India's industrial structure. This sector covers wide spectrums of industry categorized into village and small scale enterprises.

This sector has acquired a prominent place in socio-economic development of the country. According to the latest world development indicators, India is now world's fourth economy after USA, China and Japan in purchase power parity (PPP) terms measured in 1999. Indian national income in PPP terms during 1999 was $ 2.23 trillion. [2]

II. SMALL SCALE INDUSTRIES INTERNATIONAL PERSPECTIVE

Performance of small scale industries in the globalization era is vital all over the world because of its role in enriching employment and capital formation [3]. In this reference the status of small scale industry in Jordan is worth consideration which may contribute to a greater extent to its growth. A case of Jordan is worth mentioning here as in Table 1 which indicates the relative importance of total number of the small scale industries in Jordan with respect to all industries in Jordan. Table 2 indicates the relative importance of total number of employees in the small scale industries in Jordan with respect to all industries in Jordan. It is clear that the share of small scale industries in Jordan increased from 89% in1990 to 90% in 2006 with minor deviation in between showing a greater thrust of SSI with respect to others while it is evident from Table 2 that the relative importance remained around 19 percent for sixteen years since 1990. This clearly demonstrates a substantial contribution of small sector in alleviating the unemployment rate in Jordan [3].

Table 1: the relative importance of SSIs in Jordan.

Table (2): The relative importance of the SSIs in Jordan (1990-2006)

	No. of SSIs	No. of all industries	Relative importance
1990	52496	58984	89%
1991	54696	61388	89%
1992	64752	74369	87%
1993	66160	75920	87%
1994	66100	80363	81%
1995	74896	91520	82%
1996	76468	93432	82%
1997	78580	95856	82%
1998	81856	98768	83%
1999	92923	109532	84%
2000	106730	118271	90%
2001	103753	115820	89%
2002	105662	118492	89%
2003	130149	140338	92%
2004	126196	141944	88%
2005	132166	145699	90%
2006	135732	150222	90%

Source: Department of statistics, yearly statistical bulletins, various issues.

Table-2 Total number of employees in the small industries in Jordan.

(5): Total no. of employees in the small scale industries in Jordan (1990-2006)

Year	Small scale industries	All industries	Relative importance
1990	80134	421191	19.0%
1991	83522	455621	18.3%
1992	855121	479131	17.8%
1993	88991	482231	18.4%
1994	90115	491522	18.3%
1995	93145	501993	18.5%
1996	96244	521177	18.4%
1997	98100	539981	18.2%
1998	98541	544189	18.1%
1999	100241	563131	17.8%
2000	103788	585140	17.7%
2001	93269	575930	16.2%
2002	82008	573243	14.3%
2003	123248	617466	20.0%
2004	97340	705838	14.0%
2005	100541	706111	14.0%
2006	105211	706233	15.2%

Source: Department of statistics, yearly statistical bulletins, various issues.

III. THE INDIAN PERSPECTIVE

In India small scale industries also have a rapid growth rate as given in Table 3.

Table 3: Growth rate of SSIs in India.

Year	Growth rate of SSI sector (%)	Growth rate of overall industrial sector
1993-94	5.7	6.0
1994-95	10.0	9.1
1995-96	11.5	13.0
1996-97	11.3	6.1
1997-98	9.2	6.7
1998-99	7.8	4.1
1999-2000	7.1	6.7
2000-01	8.0	5.0
2001-02	6.1	2.7
2002-03	7.7	5.7
2003-04	8.6	6.9
2004-05	9.96	8.4

Further, it is evident from Table 4 that small scale industries in India contribute substantial share to "total industrial production" and the "gross domestic product" [4].

Table-4 Contribution of SSI to GDP

Year	Contribution of SSI (%)	
	Total industrial production	Gross Domestic Product (GDP)
1997-98	39.70	7.02
1998-99	39.94	6.81
1999-00	40.02	6.69
2000-01	39.91	6.86
2001-02	39.63	6.67
2002-03	39.48	6.82
2003-04	39.42	6.71

Importance of manufacturing and its effects on the general standard of living cannot be denied. Manufacturing industries generate a substantial proportion of wealth in most of the developed or developing countries and employ a major part of their population. [5]

The manufacturing sectors of many countries also play a key role in international trade because manufactured exports provide a major source of international currency to purchase goods and raw material from countries better endowed with natural resources. Exports also provide the means of affording luxury products and technologies of other nations. These are the two factors that play an ever increasing part in motivating less developed countries to expand their own manufacturing capabilities. [6]

However, Industrial enterprise is a social system functioning in a social environment consisting of various other social systems with whom it interacts in terms of specific role relationships. [6]

IV. IDENTIFICATION OF ELEMENTS

Six domain experts from Benara Auto Private Limited, Agra and academia were selected for the purpose of identifying key elements that affect Production system. These experts were asked to give five important elements that they consider would play vital role in production system. They were asked to prioritize these elements also. The list was compiled and collated to give eight elements which bore more then one vote. Those having one vote were discarded. The list of selected elements and their votes is given in Table 5.

Table: 5 List of Potent Element and their Votes.

Elements	Vote	Representation

1. Production planning level	5	PPL
2. Level of plant operation	3	LPO
3. Operational status of plant	4	OSP
4. Production level	2	PL
5. Inventory of finished good	2	IFG
6. Pressure for quality control	2	PQC
7. Rejection level	2	RL
8. Rate of acquisition of new capacity	4	RANC

V. DEVELOPMENT OF FUZZY COGNITIVE MAP:

A graphical representation is often simpler and more lucid than quantitative approaches that are generally inadequate to deal with causal interrelationships [7].

A cognitive map is a graphical representation of mental beliefs of an underlined entity by an expert following a set of discursive representations from his own cognitive representation on a particular subject to help him to think about the presented reality [8, 9].

Fuzzy Cognitive Maps have become an important means for describing a particular domain showing the concepts and the relationship between them. They have been used for several tasks like simulation processes, forecasting or decision support. In general, the task of creating FCM is made by experts in a certain domain but it is very promising the automatic creation of Fuzzy Conceptual Maps from raw data [10].

FCMs are fuzzy-signed digraphs with feedback. Nodes in the graph that are Fuzzy Sets representing concepts. Directed edges (arrows) represent causal-effect relations between the concepts as in the case of generic Cognitive Maps. Arrows can have positive or negative values; a positive value shows a positive causal connection [11]. As is depicted in Fig.1 the value of concept B increases or decreases as concept A increases or decreases. Whereas in Fig.2 a negative causal connection causes the value of the concept B to decrease when the value of concept A increases, and also a negative causal connection causes the value of concept B to increase when the value of concept A decreases [12].

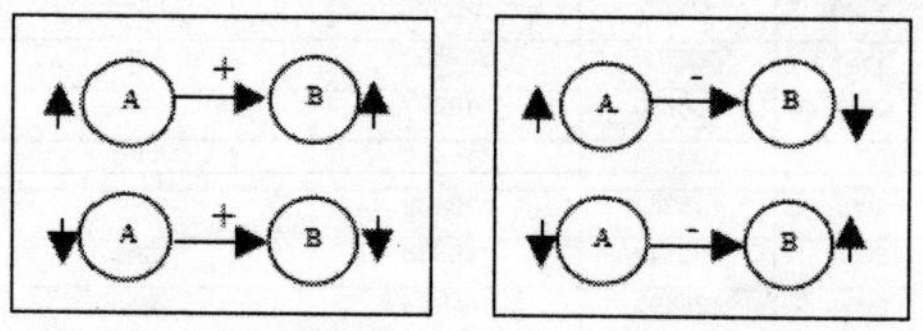

fig1.Positive causal weight   fig2. F[?]ative causal weight

Once the potent elements have been selected, the fuzzy cognitive map is developed for the production system of small scale industrial unit. For the development of the FCM the experts were asked to give their opinion to show the relationship among the different selected elements and to show the impact of one element on others which may be positive or negative. The developed cognitive map for the selected eight elements is shown in Fig.3.

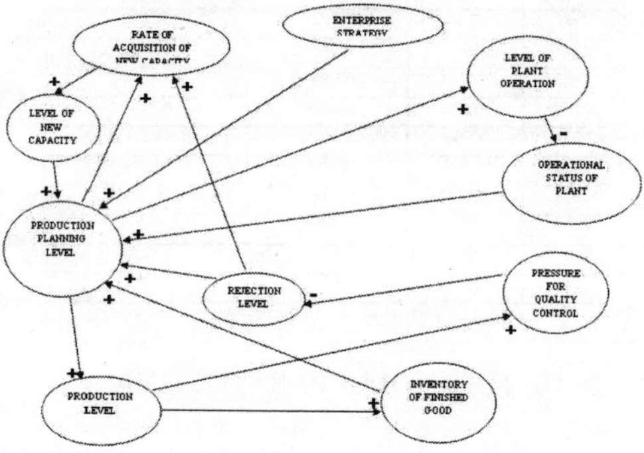

Fig. 3 Causal Loop Diagram of Production System

VI. KNOWLEDGE ACQUISITION AND DATA ANALYSIS

A FCM integrates the accumulated experience and knowledge using human experts. FCM is a signed directed graph with feedback, consisting of nodes and weighted arcs. Nodes of the graph stand for the concepts / elements, used to describe the behavior of the system. They are connected by signed and weighted arcs representing the causal relationships that exist between the concepts (Figure 4).

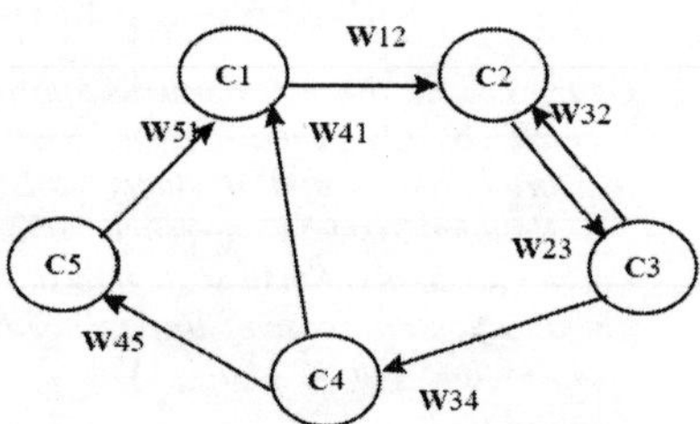

Fig.4, A simple Fuzzy Cognitive Map

It is noticed that all the values in the graph are fuzzy, so concepts take values in the range between [0, 1] and the weights of the arcs are in the interval [-1, 1]. Each concept is characterized by a number Ai that represents its value and it results from the transformation of the real value of the system's variable, for which this concept stands, in the interval [-1,1].

There are three possible types of causal relationships

 (a). Concept C_i and concept C_j could be positively related if $W_{ij} > 0$ which means that, an increase in the value of concept Ci leads to the increase of the value of concept C_j , and a decrease in the value of concept Ci leads to the decrease of the value of concept C_j .

 (b). There is negative causality between C_i & C_j if $W_{ij} < 0$ which means that, an increase in the value of concept C_i leads to the decrease of the value of concept C_j and vice versa.

 (c). There is no causality between C_i & C_j implying $W_{ij}=0$.

The value of each concept is influenced by the values of the connected concepts with the appropriate weights and by its previous value. So the value A_i for each concept Ci is calculated by the following rule

$$C_{t+1} = f(C_o W) \qquad (1)$$

Where C_{t+1} is the activation level of concept element at time t+1, W is the weight matrix of the interconnection between concepts elements, $_o$ multiplication operator and f is a threshold function. In the present paper the thresh hold function has been taken as

$f(x) = x$ for x $\square$ (0, 1),

$f(x) = 1$ for x ≥ 1, and

$f(x) = 0$ for x ≤ 0.

The new vector shows the effect of the change in the value of one concept in the Fuzzy Cognitive Map. The equation (1) can be utilized to iterate the process to get new vectors till stabilization occurs.

In the knowledge acquisition process the fuzzy information about the different elements of the developed causal map were obtained from the experts. The experts were asked to give their views about the strength of effect between the elements, on a linguistic scale quantified between .1 and 0.9. Here 0.1 shows very poor strength of effect, 0.2 for moderately poor, 0.3 for poor, 0.5 for average, 0.6 for moderately good,0.7 for good, 0.8 for moderately very good and 0.9 for very good strength of effect between the elements.
It is observed from the causalities that

 • If the Production planning level (PPL) is high, then the Level of plant operation (LPO) will increase.

 • If the Level of Level of plant operation (LPO) high, then OSP will be low.

 • If the Pressure for quality control (PQC) is high, then RL will decrease.

The weighted matrix (W) of the knowledge base is derived from the questionnaire filled by the experts based on the consensus the weight matrix is given in the Appendix I.

VII. RESULT AND CONCLUSION:

For the production system of small scale industrial unit numerous of situations have been studied. Only few of them are described here.

Situation 1: In this case, it has been assumed that only Production planning level is active on an average level and rests are not active. This is designated by switch vector $C_{1, t}$ = [.5 0 0 0 0 0 0 0]. This gives rise to a new state $C_{1, t+1}$ = [.5 .35 0 .4 0 0 0 .3]. Finally it stabilizes (Fig.5) after second iteration to give the state vector as [.5 .35 .21 .40 .32 .28 0 .3].
This implies that when Production planning level is active by an average then it results the Production level is below average, Rate of rejection is zero and other elements in poor condition.

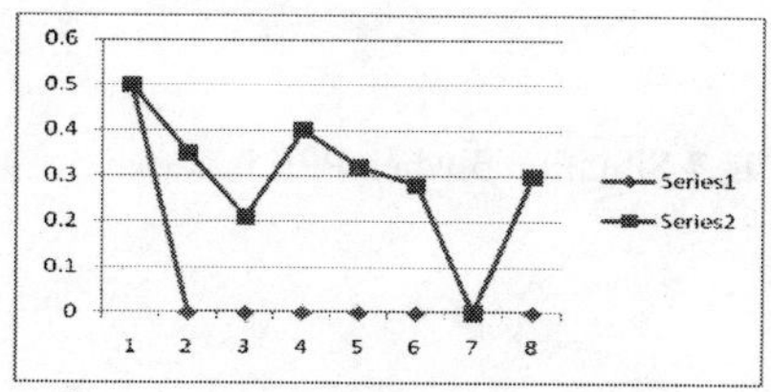

Fig.5 Situation 1 when PPL is active by average value.

Situation 2: Here the Production planning level is raised from an average level to good level of amounting .7 and rests are not active. This is designated by switch vector $C_{1, t}$ = [.7 0 0 0 0 0 0

0]. This gives rise to a new state $C_{1, t+1}$ = [.7 .49 0 .56 0 0 0 .42]. Finally it stabilizes (Fig.6) after second iteration to give the state vector as [.7 .49 .29 .56 .49 .39 0 .42].

This implies that when Production planning level raise from an average level then it results the Production level also raises from below average to moderately good.

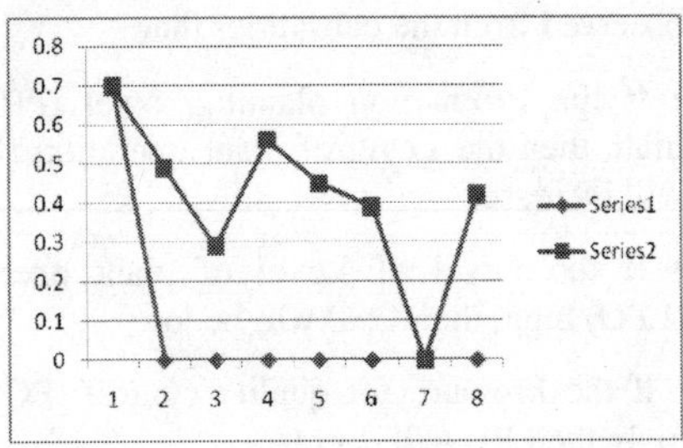

Fig.6 Situation 2 when PPL is active by .7.

Situation 3: In this situation, the Production planning level is raised from good to very good and rests are not active. This is designated by switch vector $C_{1, t}$ = [.9 0 0 0 0 0 0 0]. This gives rise to a new state $C_{1, t+1}$ = [.9 .63 0 .72 0 0 0 .54]. Finally it stabilizes (Fig.7) after second iteration to give the state vector as [.9 .63 .38 .72 .58 .50 0 .54].

This implies that when Production planning level is raised from .7 to .9 then it results the raise in Production level from moderately good to good, Level of plant operation raises from average to moderately good, Inventory of finished goods raises from average to moderately good and Rate of rejection is zero.

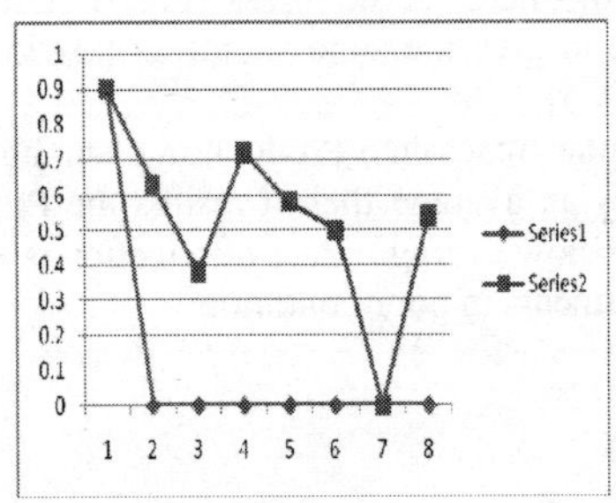

Fig.7 Situation 3 when PPL is active by .9.

VIII. REFERENCES:

[1]   Gaur. S.K et. al "A Cybernetic Approach for demand based Scenarios of Small scale industrial system and its fuzzy satisfaction" 21st National System Conference, 2007.

[2]   Basem. M. et. al "Small scale industries in the globalization era: case of Jordan", Journal of business and public affairs, Vol. 2, issue 1, 2008

[3]   Bencivenga. V., Bruce. S. and Starr. R., "Report on SSIs in Jordan", 1995.

[4]   "Comprehensive Policy Package on SSI and Tiny Sector"- announced by the Prime Minister on 30th August 2000.

[5]   Mare. R.F. "Manufacturing system economics", Holt, Rinehart and Winston Ltd, N.Y.1982.

[6]   Saini. Ram Rattan, Kumar. Parmod, Kumar. Mohan, "Sickness in small-scale industries in India".

[7]   Axelrod. R., "Structure of Decision: the Cognitive Maps of Political Elites", Princeton University Press, New Jerssey, 1976.

[8]   Kosko. B., "Neural Network and Fuzzy Systems", Prentice-Hall, 1992.

[9]   Kosko. B., "Fuzzy Cognitive Maps", International Journal of Man-Machine Studies, Vol. 24, pp 65-75, 1986.

[10]   Taber. R., "Knowledge Processing with Fuzzy Cognitive Maps", Expert Systems with Applications, Vol.2, No.1, pp 83-87, 1991.

[11]   Zhang. W.R. et. al "A Logical Architecture for Cognitive Maps", IEEE International Conference on Neural Networks, 1988.

[12]   Kumar. Bhupesh, et. al "Modeling Small Industrial System using Fuzzy Cognitive Map", Proceedings of National Systems Conference, NSC-2007.

Appendix I

Table: 6 Weight matrix for considered elements

	Production planning level	Level of plant operation	Operational status of plant	Production level	Inventory of finished good	Pressure for quality control	Rejection level	Rate of acquisition of new capacity
Production planning level	0	.7	0	.8	0	0	0	.6
Level of plant operation	0	0	.6	0	0	0	0	0
Operational status of plant	.8	0	0	0	0	0	0	0
Production level	0	0	0	0	.8	.7	0	0
Inventory of finished good	.6	0	0	0	0	0	0	0
Pressure for quality control	0	0	0	0	0	0	-.6	0
Rejection level	.8	0	0	0	0	0	0	.8
Rate of acquisition of new capacity	.8	0	0	0	0	0	0	0

The Finite Element Modelling and Analysis of Square Plate using Abaqus

Abdullah Aamir Hayat, Adnan Akhlaque and M. Naushad Alam

Mechanical Engineering Deptt, ZHCT AMU, Aligarh,

*Abstract-***The study of the influence of loading and different thickness of plate on deflection and stress pattern is very importat.A simple quadrilateral eight noded element S8R with 24 DOF is considered. Reissner–Mindlin plate theory is used forformulation. The comparision of stress distribution and bending among thick and thin plates is presented in this paper. The method involves the fundamental equations pertaining to displacement components, strain components, stresses and total potential energy due to externally applied loads. Uniformly loaded square plates with different boundary condition namely all sides fixed and all sides simply supported are analyzed using ABAQUS, with different thickness to plate-length ratios.**

Keywords: Mindlin-Reissner theory, Thin Plates, Thick Plates, Abaqus, Finite Element Analysis.

I. INTRODUCTION

Plates are defined as plane structural elements with a small thickness compared to the planar dimensions. A flat plate are extensively used in many engineering applications like roof and floor of buildings, deck slab of bridges, foundation footings, water tanks, bulk heads, turbine disks etc. Plates may be considered similar to beams, however: Plates can bend in two directions and twist, also Plates must be flat (or else they are shells)[1].

Many researchers considered the problem of the bending of rectangular plates undergoing large lateral deflection. Levy [2, 3] solved the above equation for rectangular plates.Berger [4] presented an approximate method for the large deflection analysis of plates. An excellent bibliography on the non-linear static analysis of plates is given in the book by Sathyamoorthy [5]. Ai-Kah Soh et. al. [6] provided a simple approach to formulate a quadrilateral element for the analysis of thick and thin plates.

Of the numerous plate theories that have been developed since the late 19th century, two are widely accepted and used in engineering. These are a)The Kirchhoff-Love theory of plates (classical plate theory) b) The Mindlin-Reissner theory of plates (first-order shear plate theory). The problem of a rectangular plate clamped on four sides and carrying a uniformly distributed load is of great importance and many papers have been devoted to this subject. Many authors have calculated the deflections of uniformly loaded rectangular plates with clamped edges using different methods. Vasiliev [9] presented the history of plate theories and the present state of the art of the sixth-order plate theories originally developed by Reissner, Hencky and Bolle.

The finite element method (FEM) is a computer-aided mathematical technique for obtaining approximate numerical solutions to the abstract equations of calculus that predict the response of physical systems subjected to external influences (Burnett 1988)[8]. ABAQUS is a powerful finite element software package. It is used in many different engineering fields throughout the world. ABAQUS performs static and/or dynamic analysis and simulation on structures. ABAQUS is developed and supported by Hibbitt, Karlsson & Sorensen, Inc. (HKS). ABAQUS includes four functional components:

Analysis Modules, Pre-processing Module, Post processing and Module Utilities. Singh and Elaghabash [9] reported a numerical (Ritz type) formulation procedure for the analysis of non-linear plate bending problems using the "Reissner–Mindlin plate theory". This paper analysis is based on Mindlin plate elements, which have independent displacement w and rotations θx and θy based on quadratic shape functions is briefly reviewed for analysis of both moderately thick and thin plates. Finite element procedure for the formulation of Mindlin plate element is used for the analysis of plate bending problem.

II. ANALYSIS/ FORMULATION

Finite Element Formulation Of Mindlin Plate

An 8-noded rectangular plate element with quadratic shape functions is shown in Fig. 2 is used to examine with the formulation of the Mindlin plate element. The shape functions in the natural coordinates for the quadratic rectangular element are given as

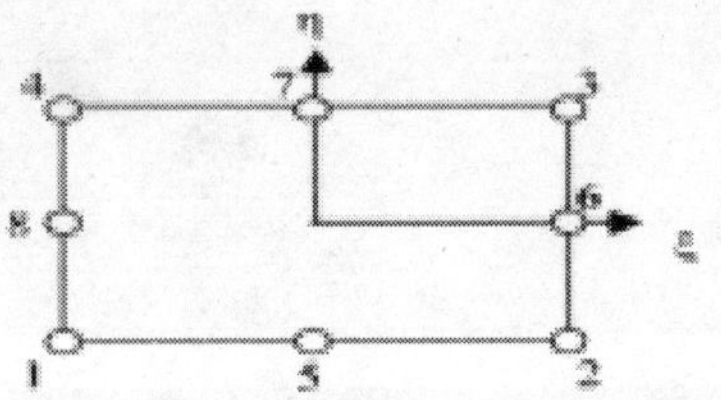

$$N_i = \frac{1}{4}(1 + \xi_i\xi)(1 + \eta_i\eta)(\xi_i\xi + \eta_i\eta - 1)$$

for corner nodes i = 1, 2, 3, 4

$$N_i = \frac{1}{2}(1 - \xi^2)(1 + \eta_i\eta)$$

for mid-side nodes i = 5, 7

$$N_i = \frac{1}{2}(1 + \xi_i\xi)(1 - \eta^2)$$

for mid-side nodes i = 6, 8

Fig. 1: A quadratic rectangular plate element

Fundamental Relations

At a typical point in a Mindlin plate, the displacement components may be represented as

$U = z\theta_x(x,y)$
$V = z\theta_y(x,y)$ (1)
$W = w(x,y)$

Where U, V, and W are the displacement components in the x, y, and z directions, respectively, w is the transverse displacement, and θ_x and θ_y are the normal rotations of the mid-plane, respectively. The rotations θ_x and θ_y can be expressed in the form

$$\theta_x = \frac{\partial w}{\partial x} - \gamma_{xz}$$
$$\theta_y = \frac{\partial w}{\partial y} - \gamma_{yz} \qquad (2)$$

where $\partial w/\partial x$ and $\partial w/\partial y$ are the slopes of the deformed median surface in the x and y directions and γ_{xz} and γ_{yz} are the shear strains.

For a displacement-based, n-noded Mindlin plate element, the displacement and normal rotations may be represented by the expressions

$$w = \sum_{i=1}^{n} N_i w_i, \qquad \theta_x = \sum_{i=1}^{n} N_i \theta_{xi}, \qquad \theta_y = \sum_{i=1}^{n} N_i \theta_{yi} \qquad (3)$$

where w_i, θ_{xi} and θ_{yi} are the displacement and rotation values at node i and Ni is the shape function with node i.

Strain–Displacement Relations

For the Mindlin plate theory, the strain components may be written in terms of the displacements of the middle surface as follows:

$$\varepsilon_x = z\frac{\partial \theta_x}{\partial x}, \qquad \varepsilon_y = z\frac{\partial \theta_y}{\partial y}$$

$$\gamma_{xy} = z\left(\frac{\partial \theta_y}{\partial x} + \frac{\partial \theta_x}{\partial y}\right) \qquad (4)$$

$$\gamma_{xz} = \frac{\partial \omega}{\partial x} - \theta_x, \qquad \gamma_{yz} = \frac{\partial \omega}{\partial y} - \theta_y$$

Curvature–displacement relations and shear strain–displacement relations are then written in the matrix form

$$[\varepsilon] = \begin{bmatrix} \varepsilon_b \\ \varepsilon_s \end{bmatrix} = \sum_{i=1}^{n} \begin{bmatrix} B_{b_i} \\ B_{s_i} \end{bmatrix} [d_i] \qquad (5)$$

in which the curvatures [ε_b], the shear strains [ε_s], and the unknowns at node i [d_i] are expressed as

$$[\varepsilon_b] = \begin{bmatrix} \varepsilon_x & \varepsilon_y & \gamma_{xy} \end{bmatrix}^T$$
$$[\varepsilon_s] = \begin{bmatrix} \gamma_{xz} & \gamma_{yz} \end{bmatrix}^T \qquad (6)$$
$$[d_i] = \begin{bmatrix} \omega_i & \theta_{xi} & \theta_{yi} \end{bmatrix}^T$$

where T shows the transposel.

The curvature–displacement matrix [B_{bi}] and the shear strain–displacement matrix [B_{si}] associated with node i may be written as follows:

$$[B_{b_i}] = \begin{bmatrix} 0 & N_{i,x} & 0 \\ 0 & 0 & N_{i,y} \\ 0 & N_{i,y} & N_{i,x} \end{bmatrix} \qquad (7)$$

$$[B_{s_i}] = \begin{bmatrix} N_{i,x} & -N_i & 0 \\ N_{i,y} & 0 & -N_i \end{bmatrix}$$

Stress–Strain Relations

The bending stress - strain relationships and the bending stresses are

$$[\sigma_b] = [D_b][\varepsilon_b], \qquad [\sigma_b] = \begin{bmatrix} \sigma_x & \sigma_y & \tau_{xy} \end{bmatrix}^T \qquad (8)$$

and the matrix of flexural rigidities may be expressed for an isotropic material as follows:

$$[D_b] = \frac{Eh^3}{12(1-v^2)} \begin{bmatrix} 1 & v & 0 \\ v & 1 & 0 \\ 0 & 0 & 1-v/2 \end{bmatrix} \qquad (9)$$

where E is Young's modulus, v is Poisson's ratio and h is the plate thickness.

The shear stress - strain relationships and the shear stresses are given as

$$[\sigma_s]=[D_s][\varepsilon_s], \qquad [\sigma_s]=\left[\tau_{xz} \quad \tau_{yz}\right]^T \quad (10)$$

where the matrix of shear rigidities for an isotropic material may be expressed as

$$[D_s]=\frac{kEh}{2(1+\nu)}\begin{bmatrix}1 & 0\\ 0 & 1\end{bmatrix} \quad (11)$$

Here, k is the shear modification factor and is normally set equal to 5/6 for homogeneous isotropic plates.

The Total Potential Energy

The total potential energy, π, can be written as

$$\pi=\frac{1}{2}\int_v \sigma_x\varepsilon_x+\sigma_y\varepsilon_y+\tau_{xy}\gamma_{xy}+\tau_{xz}\gamma_{xz}+\tau_{yz}\gamma_{yz})dv-W \quad (12)$$

where W is the work done by the applied loads. Considering the strains and inserting in the above expression, the equation can be modified as

$$\pi=\frac{1}{2}\int_A[\varepsilon_b]^T[D_b][\varepsilon_b]dA+\frac{1}{2}\int_A[\varepsilon_s]^T[D_s][\varepsilon_s]dA-W \quad (13)$$

The total potential energy for moderately thick plates is derived in the matrix form. By substituting the displacement functions in Eq. (13) and integrating over the element fields, the following equations are obtained for an element

$$[K]^e\{d\}^e=\{f\}^e \quad (14)$$

where $[K]^e$, $\{d\}^e$ and $\{f\}^e$ are the element stiffness matrix, displacement vector and the element force vector, respectively.

The Stiffness matrix k_e, can be obtained using the following relation

$$k_e=\int_{A_e}\frac{h^3}{12}\left[B_{bi}\right]^T\left[B_{bi}\right]dA+\int_{A_e}kh\left[B_{si}\right]^T\left[B_{si}\right]dA$$

The force vector f_e can be obtained using the following relation

$$f_e=\int_{A_e} N^T\{f_z,0,0\}^T dA$$

If the load is uniformly distributed in the element i.e. f_z is constant then the above equation becomes

$$F_e^T=abf_z\{1\,0\,0\quad 1\,0\,0\quad 1\,0\,0\quad 1\,0\,0\}$$

III. MODELING OF PLATE

A square plate which is subjected to a uniformly distributed load is modeled with two different boundary conditions, i.e. either a simply supported or a clamped boundary on all four edges shown in the figure below.

Fig.2. Boundary conditions for square plates.

The geometric and material properties considered, are E = 106 kN/m2, ν = 0.30, a = 8 m, q = 1 kN/m2, and k = 5/6, where q is the uniformly distributed load, ν is the Poisson's ratio of the plate material, E is the modulus of elasticity of the plate material, k is the shear modification factor, h is the plate thickness and a is the length of the plate. The objective of the analysis is to determine the maximum deflection and stresses in the plate for the two different end conditions. Square plate with thickness to length ratio of 0.001, 0.01 and 0.1 with all sides simply supported are also analysed.

IV. RESULTS AND DISCUSSION

For the **uniformly loaded square plate** the boundary condition of **simply supported** which is u = v = w = 0 at all the edges is applied and the results for centre deflection and centre stress are obtained.

Table:1.

Load (Q)	Centre Deflection (w/h)			
	h/a			ANSYS [9]
	0.1	0.01	0.001	
0	0	0	0	0
17.8	0.579	0.556	0.549	0.5494
38.3	0.86	0.841	0.834	0.8336
63.4	1.074	1.057	1.048	1.048
95	1.265	1.243	1.238	1.238
134.9	1.448	1.429	1.418	1.418

184	1.628	1.602	1.591	1.592

Load (Q)	Centre Deflection (w/h)			ANSYS [9]
	h/a			
	0.1	0.01	0.001	
0	0	0	0	0
17.8	0.2801	0.2382	0.2387	0.2391
38.3	0.54	0.468	0.473	0.4739
63.4	0.777	0.691	0.696	0.6973
95	0.9985	0.9128	0.9086	0.9101
134.9	1.207	1.126	1.109	1.115
184	1.408	1.321	1.308	1.311
245	1.615	1.534	1.503	1.5045
318	1.83	1.6892	1.6875	1.692
402	2.035	1.8688	1.8675	1.8706
245	1.81	1.779	1.765	1.766
318	1.991	1.943	1.935	1.938
402	2.168	2.117	2.103	2.106

Table:2

From table1 it is seen that the initial rate of deflection is much higher as compared to higher values of load. It is also seen that the values of centre deflection for different thickness and that the centre stress for the three thickness to length ratios shows good agreement to the analysis done in [9] at lower values of load. The rate of deflection decreases with the load and becomes nearly constant beyond the value of Q = 63.4.

For the **clamped uniformly loaded square plate** the boundary condition is u = v = w = 0 , also slope $\beta_1=\beta_2=0$, which is applied at all the edges and the results for centre deflection, centre stress and mid-side stress are obtained.

Table:3

Load (Q)	Centre Deflection (w/h)			ANSYS [9]
	h/a			
	0.1	0.01	0.001	
0	0	0	0	0
17.8	0.2801	0.2382	0.2387	0.2391
38.3	0.54	0.468	0.473	0.4739
63.4	0.777	0.691	0.696	0.6973
95	0.9985	0.9128	0.9086	0.9101
134.9	1.207	1.126	1.109	1.115
184	1.408	1.321	1.308	1.311
245	1.615	1.534	1.503	1.5045
318	1.83	1.6892	1.6875	1.692
402	2.035	1.8688	1.8675	1.8706

Table:4

Load (Q)	Centre Stress ($\sigma a^2/Eh^2$)			ANSYS [9]
	h/a			
	0.1	0.01	0.001	
0	0	0	0	0
17.8	2.53	2.556	2.557	2.5511
38.3	5.246	5.2934	5.3071	5.2973
63.4	7.89	7.98813	8.034	8.039
95	10.652	10.712	10.73	10.726
134.9	13.247	13.317	13.34	13.3575
184	16.157	15.992	15.952	15.9887
245	19.198	18.275	18.157	18.67
318	21.43	21.391	21.36	21.402
402	26.083	24.817	24.45	24.1627

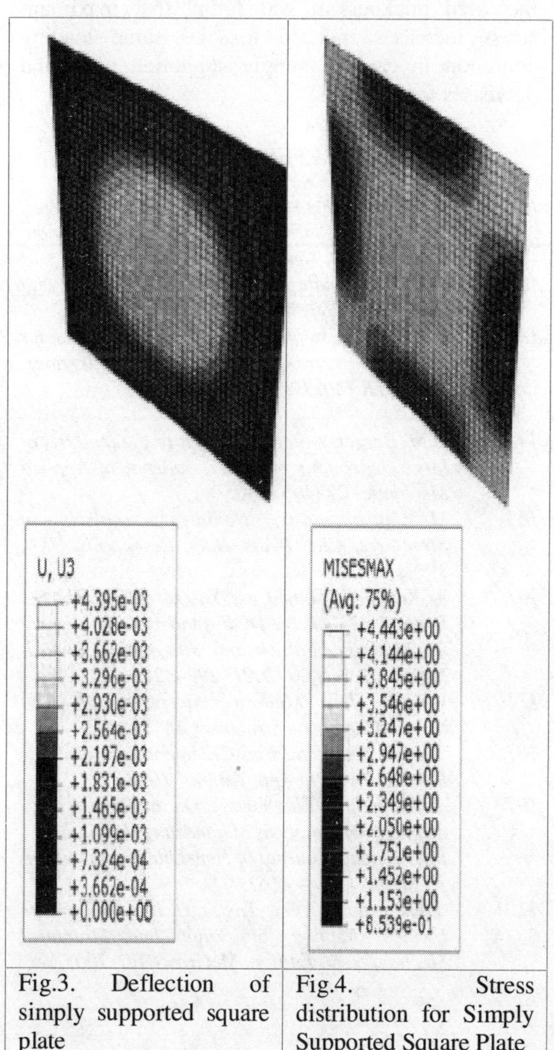

Fig.3. Deflection of simply supported square plate

Fig.4. Stress distribution for Simply Supported Square Plate

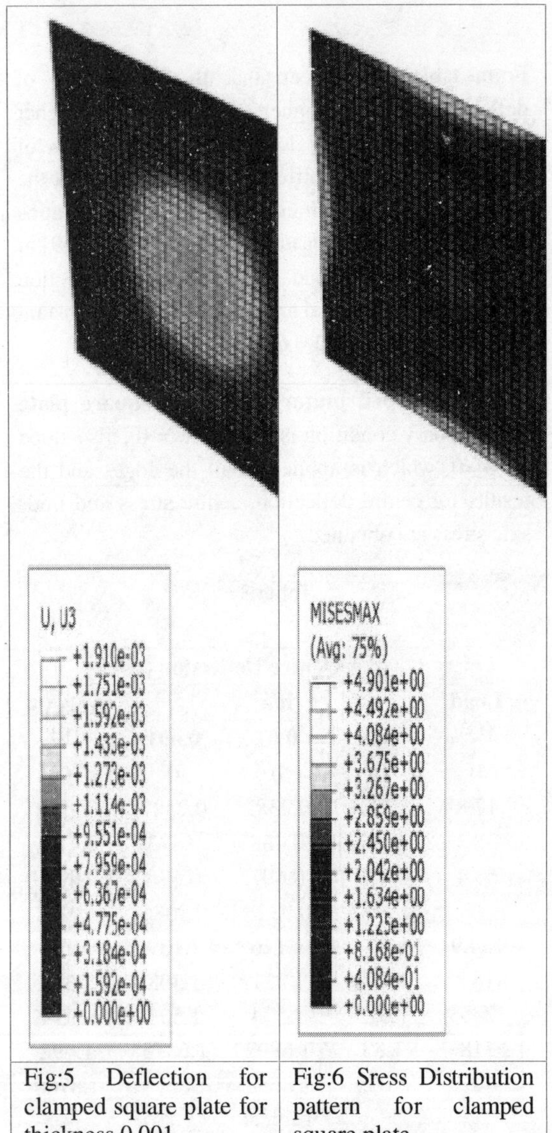

Fig:5 Deflection for clamped square plate for thickness 0.001.

Fig:6 Sress Distribution pattern for clamped square plate.

Numerical results of table 3 show that the rate of deflection is almost constant throughout. The result for moderately thick plate is not in good agreement to [9] which shows that the accuracy of the result decreases with increase in thickness and load.

Numerical results of table 4 that there is not a good agreement of result for all the thickness to length ratio, the values of mid-side stress for different values of thickness to length ratio are lower than the previous investigation done in [9]. It is also seen that the variation in the values of stress is less the plate become thinner.

V. CONCLUSION

In this paper the finite element approach for an eight node Mindlin plate bending element is presented. The geometry of eight node plate element is defined by a plane-region bounded by four straight edges and subsequently the region is mapped using the natural coordinates (ξ, η). The stiffness matrix and load vector are derived using the fundamental equations pertaining to displacement components, strain components, stresses and total potential energy due to externally applied loads. The modelling for the finite element analysis is done using a commercially available software ABAQUS. A very good comparison of the results of analysis for uniformly loaded square plate is presented. The results of deflection and stresses for different thickness to plate-length ratios in the dimensionless form are presented in this paper. For the two boundary condition the square plates having thickness 0.001 results in match with the published data from the paper by Singh and Elaghabash [9]. But when the analysis in this paper is done with increased thickness it was found that maximum stress increases induced for the same loading condition in case of simply supported plate and decreases for clamped.

VI. REFERENCES

[1]   S.P. Timoshenko, S. Woinowsky-Krieger, Theory of Plates and Shells, 2nd Edition, McGraw-Hill, New York, 1959.

[2]   S. Levy, Bending of rectangular plates with large deflections, NACA TR 737, 1942.

[3]   S. Levy, Square plate with clamped edges under normal pressure producing large deflections, NACA TR 740, 1942.

[4]   H.M. Berger, A new approach to the analysis of large deflections of plates, Journal of Applied Mechanics 22 (1955) 465.

[5]   M. Satyamoorthy, Nonlinear Analysis of Structures, CRC Press, Boca Raton, FL, USA, 1997.

[6]   Ai-Kah Soh, Song Cen, Yu-Qiu Long, Zhi-Fei Long, A new twelve DOF quadrilateral element for analysis of thick and thin plates, Eur. J. Mech. A/Solids 20 (2001) 299–326.

[7]   V.V. Vasiliev, Modern conceptions of plate theory, Composite Structures 48 (2000) 39–48.

[8]   Burnett, David S.: Finite Element Analysis From Concept to Applications, 1988.

[9]   A.V. Singh, Y. Elaghabash, On the finite displacement analysis of quadrangular plates, International Journal of Non-Linear Mechanics 38 (2003) 1149 – 1162.

[10]   Zienkiewicz O.C., Taylor R.L., The Finite Element Method for Solids and Structural Mechanics, 6th Edition, McGraw-Hill, 2005.

Analysis of Hand Transmitted Vibration through Steering of Tractor

ZakaUllah, Kashif Irshad, Abid Ali Khan, M. Muzammil

ZakaUllah. Kashif Irshad, Abid Ali Khan, M. Muzammil, Dept. of Mech. Engg, ZHCET AMU Aligarh. Email: zakaamech.amu@gmail.com

Abstract- There are several occupational illnesses related with vibration transmission from tractor to hand–arm system. In this study transmission of vibration from the steering wheel of the tractor without any trailer attached with it were discussed. The vibration measurements were carried out on the tractor randomly chosen.

An investigation was conducted to determine the transmission of vibration from the steering wheel of the hand tractor to the wrist and upper arm of the operator under actual field conditions during ploughing field. The vibration level on the steering wheel was measured and analyzed and the frequency spectra for the chosen working conditions were obtained. The results indicate that the maximum transmissibility was observed in the first two frequency interval (in Hz) i.e. 1-20 and 20-40, which may harm the operator. The frequency interval was 1-20 (target-wrist), 1-20 (target-upper arm), 20-40 (base-steering), 20-40 (base-steering) and frequency zone was 0.022, 0.3974, 0.2066 and 0.1531 respectively.

Keywords: LabView, Power spectrum density, Tractor, Vibration

I. INTRODUCTION

The term 'hand-arm vibration', is frequently used to refer to vibration from power tools, but it does not clearly indicate whether the hand and arm are the origin of the vibration or the limits of its effects. The expression 'local vibration' suggests that the effects are localized near to the point of contact with a source of vibration. While some effect can, by definition, only occur in the fingers or hand, the vibration is transmitted further into the body and the effects it produces there may be of interest. it is therefore more satisfactory to refer to hand transmitted vibration, meaning vibration entering the body at the hand.(Griffin,2004)

The reviews on the effect of vibration on human health have shown serious evidence of operator ill health that may be attributed to tractor drivers. The frequency range of 2-6 Hz has been observed to be the most harmful for the human operator because resonance occurs within this frequency range (Prasad et al.,1996)

The vibration transmitted from the steering wheel of the small tractor with a 4-wheel drive to the driver's hands were carried out on the tractor which is randomly chosen from the producer's store-house. The vibration level has been measured at idling and full load. The vibration level on the steering wheel were measured and analyzed and the frequency spectra for the chosen working condition were obtained. (Goglia et al., 2002). The vibration, which is transmitted from the handle to the hands, arms and shoulders, causes discomfort to the operator and results in early fatigue. An investigation was conducted to determine the transmission of vibration from the handle of the hand tractor

to the metacarpal, wrist, elbow and acromion of the operators under actual field conditions during transportation on a tarmacadam road, rota-tilling in a dry land and rota-puddling in a wet land condition. The maximum transmissibility was observed during the rota-tilling operation. The work related body pain (WRBP) was maximum during the rota-tilling operation, followed by transportation and rota-puddling. (Dewangan and Tewari , 2008).

This study compares the prevalence of symptoms of Hand-arm vibration syndrome and musculoskeletal symptoms in the neck and the upper limbs, between professional drivers of terrain vehicles and a referent group. They also gave information about their lifetime exposure duration driving terrain vehicles and their nicotine use. Prevalence odds ratios were determined and adjusted for age and nicotine use. Results show that there is a relation between exposure to driving terrain vehicles and some of the symptoms of Hand Arm Vibration Syndrome. Increased odds of musculoskeletal symptoms in neck, shoulders and wrists were also found and it seemed to be exposure time (Astrom et al., 2005).

II. METHODOLOGY

In this study level of vibration transmitted to hand of tractor driver while on ploughing field by using labVIEW code and Two integrated electronic piezoelectric (IEPE) sensor connected to the NI USB-9233 data acquisition device interfaced were recorded with a laptop. The recorded data auto stored in text/excels files in the laptop. The recorded data analyzed by using MATLAB programme

Experiment

A LabVIEW code was written to design the instrument for the recording of vibration levels is shown in figure (1). The data acquisition was made possible using tri axial transducer (model no. SEN041F was made by PCB piezoelectronics, NEW YORK,USA; having 10.23 mV/g,10.66mV/g and 10.41Mv/g sensitives in x, y, z direction respectively, the certificate is enclosed as appendix A)that was connected to NI card (Model No. NI9234 made by National instruments)using lead and the card was interfaced with a Acer laptop (specifications P6000 PROCESSOR ,3 GB RAM).The setup was supportive to the sampling rate of 26,400 per second .However the mean values were only recorded .The

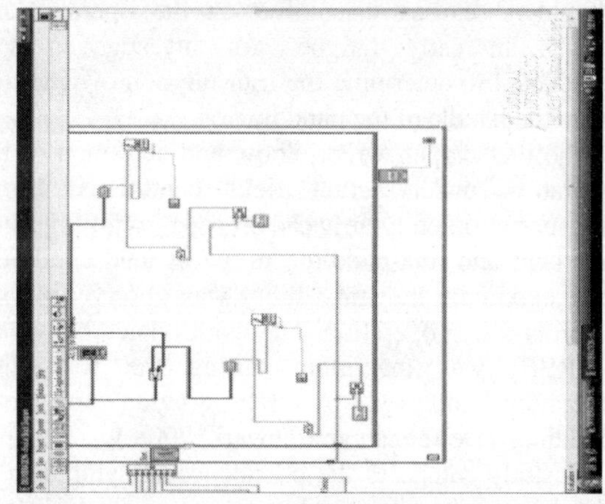

recorded data was auto stored in text/excel files in the laptop. The items used in the experimental setup are shown in figure (2). Procedure to measure the vibration on tractor steering is very much standardized. Vibration measurement on tractor steering have been performed, vibration are measured along z- axis vertical axis. We are using NI USB-9233 data acquisition device. The USB-9233 consists of two components: an NI 9233 module and aUSB-9162 USB carrier, for vibration measurement, NI USB-9233 connected with personal computer through lead.

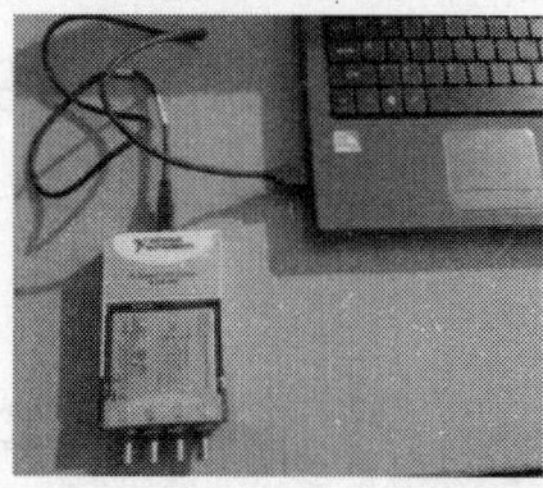

Figure.1 The Block diagram of LABVIEW code used for the recording of Vibration levels (in g)

Figure.2 The items used in data acquisition process (Triaxil transducer, Adaptor to hold transducer, data acquisition card and interfacing with Laptop)

Two integrated electronic piezoelectric (IEPE) sensor connected to BNC connector .One of the sensor are attached to steering (z-base) and second one sensor are attached to driver forearm and upperarm (near shoulder) (z-target) ,so we get The collected data was processed and analyzed with LabView TM and by using MATLAB programme for each test .

Procedure

First and foremost subject of the test was given written information about the experiment, which included the purpose of the study. The tractor that are used in experiment was in working condition, tyres of the tractor for the test were of standard size. Tests are carried out on FARMTRAC 50 model on ploughing field which is shown in figure (3).

Figure.3 a FARMTRAC 50 model on ploughing field, b) Sensors position on steering and lower arm (wrist), c) Sensors position on steering and upper arm

Farmtrac 50 from Escorts Ltd. is a 4 stroke, direct injection diesel run tractor with a capacity of 2868 cc. The vehicle comprises a mechanical constant mesh gearbox with 8 forward and 2 reverse gears. Farmtrac 50 is equipped with a recirculation ball type, worm and nut with double drop arms steering and also has an option of a power steering. The tractor is featured with drum (internal expanding shoe) brakes along with a hand operated parking brake.For measuring vibration, the tractors were moving on a specified field with certain speed. The test is carried out on tractor without any trailer attached with it but load attached to it like ploughing equipment. Subjects were considered to be healthy with no signs of musculo-skeletal system disorders. I am using two sensor for measurement of vibration in z direction, first sensor is SEN041F triaxial shear icp accelerometer, this recognized as a z-base sensor, it is attached to tractor steering of Farmtrac -50 model tractor to measure its vibration at steering, second sensor used for test is 353B18 SN 140184, it is recognized as a target sensor, this sensor attached to subject forearm and upper arm to measure the vibration transmitted to the hand of the subject.

First sensor attached to DAQ via blue cable at terminal A0-1 and second sensor is attached to DAQ via white cable at terminal AI-1. Figure (4) Shows the screen shot of the LABVIEW code for recording vibration levels in Z- direction.

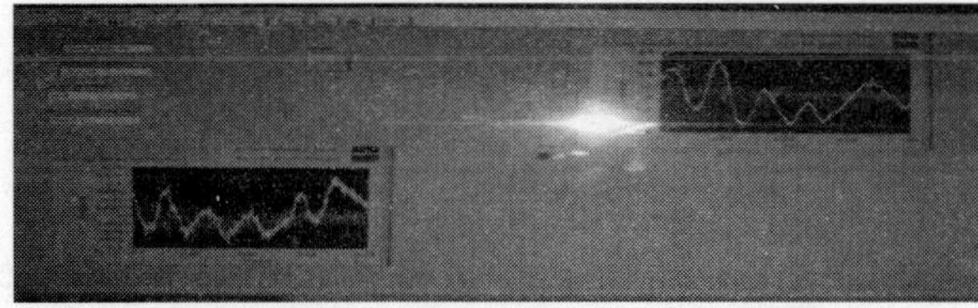

Figure.4 Screen shot of the LABVIEW code for recording vibration levels in Z direction.

III.   RESULTS

The tractors that are used in experiment were in working condition. Tests are carried out on FARMTRAC 50 model on ploughing field. The test is carried out on tractor without any trailer attached with it but load attached to it like ploughing equipment. The vibration transmitted to hand through steering of tractor has taken by using LABVIEW code with the help of laptop. The recorded data was auto stored in text/excel files in the laptop. So we get the collected data was processed and analyzed with LabView TM and by using MATLAB programme for each test. Table-1 Shows the power spectral density at different frequency zone for z-base on steering.

Table.1 Power spectral density at different frequency zone for z-base on steering.

Serial no.	Sensor position	Frequency zone(Hz)	Power spectral density
1	Base(steering)	1-20	0.1005
2	„	20-40	0.2066
3	„	40-60	0.1166
4	„	60-80	0.0373
5	„	80-100	0.0060
6	„	100-120	0.0004
7	„	120-140	0.0000
8	„	140-160	0.0000
9	„	160-180	0.0000
10	„	180-200	0.0000
11	„	200-220	0.0000
12	„	220-240	0.0000
13	„	240-260	0.0000
14	„	260-280	0.0000
15	„	280-300	0.0000
16	„	300-320	0.0000
17	„	320-340	0.0000
18	„	340-360	0.0000
19	„	360-380	0.0000
20	„	380-400	0.0000
21	„	400-420	0.0000
22	„	420-440	0.0000
23	„	440-460	0.0000
24	„	460-480	0.0000
25	„	480-500	0.0000

Table.2 Shows the power spectral density at different frequency zone for z-target on wrist.

Table.2 Power spectral density at different frequency zone for z-target on wrist.

Serial no.	Sensor position	Frequency zone(Hz)	Power spectral density
1	Target(wrist)	1-20	0.0022
2	„	20-40	0.0000
3	„	40-60	0.0000
4	„	60-80	0.0000
5	„	80-100	0.0000
6	„	100-120	0.0000
7	„	120-140	0.0000
8	„	140-160	0.0000
9	„	160-180	0.0000
10	„	180-200	0.0000
11	„	200-220	0.0000

12	,,	220-240	0.0000
13	,,	240-260	0.0000
14	,,	260-280	0.0000
15	,,	280-300	0.0000
16	,,	300-320	0.0000
17	,,	320-340	0.0000
18	,,	340-360	0.0000
19	,,	360-380	0.0000
20	,,	380-400	0.0000
21	,,	400-420	0.0000
22	,,	420-440	0.0000
23	,,	440-460	0.0000
24	,,	460-480	0.0000
25	,,	480-500	0.0000

Table.3 Shows the power spectral density at different frequency zone for z-base on steering.

Table.3 Power spectral density at different frequency zone for z-base on steering.

Serial no.	Sensor position	Frequency zone(Hz)	Power spectral density
1	Base(steering)	1-20	0.1072
2	,,	20-40	0.1531
3	,,	40-60	0.0470
4	,,	60-80	0.0382
5	,,	80-100	0.0043
6	,,	100-120	0.0003
7	,,	120-140	0.0000
8	,,	140-160	0.0000
9	,,	160-180	0.0000
10	,,	180-200	0.0000
11	,,	200-220	0.0000
12	,,	220-240	0.0000
13	,,	240-260	0.0000
14	,,	260-280	0.0000
15	,,	280-300	0.0000
16	,,	300-320	0.0000
17	,,	320-340	0.0000
18	,,	340-360	0.0000
19	,,	360-380	0.0000
20	,,	380-400	0.0000
21	,,	400-420	0.0000
22	,,	420-440	0.0000
23	,,	440-460	0.0000
24	,,	460-480	0.0000
25	,,	480-500	0.0000

Table.4 Shows the power spectral density at different frequecy zone for z-target on upper arm (shoulder).

Table.4 Power spectral density at different frequency zone for z-target on upper arm

Serial no.	Sensor position	Frequency zone(Hz)	Power spectral density
1	Target(upperarm)	1-20	0.3974
2	,,	20-40	0.0011
3	,,	40-60	0.0006
4	,,	60-80	0.0005
5	,,	80-100	0.0004
6	,,	100-120	0.0004
7	,,	120-140	0.0004
8	,,	140-160	0.0004
9	,,	160-180	0.0004
10	,,	180-200	0.0004
11	,,	200-220	0.0004
12	,,	220-240	0.0004
13	,,	240-260	0.0004
14	,,	260-280	0.0004
15	,,	280-300	0.0004
16	,,	300-320	0.0004
17	,,	320-340	0.0004
18	,,	340-360	0.0004
19	,,	360-380	0.0004
20	,,	380-400	0.0004
21	,,	400-420	0.0004
22	,,	420-440	0.0004
23	,,	440-460	0.0004
24	,,	460-480	0.0004
25	,,	480-500	0.0004

Table .5 Maximum power spectral density in accordance with the frequency zone (in Hz)

Serial number	Sensor Position (in Hz)	Frequency zone (in Hz)	Power spectral density (maximum)
1.	Base (on steering)	20 - 40	0.2066
2.	Target(wrist)	1 - 20	0.022
3.	Base(on steering)	20 - 40	0.1531
4.	Target (upper arm)	1 - 20	0.3974

IV. DISCUSSION

The vibration transmitted to hand through steering of tractor has taken by using LABVIEW code with the help of laptop. The recorded data was auto stored in text/excel files in the laptop. So we get the collected data was processed and analyzed with LabView TM and by using MATLAB programme for each test. The results we get in the form of tables- 1, 2, 3 and 4.Table-1 shows the vibration level on steering and table-2 shows the vibration level on wrist. Likewise, table-3 and table-4 shows the vibration level on steering and upper arm. The vibration level on the steering and target point wrist and upper arm has measured and analyzed and the frequency spectra for the chosen working conditions were obtained. The results indicate that the maximum transmissibility was observed in the first two frequency interval (in Hz) i.e. 1-25 and 25-50.the frequency interval was 1-25 (target wrist),1-25 (target upper arm),25-50 (base steering),25-50 (base steering) and frequency zone was 0.022,0.3974,0.2066,0.1531 respectively.

Above result table- 5 shows that the maximum power spectral density in accordance with the frequency zone (in Hz).it seems that the power spectral density is maximum at the target point shoulder i.e. 0.3974 in comparison to the target point wrist i.e.0.022. The vibration is transmitting at target point shoulder is more than the target point wrist.

IV. CONCLUSION

The following major conclusions can be drawn from the present investigation.

- The frequency zone 1-20 and 20-40 has given the maximum power spectral density.
- In the frequency zone 1-20 for target (wrist) and for target (upper arm) has given the power spectral density 0.022 and 0.3974 respectively.
- In the frequency zone 20-40 for base (steering) has given the power spectral density 0.1531 and 0.2066.
- The frequency zone 1-20 and 20-40 are the most harmful for the tractor driver.
- Prasad et al.,(1996) has also investigated that the effect of vibration on tractor driver and get the frequency zone 2-6 Hz has been the most harmful for the operator because resonance occurs within this frequency zone.likewise, Adewusi et al.,(2010) has also suggested that the operators of power tools with frequencies below 25 Hz may experience greater muscles/tissues fatigue and symptoms of musculoskeletol disorder when working with extended arm posture.

Therefore, the vibration level at the operator's seat needs to be attenuated within   acceptable limits in this frequency range. This study helped in understanding the effect of vibration level on tractor driver. The results indicated the value of research work aimed at reducing vibration levels. Ergonomic studies are recommended to determine the vibration response in agricultural production activities.

VI.   REFERENCES

[1] Astrom,C., Rehn,B., Lundstrom,R., Nisson,T, Burstrom,L. and Sundelin,G. ,2006, 'Hand arm vibration syndrome(HAVS) and musculoskeletal symptoms in the neck and the upper limbs in professional drivers of terrain vehicles-A cross sedtional study',Applied Ergonomics , volume 37 , issue 6 , pp 793-799.

[2] Dewangan, K.N., and Tewari, V.K.,2009, 'Characteristics of hand transmitted vibrationof a hand tractor used in three operational modes', International Journal of Industrial Ergonomics, volume 39, issue 1, pp 239-245.

[3] Dewangan,K.N. and Tewari, V. K . 2009, 'Vibration energy absorption in the hand arm system of hand tractor operator', Biosystems Engineering, volume 103, issue 4 , pp 445- 454.

[4] Dewangan, K.N., and Tewari,V.K.,2008, 'Characteristics of vibration transmission in the hand arm system and subjective response during field operation of a hand tractor' , Biosystem Engineering , volume 100, issue 4, pp 535-546.

[5] Goglia,V., Gospodaric,Z., Kosutic,S.,and Filipovic,D.,2002, 'Hand transmitted vibration from the stearing wheel to drivers of a small four-wheel drive tractor', Applied Ergonomics, volume 34, issue1, pp 45-49.

[6] Griffin,M.J. 'HANDBOOK OF HUMAN VIBRATION', Reprinted (2003-2004) 530-531.

[7] Goglia, V., Gospodariĉ, Z., 2002, 'Driver's exposure to hand-transmitted vibration from the steering wheel of the LPKT 40 tractor, The 30th International Symposium on Agricultural Engineering', Opatija.

[8] ISO 5349-2-2001, 2001. EN ISO 5349-2, 2001. Mechanical vibration—measurement and Evaluation of Human Exposure to Hand-Transmitted Vibration—Part 2: practical guidance for measurement at the workplace.

[9] ISO 5349-1, 2001. ISO 5349-1-, 2001. Mechanical vibration—measurement and evaluation of human exposure to hand-transmitted vibration—Part 1: general Requirements, ISO, Geneva.

[10] Kacian N. ,1997, 'A statistical report on occupational diseases in Croatia. Work and Safety' 2 , p. 271.

[11] Wasserman, D.E., 1987, 'Human vibration and Biomedical Engineering Consultant, Cincinnalu' , OH 45242, USA.

[12] Adewusi,S.A., Rakheja, S., Marcotte,P. and Boutin,J., 2010, ' Vibration transmissibility characteristics of the human hand–arm system under different postures, hand forces and excitation levels', Journal of Sound and Vibration ,Volume 329, Issue 14, PP. 2953-2971

[13] Prasad,N., Tewari,V.K.,and Yadav,R.,1996, 'Tractor ride vibration--a review' , Journal of Terrarnechanics , Vol. 32. No. 4, pp. 205-219

Barriers to Geothermal Energy Development

Ravi Kumar, M.P. Sharma
Indian Institute of Technology, Roorkee Uttarakhand
E-mail: er.ravi.49@gmail.com

Abstract- India is developing nation. Its per capita energy consumption is low. To achieve Economic Growth, we have to use more and more energy to increase the pace of development. It is estimated that Industrial energy use in developing countries constitutes about 45-50 % of the total commercial energy consumption. Much of this energy is converted from imported oil, the price of which has increased tremendously so the renewable energy sources are becoming attractive solution for clean and sustainable energy in developing country like India. Geothermal energy is the energy contained as heat in the earth's interior. The aim of this paper is the identification and analysis barriers which could hinder the wider use of geothermal energy in India. There are two big barriers in geothermal development i.e. initial investment and resources development risk. Then the problems, which could arise due to legal and administrative reasons and affect the construction and operation of a geothermal energy provision plant, are identified.
Keywords: geothermal energy, Non-technical Barriers, Financial Barriers, administrative, organizational and infrastructural challenges.

I. INTRODUCTION

Geothermal energy is the earth's natural heat available inside the earth. This thermal energy contained in the rock and fluid that filled up fractures and pores in the earth's crust can profitably be used for various purposes. Heat from the Earth, or geothermal — Geo (Earth) + thermal (heat) — energy can be and is accessed by drilling water or steam wells in a process similar to drilling for oil. Geothermal energy is an enormous, underused heat and power resource that is clean, reliable and homegrown. India has reasonably good potential for geothermal; the potential geothermal provinces can produce 10,600 MW of power. But yet geothermal power projects has not been exploited at all, owing to a variety of reasons, All partners realized that deep geothermal energy has high potential, but that its exploitation is very difficult. This is especially due to non-technical (financial) barriers and a lack of awareness among decision makers which hinders investments in geothermal power plants and industrial geothermal applications. Objectives to reach its goal of fostering a deep geothermal energy sector across India. [1]

- Reducing financial barriers hindering the initial stages of geothermal energy projects through the development and proposal of appropriate financing and funding schemes. This helps to boost investments in geothermal energy.

- Promoting and raising the awareness of emerging industrial geothermal applications. These awareness-raising activities are mainly directed toward local and regional decision makers and other potential stakeholders.

- Increasing consideration of geothermal energy as a sustainable, competitive and secure energy source within India.

- Improving the knowledge and awareness among regional decision makers by exchanging experience among the target groups.

- Spreading all findings and results as widely as possible in order to ensure that the maximum number of local and regional decision makers receive valuable information for evaluating the benefits of geothermal energy to the local/regional economy. [2]

II. VARIOUS NON-TECHNICALISSUES

Besides the technical challenges which heat and rather power production from low enthalpy fields are facing, also non-technical barriers hinder the wider use of such geothermal resources within India. One of the most important aspects hereby is the higher costs of the provision of heat and/or electricity from low enthalpy geothermal energy compared to energy production from fossil fuel energy. Additionally quite a lot of economic promising geothermal projects have been delayed, modified or failed due to various non-technical issues. The following reasons can be analyses for such failures: [2]

- High initial capital costs, and resource development risk
- Failure of drilling wells, significant depth requirements, insufficient productivity and accessibility of the reservoir.
- Low awareness and limited information about geothermal energy
- Lack of incentive schemes and uncertainty about the future of such schemes
- Organizational difficulties,
- Insufficient perception and acceptance.

Two Big Barriers

There are Two BIG BARRIERS which hinder smooth geothermal development in developing countries. They are, [4]

- Capital Investment Burden,
- Resources development risk

Financing geothermal energy projects

The financing of geothermal energy projects is rather difficult as the project phases of geological research (feasibility study, drilling, investment, operation and maintenance) are all associated with different risks and costs, but are interoperable. The stakeholders of geothermal energy projects are manifold: consumers, investors, suppliers, developers, governments, operators, and financial institutions. They all contribute to defining the economical and financial terms of the project and all have multiple interests of what the project's benefits means to them. The financial institution focuses on the viability and the risks of the project and will expect repayment of the debt, which is usually expressed in the Debt Service Cover Ratio (DSCR). The investor will focus on the return on equity of the project. The considerations of all stakeholders have to be taken into account to assess the viability of the project in a very early phase of the geothermal energy project.

Drilling

Considerations about investment costs and risks underline that the financing of the exploration and (pre)feasibility studies are an important barrier. Unsuccessful drilling is an important risk that has to be taken into account, as drilling costs are consequent and can represent a significant part of overall project costs. However, this risk must be assumed. Financial institutions often argue that for traditional financing, the risk is too high to provide a loan or other financial sources at this early stage of a project. The financial institutions will wait to provide loans to a geothermal project until the existence and the quality of the resource has been proven. Therefore, equity from one's own resources or grants from public bodies could be needed to cover these expenditures in the merely phases of the project. Private equity investors expect a high rate of return at these initial stages of the project due to the risk their investment faces. Finally, classic project finance schemes can be used at a very late project phase.

Investment

Grants can reduce the amount of investments and feed-in tariffs regulated by law and fix revenues during the operational phase. Therefore, a combination of financing schemes and incentives can be a key point for the economic success of projects and should be analyzed. Proposals of funding schemes (equity, loan, subsidies, grants, etc.) for geothermal energy are presented and analyzed in this paper for each target country, at national and international levels. A special focus of each geothermal project must be set on the geological risk insurance mechanisms that guarantee the presence and the quality of the resources. This is a necessary requirement for each project to overcome these difficulties. [2]

Main financial barriers for geothermal energy projects

To understand the necessity of public or semi-public funding instruments, it is important to get insight regarding the main financial barriers of deep geothermal projects. This is a necessary condition before analyzing the financial instruments that have been installed by the national governments or Indian institutions solution to overcome these barriers. The regulation barriers affecting deep geothermal projects have recently been analyzed. Some financial barriers have been identified. It is common knowledge that especially in India main financial barriers that hinder investments in deep geothermal energy projects are

- The high up-front costs (exploration, like seismic investigations, reprocessing and finally drilling wells) of deep geothermal energy projects.
- The high drilling risk combined with high drilling costs. These main barriers lead to two types of problems.
- The profitability of geothermal plants is far behind the profitability of other renewable energy sources because of high investment costs. The interest of investors in geothermal energy is thus far weaker than in other renewable energies.
- Because of high costs in combination with high risk, institutional investors often tend not to lend money to these projects, because they fear the risk. That makes it difficult to realize the investment because venture capital must to be attracted or one's own capital must be used.

Except in the rare case that one can provide their own funding, the search for a risk-taking investor can be a costly alternative. These barriers affect the investment in geothermal energy and its profitability. On national, European and international levels, a bundle of instruments has been developed to overcome the main barriers that hinder investment in geothermal energy. [2]

Table 1: Financial instruments to overcome the main financial barriers (Source: GEOFAR 2009)

Main barriers to overcome		Financial Instruments that exist actually
High up-front costs		Public Subsidies
		Venture loan
Geological Risk		Subsidized Insurance Mechanism
Long pay back period	High investment costs	Bank Facilities (e.g. subsidized interest loans)
		Feed-in Tariff for electricity production
	Low outcomes	Tax reduction

Table 2: Strengths and weaknesses of the offered financial instruments (Source: GEOFAR 2009)

Instruments offered	Main strengths	Main Weakness
Feed-in Tariffs	Secure income over a long term period	Acts at a very late project stage
Tax reduction	Promote increased capital investment	Affects mainly operational phase when revenues are generated
Grants	Substitute equity	Management of public money
Bank Facilities	Possibility to finance projects with high investment volumes	Difficult to apply in the stage of exploration without insurance mechanism
Tradable certificates/Quota systems	Deals with very few public money	No long term secured and fixed income

Capital Investment Burden

- Capital Investment Burden worsens the generation cost in the short term period.
- Although the generation cost is very attractive in the long term period.

Private investors are unlikely to choose geothermal power in socially and economically unstable countries, even though it is the least cost energy in the long run. [4]

III. RESOURCE DEVELOPMENT RISKS

Geothermal development has a lot of Resource Risks; Unexpected cost over-run due to

- Difficulty of Access to the Field,
- Failure of Drilling Wells,
- Depth of Reservoir,
- Poor Productivity of Reservoir,
- Decline of Steam, etc.

Depth of Reservoir varies field by field, and it affects largely the cost of drilling wells of the Project. Where as Productivity of Reservoir also varies field by field, and it affects largely the cost of generation. [4]

IV. GOVERNMENTAL SUPPORT

Currently geothermal plants participating in R&D-programmers can get funding from the national governments. However, numerous forms have to be filled to get funding and the conditions for funding are not always very clear. Sometimes the funding conditions can change within the duration of the funding or even the application period. Long and complicated approval procedures can lead to a significant time lag between an application for funding and the final decision. Such time delays can make projects even more expensive and hence less economic. To improve the procedure of deciding whether or not to fund a specific project it seems advisable

- To make the handling of the support regulations more simple and easy,
- To guarantee a short and defined time for decisions on funding applications, and
- To further improve the co-ordination between European, national and local funding institutions.

If a geothermal project seems to be profitable without subsidies an important obstacle on the way to realize such a project is often to get funding from a private bank. This situation could be changed with the help of public funds and special funding conditions or with the help of sureties and guarantees by the government. The confidence of a private bank could also be increased by long lasting arrangements on special reimbursements for the feed-in of geothermal energy guaranteed by law or by the utilities. In order to realize more geothermal projects and to increase private investments in geothermal plants the following governmental instruments and measures could be applied:

- Improvement of the funding conditions for specific geothermal demonstration projects with challenging or unproven technologies (i.e. a higher percentage of funding related to the overall costs for technologies which are still in the stage of development and demonstration such as low enthalpy electricity generation systems).
- Sufficiently high funding and flexible handling of the annual budgets for geothermal projects depending on the conditions and requests of the respective project.
- Tax incentives, e.g. a high difference between the energy taxes on the energy generation from fossil fuel energy on the one hand and from low enthalpy geothermal energy and other renewable sources on the other.
- Tax allowances (i.e. preferential conditions of depreciation) for investments in low enthalpy geothermal projects and tax reductions on the interest rates gained from investments in such projects.

- Governmental supported investment credits with especially low interest rates and a very long lifetime.
- Using geothermal energy for public buildings and paying higher energy prices for it. [3]

V. ADMINISTRATIVE CHALLENGES

In order to construct and operate a geothermal power and/or CHP plant, various administrative permissions are needed e.g. for

- The exploitation of the underground,
- The long term operation of the geothermal resource,
- The technical processes involved,
- The construction of the geothermal plant and
- The long term operation of the power and/or CHP plant.

The implementation of new geothermal projects could be hindered by e.g.

- Absence of a clear und understandable regulation framework designed for the use of geothermal energy,
- Unclear interpretation of frameworks which do not contain specific regulations or guidelines for geothermal systems,
- Existing frameworks for similar technologies which do not fit or which hinder the implementation of geothermal projects, or
- Legal frameworks which are designed for proven technology and not for technology which is still within the developing phase.

Therefore feasible ways to enhance the existing regulatory framework are needed. The existing legal framework should also be more flexible to take care of new technologies and innovative processes which are still within a demonstration phase. But this is a complex task because different policy areas as well as various policy levels have to be addressed. And this has to be realized in a similar way throughout Europe. In order to develop and establish specific and well-defined permission procedures and regulations for geothermal plants concerted political action across different countries, institutions and authorities is needed. [3]

VI. ORGANIZATIONAL AND INFRASTRUCTURAL CHALLENGES

Especially planning and realization but also operation of geothermal power and/or CHP plants for the provision of heat and/or electricity is a demanding task because it comprises various technical fields and hence involves the participation of different persons, groups, companies and institutions.

The participants involved in the implementation of such projects can be

- Decision-makers from very different areas e.g. geology, engineering science, energy economics,
- European, national and regional funding institutions,
- Private and public financiers,
- Insurance companies,
- Administrative and local permission authorities,
- Planning offices, manufacturing, engineering and construction companies,
- Plant owners and operators,
- Customers for the geothermal energy maybe accepting to pay higher prices, and
- The public, especially the local people personally affected by the plant. [3]

VII. CONCLUSION

The importance of non-technical challenges for a wider use of geothermal energy could be much less than today if the economics of the provision of geothermal heat and/or electricity were to be substantially improved. Funding activities that address non-technical or semi-technical issues (such as quality control) can decrease the need for subsidies significantly without compromising the goal of introducing geothermal energy to the market. Facing the economic uncertainties and market risks on the one side and the higher costs for the provision of geothermal energy compared to the energy costs from fossil fuel energy on the other, a broad and long lasting market introduction programme is required consisting of R&D measures on the one side and market introduction measures on the other. To finance such a programme, funds with a sufficient finance over a significantly long period of time must be available. The funds needed for such a programme could be provided by the government through different financial support measures taking the environmental and other benefits of an increased use of geothermal energy into consideration. Based on such a programme it might be possible to create a market for plants to produce geothermal heat and/or electricity which could be economically viable over the long term.

VIII. REFERENCES

[1]   *Harinarayana, "Geothermal energy scenario in India". National Geophysical Research Institute, Hyderabad (Council of scientific and industrial research). pp. 1-8.*

[2]   *The GEOFAR Project, Report "Financial instruments as support for the exploitation of geothermal energy", 2009.*

[3]   *Questionnaires of work package 5, "ENGINE-project Non-technical barriers and benefits of geothermal energy production", November 2006.*

[4]   *Business Perspectives on Encouraging Renewable Energy Policy and Technology Transfer-In a Case of Geothermal Development, West Japan Engineering Consultants, Inc.*

Effect of Vibration on Tractor Driver under Different Surface Conditions

Jiyaul Mustafa, Nidhi Singh, Mohd Farooq, Abid Ali Khan and M. Muzammil

Department of Mechanical Engineering, ZHCET, AMU, Aligarh, 202002, India

*Abstract:*Exposure to vibration is the cause of some occupational injuries and diseases. The objective of this study was determination of transmission of vibration which transmitted to tractor drivers from the seat operator system and breaking system. An investigation was conducted to determine the transmission of vibration from the seat operator system & breaking system to the neck, low back & feet under the different conditions i.e., farm field, tar-macadam road & bricks road. The vibration level was measured and analyzed and the frequency spectra for the chosen working conditions were obtained. The results indicate that the maximum transmissibility was observed in the first two frequency interval (in Hz) i.e. 1-20 and 20-40.

Keywords: Frequency, LabView, Power spectrum density, Tractor, Vibration

I. INTRODUCTION

The problems in human body, due to agriculture tractor vibrations are seating discomfort and muscle fatigue. The adverse health affects from whole-body vibration (WBV) deals with the musculoskeletal system, in particular, the lumbar spine. There is strong epidemiological evidence that occupational exposure to WBV is associated with an increased risk of low back pain (LBP), sciatic pain, and degenerative changes in the spinal system, including lumbar intervertebral disc disorders [1-2].The mainly problems in human body, due to agriculture tractor vibrations are seating discomfort and muscle fatigue. Seating discomfort led to various parameters such as the body pressure distributed under and supporting   the buttocks, thighs and the back of an operator, control of posture in static or dynamic condition [3]. The biomechanical and engineering factors like ride vibration, pressure

distribution at the seat operator interface and the body posture play an important role in a tractor seat-design. A well designed seat should be able to accommodate all sizes and shapes of users body is not well supported, several muscle groups act and should provide adequate body support. When the together to restore stability, contributing to static loading resulting in discomfort [4,5].

A large number of studies have been conducted and suggested several criteria for evaluating discomfort and the suitability of a tractor seat in a given working condition. This information available in this regard and attempts to set the most appropriate procedure for assessment of seating discomfort during tractor driving [6, 7].

India is the largest manufacturer of tractors in the world. It is the largest world market for tractors below 37 kW, due to the small size of land holdings. The current population of tractors in India is around 1.5 million and more than 0.15 million tractors are added to Indian agriculture every year [8, 9]. They are being mainly used for primary and secondary tillage operations and as a means of transportation to haul goods, people and even animals. Vibration in tractor driving can cause deafness and disorders of the spinal column and stomach. The effect of implements on tractor ride is not well understood in India [10].

Agricultural tractors are extensively used for on-road and off-road transportation and for different field operations. It is widely recognized that tractor operators are exposed to high levels of whole-body vibration during farm operations [11, 12]. In order to predict the ride vibration of an agricultural tractor, the vibrations transmitted to the driver's seat of an agricultural tractor, an analytical model has been constructed through the assembly of tyre model & cab and seat suspension models [13, 14].

II. METHOD

Materials

Mahindra Bhoomiputra 265 DI Model tractor was carried under working condition. It has a horse power of 30 H.P categories with bore and stroke of 88.9 x 96 mm. It is couched with three cylinders whereas the cubic capacity of 1788 cc. The make and model of this variant is Mahindra MDI- 1785 and its total weight somewhere comes to around 1735 kgs.

Approach

It was tried to study the effect of transfer of vibration over whole body system in tractor seat operator system and breaking system.

Participants

In the present experiment, two healthy male participants were participated.

Apparatus

A Lab VIEW code was written to design the instrument for the recording of vibration level. The data acquisition was made possible using tri axial transducer (model no. SEN041F was made by PCB piezoelectronics, NEW YORK, USA; having 10.23 mV/g,10.66mV/g and 10.41Mv/g sensitives in x, y, z direction respectively, the certificate is enclosed as appendix A) that was connected to NI card (Model No. NI9234 made by National instruments) using lead and the card was interfaced with a Wipro laptop (specifications P6000 PROCESSOR, 3 GB RAM).The setup was supportive to the sampling rate of 26,400 per second.

Figure 1: The items used in data acquisition process (Triaxil transducer, Adaptor to hold transducer, data acquisition card and interfacing with Laptop).

However, the mean values were only recorded .The recorded data was auto stored in text/excel files in the laptop. The items used in the experimental setup are shown in figure 1. Procedure to measure the vibration on tractor steering is very much standardized. Vibration measurement on tractor seat pad & breaking system have been performed, vibration are measured along z- axis vertical axis. We were used NI USB-9233 data acquisition device. The USB-9233

consists of two components: an NI 9233 module and aUSB-9162 USB carrier, for vibration measurement, NI USB-9233 connected with personal computer through lead.

Procedure

For vibration measurement from the seat operator system & breaking system to the neck, low back & feet of driver, the tractor was carried out under the different conditions i.e., farm field, tar-macadam road & bricks road. Subjects were considered to be healthy with no signs of musculo-skeletal system disorders. It was used two integrated electronic piezoelectric (IEPE) sensor connected to BNC connector. For the measurement of vibration in z direction, first sensor is SEN041F triaxial shear icp accelerometer, this recognized as a z-base sensor, it is attached to tractor seat pad & breaking system. The first set of experiment was carried out by one of the sensor are attached to seat pad (z-base) and second one sensor are attached to driver neck and low back (z-target) & second set of experiment was carried out by attached the one sensor to breaking system (z-base) & second one sensor were attached to driver feet (z-target) as shown in figure 2,so we get The collected data was processed and analyzed with Lab View TM and by using MATLAB programme for each test.

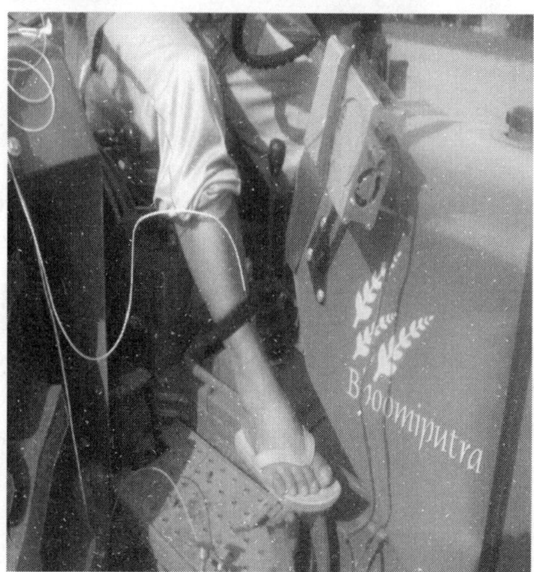

Figure2: Mahindra Bhoomiputra 265 DI Model tractor on farm field conditions **(a)** Breaking system arrangement **(b)** Seat pad arrangement.

The first set of experiment was carried out by one of the sensor are attached to seat pad (z-base) and second one sensor are attached to driver neck and low back (z-target) & second set of experiment was carried out by attached the one sensor to breaking system (z-base) & second one sensor were attached to driver feet (z-target) as shown in figure 2, So we get the collected data was processed and analyzed with Lab View TM and by using MATLAB programme for each test.

III. RESULTS & DISCUSSION

The analysis for vibration was done through the MATLAB. A program was written in MATLAB for that purpose. In the analysis, some graphs were obtained between the different variables for different condition of the sensor and for each participant.

In the analysis, graphs were obtained between the One-sided Power Spectrum Density (PSD) (dB/rad/sample) and the Frequency (Hz). A sample graph for the PSD was shown in the Figure3. The graphs were also obtained between the Frequency of Occurrence and the Vibration Amplitude, between the RMS value and the Number of Frames and for the Amplitude of Vibration and the Time (in seconds). The sample graphs for these variables were shown in the Figure 3, 4, 5 & 6 respectively.

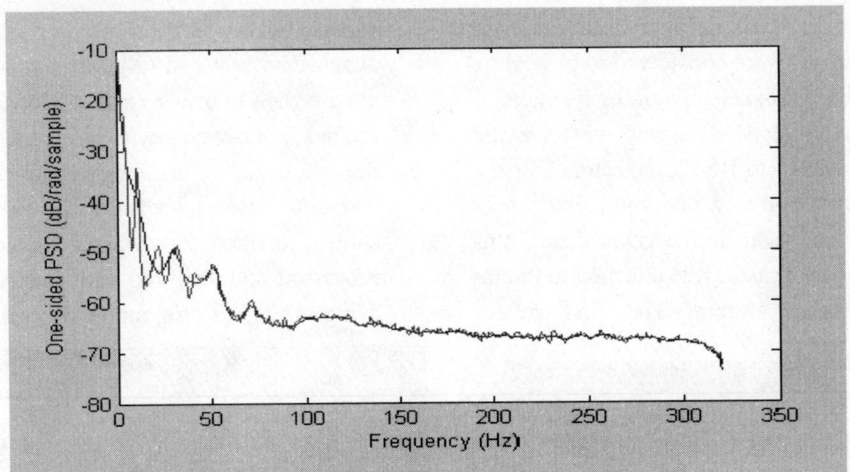

Figure 3: The graph between the One-sided Power Spectrum Density (PSD) (dB/rad/sample) and the Frequency (Hz) .

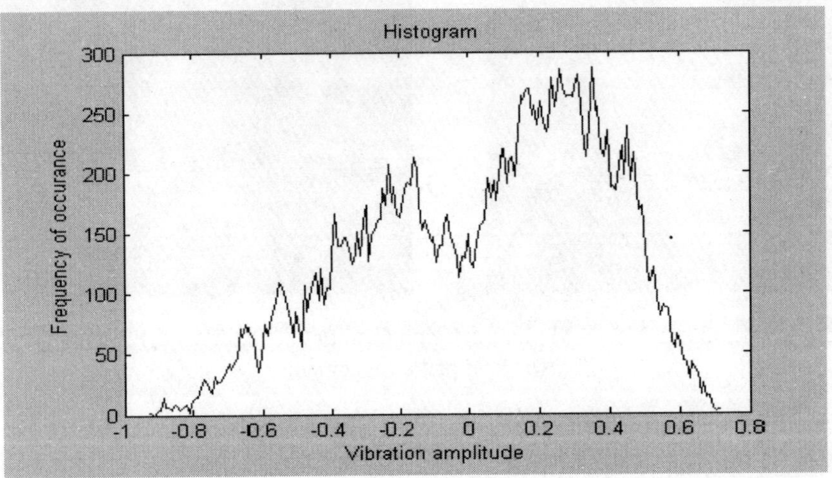

Figure 4: The histogram between the Frequency of Occurrence and the Vibration Amplitude.

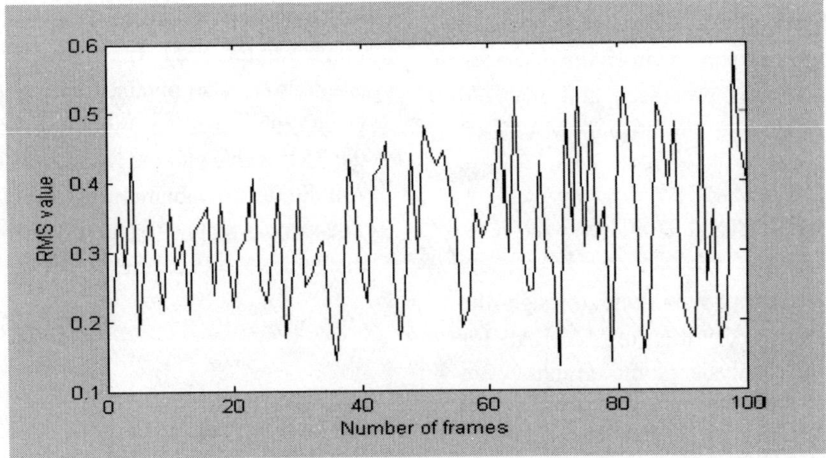

Figure 5: The graph between the RMS value and the Number of Frames.

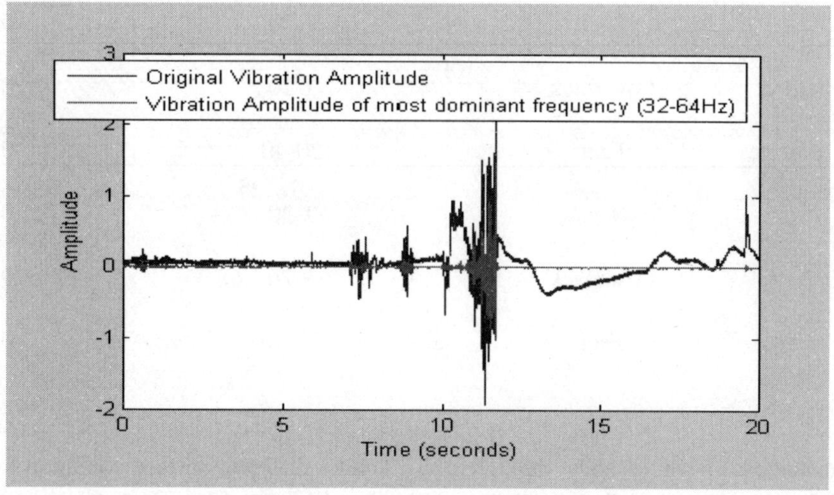

Figure 6: The graph between the Amplitude of Vibration and the Time (in seconds).

On the basis of the analysis through MATLAB, summary tables were made for participant 1 and 2. The summary table was showing the Surface conditions, Position of the sensor, Maximum Power Density and the Frequency zone.

TABLE-1: Power spectral density with frequency zone for different surface condition for Participant 1.

Surface Conditions	Position of Sensor	Frequency Zone (Hz)	Power spectral density
Farm Field	Neck	1-20	0.0840
	Low Back	1-20	0.1044
	Feet	20-40	0.4967
Tar-Macadam Road	Neck	1-20	0.1519
	Low Back	1-20	0.1899
	Feet	20-40	0.2496
Bricks Road	Neck	1-20	0.0733
	Low Back	1-20	0.0792
	Feet	20-40	0.6472

TABLE-2: Power spectral density with frequency zone for different surface condition for Participant 2.

Surface Conditions	Position of Sensor	Frequency Zone (Hz)	Power spectral density
Farm Field	Neck	1-20	0.1940
	Low Back	1-20	0.0789
	Feet	20-40	0.5997

Tar-Macadam Road	Neck	1-20	0.1258
	Low Back	1-20	0.2154
:	Feet	20-40	0.1958
Bricks Road	Neck	1-20	0.0563
	Low Back	1-20	0.0542
	Feet	20-40	0.8479

The vibrations transmitted to whole body through seat pad system & breaking system of tractor have been analyzed with Lab View TM and by using MATLAB programme for each test and the results we get in the form of tables- 1and 2. Table-1 & 2 is shows the vibration level on neck, low back & feet under different surface condition far participant 1 & 2. The results indicate that the maximum transmissibility was observed in the first two frequency interval (in Hz) i.e. 1-20 and 20-40. The frequency interval under different surface conditions was 1-20 (neck), 1-20 (low back), and 20-40 (feet).

IV. CONCLUSION & FUTURE STUDY

The results show that the vibration level of tractor driver from the seat operator system & breaking system under the different surface conditions (i.e., farm field, tar-macadam road & bricks road) have given the frequency zone 1-20 and 20-40 was the maximum power spectral density. The frequency zone 1-20 and 20-40 is the most harmful for the tractor driver.

Prasad et. at. [10] have suggested that the frequency zone 2-6 Hz has been the most harmful for the tractor driver because resonance occurs within this frequency.

We have found most harmful range of frequency zone for tractor driver; the next challenge will be evaluate the analytical model for tractor seat operator system and also, will be find spring constant & damping co-efficient of suspension system for the tractor.

V.REFERENCE

[1] Mehta,C.R., Shyam,M., Singh,P. and Verma,R.N.,2000, 'Ride vibration on tractor-implement system', Applied Ergonomics, Volume 31, pp.323-328.

[2] Ahmed, G. and Goupillon, P., 2005, 'Predicting the ride vibration of an agricultural tractor', Journal of Terramechanics, Vol. 34. No. 1, pp. 1-11.

[3] Mehta, C.R. and Tewari, V.K., 2007'Seating discomfort for tractor operators' Journal of Terramechanics, Vol. 33. No. 3, pp. 208-219.

[4] Tiwari, V.K. and Prasad, N., 1999, 'Three-DOF modeling of tractor seat-operator system', Journal of Terramechanics, Vol. 36, pp. 207-218.

[5] Dewangan, K.N., and Tewari, V.K.,2009, 'Characteristics of hand transmitted vibration of a hand tractor used in three operational modes' , International Journal of Industrial Ergonomics , volume 39, issue 1, pp 239-245.

[6] Scarlett, S.S. and Goupillon, P., 2001, 'Whole-body vibration: evaluation of emission and exposure levels arising from agricultural tractors', International Journal of Industrial Ergonomics , volume 29, pp. 209-215.

[7] Dewangan, K.N., and Tewari, V.K., 2008, 'Characteristics of vibration transmission in the hand arm system and subjective response during field operation of a hand tractor', Biosystem Engineering, volume 100, issue 4, pp 535-546.

[8] Clijmans, L., Ramon, H., Langenakens, J. and Baerdemaeker, J., (1996). 'The influence of tyres on the dynamic behavior of a lawn mower', Journal of Terramechanics, 33(4), 195–208.

[9] Crolla, D. A. and Horton, D. N. L.,(1984) 'Factors affecting the dynamic behavior of higher speed agricultural vehicles'. Journal of Agricultural Engineering Research, 30, 277–288.

[10]Goglia, V., Gospodaric, Z., Kosutic, S., and Filipovic, D., 2002, 'Hand transmitted vibration from

[11]the steering wheel to drivers of a small four-wheel drive tractor', Applied ergonomics, volume 34, issue1, pp 45-49.

[12]Adewusi,S.A., Rakheja, S., Marcotte,P. and Boutin,J., 2010, 'Vibration transmissibility characteristics of the human hand–arm system under different postures, hand forces and excitation levels', Journal of Sound and Vibration ,Volume 329, Issue 14, pp.. 2953-2971.

[13] Prasad, N., Tewari, V.K., and Yadav, R., 1996, 'Tractor ride vibration--a review', Journal of Terramechanics, Vol. 32. No. 4, pp. 205-219.

[14]Deprez, K., Moshou, D., Anthonis J., Baerdemaeker, J. D. and Ramon, H., (2005b) 'Improvement of vibrational comfort by passive andsemi-active cabin suspensions', Computers and Electronics in Agriculture, 49, 431–440.

[15]Kennes, P., Anthonis, J., Clijmans, L. and Ramon, H., (1999) 'Construction of a portable test rig to perform experimental modal analysis on mobile agricultural machinery', Journal of Sound and Vibration, 228(2), 421–441.

[16] Bouazara, M., Richard, M. J. and Rakheja, S., 2006, 'Safety and comfort analysis of a 3-D vehicle model with optimal non-linear active seat suspension', Journal of Terramechanics, 43, 97–118.

[17] Clijmans, L., Ramon, H. and Baerdemaeker, J., 1998, 'Structural modification effects on the dynamic behavior of an agricultural tractor', Transactions of the ASAE, 41(1), 5–10.

[18] Crolla, D. A., Horton, D. N. L. and Stayner, R. M., (1990) 'Effect of tyre modeling on tractor ride vibration prediction', Journal of Agricultural Engineering Research, 47, 55–77.

[19] Deprez, K., Moshou, D. and Ramon, H., (2005a) 'Comfort improvement of a non-linear suspension using global optimization' Journal of Sound and Vibration, 284, 1003 1014.

[20] Kumar, A., Mahaian, P., Mohan, D. and Varghese, M., (2001) 'Tractor vibration severity an driver healt: a study from rural India', Journal of Agricultural Engineering Research, 80(4), 313–328.

Effect of Wind Speed on Daily Yield of a Solar Still Operated at Indoor and Outdoor Conditions

Mehmood A[1] and Singh Kishan Pal[2]
[1]Department of Mechanical Engineering,Aligarh Muslim University, Aligarh (U.P.)
[2]B.S.A.College of Engineering & Technology, Mathura (U.P.)
E-mail: ayaz.amu09@gmail.com

Abstract- **The effect of wind speed on the daily productivity of a solar still operated at indoor and outdoor conditions is studied. Numerical calculations have been carried out on different days in a year in order to see the effect of speed of wind for same masses of basin water m_w for the basin type still. It is found that for the basin type still, Productivity increases with the increase of wind speed up to a crtical velocity V_c beyond which the increase in Productivity becomes insignificant for both the Indoor and Outdoor conditions.**
Keywords: Solar energy; passive cooling; Wind speed; Productivity.

I. INTRODUCTION

One major problem for the whole world and, in particular, in the third world is the availability of pure clean and healthy water especially in remote areas. Desalination systems using solar energy have been used in many countries to produce fresh water. The working of solar distillation units has been broadly classified into active and passive modes of operation. It is reported that the overall thermal efficiency of a passive distiller is higher than of an active one due to the lower operating temperature range [1]. Several researchers [2, 3, 4, 5,] have investigated the effects of climatic, operational and design parameters on the performance of single, double and multi-effect active and passive solar stills. It has been concluded that the productivity of the solar stills increases with the increase of solar radiation and ambient temperature [6] and [7]. However, it should be pointed out that there are conflicting results about the effect of wind speed on solar still productivity. Garg and Mann [6], Cooper [8], Rajvanshi [9], Soliman [10] and Malik and Tran

[11] have concluded that the increase in wind speed causes an increase in productivity; while Eibling et al. [7], Hollands [12] and Yeh and Chen [13 and 14] indicated that an increase in wind speed causes a decrease in productivity. It has also been reported by Morse and Read [15] that the wind speed has no significant effect on productivity.

Nomenclature

A_c collector area,

S solar radiation absorbed by absorbing plate

t_a ambient temperature

h_{fg} latent heat of vaporization of water, J/kg

P_{ci} partial pressure of water vapour at the glass temperature, N/m^2

P_d Yield or productivity

η_s efficiency of solar still

I_g total (global) solar radiation, W/m^2

I_{sc} extra-terrestrial radiation on a horizontal surface, W/m^2

I_b beam component of the total (global) solar radiation, W/m^2

I_d diffuse beam component of the total (global) solar radiation, W/m^2

C_p specific heat at constant pressure, J/kg $^\circ$C

S solar flux reaching the absorber plate, W/m^2

$I(t)$ solar radiation on horizontal surface, W/m^2

k   thermal conductivity, W/m °C

I_{tb} beam components of the total solar radiation

on tilted surface, W/m^2.

I_{td} diffused components of the total solar radiation on

tilted surface, W/m^2.

t   time, s

v   wind velocity (S), m/s

V_c critical velocity, m/s

k_t hourly clearness index

n   day of the year

U   heat loss coefficient, W/m^2 °C

n_2 / n_1 refractive index of air and

glass cover, respectively.

R   reflectivity

R_b beam radiation on the tilted

surface to that on the

horizontal surface.

Greek letters

α   absorptivity

β   surface tilted from the horizontal

δ   declination angle

ε   emissivity

ρ   reflectance

ω   hour angle

τ   transmittivity of the cover system of a

collector

τ_r transmittivity obtained by considering

only reflection and refraction

τ_α transmittivity obtained by considering

only absorption

θ_1 angle of incident

θ_2 angle of refraction

Φ   latitude angle

δ_e glass cover thickness

Subscripts

a   ambient air

Glass inr   glass inner surface

Glass otr   glass outer surface

II. EXPERIMENTAL SET-UP

There is a solar still which is made-up of copper sheet covered insulation, so that there is no flow of heat from still to outside then minimize the heat loss. The sheet of copper has black painted for increasing the absorbing property of heat of copper metal. In this stil we employed two numbers of U-shape heater, whose longitudinal length was 1.5 feet and heating element capacity 1500 watt each for proper heating of water. The basin type still is closed with the help of glass sheet of thickness 0.005m which is used for condensation of vapour of water at the top of still and this glass have a slope of 11.5° for installation of condensed liquid water into drain of still from where the distilled water out from still. On above the drain we use a sharp edge of glass strip to stop the droplets of condensed water and fall down into the drain. For measuring inner temperature of basin, we employed 14 thermocouples (copper+constanton) mounted on fiber beeding. First 9 thermocouples starting from basin plate are situated 1 cm apart from each other and remaining five thermocouples 2cm apart from each other and last one is 5cm apart from second last one which give the glass inner temperature. For measuring glass outer and ambient temperature we used a thermocouple attached outside sheet, for reading the temperature of thermocouples we used a digital temperature indicator which convert heat signal into electrical signal and display digitally on the screen. This solar still is kept on a iron made stand so that we used space under the still as our requirements and also this stands prevents the heat loss from solar still to the ground. The schematic diagram is shown below in Fig. (a) & the photograph of solar still shown below in Fig. (b). Now the heater removed from the solar still and for heating water, solar energy used at outdoor condition.

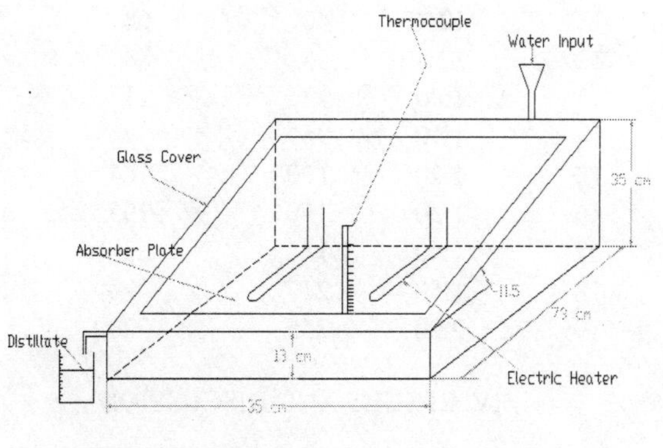

Fig. (a)

Fig. (b)

III. COMPUTATIONAL PROCEDURES

3.1. Solar still efficiency (Outdoor -Conditions)

The efficiency of solar still is calculated on the basis of solar heat used for distillation and the incident solar radiation. It is defined as ratio of the two given below:

*Efficiency of solar still (η_s) = Distillate output (kg/s) * Latent heat of vaporization of water (J/kg) Solar radiation (W/m^2) *Area of basin of still (m^2).*

3.2. Solar still efficiency (Indoor- Conditions)

The efficiency of solar still is calculated on the basis of solar heat used for distillation and the incident solar radiation. It is defined as ratio of the two given below:

Efficiency of solar still (η_s)
*= Distillate output (kg/s) * Latent heat of vaporization of water (J/kg)   Solar radiation (W/m^2) *Area of basin of still (m^2).*

Since distillate output in ml/h, to convert it into kg/s, **Experimental results are shown in Table A** ml/h = (10^{-6} m^3*10^3 kg/m^3) / 3600 s = 10^{-3}/ 3600 kg/s
Partial Pressure of vapour calculate by the equation given by Dunkle's [36]

$$P_{ci} = \exp\left[25.317 - \frac{5144}{273+T_{ci}}\right] \text{ (pascal)}$$

Experimental results are shown in Table A

Table A

Water depth in linar = 0.04 m

		Indoor Operated Still			Outdoor Operated Still		
		Distillate output			Distillate output		
S.No.	Time	S = 0 m/s	S = 36 m/s	S = 39 m/s	S = 7 m/s	S = 8 m/s	S = 10 m/s
1	11:00	0	0	0	0	0	0
2	11:10	0	0	0	0	0	0
3	11:20	0	0	0	0	0	0
4	11:30	0	0	0	0	0	0
5	11:40	0	0	0	0	0	0
6	11:50	0	2	4	0	0	0
7	12:00	0	12	10	0	0	0
8	12:10	5	23	25	0	0	0
9	12:20	12	34	37	5	7	4
10	12:30	25	48	48	10	12	8
11	12:40	40	68	68	20	23	17
12	12:50	56	88	88	30	34	28
13	1:00	73	117	108	45	50	40
14	1:10	95	143	132	65	70	60
15	1:20	120	164	150	85	90	80
16	1:30	150	195	180	105	110	100
17	1:40	182	230	207	125	132	120
18	1:50	217	270	240	145	155	140
19	2:00	255	320	280	165	185	160

IV. RESULT AND DISCUSSION

The distillate outputs measured at different times from the start-up period are plotted in Figs. 3.1 & 3.2 for different conditions. There is no output of distillate for nearly 30 to 50 minutes from the start-up period. The effect of cooling rate of the glass cover is shown in Fig. 3.1. The distillate output, at 775 W (constant heat input) of heat input for 3 hours of heating, are 255 ml, 320 ml and 280 ml with natural cooling, wind speed of 36 m/minute and wind speed of 39 m/minute respectively, showing more output at a wind speed of 36 m/minute. This was also happen in the similar fashion for out door operated still. The distillate output for 3 hours of heating, are 165 ml, 185 ml and 160 ml with wind speed of 7 m/minute, wind speed of

8 m/minute and wind speed of 10 m/minute respectively, showing more output at a wind speed of 8 m/minute.

Figures 3.3 to 3.4 show efficiency of the still for different conditions. Higher values of efficiency are achieved at moderate cooling rate of the glass cover. The efficiency of the still after three hours of heating at the wind speeds of natural, 36 m/minute and 39 m/minute, the efficiency becomes 21.17%, 27.57% and 24.19%. This was also happen in the similar fashion for out door operated still. The efficiency of the still after three hours of heating at the wind speeds of 7 m/minute, 8 m/minute and 10 m/minute, the efficiency becomes 12.15%, 12.43% and 11.94%.

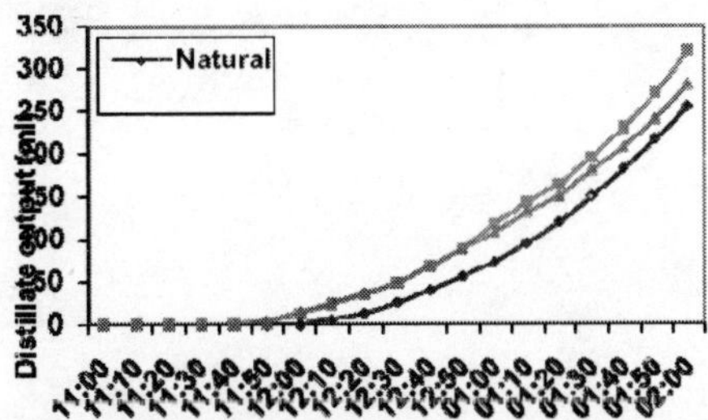

Fig. 3.1 Variation in the distillate output due to wind speed blowing past the glass cover of the Indoor distillation plant (Heat input=775 W and water depth=0.04 m).

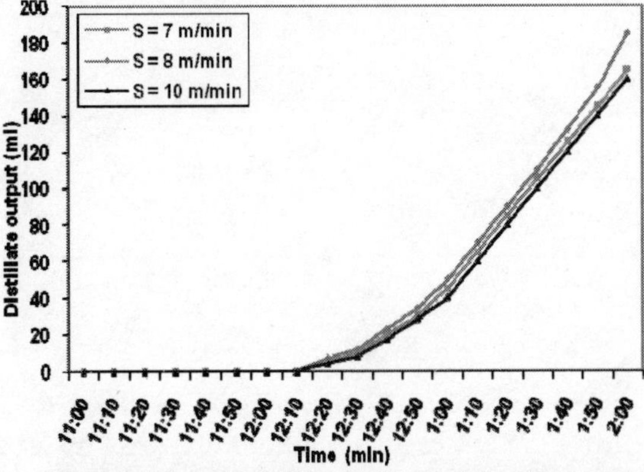

Fig. 3.2 variation in the distillate output due to wind speed blowing past the glass cover of the outdoor distillation plant (Water depth = 0.04 m).

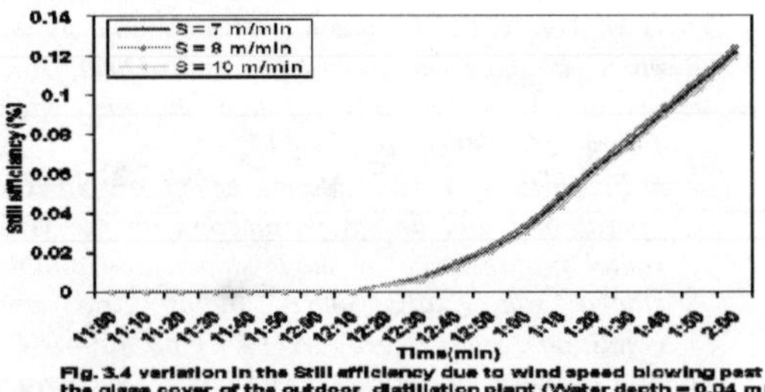

Fig. 3.4 variation in the Still efficiency due to wind speed blowing past the glass cover of the outdoor distillation plant (Water depth = 0.04 m).

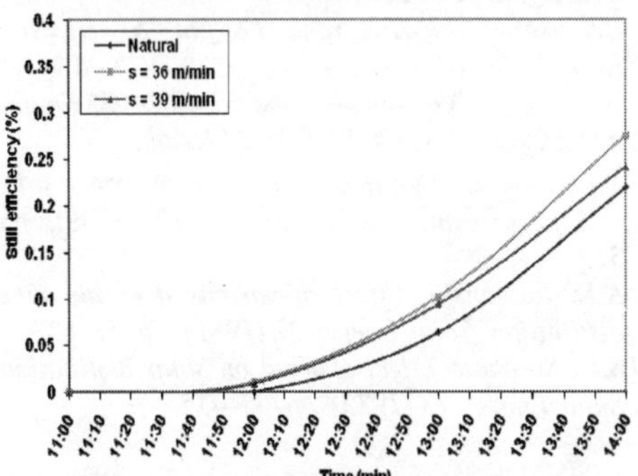

Fig. 4.32 Variation in the still efficiency due to wind speed blowing past the glass cover of the indoor distiillation plant (Heat input=775 W and water depth=0.04 m).

V. CONCLUSION

On the basis of the results obtained for the various operative conditions of the solar stills, the following conclusions may be drawn: (i) The daily productivities P_d of the basin type solar stills increase as wind speed increases up to the crtical velocity V_c. (ii) After the typical velocity V_c, the change in P_d becomes insignificant. (iii) The value of V_c is independent on the still shape, mode of operation (active or passive) and the heat capacity of brine, but it shows some seasonal dependence. V_c is found to be 36 m/s for indoor and 8 m/s for outdoor conditions respectively. (iv) The results which have been achieved in the present study may be used to explain the conflicting results which are reported in the previous studies [10, 12, 13, 14] about the effect of wind speed on the daily productivity of the solar stills.

VI. REFERENCES

[1] G.N. Tiwari, *Recent advances in solar distillation. In: Raj Kamal Solar Energy and Energy Conversion, Wiley Eastern, New Delhi (1992), pp. 32–149 (Chapter II).*

[2] S. Kumar and G.N. Tiwari, *Optimization of daily yield from an active double effect distillation with water flow. Energy Convers. Mgmt. 40 (1999), pp. 703–715.*

[3] B.A. Jubran, M.I. Ahmed, A.F. Ismail and Y.A. Abakar, *Numerical modeling of a multi-stage solar still. Energy Convers. Mgmt. 41 (2000), pp. 1107–1121.*

[4] S. Suneja and G.N. Tiwari, *Effect of water depth on the performance of an inverted absorber double basin solar still. Energy Convers. Mgmt. 40 (1999), pp. 1885–1897.*

[5] *H.M. Yeh, C.D. Ho and K.J. Hwang, Energy and mass balance in open-type multi-effect solar distiller with air flow through the last effect. Energy 24 (1999), pp. 103–115.*

[6] *H.P. Garg and H.S. Mann, Effect of climatic, operational and design parameters on the year-round performance of single-sloped and double-sloped solar stills under Indian arid zone conditions. Solar Energy 18 (1976), pp. 159–164.*

[7] *http://www.sciencedirect.com/science?_ob=Article URL&_udi=B6V2P-4B2C491-B&_user=5614075&_rdoc=1&_fmt=&_orig=search&_sort=d&view=c&_acct=C000067906&_version=1&_urlVersion=0&_userid=5614075&md5=d6a3263f0993e3809f502350261a66b3 -*

[8] *P.I. Cooper, Digital simulation of transient solar still performance. Solar Energy 12 (1969), pp. 313–331.*

[9] *A.K. Rajvanshi, Effect of various dyes on solar distillation. Solar Energy 27 (1981), pp. 51–65.*

[10] *S.H. Soliman, Effect of wind on solar distillation. Solar Energy 13 (1972), pp403–415.*

[11] http://www.sciencedirect.com/science?_ob=Article URL&_udi=B6V2P-4B2C491-B&_user=5614075&_rdoc=1&_fmt=&_orig=search&_sort=d&view=c&_acct=C000067906&_version=1&_urlVersion=0&_userid=5614075&md5=d6a3263f0993e3809f502350261a66b3 -bbib14#bbib14

[12] *K.G.T. Hollands, The regeneration of lithium chloride brine in a solar still for use in solar air conditioning. Solar Energy 7 (1963), p. 39.*

[13] *H.M. Yeh and L.C. Chen, Basin-type solar distillation with air flow through the still. Energy 10 (1985), pp. 1237–1241.*

[14] *H.M. Yeh and L.C. Chen, The effect of climatic, design and operational parameters on the performance of wick-type solar distiller. Energy Conversion Mgmt. 26 (1986), pp. 175–180.*

[15] http://www.sciencedirect.com/science?_ob=Article URL&_udi=B6V2P-4B2C491-B&_user=5614075&_rdoc=1&_fmt=&_orig=search&_sort=d&view=c&_acct=C000067906&_version=1&_urlVersion=0&_userid=5614075&md5=d6a3263f0993e3809f502350261a66b3

Information Technology Management in Supply Chain

Vikas Misra[1], M.I.Khan[2] and U.K.Singh[3]
[1] Mechanical Engineering Department, University of Petroleum & Energy Studies, Dehradun,
[2] Mechanical Engineering Department, Integral University, Lucknow.
[3] Mechanical Engineering Department. K.N.I.T. Sultanpur

Abstract: **Information is crucial to the performance of a supply chain because it provides the basis upon which supply chain managers make decisions. Information is very important factor & drives the entire Supply Chain System. Information Technology consists of the tools used to gain awareness of information, analyze this information, and act on it to improve the performance of the supply chain. The pressures of the global competition and the need for the extensive inter-organizational, collaboration is forcing industries to streamline their supply chains to make them agile, flexible and responsive. Effective use of information technology enables management to make decisions over a broad scope that crosses both functions and companies. Information Technology serves as the eyes & ears of the management in supply chain, capturing and analyzing the information necessary to make a good decision. This paper deals with integration of supply chains and specifically concentrates on the importance of distribution of information in the supply chain. This paper explores the significance of information, its uses and the technologies that enable supply chain managers to use information to make better decisions. The paper discusses how sharing and strategic utilization of information in a supply chain can radically improve execution of vital business processes and help in achieving shorter lead times, lower costs and inventory levels and finally better quality & customer satisfaction which is of utmost importance for the successfulness of the whole supply chain.**

Keywords: - Supply chain management, Information sharing, integration, internal supply chain management, business process renovation

I. INTRODUCTION:

Information is crucial to the performance of a supply chain because it provides the foundation on which supply chain processes execute transactions and managers make decisions. Without information, a manager will not know what customers want, how much inventory is in stock, and when more products should be produced and shipped. Information is the supply chain driver that serves as the glue allowing the other drivers such as inventory, transportation and facilities to work together to create an integrated, coordinated supply chain. In short, without information a manager can only make decisions blindly.

Therefore information makes the supply chain visible to a manager. With this visibility, a manager can make decisions to improve the supply chain's performance. In many ways, the information is the most important of the four supply chain drivers because without it, none of the other drivers can be used to deliver a high level of performance. Given the role of a supply chain's success, managers must understand how information is gathered and analyzed. This is Information Technology comes into play. Information Technology system consists of the hardware and software throughout a supply chain that gather, analyze and act on information. Information Technology serves as the eyes & ears (and some times a portion of the brain) of management in a supply chain, capturing and analyzing the information necessary to make good decisions.

The paper studies about the importance of distribution of information in the supply chain. It also studies about how Information Technology tools enables supply chain managers to make better decisions over a broad scope that crosses both functions and companies. [1,2]

Using Information Technology systems to capture and analyze information can have a significant impact on a firm's performance. For example a major manufacturer of computer workstations and servers found that much of the information on customer demand was not being used to set production schedules and inventory levels. The manufacturing groups lacked this information, which forced them to make inventory and production decisions blindly. By installing a supply chain software system, the company was able to gather and analyze data to produce recommended stocking levels. Using the Information Technology systems enabled the company to cut its inventory in half because managers could make decisions based on information rather than educated guesses. [3] Large impact like this underscores the importance of Information Technology as a driver of supply chain performance. Information must have following characteristics to be useful when making supply chain decisions: -

- Information must be accurate
- Information must be accessible in a timely manner.
- Information must be of the right kind.

Effective information sharing is a prerequisite for successful operation of supply chain in summary, information is crucial to making good supply chain decisions at all three levels of decision making (i.e. strategy, planning & operations) and in each of the other supply chain drivers (inventory, transportation and facilities). Information Technology enables not only the gathering of this data to create supply chain visibility, but also the analysis of this data so that the supply chain decisions made will maximize profitability.

II. SUPPLY CHAIN MANAGEMENT: -

Supply chain is a linked set of resources and processes that begins with the sourcing of raw materials and extends through the delivery of finished products to the final customer. (Bridgefeld Group ERP/ Supply Chain Glossary, 2004)

While the separation of supply chain activities among different companies enables specialization and economics of scale, there are many important issues and problems that need to be resolved for successful supply chain

operation—this is the main purpose of supply chain management.

According to the *Global Supply Chain Forum (GSFC)*, Supply Chain Management is defined as "the integration of key business processes from end user through original suppliers that provide products, services, and information that add value for customer and other stakeholders."(Chan & Qi, 2003). [1] We can only talk about Supply Chain Management, if there is a proactive relationship between a buyer and supplier and the integration is across the whole supply chain, not just first tier suppliers. (Cox, 2000). [3]

(McGuffog & Wadsley, 1999) has defined Supply Chain Management in a very comphrensive way. It states that Supply Chain Management is a set of synchronized decision & activities, utilized to effectively integrate suppliers, manufacturers, transporters, warehouses, retailers & customers so that the right product or service is distributed at the right quantities, to the right locations & at the appropriate time, in order to minimize system wide costs while satisfying customer service level requirements. It further says that Supply Chain Management is concerned with planning & coordinating activities right from procurement of raw materials to production of goods, leading finally to production of goods, with trillions of dollar traded in supply chains annually the impact of adopting better practices is tremendous.

There are several important issues in supply chain management that need to be resolved for efficient operation. Most of those problems stem either from uncertainties or inability to coordinate several activities and partners. (Turban, Mclean, & Wetherbe, 2004)

One of the most common problems in supply chains is the Bullwhip effect. Even small fluctuations in the demand or inventory levels of the final company in the chain are propagated and enlarged throughout the chain. Because each company in the chain has incomplete information about the needs of others, it has to respond with the unproportional increase in inventory levels and consequently even larger fluctuation in its demand to others down the chain. (Forrester, 1961; Forrester 1958).

Hence Supply Chain must be able to handle large numbers of events both expected & unexpected. The unexpected events, also called exceptions, typically arise because there is usually a gap between supply chain planning & execution.

III. SUPPLY CHAIN INFORMATION TECHNOLOGY FRAMEWORK: -

Given the wide range of information discussed earlier, it is important to develop a framework that helps a manager understand how this information is utilized by the various segment of Information Technology within the supply chain. Our vision of this framework is presented here. It is valuable to note that driver of Information Technology (IT) in the supply chain has increasingly been the enterprise software developed to enable processes both within and across companies. Enterprise software collects transaction data, analyzes this data to make decisions, and executes on these decisions both within an enterprise and across its supply chain. Certainly other parts of IT beyond enterprise software such as hardware, implementation services, and support are all crucial to making IT effective. Within a supplychain, however, the different capabilities provided by IT have as their most basic building block the capabilities of the supply chain's enterprise software. In many ways, software shapes the entire industry of IT as other components follow the software lead. The evolution of enterprise software provides insights not only into the future of IT, but also into what the key supply chain processes are.

The enterprise software Landscape became increasingly popular during late 2000. The unprecedented flow of venture capital into new software companies led not just to an increase in the number of software companies, but also to the proliferation of entire categories of softwares. The growth of enterprise software landscape is much more dynamic. What drives this evolution of the enterprise software landscape? Why are some categories of software companies headed for a profitable long- term future, whereas others have failed? Certainly there are a wide variety of factors affecting the natural selection of software companies. We propose, however, that three of the main drivers of the evolution taking place in enterprise software are the three major groups of supply chain processes, which we term supply chain macro processes. The successful categories of software will be those focused on macro processes. [4,5]

IV. THE SUPPLY CHAINS MACRO PROCESSES: -

The scope of supply chain management has broadened the scope across which companies make decisions. This scope has expanded from trying to optimize performance across division, to the enterprise, and now to the entire supply chain. This broadening of scope emphasizes the importance of including processes all along the supply chain when making decisions. From an enterprise's perspective, all processes within its supply chain can be categorized into three main areas: Processes focused downstream, Processes focused internally, Processes focused upstream. We use this classification to define the three Macro Supply chain Processes as follows:

Customer Relationship Management (CRM): Processes that focus on downstream interactions between the enterprise and its customer.

Internal Supply chain Management (ISCM): Processes that focus on internal operations within the enterprise.

Supplier Relationship Management (SRM): Processes that focus on upstream interactions between the enterprise and its suppliers.

WHY FOCUS ON THE MACRO PROCESSES?

As the performance of an enterprise becomes more closely linked to the performance of its supply chain, it is crucial that firms focus on these processes. After decades of focusing on the internal processes, a firm must expand the scope beyond internal processes and look at the entire supply chain to achieve breakthrough performance. The goal should be to increase the total profitability of the supply chain (also referred as supply chain surplus). Good supply chain management is a positive sum game where supply chain partners can increase their overall level of profitability by working together. Therefore, to increase the supply chain surplus most effectively, firms must expand their scope beyond their enterprise and think in terms of all three macro processes. [10]

MACRO PROCESSES APPLIED TO THE EVOLUTION OF SOFTWARE: -

As the downturn in technology spending has applied evolutionary pressure on the enterprise software landscape, we see a distinct pattern emerging. The majority of survivors have chosen to focus their products on improving their customer's macro processes. Some software firms cross over into more than one macro process, whereas others only address a small portion of macro process. But the common theme we see is that to survive, and particularly to thrive, enterprise software firms must focus on one or more of these macro processes. Almost all areas of enterprise software growth exist within CRM, ISCM, or SRM.Both new companies and larger firms within enterprise software are now targeting these three macro processes much more sharply. In the future, we see the ability to improve the three macro processes driving the winners and losers in enterprise software.

ERP software has been successful in improving data integrity within the supply chain, but by itself, data integrity provides little value. The real value from having ERP systems in place can only be obtained if these systems can be used to improve decision-making in the three macro processes. Every major ERP player has realized this and is remaking themselves into a company emphasizing products focused on the macro processes. [5]

THE SOFTWARE WINNERS WITHIN A MACRO PROCESSES:

Among software firms focused on a macro process, the following three factors determine their success;

- Functional Performance
- Integration with other macro processes
- Strength of the software firm's ecosystem

Functional performance is important to customers because it provides them with capabilities to create a competitive advantage. In addition to create raw Functional performance, we believe that the ease of use is crucial to success in this category.

The ability to integrate is important to a customer for variety of reasons. Applications that are easy to integrate are generally easier to get implemented and producing value. Integration is also crucial across different macro processes. Applications that integrate across macro processes will be able to provide the benefits if making decisions for the extended supply chain.

Finally, a firm's ecosystem – the network of software partners and, more importantly, system integrators and installed base –provides assistance in selling and implementing software. Firms that work well with implementation partners and build up large group of customers trained on their solutions have built a highly defensible position. For a customer, a strong ecosystem means a strong network to provide support both during implementation and down the road. [6]

These criteria are also important for customers of supply chain software. These criteria are the key to success for software companies precisely because they improve supply chain performance for firms. Thus companies should evaluate

software providers along these lines to find their choice of software vendor.

We now discuss each of the macro processes, what segments they consist of, who the players are, and what the future will look like.

CUSTOMER RELATIONSHIP MANAGEMENT (CRM)

The CRM macro process consists of processes that take place between an enterprise and its customers downstream in the supply chain. The goal of CRM macro process is to generate customer demand and facilitate transmission and tracking of orders. Weakness in this process results in demand being lost and a poor customer experience because orders are not processed and executed effectively. The key processes under CRM are as follows:

- Marketing
- Sell
- Order Management
- Call/ Service center

The aforementioned CRM processes are crucial to the supply chain as they cover a vast amount of interaction between an enterprise and its customers. Thus the CRM macro process is the starting point when improving supply chain performance. It is also important to note that CRM process and CRM software must be integrated with internal operations to optimize performance.

The CRM software landscape consists of three categories of companies; the best-of-breed winner, the best-of-breed startups, & the ERP players. CRM is currently dominated by Siebel Systems, the sole company in the best-of-breed winner category. However, Siebel does face serious competition from both best-of-breed startups that emphasize functional expertise as well as from the ERP players, such as SAP, Oracle, and People soft, which provide a powerful integration story and strong ecosystems. Looking foreword, Siebel, the best-of-breed winner, provides a combination of superior functionality and a strong ecosystem within SRM.[9]

4.5 INTERNAL SUPPLY CHAIN MANAGEMENT (ISCM): -

ISCM is focused on operations internal to the enterprise. ISCM includes all processes involved in planning for and fulfilling a customer order. The various processes included in ISCM are as follows:

- Strategic Planning
- Demand Planning

- Supply Planning
- Fulfillment
- Field Service

Given that the ISCM macro process aims to fulfill demand that is generated by CRM processes, their needs to be strong integration between the ISCM and CRM macro process. Similarly the ISCM processes should have strong integration with the SRM macro process.

Successful ISCM software providers have helped improve decision making within ISCM processes. Like CRM, today's ISCM landscape consists of three categories--; the best-of-breed winner, the best-of-breed startups, & the ERP players. Unlike CRM, however there is not a clear leader. There are two best of breed winners, i2 Technologies and Manugistics, which were ISCM pioneers and are currently the functional leaders.

The best-of-breed ISCM players have the leading functionality, but lack strong integration and ecosystems. These companies have been working to offer more products in the SRM and CRM space to improve their integrated offering. The ERP player's advantages are their integrated product & their ecosystems, although some ERP player's functionality is becoming more and more competitive. There are some smaller players in ISCM taking advantage of new functionality that will remain viable, especially those targeting customers in specific industries that are very dependent on advanced functionality. [8]

SUPPLIER RELATIONSHIP MANAGEMENT (SRM): -

SRM includes those processes focused on the interaction between the enterprise and the suppliers that are upstream in the supply chain. There is a very natural fit between SRM processes and the ISCM processes as integrating supplier constraints is crucial when creating plans. The major SRM processes are as follows:

- Design Collaboration
- Source
- Negotiate
- Buy
- Supply Collaboration.

Significant improvement in supply chain performance can be achieved if SRM processes are well integrated with appropriate CRM and ISCM processes.

The SRM space has four groupings of competitors. There are two best-of-breed groups that focus exclusively on SRM, one focused on design collaboration and other focused on procurement. Leading Design Collaboration firms include Agile & Matrix One while leading procurement firm are Ariba and Commerce one. The third type of player in SRM is the best-of-breed ISCM vendor that has made the natural extension of ISCM into SRM—companies such as i2 and Manugistics. Finally, the fourth category consists of the ERP players moving up into the macro processes again. SAP is the largest SRM player among the ERP player's vendors and has shown the most commitment to entering this space. SRM has already attracted all the big players from both ISCM & ERP.Therefore the future SRM landscape is likely to be dominated by one or two ISCM players and one or two ERP players. [9]

SUPPLY CHAIN INFORMATION TECHNOLOGY IN PRACTICE: -

Although there are different sets of practical suggestions for each supply chain macro process, there are several general ideas that managers need to keep in mind when making a decision regarding supply chain IT.

(1) *Select an IT system that addresses the company's key success factor:* - It is important to select supply chain IT systems that are able to give a company an advantage in the areas most crucial to the success of the business. For instance, the ability to optimally set inventory levels is crucial in the PC business where product life cycles are short and inventory becomes obsolete very quickly.

(2) *Take incremental steps and Measure values:* - One-Way to help ensure success of IT projects is to design them so that they have incremental steps. For instance, instead of installing a complete supply chain system across your company all at once, start first by getting your demand planning up and running and then move on to supply planning. Along the way, make sure that each step is adding value through improvement in the performance of the three-macro processes.

(3) *Align the level of sophistication with the need for sophistication:* - Management must consider the depth to which an IT system deals with the firm's key success factors. Therefore, it is important to consider just how much sophistication a company needs to achieves its goals & then ensure that the system chosen matches that level.

(4) *Use IT systems to support decision making, not to make decisions:* - A mistake companies generally make is installing a supply chain system & then reducing the amount of managerial effort they spend on supply chain issues. Management must keep its focus on the supply chain because as the competitive & customer landscape changes, there needs to be a corresponding change in the supply chain.

(5) *Think about future:* -Although it is more difficult to make a decision about an IT system with the future in mind than the present, it is very important that the managers include the future state of the business in the decision process. If there are trends in a company's industry indicating that significant characteristics will become crucial in the future, managers need to make sure their IT choices take these trends into account. The key here is to ensure that the software not only fits a company's current needs but also, and even more important, that it will meet the company's future needs. [6,7]

V. Conclusion

The paper shows how sharing of information is essential in making good supply chain decisions because it provides the global scope needed to make optimal decisions. Information Technology provides the tools to gather this information and analyze it to make the best supply chain decisions. Information is the factual component on which the decisions of other supply chain drivers are based. Information Technology is a large enabler for effectively integrating the main macro processes of supply chain management.

Effective and good Information Technology systems not only allow the collection of data across the supply chain, but also the analysis of decisions that maximize supply chain profitability.

VI. References

[1]. *Chan, F., Qi, HJ. (2003). An innovative performance measurement method for supply chain management. Supply Chain Management journal,8(3), 209-223.*

[2]. *Cox, A., Chicksand, L. & Ireland, P.(2001). E-Business report, Boston, MA: Earlsgate Press.*

[3]. *Cross, G.J. (2002). How e-business is transforming supply chain management. Journal of business strategy, 21 92) 36-43.*

[4]. *Drayer, Raplah, & Robert Wright.2002."Getting the most from your ERP System" Supply Chain Management Review (May-June) 44-52.*

[5]. *Escalle, Cerdic X, Mark Kotteleer, & Robert Austin.1999, ERP Technology note. Harvard Business School note 9-699-020.*

[6] *Runter, Stephen M., Brian Gibson, Kate Vitasek and Gustin, 2005. Is Technology filling the information Gap? Supply Chain Management Review (March-April) 58-64.*

[7] *Shankar, Venkatesh, and Tony Driscoll.2004. How wireless networks are reshaping the supply chain, Supply Chain Management Review, (July-August) 44-50.*

[8] *Soni, Ashok, M.V.Venkataraman and V.A.Mabert, "ERP; Common Myth vs. Evolving Reality." Business Horizons 44(3); 69-76.*

[9] *S.Chopra and P.Meindl, 2006, Coordination & Technology in Supply Chain, Supply Chain Management: An International Journal, 9s(1) 81-90.*

[10]. *Beamon, B.M. (1999) Measuring supply chain performance. International Journal of operations and Production Management, 19(3), 275-289.*

Performance of Electric Shock Absorber

Nidhi Singh, Jiyaul Mustafa, Mohd Farooq, Abid Ali Khan, M. Muzamm

Department of Mechanical , Zakir Hussain Coollege, of Engineering Technology, AMU, Aligarh

Abstract- **The electric shock absorber is a device that converts the kinetic energy of an oscillating object into electric energy. This kinetic energy is normally dumped in a form of thermal energy in a conventional, mechanical shock absorber. The electric shock absorber consists of a permanent magnet linear synchronous generator (PMLSG), a spring, and an electric energy accumulator. The designed PMLSG was studied under steady-state conditions to determine its electromechanical characteristics. For this purpose the mathematical model of the generator was proposed and a program was written in MATLAB that allowed calculating its output parameters under different operation conditions. The PMLSG operates practically in dynamic conditions within the whole system: generator – spring – controlled rectified – battery. The dynamic model of the entire system of the electric shock absorber was proposed and described using the voltage equilibrium equation for the electrical port and the force equilibrium equation for the mechanical port. The conclusion obtained indicates that the electric shock absorber is able to store part of the recovered energy in the battery.**

Keywords: Matlab, Damping force-Speed Characteristics, Efficiency and Speed Characteristics

I. INTRODUCTION

A conventional automotive shock absorber dampens suspension movement to produce a controlled action that keeps the tire firmly on the road. The shock absorber, an energy dissipating device, is used in parallel with the suspension spring to reduce the vibration excited by road irregularities or during acceleration and braking. The research about energy recovery from vehicle suspensions began more than ten years ago, first as an auxiliary power source for active suspension control, and later also as energy regenerating devices in their own accord. Suda et al

[1, 2] investigated a self-powered active suspension control system, in which one motor is used to generate the energy and another motor is used to control the vibration in another stage.

This is done by converting the kinetic energy into heat energy, which is then absorbed by the shock's oil [3, 4]. The Power-Generating Shock Absorber (PGSA) converts this kinetic energy into electricity instead of heat through the use of a linear electric motor. One important loss is the dissipation of vibration energy by shock absorbers in the vehicle suspension under the excitation of road irregularity and vehicle acceleration or deceleration. The waveform and RMS voltage of the individual coils will depend on the equilibrium position but the total energy will not. Such a regenerative shock absorber will be able to harvest 16–64 W power at 0.25–0.5 m s−1 RMS suspension velocity [5]. The electricity generated by each PGSA can then be combined with electricity from other power generation systems (e.g. regenerative braking) and stored in the vehicle's batteries. The objectives of the presenting paper, designed PGSA under steady-state conditions to determine its electromechanical characteristics.

II. CURRENTLY USED SHOCK ABSORBERS PERFORMANCE CHARACTERISTICS

The performance of shock absorbers can be described using force-speed characteristics that are shown in Fig.1. The right side characteristics show graphically how the damping force exerted on the piston of the shock absorber depends on the (maximum) speed of the piston. The left side characteristics show the damping force developed at a particular position of the piston for the operation at the particular value of the stroke. For example, at particular road conditions the piston is moving in one direction through the distance equal to the stroke. In the middle of the stroke, the piston is moving with the maximum speed, and at this speed the maximum force is exerted on it. When the piston moves in the opposite direction (negative value of speed), the negative force exerts the piston and the shape of the characteristic differs from that at positive speed. The damping force of a shock absorber depends on the speed at which the two fixing points are pulled or pushed together. The damping force-speed curve can be progressive, linear or degressive, as is shown in Fig.1. The curve shape (left side) and diagram (right side) are directly related. The smallest area and therefore the lowest mean damping, is in the diagram of the progressive curve, i.e. the actual mean damping, which is important for the springing behavior, is low. The largest area is that of the degressive curve with a rounded shape, and so it has a high mean damping.

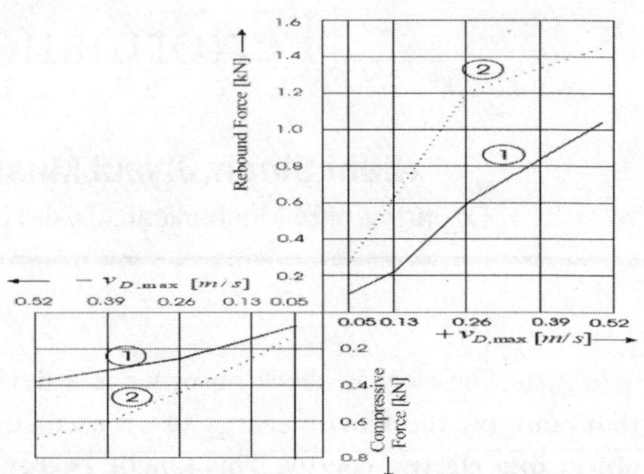

Fig.2. Rear axle damping curve: 1 is the standard setting, and 2 is for the heavy duty version

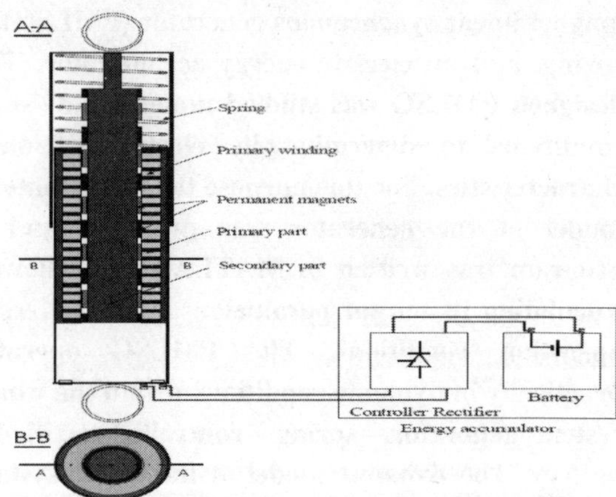

Fig.3. Diagram of electric shock absorber: A-A – axial (longitudinal) crossection, B-B perpendicular crossection [7]

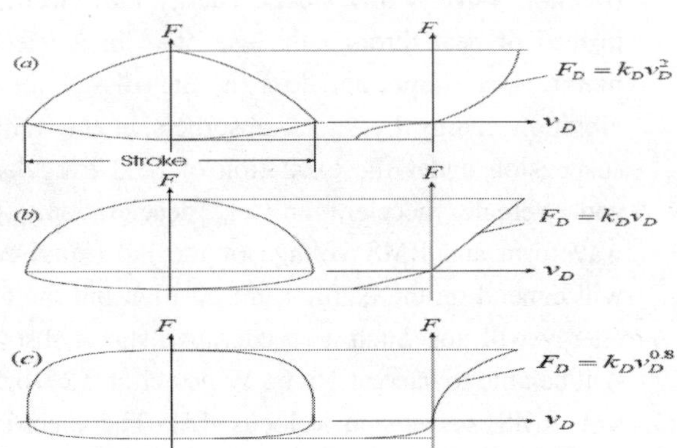

Fig.1. Damping Force-Speed curves: a) progressive, b) linear, c) degressive [6].

The exponent n can express the shape of the damping force curve F_D. In general, the force-speed characteristics can be described by the equation:

$$F_D = k_D v_D$$

Where: for $n > 1$ - progressive curve
For $n = 1$ - linear curve
For $n < 1$ - degressive curve
v_D - piston speed in m/s.
The mean piston speed $v_{D, mean}$ is:

$$v_{D, mean} = v_{D, max} / 1.62$$

As an example, Fig.2 .shows a degressive curve of the rear axle damping of a front-wheel drive vehicle. The maximum jounce force of 1.4 kN occurs at $v_{D, mean} = 0.52$m/s. However, piston speeds of 3 m/s can occur, which leads to higher force.[6]

III. ANALYSIS

Performance at Steady-state Conditions

The performance of the designed electric shock absorber should be compare with the mechanical one, and the damping force-speed characteristics (see Fig.1.) should be considered. To draw this type of characteristic for the electric shock absorber, the operation of the linear generator at constant speed should be analyzed. Thus, no spring is taken into account.

Model of Electric Shock Absorber at Steady-state Conditions

Due to the negligible role of the armature inductance, the generator model at steady state stable conditions can be regarded as the DC generator operating on the DC voltage source with the internal resistance R_s (see Fig.4)

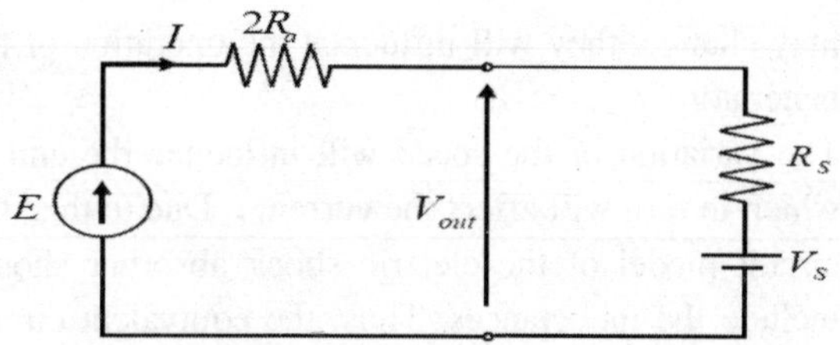

Fig.4. Simplified equivalent circuit of the electric shock absorber without inductances

In practice, the controlled rectifier connected between the generator and the source will allow adjustment to the voltage at the output terminals of the generator. The source is a battery with an internal resistance R_s and a voltage V_s.

The voltage equation for this circuit. [8]

$$E = V_s + (2R_a + R_s)I$$

Since emf

$$E = Kv$$

And damping

$$F_{em} = KI$$

The force expressed in terms of speed, the resistance, and the voltage source has the form

Steady State $\quad F_{em} = K\left(\dfrac{Kv - V_s}{2R_a - R_s}\right)$

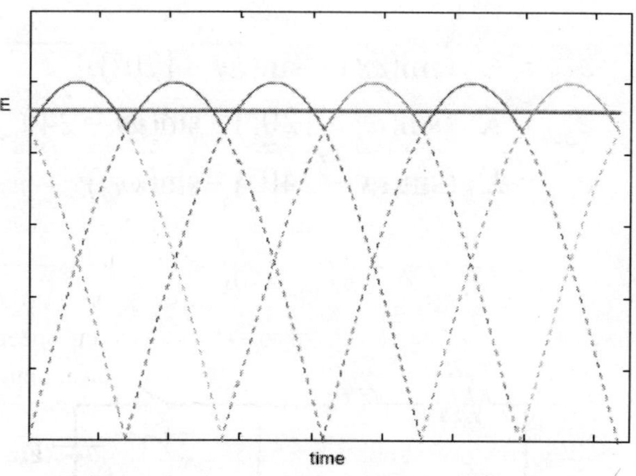

Fig. 5 Line to line voltages and output voltage in a three-phase rectifier.

Characteristics

For the purpose of this project, the calculations of the force-speed characteristics are made for three different values of the source voltage ($V_s = 0$, 12, 24 V). For each characteristic, the speed changes from 0 to 1.5 m/s, ($v = 0 - 1.5$ *m/s*). The battery resistance is set at $R_s = 0\ \Omega$. The armature resistance $R_a = 1.3\ \Omega$. The constant value is referred to the generator constant as $K_E = 119$ V.s / m [8] where $K_E = 72$ V.s/m is the voltage and force constant.

On the basis of Eqn.4, a computer program was written using MATLAB in the form of m-file. The force-speed characteristics of the electric shock absorber obtained for the parameters indicated are shown in Fig. 6.

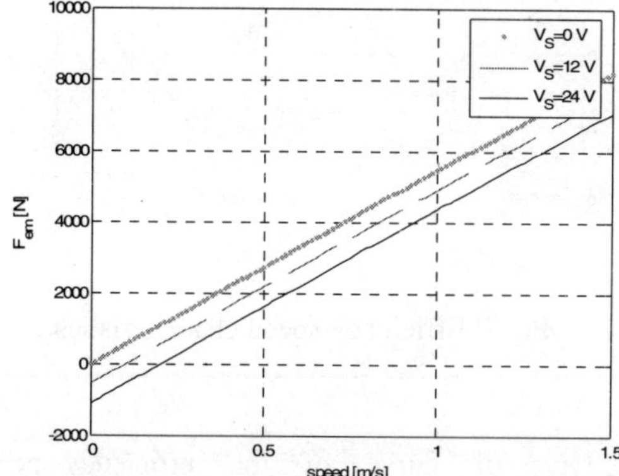

Fig. 6 Damping force-speed characteristics

This can be partially assessed by calculating the efficiency at the entire electric shock absorber system in the following way:

$$\eta = \frac{P_{out}}{P_{in}} \qquad (5)$$

Where: P_{out} - output power

$$P_{out} = V_s I \qquad (6)$$

P_{in} - input power

$$P_{in} = F_{em}v \qquad (7)$$

On the basis of Eqns. 5, 6 and 7, the computer program plots the efficiency-speed characteristic for two different values of the source voltage ($V_s = 12$ and 24 V). The efficiency-speed characteristics of the electric shock absorber obtained for the parameters indicated are shown in Fig.7.

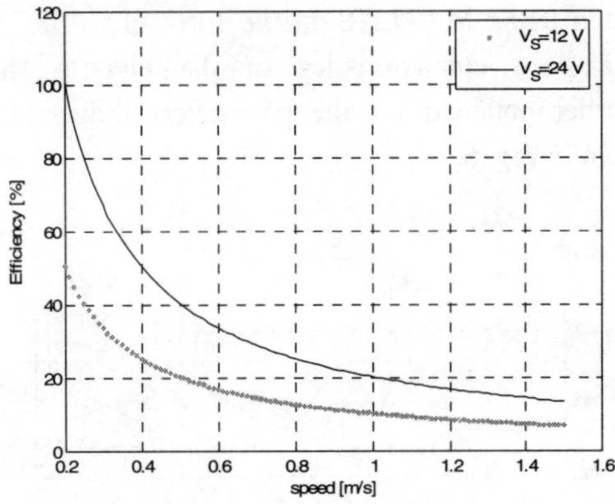

Fig.7. Efficiency-speed characteristics

The type of curve of the efficiency-speed characteristics is exponential, and the efficiency decreases inversely proportional to the speed. The greatest efficiency appears with the smallest speed. The transfer of energy from the generator to the battery occurs when $E > VS$. Since $E = Kv$, and $K = 119$ $V.s$ m, the generator starts to send the energy to the battery at speed $v = 0.2$ m/s . This explains why the efficiency-speed characteristic is drawn for speed $v \geq 0.2$ m/s. It must be pointed out that the motion losses (friction losses) were not considered in the calculation. This explains why at speed $v = 0.2$ m/s with voltage source $VS = 24V$, the efficiency tends to $\eta = 100\%$. Certainly, in practice, the efficiency is close to zero, since no power transfer occurs, and only mechanical losses exist.

Performance at Dynamic Conditions

The dynamic model of the electric shock absorber should include the spring. Since the road conditions may change, they will influence the operation of the generator.

The variation of the speed will influence the emf e, which in turn will affect the current i. Due to this; the circuit model of the electric shock absorber should include the inductances. Thus, the equivalent circuit takes the form shown in Fig.8.[8]

Mathematical Model

The voltage equation of the circuit of Fig.8

$$e = V_S + (2R_a + R_s)i + L\frac{di}{dt} \qquad (8)$$

Where R_s and V_s are the resistance and the voltage of source respectively

The total inductance

$$L = 2L_a \qquad (9)$$

Where L_a - phase inductance, which is approximately equal to the leakage inductance expressed by equation

The emf e is composed of maximum values of the line-to-line voltages that are changing in time as shown in Fig.5. The line-to-line voltages

$$e_{AB} = K_E(\sin(\omega t) - \sin(\omega t - 120°))$$
$$e_{BC} = K_E(\sin(\omega t - 120°) - \sin(\omega t - 240°))$$
$$e_{CA} = K_E(\sin(\omega t - 240°) - \sin(\omega t))$$

$$(10)$$

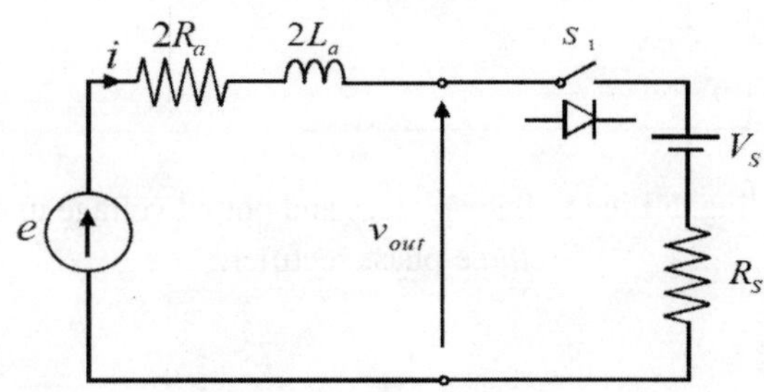

Fig.8. Simplified equivalent circuit of electric shock absorber with switch S_1 working as a diode

So that the current does not flow in the opposite direction, when the machine begins to work as a motor, the Switch 1 S is added to the circuit 8. This switch works as a diode. The switch is closed and lets the current flow only when the electromotive force e is bigger than the voltage source V_s (if $e > V_s \Rightarrow S_1$ ON).

F_{em} - electromagnetic force

$$F_{em} = K_E i \qquad (11)$$

K_E - voltage and force constant
i - output current
The instantaneous input power

$$p_{in} = ei$$

$$(12)$$

Where: e - electromotive force
The instantaneous output power

$$p_{out} = V_{out} i$$

$$(13)$$

Where: V_{out} - output voltage

IV. RESULTS AND DISCUSSIONS

For steady state, referring to the characteristics of the mechanical shock absorber in Fig.1, the type of curve of the damping force-speed characteristics is a linear one. In general, due to the controlled rectifier, the force-speed characteristic may be modified and adjusted to the particular need. The biggest force appears with the smallest voltage source ($V_s = 0$ V). If the voltage increases, the force decreases proportionally. The electromagnetic (damping) force calculated at rated conditions: voltage $V S = 24V$ and speed $v = 0.5$ *m/ s* is equal to $F_{em} = 1450N$. An important parameter is the quantity of mechanical energy converted to the electrical energy and stored

in the battery. The efficiency calculated at rated conditions: voltage $V S = 24V$ and speed $v = 0.5$ *m/s* is equal to $\eta = 40$ %. Therefore, at this speed only 40 % of the mechanical energy is converted into electrical energy and stored in the battery.

For dynamic state, the waveform of the displacement x indicates that after disturbance occurred, the oscillations damped until the electromotive force was greater than the source voltage $e > VS$, these oscillations happen if the speed of the moving part of the linear generator exceeds the certain value of the speed. Below this value, the induced voltage is equal or smaller than the source voltage ($e \leq V_s$); the current drops to 0, and no damping force $F\ em$ is produced, resulting in steady oscillations of the mechanical system. Overcome from this problem we modified the electric circuit. The modified equivalent circuit of the electric shock absorber is shown in Fig.10. The shunt resistor $R_c = 30\ \Omega$ is connected to the source voltage through the

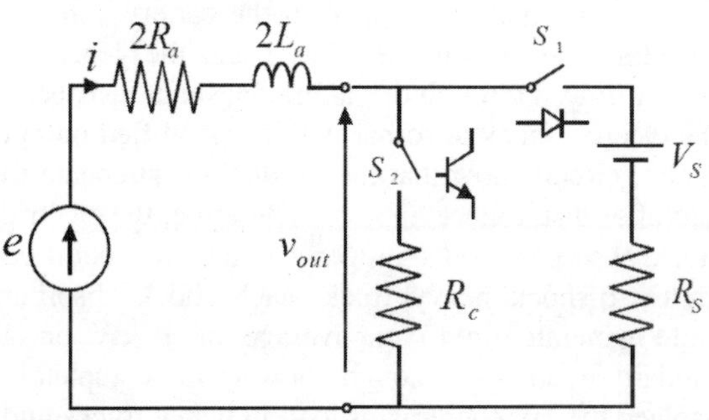

switch S_2.

Fig.9. Simplified equivalent circuit of the generator with modified electric circuit (switch S_2 works as a transistor).

During the operation

If $e > V_s \Rightarrow S_1$ ON   and S_2 OFF

If $e \leq V_s \Rightarrow S_1$ OFF   and S_2 ON   (14)

The voltage equation for the operation defined above formula is as follows

$$e = (2R_a + R_c)i + L\frac{di}{dt}$$

$$(15)$$

V. CONCLUSION

The electric shock absorber was analyzed under steady-state conditions. This allowed the determination of the performance characteristic (force-speed) that characterizes the conventional shock absorber. The results obtained from calculations indicate that the electric shock absorber has a linear characteristic. With the controller rectifier, the force-speed characteristic can be modified and adjusted to the actual road conditions. The calculation results also show that rated conditions, around 40 % of the oscillation energy can be converted into electrical energy that charges the battery. However, this parameter depends on the generator secondary part speed. If the output terminals of the rectified voltage are connected directly to the battery, then the shock absorber does not operate properly and some oscillations are not damped to zero. To make it work properly, an additional resistor was connected parallel to the battery via controlled switch (transistor). After this modification, if the generated voltage by the generator is greater than the battery voltage, the current flowed through the battery. If the generated voltage was smaller or equal to the battery voltage, the additional switch was closed and the generator current flowed through the additional shunt resistor.

The electric shock absorber with the modified output electric circuit show that the oscillations attenuate to zero after disturbance appears. Therefore, the electric shock absorber works properly under the modified circuit. 6-shock heavy truck, each shock absorber could generate up to an average of 1 kW on a standard road — enough power to completely displace the large alternator load in heavy trucks and military vehicles, and in some cases even run accessory devices such as hybrid trailer refrigeration units.

This paper is only a small portion of the growing research in this area, the simulation of the electric shock absorber will carried out in MATLAB using SIMULINK, they must be included in the future work.

VI. ACKNOWLEDGEMENT

Work in this study was partially supported by Fast Track Project entitled "Vibration induced stresses while driving tractor" Reference No. SR/FTP/ETA-64/2009 of Department of Science and Technology, New Delhi, India.

VI. REFERENCE

[1] Suda Y and Shiba T 1996 New hybrid suspension system with active control and energy regeneration Veh. Syst. Dyn. 25 (Suppl.) 641–54

[2] Suda Y, Nakadai S and Nakano K 1998 Hybrid suspension system with skyhook control and energy regeneration (Development of self-powered active suspension) Veh. Syst. Dyn. 19 (Suppl.) 619–34

[3] Reimpell, J., Stoll, H., and Betzler, J., "The automotive Chassis, Engineering Principles", Second Edition, 2001, pp. 347-385

[4] Crouse, W. and Anglin, D., "Automotive: Chassis and Body," Fifth Edition, McGraw-Hill Book Company, 1976, pp. 48-54

[5] Lei Zuo, Brian Scully, Jurgen Shestani and Yu Zhou 2010,Department of Mechanical Engineering, State University of New York at Stony Brook, Stony Brook, NY 11794, USA

[6] J A Calvo*, B Lo´pez-Boada, J L San Roma´n, and A Gauchı´a, Instituto para la Seguridad de los Vehiculos Automo´viles (ISVA), Universidad Carlos III, Madrid, Spain

[7] Mendrela, E. and Drzewoski, R., "Electric Shock Absorber for Electric Vehicles," Conference, Proc. of BASSIN' 2000, Lodz, Poland 2000.

[8] Neil Sclater and John E. Traister , Handbook of Electrical Design Details Second Edition , New York: McGraw -Hill, 2003.

Recommended Weight Limit of School Bag to Mitigate Lower Back Injury

Ajay Bangar

Maharana Pratap College of Technolgy, Gwalior

Abstract- **Material handling (MH) is an act of transferring materials (raw material, semi finished and finished product etc.) from one place to another in industry and in our daily life as well. Manual material handling (MMH) is a branch of material handling in which power and control both are executed manually to handle the material.**

Lower back injury (slip disc, hernia) due to over weight lifting is the major risk in manual material handling. Lifting, lowering and carrying of school bag by student are also a case of MMH. A survey report on school, dated April 23, 2006 in Times of India states that 30% kids have lower back pain because of their heavy school bags. Doctors said that ignoring the problem of the back pain can cause a whole damage to a child body. we have analyze recommended weight limit(RWL) of the bag that nearly all healthy students could lift or lower in a substantial period of time without an increased risk of LBP.

To calculate the RWL lifting equation (NIOSH 1991) is only one tool in a comprehensive effort to prevent work related low back pain and disability. Equation is

RWL = LC x HM x VM x DM x AM x FM x CM

Where LC is load constant, HM horizontal multiplier, VM vertical multiplier, DM distance multiplier, AM asymmetric multiplier, FM frequency multiplier, CM coupling multiplier .All multiplier vary according to existing risk factor

Analysis of RWL is based on the failure of spinal disc due to compressive force and bending moment on lower back.

Keywords - MH (Material Handling), MMH (Manual Material Handling), RWL (Recommended Weight Limit), LBP (Lower Back Pain), HAT (head+arms+torso), HERNIA.

I. OBJECTIVES

- **To emphasized the significant importance of material handling in our daily life**

- **To make aware the students about the risk of lower back injury due to the weight of bag.**

II. BIOMECHANICS

Biomechanics is the science concerned with the internal and external forces acting on the human body and the effects produced by these forces. It plays a major role to explain how different forces act on the body and why it is important to use proper techniques when lifting, carrying, pushing, and pulling during manual material handling tasks.During the standing condition the center of gravity (C.G.) of the body is located ventral to the spine this creates the need for the erector spinal muscles to be active. This creates the need for compressive force in individual vertebrae that equal the weight of body. During lifting a load position of C.G. of body changes and total weight (body+load)   produce a bending moment and compressive force on spine.

Total Bending Moment

TOTAL BENDING MOMENT $= W_H \times D_H + W_B \times D_B$,

Where, D_H is distance of HAT from joint, D_B is distance of bag from joint

W_H is weight of HAT, W_B is weight of bag

Back muscles must produce a moment of force to counteract total bending moment, Otherwise person bends forward.

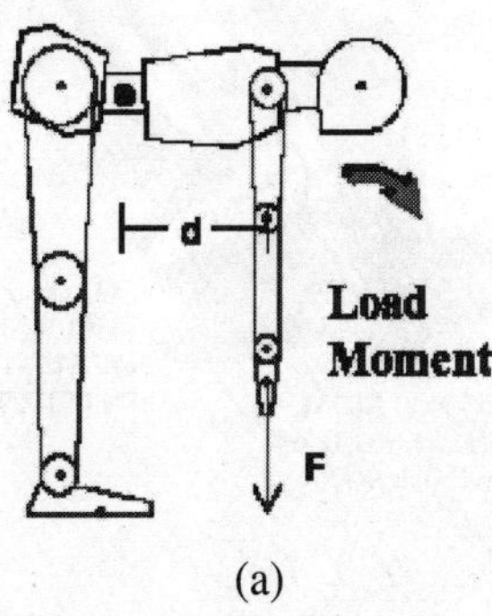

(a)

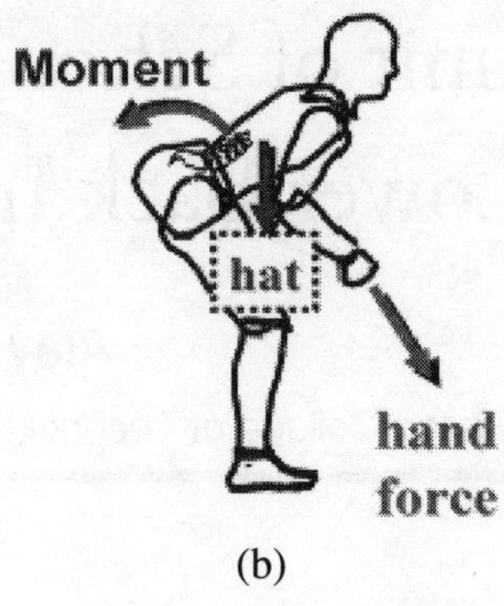

(b)

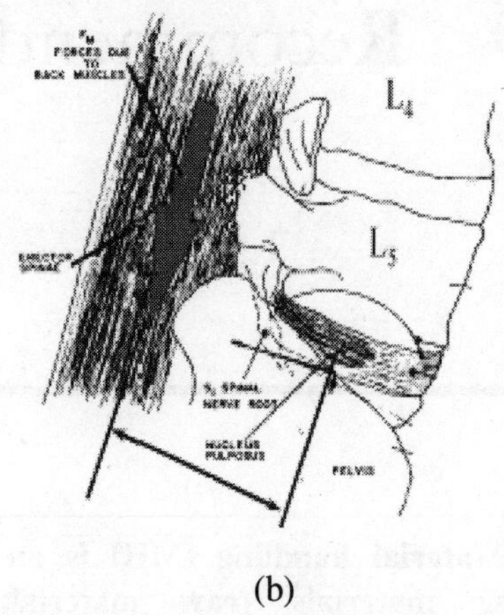

(b)

Fig 3.2 compressive force

Shear Force (F_s)

Shear Forces is defined as a force that acts parallel or tangent to a surface to create sliding of one object with respect to another. With respect to the lumbar spine, the forces due to the mass of the upper body and the forces acting on the hands create shear forces at the L4-L5 joint.

$$F_s = W_H \cos\theta + W_B \cos\theta$$

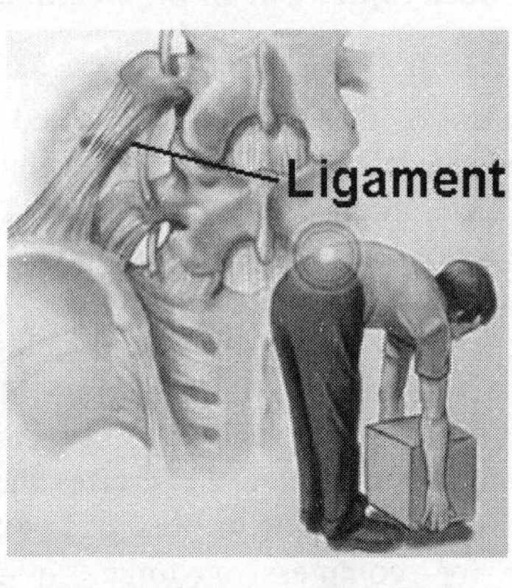

(c)

Fig.3.1 bending moment on lower back

Compressive Force On Spine (F_c)

$$F_C = F_M + W_H \sin\theta + W_B \sin\theta$$

Where

F_M is the compensative force of spinal muscle

F_M = Total bending moment/ Distance of spinal muscle from joint

(distance of spinal muscle from joint is 3 to 4 cm.)

θ is angle of trunk with the horizontal

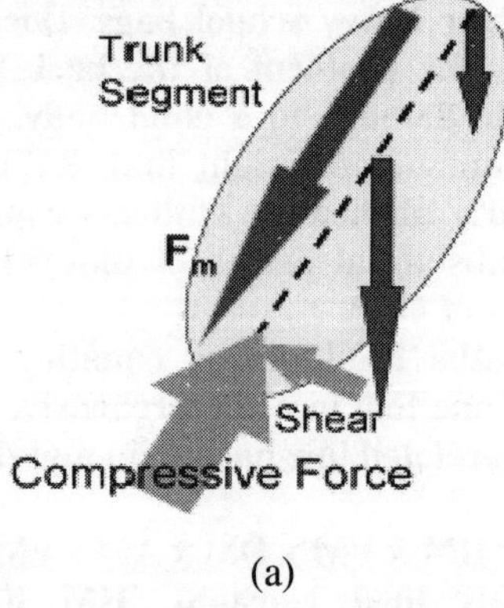

(a)

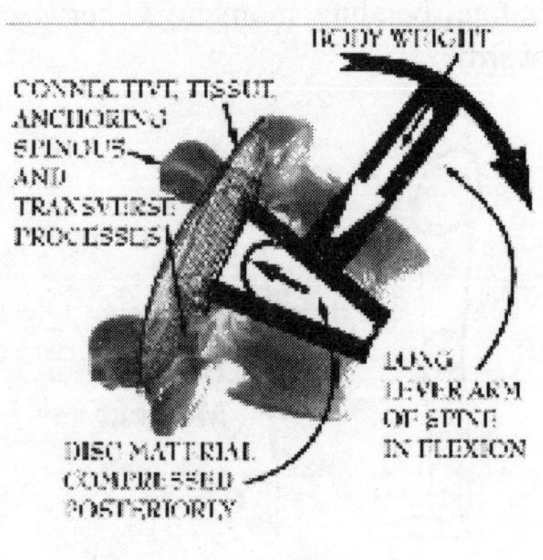

(a)

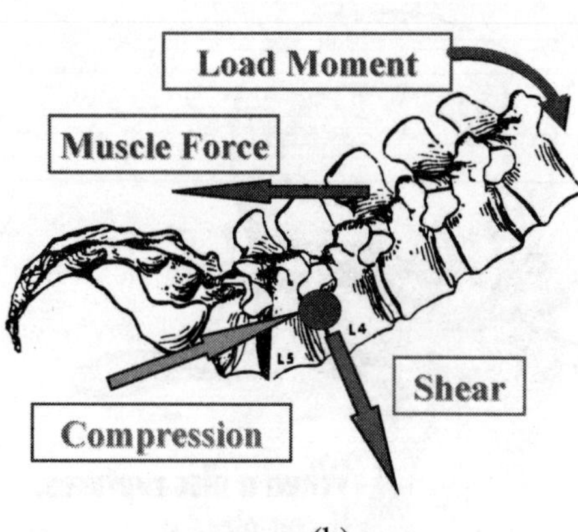

(b)

Fig3.3 Force on spine

III. ANATOMY

The spine includes vertebrae (bones), discs (cartilaginous pads or shock absorbers), the spinal cord and nerve roots (neurological wiring system), and blood vessels (nourishment). Ligaments link bones together, and tendons connect muscles to bones and discs. The ligaments, muscles, and tendons work together to handle the external forces the spine encounters during movement, such as bending forward and lifting. When back muscles encounter excessive external force, individual strands can stretch or tear between each vertebra is a "cushion" called the intervertebral disc, their function is to absorb forces and restrict excessive motion of the vertebra. The discs are composed of two regions a
gelatinous mass (nucleus pulpous) in the centre region of the disc, which allows for even distribution of pressure throughout the entire disc,

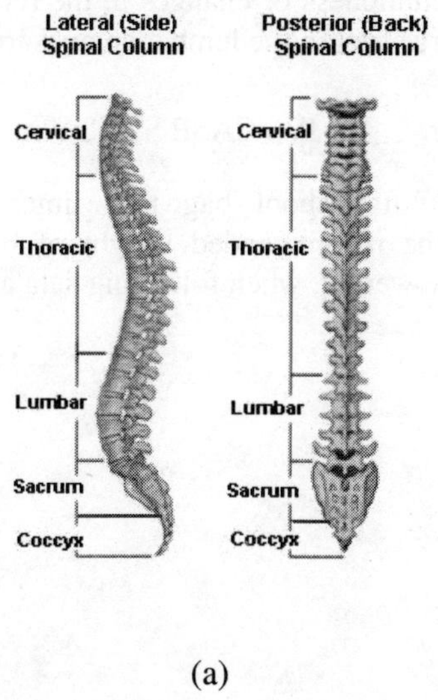

(a)

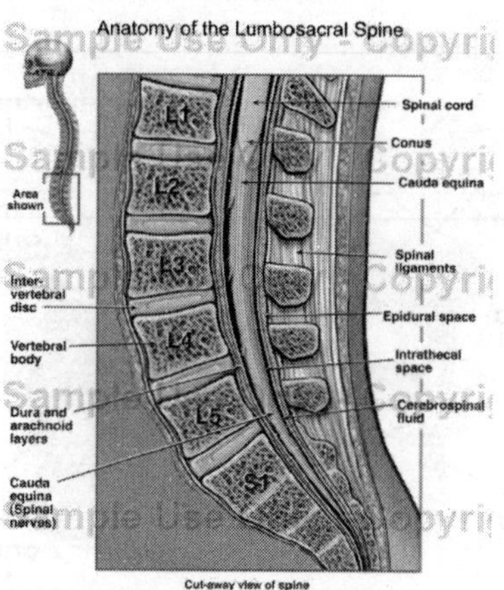

(b)

Fig.4.1 Spinal column

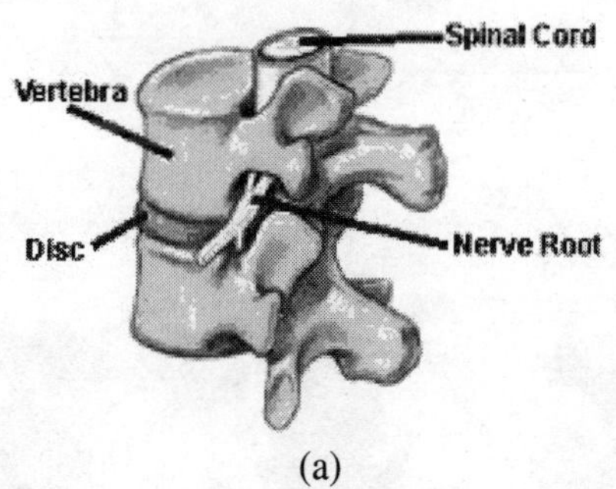

(a)

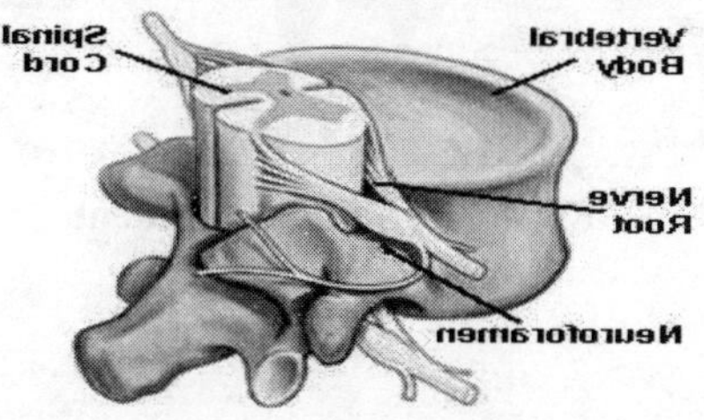

(b)

Fig.4.2 Action of disc and spinal cord

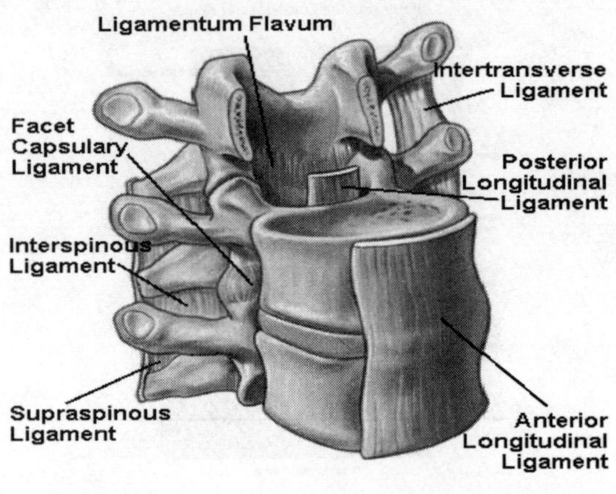

(a)

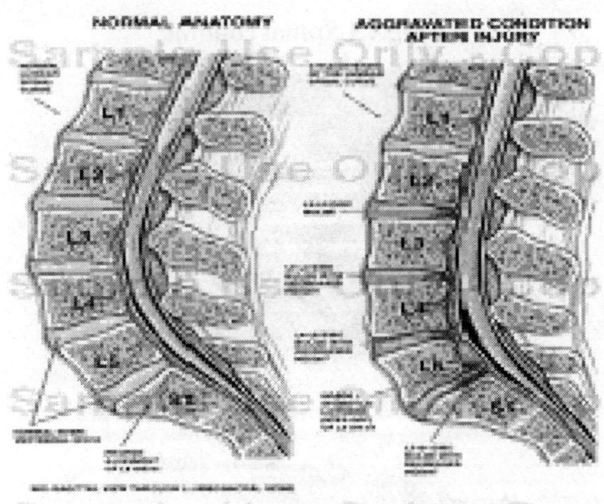

(b)

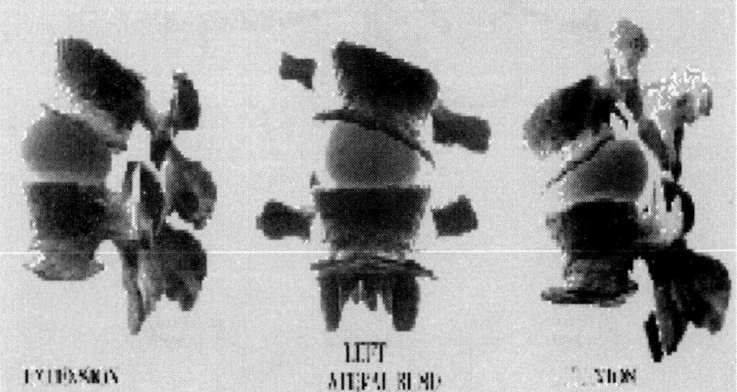

Fig 4.3 Position of ligament

(a)

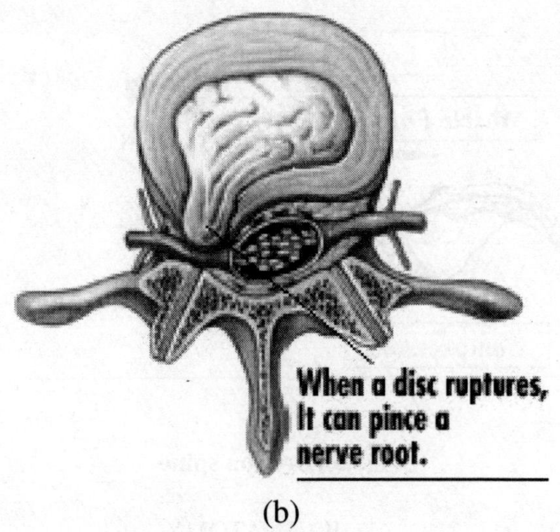

(b)

Figure 4.4 Rupture of spinal disc

These discs have a rigid outside rim, but are soft and gel-like inside. The discs can bulge and press on a nerve root, causing irritation. "Activity, stress or a mechanical problem in the spine can cause one of the discs to bulge and become misshapen just as a rubber tire might with pressure on it," Dr. Said. "When this happens, the disc may pinch or put pressure on a nerve root and the patient experiences pain. This is what frequently happens in a mild or moderate case of low back or leg pain." Occasionally, the disc will bulge to the point where it herniated or ruptures and puts even greater pressure on the nerve root. In the lower back, the nerve roots lead to the legs and irritation may cause not only back pain, but also pain that radiates down one or both of the legs. There also can be muscle weakness, numbness or changes in the reflexes in the legs if a nerve root in the lumbar spine is irritated.

IV. CASE STUDY

A student lifts school bag four times in a day. Calculate the recommended weight of bag for safe lifting and lowering, when following data are given

Table No.1 DATA SHEET

S.No.	Particular	Mean Values
1	Name	RAJKUMAR
2	Father's Name	DINESH
3	Class	8
4	Age	14
5	School	BALAK MANDIR
6	Weight (Kg.)	30 Kg.
7	Height (cm)	157
8	Waist Height (cm)	94
9	Knee Height (cm)	48
10	Arm Length (cm)	58
11	Weight of Bag (Kg.)	Is to determine
12	Way of Carrying	By both hand
13	Frequency	Four time in a day
14	Do you have back pain	Yes
15	Load Constant(Kg.)	08
16	Minimum distance of bag (cm)	20
17	Shoulder height(cm)	132

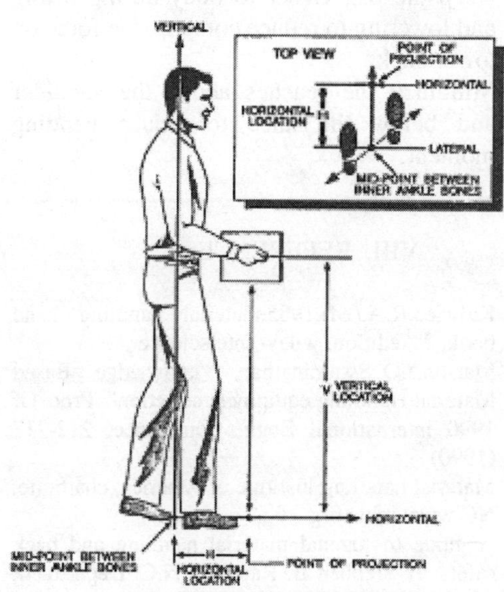

Fig.5.1 Graphical Location

V. CALCULATIONS

Recommended weight can be calculated by the lifting equation

RWL = LC x HM x BM x DM x AM x FM x CM
from data sheet(Table No.1)
LC =10 Kg.
HD = 50 cm.

VH = 0cm
VD = 132cm.
Φ = 0
F = 4 times in a day
Coupling = good
See the corresponding multiplier from the table no 2
HM = .50
VM = .78
DM = .86
AM = 1
FM = 1
CM = 1
RWL = LC x HM x VM x DM x FM x AM x CM
    = 10 x .50 x.78 x .86 x 1 x1 x 1
    = 3.35 Kg.

The weight of school bag should not be greater then 3.35 Kg. when the load constant is 10. Although the value of recommended weight varies according to load constant and height of student

Bending Moment

TOTAL BENDING MOMENT = W_B x D_{B+} W_H x D_H
W_B = 3.35 Kg.
W_H = 30/2 Kg.
D_H = .25 m.
D_B = .50 m
Total bending moment
 = (3.35x .50 + 30 / 2 x.25)9.81
 = 53.21 N-m

Compressive Force

Compressive force of lower back directly related to the bending moment about L5-S1 disk. then the compressive force (Fc) will be

$$Fc = F_M + W_H \sin\theta + W_B \sin\theta$$
$$Fc = F_M \qquad (\text{when } \theta = 0)$$

Fc = Total bending moment / distance of mussels from joint
So,   Fc = 53.21 N-m / .035
           Fc =1.520 K-N

While the normal maximum compressible force which can be withstand by the disk is 1.7 KN. So it is the safe value.

Shear Force

$Fs = W_H \cos\theta + W_B \cos\theta$
Fs = 15x9.81 cos0 + 3.35x9.81 cos0 (when θ =0)
F_S =147.15 + 32.86
F_S =180.01N

Table No. 2 Multiplying Factor

S.N.	H(cm)	HM	V(cm)	VM	D(cm)	DM	A(Φ)	AM	F(per min)	FM	C	CM
1	25	1	0	.78	25	1.00	0	1.00	.2	1	Good	1.00
2	30	.83	10	.81	40	.93	15	.95	.5	.97	Fair	.95
3	35	.70	20	.84	55	.90	30	.90	1	.94	Poor	.90
4	40	.63	30	.87	70	.88	45	.86	2	.91		
5	45	.55	40	.90	85	.87	60	.81	3	.88		
6	50	.50	50	.93	100	.87	75	.76	4	.84		
7	55	.45	60	.96	115	.86	90	.71	5	.80		
8	60	.42	70	.99	130	.86	105	.66	6	.75		
9	65	00	80	.99	145	.85	120	.62	7	.60		
10	70	00	90	.96	160	.85	135	.57	8	.52		

VI. RESULT

Recommended weight of bag should not be greater then 3.35 Kg. when it is lifted from the ground. Value of recommended weight varies according to the following aspect.

1. Height of Students.
2. Load Constant.

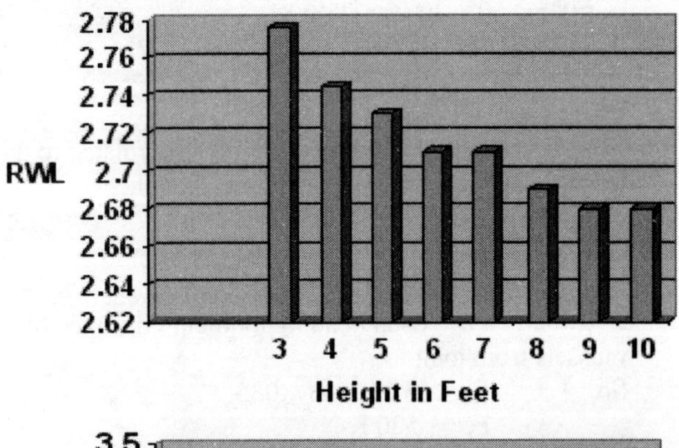

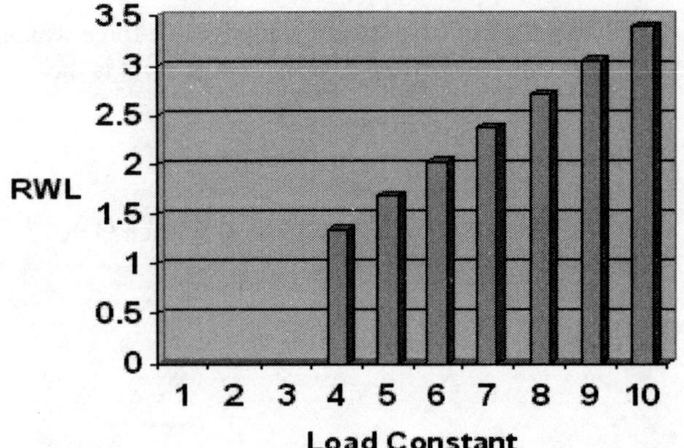

Fig 5.2 variation in RWL

VII. CONCLUSION

- Weight of the school bag should be less than 10%of body weight.
- Keep the bag closer to body during lifting and lowering to reduce compressive force on lower back.
- Minimize the reaches above the shoulder and below the knee to reduce bending moment.

VIII. REFERENCES

[1]   Kulwiec,R.A.(ed),1985,material handling hand book, 2nd edition, wiley- interscience

[2]   Matson,J.O.,Swaminathan, "knowledge Based Material Handling equipment selection" Proc. Of 1990 international Engg. Conference, 212-217 (1990)

[3]   Material handling institute of America, charbotte, NC www.mhia.org.

[4]   A guide to manual material handling and back safety by Stephen B. Randall , N.C. Department of Labor division of occupational safety and health.

[5]   OR-OSHA 206 Ergonomics of manual material handling

[6]   ThomasR.Waters, Application manual for the revised NIOSH LIFTING EQUATION., U.S. National Institute for occupational safety and health (NIOSH) 1991.

[7]   M.B. Frazer, W.P. Neumann and R.W. Norman, Ergonomics Initiative in Injury Prevention faculty of applied health sciences, university of Waterloo

[8]   Zurich Services corporation Risk Engineering, 1400 American Lane schaumbur, IL 60196

[9]   Donald S. Bloswick (Ph. D. , P.E., C.P.E.) university of Utah Dept. of Mech. Engg. Salt Lack city UT84112 for estimation of back

compressive force in manual material handling task.

[10] Canadian center for occupational health and safety.

[11] Medicine Net.com, Medical author Willium C. Shiel Jr. M.D., FACP, FACR for anatomy of lower back.

[12] Spine universe, Stewart G. Eidelson MD, spine universe founder and orthopedic surgeon, Assistant Professor university of Miami at FAU orthopedic surgery associates Boca raton USA for lumber spine.

[13] Jean-Jacques Abitbol MD, FRCSC , orthopedic surgeon California spine group MC San Diego CA USA for sprains and strains.

[14] B Davis et al.,physical Education and study of sport.

[15] Potvin,J.,Norman, R. and Wells, R., "A field method for continuous estimation of dynamic compression force on the L4/L5 Disc during the performance of Repetitive industrial risk

[16] Chaffin D.B. & G. Anderson . Occupational biomechanics. John wiley and sons 1984.

Solar Photovoltaic Automatic Peak Power Tracker by using Perturb & Observe

[1]Puneet Chaudhary, [1]Shiv Kumar Mittal, [2]Neeraj Jain

[1]Anand Engg. College, Agra [2]LIETW, Alwar

Abstract—**Renewable energy sources play an important role in electricity generation. Energy from the sun is the best option for electricity generation. Electricity from the sun can be generated through the solar photovoltaic modules (SPV).The SPV comes in various power output to meet the load requirement. Maximization of power from a solar photo voltaic module (SPV) is of special interest as the efficiency of the SPV module is very low. A peak power tracker should be used for extracting the maximum power from the SPV module. In this work Perturb & Observe (P&O) Algorithm is implemented for extracting maximum power. In the present work a photovoltaic array (PVA) simulation model is developed in MATLAB / Simulink environment. The model is developed using basic circuit equations of the solar cells including the effects of solar irradiation and temperature changes. A dc/dc converter followed by a dc/ac inverter is also implemented to connect a PV array to grid. Therefore an efficient user friendly simulation model of the photovoltaic array is developed. Test and validation studies with proper load matching circuits are simulated and results are presented.**

Keywords — Photovoltaic models, Photovoltaic power systems, Power generation, P&O Algorithm .

I.INTRODUCTION

Renewable energy sources play an important role in electric power generation. The main challenge in replacing legacy systems with newer more environmentally friendly alternatives is how to capture the maximum energy and deliver the maximum power at a minimum cost for a given load. The energy received by the earth from the sun is so enormous that the total energy consumed annually by the entire world is supplied in as short interval of time as a half hour. On a clear day the sun's radiation on the earth can be 3000 watts per square meter depending on the location. The sun is a clean and renewable energy source, which produces neither greenhouse effect gas nor nitrogen waste through its utilization. The photovoltaic process is completely solid state and self-contained. There are no moving parts and no materials are consumed or emitted. solar power cost per kilowatt-hour has to be competitive with fossil fuel energy sources. Currently, solar panels are not very efficient with only about 12 ~ 20% efficiency in their ability to convert sunlight to electrical power.

The efficiency can drop further due to other factors such as solar panel temperature and load conditions. In order to maximize the power derived from the solar panel, it is important to operate the panel at its optimal power point. To achieve this, a type of charge controller called a Maximum Power Point Tracker may be designed and implemented.

The photovoltaic (PV) modules are made up of silicon cells. The silicon solar cells give output voltage of around 0.7V under open circuit condition. When many such cells are connected in series we get a solar PV module. Normally in a module there are 36 cells which amount for an open circuit voltage of about 20V. The current rating of the modules depends on the area of the individual cells. Higher the cell area high is the current output of the cell. For obtaining higher power output, the solar PV modules are connected in series and parallel combinations forming solar PV arrays. A solar cell is a non-linear power source and its output power depends on the terminal operating voltage. The Maximum Power Point Tracker (MPPT) compensates for the varying voltage vs. current characteristics of the solar cell. The MPPT tracks the output voltage and current from the solar cell and determines the operating point that will deliver the maximum power. A properly designed MPPT should be able to accurately track the constantly varying operating point where the maximum power is delivered in order to increase the efficiency of the solar cell.

DC/DC Boost Converter

A dc/dc converter forms an integral part of any MPPT system. Without dc/dc converter no MPPT system are designed. The dc/dc converter used in the system is a step up converter in which the output voltage is higher than the input voltage. The boost converter or the step up converter has the output voltage greater than the input. The voltage transformation ratio is

$$V_o = V_i / (1-D)$$

where, D is the duty cycle,

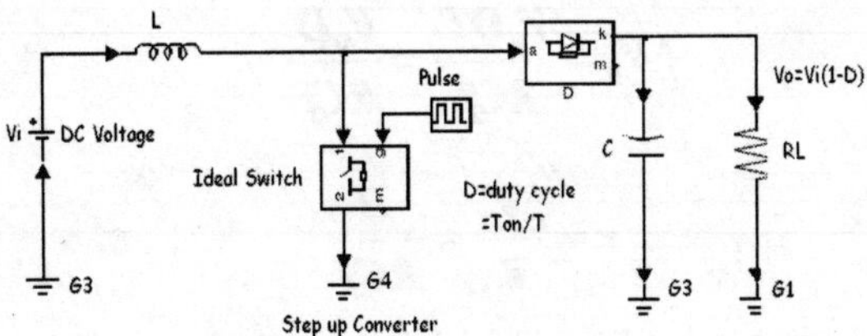

Fig.1(a) show the model of Step up or boost converter

By varying D the output voltage can be changed and it is always more than V_i. The advantage of this converter is that the input and output current both are continuous, where as in the step down converter the input current is discontinuous. The boost converter can be implemented in the MPPT system where the output voltage of the system is required to be higher than the input voltage. Generally in grid-connected systems where the MPPT system is part a boost converter is utilized which maintains a high voltage even if the array voltage falls.

Analysis of Boost converter

When Switch is closed

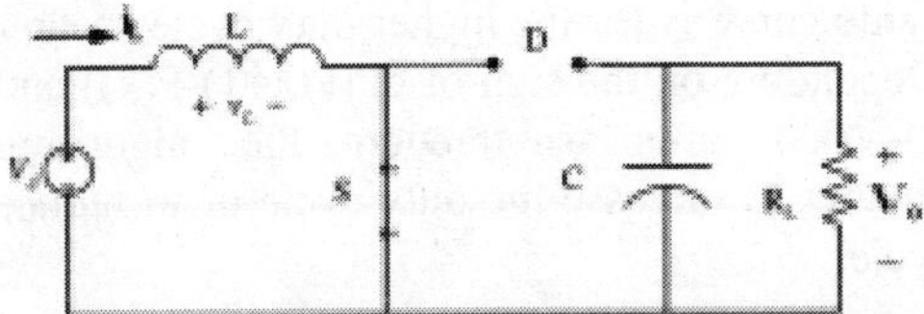

Fig.1(b) boost converter when switch (S) is closed

As shown in Fig.2, when switch (S) is closed, diode (D) is reversed. Thus output is isolated. The input supplies energy to inductor

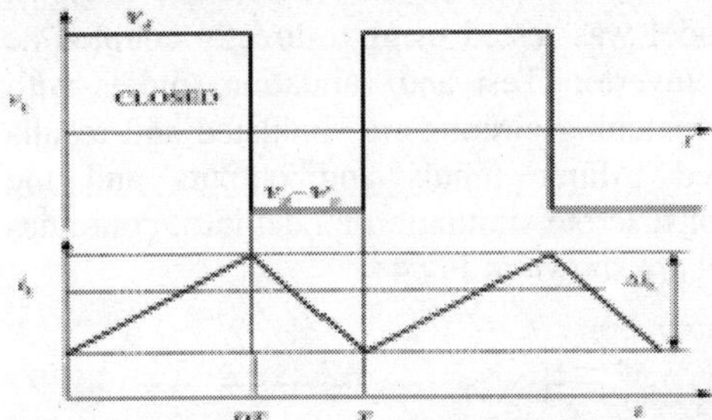

Fig.2(a) Voltage & current across inductor when Switch is closed

As shown in Fig. 2(a), when switch (S) is closed

$$v_L = V_d = L\frac{di_L}{dt}$$

$$\Rightarrow \frac{di_L}{dt} = \frac{V_d}{L}$$

$$\frac{di_L}{dt} = \frac{\Delta i_L}{\Delta t} = \frac{\Delta i_L}{DT}$$

$$\left(\Delta i_L\right)_{closed} = \frac{V_d}{L}DT$$

When switch is open

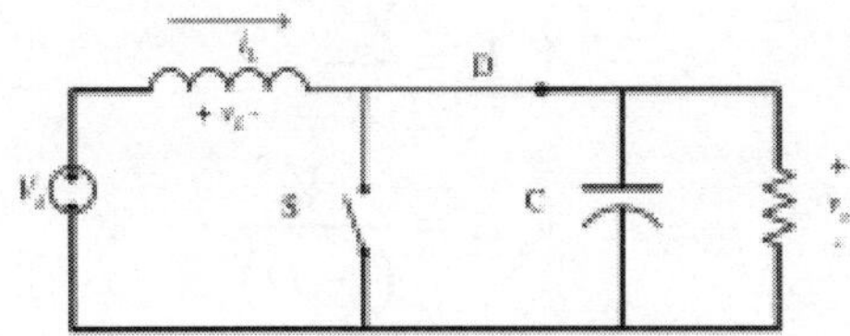

Fig.3 Boost circuit when switch is open

As shown in Fig.3 when switch (S) is opened, the diode (D) will be in forward bias and output stage receives energy from the input as well as from the inductor. Hence output is large.

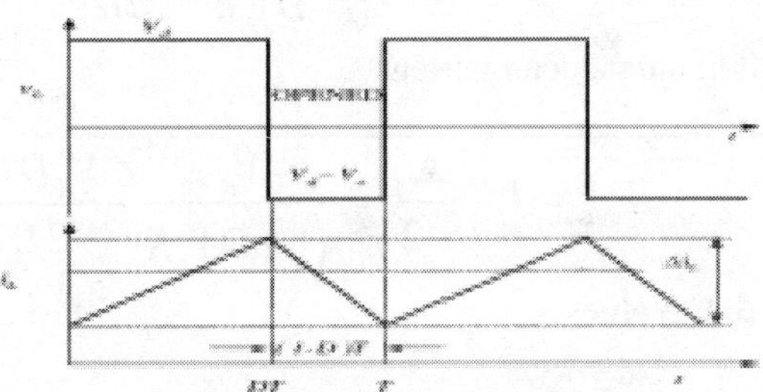

Fig.3(a) Voltage & current across inductor when switch is open

$$v_L = V_d - V_o = L\frac{di_L}{dt}$$

$$\Rightarrow \frac{di_L}{dt} = \frac{\left(V_d - V_o\right)}{L}$$

$$\frac{di_L}{dt} = \frac{\Delta i_L}{\Delta t} = \frac{\Delta i_L}{\left[(1-D)T\right]}$$

$$\left(\Delta i_L\right)_{opened} = \frac{\left[(V_d - V_o)(1-D)T\right]}{L}$$

Steady - state operation

Steady - state operation requires that i_L at the end of switching cycle is the same at the beginning of the next cycle. That is the change of i_L over one period is zero, i.e.

$$\left(\Delta i_L\right)_{closed} + \left(\Delta i_L\right)_{opened} = 0$$

$$V_o = \frac{V_d}{\left(1-D\right)}$$

From equation (4.9), If input v_d is constant than by varying duty cycle, D output voltage can be modified by boost converter.

Average, Maximum, Minimum Inductor Current

Input Power = Output Power

$$V_d I_d = \frac{V_o^2}{R}$$

$$V_d I_L = \frac{V_d^2}{\left(1-D\right)^2 R}$$

Average inductor current

$$I_L = \frac{V_d}{\left(1-D\right)^2 R}$$

Maximum inductor current

$$I_{max} = I_L + \frac{\Delta i_L}{2} = \frac{V_d}{\left(1-D\right)^2 R} + \frac{V_d DT}{2L}$$

Minimum inductor current

$$I_{min} = I_L - \frac{\Delta i_L}{2} = \frac{V_d}{\left(1-D\right)^2 R} - \frac{V_d DT}{2L}$$

L & C Values

For Continuous current mode operation

Minimum inductor current

$$I_{min} \geq 0$$

$$\frac{V_d}{\left(1-D\right)^2 R} - \frac{V_d DT}{2L} \geq 0$$

$$L_{min} = \frac{D(1-D)^2 TR}{2} = \frac{D(1-D)^2 R}{2f}$$

L_{min} is the minimum inductor current to ensure continuous mode of operation.

Output voltage ripple

The charge can be written as

$$Q = CV_0$$

$$\triangle Q = C \triangle V_0 = \left(\frac{V_0}{R}\right) DT$$

$$\Delta V_0 = \frac{V_0 DT}{RCf} = \frac{V_0 D}{RCf}$$

$$r = \frac{\Delta V_0}{V} = \frac{D}{RCf}$$

Thus, ripple can be reduced by Increasing switching frequency and by increasing capacitor size.

Perturb & Observe Algorithm

Among different maximum power point algorithms, the perturb and observe algorithm is simple and also gives good results. The algorithm reads the value of voltage and current from the solar PV module. Power is calculated from the measured voltage and current. The value of voltage and power at k^{th} instant are stored. Then next values at $(k+1)^{th}$ instant are measured again and power is calculated from the measured values. The power and voltage at $(k+1)^{th}$ instant are subtracted with the values from k^{th} instant .If we observe the power voltage curve of the solar PV module we see that in the right hand side curve where the voltage is almost constant the slope of power voltage is negative where as in the left hand side the slope is positive. The right side curve is for the lower duty cycle (nearer to zero) where as the left side curve is for the higher duty cycle (nearer to unity).Depending on the sign of dP(P(k+1)-P(k)) and dV(V(k+1)-V(k)) after subtraction the algorithm decides whether to increase the duty cycle or to reduce the duty cycle.

II. ANALYSIS AND SIMULATION OF MPPT

A photovoltaic array (PVA) simulation model to be used in MATLAB-Simulink GUI environment is developed. The model is developed using basic circuit equations of the photovoltaic (PV) solar cells including the effects of solar irradiation and temperature changes. The new model was tested using a directly coupled ac load via an inverter. Test and validation studies with proper load matching circuits are simulated and results are presented .Major inputs and outputs and the conversion of discrete simulation model into, continues circuit model are shown in Fig.4

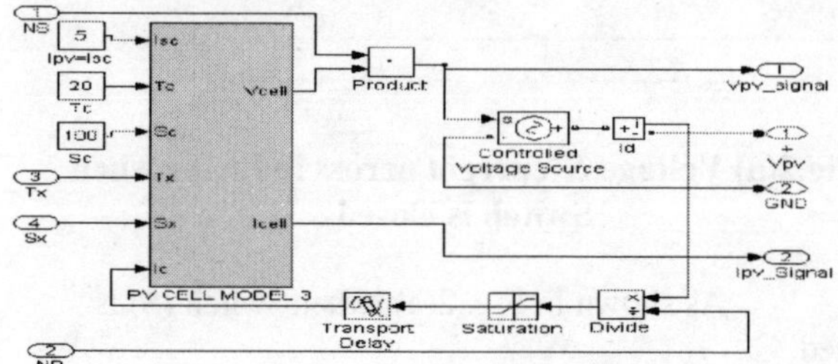

Fig.4 Show the Basic O/P Voltage equation of PV Cell

The simulink setup of the MPPT system is shown below. The logic is developed according to the flow chart. Direct duty cycle control method is implemented. The algorithm outputs a signal which has a value between 0&1.

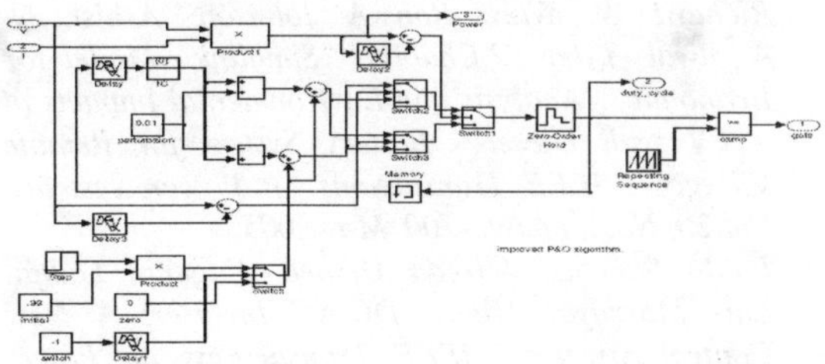

Fig .5 shows the Perturb & Observe Algorithm Simulated Results

Boost Converter PV Cell output with Fixed Temperature and Irradiance level

Number of cell connected in series = 36

Temperature of cell $= 25^0$C

Irradiance of cell $= 1000$W/m^2

From Fig.6 it is observed that output of boost converter increases voltage level from 18V to 20V and current is about 3ampere.

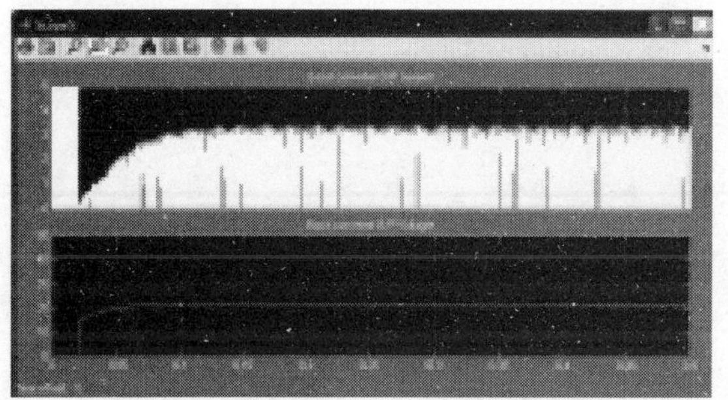

Fig.6 shows the Boost converter O/P Voltage & Current

It is observed from above fig. that THD is reduced up to 18.20 percent using LC filter and output active power is about 80watt.When Irradiance level changes from 1000 to 500 W/m^2 and temperature is 25^0 C, output voltage changes from 15.2V to 14.2 V and output power is changes from 80W to 75W as shown in above fig.7

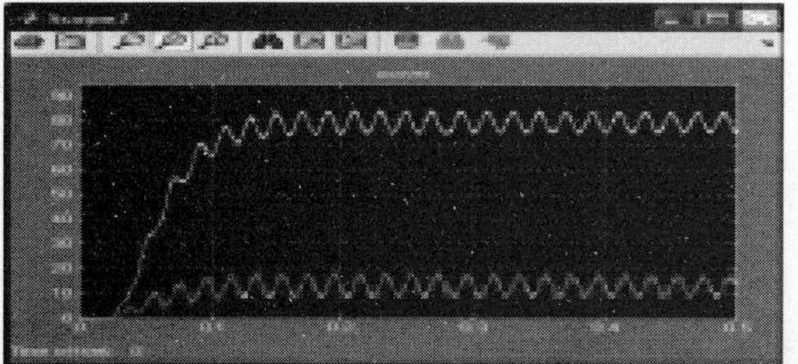

Fig.7 Shows the change in output power with change in Irradiance level

When Irradiance level is 1000 W/m^2 and temperature changes from 25 to 50^0C output power remains about 80 Watt.

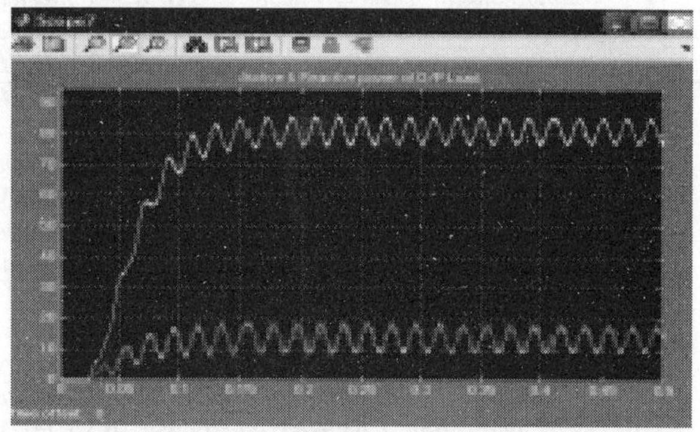

Fig .8 Shows the change in output power with change in temperature

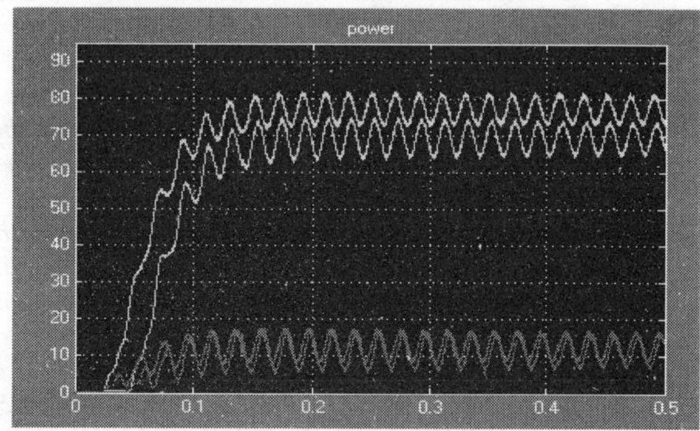

Fig .9 Shows the combined effect of the change in temperature & Irradiance level

III. CONCLUSION

The PVA model is designed in a Matlab/Simulink GUI environment to analysis the maximum power point search and a dynamic perturb & observe algorithm are used to detect the maximum power point.   When irradiance level changes from 1000 W/m^2 to 500 W/m^2 power output is decreases from 80W to 75W .When temperature is changes from 25^0C to 50^0C, power output is not too much change it is about 80 W .It shows that with changes in irradiance level and temperature output power is almost constant.

Future Scope

Development of microcontroller based dedicated MPPT controller for solar PV module based on P& O algorithm. This can be low cost embedded controller. A whole stand alone system including the MPPT system and inverter system can be developed using dSPACE technology.

IV References

1. Wu Libo, Zhao Zhengming, Liu Jianzheng, "A Single-Stage Three- Phase Grid Connected Photovoltaic System with Modified MPPT Method and Reactive Power Compensation" IEEE Transactions On Energy Conversion, Vol. 22, No.4, pp.881-886, December 2007.

2. Zhenhua Jiang , Roger A Dougal, "Multiobjective MPPT / Charging Controller for Standalone PV Power Systems under Different Irradiance and Load Conditions" IEEE Transactions on Industrial Electronics, pp.1154-1160, May, 2004

3. I.H. Altas, A.M.Sharaf, "A Novel Photovoltaic On-Line Search Algorithm for Maximum Energy Utilization", the International Conference on Communication, Computer and Power (ICCCP07), Oman, February 19-21, 2007.

4. I.H. Altas, A.M.Sharaf, "A photovoltaic Array Simulation Model for Matlab Simulink GUI Environment" IEEE Transactions on Power Electronics, pp. 341-345, March, 2007.

5. Magid Nikraj, Hooman Dehbonei , Chem Nayar, "A DSP-Controlled Photovoltaic System with Maximum Power Point Tracking" Centre for Renewable Energy and Sustainable Technologies, Australia,March,2006.

6. Krisztina Leban, Ewen Ritchie, "Selecting the accurate solar panel simulation model", Nordic workshop on Power and Industrial Electronics, June 9-11, 2008.

7. Mohamed Azab, "Improved circuit model of photovoltaic Array", PWASET, Vol.34, pp.56-59 October, 2008.

8. Edro Gomes Barbosa, Henrique, Antonio Carvalho Braga, Marcio do Carmo Barbosa Rodrigues, "Boost Current Multilevel Inverter and Its Application on Single Phase Grid-Connected Photovoltaic Systems" IEEE Transactions on Power Electronics, Vol.21, No.4, pp. 1116-1124, July 2006.

9. Richard W. Wies, Ron A Johnson, Ashish N. Agrawal ,Tyler J.Chubb, "Simulink Model for Economic Analysis and Environmental Impacts of a PV with Diesel- Battery System for Remote Villages", IEEE Transactions on Power Systems, Vol.20, No.2, pp.692-700, May 2005.

10. Pablo Sanchis, Alfredo Ursaea, Eugenio Gubfa, Luis Marroyo, "Boost DC-AC Inverter: A New Control Strategy" IEEE Transactions on Power Electronics, Vol.20, No.2, pp. 343-353, March 2005.

11. E Ozdemir, S.Ozdemir, L.M.Tolbert, B.Ozpineci, "Fundamental Frequency Modulated Multilevel Inverter for Three-Phase Stand-Alone Photovoltaic Application" IEEE Transactions on Power Electronics, Vol. 15, No.3, pp.148-153, Sept. 2008.

12. R. B. Darla, "Development of Maximum Power Point Tracker for PV Panel Using SEPIC Converter" IEEE Transactions On Power Electronics, , pp. 650- 655, August,2007.

13. M. Saraf , E.Elbakush, I.H.Altas, "A novel dynamic error driven tracking controller for PV powered PMDC motor drives" University of New Brunswick Federation,Canada.Korea, April, 2008.

Turnaround Planning for Finding Optimum Maintenance Time in Steel Rolling Mill

Manish Rawat [#1], **V. K. Khare** [*2], **Manish Vishwakarama** [#3]

[1]Department of Mechanical Engineering CSE JHANSI
[2]Department of Mechanical Engineering MANIT BHOPAL
[3]Department of Mechanical Engineering MANIT BHOPAL

Abstract— Hundreds of articles and dozens of paper have been written to optimize the Maintenance time and planning in shutdown. Despite this, there are some key shortcomings of the existing maintenance procedure and planning. In major scheduled plant shutdown of long duration and high intensity, it can consume an equivalent cost of a yearly maintenance budget in four to five weeks. It also requires the greatest percentage of the yearly process outage days. Without scheduled maintenance outages, equipment will fail, and an unscheduled outage is up to ten times more expensive than a scheduled outage. The cost is much higher again if the outage is due to a catastrophic failure. These drawbacks need further research.

The main objective of this dissertation is to present an application of critical path analysis of maintenance activity in steel rolling mill so that it allows reducing the maintenance time in rolling unit. Under the main research objective are to reduce the overall turnaround maintenance time ,efficient utilization of the man, material, money during the shutdown phase ,to reduce the maintenance cost and timing of the shutdown is critical to Marketing, Sales, and Accounting. When the plant is shutdown, there may be a negative cash flow on the revenue side.

Key words: Maintenance, CPM, Scheduling, cost.

INTRODUCTION

It is perhaps an obvious statement to make that given the complexity of the technology currently employed in the design and building of manufacturing plant. The limitation of the technology used to maintain it and the ever increasing problem of it ageing there will always be a need for maintenance work on the physical assets of the manufacturing activity. Further more in an intensely competitive global market, characterized by increasing scales of production, the effective planning and management of that maintenance activity is coming to be seen as an ever more critical business process-one which is capable of differentiating the excellent performers from those who are merely capable and from those who are less capable. Owners of commercial facilities, manufacturing processes, and industrial plants recognize that maintenance of their equipment assets is a reality. Most inspection, repair, replacement, alteration, and minor maintenance work can be done while the plant is in operation or "on line." However, there will come a time when a plant has to undergo a scheduled process outage for the major maintenance work. This outage is referred to as a "plant shutdown."

The management of a plant shutdown is known as the **"plant turnaround."** Turnaround is defines as-

"An engineering event during which new plant is installed Existing plant overhauled and redundant plant removed."

Concept of Turnaround

Plant turnaround can be defined as a large number of people and equipment working in a small area against a tight schedule and even tighter budget. In order to meet its goals, a turnaround requires a high degree of planning, scheduling, and coordination so that the right equipment and personnel are available at the right time. However, many turnaround planners focus on scheduling the mechanical and maintenance aspects of the project and neglect to include critical support services that can help keep a turnaround on track. The plant turnaround procedure is a continuous process from one major scheduled maintenance outage to the next. It starts well before the plant is taken off-line and continues for a period of time after the scheduled major maintenance work has been completed. The plant shutdown is part of the plant turnaround procedure called the "execution phase." Plant shutdowns for scheduled major maintenance work are the most expensive and time-consuming of maintenance projects because of the loss of production and the expense of the turnaround itself. They being complex proves to be costlier and difficult to manage. A plant shutdown always has a negative financial impact. This negative impact is due to both loss of production revenue and a major cash outlay for the plant turnaround and shutdown expenses. The positive side is not as obvious; therefore, it is often over looked. The positive impacts are an increase in equipment asset reliability, continued production integrity, and a reduction in the risk of unscheduled outages or catastrophic failure. a major scheduled plant shutdown is of short duration and high intensity. It can consume an equivalent cost of a yearly maintenance budget in four to five weeks. It also requires the greatest percentage of the yearly process outage days. As the plant shutdown is the major component of plant downtime and maintenance costs, proper plant turnaround management will have a significant impact on the bottom line.

Turnaround Planning

The oil refining and petrochemical processing industries have their own nomenclature for maintenance projects. For the purposes of this tutorial, "turnaround" is intended to encompass all types of industrial projects for existing process plants including I&Ts (Inspection & Testing), shutdowns, emergency outages, debottlenecking projects, revamps, catalyst regeneration, etc. where an operating plant must be shut down until the work is completed and then restarted - thus "turning around" the unit/plant. Turnaround project planning and scheduling is an important function that has a direct and dramatic impact on maintenance costs and bottom line profitability of a process plant. Maintenance costs are the result of the expenditure of menpower, equipment and materials. Keeping menpower and equipment usage efficient turnaround costs can be controlled precisely. Through judicious planning and scheduling, a maintenance planner / scheduler can help his organization save on menpower costs, ensure the shortest possible downtime, and achieve the most efficient use of equipment. The secret to achieving the most efficient plan is to remove all wasted motion, all unnecessary movements or transports, and minimize crew and equipment redeployment.

Phases of Turnaround planning

- Pre-shut down phase
- Shutdown phase.
- Post shutdown phase.

Steps of Turnaround planning

- **Logic**- In the first steps, all the jobs that make up the project are arranged in their correct and sequential order. Here no consideration is given as to how long the jobs will take or what resources will be required.

- **Timing**- Estimation of the time that each job will take and placing of this information against each job on the network.

- **Analysis**- This determines the time of the project will take for completion and identify the critical path (critical jobs).

- **Scheduling**- Scheduling means deciding the starting and finishing dates of all remaining non critical jobs.

- **Controlling** – This involves the physical progress of the project against the schedule, and taking corrective actions when necessary

LITERATURE REVIEW

Shet K.C et. al. [1] in their work estimating and predicting of Turnaround Time for Incidents in Application service maintenance Projects. Authors studied only on Estimating and Predicting Turnaround Time for Incidents. Problems and Enhancements are not in the scope of this research work.

Shaligram Pokharel et. al.[5] in their work shows that, if project management practice and involvement of external experts and parties are allowed in the maintenance projects, then issues in maintenance projects can be addressed more clearly and the cost and schedule for such a maintenance project can also be optimized. The use of information technology in the whole process can be facilitated not only during the planning phase, but also during the execution and review process. The paper should help one to understand the implications of starting a turn-around maintenance project and the issues built therein. The case study highlights that collaborative planning and execution of TAM are useful. Collaboration could be in terms of internal parties, such as decision makers and managers, or of external parties, such as external experts and contractors.

Salih O. Duffuaa et.al. [2] In their studied about Turnaround maintenance (TAM) is an essential activity in process industry. It plays an important role in maintaining consistent productive capacity. It is a major project that requires sound planning, execution and control. In this paper guidelines for a structured approach for managing TAM is outlined and the current practice of TAM in petrochemical industry is assessed. The assessment of the current practice is conducted through a structured questionnaire that covers all the essential phases and activities of TAM. Suggestions for improving the practice of TAM are also provided in the paper. This paper provided a structured approach and guidelines for initiating, planning, executing and terminating TAM. The guidelines are expected to evolve into a manual that can be utilized by process industry for managing TAM.

Chen L et. al. [8, 9] in their work studied Maintenance scheduling involving both generating units and transmission equipment has not received any attention in the literature. In fact, only two papers mention transmission network as constraints when discussing scheduling of generating units for multi-area systems. The approach that we adopted is aimed at answering a question how a certain chronological sequence of scheduled outages for equipment maintenance can impact system reliability. The economic implications are also studied. As a result of our investigations,

A. Crespo Marquez et. al. [3] in their work offers a practical vision of the set of activities composing each management block, and the result of the paper is a classification of the different maintenance engineering tools. The discussion of the different tools can also classify them as qualitative or quantitative. At the same time, some tools will be very analytical tools while others will be highly empirical. The paper also discusses the proper use of each tool or technique according to the volume of data/information available

By Rod Oliver et. al.[6] in their work suggests that the organization must measure turnaround performance and observe trends. As with all measurements, a single indicator can mislead. It is, therefore, necessary to design a number of criteria to provide a balanced indication of performance. Having a work process does not guarantee a successful turnaround, but benchmarking considerably reduces the likelihood of failure. Organizations that complete turnarounds on time, on budget, and without surprises invariably have a defined work process and adhere to it. Following the process is the key; today's technology provides tools that allow organizations to specify the process, define the tasks, and measure adherence. Web-based technology is an ideal medium to organize and control the multitude of

tasks, information, and issues that are critical to the successful completion of turnarounds.

S.O. Duffuaa et. al.[7] in their work addresses the problem of maintenance planning and scheduling and reviews pertinent literature. discusses the characteristics and the complexity of the problem. advocates mathematical programming approaches for addressing the maintenance scheduling problem. Gives examples to demonstrate the utility of these approaches Proposes expansion of the state-of-the-art maintenance management information system to utilize the mathematical programming approaches and to have better control over the maintenance scheduling problem.

Michael Harker [10] in this paper author States that, despite the rhetoric in the literature concerning the importance of information in the market-planning process, there is scant attention to the "life blood of the planning process" in the published work on company turnarounds. Addresses this problem through a study of companies undergoing turnaround in the mature engineering industry. Compares the practices in three companies which were successfully turned around with a company in which turnaround attempts did not come to fruition. finds that successful turnaround managers go through an elaborate process, "destiny development", to establish the foundations of a turnaround plan. The turnaround plan relies heavily on sound and innovatively generated information. explores the destiny development concept for theoretical and practical purposes.

M. Ben-Daya et. al. [11] in their work Focuses on the relationship between maintenance and quality. Then briefly reviews models relating production and quality on the one hand, and production and maintenance on the other. Using concepts from predictive maintenance and the measure of equipment effectiveness from total productive maintenance (TPM), identifies explicit links between maintenance and quality. emphasizes the need to develop adequate models relating maintenance and quality. presents two approaches for modeling the link between maintenance and quality. The first approach is based on the concept of imperfect maintenance and the second one uses the Taguchi approach to quality and maintenance.

RESEARCH APPROACH AND METHODOLOGY

Turnaround Planning - Critical Path Scheduling

After all work orders have been prepared and reviewed (approved), you will be ready to prepare a schedule. If more work orders are issued after you create the schedule, you can and should incorporate them into the schedule. This is a constant process, as you will get additional work orders for repairs arising from inspections. Remember the importance of maintaining the schedule constantly, as the number of changes to the work scope, progress or the lack of progress could otherwise render the schedule obsolete. The schedule must be updated at the end of every shift. This is usually twice a day. Failure to update the schedule with this frequency will impair the ability to make critical decisions, such as adding, maintaining or reducing manpower, reassigning crews, call on specialty contractors, etc.[18]

Critical Path Method

Critical path scheduling is the act of applying a logical sequence (by defining constraints) to the activities defined in the work orders. Most project management software employs a PDM (Precedence Diagramming Method) interface for defining the logic network. The sequence of activities which have no float or slack (Float = 0 hours) is called the **critical path**. It determines the remaining duration of the turnaround. The first step to turnaround scheduling is to define all 'hard' constraints. These are constraints that must be honored. For example, you cannot inspect the interior of a vessel until the man ways have been opened. It is not necessary (although it is not detrimental) to add redundant constraints such as:

- **A --> B**
- **B --> C**
- **A --> C** (this is redundant and unnecessary)

Activities can have multiple predecessors and/or successors. Activities can be started as soon as all of their predecessors are completed.

- **Predecessors** - the activities that must be completed before the next one can start
- **Successors** - all activities that follow a specific task.

Activities can start as early as desired, or can be delayed until they run out of float or slack, thus becoming critical. At that point they are identified as the **critical path**. Any delay of the critical path activities will cause an equal delay for the entire schedule. Most activities will have float or slack, which is the amount of time they can be delayed until they become critical (Float = 0 hours) and impact the unit's start-up date. Realistically, activities that have very little float or slack should be treated as critical simply because there may be a degree of error in the estimates. A sequence of activities with float = 5 hours could easily be critical if their combined durations were underestimated by five hours (or the critical path was similarly overestimated).

Priority

Be sure to schedule all equipment inspections early. This is very important, because some findings could require major repair work that might impact the schedule. All high man-hour work orders should be started as soon as possible. Some equipment will merit a lowered priority, if the past experience indicates little or no repair work will be required. Consult the inspection reports to identify the extent of the repairs during past turnarounds. Low priority work is usually classified as "fill-in" work. It usually includes all kind of small jobs - mainly piping and valve work. You can spread out this work over the duration of the turnaround, to help smooth out the manpower requirements. The scope of these small jobs seldom grows into a larger one, and has no probability of showing up as the critical path.

They may, however, in the aggregation of several jobs, result in a critical mass of work (that cannot be finished with available resources within the current critical path timeframe) and therefore eventually cause a delay in the schedule (overtaking the critical path). Critical mass develops when the rate of progress is insufficient to complete the work before the critical path end date. It is usually due to insufficient manpower. This is the reason for keeping a close watch on the actual number of workers, every shift, and comparing it with the schedule requirements.

Sequencing the Work

After the basic schedule has been created, and the work prioritized (sequenced) according to an Operation / Production equipment availability schedule and the other considerations discussed earlier, you should sequence the work in such a manner as to enhance the utilization of manpower, tools and equipment. In sequencing the work, we have to consider the type of job, the resources or skills involved and the physical layout of the unit or plant. The first step is to determine the number of crews. We do this by reviewing a resource histogram (utilization) report for all resources and record the peak leveled number of craftsmen. So, we divide by two to arrive at the peak leveled number of crews, and add ten or twenty percent. This is a good rule of thumb for preliminary manpower planning. The reason you need to hire more men than scheduled is to compensate for absenteeism, dismissals, and additional work arising from inspection.

Every time you tie or restrain activities to sequence manpower, check to see if that action resulted in making the activities critical (or near critical). Near critical activities have very little floats or slack? If the activities have become critical, then it is best to undo the tie or restraint, otherwise you may be scheduling too tightly - increasing the probability for an overrun. This is a trial-and-error method, but it is not too difficult to achieve, and the result will be a workable schedule with realistic manpower utilization.

DATA COLLECTION AND ANALYSIS

For the data collection I visit in **Bhusan power & steel limited** in a process of setting up an integrated steel plant, including a state of the art CSP mill, at rengali sambalpur, Orissa. It is a fully integrated 1.5 Million TPA Steel making Company with turnover of INR3873.

PRE SHUTDOWN ACTIVITY FOR IMPLEMENTING THE TRUNAROUND PLANNING
1 PERMISSION FROM PRODUCTION DEPARTMENT FOR THE SHUTDOWN
2. SPARES MANAGEMENT
3. MTERIAL REQUIREMENT PLANNING
4.BILL OF MATERIAL
5. PRODUCT STRUCTURE
6. CALLING OF INDIGENOUS MANUFACTURE
7. ASSIGNING THE DUTIES FOR WORK ODERS

Roll maintenance Activity Time

ACTIVITY	TIME (HRS.)
1. DISCHARGE OF THE ROLLS FROM ROLL ASSEMBLY [A]	3
2. MISALIGNMENT CHECKING OF THE ROLL SHAFT [B]	5
3. MISALIGNMENT CHECKING OF BEARING HOUSING [C]	5
4. LUBRICATION INSPECTION OF ROLL SHAFT AND BEARING HOUSING [D]	2
5 COOLING OF THE ROLLS [E]	6
6 MAINTENANCE OF BULL GEAR HOUSING AND BEARING HOUSING [F]	7
7. INSPECTION OF THE SURFACE ROUGHNESS OF THE ROLLS [G]	2
8. SURFACE ROUGHNESS FINISHING OF THE ROLLS [H]	2

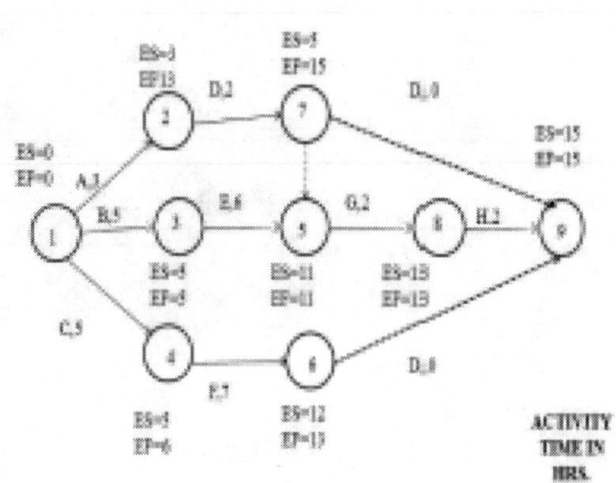

Bull Gear and Bearing Housing Maintenance Activity Time

ACTIVITY	TIME (HRS.)
1.INSPECTION OF TOP OIL LEVEL AND OIL QUALITY [A]	2
2.CHECKING OF GEAR BOX ALIGNMENT AND TEETH MESHING [B]	3
3.ALIGNMENT OF THE GEAR BOX SHAFT AND MOTOR SHAFT [C]	4
4.VIBRATION CHECKING OF GEAR (BULL TYPE) [D]	1
5.CHECKING AND REPLACEMENT OF THE ROLLING ELEMENT IN ROLLING BEARING [E]	2
6.ALIGNMENT CHECKING FOR BEARING [F]	2
7.BEARING HOUSING RIGIDITY CHECKING [G]	2
8.VIBRATION CHECKINH OF THE BEARING [H]	1
9.FIT ALL COMPONET IN THE GEAR BOX [I]	6
10. FIT ALL COMPONENT IN BEARING IN PROPER POSITION [J]	12

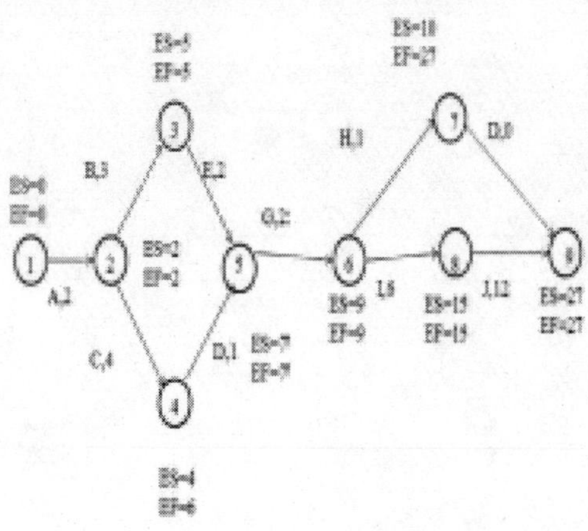

BULL GEAR AND BEARING MAINTENANCE

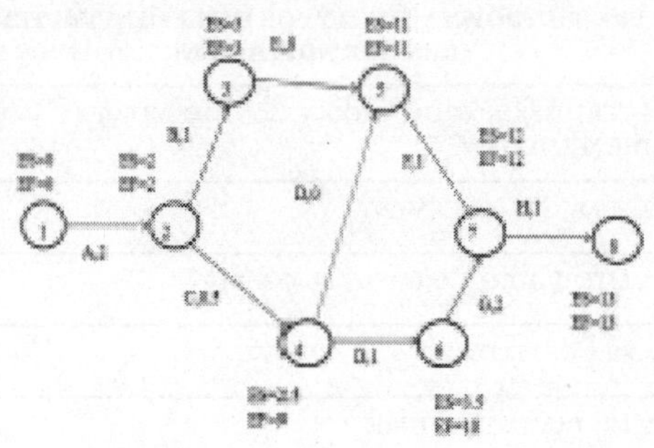

MAINTENANCE OF THE HYDRUALIC SYSTEM ACTIVITY

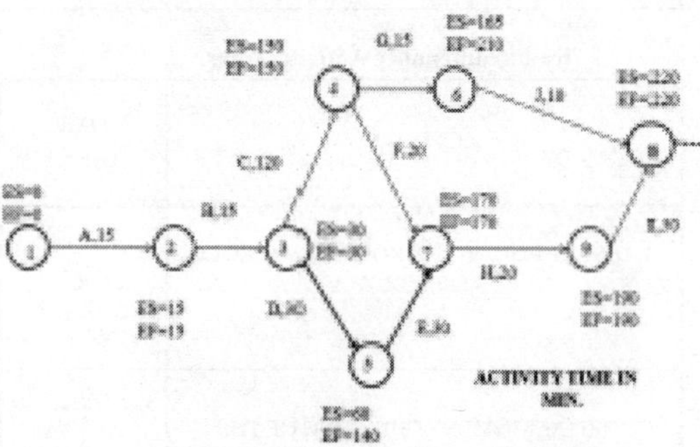

POST SHUTDOWN ACTIVITY CHART

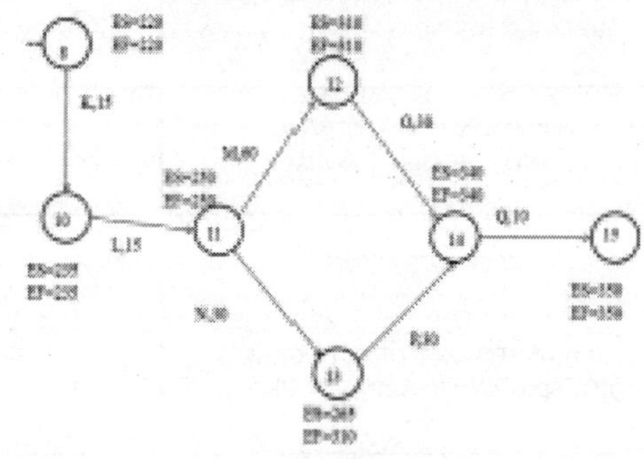

POST SHUTDOWN ACTIVITY CHART

ACTIVITY	TIME (HRS.)
1. CLEANING OF EACH CHAMBER (RETURN AND SUCTION OIL LINE) [A]	2
2. INSPECTION OF LEKAGE AND REPLACEMENT OF HOUSE PIPE [B]	1
3. OIL LEVEL INSPECTION AND TEMPERATURE INSPECTION [C]	0.5
4. OIL FILTER INSPECTION AND REPLACEMENT [D]	1
5. HEAT EXCHANGER MAINTENANCE AND DESCALING [E]	8
6. SOLENOID VALVE MAINTENANCE [F]	1
7. CLEANING OF STRAINER IN WATER SUPPLY LINE [G]	2
8. CIRCULATION PUMP MAINTENANCE [H]	1

Maintenance of hydraulic system activity time

ACTIVITY	TIME (MIN.)
1. CHECK / TIGHTEN ALL TH E FOUNDATION BOLTS OF GEAR BOX [A]	15
2. TIGHTEN ALL THE COUPLING BOLTS [B]	15
3. CHECK GEAR COUPLING (MOTOR TO GEAR BOX) TEETH CONDITION [C]	2 HR.

	TIME
4. CHEKING THE ALIGNMENT OF MOTOR AND GEAR BOX [D]	30
5. CHECK FOR PROPER GREASING IN ALL GREASING POINTS & DO GREASE WHERE REQUIRED [E]	30
6.CHEKS FOR ANY ABNORMAL SOUND / VIBRATION FROM GEAR BOX [F]	20
7.CHECKSFOR ANY ABNORMAL SOUND FROM COUPLING [G]	15
8. CHECK / ARREST ANY LUB. OIL LEAKAGE FROM GEAR BOX [H]	20
9. CHECK / ARREST ANY LUB. OIL LEAKAGE FROM PIPE LINE [I]	30
10. CHECKS / ARREST ANY HYD. OIL LEAKAGE FROM THE ACTUATOR [J]	10
11. CHECKS / ARREST ANY HYD. OIL LEAKAGE FROM THE PIPE LINE/ JOINTS [K]	15
12.GEAR&PINION NOISE CHECKING [L]	15
13.CHECK TEMPERATURE OF BEARING POINTS [M]	60
14.CHECKS AXIAL AND RADIAL PLAY OF THE SHAFT [N]	30
15.CHECK THE PINION SHAFT TOP THREAD PORTION FOR HYDRUALIC JACK MOUNTINGS [O]	30
16.CHECKS THE PINION SHAFT CHROMPLATING OF TAPER SLEEVE MOUNTING PORTION [P]	30
17. CHECK THE QUALITY OF LUB. OIL (WATER INCLUSION) [Q]	10
18. CHECK FOR PROPER OPERATION OF THE WHOLE UNIT [R]	7 HR.

SHUTDOWN ACTIVITY	TIME (HR.)
OVERALL ROLL MAINTENANCE [A]	15
OVERALL BULL GEAR AND BEARING MAINTENANCE [B]	27
OVERALL HYDRAULIC SYSTEM MAINTENANCE [C]	13
OVERHAULING OF NOZZEL IN COOLING SYSTEM [D]	2
OVERHAULING OF COIL CONVEYOR [E]	3
OVERHAULING OF DISCHARGE ROLLER TABLE [F]	5
OVERHAULING OF WALKING BEAM [G]	1
OVERHUALINH OF STEPPING MACHINE [H]	1
OVERHAULING OF LOOPER SYSTEM [I]	2
OVERHAULING OF APPROACH ROLLER TABLE [J]	3

Master network chart

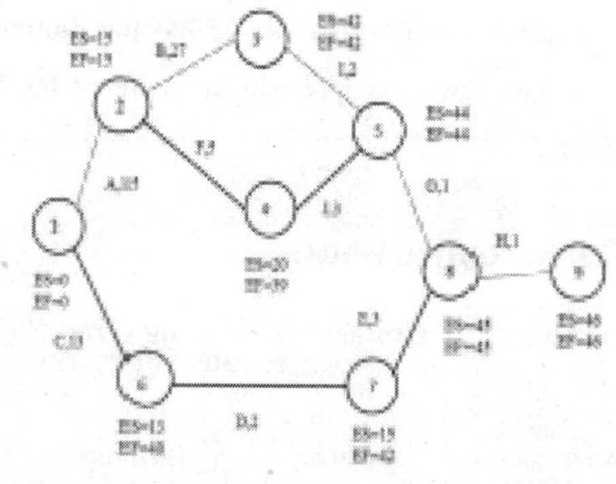

Post Shutdown Activity Time

Overall Shutdown Activities Time and Master Network Chart

In the above table time of the activities A, B, C is all ready reduced time with help of CPM and time of remaining activities is predefine by the company and this is the complete overall shutdown time in shutdown phase of turnaround planning and with help of these activities time we construct the overall shutdown network chart.

Result and Discussion

The subject of turnaround planning is very new for maintenance and it is only used in petrochemical & oil industries for planning, scheduling of the maintenance activity. It is evident that maintenance planning plays a key role in the plant productivity and operation. A good maintenance planning is well suited to the total production duration on facilities to reduce the down time of the machine and cost of the maintenance in major scheduled plant shut down of long and high intensity. In rolling mill unit, roll maintenance, bull gear and bearing maintenance and maintenance of hydraulic system is required for the effective maintenance during the shutdown.

1. With the help of **CPM** all the activities have been scheduled in the rolling unit. With the reduction in overall maintenance time the machine availability has increased significantly. In this reported work, during the analysis the critical activity has been identified giving more attention on non critical once. With the help of **turnaround planning** we reduced the overall turnaround time from **82 hours to 51.8 hours**. That means total time save during shutdown and post shutdown **30.2 hours**.

2. Statistical data of the concerned organization revealed that we have save 30.2hours in terms of down time. During this restoration work the organization lost profit of **Rs 25000 per hours**. So we have save the approximate profit of **Rs.7, 55,000.**

Time comparisons:-

ACTIVITY	TIME BEFORE MPLEMENTATION	TIME AFTER IMPLEMENTATION
ROLL MAINTENANCE ACTIVITY	20 HOURS	15 HOURS
BULL GEAR AND BEARING MAINTENANCE ACTIVITY	33 HOURS	27 HOURS
HYDRAULIC SYSTEM MAINTENACE ACTIVITY	18 HOURS	13 HOURS
TOTAL OVERALL TIME DURING SHUTDOWN PHASE	20+33+18+17(pre defined) = 72 HOURS	46 HOURS
POST SHUTDOWN ACTIVITY	10 HOURS	5.8 HOURS
TOTAL TIME FOR OVER ALL TURNAROUND	82 HOURS	51.8 HOURS

3. This table 1.8 shows the effect on the overall turnaround time during turnaround phases after implementation the turnaround planning. time before implementing the turnaround planning shutdown phase was takes **72 hours** which is reduced by CPM to **46 hours** and post shutdown phase was takes **10 hours** this is reduced to 5.8 hours.

ACTIVITIES	TIME BEFORE IMPLEMENTATION OF TURNAROUND	TIME AFTER IMPLEMENTATION OF TRUNAROUND
SHUTDOWN PHASE	72 HOURS	46 HOURS
POST SHUTDOWN PHASE	10 HOURS	5.8 HOURS
TOTAL OVERALL TIME	82 HOURS	51.8 HOURS

Table 1.8

Cost comparision

This table shows the cost saving after implementation of Turnaround planning.

		PRODUCTION COST PER HOURS	TOTAL COST
TIME BEFORE IMPLEMENTATION	82 HOURS	Rs 25,000	Rs 20,50,000
TIME AFTER IMPLEMENTATION	51.8 HOURS	Rs 25,000	Rs 12,95,000
TIME AND COST SAVED AFTER IMPLEMENTATION	30.2 HOURS	Rs 25,000	**Rs 7,55,000**

Conclusions:-

Turnaround impacts the business in a particular way. It is financed from company profits and because it is typically an expensive event it can have a major impact on those profits in the year in which it is performed. The cost of turnaround performed at intervals greater than one year may be spread over a no. of year to lessen the apparent impact on one year's profit, but loss of profit remains same.

1. Down time losses have been reduced from 82 hours to 51.8 hours all total charges rendered for over all restoration work reported is around Rs. 2000 per hr. (excluding replacement spare parts charges) this means around Rs 25000 is saved for one particular down time.

2. Even with the stand by unit usage around 2 hours time in terms of productivity a to retain the production and Rs 50,000 in terms of profit

3. In the reported work it has been analyzed that the down time losses have been reduced to the significant level thus reducing the maintenance and production losses.

Reference:-

1. **"Estimating and Prediction of Turnaround Time for Incidents in Application Service Maintenance Projects".** Perot Systems, EPIP Phase II, Whitefield Industrial Area, Computer Department, National Institute of Technology Karnataka,

2. **Turnaround maintenance in petrochemical industry: practices and suggested improvements** Author(s): Salih O. Duffuaa, M.A. Ben Daya Journal: Journal of Quality in Maintenance Engineering Year: 2004 Volume: 10 Issue: 3 Page: 184 – 190

3. **The maintenance management framework: A practical view to maintenance management** Author(s): **A. Crespo Marquez, P. Moreu de Leon, J.F. Gomez Fernandez, C. Parra Marquez, M. Lopez Campos Journal: Journal of Quality in Maintenance Engineering** Year: 2009 Volume: 15 Issue: 2 Page: 167 – 178

4. **Optimal outage scheduling—example of application to a large power system** George Anders*, Gomaa Hamoud Armando M. Leite da Silva Luiz A. da Fonseca Manso

5. **Turn-around maintenance management in a processing industry: A case study Author(s): Shaligram Pokharel, Jianxin (Roger) Jiao Journal: Journal of Quality in MaintenanceEngineeringYear: 2008 Volume: 14 Issue: 2 Page: 109 – 122**

6. **Complete Planning for Maintenance Turnarounds Will Ensure Success** By Rod Oliver, Meridium, Inc. *From Oil & Gas Journal, April 29, 2002*

7. **Mathematical programming approaches for the management of maintenance planning and scheduling** Author(s): S.O. Duffuaa, K.S. Al-Sultan Journal: Engineering Year: 1997 Volume: 3 Issue: 3 Page: 163 - 176

10. **Managing company turnarounds: how to develop "destiny"** Author(s): Michael HarkerJ journal: Marketing Intelligence & Planning, Year: 1996 Volume: 14 Issue: 3 P age: 5 – 10

11. **Maintenance and quality: the missing link Author(s): M. Ben-Daya, S.O. DuffuaaJournal: Engineering Year: 1995 Volume: 1 Issue: 1 Page: 20 - 26**

12. **Planning the Turnaround** Turnaround, *Shutdown and Outage Management, 2005, Pages 74-102*

13. **Excuting the turnaround** *Turnaround Management, 1999, Pages 154-173*

14. **Case study 5—Logistics** *Turnaround, Shutdown and Outage Management, 2005, Pages 236-241*

15. **Management of plant turnarounds – Part 1: Turnaround methodology** *Plant Maintenance Management Set, 2006, Pages 109-133*

16. **Management of plant turnarounds – Part 2: Turnaround methodology** *Plant Maintenance Management Set, 2006, Pages 109-133* Anthony Kelly

17. **Developing a turnaround strategy—A case study approach** *Omega, Volume 20, Issue 3, May 1992, Pages 345-352* S Chakraborty, S Dixit

Variation Reduction via Taguchi Technique "A Case Study of Quality"

[1]Sanjeev Pratap Singh, [1]Saurabh Dhaked, [2]Praval Pratap

[1]Department of Mech. Engg., MNIT, Jaipur, 302017, India,
[2] Sharda University., India,

Abstract – **In the current scenario leading manufacturing industries are striving for quality and cost reduction to meet customer satisfaction, market demand, better quality at a lower cost and increasing productivity. The Taguchi method involves reducing the variation in a process through robust design of experiments. The overall objective of the method is to produce high quality product at low cost to the manufacturer. The Taguchi method was developed by Dr. Genichi Taguchi of Japan who maintained that variation. The experimental design proposed by Taguchi involves orthogonal arrays to organize the parameters affecting the process and the levels at which these should vary. In this paper, the specific steps involved in the application of the Taguchi method will be described of using the Taguchi method to design experiments is given.**
Keywords: Taguchi Techniques, Robust Design, Orthogonal arrays, DOE.

I. INTRODUCTION

In industry, designed experiments can be used to systematically investigate the process or product variables that influence product quality. After you identify the process conditions and product components that influence product quality, you can direct improvement efforts to enhance a product's manufacturability, reliability, quality, and field performance.

Because resources are limited, it is very important to get the most information from each experiment you perform. Well-designed experiments can produce significantly more information and often require fewer runs than haphazard or unplanned experiments. In addition, a well-designed experiment will ensure that you can evaluate the effects that you have identified as important. For example, if you believe that there is an interaction between two input variables, be sure to include both variables in your design rather than doing a "one factor at a time" experiment. An interaction occurs when the effect of one input variable is influenced by the level of another input variable.

To this end, Taguchi (1986) introduced a systematic method for applying experimental design, so called robust design. The experimental design proposed by Taguchi involves using orthogonal arrays to organize the parameters affecting the
process and the levels at which they should be varies. Instead of having to test all possible combinations like the factorial design, the Taguchi method tests pairs of combinations. This allows for the collection of the necessary data to determine which factors most affect product quality with a minimum amount of experimentation, thus saving time and resources.

In this paper, the specific steps involved in the Taguchi method will be described and case study of using the Taguchi method to design experiments will be given.

II. LITERATURE SURVEY

Techniques of controlling/ reducing process variations

There are four major techniques for reducing the process variations namely,

- Statistical Process Control (SPC) approach, these are
- Six-Sigma approach,
- Shainin-DOE approach and
- Taguchi Loss Function approach.

These are briefly described below.

Statistical Process Control (SPC) Approach

Statistical process control was pioneered by Walter A. Shewhart in the early 1920s. The main goal of SPC is to achieve product quality by monitoring the stability of the underlying processes. In SPC, a stable process is defined as a process with only chance variation, resulting in a stable process-mean and a stable spread around this process-mean. Samples from the production process are checked at regular intervals to monitor the process. Statistical tools such as control charts are used to determine whether the process location and variation are stable or not. SPC is a very effective tool for reducing variations.

Six-Sigma Approach

Six Sigma is a business management strategy originally developed by Motorola in 1987. The term "Six Sigma" comes from a field of statistics known as process capability studies. Six Sigma seeks to improve the quality of process outputs by identifying and removing the causes of defects and minimizing variability in manufacturing and business processes. Continuous efforts to achieve stable and predictable process results and reduce process variation are of vital importance to business success.

Shainin Approach

The Shainin System is the name given to a problem solving system developed by Dorian Shainin, who established his own consulting practice: Shainin LLC in 1975. Shainin described his colourful method as the American approach to problem solving, with the same goals as that of the Taguchi approach. Dorian Shainin put several techniques – both known and newly invented – in a coherent stepwise strategy for problem solving in a manufacturing environment. This strategy is called the Shainin system. Shainin approach has no complexity, no complex mathematics and is intended for use in the on line production. Shainin approaches are effective in reducing the process variation and improving the process.

Taguchi Approach

Taguchi Approaches are statistical methods developed by Genichi Taguchi to improve the quality of manufactured goods, and more recently also applied to engineering Taguchi used and promoted statistical techniques for quality from an engineering perspective rather than from a statistical perspective. Although Taguchi has played an important role in popularising Design of Experiments (DOE), it would be wrong to consider Taguchi Methods as just another way to perform DOE. Taguchi methods have been used to evaluate factor effects and to specify design parameters. The Taguchi methodology is popular in the design stage, but also applicable during manufacturing stage for improving products and processes.

III. TAGUCHI ROBUST DESIGN

Robust Design method, also called the Taguchi Method, pioneered by Dr. Genichi Taguchi, greatly improves engineering productivity. By consciously considering the noise factors (environmental variation during the product's usage, manufacturing variation, and component deterioration) and the cost of failure in the field the Robust Design method helps ensure customer satisfaction. Robust Design focuses on improving the fundamental function of the product or process, thus facilitating flexible designs and concurrent engineering. Indeed, it is the most powerful method available to reduce product cost, improve quality, and simultaneously reduce development interval.

Taguchi on Quality

Quality has been defined by many as, "being within specifications", "zero defects", or "customer satisfaction". However, these definitions do not offer a method of obtaining quality or a means of relating quality to cost. Taguchi proposes a holistic view of quality which relates quality to cost, not just to the manufacture at the time of production but to the customer and society as whole. Taguchi defines quality as," The quality of a product is the (minimum) loss imparted by the product to society from the time product is shipped". This economic loss is associated with losses due to rework, waste of resources during manufacture, warranty costs, customer complaints and dissatisfaction, time and money spent by customer on falling products and eventual loss of market share.

Classical Approach to Quality Control: If you ship everything between the spec limits, can you claim "perfect quality"?

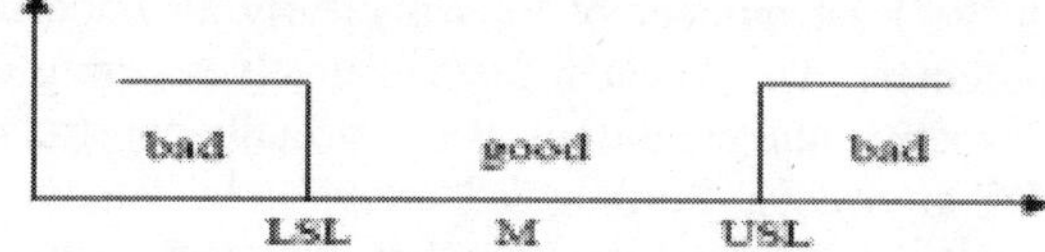

Fig. 1: Classical Approach to Quality Control

There is no difference between a student who passes with 60% and a student who fails with 59%, they are both lousy.

Need a Better Way to Define Quality: Quality loss is a financial loss.

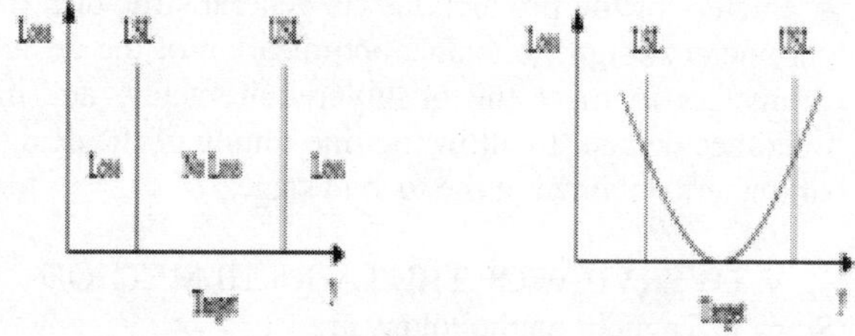

Fig. 2: Classical Quality loss and Taguchi Uses Quadratic Loss

Taguchi's quadratic loss function is the first operational joining of cost of quality and variability of product that allows design engineers to actually calculate the optimum design based on cost analysis and experimentation with the design.

Different Loss Functions Depending On Objective:

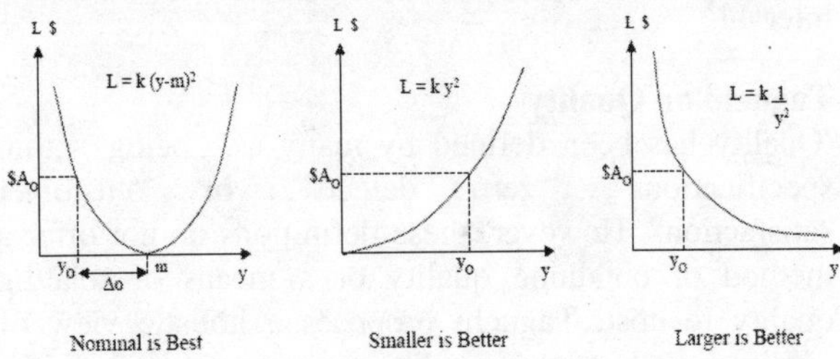

Fig. 3: Different Loss Functions Depending On Objective

IV. TAGUCHI METHODOLOGY

Taguchi (DOE) methods are used for maximizing robustness of products and processes, thereby achieving high quality at a low cost. (Antony et al.). The method was originally proposed as a means of improving the quality of products through the application of statistical and engineering concepts. Since experimental procedures are generally expensive and time consuming, the need to satisfy the design objectives with the least number of tests is clearly an important requirement. The Taguchi method involves laying out the experimental conditions using specially constructed tables known as "orthogonal arrays". In orthogonal array experiments, the number of test runs is minimized, while keeping the pair-wise balancing property (Byrne, Quality Progress 1987, 20, 19–26). The use of these tables ensures that the experimental design is both straightforward and consistent.

The methodology of Taguchi, the "father of robust design "who envisioned a three-stage design methodology (Swan, D. A., Savage, Qual. Reliable. Eng. Int. 1998, 14, 29–41.) including: (i) 'System design' to help determine the basic performance parameters of the product and its general structure; (ii) 'parameter design' to enable optimization of the design parameters to meet the quality requirements; and(iii) 'tolerance design' to allow the fine-tuning of the design parameters obtained in the second stage.

V. OVERVIEW OF THE TAGUCHI METHOD

Steps in Taguchi methodology are:
1. Determine the Quality Characteristic to be Optimized
2. Identify the Noise Factors and Test Conditions
3. Identify the Control Parameters and Their Alternative Levels
4. Design the Matrix Experiment and Define the Data Analysis Procedure
5. Conduct the Matrix Experiment
6. Analyze the Data and Determine the Optimum Levels
7. Predict the Performance at these Levels

VI. WHAT IS A TAGUCHI DESIGN?

A Taguchi design, or an orthogonal array, is a method of designing experiments that usually requires only a fraction of the full factorial combinations. In a Taguchi design, the array is orthogonal, which means the design is balanced so that factor levels are weighted equally. Because of this, an orthogonal array is one in which each factor can be evaluated independently of all the other factors.

In robust parameter design, you first choose control factors and their levels and choose an orthogonal array appropriate for these control factors. The control factors comprise the inner array. At the same time, you determine a set of noise factors, along with an experimental design for this set of factors. The noise factors comprise the outer array.

The experiment is carried out by running the complete set of noise factor settings at each combination of control factor settings (at each run). The response data from each run of the noise factors in the outer array are usually aligned in a row, next to the factors settings for that run of the control factors in the inner array. Each column in the orthogonal array represents a specific factor with two or more levels. Each row represents a run; the cell values indicate the factor settings for the run. By default, Minitab's orthogonal array designs use the integers 1, 2, 3... to represent factor levels. If you enter factor levels, the integers 1, 2, 3... Will be the coded levels for the design.

The following table displays the L8 (2**7) Taguchi design (orthogonal array). L8 means 8 runs. 2**7 means 7 factors with 2 levels each. If the full factorial design were used, it would have 2**7 = 128 runs.

The L8 (2**7) array requires only 8 runs - a fraction of the full factorial design. This array is orthogonal; factor levels are weighted equally across the entire design. The table columns represent the control factors, the table rows represent the runs (combination of factor levels), and each table cell represents the factor level for that run

L8 (27) Taguchi Design**

Run	A	B	C	D	E	F	G
1	1	1	1	1	1	1	1
2	1	1	1	2	2	2	2
3	1	2	2	1	1	2	2
4	1	2	2	2	2	1	1
5	2	1	2	1	2	1	2
6	2	1	2	2	1	2	1
7	2	2	1	1	2	2	1
8	2	2	1	2	1	1	2

In the above example, levels 1 and 2 occur 4 times in each factor in the array.

If you compare the levels in factor A with the levels in factor B, you will see that B1 and B2 each occur 2 times in conjunction with A1 and 2 times in conjunction with A2. Each pair of factors is balanced in this manner, allowing factors to be evaluated independently.

VII. EXPERIMENTS AND ANALYSIS: CASE STUDY

A company is having difficulty with its current yields. Silicon processors are made on a large die, cut into pieces, and each one is tested to match specifications. The company has requested that you run experiments to increase processor yield. The factors that affect processor yields are temperature, pressure, doping amount, and deposition rate.

The operating conditions for each parameter and level are listed below:

Table: 1 Parameters

FACTOR	STRATEGY PARAMETER
A	**Temperature(°C)**
B	**Pressure(psi)**
C	**Doping Amount (%)**
D	**Deposition Rate(mg/s)**

Table: 2 Operating parameters

Factor	Level 1	Level 2	Level 3
A	100	150(current)	200
B	2	5 (current)	8
C	4	6 (current)	8
D	0.1	0.2 (current)	0.3

Solution

The L9 orthogonal array should be used. The filled in orthogonal array should look like this: For our experiment, we considered four parameters. These five strategy parameters or factors for convenience are represented by the letters A~D. The factors (A~D) and corresponding parameters are listed in Table1. The levells of each operating parameters are listed in table 2. For our experiments, L9 (3^4) OA was selected, which represents 9 experiments with three level and four factors. The selected Oa is represented in Table 3. On the other hand, the full factorial design, which

represents the traditional or classical approach, considers all possible combinations of input parameters and for the given situation would require 81(3^4) experiments to be performed.Table 1 Factors and the correponding parameters.

Table: 3 The full factorial design

Experiment number	Temperature	Pressure	Doping Amount	Deposit ion Rate
1	100	2	4	0.1
2	100	5	6	0.2
3	100	8	8	0.3
4	150	2	6	0.3
5	150	5	8	0,1
6	150	8	4	0.2
7	200	2	8	0.2
8	200	5	4	0.3
9	200	8	6	0.1

This setup allows the testing of all four variables without having to run 81 [=3^4=(3 Temperatures) (3 Pressures) (3 Doping Amounts) (3 Deposition rates)] separate trials.

b) Conducting three trials for each experiment, the data below was collected. Compute the SN ratio for each experiment for the target value case, create a response chart, and determine the parameters that have the highest and lowest effect on the processor yield.

Shown below is the calculation and tabulation of the SN ratio.

$$S_{m1} = (87.3 + 82.3 + 70.7)^2/3 = 19248.0$$
$$S_{T1} = 87.3^2 + 82.3^2 + 70.7^2 = 19393.1$$
$$S_{e1} = S_{T1} - S_{m1} = 19393.1 - 19248.0 = 145.1$$

$$V_{e1} = S_{e1} / N - 1 = 145.1 / 2 = 72.5$$

$$SN_1 = 10 \log (1/N) (S_{m1} - V_{e1}) / V_{e1}$$

Where N = no. of trail (3) in this case.

Table: 4 Conducting three trials for each experiment & there mean

Ex. No.	A	B	C	D	Trail 1	Trail 2	Trial 3	Mean
1	100	2	4	0.1	87.3	82.3	70.7	**80.1**
2	100	5	6	0.2	74.8	70.7	63.2	**69.6**
3	100	8	8	0.3	56.5	54.0	45.7	**52.1**

4	150	2	6	0.3	79.8	78.2	62.3	**73.4**
5	150	5	8	0.1	77.3	76.5	54.0	**69.3**
6	150	8	4	0.2	89.0	87.3	83.2	**86.5**
7	200	2	8	0.2	64.8	62.3	55.7	**60.9**
8	200	5	4	0.3	99.0	93.2	87.3	**93.2**
9	200	8	6	0.1	75.7	74.0	63.2	**71.0**

Table: 5 Conducting three trials for each experiment & there SN

Ex. No.	A	B	C	D	Trail 1	Trail 2	Trial 3	SN
1	1	1	1	1	87.3	82.3	70.7	19.5
2	1	2	2	2	74.8	70.7	63.2	21.4
3	1	3	3	3	56.5	54.0	45.7	19.3
4	2	1	2	3	79.8	78.2	62.3	17.6
5	2	2	3	1	77.3	76.5	54.0	14.3
6	2	3	1	2	89.0	87.3	83.2	29.2
7	3	1	3	2	64.8	62.3	55.7	22.2
8	3	2	1	3	99.0	93.2	87.3	24.0
9	3	3	2	1	75.7	74.0	63.2	20.4

Shown below is the response table. This table was created by calculating an average SN value for each factor. A sample calculation is shown for Factor B (pressure):

Table: 6. Average SN value for each factor

Experiment Number	A(Temp)	B(pres)	C(dop.)	D(Dep)	SN
1	1	1	1	1	19.5
2	1	2	2	2	21.4
3	1	3	3	3	19.3
4	2	1	2	3	17.6
5	2	2	3	1	14.3
6	2	3	1	2	29.2
7	3	1	3	2	22.2
8	3	2	1	3	24.0
9	3	3	2	1	20.4

The effect of this factor is then calculated by determining the range:

$$\triangle = Max. - Min. = 23.0 - 19.8 = 3.2$$

Table: 7 Rank of parameters

Level	A (Temp.)	B (Press.)	C (Dop.)	D (Dep.)
1	20.1	19.8	24.2	18.1
2	20.4	19.9	19.8	24.3
3	22.2	23.0	18.6	20.3
$\triangle$	2.2	3.2	5.6	6.3
Rank	4	3	2	1

VIII. CONCLUSION

In this Case Study, we have seen that higher rank is allocated to deposition rate and lower rank is allocated to temperature and doping rate and pressure got rank 1 and 3. The deposition rate has largest effect on the processor yield because is has higher rank and that temperature has the smallest effect on the processor yield because it has lower rank. So we can easily analysis these type of problem by using design of experiment via taguchi method.

IX. REFERENCES

[1]. Antony, J., Kaye, M., Frangou, A., A strategic methodology to the use of advanced statistical quality improvement techniques. TQM Magazine 1998, 10, 169–176.

[2]. Byrne, D.M., Taguchi, S., The Taguchi approach to parameter design. Quality Progress 1987, 20, 19–26.

[3]. De mast, J., Shippers, W.A.J., Does, R.J.M.M. and Van Den Heuvel, Edwin R., (2000), "Steps and Strategies in Process improvement", Quality and Reliability Engineering International, 16: p.301-11.

[4]. Roy, R. K., a primer on the taguchi method, van no strand Reinhold, New York 1990.

[5]. Rosa, Jorge Luiz, Robin, Alain, Silva, M. B., Baldan, Carlos Alberto, Peres, Mauro Pedro (2009) "Electrode position of copper on titanium wires: Taguchi experimental design approach." Journal of Materials Processing Technology, vol. 209, pp. 1181-1188.

[6]. Taguchi, G.(1986).Introduction to quality engineering. White Plains, NY: Asian Productivity Organization.

[7]. Taguchi, G., & Wu, Y. (1980). Introduction to off-line quality control. Nagoya, Japan: Central Japan Quality Control Association.

[8]. Tay, K.-M. and C. Butler, (1999), "Methodologies for experimental design: A survey, comparison and future predictions.", Quality Engineering, 1999. 11(3): p. 343-356.

[9]. Tanco, Martin, Viles, Elizabeth and Pozueta, Lourdes, (2008), "Are All Designs of Experiments Approaches Suitable for Your Company?", Proc. World Congress on Engineering 2008, July 2-4, London, U.K., Vol II.

[10]. Swan, D. A., Savage, G. J., Continuous Taguchi-a model based approach to Taguchi's "quality by design" with arbitrary distributions. Qual. Reliab. Eng. Int. 1998, 14, 29–41.

Recent Trends in Manufacturing Waste Management Programme in a Process Industry

Ashish Shastri, Sanjay Goyal, Ajay Bangar

Department of Mechanical Engineering MPCT Gwalior

Abstract–**The principles of waste management are well developed and form a practical standpoint, thoroughly implemented in the process industries. This paper presents the concept and practical approaches to implement a "Waste Management Program". Such a program increase yields, recoveries and output while it decreases unit costs. In fact such a waste management program incorporating statistical process control will be an essential ingredient to effectively complete in the future. The waste management program aims at developing a strategy known as 5S philosophy which helps a company to stand when environment changes abruptly rather than a change or perish situation. The 5S comprises the basic civilization that protects corporate organisms from the ravages of the changing environment. The philosophy suggests factories, that to survive factories must learn from environment; they must destroy their former organizational concepts and customs that no longer apply and build up more appropriate and stronger organizations. 5S is a shop floor method to make every scrap count while wasting nothing. The 5S may be defined as:**

The first S (Seiri) is for "Sort"- This is about removing from our selected area anything that simply does NOT belong in it: broken tools or parts, trash, remains of pipe, wire, paint, brushes, oil canst process, boxes, bags, etc.

The second S (Seiton) is for "Set in place"-The goal is for anyone (when this step is finished), to find anything they may need in just a few seconds. The "place" assignment must be logical and determined by the direct user of the workstation.

The third S (Seiso) is for "Shine" or "Super-Cleaning" -This step is very important and focuses to find the root cause(s) or the origin(s) of contamination

The fourth S (Seiketsu): "Standardize" - Standardization is the assimilation of the new way of having our workstations looking, feeling, and functioning.

Finally the fifth S (Shitsuke): "Sustain"- Like any progress we achieve in life, it is critical to do all that it may take to preserve it.

I. INTRODUCTION

A Company is like a living organism. When its environment changes abruptly, its must also change- or perish. With the future challenged in the mind, many companies are gearing up by promoting ambitious companywide "rationalization" efforts. What kind of rationalization policy should a company adopt? The answer is 5' S (the Japanese "s" words that we call the five pillars of the visual workplace). The 5-S comprises the basic civilization that protects corporate organisms from the ravages of the changing environment. Not only are the 5-S the foundation upon which a company must stand to survive, they are also part of a company's unique corporate culture. No matter how it envisions it –companies truly intent on finishing this century ahead of competition must take the 5-S their first step.

Every ounce of this far is pure waste, and overweight companies are the least likely to win the marathon race for survival in to the next century.

"Good workplace develops beginning with the 5-S. Bad workshops fall apart from beginning with the 5-S."

Products and processes today involved very close tolerances. Variability must be fully controlled. The physical work environment is critical for drive high quality, low cost & speedy delivery.Ever-growing consumer demand for quality products is forcing people from other industries to rethink their workplaces. In the industry the Organization, Orderliness, Cleanliness, Standardized cleanup, and Discipline are needed. 5-S is to be the most important step for productivity improvement and safety today. In short 5-S is the simple activity that can be difficult to implement. And this is what the companies to survive in the years ahead.

II. PROBLEM DISCUSSION

Competitiveness is one of the strategic goals for all companies. To be more competitive one of the way is to meet customer demand promptly. To respond customer faster, production should be accomplished for each demand quickly so the faster a company is, the more competitive it is. For this reason, production time becomes important. Also, efficiency is an effect on competitiveness because efficiency level affects production time and much more elements of a production process. Employee satisfaction is important for increasing the level of efficiency in working process since satisfied employees work with more motivation and performance which conduces efficiency in production.

In manufacturing area employee satisfaction is disregarded while all the attention is given to production but it should not be forgotten that without considering employee satisfaction it is not easy to gather a good result or useful solution to enhance production like reducing the production time.5-S is a useful key concept for the companies for achieving strategic goals. 5-S provides the chance to satisfy the employee as well as having effective way of production.

So we can apply 5-S in order to eliminate waste and to reduce production lead time while keeping the customer satisfaction level high in terms of money. Moreover methods can be applied in an improved way, different than normal process. Thus we can determine the current state, identify problems, and generate suggestions on the production time and employee satisfaction. The elimination of different waste like overproduction, transportation & waiting is also the goal of this dissertation work. Thesis work focused on the elimination of the wastes and consequently increases the productivity of the organization. This also increases the morale of the workers.

III. THE CONCEPT

5-S Concept

The 5-S is a philosophy, which focuses on effective workplace organization and standardized work procedures. It is based on five Japanese words that begin with S.

The 5-S philosophy originated in the post World War II era (probably in the mid 1950s) in Japan. At that time, Japanese manufacturing companies were forced to produce with a very few resources, so they developed a shop floor method to make every scrap count while wasting nothing. Originally there were only four activities in the Japanese system, each beginning with the letter S. They were –

- **SEIRI** - Sorting
- **SEITON** - Setting in order , straightening, simplifying
- **SEISO** - Sweeping, Shining, Systematic Cleaning
- **SEIKETSU** - Standardizing

Later, a fifth activity was added, called **SHITSUKE** (sustaining), which completed the five S elements known as **5-S.** Today, the 5-S system retains its fundamental power to change the workplace & involve everyone in the improvement process. It is a system, to reduce waste and optimize productivity through maintaining an orderly Workplace to achieve more consistent operational results. It may be applied to any workplace for a short period of time due to its simple nature. The brief description of 5-S is as follows: [8]

The 'S'	5-S Element	What is Involved?	Objective
1-S	SEIRI	• Segregate necessary from unnecessary • Remove what is not required • Decide on frequency of sorting	Saving & Recovering Space
2-S	SEITON	• Arranging in order • A place for everything and every useful thing in its place.	• Minimizing search time • De-cluttering the workplace
3-S	SEISO	• Cleaning the workplace/ equipment • Ensuring Tip Top condition	• Inspecting the problems. • Taking corrective actions faster.
4-S	SEIKETSU	Working Methodology (procedures and work instructions)	Achieving higher productivity and better quality.
5-S	SHITSUKE	• Forming the habit • Training and discipline	Doing it first time and every time.

Resistance to 5-S

Any company introducing 5-S is likely to encounter various kinds of resistance, whether from the shop-floor or clerical staff which is as follows:

- *What's so great about Organization and orderliness?*
- *Why clean when it Just Gets Dirty Again?*
- *Implementing, Organization & Orderliness: Will Not Boost output!!*
- *Why Concern Ourselves with Triviality?*
- *We already Implemented Organization and Orderliness/*
- *My filing System is a Mess – but I Know My Way around It!*
- *We did the 5-S years ago*
- *We're Too Busy to Spend Time on Organization and Orderliness*
- *Who Are They to Tell Me What to Do?*
- *We don't need the 5-S – We're making money, so just let us do our work*

It can be difficult to implement the 5-S or other improvement techniques at companies that are currently profitable." Generally, such people fail to recognize how many processes are involved in making a product instead of emphasizing the product.

Objective of implementation of 5-S in the company

1. Increased Efficiency (Company)

It affects deeply the efficiency of the operation, whether it refer to on plant floor or office environment.

2. Better Morale (Employees)

The people feel that they are valued and that their work is more appreciated as the work conditions have improved. Less frustration is visible when people don't have to waste their time looking for things,

Satisfied Client (Customer)

By implementing 5-S, it can be can rest assured that everything is in control and have best interest at heart .Therefore; it also improves the relationship between the interests of the customer, the company shareholders and the employees.

Benefits of 5-S

- Greater efficiency in achieving goals, greater readiness for new tasks.
- Fewer hazards less spending on replacing lost or damaged items.
- Less stress Greater self-esteem Decrease down time.
- Raise employee morale, Identify problems more quickly.
- Develop control through visibility, Establish convenient work practices.
- Improved quality, achieve work standardization.
- Decreased changeover time, improved safety.

IV. REFRENCES

[1]   Becker, J.E. (2001), "Implementing 5-S to promote safety and housekeeping", Professional Safety, Vol. 46 No.8, pp.29-31.

[2]   Douglas, A. (2002), "Improving manufacturing performance", Quality Congress. ASQ's Annual Quality Congress Proceedings, Vol. 56 pp.725-32

[3]   Eckhardt, B. (2001), "The 5-S housekeeping program aids production", Concrete Products, Vol. 104 No.11, pp.56.

[4]   Zutshi, A., Sohal, A.S. (2000), "Visual management system: the experiences of Australian organisations", Journal of Manufacturing Technology Management, Vol. 16 No.2, pp.211.

[5]   Tice, J., Ahouse, L., Larson, T. (2005), "Lean production: aligning environmental management with business priorities", Environmental Quality Management, Vol. 15 No.2, pp.1-12.

Performance Measurement System: Some Issues and Gaps

[1]Bhagwan Das Gidwani, [2]Milind Kumar Sharma, [3]G. S. Dangayach

[1]Deptt. of Mech. Engg. Govt. Engg. College, Ajmer (Raj.).
[2]Deptt. of Prod. & Indl. Engg. MBA Engg. College JNV Univ., Jodhpur (Raj.)
[3]Deptt. of Mech. Engg. Malaviya National Institute of Tech., Jaipur (Raj.)

Abstract—**Performance measurement is process of quantifying action, where measurement is the process of quantification and action leads to performance. The design, implementation and use of adequate performance measurement and management frameworks can play an important role, if organizations are to succeed in an increasingly complex, interdependent and changing world.**

The adoption of Performance Measurement system (PMS) is a major endeavor and a complex project with many potential difficulties concerning budget overruns and benefits reached. Effective PMS has become a potentially valuable way of securing competitive advantage and improving organizational performance. To operate in an extremely volatile and unpredictable global business environment, availability of having the right kind of performance evaluation system in the industry has become a prerequisite for successful operations. For the same purpose, Industries must apply contemporary performance measurement methodologies which shall ensure improved efficiency and effectiveness in their overall business.

Industries require a PMS, which can be customized to their regional needs and to compete in globalised market. In our study, performance measures literature have been reviewed and gaps for future research have been identified. Findings are of use for industry in optimizing business performance measurement decisions. Emphasis is placed on introducing the PMS followed by a description of methodologies adopted. The current research work being undertaken by various authors is also highlighted.

I. INTRODUCTION

Measurement is a prerequisite to management in any firm. Many manufacturing and service organizations have used measures and measurement systems to determine their performance [1]. Performance Measurement System (PMS) is an emerging and evolving field in the era of globalization. PMS has been the dominant research paradigm of the last decade in the European countries but underdeveloped and under researched in the developing nations such as India. The objective of this study is to examine systematically the literature dealing with the different facets of PMS. In the process, issues relevant to the advancement of the theory and practice of PMS are identified and discussed.

II. THE CONTEXT OF THE STUDY

Performance measurement is the process of quantifying action, where measurement is the process of quantification and action leads to performance [2]. The terms efficiency and effectiveness are used precisely in this context. Effectiveness refers to the extent to which customer requirements are met, while efficiency is a measure of how economically the firm's resources are utilized when providing a given level of customer satisfaction [3]. The level of performance a business attains is a function of the efficiency and effectiveness of the actions it undertakes, and thus:

- *Performance measurement* can be defined as the process of quantifying the efficiency and effectiveness of action.
- A *performance measure* can be defined as a metric used to quantify the efficiency and/or effectiveness of an action.
- A *performance measurement system* can be defined as the set of metrics used to quantify both the efficiency and effectiveness of actions [4].

Measurement may be the "process of quantification", but its affect is to stimulate action, and as Mintzberg [5] has pointed out, it is only through consistency of action that strategies are realized. There are various ways in which these performance measures can be categorized, ranging from Fitzgerald *et al.* [6] framework of results and determinants through to Kaplan and Norton's [7] balanced scorecard.

Neely *et al.* [8] focused on the issues associated with the design of performance measurement systems, rather than the detail of specific measures. Their article is structured around the framework as shown in Figure 1, which highlights the fact that a performance measurement system can be examined at three different levels:

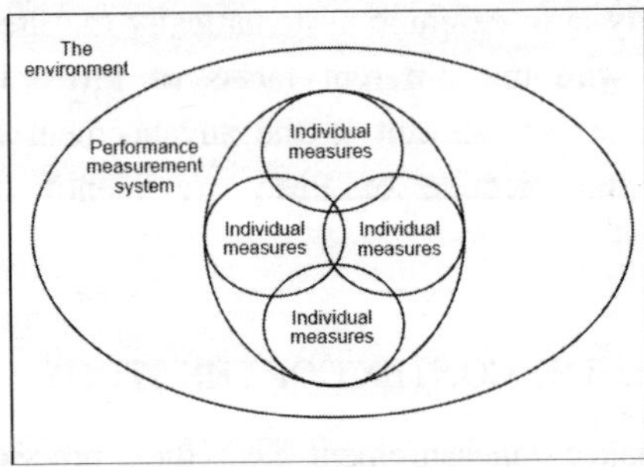

Fig.1 A framework for PMS design given by Neely *et al.*[8]

- The individual performance measures;
- The set of performance measures – the PMS as an entity;
- The relationship between PMS and the environment within which it operates.

At the level of the individual measure, this "performance measurement system" can be analyzed by asking questions such as:

- What performance measures are used?
- What are they used for?
- How much do they cost and
- What benefit do they provide?

Bititci *et al.* [9] during their research observed that, in recognition of the need for more relevant, better structured and integrated performance measurement systems, a number of frameworks and models for performance measurement have been developed, such as:

- Balanced scorecard (Kaplan and Norton, 1996);
- SMART (Strategic Measurement Analysis and Reporting Technique) (Cross and Lynch, 1988-1989);
- Performance measurement for world class manufacturer (Maskel, 1989);
- Performance measurement questionnaire (Dixon et al., 1990);
- Cambridge performance measurement design process (Neely et al., 1995;1996); and
- Integrated performance measurement systems reference model (Bititci and Carrie, 1998; Bititci et al., 1998a).

After reviewing all above frameworks a model for integrated and dynamic performance measurement systems was developed and illustrated by Bititci *et al.* [9]. The issues of designing, implementing, using and continuously updating performance measurement systems in manufacturing companies were addressed by Bourne *et al.* [10]. According to them specific processes are required to continuously align the performance measurement system with strategy. Kennerley and Neely [11] presented a framework of the factors affecting the evolution of performance measurement systems which provides a structured view of the factors affecting the evolution of performance measures and measurement systems. Model for measuring and managing performance in extended enterprises which includes intrinsic and extrinsic inter-enterprise coordinating measures was proposed by Bititci *et al.* [12].

Yilmaz and Bititci [13] compared Performance measurement in the value chain in contrast with manufacturing v/s. tourism and demonstrated the usability of Supply Chain Operations Reference (SCOR) Model-like frameworks in the tourism industry to manage and measure the value chain processes. Tzelepis and Skuras [14] presented link between Strategic performance measurement and the use of capital subsidies. They concluded that capital subsidies have a positive impact on firms' long-term strategic orientations such as the firms' net market growth and the optimal scale of operation. An empirical test of customer-based performance measures of Interaction Fluency was developed by Harold and Douglas [15] to contribute to the assessment of customer service performance across multiple contact points upon the implementation of relationship management practices and technologies. One more empirical study was carried out by Garengo and Bititci[16] in Scottish SMEs to explain contingency approach to performance measurement. A relationship between the contingency factors and performance measurement was formalized by them.

Juan *et al.* [17] presented Performance measurement system for enterprise networks. They developed a PMS that overcomes difficulties and, at the same time, provide enterprises with a simple, efficient, robust and useful framework. Olsen *et al.* [18] presented a article on PMS and relationships with performance results using a case analysis of a continuous improvement approach to PMS design. Framework presented by them for evaluating the effectiveness of a PMS was based on three criteria – i.e. causality, continuous improvement and process control.

Bhagwat and Sharma [19] have developed balanced scorecard approach which provides a useful guidance for managers in evaluation and measuring of Supply Chain Management in a balanced way. Many other researchers have also attempted to summarize some of the most appropriate performance metrics and measures of SCM. According to Mason-Jones and Towill [20], Proper control of the order is possible, provided that the order entry method is capable of providing timely, accurate and usable data at various entry levels, and hence, can be used as a metric of performance measure. Gunasekaran *et al.*[21], proposed that reduction in the order cycle time leads to a reduction in the supply chain response time. According to Chen [22] Performance-measurement systems that are based on traditional cost-accounting systems, do not capture the relevant performance issues for today's manufacturing environment. A variety of integrated systems have been proposed to overcome the limitations of the traditional performance-measurement systems. Johnston and Pongatichat [23] concluded that recent studies of

world-class manufacturing organizations have found evidence of a consistency between performance measures and systems with manufacturing and organizational objectives.

Meekings *et al.* [24] have argued that without appropriate plumbing, performance management systems do not drive organizational change and improvement. According to them key elements of a plumbed-in performance management system is: performance architecture; performance insights; performance focus; and performance action. Andre *et al.* [25] have examined the relationship between the level of completeness of a strategic performance management (SPM) system implementation and the advantages and disadvantages an organization experiences from this system. They concluded that organizations that have fully completed the SPM implementation gain more financial and non-financial advantages. Lehtinen and Ahola [26] have assessed compatibility of existing PMS's with the central features of extended enterprises and concluded that the common underlying reasons for measuring performance are highly valid also in the context of an extended enterprise.

III.   THE LITERATURE GAP

Performance measurement is not and never can be a field of academic study because of its diversity. Certainly, as eluded by Neely [27], even the most widely cited authors in the field come from a variety of different disciplinary backgrounds – accounting, information systems, operations management and operations research. The characteristics of modern performance measurement systems as explained by various researchers are:

- must reflect relevant non-financial information based on key success factors of each business (Clarke, 1995);
- should be implemented as a means of articulating strategy and monitoring business results (Grady, 1991);
- should be based on organizational objectives, critical success factors, and customer needs and should monitor both financial and nonfinancial aspects (Manoochehri, 1999);
- must accordingly change dynamically with the strategy (Bhimani, 1993);

- must meet the needs of specific situations in manufacturing operations and should be longterm oriented as well as simple to understand and implement (Santori and Anderson, 1987);
- must make a link to reward systems (Tsang et al., 1999); and
- financial and non-financial measures must be aligned and fit within a strategic framework (Drucker, 1990; McNair and Mosconi, 1987).

While there is evidence that some firms are increasingly utilizing non-financial performance measures in their day-to-day decision-making processes [28], there appears to be very little evidence that these measures are formal and directly linked to the firm's strategy and effectiveness. Some of the reasons mentioned by Hudson *et al.*[29] in the literature for not using PMS in industries include shortage of human and capital resources, lack of strategic planning, misconception of the benefits of performance measurement and an overall technical orientation. Traditional mindset does not allow industrialists to invest their resources much in PMS.

In order to respond proactively to emerging challenges, management requires up-to-date and accurate information on its performance. It is natural that organizations measure their performance in order to direct organization's resources towards important organizational goals and in designing strategy. Investigations of the performance impact of the balanced scorecard have delivered mixed results [30-32]. Since performance measuring can profoundly affect the motivation of individuals [33], the PMS must take the human factor into consideration.

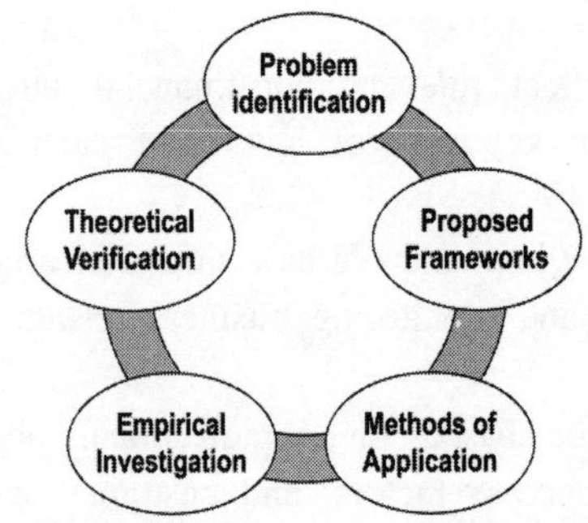

Fig.2 The evolution of the field of performance measurement [27]

IV. CONCLUSIONS

A comprehensive but brief survey of methods applied to PMS has been presented here to provide background information regarding the present research and development trends in this area. This paper has taken into account 33 papers, using various methods and strategies discussed above. Only some salient features and significant developments have been highlighted here and for proper design and evaluation for a particular application, one has to study in depth many trade off considerations and relevant texts given in references.

Based on the literature, performance measurement must be grounded on information availability, reliability and responsibility. As such, in a systematic PMS information is not only tracked and stored; rather it is made readily available to management and external entities. The reliability of the information flow is obtained from the shop-floor, as well as from the market place. Responsibility means the responsibility of all elements of the organization, as they attempt to execute their tasks with efficiency within the context of strategic effectiveness guidelines. Thus, the performance measurement organizational effort must be viewed as a complete organizational system rather than a collection of standalone models and tools.

V. FUTURE SCOPE

The performance measurement, measures and systems must be viewed from a continuous improvement perspective. In this context, the PMS must be viewed as a collection of procedures, techniques, processes. Perhaps future research should examine whether this systematic approach to performance measurement and measures should be based on a top-down or a bottom-up orientation. In the context of performance measurement, practitioners and scholars need to combine their efforts in order to validate what is known and build on it. The results of this joint effort may lead to the deployment of practical philosophies, systems and measures, which the practice of performance measurement badly need. This research is a small step toward that end.

The future scope of performance measurement, measures and systems must be viewed from a continuous improvement perspective. In this context, the PMS must be viewed as a collection of procedures, techniques, processes, and more importantly, people working together toward continuously improving the multifacets of manufacturing performance and measurement. Future research should also empirically examine the effectiveness of the integrated approach to these efforts versus the measure-specific approach. In the context of manufacturing performance measurement, practitioners and scholars need to combine their efforts in order to validate what is known and build on it.

VI. REFERENCES

[1]   Mettanen, P., 2005, "Design and implementation of a performance measurement system for a research organisation", *Production Planning and Control*, Vol. 16, No. 2, pp.178–188.

[2]   Kotler, P., 1998. *Marketing Management Analysis, Planning and Control,* Prentice-Hall, Englewood Cliffs, NJ.

[3]   Slack, N., 1991. *The Manufacturing Advantage: Achieving Competitive Manufacturing Operations,* Mercury, London.

[4]   Neely, A.D., 1994. "Performance measurement system design – third phase", draft of the fourth section of the "Performance Measurement System Design Workbook".

[5]   Mintzberg, H., 1978. "Patterns in strategy formulation", *Management Science*, 24(9), pp. 934-48.

[6]   Fitzgerald, L., Johnston, R., Brignall, S., Silvestro, R. and Voss, C., 1991. *Performance Measurement in Service Business,* CIMA, London.

[7]   Kaplan, R.S. and Norton, D.P., 1992. "The balanced scorecard – measures that drive performance", *Harvard Business Review,* pp. 71-79.

[8]   Andy Neely, Mike Gregory and Ken Platts, 1995. "Performance measurement system design- A literature review and research agenda", *International Journal of Operations & Production Management*, 15(4), pp. 80-116.

[9]   Umit S. Bititci, Trevor Turner and Carsten Begemann, 2000. "Dynamics of performance measurement systems", *International Journal of Operations & Production Management*, 20(6), pp.692-704.

[10] Mike Bourne, John Mills, Mark Wilcox, Andy Neely and Ken Platts, 2000. "Designing, implementing and updating performance measurement systems", *International Journal of Operations & Production Management*, 20(7), pp. 754-771.

[11] Mike Kennerley and Andy Neely, 2002. "A framework of the factors affecting the evolution of performance measurement systems", *International Journal of Operations & Production Management*, 22(11), pp. 1222-1245.

[12] Umit S. Bititci, Kepa Mendibil, Veronica Martinez and Pavel Albores, 2005. "Measuring and managing performance in extended Enterprises", *International Journal of Operations & Production Management*, 25(4), pp. 333-353.

[13] Yildirim Yilmaz and Umit Bititci, 2006. "Performance measurement in the value chain: manufacturing v. tourism", *International Journal of Productivity and Performance Management*, 55(5), pp. 371-389.

[14] Dimitris Tzelepis and Dimitris Skuras, 2006 "Strategic performance measurement and the use of capital subsidies", *International Journal of Productivity and Performance Management*, 55(7), pp. 527-538.

[15] Harold Cassab and Douglas L. MacLachlan, 2006. "Interaction fluency: a customer performance measure of multichannel service", *International Journal of Productivity and Performance Management*, 55(7), pp. 555-568.

[16] Patrizia Garengo and Umit Bititci, 2007. "Towards a contingency approach to performance measurement: an empirical study in Scottish SMEs", *International Journal of Operations & Production Management*, 27(8), pp. 802-825.

[17] Juan José Alfaro Saiz, Angel Ortiz Bas and Raúl Rodriguez Rodriguez, 2007. "Performance measurement system for enterprise networks", *International Journal of Productivity and Performance Management*, 56(4), pp. 305-334.

[18] Eric O. Olsen, Honggeng Zhou, Denis M.S. Lee, Yoke-Eng Ng, Chow Chewn Chong And Pean Padunchwit, 2007. "Performance measurement system and relationships with performance results A case analysis of a continuous improvement approach to PMS design", *International Journal of Productivity and Performance Management,* 56(7), pp. 559-582.

[19] Rajat Bhagwat and Milind Kumar Sharma, 2007. "Performance measurement of supply chain management: A balanced scorecard approach**",** *Computers and Industrial Engineering*, 53(1), pp. 43-62.

[20] Mason-Jones, R., & Towill, D. R., 1997, "Enlightening supplies", *Manufacturing Engineering*, 3, pp.156–160.

[21] Gunasekaran, A., Patel, C., & Tirtiroglu, E., 2001, "Performance measures and metrics in a supply chain environment", *International Journal of*

Production and Operations Management, 21(1/2), pp.71–87.

[22]   Chen, C. C., 2008. "An objective-oriented and product-line-based manufacturing performance measurement", *International Journal of Production Economics*, 112, pp. 380-390.

[23]   Johnston R. and Pongatichat P., 2008. "Managing the tension between performance measurement and strategy: coping strategies", *International Journal of Operations & Production Management*, 28(10), pp. 941-967.

[24]   Alan Meekings, Simon Povey and Andy Neely, 2009. "Performance plumbing: installing performance management systems to deliver lasting value", *Measuring Bussiness Excellence*, 13(3), pp.13-19.

[25]   Andre de Waal, Karima Kourtit and Peter Nijkamp, 2009. "The relationship between the level of completeness of a strategic performance management system and perceived advantages and disadvantages", *International Journal of Operations & Production Management*, 29(12), pp. 1242-1265.

[26]   Jussi Lehtinen and Tuomas Ahola, 2010. "Is performance measurement suitable for an extended enterprise?", *International Journal of Operations & Production Management*, 30(2), pp181-204.

[27]   Andy Neely, 2005, "The evolution of performance measurement research Developments in the last decade and a research agenda for the next", *International Journal of Operations & Production Management,* Vol. 25 No. 12, pp. 1264-1277.

[28]   Burns, J., Scapens, R. and Turley, S. (1997), "The crunch for numbers", Accountancy, Vol. 119 No. 1245, pp. 86-7.

[29]   Hudson, M., Smart, P.A. and Bourne, M., 2001, "Theory and practice in SME performance measurement systems", *International Journal of Operations & Production Management*, Vol. 21, No. 8, pp 1096-1115.

[30]   Banker, R.D., Potter, G. and Srinivasan, D. (2000), "An empirical investigation of an incentive plan that includes nonfinancial performance measures", The Accounting Review, Vol. 75 No. 1, pp. 65-92.

[31]   Ittner, C.D. and Larcker, D.F. (2003), "Coming up short on non-financial performance measurement", Harvard Business Review, November, pp. 88-95

[32]   Neely, A.D., Kennerley, M. and Martinez, V. (2004), "Does the balanced scorecard work: an empirical investigation", Proceedings of the 4th International Conference on Performance Measurement, Edinburgh.

[33]   Sinclair, D. and Zairi, M. (1995a), "Effective process management through performance measurement. Part I: applications of total quality-based performance measurement", Business Process Re-engineering & Management Journal, Vol. 1 No. 1, pp. 75-88.

Biodiesel Production for Sustainable Development and Indian Perspective

Yogesh Dhote, Manish Sharma, Yashvir Singh and Nishant Singh

Department of Mechanical Engineering, Hindustan College of Science & Technology,
Farah, Mathura

Abstract---- **Today's world is heavily dependent on fuels, which are almost entirely derived from fossil fuels. For almost last three decades, since the increase in the petroleum products prices, various other alternatives have been explored with the goal of replacing conventional petroleum supplies at least partially. This need has been supplemented by the concern for the increasing pollution owing to the fossil fuel combustion. Thus, the need of the hour is to utilize renewable energy sources. Biodiesel refers to a vegetable oil or animal fat based diesel fuel consisting of long-chain alkyl (methyl, propyl or ethyl) esters. Biodiesel is typically made by chemically reacting lipids (e.g., vegetable oil, animal fat (tallow)) with an alcohol. Biofuels offer an attractive alternative to fossil fuels, but a consistent scientific framework is needed to ensure policies that maximize the positive and minimize the negative aspects of biofuels. Numerous countries are moving towards the partial and gradual replacement of fossil fuels with biofuels, mainly ethanol and biodiesel. The increased move towards biofuels is spurred by global political, economical and environmental events, especially rising crude oil prices.**

Keywords: *Biodiesel, Jatropa, Fuel, Blends etc.*

I. INTRODUCTION

The use of vegetable oils for engine fuels may seem insignificant today. But such oils may in the course of time become as important as petroleum and the coal tar products of present time." - Rudolf Diesel (1912). Bio-fuel is a renewable source of combustible material whose energy content can be beneficially utilized for transportation purposes. In 1895, Rudolf Diesel developed a new engine with the intention that it could use a variety of fuels, including vegetable oil. When he showcased it to the public at the 1900 Paris World's Fair, he had the engine run on peanut oil. As the diesel engine became more widely adopted in subsequent years, however, petroleum-based diesel fuel proved to be less expensive and became the fuel of choice. Biodiesel's definition has been a work in progress. Since the early 1900s, biodiesel has been defined as an alternative form of diesel fuel made from vegetable oils or animal fats and alcohol.

All the bio-fuels stem from agricultural sources in a closed carbon dioxide cycle. In India there is significance for searching the replacement of fossil fuels. Almost 80% of our fuel requirement of diesel and petroleum is presently imported. The nation is heavily dependent on this input straining its foreign exchange resources. This is particularly a major weakness from defense point, in case of any aggression, war or international emergency our entire activity can get affected if the import of fuel or raw oil stops even for a small period.

We are fortunate to have ample renewable resources, which can substantially replace the imported oil requirements strengthening the economy of the nation and also making it more self-reliant to face any emergency. Following are major advantages of replacing present fossil diesel oil by bio diesel.

a) Bio diesel is a renewable resource and thus there is no fear like exhausting of present fossil stocks.

b) Bio diesel contains almost no sulphur and hence the exhaust gas emission is far better in quality for controlling pollution caused by fossil fuel combustion.

c) Bio diesel is produced from seed oil of plants. Increasing the vegetation or plantation will help consuming the carbon dioxide produced by combustion and releasing oxygen maintaining the ozone in the atmosphere.

d) Decentralized production of bio diesel globally will be more economic to feed local requirements reducing the transport cost of basic fossil fuels.

e) Non-edible oils can prove viable alternatives to edible oils for bio diesel production.

II. Properties of Biodiesel

a) Biodiesel has better lubricating properties and much higher cetane ratings than today's lower sulfur diesel fuels. Biodiesel addition reduces fuel system wear in low levels.

b) In high pressure systems biodiesel have good properties to increase the life of the fuel injection equipment that relies on the fuel for its lubrication.

c) The calorific value of biodiesel is about 37.27 MJ/L which is 9% lower than regular number to petro diesel.

d) Variations in biodiesel energy density are more dependent on the feedstock used than the production process. Still these variations are less than for petrodiesel.

e) It has been claimed biodiesel gives better lubricity and more complete combustion thus increasing the engine energy output and partially compensating for the higher energy density of petrodiesel.

f) Biodiesel is a liquid which varies in color between golden and dark brown depending on the production feedstock.

g) It is immiscible with water, has a high boiling point and low vapor pressure.

h) The flash point of biodiesel (>130 °C) is significantly higher than that of petroleum diesel (64 °C) or gasoline (−45 °C). Biodiesel has a density of ~ 0.88 g/cm³, higher than petrodiesel (~ 0.85 g/cm³).

i) Biodiesel has virtually no sulfur content, and it is often used as an additive to Ultra-Low Sulphur Diesel (ULSD) fuel to aid with lubrication, as the sulfur compounds in petrodiesel provide much of the lubricity.

III. MATERIAL COMPATIBILITY

a) Plastics: High density polyethylene (HDPE) is compatible but polyvinyl chloride (PVC) is slowly degraded. Polystyrenes are dissolved on contact with biodiesel.

b) Metals: Biodiesel has an effect on copper-based materials (e.g. brass), and it also affects zinc, tin, lead, and cast iron stainless steels (316 and 304) and aluminum are unaffected.

c) Rubber: Biodiesel also affects types of natural rubbers found in some older engine components. Commonly used synthetic rubbers FKM- GBL-S and FKM- GF-S found in modern vehicles were found to handle biodiesel in all conditions.

IV. LOW TEMPERATURE GELLING

When biodiesel is cooled below a certain point some of the molecules aggregated and form crystals. The fuel starts to appear cloudy once the crystals become larger than one quarter of the wavelengths of visible light; this is the cloud point (CP). As the fuel is cooled further these crystals become larger. The lowest temperature at which fuel can pass through a 45 micrometer filter is the cold filter plugging point (CFPP). As biodiesel is cooled further it will gel and then solidify.

V. CONTAMINATION BY WATER

Biodiesel may contain small but problematic quantities of water. Although it is not miscible with water, it absorbs water from atmospheric moisture. Water contamination is also a potential problem when using certain chemical catalysts involved in the production process, substantially reducing catalytic efficiency of base (high pH) catalysts such as potassium hydroxide. The presence of water is a problem because:

a) Water reduces the heat of combustion of the bulk fuel. This means more smoke, harder starting, less power.

b) Water causes corrosion of vital fuel system components: fuel pumps, injector pumps, fuel lines, etc.

c) Water & microbes cause the paper element filters in the system to fail (rot), which in turn results in premature failure of the fuel pump due to ingestion of large particles.

d) Water freezes to form ice crystals near 0 °C (32 °F). These crystals provide sites for nucleation and accelerate the gelling of the residual fuel.

e) Water accelerates the growth of microbe colonies, which can plug up a fuel system. Biodiesel users who have heated fuel tanks therefore face a year-round microbe problem.

f) Additionally, water can cause pitting in the pistons on a diesel engine.

VI. EFFICIENCY AND ECONOMICS

Average crops of rapeseed produce oil at an average rate of 1,029 L/ha, and high-yield rapeseed fields produce about 1,356 L/ha. The ratio of input to output in these cases is roughly 1:12.5 and 1:16.5. Biodiesel is less costly to deploy (solar cells cost approximately Rs. 50,000 per square meter) and transport (electric vehicles require batteries which currently have a much lower energy density than liquid fuels). When straw was left in the field, biodiesel production was strongly energy positive, yielding 1 GJ biodiesel for every 0.561 GJ of energy input (a yield/cost ratio of 1.78).

When straw was burned as fuel and oilseed rape meal was used as a fertilizer, the yield/cost ratio for biodiesel production was even better (3.71). In other words, for every unit of energy input to produce biodiesel, the output was 3.71 units (the difference of 2.71 units would be from solar energy)

VII. ENERGY SECURITY

One of the main drivers for adoption of biodiesel is energy security. This means that a nation's dependence on oil is reduced, and substituted with use of locally available sources, such as coal, gas, or renewable sources. Thus a country can benefit from adoption of biofuels, without a reduction in greenhouse gas emissions. The US National Renewable Energy Laboratory (NREL) states that energy security is the number one driving force behind the US biofuels programme, and a White House "Energy Security for the 21st Century" paper makes it clear that energy security is a major reason for promoting biodiesel. The EU commission president, Jose Manuel Barroso, speaking at a recent EU biofuels conference, stressed that properly managed biofuels have the potential to reinforce the EU's security of supply through diversification of energy sources.

VIII. ENVIRONMENTAL EFFECT

The surge of interest in biodiesels has highlighted a number of environmental effects associated with its use. These potentially include reductions in greenhouse gas emissions, deforestation, pollution and the rate of biodegradation.

According to the EPA's Renewable Fuel Standards Program Regulatory Impact Analysis, released in February 2010, biodiesel from soya oil results, on average, in a 57% reduction in greenhouse gases compared to fossil diesel, and biodiesel produced from waste grease results in an 86% reduction.

IX. FOOD, LAND AND WATER VS FUEL

In some poor countries the rising price of vegetable oil is causing problems. Some propose that fuel only be made from non-edible vegetable oils such as camelina, jatropha or seashore mallow which can thrive on marginal agricultural land where many trees and crops will not grow, or would produce only low yields. Others argue that the problem is more fundamental. Farmers may switch from producing food crops to producing biofuel crops to make more money, even if the new crops are not edible. The law of supply and demand predicts that if fewer farmers are producing food the price of food will rise. It may take some time, as farmers can take some time to change which things they are growing, but increasing demand for first generation biofuels is likely to result in price increases for many kinds of food. Some have pointed out that there are poor farmers and poor countries who are making more money because of the higher price of vegetable oil.[76]

Biodiesel from sea algae would not necessarily displace terrestrial land currently used for food production and new alga-culture jobs could be created.

X. PRODUCTION

Biodiesel is derived from biological sources, such as vegetable oils or fats, and alcohol. Commonly used feedstocks are as given below.

Vegetable Oils:
   a)   Soybeans
   b)   Rapeseed
   c)   Canola Oil (a modified version of rapeseed)
   d)   Safflower Oil
   e)   Sunflower Seeds
   f)   Yellow Mustard Seed

Animal Fats:
   a)   Lard
   b)   Tallow
   c)   Poultry Fat

Other Sources:
   a)   Recycled Restaurant
   b)   Cooking Oil (Yellow Grease)

XI. COMPARISON OF FUEL PROPERTIES

As the depletion of fossil fuel has already been started and it is expected that diesel prizes will never came down, there is a need to think of other substitute fuels. In this reference biodiesel and hydrogen are expected to be the future fuel. Comparison of fuel properties have been shown for fossil diesel vs biodiesel (Refer Table 1) and also between hydrogen and the biodiesel (Refer Table 2).

Table 1 Comparison of Fuel Properties between Diesel and Biodiesel

Fuel Property	Diesel	Biodiesel
Fuel Standard	ASTM D975	ASTM PS 121
Fuel composition	C10-C21 HC	C12-C22 FAME
Lower Heating Value, Btu/gal	131,295	117,093
Kin. Viscosity, @ 40° C	1.3-4.1	1.9-6.0
Specific Gravity kg/l @ 60° F	0.85	0.88
Density, lb/gal @ 15° C	7.079	7.328
Water, ppm by wt	161	.05% max
Carbon, wt %	87	77
Hydrogen, wt %	13	12
Oxygen, by dif. wt %	0	11
Sulfur, wt %	.05 max	0.0 - 0.0024
Boiling Point (°C)	188-343	182-338
Flash Point (°C)	60-80	100-170
Cloud Point (°C)	-15 to 5	-3 to 12
Pour Point (°C)	-35 to -15	-15 to 10
Cetane Number	40-55	48-65
Stoichiometric Air/Fuel Ratio wt./wt.	15	13.8
BOCLE Scuff, grams	3,600	>7,000
HFRR, microns	685	314

Table 2 Comparison of Fuel Properties between Biodiesel and hydrogen as fuels of the future

	Biodiesel	Hydrogen
Technological Readiness	Can be used in existing diesel engines, which have already been in use for 100 years	Electrolyzing water (most likely using fossil fuel energy) or reforming fossil fuels. Most likely non-renewable methods with large net CO2 emissions.
Fuel Distribution System	Can be distributed with existing filling stations with no changes	No system currently exists, would take decades to develop. Would cost $176 billion to put one hydrogen pump at each of the filling stations in the US.
Fossil Energy Balance (higher is better)	3.2 units (soy) 4.3 units (rapeseed)	0.66 units (steam reforming of natural gas)
Large scale fuel development cost analysis	For an estimated $169* billion, enough algae farms could be built to completely replace petroleum transportation fuels with biodiesel	To produce enough clean hydrogen for our transportation needs would cost $2.5 trillion (wind power) or $25 trillion (solar)
Safety	Flash point over 300° F (considered "not flammable")	Highly flammable, high pressure storage tanks pose a large risk due to stored mechanical energy, as well as flammability explosiveness
Time scale for wide scale use	5-15 years	30-70 years optimistic assumption
Cost of engines	Comparable to existing vehicles	Currently 50-100 times as expensive as existing engines
Tank capacity required for 1,000 mile range in conventional sedan	20 gallons	268 gallons

XII. BLENDS

Fig.1 shows Pure Biodiesel (B-100) from Soyabean. The blends of biodiesel and conventional hydrocarbon-based diesel are products most commonly distributed for use in the retail diesel fuel marketplace. An important distinction needs to be made between biodiesel and biodiesel blends. Biodiesel is commonly mixed with diesel No. 2 to form a biodiesel blend.

As stated above, a mixture of biodiesel and diesel is not biodiesel, but is referred to as a biodiesel blend. Pure biodiesel, also known as neat biodiesel, is commonly noted as B100, indicating that the fuel has 100 percent biodiesel (noted by the 100) and 0 percent diesel. Blends of 20 percent biodiesel with 80 percent petroleum diesel (B20) can generally be used in unmodified diesel engines. Biodiesel can also be used in its pure form (B100), but may require certain engine modifications to avoid maintenance and performance problems.

Fig.1 Pure Biodiesel (B-100) from Soyabean

XIII. APPLICATIONS

Biodiesel can be used in pure form (B100) or may be blended with petroleum diesel at any concentration in most injection pump diesel engines. New extreme high pressure (29,000 psi) common rail engines have strict factory limits of B5 or B20 depending on manufacturer. Biodiesel has different solvent properties than petrodiesel, and will degrade natural rubber gaskets and hoses in vehicles (mostly vehicles manufactured before 1992), although these tend to wear out naturally and most likely will have already been replaced with FKM, which is nonreactive to biodiesel. Biodiesel has been known to break down deposits of residue in the fuel lines where petrodiesel has been used. As a result, fuel filters may become clogged with particulates if a quick transition to pure biodiesel is made. Therefore, it is recommended to change the fuel filters on engines and heaters shortly after first switching to a biodiesel blend. There are many success stories about biodiesel that they can be used for vehicles on roadways, railways and airways.

XIV. INDIAN SCENARIO

Biodiesel is meant to be used in standard diesel engines and is thus distinct from the vegetable and waste oils used to fuel converted diesel engines. Biodiesel can be used alone, or blended with petro-diesel. Biodiesel can also be used as a low carbon alternative to heating oil. Table 3 shows the projected demand for petrol, fossil diesel and biodiesel as proposed by Planning Commission of Government of India in the year 2003.

Table 3 Projected demand for petrol, fossil diesel and biodiesel requirements

Year	Petrol Demand Mt	Ethanol blending requirement (in metric ton)			Diesel Demand Mt	Biodiesel blending requirement (in metric ton)		
		@5%	@10%	@20%		@5%	@10%	@20%
2006-07	10.07	0.50	1.01	2.01	52.32	2.62	5.23	10.46
2011-12	11.85	0.64	1.29	2.57	66.91	3.35	6.69	13.38
2016-17	16.40	0.82	1.64	3.28	83.58	4.18	8.36	16.72

Source: Planning commission Govt. of India, 2003

India has a vast untapped potential of non-edible oil-bearing plant species. According to a survey conducted in 2002, twelve most important and abundant non-edible oil-bearing trees produce as much as 97 lakh tones of seeds per year, of which only 12% is utilized. These species of trees are Neem, Mahua, Undi, Jatropha, Castor, Kusum, Pilu, Dhupa, Nahor, Kokum and Sal. They (and by-products from their processing) are increasingly being used in modern industry for cosmetics, varnishes and paints, lubricants, resins, adhesive, dyes and inks, explosives, cellophane, pesticides, pharmaceuticals, etc. recently their potential as a biofuel has been investigated and hence they can be utilized as a feedstock for bio diesel production. Out of these, the primary focus of our project is on Jatropha (Ratanjot in Hindi).

Jatropha is a very sturdy forest shrub/tree having life of about 50 years. It is resistant to high winds and drought but is susceptible to freezing temperatures below 1 °C; the seeds contain average of 35% oil and yield about 75% cake after cold pressing of the seeds in screw type oil expeller. 1 Kg. of oil will be available from 4 kg of seeds. The plant after 2-3 years can on an average produce 2 kg of seeds per plant. This output will increase after the maturity and development of the tree. Such 1000 trees can be planted in one acre of land. This being a sturdy forest tree we can use waste lands. The tree is rain fed plant and does not require more watering after the first two the three years. It also does not require any special care as regards to fertilizers or pesticides. Thus this tree is selected considering optimum earning/acre. The plucking of pods and getting the seeds will provide excellent opportunities for rural employment. Thus, this opens up another major objective of the project. After getting 25 to 30% of the oil from the seeds, balance yield is 70 to 75% of cakes. This is a major by product of bio diesel production. Presently the cost of biofuel seeds as decided by Chhatisgarh Government is as follows.

Jatropa = 550 / Quintal; Karanji = 500 / Quintal.

India is having vast wastelands available for forest plantation. By proper propagation and cultivation of forest trees like Jatropha and utilizing the same for producing bio diesel, our nation can change this backward and poor area into very prosperous and developing areas within a decade time. India's biodiesel processing capacity is estimated at 600,000 tons per year. The government is now likely to fix a price of Rs. 34/- a liter for purchase of biodiesel by oil marketing companies.

XV. CONCLUSION AND DISCUSSION

Increased utilization of biodiesel could have a wide range of impacts on the national economy. For the agricultural sector, biodiesel would provide a new market for their products. As biodiesel fuel may cost more than conventional diesel, there could also be adverse economic impacts if costs are passed on in a variety of consumer goods. Government of India started BioFuel mission in 2003, but Indian BioFuel Policy was finally announced on 23rd Dec 2009. The Union Cabinet in its meeting gave its approval for the National Policy on Biofuel prepared by the Ministry of New and Renewable Energy, and also approved for setting up of an empowered National Biofuel Coordination Committee, headed by Prime Minister of India and a Biofuel Steering Committee headed by Cabinet Secretary. It gives a rough guideline, which was actually proposed many years back. Main stumbling blocks are still not resolved. There are no Figures or Financial commitments.

To set up a biodiesel plant, one should have own Jatropha plantation. Oil is currently not easily available. To set up 1,000 liters per day plant, you need a plantation in 200 hectares. Once you have a source of oil, you can set up a BioDiesel plant. Such a large patch of land is not available in India as one piece. Hence the plants have to collect seeds from a number of farmers.

The government needs to take confidence-building measures and clearly formulate its policy and explain to farmers that their role is vitally important in the success of the biodiesel program. Financial assistance should be given to NGOs in developing a large-scale awareness/training program for farmers. The government should arrange tours for reputable NGOs and progressive farmers to other countries/States to enable them to witness the success of biodiesel production first-hand. The National Biodiesel Program of

India is facing the challenges in reference to unavailability of sufficient validated data on agronomic practices; yield potential and water requirement of biofuel crops. There is a need to enhance production and supply of *Jatropha* and *Pogamia* and *other non-edible* oils to be used as biodiesel. The proposed strategy is to bring vast areas of degraded and low quality is to bring vast areas of degraded and low quality lands under biodiesel plantation.

XVI. REFERENCES

[1] *Amaral W. A. N. do and Pezzo C., "Mapping a Course for Biofuels: Science-informed Policy Needed, Global Change News Letter No. 70 December 2007.*

[2] *Gonsalves Joseph B., "An Assessment of the Biofuels Industry in India", United Nation's Conference on Trade and Development, 18 October 2006, Geneva.*

[3] *Raju K. U., "Biofuel in Soth Assian countries: an overview", Asian Biotechnology and Development Review, Vol. 8, No. 2, PP 1-9 (2006).*

[4] *Planning Commission, Government of India, "Report of the Committee on Development of Biofuel", 16 April 2003.*

[5] *Christopher Strong, Charlie Erickson and Deepak Shukla, "Evaluation of Biodiesel Fuel: Literature Review, January 2004.*

[6] *Michael S. Briggs, Joseph Pearson, Ihab H Farag, "Incorporating Lessons on Biodiesel into the Science Classroom", Presentation at the NH Science Teacher Association (NHSTA) Annual Conference, Session 15, March 22, 2005, Philips Exeter Academy.*

[7] *http://www.un.org/esa/sustdev/documents*

[8] *http://www.mdt.mt.gov/research/docs*

[9] *http://www.climate_leader.org*

[10] *http://unfccc.int/kyoto_protocol/items*

[11] *http://www.ipcc.ch/pub/un/syreng*

[12] *http://en.wikipedia.org/wiki/Biodiesel*

BPR, Innovation and People Management: A Creative Approach

Raghu Kumar[1], Milind Kumar Sharma[2]
[1]Defense (Indian Armed Forces) Nagtalao, Jodhpur.
[2]Department of Production and Industrial Engineering M.B.M. Engineering College, J.N.V. University
Jodhpur, Rajasthan State, India, 342001

Abstract-The Business Process Reengineering (BPR) derives its existence from different disciplines, and four major areas can be identified as being subjected to rapid changes in BPR - organization, technology, strategy, and people - where a process view is used as common framework for considering these dimensions. The involvement of two dynamic entities viz. people and technology, in any organization, calls for a deliberate analysis of all interwoven factors. When such a study is being undertaken, it is well in order to also consider the effects of another emerging trend, innovation (and its management), on the processes presently in vogue. This study addresses effects of latest trends in (Innovation Management Techniques) IMTs in the backdrop of the BPR based initiatives vis-à-vis the interactions involving human and technical elements. While trying to address issues like innovation strategy which requires strategic planning, it tries to recognize the requirements to develop goals based on the organization and the processes. This is where an assessment of the organizational agility may be determined vis-à-vis contingency. In recent years, organizations have started realizing that investing in maintenance to limit breakdowns and operation's interruption due to equipment failure, could actually become a comparative advantage in today's increasingly competitive environment. Investing in maintenance implies here not better equipment or the infrastructure but the "people management" itself. It can be shown that by following the simple principles of the innovation management and involving the employees to believe in the innovative approaches to the modern day technocrat work culture, the BPR principles get invariably involved.

Keywords: Business Process Reengineering (BPR); Innovation management Techniques (IMTs); Operational Performance Indicators (OPIs);

I. INTRODUCTION

Our day to day activities indicate a plethora of technology which has been thrust upon a group of unsuspecting people who are basically not tech savvy unlike the small fraction of senior lot, who at least behave, if not fully tech savvy. Remaining minute portion of actual tech savvy personnel do not often get the type of reckoning they deserve as they are a rare commodity and utilized mercilessly for "running-the-show"! We can borrow from above words of the writer and thus develop a system which has a bunch of subsystems which are inherently self developing without external coercion. That is to say in other words have a system which has the capability to adapt to the changing scenario of technological advancement without much effort but with an inherent responsibility and vision of its own.

II. SCOPE

Rather than practicing management of the people and the associated infrastructure and environment which tend to cause those uncontrolled number of situations to develop, the art of management need to look towards management "for the people" and "for the facility". This dictates that the existing management principles need to be scrutinized in the backdrop of the prevailing practices in the organization. Thus there is a need to develop beyond the self imposed constraints in people management, the shackles, which are based on the covenants developed in erstwhile non industrialized era of 18th century.

The word change may not produce as desired an effect on any unsuspecting individual as much as it may on someone who is engrossed with the sole aim of radical redesign of the organizational processes. At the same time it may cause some sleepless nights and nightmares to people who are working in industries 24X7 owing to the new dimension being offered with a high content of 'fear of unknown'. The Business Process Reengineering (BPR) derives its existence from different disciplines, and four major areas can be identified as being subjected to rapid changes in BPR - organization, technology, strategy, and people - where a process view is used as common framework for considering these dimensions. The reengineering is about achieving a significant improvement in processes so that contemporary customer requirements of quality, speed, innovation, customization, and services are met. A well-known BPR theorist, uses the term process innovation, which he says "encompasses the envisioning of new work strategies, the actual process design activity, and the implementation of the change in all its complex technological, human, and organizational dimensions" (Thomas H. Davenport 1993).

Whenever the feedback on processes which make up the existence of an organization is sought, be it financial / technical / human / others, striking feature of the feedback would be that of a well oiled and sound mechanism functioning with highest efficiency seemingly without any scope for improvement. But a gradual and deliberate analysis would definitely invoke the process of ideation and discover areas for improvement, though not necessarily resulting in any immediate enhancement to the throughput but definitely contributing towards an overall upliftment of the moods of change aspiring individuals.

Innovation Management

Unlike quality, which has seen widely used terms and methods rise to the status of international standards, such as the ISO Standards, alongside a substantial awareness-raising campaign among clients and suppliers alike, in the case of innovation these factors have not yet been sufficiently established. Thus the introduction of the phrase 'Change Management' at this juncture is significant. The observed practices reveal that irrespective of the advancement of the technology the ability to excel is always limited to the percentage of involvement of the people. However the principles, methods and techniques/tools for effectively assessing the potential value of a technology and its contribution to company's competitiveness and profitability needs to be evaluated before accepting an innovative process. The "vision of success" has to be created at the earliest to mend for past follies of not having committed to innovate scientifically.

The better management technique thus imply not just extracting work out of people most profitably; But it connotes how not to degrade the workforce in doing so! While our excursion points to a triad among the 5 M's- **Man**, **Machine**, Money, Media and **Management**-we find it, unsurprisingly, that there is a discord. The real technique lies in maintaining the present workforce and ensuring that they keep performing at highest levels irrespective of the odds, failing which the organization would face with the daunting task of training new workforce ever so often. This jeopardizes the safety and reliability of the system apart from breaking the cohesion which is ethereal and runs as an undercurrent within an organization. Therefore at this juncture we start following the principles of BPR. It may also be required to thoroughly evaluate the technology and its value from technical, market and consumer perspectives and reconcile the results within a valid methodology.

III. INNOVATION PRINCIPLES AND HUMAN ELEMENT

Integration of the relevant technology with the correct strategy and people to promote creativity and entrepreneurship at the local level is the need of the hour. This requires role to be established for influencing innovation management and the strategy development. In this context we can examine some basic principles of the innovation management which are:

Innovation as a Strategy

Strategy development refers to the organization's approach to the perceived changes on technological frontiers. The development of innovation faculty as a strategy in itself involves identification of basic areas required to develop great ideas which will propel the organization ahead in the future.

Specific Management Practices to Support Innovation

The five common elements of any management system can be listed out to form major heads as under:

Policy
A policy guideline is an essential backbone of any system, and more so when such a system is expected to run in a semi-independent manner, without frequent interferences from the top management.

Planning
The basic ingredients of planning for innovation would simply consist of taking a wide look around

at what's going on outside the organization and how it might effect the organization.

Implementation and Operation

Before we start operating on innovation management it is apparent that the basic definition of *innovation* be understood by one and all in the organization in the contextual sense so that there is no discrepancy and confusion in the orders that follow later or the feedback which goes to the higher echelon.

Monitoring and Measurement of Performance

It may be described as process of applying technical and administrative direction and surveillance to each constituent of a system (APIC 2006). This is required to ensure the following:

- Identify & document the physical and functional characteristics of each constituent.

- Control changes there to viz. to characteristics of the constituent.

- Record and implement the changes.

- Verify compliance to specified requirements (Base Lines).

- *Improvement* - This in the corporate parlance is also known as incremental innovation.
- *Management Review* - The review consists of the reporting of all current and historical data concerned with each entity throughout its life cycle.

Development of a "vision of success"

This provides a common set of goals with the vision of the industry in focus. Innovation is certainly occurring globally and the earlier this fact is understood the better it is for any organization.

Nurturing an entrepreneurial culture

Involvement of everyone in the organization is most important when planning to develop a culture to innovate. This should invariably involve recognition of the efforts of each individual and the prompt rewards too should follow. There also need to be an efficiency bar test required to be achieved by these individuals periodically to remain focused in their work.

IV. OPERATIONAL PERFORMANCE INDICATORS (OPIS)

While examining the above guidelines for the sake of implementing innovation strategies and developing innovation management we can see that an obvious offshoot would be a model for assessment / evaluation of the performance criteria in the form of OPIs. The performance measurement

in classic sense would involve, broadly, financial and operational measurements (Operations Management, Tata McGraw Hill 2006). However these general performance criteria could be developed to involve the following principles (APIC 2006) which invariably transpire to contain human elements:

- The overall improvement in customer satisfaction perceived vis-à-vis efforts undertaken (in terms of market share / customer retention) and interpretation of market share changes.

- Improvement trends in key operational performance indicators such as productivity, cycle time, waste reduction, new product introduction and defect levels.

- Financial benefits derived from improvements in employee safety, absenteeism and turnover.

- Benefits and costs associated with the education and training of the workforce.

- The relationship between knowledge management and innovation.

- Cost and revenue implications of employee related problems and effective problem resolution.

- Individual or aggregate measures of productivity and quality relative to competitor's performance.

- Relationships among product and service quality, operational performance indicators, and overall financial performance trends as reflected in indicators such as operating costs, revenues, asset utilization, and value added per employee.

- Allocation of resources among alternative improvement projects based on cost/benefit implications or environmental and community impact.

- Net earnings derived from quality, operational, and human resource performance improvements.

- Contributions of improvement activities to cash flow, working capital use,a nd shareholder value.

- Profit impacts of customer retention.

- Cost and revenue implications of new market entry, including global market entry or expansion.

- Cost and revenue, customer, and productivity implications of engaging in or expanding e-commerce or e-business and use of the internets and intranets.

- Market share versus profits.

- Trends in economic and market indicators of value and the impact of these trends in organizational sustainability.

V. INTEGRATING INNOVATION, TECHNOLOGY AND PEOPLE – BPR WAY

It can be seen that by using the aforesaid performance criteria it becomes easier to start following the BPR guidelines. The ability provided by the detailed performance measurement done as above will entail managers to establish the outcome based plans for specialized tasks which were hitherto performed by individual groups or entirely individuals. This gradually improves the quality and throughput of small sections or groups of workers which are now better equipped at handling challenges in terms of varying customer requirements compared to these groups which handled small tasks earlier. For example the responsibility for generating, planning and estimating, and authorizing work should be separate from the responsibility for performing work. Similarly, it is preferable for the Quality Assurance functions to be the responsibility of an autonomous organization, apart from those ordering and performing the work. This provides the system with checks and balances and free of the appearance of conflict of interest.

Parallel linking of processes

These small units of excellence should also be capable of handling external information and reacting suitably to the stimulus thus received, thereby reducing errors caused by entirely alien group handling related information and resulting in errors due to multiple contact points (Operations Management, Tata McGraw Hill 2006). Further the linking of the parallel activities should be done instead of linking their results which will invariably cause delays in the output. Work planning for each process thus should provide a sufficient volume of work sufficiently in advance of the required completion date to permit balancing the facilities maintenance workload among different processes, acquiring material, arranging timely contract support, achieving priorities, and coordinating all the elements.

Geographical diversity

Rapid advances in some 'infrastructural' technological areas such as information and communication technology, biotechnology, and advanced materials penetrate into the whole industry. Globalization also denotes the intensifying competition and increasing cooperation among countries. As the world economy has been increasingly integrated, on the one hand, it has increasingly intensified competition among countries more than ever; it has also increasingly created opportunities of new forms of cooperation among countries and/or economic agents, on the other hand. In turn, diffusion of new ideas, knowledge, and technologies has been further encouraged through these intensified competition and cooperation. Therefore a requirement exists for facilities maintenance managers to use modern maintenance systems and methods to control their work activities, account for resources they are provided and to monitor and report work execution through the full use of various industry standard metrics and other management indicators irrespective highly dispersed geographical location.

Building control into the processes

The measurement, analysis, and knowledge management would be the criteria for all key information about effectively measuring, analyzing and reviewing performance. BPR requires that the goals and expectations are clearly established, results monitored and fed back to the employees. Knowledge management at lower levels become highly difficult due to the gap between the nature of information required and that is provided. The knowledge can be explicit and tacit knowledge; The mixing of each other will prove the entire exercise of collection and collation totally futile. Thus ascertaining the right information being fed will go a long way in achieving desired process control.

VI. SALIENT ASPECTS OF THE STUDY THAT GARNER ATTENTION

It is said 'If everything is under control you are just not driving fast enough'! There is after all no way of knowing the judicious mixture of all components most suitable for any given environment in any specific industry. Though everyone understands the urgency in implementation of the factors being discussed, the changes to be brought about should be deliberate and gradual to avoid resentment among the workforce. However deeper knowledge of the various interacting features of all these independent tools will definitely provide a new and innovative way to handle the industry forces in all circumstances.

VII. IMPLICATIONS FOR FUTURE RESEARCH

The aforesaid tools and features discussed are just an attempt in understanding the interactions of the modern tools presently available to any manager in the industry. However the undercurrents developed due to the adoption of various tools in any industry are left alone for the fear of disturbing or attracting wrath of senior management officials. The outcome of such an implementation is generally left unattended and gradually dies its death over a period of time. This actually raises a question mark on the efficacy of the tool thus adopted, which was successful elsewhere.

Facts about implementation of such techniques need to be studied in depth for better understanding of all interactions when associated with human elements. The same need to be documented with additional care for future references and enhancements as and when required by the academicians and industry experts.

Biographical notes: Raghu Kumar is presently pursuing higher studies after finishing M.Tech in Aerospace Engineering from IIT, Madras. He is presently working for the Armed Forces.

Milind Kumar Sharma is Associate Professor in the Department of Production and Industrial Engineering at M.B.M. Engineering College, Faculty of Engineering, J.N.V. University, Jodhpur, Rajasthan State, India.

VIII. REFERENCES

[1] *Army Performance Improvement Criteria. (2006) US Army.*

[2] *Richard, B, Chase., F, Robert, Jacobs., Nicholas, J, Aquilano. and Nitin, K, Agarwal. (2006) [3]'Operations Management for Competitive Advantage', TATA MCGRAW – HILL, Eleventh Edition, pp 418-419, pp 769-770.*

[4] *Eleni Sefertzi, (2000) 'Creativity', Report produced for the EC funded project INNOREGIO: dissemination of innovation and knowledge management techniques.*

[5] *Simon Rogerson, (1996) 'Business Process Re-engineering', Originally published as ETHIcol in the IMIS Journal Volume 6 No 2.*

[6] *Davenport, Thomas (1993), Process Innovation: Reengineering work through information technology, Harvard Business School Press, Boston.*

[7] *Yannis, B. Vassilis, K. 'Project Management', Report produced for the EC funded project INNOREGIO: dissemination of innovation and knowledge management techniques.*

Design Features and Limitations of Electromagnetic Regenerative Shock Absorbers

[1]**Nitin V. Satpute**, [2]**Shankar Singh**, [3]**L.M. Jugulkar**

[1]Sant Longowal Institute of Engineering & Technology (SLIET), Longowal, Distt. Sangrur, Punjab.
[2]Department of Mechanical Engg.(SLiET), Longowal, Distt. Sangrur, Punjab.
[3]Rajarambapu Institute of Technology, Rajaramnagar, Sangli, Maharashtra.

Abstract—**The vehicle is subjected to constant source of vibrations due to uneven road profile, braking and corner turning. Conventional shock absorbers are designed for minimizing acceleration transmitted to the occupants and for minimizing vertical lift of tyre from ground. Study identifies that significant amount of energy is lost in Shock Absorber due to the above motions. The paper identifies various means for harvesting the energy lost in Shock Absorbers, their relative merits and demerits have been identified.**

Three types of regeneration means have been discussed these include electromagnetic damping by using linear generator, using air compressor for converting vibration energy into pressure energy and using fluid compressor for converting vibration energy into the fluid pressure energy Reverse engineering (R.E.) of mechanical component is reconstruction of the component based on geometrical information which is captured by using appropriate 3 D digitizing technique, sufficient to replicate the component by using appropriate manufacturing techniques.

I. INTRODUCTION

Nowadays automobile manufacturers are striving for reducing fuel consumption. In doing so, the energy lost in the automobile Shock Absorbers have not attracted many researchers. Even though study identifies that significant amount of energy is lost in the Shock Absorbers. Conventionally the hydraulic Shock Absorbers consist of piston and cylinder arrangement, in this type of construction vibration energy of the Automobile is converted in to heat through fluid friction. This type of construction generally provide non variable damping factor. Lee et al. [18] have described a passive Shock Absorber of which damping factor varies with respect to the stroke of the piston. More recently the automobile manufacturers are using Semi-active and Active Shock Absorbers that vary the damping factor in response to vibration velocity (in case of Semiactive suspension) and apply the active force between wheel and vehicle body for attenuating the vibrations (in case of active suspension). These two types of Shock Absorber give better performance than that of passive one, but it comes at higher cost.

The shock Absorbers are designed for vehicle safety criteria and comfort criteria [2]. In case of these at lower vibration velocities, acceleration transmitted to the occupants is minimized (comfort criteria), whereas at higher vibration velocities, displacement of the wheel from ground is minimized (comfort criteria).

Since vehicle safety and comfort criteria are conflicting optimization needs to be carried out for deciding parameters of the Shock Absorber Till date much development has been done for optimizing passive, active and semi active Shock Absorber, but no much effort has been taken on energy harvesting from Shock Absorbers. The present paper will discuss reasons behind this. An automobile while moving on the road comes across vertical acceleration of up to 1.2 m/sec2, in case of abrupt bump, this value will further increase. Vertical velocity ranges from 0 to 0.5 m/sec in normal running.

This means that maximum damping force required to attenuate this vibration is 1200N for a car. For a heavy vehicle like truck this force further increases. Also since vibration intensity for automobile is not constant, the Shock Absorber has to provide variable damping intensity to meet safety and comfort criteria.

 Not many researchers across the globe have tried to develop any regenerative damping means that will satisfy the above requirements.

Zen et al. (2010) have designed an electromagnetic Shock Absorber for a vehicle weighing 1.3 ton [4]. The constructed Shock Absorber in the present study is a linear generator. This linear generator has two columns of permanent magnets, connected with pole pieces. An air gap of 5 mm is formed in between the pole pieces with the air gap flux density of 2.7 T. Damping force offered by the Shock Absorber is maximum when voltage generated by the coil is maximum. In the presented work ANSYS static FEA analysis is used to decide optimal arrangement of the permanent magnet and pole pieces.

The authors have calculated that maximum damping force of 1145 N can be developed by the presented Shock Absorbers, when the vehicle is travelling on a road with the speed of 20Km/Hr. The calculations done by the authors show that total current in the coil will be 60 Amperes and generated Electric Power will be of 4158 W.

Detailed mathematical model have not been presented by the authors in the paper to analyze the Shock Absorber performance. For an automobile high value of damping force has to be provided by the Shock Absorber in case of extreme bump or pothole, the Shock Absorber presented in the literature will not be able to provide this much high damping force.

Zador et al. have suggested an Active Shock Absorber that is using the energy harvested form the vibrations for applying active force between wheel and vehicle body [5]. The presented system is designed for electric vehicle, which has much lower mass, and runs with lower velocities as compared to vehicles running on fossil fuels. Variable in intensity damping is not achievable with the suggested Shock Absorber. Also higher value of damping force will not be provided by the Shock Absorber as in case of bump or pothole.

Nakano et al. have suggested a similar Regenerative – Active Shock Absorber that uses DC motor for harvesting energy of the Shock Absorber [6]. This DC motor is used for converting vibration energy in to electric energy, also for applying active force between wheel and vehicle body part of the energy earlier harvested is used. This Shock Absorber suffers from same disadvantages as of the earlier one.

Gupta et al. has noticed that used of linear generator for energy harvesting has some limitations [10]. Since maximum stroke available for energy harvesting is limited to working space of the Shock Absorber, efficiency if the linear generator is limited.The authors have suggested use of DC motor for electric energy generation rather than that of linear generator. An arrangement has been done to have maximum number of rotations of the DC motor for given stroke length of the Shock Absorber. The authors have noticed significant amount of increase in efficiency of the Shock Absorber because of this arrangement.

Goldner et al. (2001) presents preliminary study of energy recovery form vehicle Shock Absorbers [9]. As vehicle comes across road irregularity, the sprung mass vibrates with some velocity. In the presented work kinetic energy of vibration is calculated based on this velocity of vibration. This energy is lost in the form of heat in conventional shock absorber. Calculations for kinetic energy lost are done for various sizes of speed bumps.

In the presented paper the authors have used permanent magnet and electromagnetic coil moving relative to each other at 20 Hz frequency. The authors have verified theoretical and experimental values of generated voltage. The authors have concluded that significant amount of energy can be generated from the Shock Absorbers. A group of students from MIT, USA have developed a new Shock Absorber that uses a hydraulic pump inside the shock Absorber. This pump harvests the energy of vibrations from the Shock Absorber and uses the same pressurizing the fluid. Detailed technical report for this is not available.

Recently Bose Electric Company have claimed to have invented an Electric Shock Absorber that converts vibration energy in to useful electricity. Detailed technical report for this is not available.

Designing Linear Generator for Shock Absorbers

Linear generator is designed with following parameters:

Maximum acceleratotion at Un sprung mass: 1.5 m/sec2, Maximum velocity = 0.5m/sec, Air gap in the Generator: 8mm, Air gap magnetic flux density: 0.8 T, Average coil ida. Of electromagnetic coil: 150mm, Number of turns: 70, Number of permanent magnets and electromagnet sets: 10,

Emf induced in each coil is calculated as:

$$E(t) = B_{gPM} * U(t) * \pi * D_{avg} * 2p * n_c * \left\{ \frac{l_{PM}}{l_{PM} + l_{stroke}} \right\}$$

B_{gPM} : Air gap flux density (0.8 T)

$U(t)$: Relative velocity between conductor and Magnet

D_{avg} : Average coil dia (150mm)

P   : number of poles (1)

n_c : Number of turns (70)

l_{PM} : Length of Permanent Magnet (25)

l_{stroke} : Stroke length of Coil relative to the magnet (20 mm)

n_c : Product of number of Turns and Current (Amp-Turns)

The winding can be arranged such that total emf will increase 10 fold times, since there are 10 permanent magnets and 10 coils. Current induced in the coil depends on internal inductance and resistance, which are given as:

$$L_y \approx \frac{1}{8} * 2p * \mu_0 * n_c^2 * \pi * D_{avg} * \frac{l_{PM} + l_{stroke}}{h_{PM} + g + h_{coil}}$$

$$R_s = \rho_\omega * \pi * D_{avg} * \frac{n_c^2 * 2p}{\underset{\text{coil}}{I_n * n_c'}}$$

g=2 mm, for the given coil configuration, Internal Resistance Rs= Ohms, Inductance Ls= Henry

It is assumed that capacitor is used to nullify the effect of inductance. Current induced in each coil is calculated using the following formula:

EMF = Current X Resistance

E = I x R

Damping effect in the Shock Absorber will take place only when the generated EMF is utilized for some purpose. If the generated emf is not used for any purpose there will be no damping (Braking) effect. If the generator is to be used for charging the battery, the generated voltage has to be higher than battery threshold voltage. If the voltage is higher than battery threshold voltage, damping effect is possible. If the generated voltage is lesser than the battery threshold voltage, no damping effect is produced. Assuming that the generated voltage is used, Damping force is computed as:

$$F_e(t) = B_{gPM} *$$

$$n_c * \left\{ \frac{l_{PM}}{l_{PM} + l_{stroke}} \right\} * \pi * D_{avg} * 2p * l(t)$$

Since this is the force on each conductor, total damping force on 10 conductors is 10 times the above force.

Energy loss in Copper wires is calculated as:

$$P_{coil} = R_s * I_m^2$$

E loss Input energy to the generator (E1) is the kinetic energy of Sprung mass

E1 = (1/2)* (vehicle mass)*(vertical velocity ^2)

Energy generated by the generator

E2 = (E*I) – (Energy loss in Copper wire)

Efficiency of the generator = (E2 - Eloss) / (E1)

For constructing Linear Generator for Shock Absorber, Permanent magnets are mounted in stationary part and the coils are connected to the movable part. This type of construction is preferred to avoid damage to the magnet due to continuous vibrations. Efficiency and maximum damping force produced by the generator depends on air gap flux density. Using the permanent magnets of rare earth metals will insure higher flux density in the air gap. Primary winding should be kept shorter to reduce cost of the construction. Also primary part should be laminated to avoid eddy current losses.

Permanent magnets are used along with pole pieces to ensure proper circulation of the magnetic flux. Static FEA can be used for deciding optimal configuration of the magnets and pole pieces. This will ensure higher magnetic flux density in the air gap of the linear generator. Transient FEA can be performed to decide optimal configuration of the coil for maximizing efficiency of the linear generator.

Figure 1 shows construction of linear generator for use in Shock Absorbers. Figure 2 shows quarter car model used for analyzing the Shock Absorber. Initial velocity is given at 'x', the performance of the Shock Absorber is decided by computing acceleration at the sprung mass and displacement of the unsprung mass.

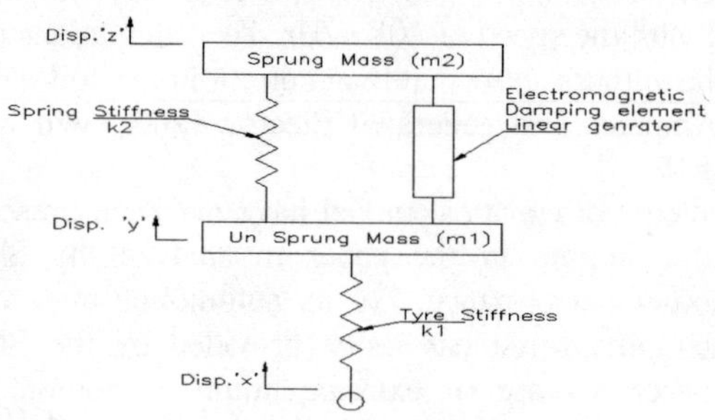

Figure1: Quarter Car Model

Differential equation for the Quarter car model are written as,

$$m_1 \ddot{y} = -k_2 (y-z) + k_1 (x-y) - F_D$$

$$m_2 \ddot{z} = k_2 (y-z) - F_D$$

Where ' FD ' is the electromagnetic damping force.

Figure 2 Shows MAT LAB Simulink model constructed for solving the above differential equation. Results of the above simulation are as shown in table 1.

Sr.No.	Velocity 'x dot' (m/sec)	EMF E (Volts)	Generator Efficiency (%)	Acceleration at sprung mass (m/sec2)	Displacement of tyre (mm)
1	0.1	2.8	0	0.2	5
2	0.3	8.63	0	0.8	8
3	0.5	14.38	2.58	1.1	18
4	0.8	23.02	12	3.3	30
5	1.0	28.77	14	4.7	46

Table 1 shows EMF, Current and Damping force generated by the linear generator for various Vibration velocities with the given coil configuration

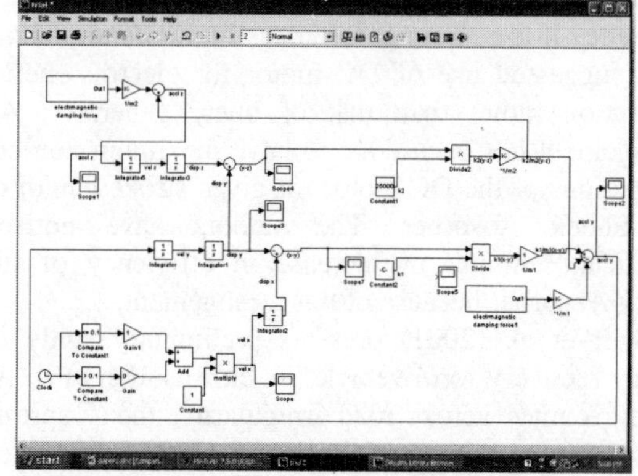

Figure 2: Simulink Model

II. CONCLUSIONS

1. At lower suspension velocities, lower emf is generated; hence damping force offered is very less. Since the generated emf is too low, it will not exceed battery threshold voltage, and hence is not useful. For creating damping effect with the lower suspension velocities, the generated emf has to be passed through external resistance.

2. The Linear Generator will perform better only at design velocity (in this case 0.5 m/sec), at this velocity the acceleration is not very high, displacement of tyre is also limited and sufficient emf is generated that will overcome battery threshold voltage.

3. At higher velocities of suspension, emf generated increases, but at the same times both acceleration at sprung mass and tyre deflection increases. Hence it can be concluded that maximum damping force generated by this Shock Absorber is limited.

4. At higher suspension velocities it is necessary to generate higher damping force, in absence of which the wheels may get lifted off the ground, giving rise to security concern.

5. To improve performance of this Shock Absorber, additional damping means may be incorporated. This damping means should be effective only at higher suspension velocities, so that the requirement of higher damping force can be met.

III. BIBLOGRAPHY

[1] John C. Dixon , "The Shock Absorber Handbook", John Wiley and Sons, 2007.

[2] Dipl.-Ing. Jörnsen Reimpell, Dipl.-Ing. Helmut Stoll, Prof. Dr.-Ing. Jürgen W. Betzler, " The Automotive Chassis Engineering Principles", Reed Elsevier and Professional Publishing Ltd, 2001.

[3] Emanuele Guglielmino ,Tudor Sireteanu,Charles W. Stammers , Gheorghe Ghita, Marius Giuclea, "Semi-active Suspension Control", Springer, 2008.

[4] Longxin Zhen and Xiaogang Wei, "Structure and Magnetic Field Analysis of Regenerative Electromagnetic Shock Absorber" (Paper No.11529550), IEEEWASE International Conference on Information Engineering, Beidaihe- Hebei, 2010.

[5] Istvan Zador, "Rear earth and high temperature superconducting permanent magnet synchronous tube motor/generator optimization for the components of the car suspension system", Ph. D. Thesis, Budapest University of technology and economics, Budapest, (2008).

[6] Nakanoa, K., Sudab,Y. and Nakadai. S, "Self powered Active Vibration Control using a Single Electric Actuator", Journal of Sound and Vibration, (2003).

[7] Lei Zuo, Brian Scully, Jurgen Shestani and Yu Zhou,"Design and characterization of an electromagnetic energy harvester for vehicle suspensions", IOP Publishing Smart materials and structures, (2010).

[8] Morteza Montazeri-Gh and Mahdi Soleymani, "Active suspension system in parallel hybrid electrical vehicles", IUST International Journal of Engineering, (2008).

[9] R. B. Goldner and P. Zerigian, "Preliminary study of energy recovery in vehicle using magnetic shock absorber",SAE Technical Paper Series, Paper number: 2001-01-2071,(2001).

[10] Glenn R. Wendel and Gary L. Stecklein, "Regenerative active suspension on rough terrain vehicles",SAE Journal of Commercial vehicles, Paper number: 940984, (1994).

Impact of Green-House Gas on Environment during Growth and Use of Transesterified Karanja Oil

Ashok Yadav[1], Onkar Singh[2], Naveen Kumar[3]
[1]Mechanical Engineering Department, Sachdeva Institute of Technology, Mathura, U.P.
[2]Mechanical Engineering Department, H.B.T.I., Kanpur, U.P.
[3]Mechanical Engineering Department, Delhi College of Engineering, Delhi.

Abstract- **Increase in greenhouse gases is a cause of concern for our environment. This article reports the replacement of the CO_2 produced during combustion of diesel fuel with that of bio-diesel. Diesel fuel ($C_{16}H_{34}$) releases 3.11 kg of CO_2 per kg of fuel used whereas karanja bio-diesel ($C_{20}H_{38}O_2$) releases 2.83 kg of CO_2 per kg of fuel used for combustion. 4301 kg of bio-diesel can replace 3,737 kg of diesel fuel and would replace 11,621 kg of carbon dioxide from petroleum fuel with 12,172 kg of CO_2 with renewable karanja seed source. Karanja bio-diesel is a renewable source of energy with little less energy content (15%) than diesel but as the CO_2 emitted was fixed up by plants recently, this fuel would reduce the greenhouse gases from the environment.**

Keywords: greenhouse gases, bio-diesel, carbon sequestration, karanja seed, photosynthesis.

I .INTRODUCTION

Extensive industrialization has resulted in increase in the greenhouse gases in our environment. A report of Global Change Research Information Office, USA says, "The consensus of most scientists worldwide is that increasing concentrations of greenhouse gases (carbon dioxide and methane, for example) will lead to significant climate warming, shifts in precipitation patterns and rising sea levels, although the magnitude, timing, and regional patterns of these changes cannot be accurately predicted at this time" [1].

Gust has described the basic principles of photosynthesis which mainly uses atmospheric CO_2 [2]. Emil and Winnett carried out a detailed study about the effect of biomass for the emission and sequestration of atmospheric carbon [3]. Excess crop material (residue) can be used to improve the post harvest soil conditions. Field reclamation principles detail various chemical and physical processes occurring in the field after harvest, including the post-harvest soil conditions where carbon plays an important role [4, 5].

Studies have shown the present and possible future uses of residual crop biomass [5]. Carbon fixed by plant may take various pathways for its emission to the environment and sequestration. For example, the 86104biomass for furniture and building. Ellington and Meo made calculations for all greenhouse gases emitted from technical system like production and then use of ethanol. A system approach (total analysis of all outputs and inputs) was used to compute the total warming related to the total lifetime useful output of the system to yield an index of performance (greenhouse warming effect index) [6].

Several authors have suggested that increment of carbon dioxide in to the atmosphere as a result of engine combustion can be reduced by replacing petroleum diesel with transesterified vegetable oil [7]. Non-edible vegetable oils such as Karanja (*Pongamia pinnata*), Neem (*Azadirachta indica*) and Jatropha (*Jatropha curcas*) can be transesterified and used as fuel in compression ignition engine. These oilseed bearing trees fix carbon dioxide from the atmosphere through photosynthesis process. So, carbon released during combustion of these oils had already been fixed up in the near past.

Figure 1 shows the interdependence of three important agricultural cycles and various processes of each cycle in which matter take part from growth to decay of the plant. Respiration, photosynthesis and microbial decomposition are the processes of carbon life cycle demonstrating carbon exchange with environment and soil.

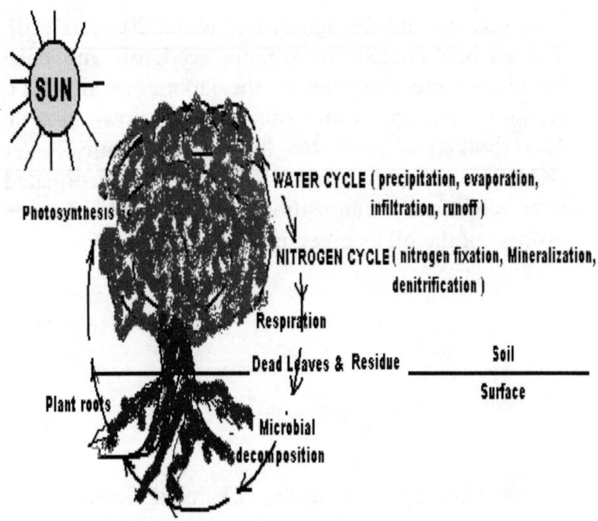

Figure 1: Interdependence of carbon, water and nitrogen cycles [5]

This study focuses on the process of fixing carbon for transesterified oil of karanja seed and effect of substitution of this oil for petroleum diesel.

Photosynthesis

Process of photosynthesis is responsible for production of energy which these plants can utilize. Carbon is incorporated into life forms through the basic process of photosynthesis which is performed in the presence of sunlight by all life forms that contain chlorophyll. This process converts carbon dioxide from the atmosphere or dissolved water in to glucose molecules. These glucose molecules are either converted into other substances or used to provide energy for the synthesis of other biologically important molecules. For the growth and development of the plant, the glucose in combination with nutrients is absorbed from the soil.

In terms of balanced chemical formula, photosynthesis equation can be written as:

$$6CO_2 + 12H_2O \xrightarrow{\text{sunlight energy}} C_6H_{12}O_6 + 6O_2 + 6H_2O$$

Above chemical equation of photosynthesis shows that whole of the carbon from atmospheric carbon dioxide is absorbed by plant which is a very important observation with regard to carbon cycle.

An evaluation of the carbon life cycle requires information about photosynthesis as well as the various processes required to convert the material in to useful form. During transesterification process, alcohol reacts with vegetable oil resulting in ester and

glycerol molecule formation. Since these alcohol and glycerin molecules contain carbon, transesterification directly affects the carbon life cycle. As seed cake leftover after oil extraction from the vegetable seed contains substantial amount of carbon, the end use of this cake for organic fertilizer, combustion or live feed stock, makes its entry into the carbon life cycle.

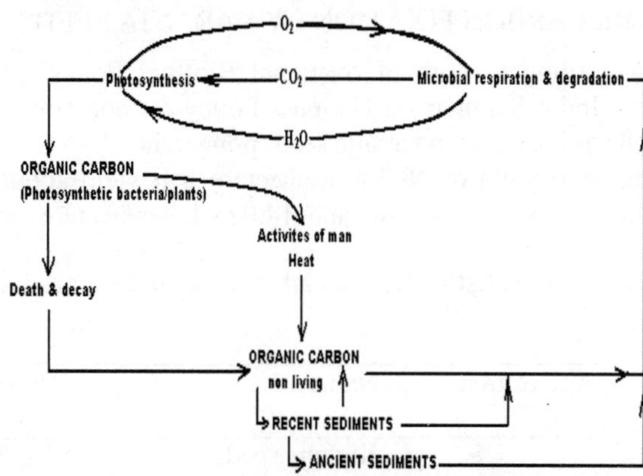

Figure 2: Micro-organisms contribution in carbon life cycle [8]

Figure 2 depicts that organic compounds such as emissions from internal combustion engine are biodegraded in various ways by micro-organisms. Bio-diesel (transesterified vegetable oil) production from karanja oil is a part of this cycle. The plant has fixed the carbon dioxide in recent or same year during its growth, which is released to the atmosphere by engine combustion. New crop of Karanja seeds for future bio-diesel production are fixing carbon during current growth cycle. Figure also explains the carbon dioxide production from petroleum fuels, as the carbon sequestered by the plants which were subjected to heat and pressure under earth crust created rich resources of petroleum. As this petroleum is utilized, the carbon dioxide is returned to the atmosphere.

Though all the carbon is fixed by plant from atmospheric carbon dioxide in photosynthesis process, yet not all of it stays within the plants, as the various studies reflect. In a study on rapeseed seedlings conducted by Shepherd & Davis, it was concluded that 17-19% of the fixed CO_2 was translocated to the roots over a period of two weeks [9]. Some part of this was released into atmosphere through respiration and some was released in to the soil. As a result, 30-40% of the total carbon translocated to the roots, remains in the roots. So, the carbon released into the soil would continue to degrade with the help of micro-organisms and might be released as CO_2 via biological activity. Therefore, it can be concluded that plant takes in more CO_2 than its accumulation in the plant biomass. Neem and Karanja would help in permanent sequestering of atmospheric CO_2, if all the wood obtained from these trees is used for non-combustion purposes like furniture, buildings, industries etc.

III. CARBON FIXATION BY KARANJA SEEDS

According to draft of 'National Bio-fuel Policy' at All India Seminar on National Policy on non-edible oils as bio-fuels, a karanja seed (pongamia pinnata) yield of 18.7 tones/hectare will yield about 4.675 tones/hectare oil and 14.025 tones/hectare of

cake considering the age of tree about 20 years [10]. The carbon content in karanja seed, oil and cake (residue) were analyzed in the laboratory at HBTI, Kanpur, India and results obtained are shown in table 1. Hypothetical formulae for karanja methyl ester (KME) and Karanja ethyl ester (KEE) were estimated from chemical composition and transesterification process of the oil as given in table 2.

Table1: Estimation of carbon content for a production of 18.7 tones/ha and 7.7 tones/ha of karanja seeds

Age of tree	Name	Production (kg/ha)	carbon content (%)	kg C/ha	kg CO_2/ha
20 years	Karanja seeds	18,700	0.6147	4653.8	42,147.9
	Karanja oil	4,675	0.8108	1534.6	13,898.4
	Karanja cake	14,025	0.4765	2705.6	24,504.0
	Karanja seed cover/pod	18,700	0.4223	3197.2	28,955.7
10 years	Karanja seeds	7,700	0.6147	1916.2	17,355.0
	Karanja oil	1,925	0.8108	631.9	5,722.9
	Karanja cake	5,775	0.4765	1114.1	10,089.8
	Karanja seed cover/pod	7,700	0.4223	1316.5	11,922.9

Table 2: Hypothetical formulae for the methyl and ethyl esters of karanja oil and heating values

Fuel	hypothetical Formulae	Molecular weight	Heating values (kJ/kg)
KME	$C_{19}H_{36}O_2$	296	36635
KEE	$C_{20}H_{38}O_2$	310	36680
Diesel	$C_{16}H_{34}$	226	42210

IV. RESULTS AND DISCUSSION

Processing

As a number of processing techniques are employed for conversion of karanja seeds in to bio-diesel, carbon life cycle becomes more complex. From oilseeds which contain about 61.5% carbon, oil is extracted initially, via mechanical, chemical or a combination of the two processes. The oil contains about 81% of carbon. Karanja cake (residue) obtained during oil extraction contains 47.6% of carbon and can be used as biomass fuel resource, organic fertilizer or feed for livestock.

The raw oil is filtered and refined through various techniques such as filtration, deguming etc. The oil is

transesterified to lower the viscosity and improve the cetane rating so that this fuel may be used in CI engine. Transesterification incorporates removal of glycerol molecule from triglyceride and addition of alcohol molecule. So, this process alters carbon content in transesterified oil and the byproduct.

Combustion

Fuel used in the CI engine to produce power, undergo combustion process. Petroleum fuel composed of carbon and hydrogen with approximately 86% carbon and 14% hydrogen by weight [11].

The combustion equation for diesel fuel ($C_{16}H_{34}$) may be written as:

$$C_{16}H_{34}+24.5O_2+92.12N \longrightarrow 16CO_2 + 17\ H_2O + 92.12\ N$$

This equation is obtained from hydrocarbon generalized combustion equation [11]:

$$C_nH_{an+b}+cO+3.76cN \longrightarrow dCO_2+eH_2O+3.76cN$$

a, b, c, d, e and n are the constants for a particular fuel.

Combustion equation for karanja ethyl ester (bio-diesel) when used as fuel in IC engine may take the following form:

$$C_{20}H_{38}O_2 + 28.5O_2 + 92.12N \longrightarrow 20CO_2 +19H_2O + 92.12N$$

Above combustion equation for diesel fuel show that 1 kg of diesel produces 3.11 kg of CO_2. Similarly, 1 kg of karanja ethyl ester when used as fuel in CI engine produces 2.83 kg of CO_2. So, diesel produces 9.9% more CO_2 per kg than bio-diesel.

In this theoretical combustion process, all the carbon content of bio-diesel is converted to CO_2 and nitrogen of air remains unaffected during the process. During actual combustion process, a part of the carbon is converted into CO, HC and aldehyde while some part of nitrogen is reacted to form NO_x. Exhaustive engine tests have shown that amount of carbon conversion into CO_2 is 99% and less than 1% in to other forms [11]. Though intermediate products (CO, HC etc) are undesirable in the environment, these do not produce any global warming effect and with time return to CO_2 through the action of microbes and various other degradation processes.

Diesel produces 9.9% more CO_2 per kg than does bio-diesel; but diesel has 15% more energy per kg than bio-diesel, requiring less fuel to do the same work. So, it can be fairly assumed that for per unit of work done, the amount of CO_2 produced would be nearly the same whether using diesel fuel or bio-diesel.

Reduction in CO_2 through use of bio-diesel

Production of CO_2 from bio-diesel and diesel can be compared on the two basis; first, replacing diesel fuel with bio-diesel and second, considering other factors such as production, processing and transportation with replacement of diesel with bio-diesel.

Using diesel fuel 9.9 % more CO_2 is produced and 15.1% more energy is available than bio-diesel which implies approximately the same amount of CO_2 would be produced for an equivalent energy content of bio-diesel. This difference is due to (a) the oxygen in the bio-diesel and (b) theoretical verses actual

process. Transesterification requires an addition of ethyl alcohol C_2H_5OH (mol. wt = 46) and removal of glycerin $C_3H_8O_3$ (mol. wt = 98). About 10% alcohol is reacted and the same amount by weight glycerin is removed, so the amount of ester produced is essentially equivalent to the oil used. Typically a hectare of karanja tree at an age of 20 years produces 4,301 kg of bio-diesel with a conversion efficiency of 92%. This bio-diesel can replace 3,737 kg of diesel fuel and would replace 11,621 kg of carbon dioxide from petroleum with 12,172 kg of CO_2 with renewable karanja seed source. Some of the carbon in bio-diesel may be fixed as free carbon and some is stored in less biodegradable forms. So, it is evident that renewable fuel removes more carbon dioxide than the simple exchange of fuel combustion equivalents would suggest.

A detailed exploration of the carbon dioxide emissions from bio-diesel compared with diesel fuel would require an analysis of the total system. As karanja bio-diesel is grown on agricultural land, it requires various inputs of energy for harvesting, tillage, crop transport, storage and labor. Processing requires energy to operate pumps, presses, production of storage vessels and labor. Researchers have estimated from 3 to 4 units of energy output for each unit of energy input for bio-diesel.

V. CONCLUSION

Energy to drive IC engine can be obtained through combustion process which in turn converts hydrocarbon fuel to carbon dioxide and water. Diesel fuel ($C_{16}H_{34}$) releases 3.11 kg of CO_2 per kg of fuel used whereas bio-diesel ($C_{20}H_{38}O_2$) releases 2.83 kg of CO_2 per kg of fuel used for combustion. A hectare of karanja tree of age 20 years produces 4,301 kg of bio-diesel with a conversion efficiency of 92%. This bio-diesel can replace 3,737 kg of diesel fuel and would replace 11,621 kg of carbon dioxide from petroleum with 12,172 kg of CO_2 with renewable karanja seed source. The carbon dioxide released by petroleum fuel increases the amount of CO_2 in environment while CO_2 released by bio-diesel was fixed by plants in a recent year and will be recycled by the forthcoming generation of crops.

VI. REFERENCES

[1]   Gibbsons JH (1995). *Our changing planet. Subcommittee on Global Change Research, Global Change and Research Program, 300D Street, S.W. Washington DC 20024.*

[2]   Gust D (1995). *Why study photosynthesis? ASU photosynthesis center. www information page.*

[3]   Emil JL, Winnett S (1995). Biomass and Global Climate Change: An overview. U.S. EPA Climate Change Division, Washington DC 20460.

[4]   Reicosky DC (1994). Crop residue management: soil, crop, climate interactions. Crop residue management, ed. JL Hatfield and BA Stewart. Lewis, Boca Raton, 1994, pp. 191-214.

[5]   Galinato G, Peppersack J, Taylor R (1987). Assessment of agricultural crop residue for energy recovery in Idaho. Idaho department of water resources, Boise, Idaho, December.

[6]   Ellington RT, Meo M (1992). Calculating the net greenhouse warming effect of renewable energy resources: methanol from biomass, Journal of Environmental Systems, 20, pp.287-301.

[7]   Sagar AD (1995) Automobiles and Global Warming: Alternative fuels and other options for carbon dioxide emission reduction. Environment Impact Assessment Review, Vol. 15. Elsevier Publishing, pp.241-274.

[8]   Cripps RE, Watkinson RJ (1978). Polycyclic aromatic hydrocarbons: metabolism and environmental aspects. In Developments in Biodegradation of Hydrocarbons-1, ed. R. J. Watkinson. Applied Science Publishers, London, pp. 113-134.

[9]   Shepherd T, Davies HV (1993). Carbon loss from the roots of forage rape (brassica napus L.) seedlings following pulse-labeling with CO_2, Annals of Botany, 72, pp.155-163.

[10]   Shrinivasa U (2003). Report on National Policy on non-edible oils as biofuels, Draft National Bio-Fuel Policy, ISC Bangalore, India, Feb.1-2.

[11]   Charles LP, Hustrulid T (1998). Carbon Cycle for Rapeseed oil Bio-diesel Fuels, Biomass and Bioenergy, Vol. 14, 2, pp.91-101.

Two Dimensional Static Human Modeling During Manual Material Handling Task

Kashif Irshad, Asmar Hussain, Zaka Ullah , Abid Ali Khan, M. Muzammil

Dept. of Mech. Engg, ZHCET, AMU, Aligarh, U.P.

Abstract-**In this study two dimensional static human model were investigate and represents a methodology to determine the effects of different loads under different lifting techniques on human body.For calculation, theoretically 5 to 100 kg of load were lifted and used to quantify the effects of a physical activity on an individual by estimating the forces and moments acting on the musculoskeletal system of the body as it performs a task. Study confirm that as the lifted load increases reaction forces acting at different joints (i.e. elbow, shoulder, hip, knee, and ankle) keeps on increasing during both stoop and squat lifting. Stooping requires more exertions in the hip, shoulder and elbow joints, whereas squatting requires higher moments at ankles and knees, as expected. The hip and ankle joint contributed to the most part of the support moment during the squat lifting, and the knee moment played an important role in stoop lifting .The ankle and Hip generated power and only the knee joint absorbed power in squat lifting .The hip generated power in the stoop lifting. By comparing static strength of the muscle with (Stobbe, 1982, Static Muscle Strength Moment data) load lifting capability of the each link of body was evaluated and found out that female has less lifting capability as compare to male during both types of lifting techniques.**

Keywords: Static muscle strength, manual material handling, body posture

I. INTRODUCTION

In this era of extensive use of automation, manual material handling (MMH) tasks are still inherent in many different jobs in industry. Due to the nature of tasks or situations that will not allow the use of mechanical handling equipment, the "man-machine" is the primary source or device to perform physical activities such as pulling, pushing, carrying and lifting. .The (Kansas Department of Human Resources, 2000) cites manual lifting as the highest classified cause (19.6%) of mishap incidence and also as the highest classified cause (31.1%) of injury on the job. The average compensable cost reported for injury resulting from lifting was $11,793. In the workplace today, it is reported that the second greatest reason for lost workdays is injuries that are due to overexertion when performing lifting tasks (Snook, 1988). It is estimated that 2% of the U.S. workforce suffers compensable back injuries for a total of over 500,000 injuries annually, and that low back cases constitute about 16% of all workers compensation cases and are responsible for 33% of the total cost (Andersson, 1998; Bernard, 1997). With Americans making nearly 20 million physician office visits per year for back complaints, back pain is the second most frequent reason for physician visits, accounting for about 2.8% of all office visits (Hart, et al., 1995; National Academy of Sciences, 2001). Each year, approximately 2% of the US work forces are compensated for back injuries (Andersson, 1998), and only 50% of that worker whose leave of absence lasts for 6 months or more eventually return to work (Valat,2005) .(Frymoyer,1989) estimates that 9.2 million Americans are presently impaired and 2.4 million are disabled by low back pain. Each year, the estimated 600,000 back injuries in the United States cost industry 10 to 14 billion dollars in workers' compensation costs and up to 149 million workdays annually (Occupational Safety and Health Administration, 1992; National Academy of Sciences, 2001).

Biomechanical models are used to quantify the effects of a physical activity on an individual by estimating the forces and moments acting on the musculoskeletal system of the body as it performs a task. Ergonomists are interested in modeling the stresses produced m MMH activities, particularly lifting. Biomechanical models have been used to compare different lifting techniques, evaluate the effects of job design and predict individual lifting capacities. The primary goal of biomechanical modeling is to reduce work-related injuries of the musculoskeletal system (Ayoub, 1992). Biomechanical models may be classified as two- or three-dimensional static or dynamic models. Biomechanical models are used to estimate the forces and moments occurring at each joint center. In static models, the stresses are a function of the load held in the hands, body segment weights and the posture of the body (Garg et al., 1982), while the effects due to acceleration are considered negligible. Dynamic models include inertial effects in addition to the static stresses.

II . METHOD

In this study Biomechanical approach was used. According to (Frankel, 1980). In order to eliminate the indeterminate nature of the models, some simplification and assumptions are necessary. In a biomechanical analysis, the human body is usually viewed as a system of linkages and joints. Each of the links is the same length as their corresponding human segments and possesses the same mass and moment of inertia as their human counterpart. The mass is considered to be concentrated at a single point on the link, the center of mass (CM). The motion of each link is performed about its corresponding joint and can be viewed as a lever system. Through the joints of these linkage systems, the surrounding muscles and tendons are used to apply the moment required to produce the necessary motions.

Force and moment acting on the body while lifting different load was calculated during both the squat and stoop lifting technique. Furthermore load lifting strength of each joint of both male and female was compared with Static Muscle Strength Moment Data (Nm) given by (Stobbe, 1982).

Two dimensional static model of the Human Body was represented as a five-segment open kinematic chain. In static models, the stresses are a function of the load held in the hands, body segment weights and the posture of the body (Garg et al., 1982), while the effects due to acceleration are considered negligible. The link lengths were derived from the over the body dimensions using empirical relations of (Drillis et.al, 1996) as described below.

Links Length

Hand grip to elbow, L1= 0.146H

Ankle to knee, L2= 0.246H

Elbow to shoulder, L3= 0.186H

Knee to the center of Hip, L4= 0.245H

The center of hip to shoulder, L5=0.475H,

Center of Mass of body

COM of Hand grip to elbow =0.430 x L1

COM of Ankle to knee = 0.433 x L2

COM of Elbow to shoulder=0.436 x L3

COM of Knee to the center of Hip=0.433 x L4

COM of The center of hip to shoulder=0.66 x L5

Link weights derived from body weight

Link weights are derived from body weight using relationships of Dempster and other researchers (Garg et.al, 1975), which are as follows:

Hand grip to elbow (each), W1= 0.025*W

Ankle to knee (each), W5= 0.046*W

Elbow to shoulder (each), W2= 0.031*W

Knee to the center of Hip (each), W4= 0.105*W

The center of hip to L5/S1 disc, W3a= 0.191*W

Head neck and Trunk above L5/S1 disc, W3b= 0.363*W

W3=W3a+W3b

Location of Center of gravity of links

For analysis it may be assumed that the link masses are concentrated at the center of gravity of the link. Location of the Centre of Gravity (CG) for link segments were also thus derived according to Dempster's relations from the link lengths (Garg et.al,1975) as described below:

Elbow to C.G. of Lower Arm, LG1= 0.430* L1

Ankle to CG of Lower Leg, LG2=0.567* L2

Shoulder to CG of Upper Arm (Each), LG3=0.436* L3

Knee to CG of Upper Leg (Each), LG4=0.567 *L4

Centre of Hip to CG of Trunk between Hip and L5/S1 Disc, LG5=0.5* L5

L5/S1 Disc to CG of Trunk above L5/S1, Neck and Head, LG6=0.4321* L5

Force component:

For each joint, the resultant force & moment should be equal to zero .Weight of each link, load & reaction force of adjacent joint were evaluated.

For each joint $\sum F = 0$

Relbow= (-W1) + (-Load)

Rshoulder = (-W2) + Relbow

Rhip = (-W3) + Rshoulder

Rknee = (-W4) + Rhip

Rankle = (-W5) + Rknee

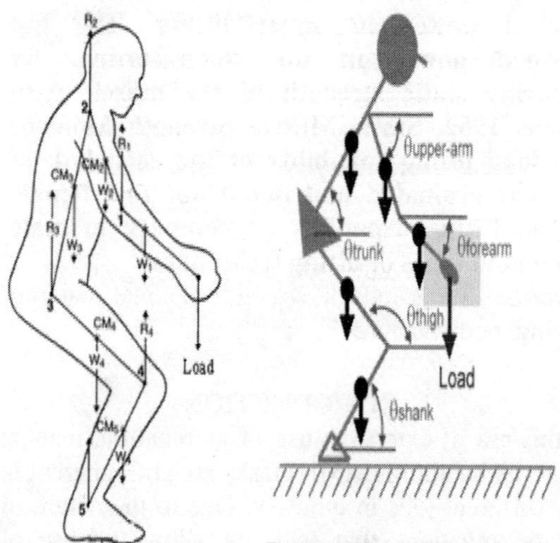

Fig 1. Show free body diagram of a two dimensional static human model during load lifting task

Moment Component

The moment produced by the weight of each segment, the moment produced by the load, & the moment produced by the reaction force of adjacent joint.

For each joint, $\sum M = 0$

Melbow= (Load) (L1) (Cos θ forearm) + (W1) (COMforearm) (Cos θ forearm) + (Relbow) (0)

Mshoulder = (Relbow) (L3) (Cos θ upper-arm) + (W2) (COMupper-arm) (Cos θ upper-arm) + (Rshoulder) (0) + Melbow

Mhip = (Rshoulder) (L5+L6) (Cos θ trunk) + (W3) (COMtrunk) (Cos θ trunk) + (Rhip) (0) + Mtrunk

Mknee = (Rhip) (L4) (Cos θ thigh) + (W4) (COMthigh) (Cos θ thigh) + (Rknee) (0) + Mhip

Mankle = (Rknee) (L2) (Cos θ shank) + (W5) (COMshank) (Cos θ shank) + (Rankle) (0) + Mknee

III. RESULTS

Biomechanical models are used to estimate the forces & moments occurring at each joint center. In this study two dimensional static human model were consider. In static models, the stresses were a function of the load held in hands, body segment weights and the posture of the body while the effect due to acceleration were considered negligible. Here forces and moments at different joints acting during lifting different loads evaluated .Table 1 shows reaction forces acting at different joints(i.e. elbow ,shoulder ,hip, knee , ankle) when different load were lifted by human being during squat lifting.

Table 1. Variation of reaction force at different joints during squat lifting

Load	Relbow	Rshoulder	Rhip	Rknee	Rankle
5	6.625	8.64	44.65	51.475	54.465
10	11.625	13.64	49.65	56.475	59.465
15	16.625	18.64	54.65	61.475	64.465
20	21.625	23.64	59.65	66.475	69.465
25	26.625	28.64	64.65	71.475	74.465
30	31.625	33.64	69.65	76.475	79.465
35	36.625	38.64	74.65	81.51	84.465
40	41.625	43.64	79.65	86.475	89.465
45	46.625	48.64	84.65	89.475	94.465
50	51.625	53.64	89.65	96.475	99.465
55	56.625	58.64	94.65	101.475	104.465
60	61.625	63.64	99.65	106.475	109.465
65	66.625	68.64	104.65	111.475	114.465
70	71.625	73.64	109.65	116.475	119.465
75	76.625	78.64	114.65	121.475	124.465
80	81.625	83.64	119.65	126.475	129.465
85	86.625	88.64	124.65	131.475	134.465
90	91.625	93.64	129.65	136.475	139.465
95	96.625	98.64	134.65	141.475	144.465
100	101.625	103.64	139.65	146.475	149.465

Table 2 shows reaction forces acting at different joints (i.e. elbow, shoulder, hip, knee, ankle) when different load were lifted by human being during stoop lifting.

Table 2. Variation of reaction force at different joints during stoop lifting

Load	Relbow	Rshoulder	Rhip	Rknee	Rankle
5	6.625	8.64	44.65	51.475	66.465
10	11.625	13.64	49.65	56.475	72.465
15	16.625	18.64	54.65	61.475	78.465
20	21.625	23.64	59.65	66.475	84.465
25	26.625	28.64	64.65	71.475	90.465
30	31.625	33.64	69.65	76.475	96.465
35	36.625	38.64	74.65	81.51	102.465
40	41.625	43.64	79.65	86.475	108.465
45	46.625	48.64	84.65	89.475	114.465
50	51.625	53.64	89.65	96.475	119.465
55	56.625	58.64	94.65	101.475	124.465
60	61.625	63.64	99.65	106.475	129.465
65	66.625	68.64	104.65	111.475	134.465
70	71.625	73.64	109.65	116.475	139.465
75	76.625	78.64	114.65	121.475	144.465
80	81.625	83.64	119.65	126.475	151.465
85	86.625	88.64	124.65	131.475	157.465
90	91.625	93.64	129.65	136.475	163.465
95	96.625	98.64	134.65	141.475	169.465
100	101.625	103.64	139.65	146.475	175.465

In a biomechanical analysis, the human body is usually viewed as a system of linkages and joints. The development of valid and useful force-motion relationships requires that the body be represented by a simple yet realistic mechanical model. Each of the links is the same length as their corresponding human segments and possesses the same mass and moment of inertia as their human counterpart. The mass is considered to be concentrated at a single point on the link, the center of mass (CM).Table 3 represent weight link length and centre of mass of segment of human body. Link weights were derived from body weight using relationships of Dempster and other researchers (Garg et.al, 1975) while link length were derived from body height using relationships of (Drill et al, 1996) and center of mass of segment were derived by using relationships of Dempster and other researchers (Garg et.al, 1975).

Table 3. Weight, Length and centre of mass of human body (link)

Link	Weight of link (N)	Length of link (m)	Centre of mass of segment (m)
Forearm	15.94	.256	.110
Upper arm	19.76	.327	.142
Thigh	66.95	.4312	.1867
Shank	29.33	.4329	.1874
Trunk	353.12	.836	.5511

Table 4. represents static muscle strength moment at different loads during squat lifting .The moment produced by the weight of each segment, the moment produced by the load, & the moment produced by the reaction force of adjacent joint is calculated by summing all the moments for each joint, ($\sum M=0$) .Weight of each segment or link, length of link & centre of mass of link is evaluated from table 3, while reaction force acting at each joint is evaluated from table 1.

Table 4. Static muscle strength moment data at different loads during Squat lifting

Load (kg)	Elbow moment (Nm)	Shoulder moment (Nm)	Hip moment (Nm)	Knee moment (Nm)	Ankle moment (Nm)
5	4.89	25.67	44.38	-54.85	.078
10	9.18	43.54	65.31	-49.329	6.067
15	13.48	62.34	86.73	-34.14	14.925
20	16.29	78.45	105.65	-26.73	26.20
25	20.21	96.64	122.07	-16.81	38.78
30	24.14	114.67	143.23	06.89	47.45
35	28.06	132.45	164.45	08.65	63.67
40	31.99	150.23	185.45	19.67	78.45
45	35.915	167.23	206.89	29.34	88.27
50	39.83	185.67	217.46	34.45	102.576
55	43.76	203.31	238.19	49..23	113.678
60	47.68	221.76	256.78	62.46	124.76
65	51.61	239.778	277.67	76.78	141.32
70	55.53	256.34	297.98	88.67	155.56
75	59.45	274.67	311.67	99.57	173.78
80	63.38	294.134	339.768	107.35	191.34
85	67.30	310.23	350.57	119.78	201.89
90	71.231	328.89	368.78	131.56	213.78
95	75.155	346.978	379.256	148.43	227.87
100	79.09	363.45	398.35	161.87	244.34

Each joint have different joint angle while lifting load. Therefore for calculation of moment of Elbow joint angle was taken to be 70 degree and for Shoulder joint angle was taken to be 30 degree and for Hip it was 86 degree and for Knee it was 120 degree and for Ankle it was taken to be 80 degree. While table 5 represents static muscle strength moment at different loads during stoop lifting. Elbow joint angle was taken to be 50 degree and for Shoulder joint angle was taken to be 30 degree and for Hip it was 65 degree and for Knee it was 90 degree and for Ankle it was taken to be 90 degree

Table 5. Static muscle strength moment data at different loads during Stoop lifting

Load (kg)	Elbow moment (Nm)	Shoulder moment (Nm)	Hip moment (Nm)	Knee moment (Nm)	Ankle moment (Nm)
5	9.19	30.22	115.54	115.54	115.54
10	17.269	52.12	139.12	139.12	139.12
15	25.34	74.34	163.199	163.199	163.199
20	33.42	96.78	187.406	187.406	187.406
25	41.23	116.23	208.62	208.62	208.62
30	49.78	134.23	230.74	230.74	230.74
35	57.34	156.87	254.19	254.19	254.19
40	67.34	175.34	275.34	275.34	275.34
45	74.56	194.45	298.53	298.53	298.53
50	86.78	211.67	322.55	322.55	322.55
55	94.23	231.98	346.04	346.04	346.04
60	103.67	257.87	367.56	367.56	367.56

65	112.34	275.87	389.92	389.92	389.92
70	123.45	295.76	414.23	414.23	414.23
75	132.78 9	310.83	434.12	434.12	434.12
80	141.34	332.58	456.22	456.22	456.22
85	150.97	357.38	478.38	478.38	478.38
90	158.23	377.75	501.23	501.23	501.23
95	166.45	393.67	524.36	524.36	524.36
100	174.48	408.78	547.34	547.34	547.34

IV. DISCUSSION

This study represents a methodology to determine the effects of different loads under different lifting techniques on human body. A two-dimensional sagittally symmetric human body model was constructed as a five rigid link mechanism for the biomechanical simulation of manual lifting tasks. These links possessed the same length, mass, and inertia properties as estimated for their human counterparts. Therefore, any movement or configuration could be described with the five generalized coordinates describing the relative orientation of each of these five links with respect to the parent body. Joints at the ankle, knee, hip, shoulder, and elbow were all treated as one-degree of freedom revolute joints. Spinal column was considered as one rigid link that includes mass of the head and neck appropriately located. The hands were also modeled as parts of the forearms, and their relative motion with respect to forearms were neglected. It was further assumed that subject did not walk with the load during the lift, i.e., foot was fixed on the ground

(Keyserling, 2000) identified key factors associated with lifting. These were:

The amount of weight lifted, and Body posture (e.g. forward bending of the trunk increases the load on the lower back).

These all factors were evaluated in our study .Study confirm that as the lifted load increases   reaction forces acting at different joints (i.e. elbow, shoulder, hip, knee, and ankle) keeps on increasing during both stoop and squat lifting. .In this case, all parameters were kept constant, except the initial angular positions of body segments. They were changed so as to have a lifting mode of a squat, a stoop, In the stoop, the ankles and knees were kept straight, whereas in the squat, the back was kept to be vertical during the lift .

Load lifting strength

Static Strength is defined as when there is no perceptible change in muscle length during an isometric effort, then the involved body segments do not move; in physical terms, all forces acting within the system are in static equilibrium, as stated by Newton's First Law. The static condition is theoretically simple and experimentally well controllable. It allows a rather easy measurement of muscular effort. (Stobbe, 1982) evaluated Static Muscle Strength Moment Data

(Nm) for 25 Men and 22 Women employed in Manual Jobs in Industry which was represented in Appendix A .By comparing our results with (Stobbe, 1982) Static Muscle Strength Moment data for 5 percentile male and female ,limits of load lifting strength for each joints of body can be easily examined. Now let us examine static muscle strength of male. During Squat lifting maximum load which elbow can bear was 35 kg & during stoop lifting was 15 kg .For shoulder during squat lifting was 15 kg & for stoop lifting was 10 kg, for hip during squat lifting was 20 kg and for stoop lifting it was 5 kg, for ankle during squat lifting was 35kg and for stoop lifting result shows moment with greater magnitude which was above the load lifting capacity of an individual and for knee during squat lifting maximum load lifted was 65 kg and for stoop lifting again result shows moment with greater magnitude which was above the load lifting capacity of an individual .For female static muscle strength during Squat lifting maximum load which elbow can bear was 10 kg & during stoop lifting was 5 kg . For shoulder during squat lifting was 5 kg & for stoop lifting was less than 5 kg, for hip during squat lifting was less than 5 kg and for stoop lifting result shows moment with greater magnitude which was above the load lifting capacity of an individual, for ankle during squat lifting was 20 kg and for stoop lifting result shows moment with greater magnitude and for knee during squat lifting it was 55 kg and during stoop lifting result shows moment with greater magnitude which was above the load lifting capacity of an individual. Thus it was seen clearly that female have less strength as compared to male and also maximum strength of joints depends upon type of lifting techniques   & changes as we change the lifting technique. Also, higher levels of moment applications are necessary to accomplish the task, especially at the start of the lift.

V. CONCLUSION AND SCOPE FOR FUTURE STUDY

Conclusion

The purpose of this study was to investigate two dimensional static human models which were used to quantify the effects of a physical activity on an individual by estimating the forces and moments acting on the musculoskeletal system of the body as it performs a task. Static model was used as it requires relatively simple logic and simpler task data. Thus with the help of this model comparison between different lifting techniques i.e. (Squat and stoop), while lifting different loads were evaluated and also muscle strength of each link of body were compared with (Stobbe, 1982) Static Muscle Strength Moment Data. From the above results, squat lift technique (i.e., knee bent and back straight) was generally considered to be safer than the stoop lift (i.e., knee straight and back bent) in bringing the load closer to the body and, hence, reducing the extra demand on back muscles while counterbalancing the moments of external loads and hence reduces effect of low back disorder. Above result shows that female has less load lifting capability as compare to male during both types of lifting techniques. Study confirm that as the lifted load increases reaction forces acting at different joints (i.e. elbow, shoulder, hip, knee, and ankle) keeps on increasing during both stoop and squat lifting. Stooping requires more exertions in the hip, shoulder and elbow joints, whereas squatting requires higher moments at ankles and knees, as expected. The hip and ankle joint contributed to the most part of the support moment during the squat lifting, and the knee moment played an important role in stoop lifting The ankle and Hip generated power and only the knee joint absorbed power in squat lifting .The hip generated power in the stoop lifting.Further, work requiring stooped postures is strongly associated with high incidence of low back disorders (LBDs). But still it is preferred as worker find easy to lift heavy load which when repetitive done cause low back disorders. Therefore, it can be advised that one should not lift a load that is sufficiently heavy to require a joint moment that exceeds the associated joint strength for the individual, which can be injurious.

VI. REFERENCE

[1] Andersson, G. B., (1997), the epidemiology of spinal disorders. In: Marras, W.S. (2000),

[2] Occupational low back disorder causation and control. Ergonomics, vol. 43, no. 7, pp. 880 - 902.

[3] Ayoub, MM. (1992). Problems and solutions in manual materials handling: die state of die art. Ergonomics. 3 5(7/8 V 713-728).

[4] Bernard, B. P. (1997). Musculoskeletal disorders and workplace factors : a critical review of epidemiologic evidence for work-related musculoskeletal disorders of the neck, upper extremity, and low back, US Department of Health and Human Services (DHHS) publication no. 97-141 (Cincinnati: National Institute for Occupational Safety and Health)

[5] Chaffin D.B., (1969) .A computerized biomechanical model—Development of and use in studying gross body action, Journal of Biomechanics, Vol. 2. 4, pp :429-441

[6] Dempster, W.T., (1955). Space requirements of the seated operator, WADC-TR-55-159, Aerospace Medical Research Laboratories: Dayton, OH.

[7] Frankel, V., and Nordin,M,(1980). Basic biomechanics of skeletal system, Henry Kempton Publishers, London.

[8] Freivalds, A., Chaffin, D.B., Garg. A., and Less, KS. (1984). A dynamic biomechanical evaluation of lifting maximum acceptable loads. Joumal of Biomechanics. 17(4). pp 251-262

[9] Frymoyer, J.W., (1989). Epidemiology. In: Frymoyer, J.W., Gordon, S.L., eds Symposium on new perspectives on low back pain. Park Ridge, IL: American Academy of Orthopaedic Surgeons, pp. 19-33.

[10] Garg, A. (1982). Lifting and back injuries. Plant Engmeering. 89-93. Yeung, S., Genaidy, A.

[11] Garg A. and Chaffin D.B., (1975). A biomechanical computerized simulation of human strength, AIIE transaction, March, pp. 1-15

[12] Hart, G.L., Deyo, R.A. and Cherkin, D.C., (1995). Physician office visits for low back pain., Spine, Vol .20, pp. 11-19.

[13] J.P. Valat, (2005) .Factors involved in progression to chronicity of mechanical low back pain, Joint Bone Spine, Vol. 72, pp. 193-195.

[14] Kansas Department of Human Resources (2000). Workers Compensation 26th Annual Report, Fiscal Year 2000.

[15] Keyserling, W., (2000), Workplace risk factors and occupational musculoskeletal disorders, part 1: a review of biomechanical and psychophysical research on risk factors associated with lowback pain. American Industrial Hygiene Association Journal, vol. 61, pp. 30 - 50.

[16] National Academy of Sciences (2001). Musculoskeletal disorders and the workplace: low back and upper extremities. National Academy Press, Washington, D.C.

[17] National Research Council, (1999), Work-related Musculoskeletal Disorders (Washington, DC: National Academy Press). In: Op De Beeck, R., and Hermans, V., 2000, Research on workrelated low back disorders. European Agency for Safety and Health at Work, ISBN 92 950007 02 06.

[18] National Institute for Occupational Safety and Health (1991). DHHS, Work Practices Guide for Manual Lifting

[19] Snook, H., (1988). Approaches to the Control of Back Pain in Industry: Job Design, Job Placement and Education/Training, Professional Safety Vol. 33. 8, 23-31.

Dynamic System Modeling of Stock Market Volatility Via Fuzzy and Neuro Fuzzy Approach: Impact of Real Economic Indicators in Indian Stock Market

Swami P Saxena[1], D.K. Chaturvedi[2], Anurag Gupta[3], Sonam Bhadauriya[1]

[1] Department of Applied Business Economics [2] Department of Electrical Engineering
Dayalbag Educational Institute ,Dayalbagh,Agra
[3] Mechanical Engineering Department, Anand Engineering College, Agra, U.P.

Abstract- The modeling of stock market volatility is one of the key areas of present financial research as stock market is the main determinant of economic development of a country. A number of researches have been investigated on the modeling of relationship between macroeconomic indicators and stock market volatility. But in the context of India not many researches can be traced in the literature, moreover the present study has used the neuro fuzzy modeling, which has not been previously used by researchers in India. The period modeled is from April 1999 to March 2010. Daily index of S&P CNX Nifty is used as indicator of stock market volatility and only real macroeconomic indicators (Gross Domestic Product, Index if Industrial Production and Inflation) are used for modeling purpose. The study found that neuro fuzzy model developed for stock price volatility is quite suited for this application and also provide opportunity for predicting stock market volatility in future by predicting the future time series of independent variables.

Keywords: Stock Market Volatility, Real Economic Indicators, Fuzzy Logic, Adaptive Neuro-Fuzzy

I. INTRODUCTION

The volatility of stock market has been studied for many years and number of stylized facts has been applied. Many researchers claim that the stock market is a chaos system as it is a nonlinear process which can't be easily expressed. Stock market is a dynamic and complicated system. So it is difficult to deal it with normal analytical methods. The existence of nonlinearity of stock market is propounded by many researchers and financial analysts. To deal with system nonlinearity, non conventional methods like artificial neural networks (ANN), fuzzy logic are considered best.

The impact of real macroeconomic indicators on stock market volatility has been studied extensively in finance literature. An illustrative list of studies includes Famma EF (1981), Ibrahim HM (1999), Ahmed MF (1999), Charkravarty S (2005), Mehr (2005), Chaudhary SSH (2006), Adam (2008) etc. These studies indentify macroeconomic factors such as gross domestic product, industrial production, inflation, trade balance, money supply and so on as being important in explaining stock market volatility.

Corradi V (2009) stated that studying the impact of real macro economy indicators on stock market volatility has important implications for both market practitioners and policy makers. Policy makers are interested in the main determinants of volatility and in its effects on real activity to conduct national macroeconomic policies without the fear of influencing capital formation and the stock trade process. Moreover, economic theory suggests that stock prices should reflect expectations about future corporate performance, and corporate profits generally reflect the level of economic activities. If stock prices accurately reflect the underlying fundamentals, then the stock prices should be employed as leading indicators of future economic activities, and not the other way around. Market practitioners are mainly interested in the direct effects time-varying volatility exerts on the pricing and hedging of plain vanilla options and more exotic derivatives. Maysami RC (2004) concluded that the causal relations and dynamic interactions among macroeconomic determinants of the economy and stock prices are important in the formulation of the nation's macroeconomic policy.

This paper consists of seven sections. First the introduction states the motivation and goal of the research by describing the importance of nonlinear modeling of stock market volatility. In the second section, specific studies conducted on impact of macroeconomic variables on stock market volatility are presented. Then a brief introduction of selected real economic indicators is described in the next section. The fourth section deals with the aspects concerned with database and the neuro fuzzy modeling. The fifth section is the model development phase & sixth is the simulation phase. In the last section conclusions are derived on the basis of developed model.

II. LITERATURE REVIEW

The relationship between macroeconomic variables and stock market returns is, by now, well-documented in the literature. Maysami RC (2004) examined the long-term equilibrium relationships between selected macroeconomic variables and the Singapore stock market index. Leblebicioglu K (2004) implemented a rule based fuzzy logic model to forecast the monthly return of the ISE100 Index by combining technical analysis, financial analysis and macroeconomic analysis.

Chowdhury SSH (2006) examined how the macroeconomic risk associated with industrial production, inflation, and exchange rate is related reflected to the stock market return in the context of Bangladesh and concluded that there is relation between stock market volatility and macroeconomic volatility. Engle RF (2006) developed a model allows long horizon forecasts of volatility to depend on macroeconomic developments, and delivers estimates of the volatility to be anticipated in a newly opened market. Aguiar RA (2006) presented empirical tests for the overreaction and under reaction hypothesis for petrol/petrochemical and textile firms in the Brazilian stock market.

Humpe A (2007) examined whether a number of macroeconomic variables influence stock prices in the US and Japan. Adam (2008) in the conclusion of their study established that there is co-integration between macroeconomic variable and Stock prices in Ghana indicating long run relationship.

III. REAL ECONOMY INDICATORS

Real economy is the physical side of the economy dealing with goods, services and resources. This side is concerned with using resources to produce the goods and services that make the satisfaction of wants and needs possible. This should be contrasted with the paper economy, or financial side of the economy. Main real economic indicators selected for the study purpose are:

1. **Gross Domestic Product:** GDP is the value of all goods and services produced in the economy in the given period.
2. **Industrial Production:** Index of Industrial Production is a government report that provides a fixed weight measure of the

physical output of the nation's factories, mines and utilities.

Inflation Rate

It is the percentage rate of increase of the level of prices during a given period, usually measured by the Consumer Price Index (CPI) and the Wholesale Price Index (WPI). WPI is considered as much better indicator of inflation rate because CPI measures monthly change in the prices of a defined basket of consumer goods and WPI measures changes in prices in the manufacturing and distribution sector of the economy tends to lead the consumer price index by 60 to 90 days.

IV. DATABASE

It is rightly said that data selection must be performed judiciously to avoid the "garbage-in, garbage-out" syndrome often associated with computers. Performance of fuzzy logic is highly dependent on the quality & appropriateness of its input data. If relevant data inputs are not included, the performance will suffer needlessly.

The daily S&P CNX Nifty index is taken as the dependent variable for the modeling. The data for the study are collected from the website of the National stock Exchange (NSE) (www.nseindia.com). S&P CNX Nifty is a well-diversified 50-stock index accounting for 23 sectors of the economy. Three real macro economic variables Gross Domestic Product (GDP), Index of Industrial Production (IIP) & Wholesale Price Index (WPI) are taken as the independent variables and the related data are collected from the website of Reserve Bank of India (www.rbi.org). In order to measure the impact of selected macroeconomic indicators on volatility of S&P CNX Nifty, eleven year data from 1999-00 to 2009-10 is taken for modeling purpose.

V. NEURO FUZZY MODELING

Keeping in mind the weaknesses of conventional analytical techniques, researcher apply neuro-fuzzy based model as a forecasting tool for predicting the behavior of the stock market. The study also presents a comparison between the results of fuzzy and neuro fuzzy logic based model.

Fuzzy Logic was initiated in 1965 by Zadeh Lotfi, professor for computer science at the University of California in Berkeley. Basically, Fuzzy Logic is a multi-valued logic that allows intermediate values to be defined between conventional evaluations like true/false, yes/no, high/low, etc. Fuzzy logic is a superset of conventional (Boolean) logic that has been extended to handle the concept of partial truth - truth values between "completely true" and "completely false". But, it requires a sufficient expert knowledge for the formulation of the rule base, the combination of the sets and the defuzzification. In General, the employment of fuzzy logic might be helpful, for very complex processes, when there is no simple mathematical model, for highly nonlinear processes or if the processing of expert knowledge is to be performed.

Chaturvedi DK (2005) stated that neural networks and fuzzy systems each have their own advantages and disadvantages. Both the systems are very different but they have a close relationship because both of them can work with imprecision in a space that is not defined by crisp, deterministic boundaries. The shortcomings of neural networks and of fuzzy systems may be overcome if both operate competitively and co-operately. Fuzzy model requires large power in representing linguistic and structured knowledge by fuzzy sets and usually rely on the domain experts to provide the required knowledge for a specific problem. On the other hand, artificial neural networks models are particularly good for non linear mappings because these models are developed via training. Combination of fuzzy systems and neural networks is known as neuro fuzzy model. So, the neuro fuzzy approach of modeling has also been applied by the researcher to make the study more strengthened. For developing fuzzy and neuro-fuzzy model of stock market volatility, Fuzzy toolbox of MATLAB (7.0) is used.

VI. MODEL DEVELOPMENT PHASE

In this phase of the study researcher developed fuzzy logic based and neuro fuzzy logic based models for prediction of stock market behavior by following some steps. These steps includes indentification of key variables and their ranges, development of fuzzy knowledge base, development of fuzzy rule base and development of nuero fuzzy logic based model.

Step 1: The first step in model development is the identification of key variables and their ranges. The key variables identified for the Stock Market Volatility are S&P CNX Nifty, GDP, IIP and WPI: The range of these variables is given in table 1:

Table 1: variables and their Range

S. No.	Variable Name	Variable Type	Range
1.	S&P CNX Nifty	Dependent	830 – 6310
2.	GDP	Independent	386 – 1546000 Crore Rupees
3.	IIP	Independent	128-368
4.	WPI	Independent	116-276

Step 2: Second step in model development is development of fuzzy knowledge base. For this purpose, ranges of above variables have been divided into nine membership functions. Linguistic terms for these membership functions are as Extremely Low, Very Low, Low, Lowly Medium, Medium, Highly Medium, High, Very High and Extremely High. Some membership values are

given to all membership functions as shown in box 1.

Sugeno method of fuzzy inference has been used for fuzzy modeling. Sugeno method is ideal for acting as an interpolating supervisor of multiple linear controllers that are to be applied, respectively to differed operating conditions of a dynamic nonlinear system because of the linear dependence of each rule on the input variable of the system.

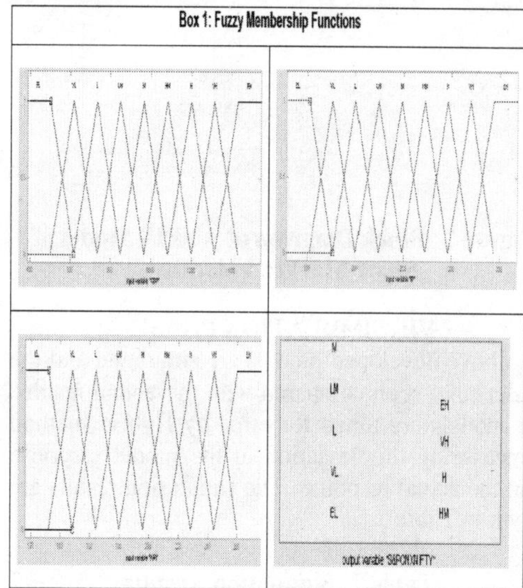

Step 3: For the development of fuzzy rule base model developer established 93 rules as the stock market is highly volatile system that's why researcher has to define a large number of rules. These rules are defined on the basis of causal relationship between identified variables. For this purpose, the causal loop diagram has been drawn as below.

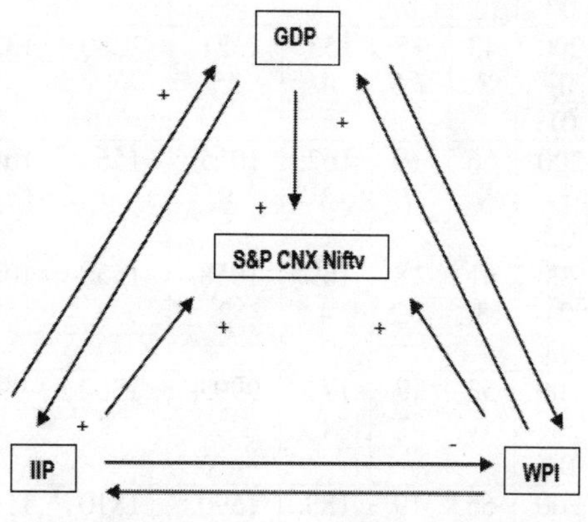

Figure 1: Causal Loop Diagram for Stock Market Volatility

Step 4: The fuzzy knowledge base developed in previous is fixed and does not have adaptability. Hence, Nuero-Fuzzy approach has been used to make it more flexible and adaptive. The Adaptive neuro fuzzy logic inference system (ANFIS) model is developed by Subtractive Clustering Method in the fuzzy logic toolbox of MATLAB.

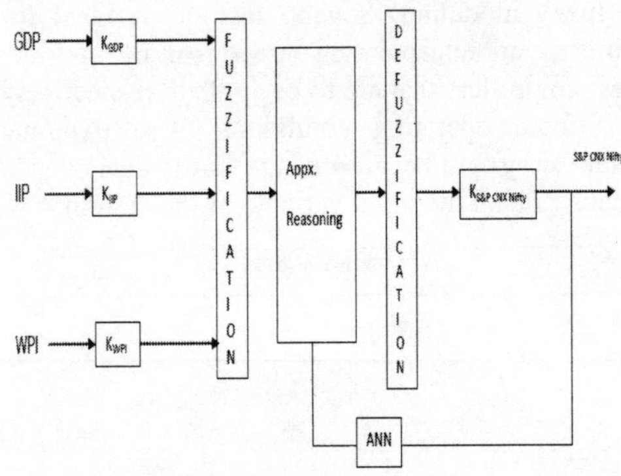

Figure 2: Block Diagram of ANFIS Model of Stock Market Volatility

VII. SIMULATION PHASE

The above developed models are simulated and the results have been compared with the actual results. The models are tuned for error by Hybrid method to measuring the deviation in the model response from the actual response. The simulation results are shown in Table 2.

Table 2: Simulation Results

S.N.	Year	Crisp Values of Independent Variables			S&P CNX Nifty		
		GDP	IIP	WPI	Actual	Fuzzy Model	ANFIS Model
1.	1999-00	415	151	142.8	1148.1	1280	1150
2.	2000-01	433	157.5	153.4	1321.25	1280	1320
3.	2001-02	569	167	162.3	1015.8	1560	1010
4.	2002-03	613	182.2	169.4	1058.2	1550	1060
5.	2003-04	580	174	173.1	999.4	1560	999
6.	2004-05	666	197.8	188.4	1590.35	1810	1590
7.	2005-06	884	232.5	197.2	2778.55	3110	2780
8.	2006-07	1040	265.5	208.8	4109.05	3650	4110
9.	2007-08	993	263.1	212.3	4260.9	3570	4260
10.	2008-09	1175	276.2	241.5	4290.3	4090	4290
11.	2009-10	1480	334.3	246.5	5117.3	5040	5120

Note: Average testing errors for fuzzy and ANFIS Models are respectively 449.85 and 0.1131.

VIII. CONCLUSION

The study shows that there is a positive causal inks between the real economic determinants and stock market prices, which concludes that macro real economic fundamental news can be used to predict stock market prices in India. Moreover, stock exchange performance can also represent the macroeconomic movement in India. Effects of gross domestic product, index of industrial production, inflation rate on S&P CNX Nifty are shown in fig 3 to 5.

The study has developed fuzzy and neuro fuzzy based models for stock market volatility due to real economic indicators. The results show that both the models are quite suited for this application. But the adaptive neuro fuzzy model is more successful in predicting the stock market prices, as the deviation of model response to actual response is zero percent in this model.

On the basis of simulation results researcher want to state that both the modeling approaches are perfect for complex problems, which are highly non linear in nature as forecasting stock market volatility. Both the modeling techniques have their advantages and limitations. Fuzzy model posses large brain power and time in representing linguistic and structured knowledge by fuzzy sets and membership functions. Further the compensatory operators in the fuzzy models as connectives are found quite suitable and give results very close to the actual result. On the other hand, neural network models are particularly good for non-linear mappings and for providing parallel processing facility to simulate complex system because these models are developed via training. Furthermore, while the behavior of fuzzy models can be understood easily due to their logical structure and step-by- step inference procedures, a neural network normally acts as a black box, without providing explanation facility. From these investigations, it is quite natural to consider the possibilities of integrating the two paradigms, in order to utilize the desired strength of both types of models to produce improved results.

APPENDICES

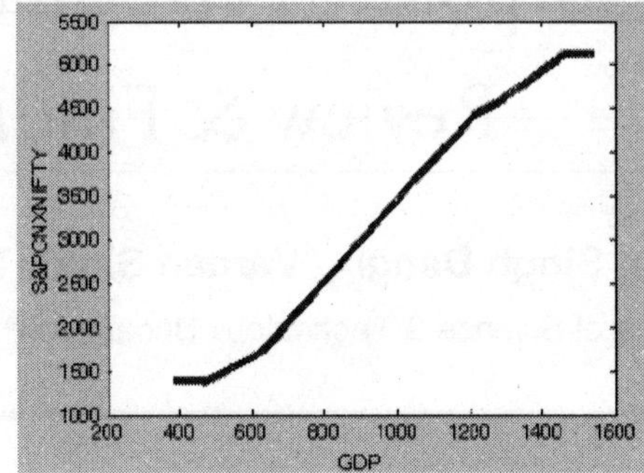

Figure 3: Effect of GDP on Stock Market

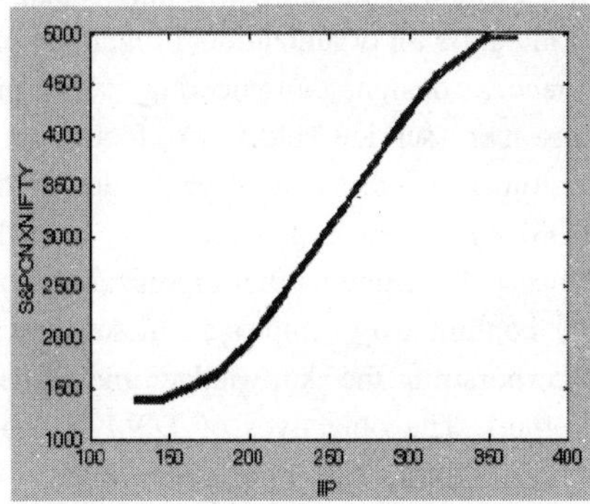

Figure 4: Effect of IIP on Stock Market

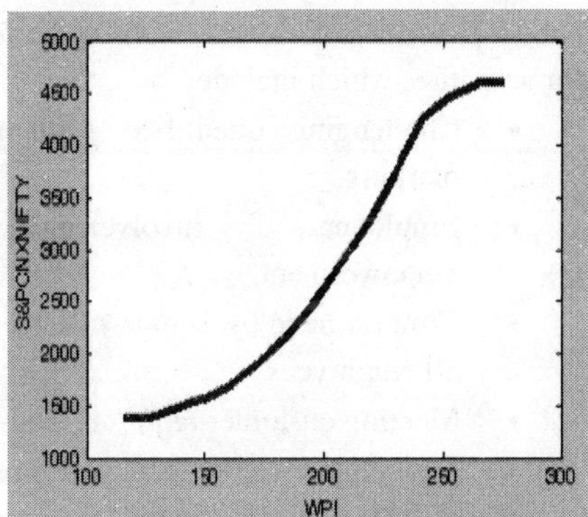

Figure 5: Effect of WPI on Stock Market

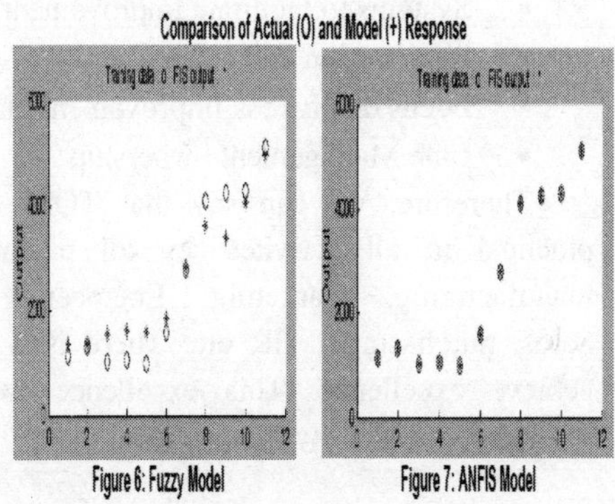

Figure 6: Fuzzy Model   Figure 7: ANFIS Model

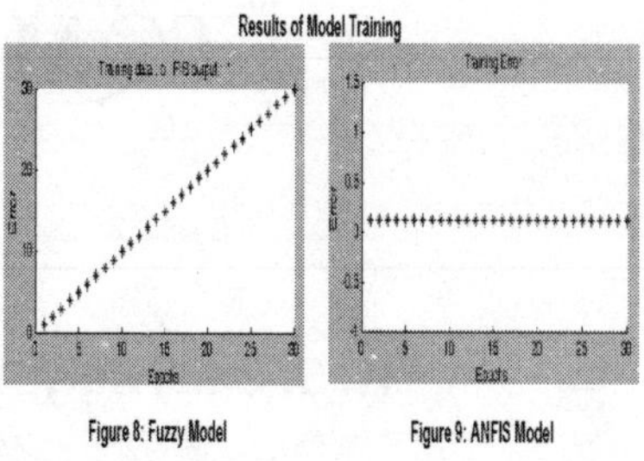

Figure 8: Fuzzy Model   Figure 9: ANFIS Model

IX. REFERENCES

[1] Adam, Anokye M and Tweneboah , George (2008), Do macroeconomic variables play any role in the stock market movement in Ghana?, School of Management , University of Leicester, UK

[2] Aguiar RA, Sales RM and Sousa LA (2006), A behavioral fuzzy model for analysis of overreaction and under reaction in the Brazilian stock market

[3] Ahmed MF (1999), Stock market, macroeconomic variables and causality: the Bangladesh case, savings and development

[4] Chaerkravarty S (2005), Stock market and macroeconomic behavior in India, Institute of Economic growth, Delhi

[5] Chaturvedi DK (2005), Dynamic model of HIV/Aids population of Agra region, Int. J. of Environmental Research and Public Health

[6] Chowdhury SS, Mollik AT and Akhter MS (2006), Does predicted macroeconomic volatility influence stock market volatility? evidence from the Bangladesh capital market, University of Rajshahi, Bangladesh

[7] Corradi V and Distaso W (2009), Macroeconomic determinants of stock market volatility and volatility risk-premia, London School of Economics, London

[8] Engle RF and Rangel JG (2006), The Spline-GARCH model for low frequency volatility and its global macroeconomic causes, New York University, US

[9] Famma EF (1981), Stock prices, inflation, real activity and money. Am. Economic Review

[10] Humpe A and MP (2007), Can macroeconomic variables explain long term stock market movements? A comparison of the US and Japan", University of St Andrews, US

[11] Ibrahim HM (1999), Macroeconomic indicators and stock prices in Malaysia; An empirical analysis, Asian Economics Journal, 13(2)

[12] Leblebicioglu K and Aksoy H (2004), Modeling ISE100 index through fuzzy logic, Eleventh Annual MFS Conference Proceedings, Turkey

[13] Maysami RC, Lee CH and Mohamad AH (2004), Relationship between macroeconomic variables and stock market indices: cointegration evidence from stock exchange of Singapore's All-S sector indices, Jurnal Pengurusan

[14] Mehr (2005), Stock market consequences of macroeconomic fundamentals, institute of Business and Technology, Karachi, Pakistan

TQM Impact on Corporate Performance Review & Findings

[1]R. S. Sikarwer, [1]Y. S. Tomar, [2]Raman Singh Dangi , [1]Vardan Singh Tomar

[1]Vidhya Peeth Insitute of Science & Technology Bhoapl, M.P. (India)

[2] TIT, Bhopal

Abstract – **This paper represents a review of the literature on the Impact of total Quality Management (TQM) on corporate performance and supported by various philosophies, models and framework of TQM. This research presents new data and empirical insights into the relationship between TQM implementation and corporate performance. The main purpose of this research is to determine the impact of TQM implementation on corporate performance in the context of India manufacturing industry. As very few research on TQM has been done in the developing countries, so our aim is to analyze the status of the Indian Manufacturing Industry for TQM implementation, as India is becoming a major sourcing base for the world and there is a paucity of such research. The main implication of the findings for managers is that a typical manufacturing organization is more likely to achieve better performance in employee relations, customer satisfaction, operational performance and business performance with TQM than without TQM. There are very few work which gives the idea of the impact of TQM implementation on organizational performance. Therefore going for this kind of research will be a boon for the Indian Companies.**

Key-Words: Corporate, Implementation, TQM, Review, Empirical.

I. INTRODUCTION

Total Quality is a description of the culture, attitude and organization of a company and strives to provide customers with products and services that satisfy their needs.

Total Quality Management, TQM is a method by which management and employees can become involved in the continuous improvement of the production of goods and services. It is a combination of quality and management tools aimed at increasing business and reducing losses due to wasteful practices.

TQM is a management philosophy that seeks to integrate all organizational functions (marketing, finance, design, engineering and production, customer service etc.) to focus on meeting customer needs and organizational objectives. TQM views an organization as a collection of process. It maintains that organizations must strive to continuously improve these processes by incorporating the knowledge and experience of workers. The objectives of TQM is "do the right things, right the first time, every time."

TQM is now becoming recognized as a generic management tools, just as applicable in service and public sector organizations. TQM is the foundation for activities, which include:

- Challenging quantified goals and bench marking.
- Employee involvement and empowerment.
- Commitment by senior management and all employees.
- Meeting customer requirements.
- Reducing development cycle times.
- IIT/ Demand flow manufacturing.
- Improvement teams.
- Reducing product and service costs.
- Systems to facilitate improvement.
- Recognition and celebration.
- Focus on process/Improvement plans.
- Line Management ownership.

Therefore, we can say that TQM must be practical in all activites, by all personnel, in manufacturing, marketing, Engineering, R&D Sales, purchasing , HR etc. There is a need to achieve excellence. This excellence can come through Total Quality Management.

II. EFFECTS OF TQM IMPLEMENTATION:

Many firms have arrived at the conclusion that effective TQM implementation can improve their competitive abilities and provide strategic advantages in the market place and several studies have shown that the adoption of TQM practices can allow firms to compete globally several researches also reported that TQM implementation has led to improvements in quality., productivity and competitiveness in only 20-30% of the firms that have implemented it. A study shows that a 90% improvement rate in employer relations, operating procedures, customer satisfaction, and financial performance is achieved due to TQM implementation. But some of the scientist reported that TQM implementation has uncertain or even negative effects on performance. Also some of the scientist reported that TQM implementations are highly dependent on top management support. Thus conflicting research findings have been reported surroundings effects of TQM implementation on organizational performance.

III. CONCEPT OF TQM

Some of the Scientist i.e. gurus like Deming etc. The primary authorities of total quality management have developed certain propositions in the field of TQM, which have gained significant acceptance throughout the world. After careful study of their work, it has been found that these quality gurus have different views about TQM, although some similarities can be found. So far, TQM has come to mean different things to different people (Hack man and Wageman, 1995).

Towards TQM through quality circle:

Quality Circle is important in the following aspects:

- Improvement in human relations and works-area morale.
- Promotion of participate culture.
- Enhancement of job interest.
- More effective team work.
- Reducing defects and improving quality.
- Improving housekeeping cost effectiveness, safety etc.
- Improvement of productivity.
- Enhancing problem solving capability.
- Encouraging on attitude of problem prevention.
- Improving communication.
- Promotion of personal and leadership development.
- Catalyzing attitudinal changes.

Thus quality circle is a mean towards total quality management.

IV. TOP MANAGEMENT

The primary role of the top management is to convey to all the employees of the organization and its commitment to TQM and participative concept of quality circles. External visible support to the movement by attending the major function such as annual conventions, messages in newsletters, etc. Formulates quality policy and circulate it to all employees. They form quality council and preside over its meeting for constantly overseeing the progress of the movement, and taking decisions of policy, once a quarter sanction necessary funds for the movement. So the Chairman of Directors of any large corporation do not fall within the formal structure of quality circles, it plays an important role to ensure success of implementation of the concept in the organization. For the purpose, a quality council should be formed, reporting directly to Managing Director which has two wings, one for overseeing TQM as a whole and other to take an overview of quality circles movement in the organization.

V. DOCUMENT SHOULD COVER THE FOLLOWING ASPECT OF TQM

- Company's quality policy and objectives.
- Quality organization and responsibilities of the Quality Manager.
- Procedure for preparing the product design and its evaluation.
- Vendor relations and the Control of incoming materials.
- Manufacturing Control procedures, with clear instructions, regarding responsibility for planning and operation of process controls, and authority for approving or stopping the process.
- Planning, manufacturing and procurement of test equipment and the running of the test laboratory.
- Product acceptance procedures.
- Packing and shipment control.
- Investigation of customer complaints on quality.
- Quality motivation and training.
- Quality Audit.

VI. REWIEW OF QUALITY AWARD MODELS

There are several quality Awards, such as the Deming prize in Japan (1996). The Malcolm

Baldrige National Quality Award in the United States of America (1999). The broad aims of these awards are described follows (Ghobadian and WOO, 1996). The European Quality award in Europe (1994).

These awards encourage firms to introduce a continuous improvement process, Promote understanding of the requirements for the attainment of quality excellence and successful deployment of TQM etc.

VII. CONCLUSION

After review of various literatures it has been found that there is very little work which gives the idea of the impact of TQM implementation on organization performance. Also different researchers have different kind of finding according to the environment in which they have done the study, which has very less applicability for the Indian companies. Therefore going for this kind of research will be a boon for the Indian Companies.

VIII. REFERENCE

[1] Crosby Philips B. (1992). Complete new Quality for 21st Century, Duttam, Canada.

[2] Demig, (1982). Quality Productivity and Competitive positioning, Cambridge M.A.N.I.T Press, Cambridge.

[3] Juran, (1974), Quality Control Hand book, 3rd Edition, MC Graw Hill Book Company, New York.

[4] Ambroz, M. (2004). Total quality System as a product of the eimpowered Corporate Cultute. The TQM Magazine, Vol. 16. No. 2, PP. 93-104.

[5] Jha, U.C and Sunand Kumar (2008) "Impact of TQM on firm's performance; An Empirical Analysis of Indian Manufacturing Industry" at 12th International SOM conference, IIT, Kanpur.

CNT -Strength behind Composite- A Review

[1]**Kamal Sharma,**[1]**Kuwar Mausam,**[2]**Joyti Vimal Srivastava**
[1]Department of Mechanical Engineering, GLA UNIVERSITY, Mathura
[2]Department of Mechanical Engineering,MITS Gwalior,

Abstract — Superlative mechanical properties of carbon nanotubes (CNT's) make them the filler material of choice for composite reinforcement. In this work we review the progress to date in the field of mechanical reinforcement composites are introduced and the prerequisites concern to the properties of CNT's for successful reinforcement discussed. Various nanofillers have been discussed and the properties of different composites being tabulated. The effectiveness of different properties of CNT's like electrical conductivity, optical properties, chemical reactivity, mechanical strength, growth mechanism etc. In addition we discuss the levels of fabrication methods of CNT's for both multi walled CNT's and Single walled CNT's (MWCNT and SWCNT) that have actually been achieved. Finally we compare and tabulate these results. This leads to a discussion of the most promising processing methods for mechanical reinforcement and the outlook for the future.

Key-words— Carbon nanotubes, mechanical strength, optical properties, chemical reactivity

I. INTRODUCTION

Properties of Nanotubes

Different from other carbon materials, such as graphite, diamond and fullerene (C60, C70, etc.), CNTs are one dimensional carbon materials which can have an aspect ratio greater than 1000. [1] They can be envisioned as cylinders composed of rolled-up graphite planes with diameters in nanometer scale. The cylindrical nanotube usually has at least one end capped with a hemisphere of fullerene structure.[1] Depending on the process for CNT fabrication, there are two types of CNTs : single-walled CNTs (SWCNTs) and multiwalled CNTs (MWCNTs). SWCNTs consist of a single graphene layer

(**Fig. 1A**) rolled up into a seamless cylinder whereas MWCNTs consist of two or more concentric cylindrical shells of graphene sheets[1-2]

(**Fig. 1B**) [1-2] coaxially arranged around a central hollow core with vander Waals forces between adjacent layers.

According to the rolling angle of the graphene sheet, CNTs have three chiralities: armchair, zigzag and chiral one. The tube chirality is defined by the chiral vector, Ch = na1 + ma2 (**Fig. 2**),[2] where the integers (n, m) are the number of steps along the unit vectors (a1 and a2) of the hexagonal lattice . Using this (n, m) naming scheme, the three types of orientation of the carbon atoms around the nanotube circumference are specified. If n = m, the nanotubes are called "armchair". If m = 0, the nanotubes are called "zigzag". Otherwise, they are called "chiral". The chirality of nanotubes has significant impact on their transport properties, particularly the electronic properties. For a given (n, m) nanotube, if (2n + m) is a multiple of 3, then the nanotube is metallic, otherwise the nanotube is a semiconductor.

`Each MWCNT contains a multi-layer of graphene, and each layer can have different chiralities, so the prediction of its physical properties is more complicated than that of SWCNT.

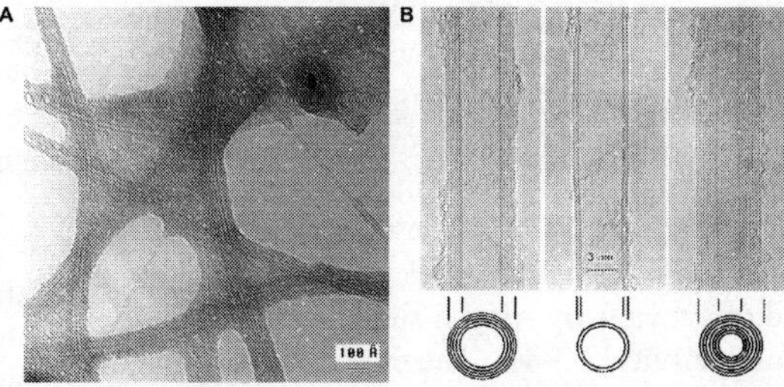

Fig 1. TEM images of different CNTs (A: SWCNTs; B: MWCNTs with different layers of 5, 2 and 7). . [1]

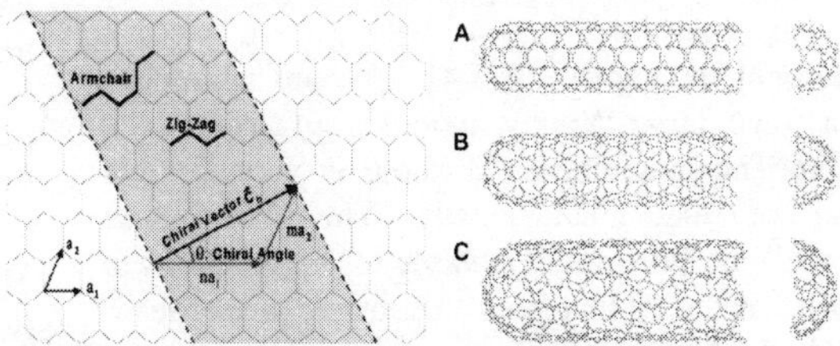

Fig 2. Schematic diagram showing how a hexagonal sheet of graphene is rolled to form a

CNT with different chiralities (A: armchair; B: zigzag; C: chiral). [1]

Special properties of carbon nanotubes [4-5]
Electronic, molecular and structural properties of carbon nanotubes are determined to a large extent by their nearly one dimensional structure. The most important properties of CNTs and their molecular background are stated below.

Chemical reactivity. [4-5] The chemical reactivity of a CNT is, compared with a graphene sheet, enhanced as a direct result of the curvature of the CNT surface. Carbon nanotube reactivity is directly related to the pi-orbital mismatch caused by an increased curvature. Therefore, a distinction must be made between the sidewall and the end caps of a nanotube. For the same reason, a smaller nanotube diameter results in increased reactivity. Covalent chemical modification of either sidewalls or end caps has shown to be possible. For example, the solubility of CNTs in different solvents can be controlled this way. Though, direct investigation of chemical modifications on nanotube behaviour is difficult as the crude nanotube samples are still not pure enough.[3]

Electrical conductivity [2-3-4]. Depending on their chiral vector, carbon nanotubes with a small diameter are either semi-conducting or metallic. The differences in conducting properties are caused by the molecular structure that results in a different band structure and thus a different band gap. The differences in conductivity can easily be derived from the graphene sheet properties. It was shown that a (n,m) nanotube is metallic as accounts that: $n=m$ or $(n-m) = 3i$, where i is an integer and n and m are defining the nanotube. The resistance to conduction is determined by quantum mechanical aspects and was proved to be independent of the nanotube length. For more, general information on electron conductivity is referred to a review by Ajayan and Ebbesen.

Optical activity[2-3-4]. Theoretical studies have revealed that the optical activity of chiral nanotubes disappears if the nanotubes become larger.Therefore, it is expected that other physical properties are influenced by these parameters too. Use of the optical activity might result in optical devices in which CNTs play an important role.

Mechanical strength. [2-3-4] Carbon nanotubes have a very large Young modulus in their axial direction. The nanotube as a whole is very flexible because of the great length. Therefore, these compounds are potentially suitable for applications in composite materials that need anisotropic properties.

Growth mechanism [5]
The way in which nanotubes are formed is not exactly known. The growth mechanism is still a subject of controversy, and more than one mechanism might be operative during the formation of CNTs. One of the mechanisms consists out of three steps. First a precursor to the formation of nanotubes and fullerenes, C2, is formed on the surface of the metal catalyst particle. From this metastable carbide particle, a rodlike carbon is formed rapidly. Secondly there is a slow graphitisation of its wall. This mechanism is based on in-situ TEM observations. The exact atmospheric conditions depend on the technique used, later on; these will be explained for each technique as they are specific for a technique. The actual growth of the nanotube seems to be the same for all techniques mentioned.

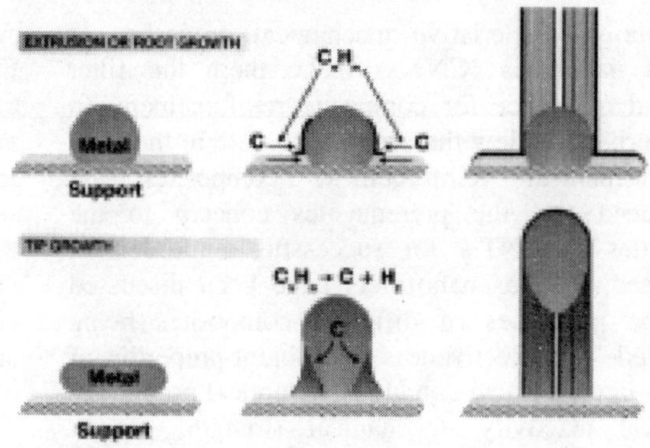

Fig 3. Visualisation of a possible carbon nanotube growth mechanism.

There are several theories on the exact growth mechanism for nanotubes. One theory postulates that metal catalyst particles are floating or are supported on graphite or another substrate. It presumes that the catalyst particles are spherical or pear-shaped, in which case the deposition will take place on only one half of the surface (this is the lower curvature side for the pear shaped particles). The carbon diffuses along the concentration gradient and precipitates on the opposite half, around and below the bisecting diameter. However, it does not precipitate from the apex of the hemisphere, which accounts for the hollow core that is characteristic of these filaments. For supported metals, filaments can form either by 'extrusion (also known as base growth)' in which the nanotube grows upwards from the metal particles that remain attached to the substrate, or the particles detach and move at the head of the growing nanotube, labelled 'tip-growth'. Depending on the size of the catalyst particles, SWNT or MWNT are grown. In arc discharge, if no catalyst is present in the graphite, MWNT will be grown on the C2-particles that are formed in the plasma.

Synthesis of SWNT [6]
If SWNTs are preferable, the anode has to be doped with metal catalyst, such as Fe, Co, Ni, Y or Mo.A lot of elements and mixtures of elements have been tested by various authors16 and it is noted thatthe results vary a lot, even though they use the same elements. This is not surprising as experimental conditions differ. The quantity and quality of the nanotubes obtained depend on various parameters such as the metal concentration, inert gas pressure,

kind of gas, the current and system geometry.Usually the diameter is in the range of 1.2 to 1.4 nm. A couple of ways to improve the process of arc discharge are stated below.

- Inert gas
- Optical plasma control
- Catalyst

d) Improvement of oxidation resistance

e) Open air synthesis with welding arc torch

Synthesis of MWNT [6]

If both electrodes are graphite, the main product will be MWNTs. But next to MWNTs a lot of side products are formed such as fullerenes, amorphous carbon, and some graphite sheets. Purifying the MWNTs, means loss of structure and disorders on the walls. However scientists are developing ways to gain pure MWNTs in a large-scale process without purification. Typical sizes for MWNTs are an inner diameter of 1-3 nm and an outer diameter of approximately 10 nm. Because no catalyst is involved in this process, there is no need for a heavy acidic purification step. This means, the MWNT, can be synthesized with a low amount of defects.

- Synthesis in liquid nitrogen
- Magnetic field synthesis
- Plasma rotating arc discharge

Major production methods and their efficiency [9-10-23]

Method	Arc discharge method	Chemical vapour deposition	Laser ablation (vaporization)
Who	Ebbesen and Ajayan, NEC, Japan 1992	Endo, Shinshu University, Nagano, Japan	Smalley, Rice, 199514
How	Connect two graphite rods to a power supply, place them a few millimeters apart, and throw the switch. At 100 amps, carbon vaporizes and forms hot plasma.	Place substrate in oven, heat to 600 oC, and slowly add a carbon-bearing gas such as methane. As gas decomposes it frees up carbon atoms, which recombine in the form of NTs	Blast graphite with intense laser pulses; use the laser pulses rather than electricity to generate carbon gas from which the NTs form; try various conditions until hit on one that produces prodigious amounts of SWNTs
Typical yield	30 to 90%	20 to 100 %	Up to 70%
SWNT	Short tubes with diameters of 0.6 - 1.4 nm	Long tubes with diameters ranging from 0.6-4 nm	Long bundles of tubes (5-20 microns), with individual diameter from 1-2 nm.
MWNT	Short tubes with inner diameter of 1-3 nm and outer diameter of approximately 10 nm	Long tubes with diameter ranging from 10-240 nm	Not very much interest in this technique, as it is too expensive, but MWNT synthesis is possible.
Pro	Can easily produce SWNT, MWNTs. SWNTs have few structural defects; MWNTs without catalyst, not too expensive, open air synthesis possible	Easiest to scale up to industrial production; long length, simple process, SWNT diameter controllable, quite pure	Primarily SWNTs, with good diameter control and few defects. The reaction product is quite pure.
Con	Tubes tend to be short with random sizes and directions; often needs a lot of purification	NTs are usually MWNTs and often riddled with defects	Costly technique, because it requires expensive lasers and high power requirement, but is improving

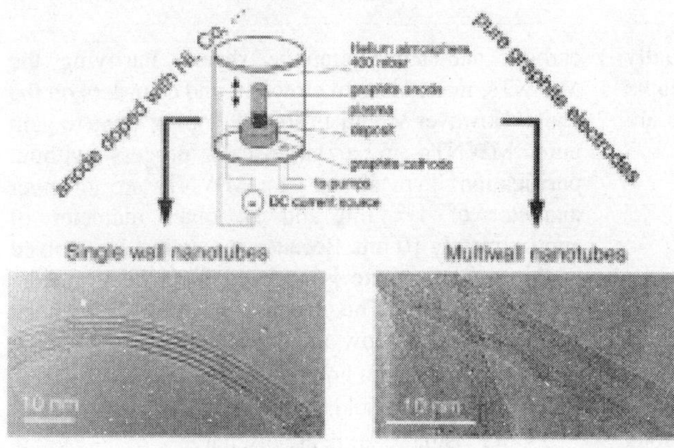

Fig 4. Experimental set-up of an arc discharge apparatus.

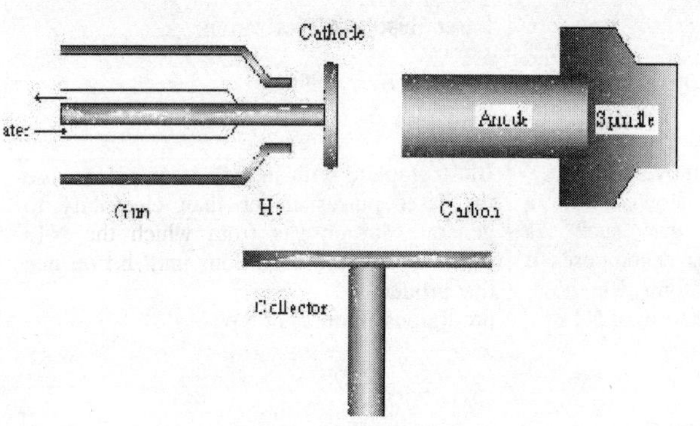

Fig 5. Schematic diagram of plasma rotating electrode system.

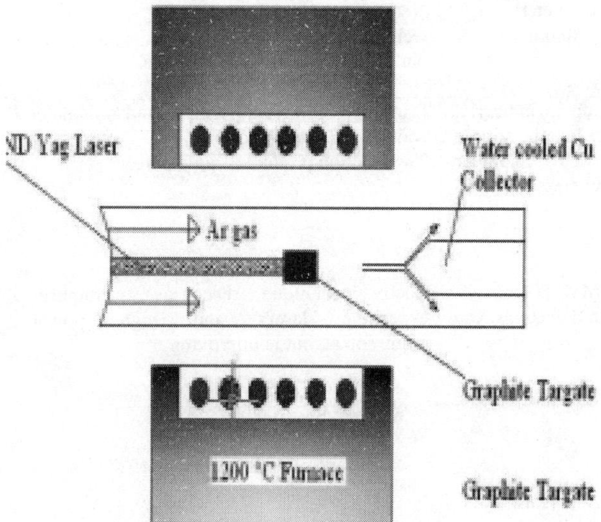

Fig 6. Schematic drawing of a laser ablation apparatus.

SWNT versus MWNT [14]

The condensates obtained by laser ablation are contaminated with carbon nanotubes and carbon nanoparticles. In the case of pure graphite electrodes, MWNTs would be synthesised, but uniform SWNTs could be synthesised if a mixture of graphite with Co, Ni, Fe or Y was used instead of pure graphite. SWNTs synthesised this way exist as 'ropes', see Figure 2-10 28,30. Laser vaporisation results in a higher yield for SWNT synthesis and the nanotubes have better properties and a narrower size distribution than SWNTs produced by arc-discharge. Nanotubes produced by laser ablation are purer (up to about 90 % purity) than those produced in the arc discharge process. The Ni/Y mixture catalyst (Ni/Y is 4.2/1) gave the best yield.

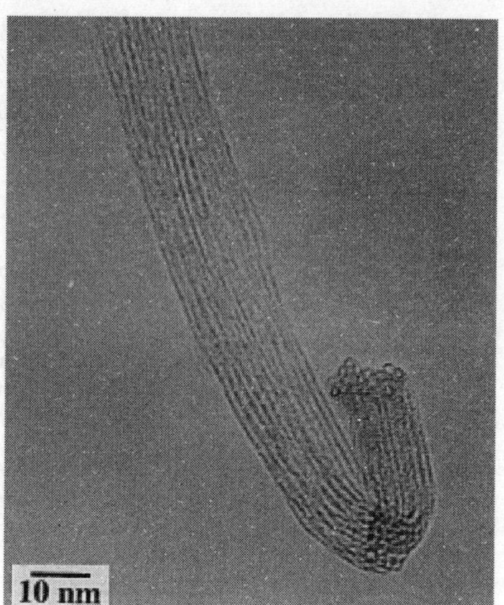

Fig7 TEM images of a bundle of SWNTs catalysed by Ni/Y (2:0.5 at. %) mixture, produced with a continuous laser.

The size of the SWNTs ranges from 1-2 nm, for example the Ni/Co catalyst with a pulsed laser at 1470 °C gives SWNTs with a diameter of 1.3-1.4 nm26. In case of a continuous laser at 1200 °C and Ni/Y catalyst (Ni/Y is 2:0.5 at. %), SWNTs with an average diameter of 1.4 nm were formed with 20-30 % yield.

NANOFILLER- [16]

Nanofiller are classified into three categories depending on their geometry, as shown in **Fig. 8**

1. Nanoparticles: When the three dimensions of particulates are in the order of nanometers, they are referred as equi-axed (isodimensional) nanoparticles or nanogranules or nanocrystals.

2. Nanotubes: When two dimensions are in the nanometer scale and the third is larger, forming an elongated structure, they are generally referred as 'nanotubes' or nanofibers/whiskers/nanorods.

3. Nanolayers: The particulates which are characterized by only one dimension in nanometer scale are nanolayers/nanoclays/nanosheets/nanoplatelets. These particulate is present in the form of sheets of one to a few nanometer thick to hundreds to thousands nanometers long.

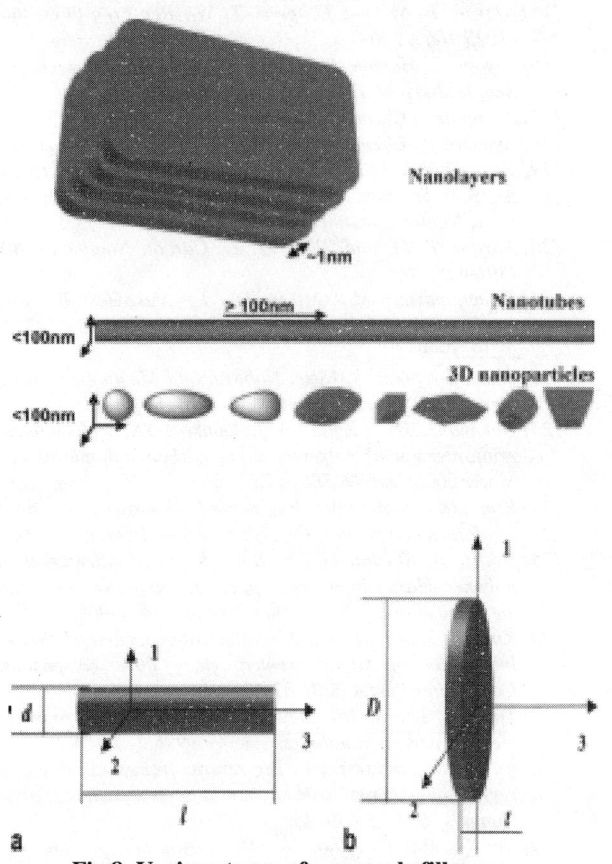

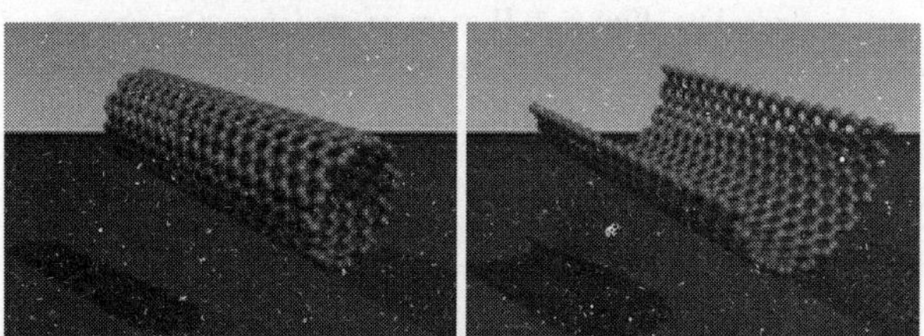

Fig 8. Various types of nanoscale filler

6.1 Types of carbon nanotubes and related structures [30]

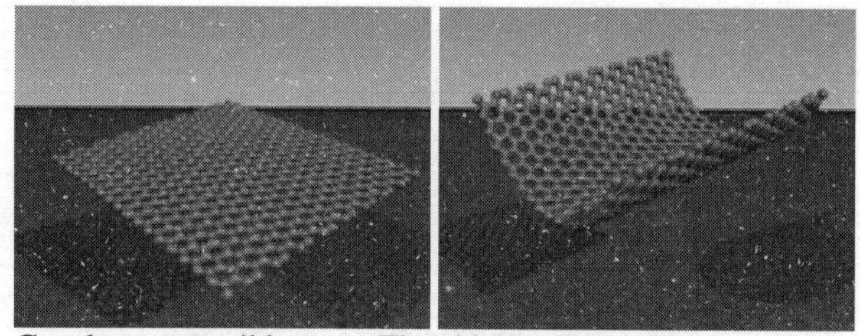

Armchair   The chiral vector is bent, while the translation vector stays straight

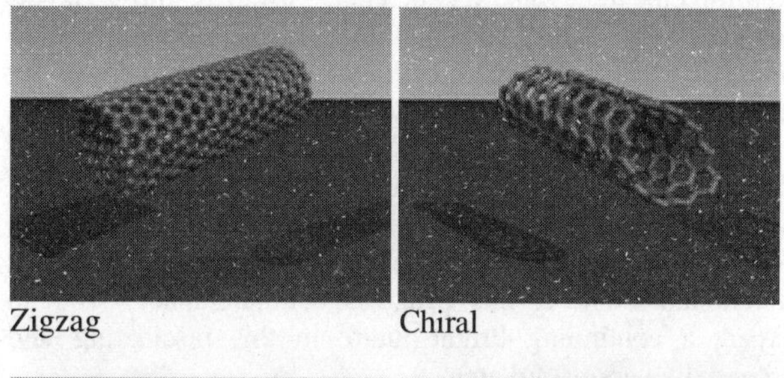

Graphene nanoribbon   The chiral vector is bent, while the translation vector stay straight

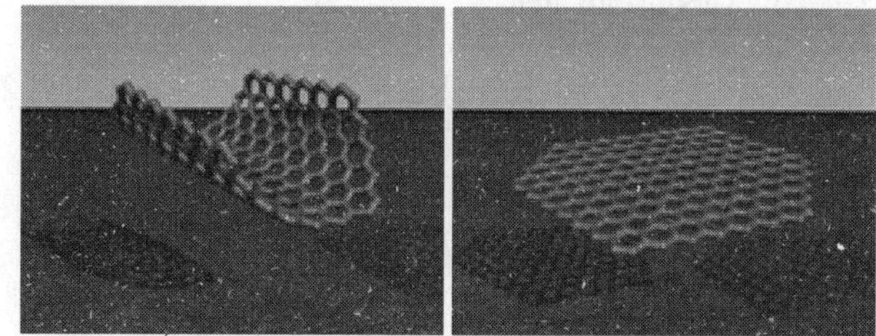

Zigzag   Chiral

n and m can be counted   Graphene nanoribbon at the end of the tube

Fig 9. Various types of nanotube structure

Comparison of mechanical properties of different nanotube structure:-

Material	Young's modulus (TPa)	Tensile strength (GPa)	Elongation at break (%)
SWNT	~1 (from 1 to 5	13–53	16
Armchair SWNT	0.94	126.2	23.1
Zigzag SWNT	0.94	94.5	15.6–17.5
Chiral SWNT	0.92		
MWNT	0.27–0.8–0.95	11–63–150	

SS	0.186–0.214	0.38–1.55	15–50
Kevlar–29&149	0.06–0.18	3.6–3.8	~2

VI CONCLUSION

In conclusion, much progress has been made over the last few years in composite reinforcement using carbon nanotubes. Reinforcement levels are creeping towards those levels predicted by theory. In addition, many researchers are branching out with the fabrication of novel structures by innovative processing techniques. Some of these materials, such as coagulation spun fibres have even challenged state of the art materials such as Kevlar. Most importantly, the overriding problems of dispersion and stress transfer are being addressed using chemical techniques such as functionalisation. If this progress continuesapace, we can expect a continuing bright future in this fascinating and potentially very useful area.

VII. REFERENCES

[1] Iijima S. Helical microtubules of graphitic carbon. Nature 1991;354:56–8.

[2] Bethune DS, Klang CH, De Vries MS, et al. Cobalt-catalysed growth of carbon nanotubes with single-atomic-layer walls. Nature 1993;363:605–7.

[3] Dresselhaus MS, Dresselhaus G, Saito R. Physics of carbon nanotubes. Carbon 1995;33:883–91.

[4] Thostenson ET, Ren ZF, Chou TW. Advances in the science and technology of CNTs and their composites: a review. Compos Sci Technol 2001;61:1899–912.

[5] Yakobson BI, Avouris P. Mechanical properties of carbon nanotubes. Top Appl Phys 2001;80:287–327.

[6] Qian D, Wagner GJ, Liu WK, Yu MF, Ruoff RS. Mechanics of carbon nanotubes. Appl Mech Rev 2002;55:495–533.

[7] Reich S, Thomsen C, Maultzsch J. Carbon nanotubes: basic concepts and physical properties. New York: Wiley-VCH; 2004. p. 31–40.

[8] Koo JH. Polymer nanocomposites: processing, characterization, and applications. New York: McGraw-Hill; 2006. p. 1–8.

[9] Ajayan PM, Schadler LS, Braun PV. Nanocomposite science and technology. Weinheim: Wiley-VCH; 2003. p. 77–80.

[10] D.A.Bochvar and E.G.Gal'pern, Dokl.Akad.Nauk.USSR, 209, (610, 1973

[11] I.V.Stankevich, M.V.Nikerov, and D.A.Bochvar, Russ.Chem.Rev., 53, (640, 1984

[12] H.W.Kroto, J.R.Heath, S.C.O'Brien, R.F.Curl, and R.E.Smalley, Nature, 318, (162, 1985

[13] Iijima, Sumio, Nature (London, United Kingdom), 354, (6348), 1991

[14] .Dresselhaus, M. S., Dresselhaus, G., and Eklund, P. C., 1996

[15] Ajayan, P. M. and Ebbesen, T. W., Rep.Prog.Phys., 60, (1025-1065, 2003

[16] Niyogi, S., Hamon, M. A., Hu, H., Zhao, B., Bhowmik, P., Sen, R., Itkis, M. E., and Haddon, R. C.,

[17] Accounts of Chemical Research, 35, (12), 2002

[18] Avouris, P., Chemical Physics, 281, (2-3), 429-445, 2002

[19] Tans, Sander J., Devoret, Michel H., Dal, Hongjie, Thess, Andreas, Smalley, Richard E., Geerligs, L. J., and Dekker, Cees, Nature (London), 386, (6624), 1997

[20] Ajayan, P. M. and Zhou, O. Z., Carbon Nanotubes, 80, (391-425, 2001

[21] Damnjanovic, M., Milosevic, I., Vukovic, T., and Sredanovic, R., Physical Review B, 60, (4), 2728-2739, 1999

[22] Yasuda, Ayumu, Kawase, Noboru, and Mizutani, Wataru, Journal of Physical Chemistry B, 106, (51),2002

[23] Coleman JN, Khan U, Gunko YK. Mechanical reinforcement of polymers using carbon nanotubes. Adv Mater 2006;18:689–706.

[24] Kim JK, Mai YW. Engineered interfaces in fiber reinforced composites. Oxford: Elsevier; 1998. p. 1–100.

[25] Hodzic A, Stachurski ZH, Kim JK. Nano-indentation of polymer–glass interfaces part I: experimental and mechanical analysis. Polymer 2000;41:6895–905.

[26] Kim JK, Sham ML, Wu JS. Nanoscale characterisation of interphase in silane treated glass fibre composites. Composites Part A 2001;32:607–18.

[27] Thess A, Lee R, Nikolaev P, et al. Crystalline ropes of metallic carbon nanotubes. Science 1996;273:483–8.

[28] Breuer O, Sundararaj U. Big returns from small fibers: a review of polymer/ carbon nanotube composites. Polym Compos 2004;25:630–45.

[29] Xie XL, Mai YW, Zhou XP. Dispersion and alignment of carbon nanotubes in polymer matrix: a review. Mater Sci Eng R 2005;49:89–112. en.wikipedia.org/wiki/Carbon_nanotube(accesson 01/02/2011)

[30] Peng-Cheng Maa, Naveed A. Siddiqui a, Gad Marom b, Jang-Kyo Kim a,* Dispersion and functionalization of carbon nanotubes for polymer-based nanocomposites: A review Composites: Part A 41 (2010) 1345–1367

[31] Nanda Gopal Sahooa, Sravendra Ranab, Jae Whan Chob,□, Lin Li a,□□, Siew Hwa Chana Polymer nanocomposites based on functionalized carbon nanotubes Progress in Polymer Science 35 (2010) 837–86

Experimental investigations of Mechanical Properties of Alumina Particle Reinforced in Metal Matrix Composite of Aluminum

Lokesh Upadhyay[1], Puneet Bansal[1] and Mohd.Arif[2]

[1]Department of Mechanical Engineering, Sachdeva Institute of Technology,Mathura,U.P.
[2]Department of Mechanical Engineering, Anand Engineering College, Agra,U.P.

*Abstract-***Metal matrix compositions (MMC) have become a leading materials and particles reinforced aluminum MMCs have received considerable attention due to their excellent mechanical properties like high hardness, high tensile strength etc. These materials difficult to machine because of high hardness and abrasive nature of reinforcing elements like alumina particles. In this study, homogenized (2%, 4%, and 6%) by weight of alumina aluminum metal matrix composite materials were fabricated and selected as work piece for experimental investigations of Tensile Strength, EDX SEM, and Hardness. The microstructures and mechanical properties of produced composite specimens have been investigated. It has been observed that increase of reinforcement element produced better mechanical properties such as hardness and tensile strength. The main effects have been discussed over concentration effect on Tensile Strength, EDX, SEM and Hardness have been determined**

Keywords: Metal matrix composites; SEM ,Hardness , EDAX

1. INTRODUCTION

There have been huge work in engineering materials since 1950. Several Super Alloys, Heat Resisting Materials, Smart materials have been developed for various industrial applications e.g. Aerospace, Automotives, Defense, Medical equipments. A new generation materials are developing of low density and high mechanical properties one of this is Composite Materials. (Matthew and Rawling, 1994)

Composite materials are important materials since they having superior mechanical properties. Metal Matrix Composite (MMC) are widely used because of their great mechanical properties such as high strength, Hardness, Stiffness, Wear, Corrosion resistance. Aluminum based MMCs are widely used due to their economical production. (Beder and Ögel, 2004)

As the matrix element, aluminum, titanium and magnesium alloy are used, while the popular reinforcements are silicon carbide (SiC) and alumina (Al_2O_3). Aluminum-based Al2O3 particle reinforced MMC materials have become so useful engineering materials due to their properties such as low weight, heat-resistant, wear-resistant and low cost. (N.P.Hung and K.P.ng 1994)

Embury (1985) reported that surface tension in MMCs decreases Ductility of Matrix element of MMCs. Ejiofor and Reddy in 1997 has explained that strengthening mechanism of discontinues reinforced metal and intermetallic matrix composite with constipation of matrix and strength of basic structure. Selezenev et al (1998) has observed that there was 10% increase in Tensile strength and no change in fracture toughness by addition of 1.5% of mg but silicon added to enhanced the mechanical properties.

Increase in the industrial applications of composite materials brings the matter of shaping of these materials. From this point of view, it is natural that the methods of production by machining are considered initially, for why machining has a prevalence of approximately 70% among all production meth- ods. The processing parameters affecting machinability of a material are the values of cutting speed according to selected set of material properties of work piece and machining

In this study SEM, EDX, Tensile Strength and Hardness of a Metal Matrix Composite (MMCs) of Alumina particle reinforced into Aluminum is analyses and thus results are find out that how the concentration of the Alumina particles effects these SEM, EDX, Tensile Strength, Hardness

II. EXPERIMETAL

The manufacturing of the composite material we select two materials. The Al 2024 alloy is used as base metal and α- Al_2O_3 is used as reinforcement material for manufacturing. The range of particle size of alumina is 16-30 μm and average particle size is 25μm.The general properties of aluminum are non toxic, impervious, non magnetic, non sparking, easily machined, low density, corrosion resistant, electricity conductor, non-magnetic, malleable. Alumina is the most cost effective and widely used material in the family of engineering ceramics. It has many desirable characteristics like hard, wear resistant, good thermal conductivity, high strength and stiffness, excellent size and shape capability.

The Al_2O_3 reinforced Al composite was produced by sand casting in foundry shop The dimensions of final product were 45mm in diameter and 300mm length. In order to obtain matrix material at the beginning phase of the production, 99.9% pure aluminum was melted in the crucible at 700°C on muffle furnace. Then the alumina was added in the crucible and steered continuously. In order to increase wetting capability of reinforcement, 2% of Mg was added. In our experiment three types of alumina particles reinforced metal matrix composites were casted. In the first type 10% by weight alumina and remaining aluminum and could able to mix 2% by weight alumina in the final cast product In type second 20% by weight alumina and remaining aluminum and could able to mix 4% by weight alumina in the final cast product. In third type 30% by weight alumina and remaining aluminum and could able to mix 6% by weight alumina in the final cast product which contained 2% alumina; another contained 4%

alumina and the remaining were of 6% alumina by weight and one piece of 100% aluminum by weight.

Hardness, tensile strength, scanning electron microscope images and EDAX have been characterized of work piece material. Hardness of composite is tested on the Rockwell hardness testing machine .The model of machine is Fine Engineering industries, S No. NR S. and pressure imposed capacity is 50kgf, 100kgf, 150kgf. We had applied 100kgf at B scale1/16'' ball penetration, of pressure on our three specimens of 2%, 4% & 6% by weight in alumina particles reinforced aluminum metal matrix composite. The maximum capacity of UTM is 40,000Kgf. When the composite specimen is obtained from casting, their micro-Structure of 2%, 4%, and 6% will examined with scanning electron microscope (SEM). Energy dispersive X-rays Spectroscope (EDAX) gives the dispersive energy of Materials in terms of percentage workspace material.

III. RESULT AND DISCUSSIONS

Hardness

it has been observed that hardness of composite increases with increasing the reinforcement ratio. This is due to the fact that alumina is very hard and brittle material, so reinforcement of alumina increases the hardness of the composite material.

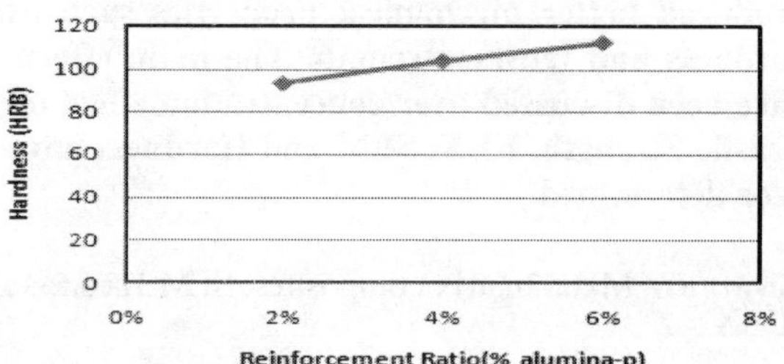

Tensile strength

From main effect plot of tensile strength (figure 3.4), it has been observed that there is not so much effect of reinforcement on the tensile strength of the composite material. It is also increase with increasing the reinforcement but at slower rate.

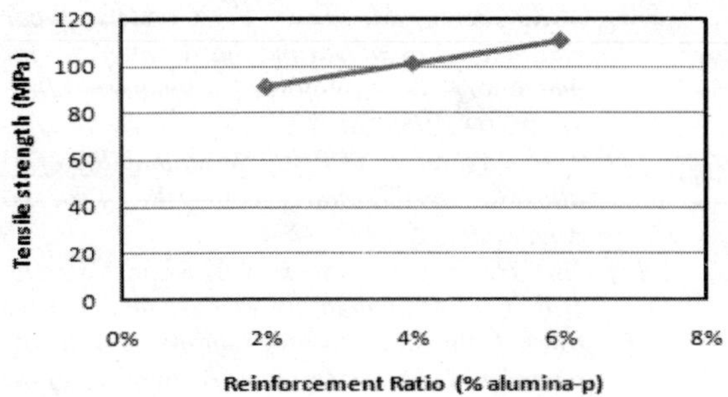

Energy –dispersive X- ray spectroscopy (EDAX)

The EDAX analysis stands for energy dispersive X-ray spectroscopy. It is a technique used for identifying the elemental composition of the specimen. The EDAX of workpeice material of 6% reinforcement of alumina shown in fig. (3.6).

Element	Wt.(%)	Error
Aluminum	92.49	4.4
Oxygen	8.95	1.8

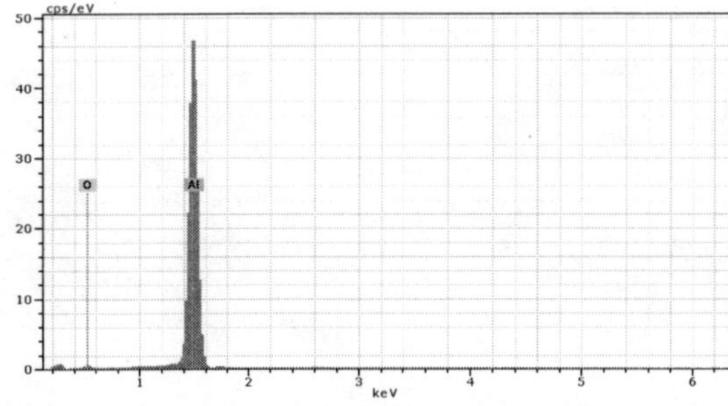

SEM of composite specimens

The SEM images of work piece material of 2%, 4%, and 6% reinforcement of alumina shown in fig (3.5)

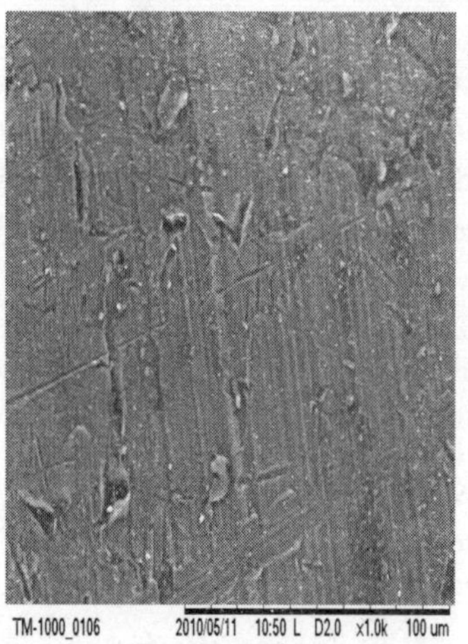

Fig (a)

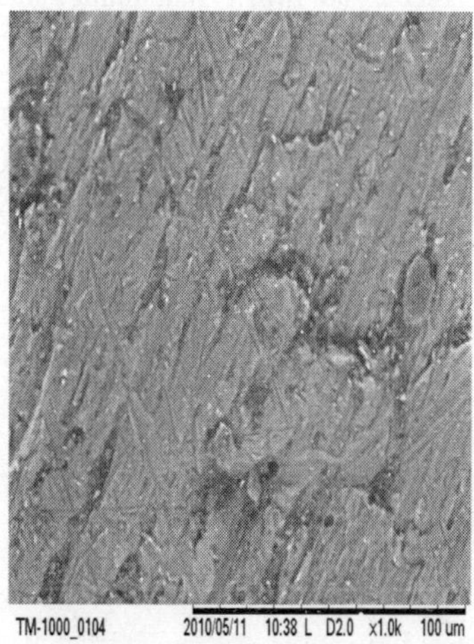

Fig (b)

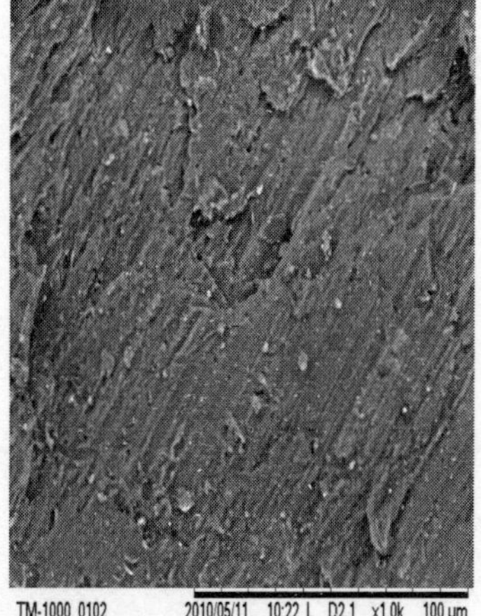

Fig (c)

Fig 3.5: (a,b,c) SEM images of composite specimens, 2%, 4%, and 6% Cons. of alumina.

IV. CONCLUSION

In this study mechanical properties of the alumina reinforced aluminum composite material has examined. The effect of Tensile strength, SEM, EDX, Hardness finds out thus some conclusions are takes.

a)   EDAX is analyses as per the Process.
b)   SEM shows the orientations of the Alumina Particles arranged into the MMCs of Aluminum.
c)   With increase in reinforcement ratio, the hardness and tensile strength of composite material found to be increase.

V REFERENCES

[1]   K. Winert, A consideration of tool wears mechanism when machining metal matrix composites, Ann. CIRP 42 (1) ,1993, 95–98.

[2]   Kilickap, Orhan Cakir 2008.Investigate the mechanical and machinability properties of SiC-p reinforced Al-MMC. Journal of materials processing technology 198 ,2008, 220-225.

[3]   M.K. Brun, M. Lee, F. Gorsler, Wear characteristics of various hard materials for machining SiCp reinforced aluminium alloy, Wear 104 ,1985, 21–29.

[4]   J. Monaghan, P. O'Reilly, Machinability of Al alloy/SiC metal matrix composite, Process. Adv. Mater. 2 ,1992, 37–46.

[5]   N.P. Hung, F.Y.C. Boey, K.A. Khor, C.A. Oh, H.F. Lee, Machinability of cast and powder formed aluminium alloys reinforced with SiC particles, J. Mater. Process. Technol. 48 (1–4) ,1995, 291–297.

[6]   N. Raghunath, P.M. Pandey , Improving Accuracy through shrinkage modeling by using Taguchi method in SLS, International Journal of Machine Tools and Manufacture, 2007, 47, pp 985-995

[7]   W.H. Yang, Y.S. Tarng, Design optimization of cutting parameters for turning operations based on the Taguchi method, Journal of Materials Processing Technology 84 ,1998,122–129.

Mechanical Properties of Dissimilar Joints

[1]**Kapil Gupta**, [2]**Rahul Chhibber**

[1]Deptt. of Mechanical Engineering, Hindustan College of Science & Technology, Farah Mathura.
[2]Deptt. of Mechanical Engineering, Thapar University, Patiala, Punjab.

Abstract- **The mechanical properties of dissimilar weld joints were examined. A dissimilar weld joint were formed by joining ferritic steel and an austenitic stainless steel by using gas tungsten arc weld (GTAW) process. Mechanical properties of the specimens were obtained by means of microhardness testing, tensile testing and bending fatigue testing,. The highest microhardness values were recorded on the ferritic–austenitic dissimilar weld joint, whereas the highest tensile strength and bending fatigue life were obtained with the austenitic–austenitic joints.**
Keywords: Dissimilar Weld, Gas Tungsten Arc elding, Ferrite Steel, Austenite Steel, Filler Material.

I. INTRODUCTION

Dissimilar joints between austenitic stainless steel and ferritic stainless steel are extensively utilized in many high-temperature applications in energy conversion systems. In central power stations, the parts of the boilers that are subjected to lower temperature are made of ferritic steel for economic reasons. The other parts, operating at higher temperatures, are constructed with austenitic stainless steel. Therefore, the transition welds are needed between the two materials. In addition, dissimilar joints are used in the main steam lines of power plants, in nuclear reactors, and in petrochemical plants. Dissimilar metal welds are also widely used in the construction of vessels and heat exchangers for several industrial applications [1].
There have been are several studies of the welding of dissimilar materials because the failure in dissimilar joints can occur before the components reach their expected design life [2]. The investigations have show that large thermal stresses can occur in dissimilar joints due to the difference in thermal expansion during temperature fluctuations [3]. Klueh et al. [4] pointed out that the thermal stresses are caused by metallurgical changes, including carbon migration and carbide precipitation at the grain boundaries [5]. In addition, there have been studies of the effect of hardness changes on mechanical properties of the dissimilar welds [6–8]. Joining dissimilar metals using new methods has also been a subject of interest in recent years [9,10]. The high energy density-beam welding process has been especially developed and success- fully used in many industrial applications, for which more information can be found in [11].

The morphology of the transition region in dissimilar austenitic–ferritic steel welds has been investigated, and Pan et al. [12] pointed out that the microstructure variations are dependent on the carbon content, the cooling rate, and the segregation of alloying elements. When joining ferritic steel to austenitic steel, the most important requirement is the avoidance of solidification cracking. Consequently, high cooling rates are required on the austenitic steel side of the joint [13]. The present work includes mechanical and structural property determinations on dissimilar weld joints, to investigate why failures take place before such joints reach their expected design life.

II. EXPERIMENTAL DETAILS

Test Materials

The tests were performed using plates of ferritic steel (type 508) and austenitic steel (type 316) welded by the gas tungsten arc weld (GTAW) process using an austenitic filler metal (type 309 L). The chemical compositions of the two base materials and the filler metal are given in Table 1.
Also the material properties are given in table 2.

	% Wt									
Type	C	Mn	Si	P	S	Cu	Ni	Cr	Mo	Fe
508	0.25	1.21	0.69	0.012	0.03	0.03	0.82	0.16	0.47	Rest
316	0.03	0.77	1	0.027	0.003		13.25	18	2.25	Rest
309 L	0.2	2	0.75	0.045	0.03		13	22		Rest

Table 1: Chemical Composition of the base material and filler material

	508	316	309 L
Ultimate Tensile Strength in MPa	595	545	644
Yield Strength in MPa	424	205	489
Poison's Ratio	0.29	0.27-0.30	0.27-0.30
Young Modulus in GPa	193	205	190
% Elongation	29	40	25

Table. 2: Material Properties

Material Combination:

Combination of two similar welds and one dissimilar weld were made, the joint configurations being shown schematically in Tab.3.

Sample	Material Combination
A	Ferritic-Ferritic (316-316)
B	Austenite-Austenite(508-508)
C	Ferrritic-Austenite(316-508)

Welding Parameters

Welding Process: GTAW
Current Polarity: DCRP
Welding Speed: 100 mm/min
Welding current: 140–160 A
Welding Voltage: 24 V.

Test Performed

The experimental work included

- Microhardness testing
- Tension testing
- Fatigue testing

All the weld specimens for each experiment were cut, machined, and examined to de fine their mechanical and structural properties.

Polished specimens were prepared for the mechanical property testing. The geometries of the specimens for the tension tests and fatigue tests are shown in Fig. 1.

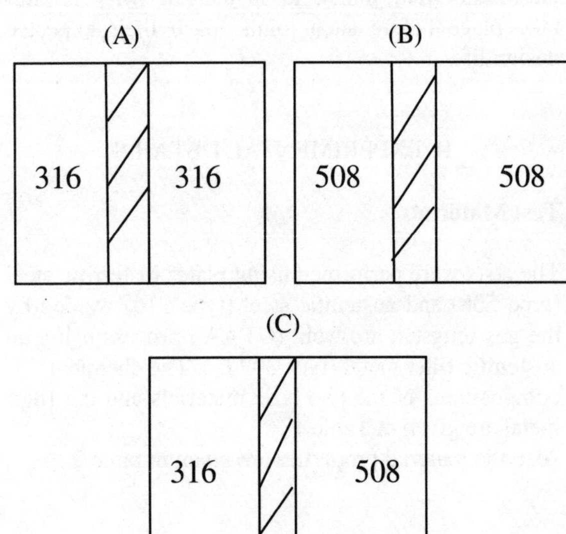

Tab.3 : Material Combination

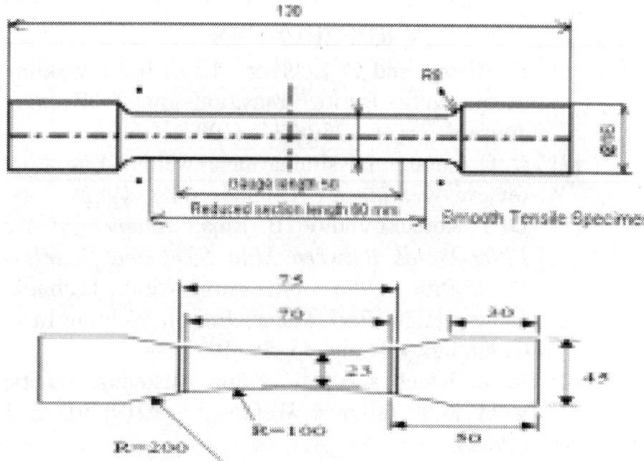

Fig.1. Geometry of the specimen (a) Tensile (b) Fatigue

Tension tests were conducted using universal test machine. The strain rate was 20 mm/minute. At least three specimens were tested for each material and for each similar or dissimilar joint.

Fatigue lives were determined by using a bending fatigue machine. Between 12 and 45 specimens were used to obtain the fatigue strength for each material. At least three specimens were tested at each stress level and the average value calculated for each level. The cycling rate was 2815 rpm (about 50Hz), and tests were run out to a maximum of 10^7 cycles, at which point they were terminated if no failure had occurred. Statistical analysis using four or five stress levels was done after the fatigue test results. The staircase method was employed to determine fatigue strength.

III. RESULTS AND DISCUSSION

Tensile Properties and Microhardness Distribution

The measured tensile properties of the ferritic steel, austenitic steel and the three different welds are reported in Table 4. The tensile strength values of the ferritic–austenitic dissimilar joints were intermediate between those of ferritic–ferritic and austenitic–austenitic welds. The ductility of the dissimilar weld was much higher, at 60%, than that of either of the two similar welds, around 40%.

Table 4 also contains the results of the microhardness tests. These were taken along a line in the middle of the thickness and include base metal, the heat affected zones (HAZ), and the weld area.

The distributions of microhardneses are graphically depicted in Fig. 2. The higher hardness values were generally obtained in the HAZ regions of the welds. The hardness values recorded in the HAZ of the ferritic–ferritic and austenitic–austenitic similar metal joints were found to be around 135–213 and 195–230HV, respectively. Those recorded in the HAZ of the dissimilar weld depended on which side of the joint was being examined, and reflected the values recorded with the corresponding base metal The highest values were seen in the weld zone of the dissimilar joint, being in the range 220–275HV. The hardness values across the all-austenitic joint changed very little, because the phase structure of Austenite steel was not transformed by the temperature fluctuations that occur during welding. By contrast, the all-ferritic joint showed much higher hardness in the weld zone and the HAZ, reflecting the fact that it had undergone a phase transformation during the welding process.

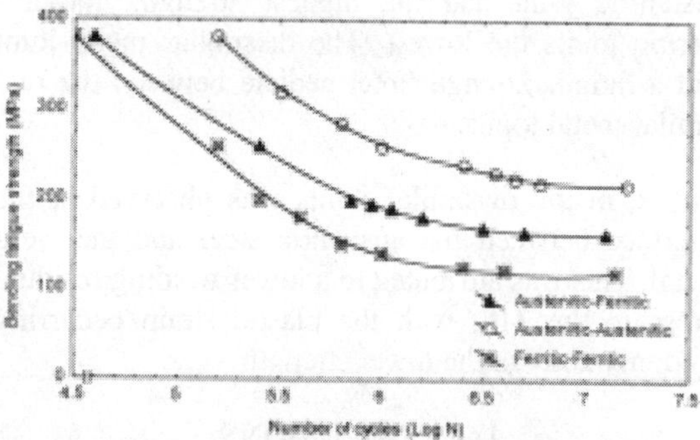

Fig. 2. Hardness profiles of the similar and dissimilar welds

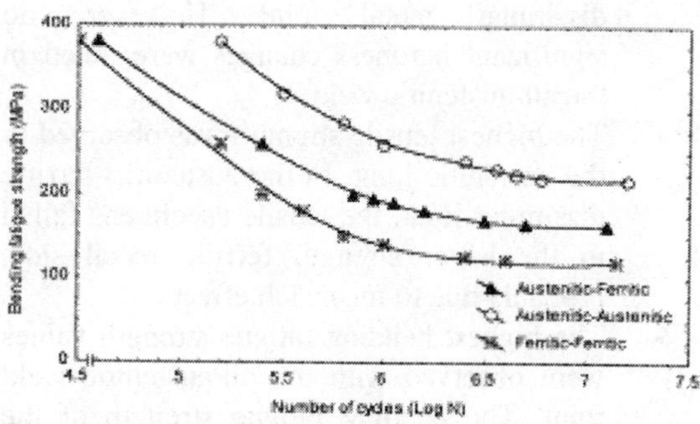

Fig. 3. Bending fatigue life of similar and dissimilar welds

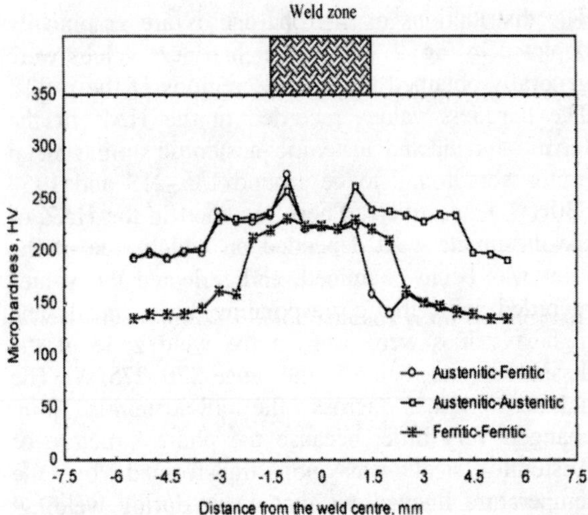

The specimen configuration for both the tension test and the bending fatigue test is shown in Fig. 1. The bending fatigue curves are shown in Fig. 3. The all-austenitic joint had the highest strength, the all-ferritic joints the lowest. The dissimilar metal joint had a fatigue strength intermediate between the two similar metal joints.

Failure in the dissimilar joints was observed at the interface between the austenitic steel and the weld metal. This was attributed to a lower welding residual stress in the 316, with the plastic strain occurring predominantly in the lower strength.

IV CONCLUSIONS

The following conclusions can be made from the results of this work.

- Extreme hardness increases were observed in the HAZ regions of the austenitic–ferritic dissimilar metal joint. However, no significant hardness changes were noted in the all-austenitic weld.
- The highest tensile strength was observed in the austenitic joint. In the austenitic–ferritic dissimilar joint, the tensile specimens failed in the lower strength ferritic metal side, probably due to mismatch effects.
- The highest bending fatigue strength values were observed with the all-austenitic weld joint. The bending fatigue strength of the dissimilar metal joint was intermediate between that of the all austenitic joint and that of the all-ferritic joint.
- Fatigue failure took place at the interface between the austenitic steel and the weld metal in the dissimilar joints. The predominant fracture mode was a transgranular ductile fracture.

V REFERENCES

[1] S. Missori and C. Koerber: "Laser beam welding of austenitic–ferritic transition joints". *Welding Journal.* 76(3): 125s–134s (1997).

[2] C. D. Lundin: Dissimilar metal welds—Literature review. *Welding J.* 61(2):585s–635s (1982).

[3] D. J. Kotecki and V. B. Rajan: *Submerged Arc Fillet Welds Between Mild Steel and Stainless Steel.* Proc. 1996 Conf. Adv. Weld. Technol.: Joining High- Perf. Mater., Edison Welding Inst., Columbus, OH, pp. 33–58 (1997).

[4] R. L. Klueh and J. F. King: Austenitic–ferritic weld joint failures. *Welding J.* 61(9):302–311 (1982).

[5] S. K. Albert, T. P. S. Gill, A. K. Tyagi, S. L. Mannan, S. D. Kulkami, and P. Rodriguez: Soft zone formation in dissimilar welds between two Cr-Mo steels. *Welding J.* 76(3):135s–142s (1997).

[6] *C. D. Lundin, K. K. Khan, and D. Yang: Effect of Carbon Migration on the Metallurgical Structure and Mechanical Properties of Cr-Mo Weldments. Proc. Conf. Recent Trends in Welding Science and Tech., Gatlinburg, TN (May 14–18, 1989), ASM Intl Materials Park, OH, pp. 291–296 (1990).*

[7] *E. Zumelzu and C. Cabetas: Study of welding such dissimilar materials as AISI 304 stainless steel and DHP copper in a sea water environment: Influence of weld metal on corrosion. J. Mater. Process. Technol. 57(3):246–252 (1996).*

[8] *R. H. Ryder and C. F. Dahms: Design criteria for dissimilar metal welds. Welding Res. Council Bull. 350:1–11 (1990).*

[9] *Z. Sun and R. Karppi: Application of electron beam welding for the joining of dissimilar metals.J. Mater. Process. Technol. 59(3):257–267 (1996).*

[10] *R. Wise: New technique for joining dissimilar ma- terials. Welding Rev. Int. 12(1):40–42 (1993).*

[11] *Z. Sun: Feasibility of producing ferritic/austenitic dissimilar metal joints by high energy density laser beam process.Int. J. Pressure Vessels Piping68(2):153–160 (1996).*

[12] *C. Pan and E. Zhang: Morphologies of the transition in dissimilar austenite- ferritic welds Mater.Char 36(1),5-10 (1996)*

[13] *C. Yeni, S. Erim, G. Cam, and M. Kocak: Microstructural Features and Fracture Behavior of Laser Welded Similar and Dissimilar Steel Joints. Int. Symp. Welding Technol. 96, Turkey (May 12–15, 1996).*

[14] *F. Gauzzi and S. Missori: Microstructural transformation in austenitic-ferritic transition joints , Mater Sci 23, 782-789, (1998)*

[15] *S. K. Bhambri, V. Singh, and K. Rajanna: Microstructure stage II fatigue crack growth rates and dynamic fracture toughness—A correlation.Eng. Fract. Mech. 31(5):83–792 (1988)*

Shape Memory Alloys

Rahul Jain , Roopam Bhattacharya

Department of Mechanical Engineering, FET Agra College Agra

Abstract - **Shape memory alloys (SMA) are materials that have the ability to return to a former shape when subjected to an appropriate thermomechanical procedure. Shape memory materials are also super elastic, namely they are able to sustain a large deformation at a constant temperature, and when the deforming force is released they return to their original undeformed shape. Pseudoelastic and shape memory effects are some of the behaviors presented by these alloys. The unique properties concerning these alloys have encouraged many investigators to look for applications of SMA in different fields of human knowledge. Some shape memory alloys like NiTi show noticeable high damping property in pseudoelastic range. Due to its unique characteristics, a NiTi alloy is commonly used for passive damping applications, in which the energy may be dissipated by the conversion from mechanical to thermal energy. The purpose of this paper is to present a brief discussion on the behavior of SMA and to describe their most promising applications in the biomedical area. These include cardiovascular and orthopedic uses, and surgical instruments.**

Shape recovery in Shape memory alloys is due to solid-to-solid phase transformation from martensite to austenite. SMA is typically formed above 500-700°C where the material exhibits highly ordered austenitic crystalline structure. Upon cooling below transition temperature, unstrained SMA will have twinned martensitic structure. SMA deforms easily under stress due to twin boundaries propagating throughout the structures in the direction of stress. Austenite crystalline structure is regained upon heating, returning to its original shape.

Keywords: Shape memory alloys, biomaterials, pseudoelasticity

I. INTRODUCTION

Metals are characterized by physical qualities as tensile strength, malleability and conductivity. In the case of shape memory alloys, we can add the anthropomorphic qualities of memory and trainability. Shape memory alloys exhibit what is called the shape memory effect. If such alloys are plastically deformed at one temperature, they will completely recover their original shape on being raised to a higher temperature. In recovering their shape the alloys can produce a displacement or a force as a function of temperature. In many alloys combination of both is possible. We can make metals change shape, change position, pull, compress, expand, bend or turn, with heat as the only activator. Key features of products that possess this shape memory property include: high force during shape change; large movement with small temperature change; a high permanent strength; simple application, because no special tools are required; many possible shapes and configurations; and easy to use - just heat. Because of these properties shape memory alloys are helping to solve a wide variety of problems. In one well–developed application shape memory alloys provide simple and virtually leakproof couplings for pneumatic or hydraulic lines. The alloys have also been exploited in mechanical and electromechanical control systems to provide, for example, a precise mechanical response to small and repeated temperature changes. Shape memory alloys are also used in a wide range of medical and dental applications (healing broken bones, misaligned teeth)

II.HISTORY

First observations of shape memory behaviour were in 1932 by Ölander in his study of "rubber like effect" in samples of gold–cadmium and in 1938 by Greninger and Mooradian in their study of brass alloys (copper–zinc). Many years later (1951) Chang and Read first reported the term "shape recovery". They were also working on gold–cadmium alloys. In 1962 William   J. Buehler and his co–workers at the Naval Ordnance Laboratory discovered shape memory effect in an alloy of nickel and titanium. He named it NiTiNOL (for nickel–titanium Naval Ordnance Laboratory). Buehler's original task was finding a metal with a high melting point and high impact resistant properties for the nose cone of the Navy's missile SUBROC. From among sixty compounds, Buehler selected twelve candidates to measure their impact resistance by hitting them with hammer. He noted that a nickel–titanium alloy seemed to exhibit the greatest resistance to impact in addition to satisfactory properties of elasticity, malleability and fatigue. One day he took some NiTiNOL bars from melting furnace and laid them out on a table to cool. He intentionally dropped one on the floor out of curiosity. The bar produced a bell–like quality sound, then he ran to the fountain with cold water and chilled the warm bar. The bar was once again dropped on the floor. On his amazement it exhibited the leaden–like acoustic response. Buehler knew that acoustic damping signalled a change in atomic structure that can be turned off and on by simple heating and cooling near room temperature, but he did not yet know that this rearrangement in the atomic structure would lead to shape memory effect. It was in 1960 when Raymond Wiley joined Buehler's research group. He worked on failure analysis of various metals. He demonstrated to his management the fatigue resistance of a NiTiNOL wire by flexing it. The directors who were present at this meeting passed the strip around the table, repeatedly flexing and unflexing it and were impressed with how well it held up. One of them, David Muzzey, decided to see how it would behave under heat. He was a pipe smoker, so he held the compressed NiTiNOL strip in the flame of his lighter. To the great amazement of all, it has stretched out completely. When Buehler heard about that, he realized that it had to be related to the acoustic behaviour he had noted earlier. After this moment, NiTi alloys increased interest of developing applications based on a shape memory alloys.

III. PSEUDO-ELASTICITY

One of the commercial uses of the shape memory alloy involves using the pseudo-elastic properties of the metal during the high temperature (austentic) phase. Pseudoelasticity occurs whenever an SMA sample is at a temperature above A_F (the temperature above which only the austenitic phase is stable for a stress-free specimen Braz J Med Biol Res 36(6) 2003 Medical applications of shape memory alloys men). Thus, one can consider an SMA sample subjected to a mechanical loading at a constant temperature above A_F. The stress-strain curve (σ-ε) in Figure 1, below, illustrates the macroscopic behavior of SMA, showing the pseudoelastic phenomenon.

A mechanical loading causes an elastic response until a critical value is reached, point A, when the martensitic transformation (austenite $\rightarrow$ martensite) arises, ending at point B. At this point, the crystal structure of the sample is totally composed of detwinned martensite. For higher stress values, SMA presents a linear response. During the unloading

process, the sample presents an elastic recovery (B $\rightarrow$ C). From point C to D one can note the reverse martensitic transformation (martensite $\rightarrow$ austenite). From point D on, the sample presents an elastic discharge. When the loading-unloading process is finished, SMA have no residual strain. However, since the path of the forward martensitic transformation does not coincide with the reverse transformation path, there is a hysteresis loop associated with energy dissipation. Another way to observe the pseudoelastic effect is indicated on the upper side of Figure 1. First, let us consider an SMA at a temperature above A_F. At this temperature, there is only one phase, i.e., austenite. At a constant temperature, a mechanical loading is applied promoting the appearance of the detwinned martensite. During the unloading process, reverse transformation takes place (detwinned martensite $\rightarrow$ austenite) and when load vanishes, the sample presents no residual strain.

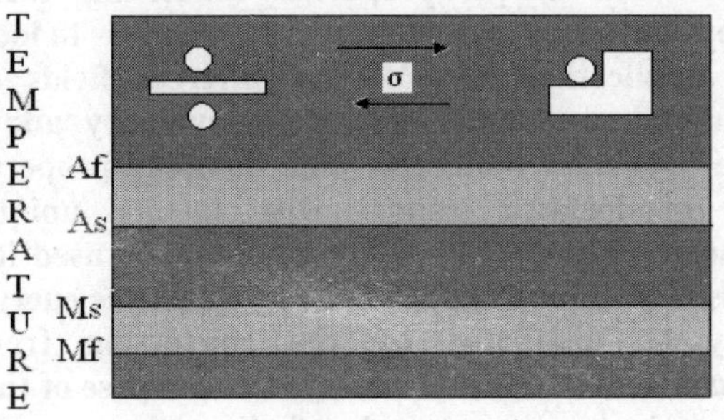

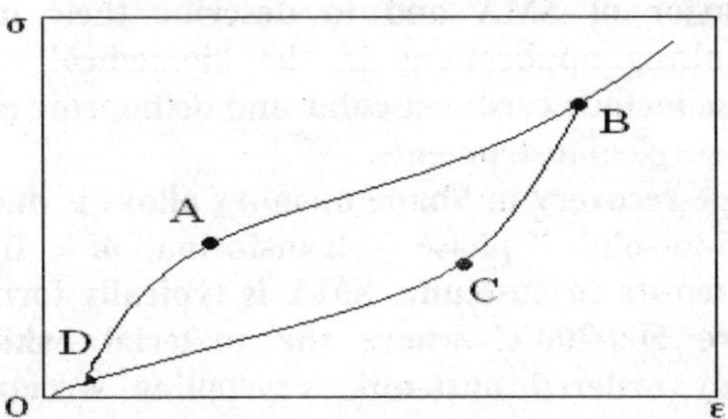

Figure 1: Pseudoelasticity. A_S, A_F and M_S and M_F = temperature at which the formation of austenite and martensite starts and ends, respectively. σ-ε = stress-strain curve.

IV. ALLOY TYPES

Since the discovery of Ni-Ti, at least fifteen different binary, ternary and quaternary alloy types have been discovered that exhibit shape changes and unusual elastic properties consequent to deformation. Some of these alloy types and variants are shown in table 1.

The original nickel-titanium alloy has some of the most useful characteristics in terms of its active temperature range, cyclic performance, recoverable strain energy and relatively simple thermal processing. Ni-Ti and other alloys have two generic properties thermally induced shape recovery and super- or pseudo-elasticity. The latter means that an SMA in its elastic form can undergo a deformation approximately ten times greater than that of a spring-steel equivalent, and full elastic recovery to the original geometry may be expected. This may be possible through several million cycles. The energy density of the alloy can be used to good effect to make high-force actuators - a modern DC brushless electric motor has a mass of 5-10 times that of a thermally activated Ni-Ti alloy, to do the same work. The superelastic Ni-Ti alloys are "stressed" by simply working the alloy. These stresses can be removed, just as with many other alloys, by an annealing process. The stressed condition is termed stress-induced martensite, which is the equivalent of being cold/hot worked.

Table 1. Shape memory alloy types.

· Titanium-palladium-nickel	· Iron-manganese-silicon
· Nickel-titanium-copper	· Nickel-titanium
· Gold-cadmium	· Nickel-iron-zinc-aluminium
· Iron-zinc-copper-aluminium	· Copper-aluminium-iron
· Titanium-niobium-aluminium	· Titanium-niobium
	·Zirconium-copper-zinc
· Uranium-niobium	·Nickel-zirconium-titanium
· Hafnium-titanium-nickel	

V. APPLICATIONS OF SHAPE MEMORY ALLOYS

Shape memory alloys have wide range of use. That is why a large number of categories of application have been introduced.

Free recovery

This category of shape memory alloys is deformed while in a martensitic phase. The only function required is that they return to previous (parent) shape upon heating. A prime application is a blood–clot filter. The filters are constructed from NiTi wires and are used in one of the outer hearth chambers to trap

blood clots, which potentially could cause serious health problems if travelling around the blood circulation system. The blood–clot filter is introduced in a compact cylindrical form about 2.0-2.5 mm in diameter. The body heat causes transformation to its functional shape.The wire is shaped to anchor itself in a vein and catch passing clots.

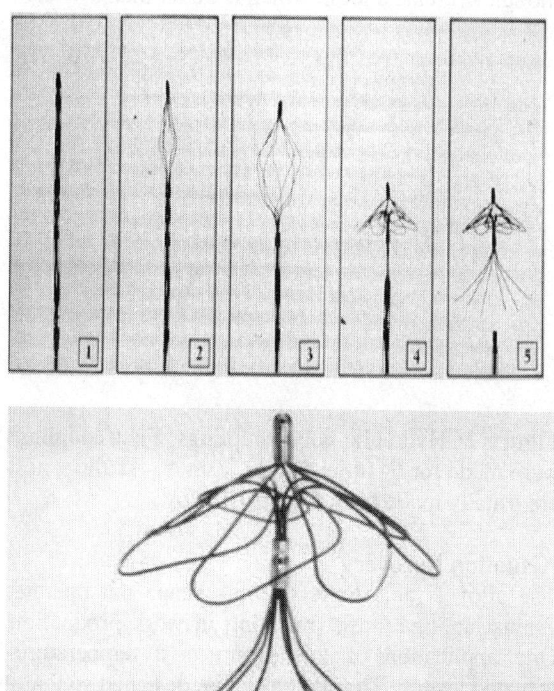

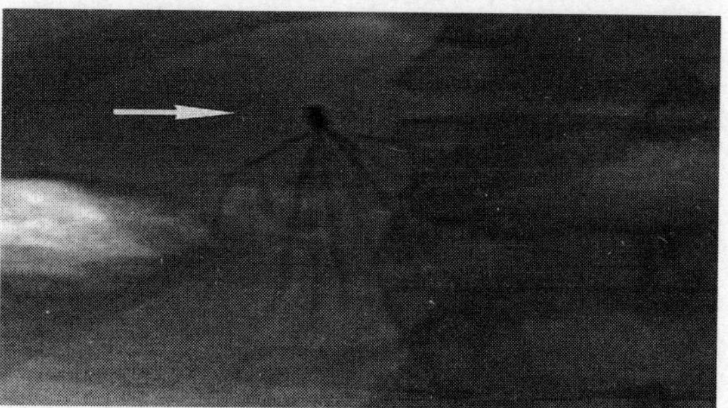

Figure 2: The blood–clot filter

Constrained Recovery

The alloy is prevented from full shape recovery. This means that the alloy generates stress on the constraining element. The most successful example of this type of product are hydraulic–tube couplings. This kind of couplings was on the first place made for F–14 jet fighter. Coupling is machined at normal temperature to have inner diameter 4 % less than the outer diameter of the tubes to be joined. After cooling the coupling under the Mf temperature, the diameters are expanded to have greater diameter than the tubes. When the coupling is warmed to austenite phase, it shrinks in diameter and strongly holds the tube end together. The tubes prevent the coupling to recover its original shape. This creates stresses powerful enough to create a joint, which is better than a weld.

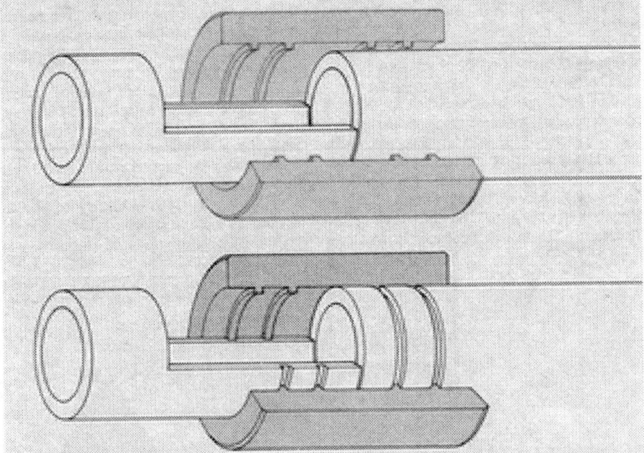

Figure 2: Hydraulic–tube couplings. First couplings were made for F–14 jet fighter from Ni–Ti alloy, now are usually made from Cu-Zn-Al alloy

Actuation Recovery

The alloy is able to recover its shape but operates against applied stress, resulting in work production. One application of this group is a temperature–actuated switch. The switch can be designed so that it opens or closes above particular temperature. Those switches are usually applied in instruments for safety purposes. One such example is a fire safety switch, which is designed to shut off the electricity or flammable gas flow when fire occurs.

INITIAL SHAPE	ADDITIONAL COLD OR HOT SHAPING	SHAPE AFTER BETATIZING AND QUENCHING	POSITION AT ROOM TEMPERATURE	'REMEMBERED' POSITION (ABOVE A$_f$)
	NO MARTENSITE	NOW CONTAINS MARTENSITE	MARTENSITE UNDER STRESS	NO MARTENSITE
	NONE	BEND TO CURVE		
		BEND STRAIGHT		

Figure 3: Temperature–actuated switch. If the switch is designed to close above the A$_f$ temperature (top), a straight rod of alloy in initial shape is cooled to martensite phase (red colour). Then an alloy is reshaped under stress. When the rod is heated above A$_f$ temperature, the martensite disappears and the rod straightens, closing the switch. If the switch is designed to open above A$_f$ temperature, the rod must

be bent before cooling (bottom). The rod is then straightened out before it is placed in the switch.

Superelastic Recovery

The only isothermal application of the memory effect, superelastic recovery involves the storage of potential energy through comparatively large but recoverable strains. This is commercially used in orthodontic applications, especially in correcting misaligned teeth. Superelastic NiTi wires are used for making continuous, gentle, corrective force to give large rapid movement of teeth. Based on clinical observations the optimal forces for tooth movement are 0.5–1.25N. The forces at the lower end of the range correspond to the smaller rooted teeth and the forces at the upper end of range are optimal for the larger rooted teeth. Optimal wires are from NiTi alloy with diameter from 0.4 mm to 0.6 mm. Similar shape memory alloy devices are used for healing broken bones. Staples of the shape memory materials are attached to each part of broken bone. These staples apply a constant, well–defined force to pull the two pieces together. This force helps to heal up the two pieces of bone together.

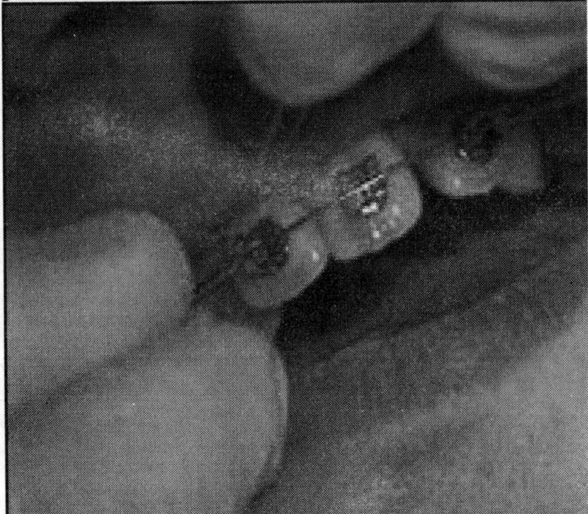

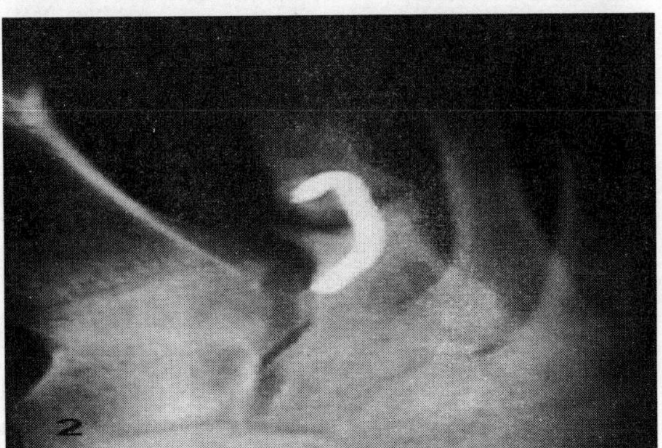

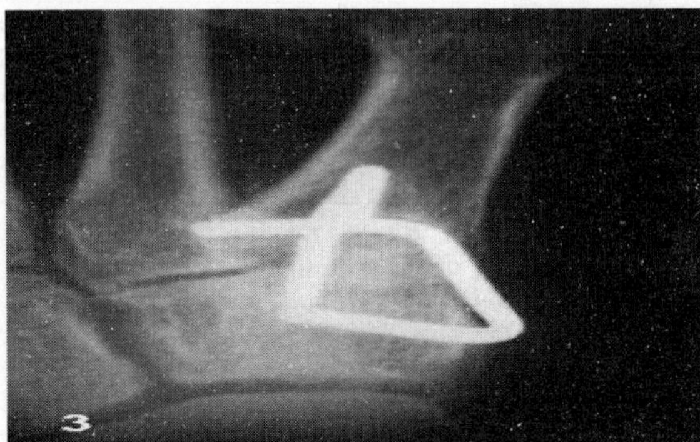

Figure 4: Arch wires for orthodontic correction of misaligned teeth (up) and TiNi shape memory clamps for healing broken bones (left, bottom)

Medical Applications

The variety of forms and the properties of SMAs make them extremely useful for a range of medical applications. For example, a wire that in its "deformed" shape has a small cross-section can be introduced into a body cavity or an artery with reduced chance of causing trauma. Once in place and after it is released from a constraining catheter the device is triggered by heat from the body and will return to its original "memorised" shape.

Increasing a device's volume by direct contact or remote heat input has allowed the development of new techniques for keyhole or minimally invasive surgery. This includes instruments that have dynamic properties, such as miniature forceps, clamps and manipulators. SMA-based devices that can dilate, constrict, pull together, push apart and so on have enabled difficult or problematic tasks in surgery to become quite feasible.

Table 2. Current examples of applications of shape memory alloys.

Aids for disabled	Micro-actuators
Aircraft flap/slat adjusters	Mobile phone antennas
Anti-scald devices	Orthodontic archwires
Arterial clips	Penile implant
Automotive thermostats	Pipe couplings
Braille print punch	Robot actuators
Catheter guide wires	Rock splitting
Cold start vehicle actuators	Root canal drills
Contraceptive devices	Satellite antenna deployment
Electrical circuit breakers	Scoliosis correction
Fibre-optic coupling	Solar actuators
Filter struts	Spectacle frames
	Steam valves
	Stents

Fire dampers	Switch vibration damper
Fire sprinklers	Thermostats
Gas discharge	Underwired bras
Graft stents	Vibration dampers
Intraocular lens mount	ZIF connectors
Kettle switches	
Keyhole instruments	
Key-hole surgery instruments	

VI. REFERENCES

[1]   Hodgson DE, Wu MH & Biermann RJ (1990). Shape Memory Alloys,Metals Handbook. Vol. 2. ASM International, Ohio, 897-902.

[2]   Mantovani D (2000). Shape memory alloys: Properties and biomedical applications. Journal of the Minerals, Metals and Materials Society, 52: 36-44.

[3]   Shape Memory Alloy Research Team (Smart) (2001). http://smart.tamu.edu

[4]   Otsuka K & Ren X (1999). Recent developments on the research of shape memory alloys. Intermetallics, 7: 511-528.

[5]   Wu SK & Lin HC (2000). Recent development of TiNi-based shape memory alloys in Twain. Materials Chemistry and Physics, 64: 81-92.

[6]   Funakubo H (1987). Shape Memory Alloys. Gordon & Bleach, New York, NY, USA.

[7]   Shape Memory Applications, Inc. (2001). http://www.sma-inc.com

[8]   van Humbeeck J (1997). Shape memory materials: state of art and requirements for future applications. Journal de Physique IV, 7: 3-12.

[9]   van Humbeeck J (1999). Non-medical applications of shape memory alloys. Materials Science and Engineering A, 273-275: 134-148.

[10]   Schetky LMcD (2000). The industrial applications of shape memory alloys in North America. Materials Science Forum, 327-328: 9-16.

[11]   Denoyer KK, Erwin RS & Ninneman RR (2000). Advanced smart structures flight experiments for precision spacecraft. Acta Astronautica, 47: 389-397.

[12]   Pacheco PMCL & Savi MA (2000). Modeling and simulation of a shape memory release device for aerospace applications. Revista de Engenharia e Ciências Aplicadas.

[13]   Webb G, Wilson L, Lagoudas DC & Rediniotis O (2000). Adaptive control of shape memory alloy actuators for underwater biomimetic applications. AIAA Journal, 38: 325-334.

[14]   Rogers CA (1995). Intelligent materials. Scientific American, September: 122-127.

[15]   Birman V (1997). Theory and comparison of the effect of composite and shape memory alloy

stiffeners on stability of composite shells and plates. International Journal of Mechanical Sciences, 39: 1139- 1149.

[16] *Duerig TM, Pelton A & Stöckel D (1999). An overview of nitinol medical applications. Materials Science and Engineering A, 273-275:149-160.*

[17] *Pelton AR, Stöckel D & Duerig TW (2000). Medical uses of nitinol. Materials Science Forum, 327-328: 63-70.*

[18] *Chu Y, Dai K, Zhu M & Mi X (2000). Medical of NiTi shape memory alloy in China. Materials Science Forum, 327-328: 55-62.*

[19] *Airoldi G & Riva G (1996). Innovative materials: The NiTi alloys in orthodontics. Bio-Medical Materials and Engineering, 6: 299-305.*

[20] *Lagoudas DC, Rediniotis OK & Khan MM (1999). Applications of shape memory alloys to bioengineering and biomedical technology. Proceedings of the 4th International Workshop on Mathematical Methods in Scattering Theory and Biomedical Technology, Perdika, Greece, October 8-10, 1999.*

[21] *Jordan L, Goubaa K, Masse M & Bouquet G (1991). Comparative study of mechanical properties of various Ni-Ti based shape memory alloys in view of dental and medical applications. Journal de Physique IV, 1: 139-144.*

[22] *Savi MA, Paiva A, Baêta-Neves AP & Pacheco PMCL (2002). Phenomenological modeling and numerical simulation of shape memory alloys: A thermo-plastic-phase transformation coupled model. Journal of Intelligent Material Systems and Structures, 13: 261-273.*

[23] *Zhang XD, Rogers CA & Liang C (1992). Modeling of the two-way shape memory effect. Philosophical Magazine A, 65: 1199-1215.*

[24] *Shabalovskaya SA (1995). Biological aspects of TiNi alloys surfaces. Journal de Physique IV, 5: 1199-1204.*

[25] *Ryhänen J (1999). Biocompatibility evolution of nickel-titanium shape memory alloy. Academic Dissertation, Faculty of Medicine, University of Oulu, Oulu, Finland.*

[26] *NMT Medical, Inc. (2001). http://www.nmtmedical.com*

[27] *Medical Devicelink (2001). http://www.devicelink.com*

[28] *Raychem (2001). http://www.raychem.com*

[29] *Duerig TM, Pelton A & Stöckel D (1996). The use of superelasticity in medicine. Metall, 50: 569-574.*

Synthesis of Silver Nanoparticles by natural Polysaccharide-Gaur Gum and Their Characterization

[1]B.S.Kushwah, [2]Sharad Chandra Upadhyaya, [3]Manish Dubey, [3]Seema Bhadauria

[1]F.E.T. R.B.S. College, Agra
[2]Dau Dayal Institute of Vocational Education, Dr. B. R. Ambedkar University, Agra
[3]Microbio and NanoTech Lab, Department of Botany, R.B.S. College, Agra

Abstract—**Stable and uniform silver nanoparticle clusters have been produced by reduction of silver ions by gaurgum, which is Natural polysaccharide and works as a stabilizer. The results showed that guar gum is very effective in binding the Ag clusters and restricts their size in the nano region. The UV/Vis spectra show that an absorption peak, occurring due to Surface Plasmon Resonance (SPR), exists at 410 nm of visible region confirms the formation of Ag clusters. The samples have been characterized by X-Ray diffraction (XRD) and Transmission electron microscopy (TEM), which reveal of the nano nature of the particles. These studies infer that the particles are mostly spherical in shape and have an average size of 16 nm.**

Keywords : GaurGum, silver nanoparticles, TEM

I. INTRODUCTION

Silver has been in use since time immemorial in the form of metallic silver, silver nitrate, silver sulfadiazine for the treatment of burns, wounds and several bacterial infections. But due to the emergence of several antibiotics the use of these silver compounds has been declined remarkably. Nanoparticle synthesis and the study of their size and properties is of fundamental importance in the advancement of recent research. It is found that the optical, electronic, magnetic, and catalytic properties of metal nano particles depend on their size, shape and chemical surroundings.

Nanotechnology is gaining tremendous impetus in the present century due to its capability of modulating metals into their nanosize, which drastically changes the chemical, physical and optical properties of metals. Metallic silver in the form of silver nanoparticles has made a remarkable comeback as a potential antimicrobial agent. The use of silver nanoparticles is also important, as several pathogenic bacteria have developed resistance against various antibiotics. Hence, silver nanoparticles have emerged up with diverse medical applications ranging from silver based dressings, silver coated medicinal devices, such as nanogels, nanolotions, etc. In nanoparticle synthesis it is very important to control not only the particle size but also the particle shape and morphology as well.

The area of nanotechnology, which spans the synthesis of nanoscale matter, understanding/utilizing their exotic physicochemical and optoelectronic properties, and organization of nanoscale structures into predefined superstructures, promises to play an increasingly important role in many key technologies of the new millennium . As far as the synthesis of nanoparticle is concerned, there is an ever-growing need to develop clean, non-toxic and environmentally friendly (green nanochemistry) procedures. Consequently, researches in the field of nanoparticle preparation have been looking at biological systems for inspiration. The above factors, combined with academic curiosity, have lead to the development of biomimetic approaches for the growth of advance materials (bioactive materials).

Many organisms, both unicellular and multicellular, are known to produces inorganic materials either intra- or extracellularly [1]. Even through microbes have been used with considerable success in biotechnological applications, such as remediation of toxic metals, reports on their use in the synthesis of nanomaterials are extremely limited. Mukherjee and coworker have demonstrated the formation of gold nanoparticle through *Verticillium* fungus sp [1].

In the present investigation the synthesis of silver nanoparticles by biological route with gaur gum is discussed, which is an easy, simple and convenient route for preparing metal particles in nanometer range. The prepared silver nano particles have been examined using X-ray diffraction (XRD), Transmission Electron Microscope (TEM) and UV/Vis absorption spectroscopy. These studies reveal that the prepared nanoprticles are of an average size of 16 nm. Perhaps the most important factor in this process is that the silver nano particles prepared by this process are stable for months.

II. METHODOLOGY

Synthesis of Silver Nanoparticles

Materials. All reagents and solvents were used as received without further purification. Ultrapure grade silver nitrate ($AgNO_3$) was purchased from Sigma, gaur gum from Sigmaand Milli-Q water were used in all preparations.

Particle Synthesis. A variable amount of a silver nitrate stock solution was added to a variable amount of a stock solution of gaur gum and filled up to a controlled volume with deionized water. The pH of the reaction mixture was set at this stage by adding a controlled volume of NaOH .Both reactant solutions always contained the same weight percentage of gaur gum. Then, the silver solution was quickly added to the vigorously stirred acid solution. As a result of the size-dependent extinction of the silver particles, the color of the solution turned from transparent to reddish brown colour of prepared nanoparticles indicates nearly 100 % conversions of silver ions into nanoparticles. The preparation of silver nanoparticles with different electrolyte concentrations has been tried, but neither the samples with concentration other than the present one is found to be stable over 2 weeks nor of smaller size (more than 60 nm). Hence, we have recorded the data of the particle of optimum size and of comparatively better stability (over 4 months). The induction time (a few seconds) for these colors to appear was dependent on the reaction conditions. The solutions were stirred up to 24 h at room temperature. Subsequently, the dispersions were washed twice with deionized water after

sedimentation or centrifugation (max 200g). The silver particles were kept in water where they remained stable longer than a year.

Particle Characterization. The particle morphology was studied by transmission electron microscopy (TEM) on H-9500 300kV TEM and scanning electron microscopy (SEM) on a Hitachi VP-SEM S-3400N respectively. The particle size and polydispersity were determined from theTEM micrographs using a Nikon profile projector. UV-vis spectroscopy (Elico spectrophotometer) was performed on dilute (<10-5 vol %) suspensions of Ag particles in water. The optical path was 1.00 cm.

III. RESULTS AND DISCUSSION

Silver nanoparticles were synthesized according to the method described in the previous section, the colloidal solution turned pale brown, pale yellow and pale red indicating that the silver nanoparticles were formed. Figure 1 shows the photographs of samples obtained at different conditions. The colourless Ag+ solution containing gaur gum changed to pale red . UV-visible spectroscopy is one of the most widely used techniques for structural characterization of silver nanoparticles.

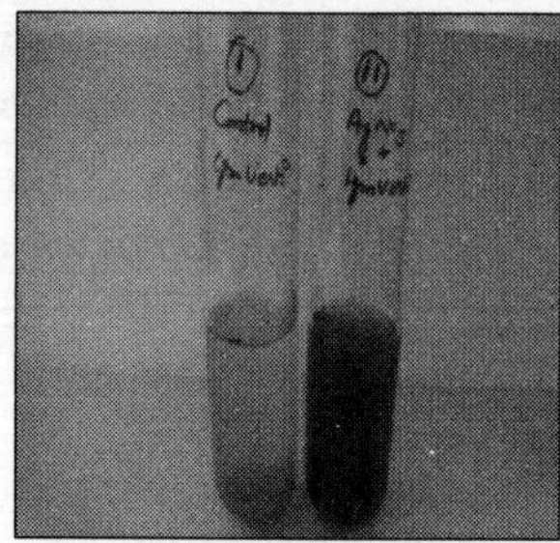

Figure 1. *Picture of tubes containing the gaur gum in AgNO₃.Note the change in colour due to nanoparticle synthesis*

The UV–Vis absorption spectrum of silver nanoparticles synthesized using gaur gum is show in figure 2. UV–Vis absorption spectrum (Fig. 2) reveals the formation of silver nanoparticles by showing surface Plasmon absorption maxima at 412 nm. The position and shape of the plasmon absorption depends on the particles size, shape and the dielectric constant of the surrounding medium. We can observe that surface plasmon absorption maximum initially at 418 nm (figure 3) is shifted to lower wavelength (figure2) with addition of gaurgum as reducing agent. The shape of the UV–Vis absorption spectrum is also noticed. In metal nano particles such as in silver, the conduction band and valence band lie very close to each other in which electrons move freely. These free electrons give rise to a surface plasmon resonance (SPR) absorption band occurring due to the collective oscillation of electrons of silver nano particles in resonance with the light wave [2].

Classically, the electric field of an incoming wave induces a polarization of the electrons with respect to much heavier ionic core of silver nanoparticles. As a result a net charge difference occurs which in turn acts as a restoring force? This creates a dipolar oscillation of all the electrons with the same phase. When the frequency of the electromagnetic field becomes resonant with the coherent electron motion, a strong absorption takes place, which is the origin of the observed colour. Here the colour of the prepared silver nanoparticles is dark reddish brown. This absorption strongly depends on the particle size, dielectric medium and chemical surroundings [3, 4]. Small spherical nano particles (< 20nm) exhibit a single surface plasmon band [5].

The UV/Vis absorption spectra of the silver nano particles are shown in the fig. 3. The absorption peak (SPR) is obtained in the visible range at 410 nm. With the above mentioned concentration. The stability of silver nanoparticles is observed for 4 months and it shows a SPR peak at the same wavelength. The dispersions of silver nanoparticles display intense colors due to the plasmon resonance absorption. The surface of a metal is like plasma, having free electrons in the conduction band and positively charged nuclei. Surface Plasmon resonance is a collective excitation of the electrons in the conduction band; near the surface of the nanoparticles.

Electrons are limited to specific vibrations modes by the particle's size and shape. Therefore, metallic nanoparticles have characteristic optical absorption spectrums in the UV-Vis region [6].

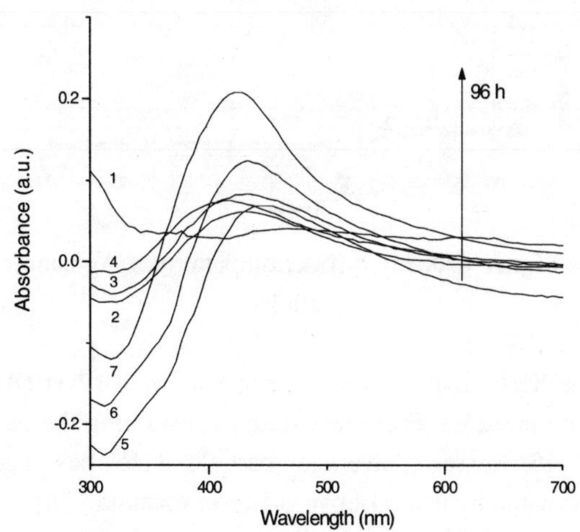

Figure 2: *UV-Vis spectra recorded as a function of time of reaction of 1 mM AgNO₃*

The average size of these particles is approximately 30 nm. The particle size histograms of silver particles (Figure 3) show that the particles range in size from 15 to 48 nm.

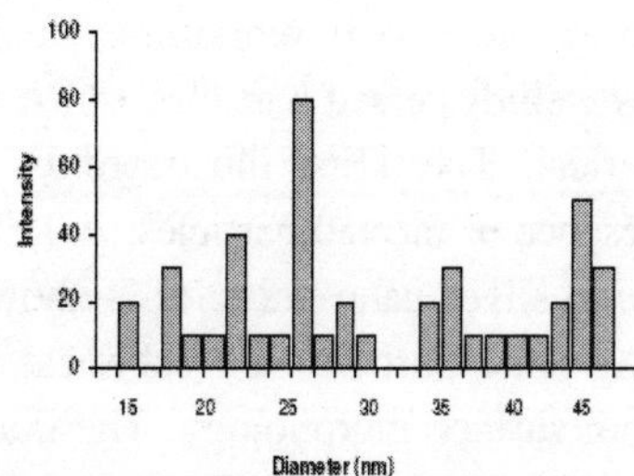

Figure 3. Particle size distributions of Ag nanoparticle

XRD Analysis

When the electron diffraction is carried out on a limited number of crystals one observes only some spots of diffraction distributed on concentric circles. The rings patterns with plane distances 2.36Å, 2.04Å, 1.45Å, 1.23Å and 0.94 Å are consistent with the plane families {111}, {200}, {220}, {311} and {331} of pure face-centred cubic (fcc) silver structure (JCPDS, File No.4-0787).

The structure of prepared silver nanoparticles has been investigated by X-ray diffraction (XRD) analysis. Typical XRD patterns of the sample, prepared by the present method are shown in the Fig.4.

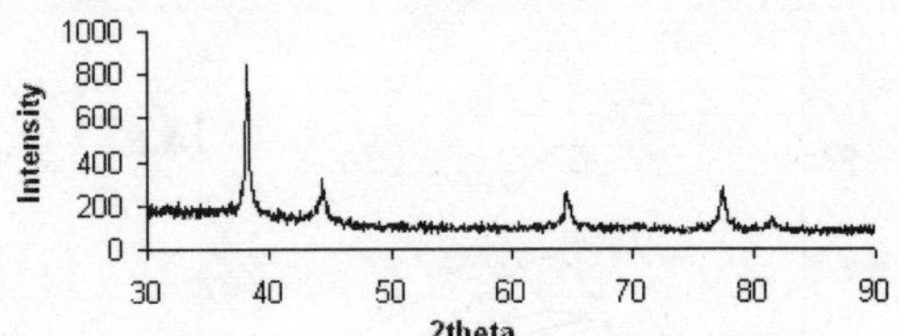

Figure 4. X-ray diffraction pattern of Ag nano particles.

The XRD study indicates the formation of silver (Ag) nano particles. From this study, considering the peak at 45 degrees, average particle size has been estimated by using Debye-Scherrer formula [7,8].

$$D = \frac{0.9\lambda}{WCos\ \theta} \qquad \ldots\ldots (1)$$

Where 'λ' is wave length of X-Ray (0.1541 nm), 'W' is FWHM (full width at half mamimum), 'θ' is the diffraction angle and 'D' is particle diameter (size). The average particle size is calculated to be around 14 nm.

TEM Analysis

The transmission electron microscopy (TEM) image of the silver synthesized is represented in Figure 5 and indicates well dispersed particles which are more or less spherical. The TEM illustrated in figure 6 show the presence of faceted particles. A TEM image of the prepared silver nano particles is shown in the fig.5. The Ag nano particles are spherical in shape with a smooth surface morphology. The diameter of the nano particles is found to be approximately 16 nm. TEM image also shows that the produced nano particles are more or less uniform in size and shape.

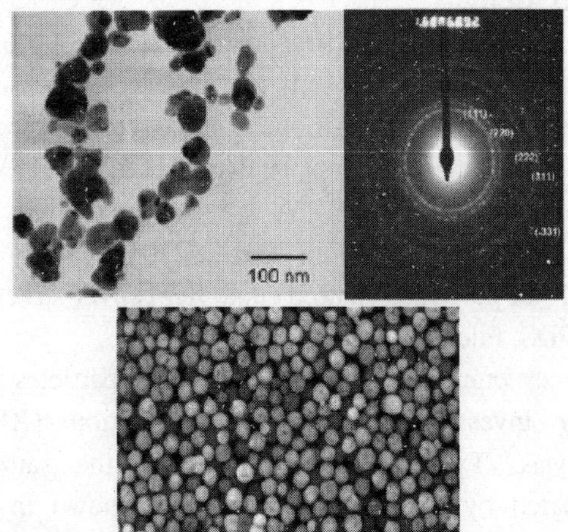

Figure 5. TEM , HEED and SEM images of silver nanoparticles synthesized

IV. CONCLUSION

Silver nanoparticles have been prepared through the reduction of silver ions by polysaccharide-gaurgum. This is one of the simplest and cheapest processes for obtaining silver nanoparticles. UV/Vis spectroscopy reveals the surface plasmon property, while XRD analysis and TEM images reveal the nano nature of the prepared samples. Average size estimated from above studies is 16 nm. The energy-dispersive spectroscopy (EDX) of the nanoparticles dispersion confirmed the presence of elemental silver signal no peaks of other impurity were detected. Highenergy electron diffraction (HEED) confirmed that formation of face-centered-cubic silver nanoparticles.

V. ACKNOWLEDGEMENTS

Authors thank to DST-NANOMISSION Programme for financial assistance and to the Principal ,R.B.S.College,Agra for providing research facilities.

VI. REFERENCES

[1] P. Mukherjee, A. Ahmed, D. Mandal, S. Scnapati, S.R Sainleem, M. I Khan, R. Ramani, R. Parischa, P. V Ajayakumar, M. Alam, M. Sastry, R. Kumar. Angew Chem. Int Ed: 40. (2001), 3585.

[2] Nath S. S., Chakdar D., and Gope G.; Synthesis of CdS and ZnS quantum dots and their applications in electronics, Nanotrends- A journal of nanotechnology and its application, 02(03), (2007).

[3] Noginov M. A., Zhu G., Bahoura M., Adegoke J., Small C., Ritzo B. A., Drachev V. P.,and Shalaev V. M.; The effect of gain and absorption on surface plasmos in metal nanoparticles, Appl. Phys. B (2006).

[4] Link S. and El-Sayed M. A.; Optical Properties and ultrafast dynamics of metallic nanocrystals, Annu. Rev. Phys. Chem. 54, 331-66 , (2003).

[5] He R., Qian X., Yin J., and Zhu Z.; Preparation of polychrome silver nanoparticles in different solvents, J. Mater. Chem. 12, 3783-3786, (2002).

[6] Wang G., Shi Ch., Zhao N., Du X. Materials Letters 61 (2007), 3795-3797.

[7] Nath S. S., Chakdar D., and Gope G.; Synthesis of CdS and ZnS quantum dots and their applications in electronics, Nanotrends- A journal of nanotechnology and its application, 02(03), (2007).

[8] Nath S. S., Chakdar D., Gope G., and Avasthi D. K.; Effect of 100 Mev Nickel Ions on Silica Coated ZnS Quantum Dot, Journal of Nanoelectronics and Optoelectronics, 3, 1-4, (2008).

Variation of Mechanical Properties, Al (6063/ Al$_2$O$_{3p}$), Metal Matrix Composite Produced By Stir Casting

Ravindra Mamgain, Kapil kr. Sharma, Arbind Prasad

Graphic Era University Dehradun

Abstract—**Composites are made up of individual materials referred to as constituent materials. There are two categories of constituent materials: matrix and reinforcement. At least one portion of each type is required. The matrix material surrounds and supports the reinforcement materials by maintaining their relative positions. The reinforcements impart their special mechanical and physical properties to enhance the matrix properties.**

The present work deals with the development of Aluminum matrix composites using aluminum alloy Al 6063 as matrix and alumina as a reinforcing material prepared by stir casting technique. The alumina amounts varied as 5, 10, and 15 percent by volume.

The mechanical properties like hardness, tensile strength, and impact strength have been investigated. The effect of addition by volume percent of alumina on these mechanical properties has been studied. The properties like hardness and impact strength vary inversely to each other with alumina content as the hardness increases with alumina content and impact strength decreases. The change in these properties is moderate for 5 percent addition of alumina and with 10 and 15 percent there are marginal changes. The tensile strength of composite decreases with the addition of alumina.

Keyword : metal matrix composite, stir casting, mechanical properties

I. INTRODUCTION

Humans have been using composite materials for thousands of years. Take mud bricks for example. A cake of dried mud is easy to break by bending, which puts a tension force on one edge, but makes a good strong wall, where all the forces are compressive. A piece of straw, on the other hand, has a lot of strength when you try to stretch it but almost none when you crumple it up. But if you embed pieces of straw in a block of mud and let it dry hard, the resulting mud brick resists both squeezing and tearing and makes an excellent building material. Put more technically, it has both good compressive strength and good tensile strength.

II. CLASSIFICATION OF COMPOSITE

Composites are of following types:

- Polymer matrix composites.
- Ceramic matrix composites.
- Metal matrix composites

Polymer matrix composites

Polymer-matrix composites consist of high-strength fibers, carbon glass, or other materials in a matrix of thermosetting or thermoplastic polymers. The fibers provide high strength at a very low weight, and the matrix holds the fibers in place. The most outstanding characteristic of polymer-matrix composites is the materials' ability to replace lightweight, high-strength metals or wood with an even lighter-weight and higher-strength alternative. [1]

Ceramic matrix composites

The class of materials known as ceramic matrix composites, or CMCs, shows considerable promise for providing fracture-toughness values similar to those for metals such as cast iron. Two kinds of damage-tolerant ceramic-ceramic composites are being developed. One incorporates a continuous reinforcing phase, such as a fiber; the other, a discontinuous reinforcement, such as whiskers. [2]

Metal matrix composites

Metal matrix composites are the engineered material having the combination of two or more materials in which the tailored properties are achieved. In the past decade, the need for lighter materials with high specific strength coupled with major advances in processing, has led to the development of numerous composite materials as a serious competitor to traditional engineering alloy of particular interest in aerospace and defence industry. [3]

The matrix alloy, the reinforcement material, the volume and shape of the reinforcement, the location of the reinforcement, and the fabrication method can all be varied to achieve required properties. Numerous metals have been used as matrices. The most important have been aluminum, titanium, magnesium and copper alloys and super alloys.

III. PROCESSING TECHNIQUES:

The fabrication of metal matrix materials may be considered in two stages: the fabrication of the composite material from base metal and fiber reinforcement and the subsequent fabrication of laminates from the composite material. In some cases, the two steps occur simultaneously depending on the final material product desired and the method of fabrication used in the process. The choice of methods used to fabricate a composite material depends on the mechanical and chemical properties of the fiber and matrix, the fiber length and size, the fiber packing, and the desired fiber configuration. A short overview of some of the methods used to fabricate aluminum matrix composites (AMCs) are discussed below.

Solid state processing

Different solid state processing techniques can be used for preparing composites. Few of these techniques are:

- Powder metallurgy technique
- Diffusion bonding
- Step pressing
- Hot-die molding
- Super plastic forming
- Hot isostatic pressing

Liquid state processing

In liquid state processing of composite, liquid metal is combined with reinforcing phase and solidified in a mould. Few of these techniques are:

- Squeeze casting
- Infiltration casting
- Investment casting
- Pressure casting
- Stir casting

IV. EXPERIMENTAL DETAIL

In the present work aluminum based alumina reinforced particulate metal matrix was prepared by stir casting. The material used and procedure for its casting is explained as follow:

Elements	Si	Mn	Mg	Cu	Fe	Ti	Al
Al 6063	0.44	0.07	0.6	0.018	0.2	0.008	Rest

Preparation of composite

The process for composite casting is shown in fig 1. The Matrix alloy used in the study is Al-Mg-Si wrought alloy matrix (6063) reinforced with Al2O3. Commercial Al-6063 (Al-98.52%, Mg-0.649, Si-0.445) alloy reinforced with 5, 10& 15 % by vol. The Matrix alloy was first melted in a graphite crucible in a electric furnace and before mixing, the Al_2O_3 particles were preheated at 300°C for 1 hour to make the surface of Al2O3 particle oxidized. The furnace temperature was first raised above the liquidus temperature to melt the alloy completely at 750°C and was then cooled down just below the liquidus temperature (700˚C) to keep the slurry in a semi solid state. The stir made of stainless steel attached with graphite blade was made to move at a rate of 200 rpm up to 15 minutes. The mixing was done for a short time period of 1 to 1.5 minutes. The composite slurry was reheated to a fully liquid state and the automatic mechanical mixing was done for about 30 minutes at stirring rate of 250 rpm. In this experiment, the molten composite was transferred from the crucible into the mould.

Fig: 1 Stir casting setup

V. RESULTS

In present work the three composites, one by addition of 5%alumina, second by 10% alumina and other by 15% addition of alumina have been cast by stir casting technique and their mechanical properties like tensile strength ,impact strength and hardness have been determined. These properties have been reported and compared with base alloy Al 6063 in the following section.

Composition	UTS Mpa	extension@ max %	extension@ break %	Impact strength	HARDNESS IN BHN
Al 6063 base alloy	117.5	11.13	11.67	1.480 Kgm/cm^2	39
Al 6063 +5%alumina (Al2O3)	109.4	2.623	2.650	0.99181 Kgm/cm^2	47
Al 6063 +10%alumina (Al2O3)	77.0	3.955	4.777	0.6176 Kgm/cm^2	61
Al 6063 +15%alumina (Al2O3)	51.0	3.038	3.636	0.18564 Kgm/cm^2	73

Tensile strength

The tensile strength of Al6063 base alloy, Al6063 +5% alumina, Al6063 +10%alumina and Al6063 +15% alumina by vol. was measured at room temperatures. The stress-strain and force-extension curves for different composition of alumina i.e. Al6063 base alloy, Al6063 +5%alumina, Al6063 +10%alumina and Al6063 +15%alumina by vol. percent are shown in fig 2.1, 2.2, 2.3, 2.4 respectively.

The UTS for Al 6063 base alloy, Al 6063 +5%alumina, Al 6063 +10%alumina and Al 6063 +15%alumina by vol. is 117.5 Mpa , 109.4Mpa, 77.0Mpa and 51.0 Mpa respectively, there is a decrease in UTS at room temperature for 5%,10% and 15% addition of alumina by vol. by 8.5 Mpa , 32.4 Mpa and 26 Mpa respectively.

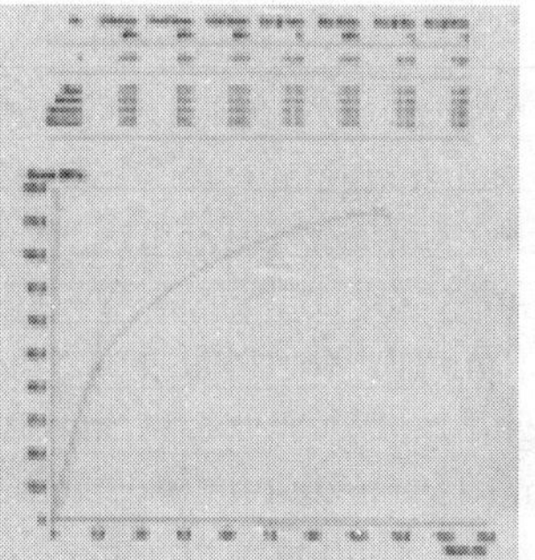

Fig 2.1: Stress-strain curve for Al6063 base alloy

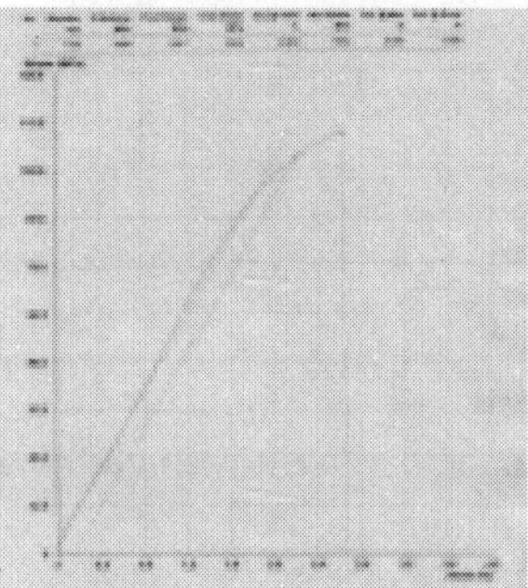

Fig 2.2: Stress-strain curve for Al6063 base alloy + 5% alumina

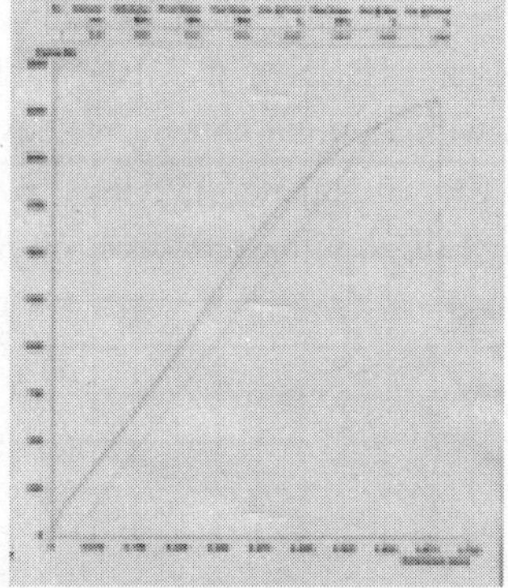

Fig 2.3: Stress-strain curve for Al6063 base alloy + 10% alumina

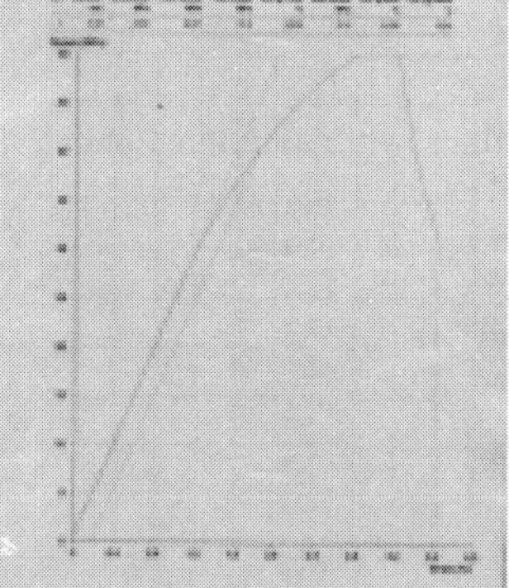

Fig 2.4: Stress-strain curve for Al6063 base alloy + 15% alumina

Impact strength

The impact strength for base alloy Al (6063) is 1.480 Kgm/cm^2. It has decreased marginally to -0.99181 Kgm/cm^2 for 5% addition, impact strength further decreases to 0.6176 Kgm/cm^2 for composite Al 6063 for 10 % and 0.18564 Kgm/cm^2 for 15%. For 5 % addition there is no significant decrease in impact strength but with 10% and 15% addition the decrease is more as compared with 5% addition of alumina.

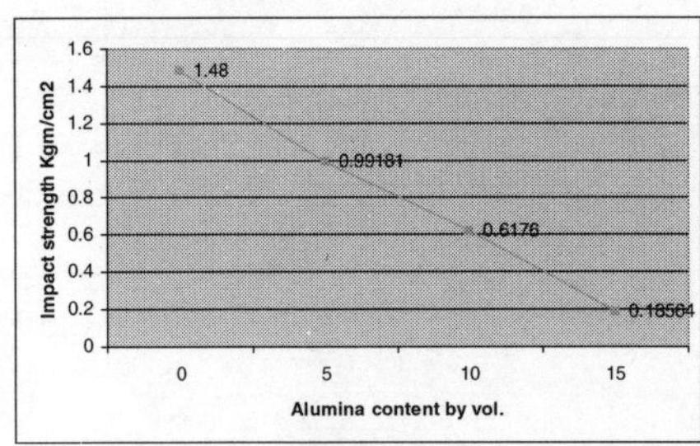

Fig 3: Variation of impact strength with alumina content.

Hardness

The hardness of Al 6063 base alloy is as low as 39 BHN and with 5% addition of alumina it increases upto 47BHN and further with 10% and 15% addition of alumina it reaches at 61BHN and 73BHN. The variation of hardness with alumina content is shown in the hardness of the composite increases with increase in vol. percent of alumina reinforced in the alloy.

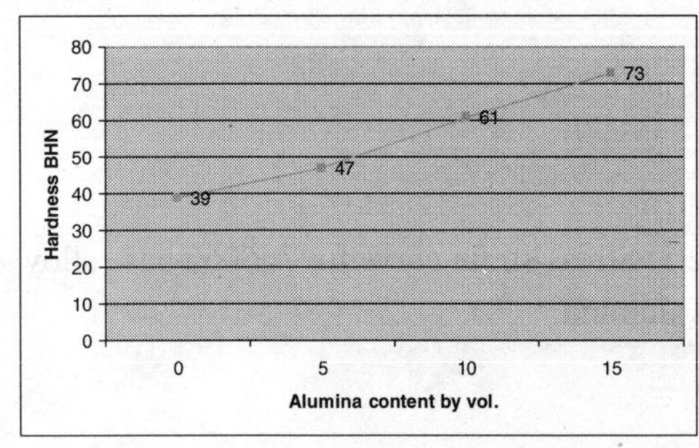

Fig 4: variation of hardness with alumina content.

Microstructure

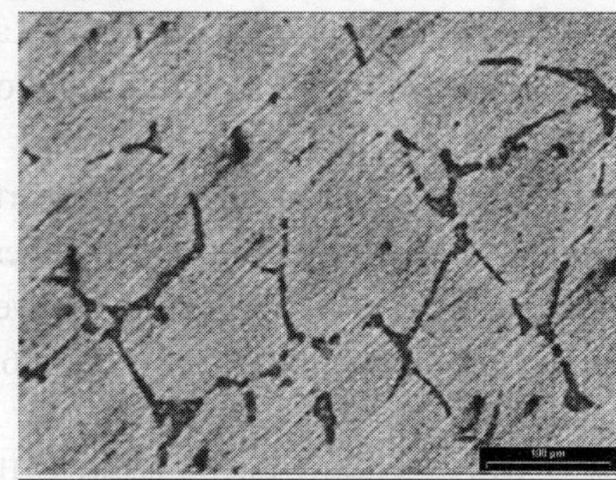

Fig 5: Optical micrograph of Al6063 base alloy + 5% alumina

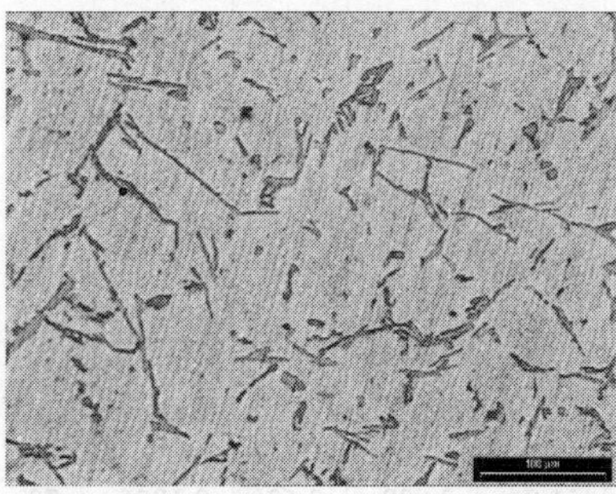

Fig6: Optical micrograph of Al6063 base alloy + 10% alumina

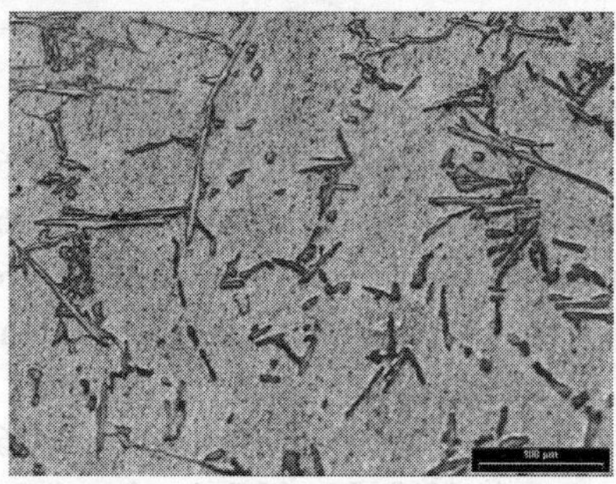

Fig 7: Optical micrograph of Al6063 base alloy + 15% alumina

VI. DISCUSSIONS

The ultimate tensile strength of the composite decreases with addition of alumina. The decrease in ultimate tensile strength may be due to the segregation of particles at some specific zones.

The second reason for decrease in tensile strength may be due to the presence of the interfacial gaps between the matrix and the reinforcement, which is unable to transfer the load from the matrix to reinforcing phase as can be seen from the optical micrograph.

The hardness of the composite increases with the addition of alumina. Hardness of the Al 6063 base alloy is 39 BHN, with the addition of 5% alumina alumina it increases to 47 BHN and with addition of 10% and 15% it increases to61 BHN and 73 BHN. The hardness of the Composite increases because hard nature of particles. With the 5% addition of particles the hardness increases by 8 BHN and with 10% and 15% it increases by 22 BHN and 34 BHN. This increase in hardness is attributed of the hard nature of particles as compared to base alloy. The result show the average value of hardness there are variations in the hardness observed for same surface of composite, this may be due to the difference in the distribution of the alumina particles as observed from optical micrograph.

The impact strength for the base alloy is 1.480 Kgm/cm^2 for the addition of 5% alumina it decreases to 0.99181 Kgm/cm^2 and with the addition of 10% and 15% it reduces to 0.6176 Kgm/cm^2 and 0.18564 Kgm/cm^2. For 5% addition there is no significant decrease in impact strength but with 10% and 15% the decrease is more as compared with 5% addition of alumina, this may be due to the brittle nature of the particles and more segregation of the particles at some specific places.

VII. CONCLUSIONS

The addition of alumina in high strength aluminum alloy (Al 6063) decreases the ultimate tensile strength.

1. Decohesion at the interface between alumina particles and ductile matrix on application of load can be one of the reason for lower tensile and impact strength.

2. The mismatch between reinforcement and matrix leads to a large stress concentration near particulate and matrix in that region fails prematurely under application of load.

3. The failure also occurs because of presence of some preexisting flaws in the reinforcement during processing.

4. With the increase in vol. fraction a strong tendency of clustering of particulates [as is evident from the optical micrograph] leads to a very inefficient load transfer mechanism causing low strain to failure.

The hardness of aluminum alloy Al 6063 is 39 BHN. There is increase in hardness from 39 to 73 BHN, on addition of 5%, 10% and 15% alumina by vol. respectively. This increase in hardness is attributed of the hard nature of particles as compared to base alloy.

VIII. REFERENCES

[1] H.Capel, S.J.HariS,P.Schulj and H.Kaufmann " Material Science technology" August 2000,Vol16, pp.765-767

[2] G.Guo, P.K.Rohatgi and S.Ray: "Aluminum composites for automotive applications" Trans. Am. Foundarymen's Soc. 1996

[3] William F.Smith "Principle of material science and engineering" Vol.2, Issue 1, 1985, pp.478-483

[4] K.J.Bhansali, And R.Mehrabian: "Advances in reinforcements for MMCs"JOM, vol.349 (1993), pp.30-40.

[5] W.L.Winterbottom, "Low cost fabrication for AMCs" Acta Metall.,15(1967),p.303

[6] Huseyin Sevik and S. Can Kurnaz: 'Properties of alumina particulate reinforced aluminum alloy produced by pressure die casting' Materials and Design 27 (2006) 676–683

Wear Behaviour of Polypropylene and its Composites

Mubbasher Ali Khan, M.Arif Siddiqui

Department of Mechanical Engineering, Aligarh Muslim University, Aligarh. U.P. (INDIA)

Abstract: **In this study the Polypropylene (PP) was taken as a base material. The PP composites were made with Talc and at various concentrations by weight. The standard specimens of the composites were moulded through Injection molding process at different injection pressures. The effect of talc on wear behavior of injected molded compacts was studied. Wear rate was decreased with the addition talc in PP. The wear rate in case of talc containing composites was insignificant at higher compaction pressures. In general, the wear rate was decreased in composites containing talc at lower injection pressure.**

I. INTRODUCTION

Polymers are widely used in engineering and medical applications. Light weight and reasonable hard polymers are of great importance in various applications of automobile industry. Costly polymers are used where wear resistance is of first priority such as in gear wheels. Inorganic mineral fillers into plastic resin improves various physical properties of the materials such as mechanical strength, modulus etc [1]. Polypropylene (PP) is, an important polymer, widely used in engineering applications due to its great recyclability, low cost etc. It has a wide range of applications in household goods, packaging, automobiles & other engineering applications such as housing for electrical systems, air ducting, headlight housing, washing machine components, housing for electric irons and toasters etc [2, 3]. Tjong and Li, studied about the mechanical properties and impact toughness of talc filled β-crystalline phase PP composites and found that the elastic modulus of the β-PP composites increases with increasing filler content but the yield strength and impact toughness decrease with increasing filler content[4]. Lapcik et. al, studied the effect of a microsize/nanosize talc filler on the physicochemical and mechanical properties of filled polypropylene. They found that the increase in filler content lead to an increase in the strength of the composite material with a simultaneous decrease in the fracture toughness. They observed that the increase in tensile strength were from 15 to 25% (the maximum tensile strength at break was found to be 22 MPa). The increase in mechanical strength simultaneously led to a higher brittleness [5]. The fillers were used to improve the properties of polymers for a variety of reasons: cost reduction, improved processing, density control, optical effects, thermal conductivity, control of thermal expansion, electrical properties, magnetic properties and improved mechanical properties, such as hardness and wear resistance.

II. MATERIALS USED

The morphology of base material, PP, and the fillers used is given in Table1:

Table1. Morphology of materials used.

S.NO.	Materials used	Morphology of additives	Supplier
1	Polypropylene		Reliance Industries Limited, Gujarat, India.
2	Talcum powder		AVARICE Laboratories Pvt. Ltd., Ghaziabad, India.

III. EXPERIMENTAL SET UP

The specimens used for the experiments were made on Injection moulding machine (De-Tech 60LNC5 made by Larsen & Tubro) in Plastic Technology Laboratory at A.M.U. Aligarh, U.P., INDIA.
The following were the specifications of the moulding machine:

Injection Unit
1) Screw diameter=35mm
2) Screw L/D ratio=20
3) Injection pressure=2024 bar
4) Shot weight in poly styrene=153g (maximum)

Clamping Unit
1) Clamping force 600KN (60 ton)
2) Moulding opening stroke =310mm
3) Total mould carrying capacity=420kg
4) Total mould weight in moving platens=280kg
5) Hydraulic ejector stroke =200mm

IV. PROCEDURE

The detail of specimen of PP and its composites used in the present study is shown in Fig.1.

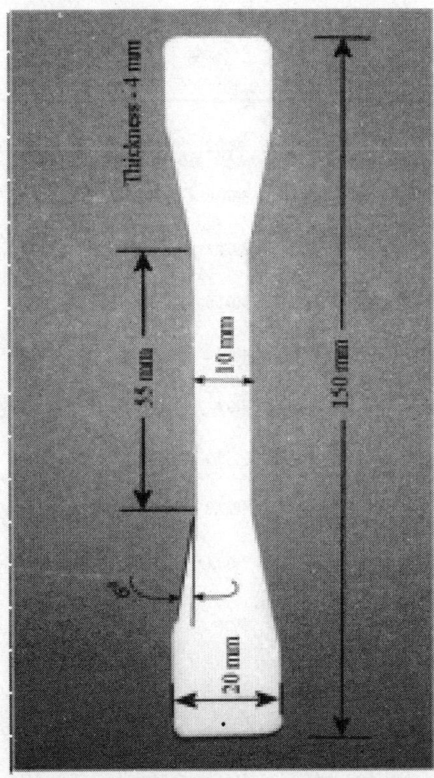

Figure1. Sample made for the analysis with dimensions

To obtain the test specimens the die was clamped onto the injection moulding machine. The PP was fed to the machine through hopper and it was entered in to the injection barrel by gravity where the PP with and without fillers were heated to a temperature of about 225°C. The material was injected into the mold by a reciprocating screw injector. The fillers were added and mixed manually and the specimens shown in Fig. 1 were made. Table 2 and 3 gives the variable parameters of the injected specimens and experiment design matrix respectively.

Table 2. Variable parameters with coding.

LEVELS	FILLERS CONCENTRATION (%)	INJECTION PRESSURE (MPa)	CODE
LOW	10	4	-1
MEDIUM	20	5	0
HIGH	30	6	+1

Table 3. Experiment design matrix.

S.No.	X_1	X_2
1	-1	-1
2	-1	0
3	-1	+1
4	0	-1
5	0	0
6	0	+1
7	+1	-1
8	+1	0
9	+1	+1

V. WEAR TEST

Wear test was carried out using self fabricated pin on disc machine as   shown in Fig.2. Samples were gripped on to the machine. The wear test was conducted at a load of 10 kg. The contact between the sample & the abrasive paper was made on the application of the load. The speed of rotatory disk was kept constant (1800rpm). The abrasive paper of 600 grit size was used in place of wear plate.

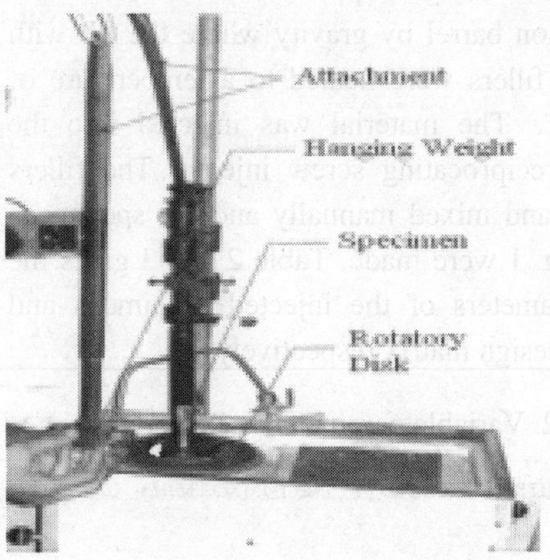

Figure 2. Wear test on the moulded specimen.

VI. RESULTS AND DISCUSSION

Table 5 shows the percent wear rate of PP composites at different concentration of talc with various injection pressure. Fig. 3 shows the wear rate of PP with and without talc powder at an injection pressure of 4 MPa. From fig. 3 it is clearly seen that there was a decrease in wear rate of PP -talc composite as compared to PP at a lower level of concentration (10%) of talc powder. But on increasing the concentration of talcum powder up to 20% there was an increase in percent wear which may be due to the creation of poor particle–polymer interactions. There was a significant decrease in wear rate at 30% talc containing PP composites and it is in agreement of the earlier work [5]. Fig. 4 shows that as the injection pressure was increased, there was an increase in percent wear which may be due to the reason that there may be the formation of layers of talcum powder in talcum powder-Polypropylene composites.

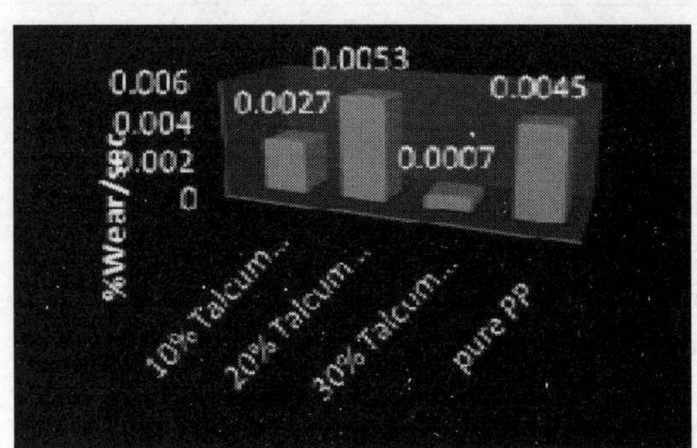

Figure 3. PP with and without Talc Powder at injection pressure of 4 MPa

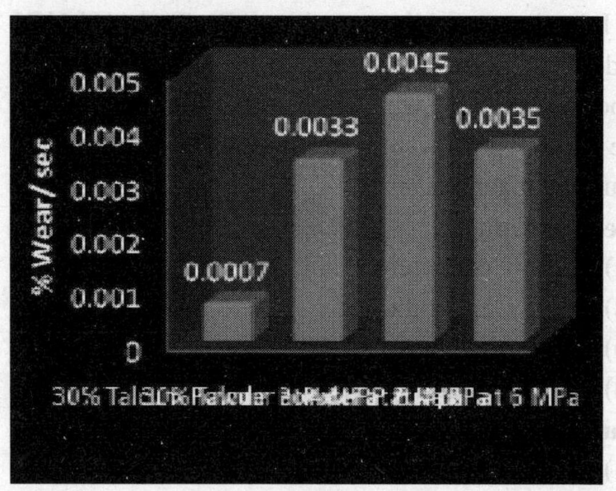

Figure 4. Percent wear rate versus PP with 30% Talcum Powder at different Injection pressure.

Table 5. Shows that the wear rate of Polypropylene (PP) and its composites.

IP: Injection pressure (MPa); V_1, V_2 Volume (mm) before and after test respectively.

* Percent by wt.

PP*	Talc*	IP	V_1	V_2	$(V_1-V_2)/V_1$ (%)	Wear/sec. (%)
90	10	4	4000	3967.2	0.82000	0.00273
		5	4013.6	3976.8	0.91688	0.00306
		6	4026	3974.4	1.28167	0.00427
80	20	4	4056.4	3991.6	1.59748	0.00532
		5	4098	4059.2	0.94680	0.00316
		6	3930	3892	0.96692	0.00322
70	30	4	3963.6	3955.2	0.21193	0.00071
		5	4048	4027.2	0.51383	0.00171
		6	3940.4	3900.4	1.01513	0.00338
		4	3978.8	3924	1.37730	0.00459
		5	3931.6	3886	1.15983	0.00387
100		6	3821.6	3780.8	1.06762	0.00356

VII. CONCLUSION

Addition of talcum powder in general reduced the percent wear rate of PP and found to be significant at 30% of talc containing PP composites where percent wear rate can be reduced upto 84%. Injection pressure did not show any significant effect on wear rate.

VIII. ACKNOWLEDGEMENT

The authors would like to express their deep gratitude to the Department of mechanical engineering and University Polytechnic of Aligarh Muslim University (AMU), for providing the laboratory facilities and financial support.

IX. REFERENCES

[1] Katz, H.S., Milevski, J.V., [1978], Handbook of Fillers and Reinforcements for Plastics,   1,1,333-457, Van Notrand Reinhold, New York.

[2] Van der Wal A. and Gaymans R.J., [1999], Polymer, 40, 6067

[3] William G.P., [1999], Polym. Eng. Sci., 39, 2445

[4] Tjong S. C. and Li R. K. Y.,[1997], Journal of vinyl and additive technology,3,1

[5] Lubomir Lapcik, Jr., Jindrova Pavlina, Lapcikova Barbora, Tamblyn Richard, Greenwood Richard, Rowson Neil, [2008],Effect of the Talc Filler Content on the Mechanical Properties of Polypropylene Composites, Wiley Inter Science, DOI 10.1002/app.28797

[6] Liang J. Z. , Tang C. Y. , Li R. K. Y. and Wong T. T. , [2007],Mechanical properties of polypropylene /CaCO$_3$ composites, Metals and Materials International, 4, 616-619

[7] Borkar S. P., Kumar V. S., & Mantha S. S., [2007], Effect of silica and calcium carbonate fillers on the properties of woven glass fibre composites, Indian Journal of Fibre & Textile Research, 32, 251-253

Analysis of Process Variables on Wire-EDM

[1]**Kapil Sharma**, [2]**Vipul Paliwal**

[1]Department of Mechanical Engineering, Graphic Era institute, Dehradun, U.K.
[2]Department of Mechanical Engineering, Anand Engineering college, Keetham Agra, U.P.
E-mail:kapilsharma9999@gmail.com

Abstract- CNC Wire Cut Electrical Discharge Machining (WEDM) is a widely accepted material removal process used to manufacture components with complicated shapes and profiles. WEDM is a unique adaptation of the conventional EDM process which uses an electrode to initialize the sparking process. WEDM utilizes a continuously traveling wire electrode made of thin copper, brass or tungsten of diameter 0.05-0.3 mm, which is capable of achieving very small corner radii. During the WEDM process, the material is eroded ahead of wire and there is no direct contact between the work piece and the wire, eliminating the mechanical stresses during machining. The effects of various process parameters of WEDM like pulse on time (T_{ON}), pulse off time (T_{OFF}), gap voltage (SV) have been investigated to reveal their impact on output parameter i.e.,Surface Roughness of high carbon high chromium steel using Response Surface Methodology.Experimental plan is performed by a standard RSM design called a BOX-BEHKEN DESIGN.The results of analysis of variance (ANOVA) indicate that the proposed mathematical model can adequately describe the performance within the limit of factors being studied.The optimal set of process parameters has also been predicted to maximize the Surface Finish.

I. INTRODUCTION

Wire Electrical discharge machining (WEDM) is a non-traditional, thermoelectric process which erodes material from the work piece by a series of discrete sparks between a work and tool electrode immersed in a liquid dielectric medium. These electrical discharges melt and vaporize minute amounts of the work material, which are then ejected and flushed away by the dielectric. A wire EDM generates spark discharges between a small wire electrode (usually less than 0.5 mm diameter) and a work piece with deionised water as the dielectric medium and erodes the work piece to produce complex two- and three dimensional shapes according to a numerically controlled (NC) path.

Wire electrical discharge machining (WEDM) is a specialized thermal machining process capable of accurately machining parts with varying hardness or complex shapes, which have sharp edges that are very difficult to be machined by the main stream machining processes. This practical technology of the WEDM process is based on the conventional EDM sparking phenomenon utilizing the widely accepted non-contact technique of material removal. Since the introduction of the process, WEDM has evolved from a simple means of making tools and dies to the best alternative of producing micro-scale parts with the highest degree of dimensional accuracy and surface finish quality.

Wire electrical discharge machining (WEDM) is a widely accepted non-traditional material removal process used to manufacture components with intricate shapes and profiles. It is considered as a unique adaptation of the conventional EDM process, which uses an electrode to initialize the sparking process. However, WEDM utilizes a continuously travelling wire electrode made of thin copper, brass or tungsten of diameter 0.05–0.3 mm, capable of achieving very small corner radii. The wire is kept in tension using a mechanical tensioning device reducing the tendency of producing inaccurate parts. During the WEDM process, the material is eroded ahead of the wire and there is no direct contact between the work piece and the wire, eliminating the mechanical stresses during machining. In addition, the WEDM process is able to machine exotic and high strength and temperature resistant (HSTR) materials and eliminate the geometrical changes occurring in the machining of heat-treated steels. WEDM was first introduced to the manufacturing industry in the late 1960s. The development of the process was the result of seeking a technique to replace the machined electrode used in EDM. in 1974.

Wire-electro discharge machining has been defined as the process of material removal of electrically conductive materials using the thermo-electric source of energy. The material removal is by controlled erosion through a series of repetitive sparks between electrodes, i.e., work piece and tool. Fig. 1.1 shows the basic features of the wire cut EDM machine.

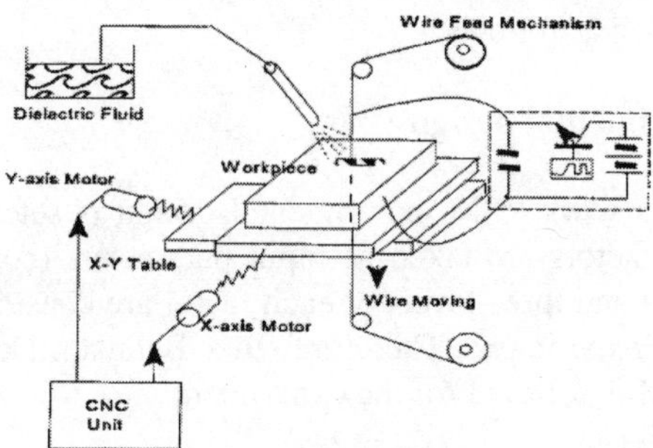

Fig. 1.1 Schematic representation of Wire EDM cutting process

The electrode is a thin wire and it is pulled through the work piece from a supply spool on to a take-up mechanism. On application of a suitable voltage, discharge occurs between the wire electrode and the work piece in the presence of a flood of deionised water of high insulation resistance.

Wire-Edm Equipment

A wire equipment machine consists of four sub systems: the position system, the wire drive system, the power supply, and dielectric system. All the four sub-systems are distinct from   conventional EDM.

RESPONSE SURFACE METHODOLOGY

Introduction to the RSM method

Many experimental programs are designed with a two-fold purpose in mind: (i) to quantify the relationship between the values of some measurable response variable(s) and those of a set of experimental factors presumed to affect the response(s) and, (ii) to find the values of the factors that produce the best value or values of the response(s). Response surface methodology (RSM) is a set of techniques designed to find the "best" value of the response. If discovering the best value or values of the response is beyond the available resources of the experiment, then response surface methods are used to at least gain a better understanding of the overall response system.

Techniques of the Response surface methodology (RSM)

The principal techniques of RSM are:
1. Setting up a series of experiments.
2. Determining a mathematical model that best fits the data collected.
3. Determining the optimal settings of the experimental factors that produce the optimum value of the response.

Terminology:
- Factors
- Response
- The response function

Factors
- Factors are processing conditions or input variables whose values or settings can be controlled
- by experimenter. Factors can be qualitative or quantitative.
- The specific factors studied in detail in this course are quantitative, and their levels are assumed to be fixed or controlled by the experimenter.
- Factors and their levels will be denoted by $X1, X2,\ldots,Xk$.

Response
- The response variable is the measured quantity whose value is assumed to be affected by
- changing the level of the factors. The true value of the response is denoted by η.
- Because of experimental error, the observed value of the response Y differs from η.
- This difference from the true value is written as $Y = \eta + \varepsilon$, where ε denotes experimental error

Response function
- When we say that the value of the true response η depends upon the levels $X1, X2,\ldots,Xk$ of k
- quantitative factors, we are saying that there exists some function of these levels, $\eta = \Phi(X1, X2,\ldots,Xk)$.
- The function Φ is called the true response function (unknown), and is assumed to be a continuous, smooth function of the Xi.

Procedure to determine optimum conditions
- This method permits to find the settings of the input variables which produce the most desirable response values. The set of values of the input variables which result in the

most desirable response values is called the set of optimum conditions.

Steps Involved

- The strategy in developing an empirical model through a sequential program of
- experimentation is as follows:
- The simplest polynomial model is fitted to a set of data collected at the points of a first-order design.
- If the fitted first-order model is adequate, the information provided by the fitted model is used to locate areas in the experimental region, or outside the experimental region, but within the boundaries of the operability region, where more desirable values of the response are suspected to be.
- In the new region, the cycle is repeated in that the first-order model is fitted and testing for adequacy of fit.
- If nonlinearity in the surface shape is detected through the test for lack of fit of the first-order model, the model is upgraded by adding cross-product terms and / or pure quadratic terms to it. The first-order design is likewise augmented with points to support the fitting of the upgraded model
- If curvature of the surface is detected and a fitted second-order model is found to be appropriate, the second-order model is used to map or describe the shape of the surface, through a contour plot, in the experimental region.
- If the optimal or most desirable response values are found to be within the boundaries of the experimental region, then locating the best values as well as the settings of the input variables that produce the best response values are determined.
- Finally, in the region where the most desirable response values are suspected to be found, additional experiments are performed to verify that this is so.

Box-Behnken Design

- The Box-Behnken design fills out a polyhedron, approximating a sphere
- For a 3 factor experiment, the 15 runs consist of three four-run, two-level factorials in two factors - with the third factor at its mid-level, and three centre points.
- Box-Behnken and face-centred cubic designs are subsets of the full three level factorial designs, except for centre points, they are complementary fractions in that no point in one design is in the other design.

II. EXPERIMENTATION

Synopsis

This portion contains the planning of step by step procedure of the experiment. The planning is based on the literature. This shows the procedure of selection of input parameters and output parameters, tool and material, machine, instrument for measurement, software for analysis. All these are described as follows:-

Experimental design

For this work RSM experimental design is selected. Three factors are taken as input parameters (control factors) and three levels of each factor are considered in this experiment. Therefore, Box Behnken Design of RSM is selected for the experiment..

Experiment parameters of study

There are various process variables of wire cut EDM affecting the machining characteristics. Based on the review of literature. The following process parameters have been selected.

- On-Time(Pulse Time(Ton))
- Off-Time(Pulse Time (Toff))
- Servo Voltage(SV)

Machine used with operational details:
Machine used in this work for experimentation is highly precise Sprint cut WEDM Electronica machine. WEDM is a four axis machine and capable to control all four axes simultaneously. The machine can perform multiplicity of operations in one set up.

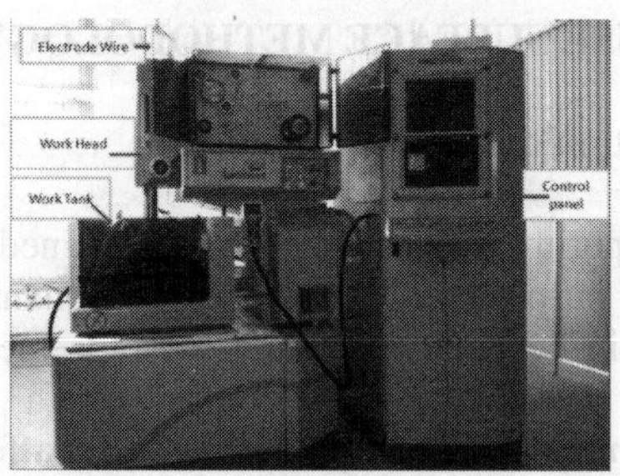

CNC Wire Cut EDM Electronica

Chemical composition of work material High Carbon High Chromium Steel

C - 2.15%
Mn - 0.415%
S - 0.049%
P - 0.055%
Si - 0.605%
Ni - 0.085%

Cr - 12.34%
Mo - 0.001%
Cu - 0.052%
Al -0.004%

Experimental Data

Table: Process parameters & their levels

Input factors	Level 1	Level 2	Level 3
Ton	105	115	125
Toff	43	53	63
SV	10	30	50

The experimental work is carried out as per the rsm design shown in table

S.NO.	TON	TOFF	S.V
1	105	43	30
2	125	43	30
3	105	63	30
4	125	63	30
5	105	53	10
6	125	53	10
7	105	53	50
8	125	53	50
9	115	43	10
10	115	63	10
11	115	43	50
12	115	63	50
13	115	53	30
14	115	53	30
15	115	53	30
16	115	53	30
17	115	53	30

Based on the experimental design as given in table , the specimen were prepared and the values of the selected machining characteristics surface roughness are reported in table

S.NO.	TON	TOFF	S.V	S.R
1	105	43	30	2.58
2	125	43	30	2.94
3	105	63	30	2.87
4	125	63	30	2.67
5	105	53	10	2.93
6	125	53	10	3.15
7	105	53	50	2.82
8	125	53	50	2.92
9	115	43	10	3.06
10	115	63	10	2.76
11	115	43	50	3.20
12	115	63	50	2.76
13	115	53	30	3.05
14	115	53	30	2.97
15	115	53	30	2.44
16	115	53	30	2.83
17	115	53	30	2.77

III. ANALYSIS AND MODELING

Synopsis

In this chapter there is discussion about the result of experiment and the individual effect of each parameter on the output characteristic.

Description of the experiment with results

The influences of the Ton, Toff and SV on machined surface roughness in wire electrical discharge machining process have been examined. Experiments have been performed on high carbon high chromium steel with a wire of diameter 0.2 mm and the obtained data has been analyzed using Response Surface Methodology.

The results of the performed experiment show that pulse on time (Ton), pulse off time (Toff) and the servo voltage (SV) influence on Surface roughness.

Effects of machining parameters on surface roughness:

1. Effects of Pulse on time on surface roughness:- It is seen from the fig.no.3(a) that machining speed increases with the pulse on time. On the contrary, surface accuracy decreases with increasing the pulse on time. This is because the discharge energy increases with the pulse on time. As a result, machining speed becomes faster with the increase of discharge energy.

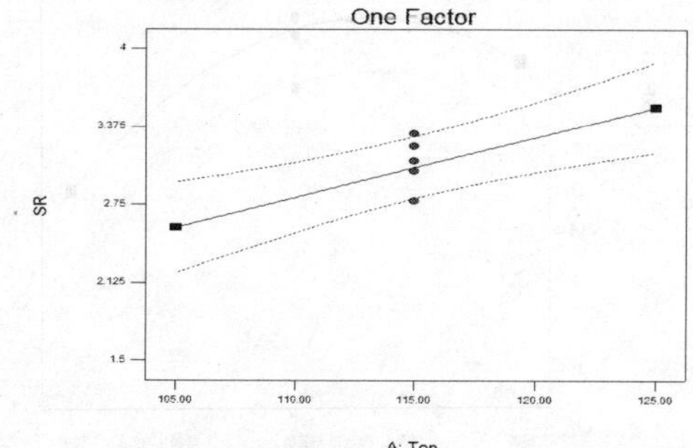

Figure 3(a)

2. Effect of Pulse off time on surface roughness: - From the fig. No.3(b) it is seen that as the pulse off time is shorter, the number of discharges within a given period becomes more. This will lead to a higher machining speed. But, surface accuracy

becomes poor because of a large number of discharge.

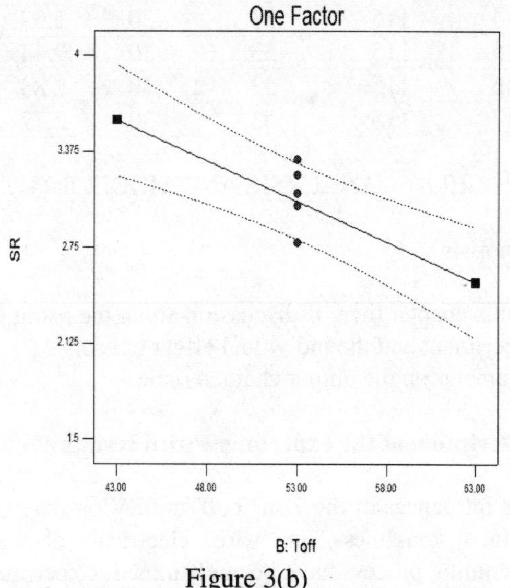

Figure 3(b)

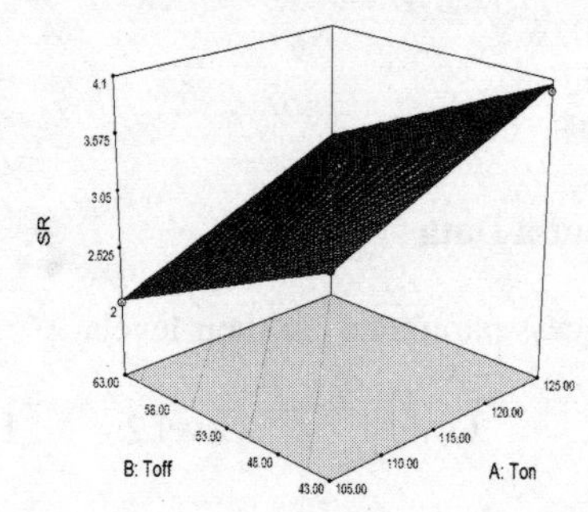

Figure 3(d)

3. Effect of Servo reference voltage on surface roughness: - From the fig.no.3(c) It is seen that Higher the servo reference voltage, the longer the discharge wait time. To obtain the longer discharge wait time, the machining speed needs to be slowed down. This will lead to a wider average discharge gap. Therefore, the discharge condition becomes more stable but the number of discharge cycles becomes within a given period. Owing to this stable machining, surface accuracy becomes better.

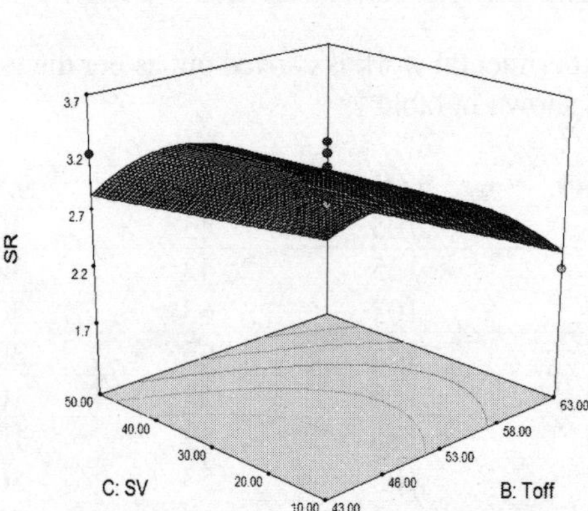

Discussion of results and achievements for individual parameters in surface roughness experiment:

The result and achievement for individual parameter is given in the response surface quadratic model

Table ANOVA for Response Surface Reduced Quadratic Model

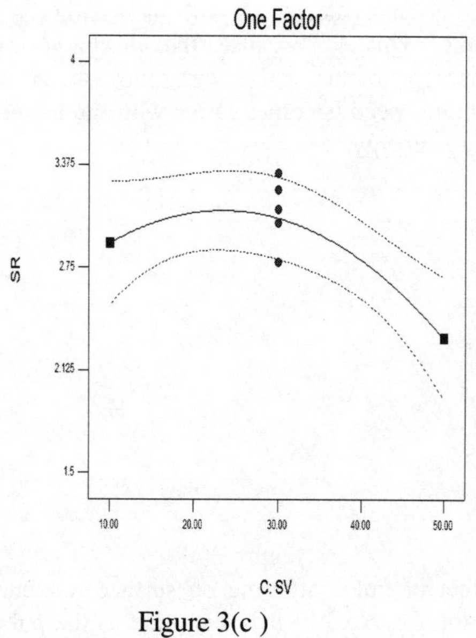

Figure 3(c)

3-D view of all the graphs

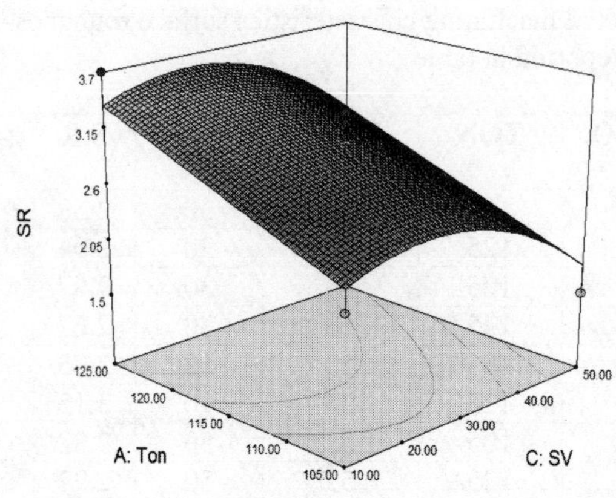

Figure 3(e)

Source	Sum Of Squares	Df	Mean square	F-value	P-value	
Model	5.59	4	1.40	11.89	0.0004	Significant
Ton (A)	1.85	1	1.85	15.77	0.0019	
Toff(B)	2.24	1	2.24	19.04	0.0009	
SV(C)	0.67	1	0.67	5.73	0.0339	
SV (C)sq.	0.83	1	0.83	7.02	0.0212	
Residual	1.41	12	0.12			
Lack Of Fit	1.24	8	0.15	3.63	0.1144	not significant
Pure Error		0.17	4	0.043		
Cor Total		7.00	16			

The Model F-value of 11.89 implies the model is significant. There is only a 0.04% chance that a "Model F-Value" this large could occur due to noise. Values of "Prob > F" less than 0.0500 indicate model terms are significant. In this case A, B, C, C^2 are significant model terms. Values greater than 0.1000 indicate the model terms are not significant. If there are many insignificant model terms (not counting those required to support hierarchy), model reduction may improve the model.

The optimal values of Ton from fig. 6.1 = 105µs

Toff from fig. 6.2 = 63 µs

SV from fig 6.3 = 50 V

Final Equation in Terms of Actual Factors:

S = -0.24124+0.048125 * Ton-0.052875 *

Toff+0.051708 * SV-1.10347E-003* SV^2

With the help of this equation the optimal value of surface roughness is 2.91 micron

IV. CONCLUSIONS

- Surface roughness increases with an increase of pulse on time, surface roughness decreases with increase of pulse off time and surface roughness increases with an increase of SV.
- In WEDM Process, use of low pulse on time (105units), high Pulse off time (63 units) and high Servo voltage (50 units) are recommended to obtained better surface finish for the specific test range in a High carbon high chromium material. The optimal value of surface roughness is 2.91 units.

V. REFERENCES

[1] Nihat Tosun and Can Cogun, "A study on kerf and material removal rate in wire electrical discharge machining based on Taguchi method", Journal of Materials Processing Technology 152 (2004) 316–322.

[2] Te-Chang Tsai and Jenn-Tsong Horng, "The effect of heterogeneous second phase on the machinability evaluation of spheroidal graphite cast irons in the WEDM process", Materials and Design 29 (2008) 1762–1767.

[3] Aminollah Mohammadi and Alireza Fadaei Tehrani, "Statistical analysis of wire electrical discharge turning on material removal rate", journal of materials processing technology 205 (2008) 283–28

[4] Jin Yuan and Kesheng Wang, "Reliable multi-objective optimization of high-speed WEDM process based on Gaussian process regression", International Journal of Machine Tools & Manufacture 48 (2008) 47–60.

[5] R.Ramakrishnan and L. Karunamoorthy, "Multi response Optimization of WEDM processes using robust design of experiment", Int J Adv Manuf Technol (2006) 29: 105–112.

[6] A.B. Puri and B. Bhattacharyya, "Modelling and analysis of white layer depth in a wire-cut . EDM process through response surface methodology", Int J Adv Manuf Technol (2005) 25: 301–307.

[7] M.S. Hewidy and T.A. El-Taweel, "Modelling the machining parameters of wire electrical discharge machining of Inconel 601 using RSM", Journal of Materials Processing Technology 169 (2005) 328–336.

[8] K.H. Ho, S.T. Newman, Rahimifard and R.D. Allen, " State of the art in wire electrical discharge machining (WEDM)", International Journal of Machine Tools & Manufacture 44 (2004) 1247–1259.

[9] Scott F. Miller and Albert J. Shih," Investigation of the spark cycle on material removal rate in wire electrical discharge machining of advanced materials", International Journal of Machine Tools & Manufacture 44 (2004) 391–400.

[10] Y.S. Liao and J.T. Huang, "A study to achieve a fine surface finish in Wire-EDM", Journal of Materials Processing Technology 149 (2004) 165–171

[11] Y.S. Liao and Y.P. Yu, "The energy aspect of material property in WEDM and its application", Journal of Materials Processing Technology 149 (2004) 77–82.

[12] T.A. Spedding and Z.Q. Wang, "Study on modelling of wire EDM process", Journal of Materials Processing Technology 69 (1997) 8- 28

[13] Y.S. Liao and J.T. Huang, "A study on the machining-parameters optimization of WEDM", Journal of Materials Processing Technology 71 (1997) 487-493.

[14] C.J.Luis and I. Puertas, "Methodology for developing technological tables used in EDM processes of conductive ceramics", Journal of Materials Processing Technology, 189(2007)301-309.

[15] J.L.LIN and C.L.LIN, The use of the orthogonal array with grey relational analysis to optimize the electrical discharge machining process with multiple performance charecteristics,International Journal of Machine Tools and manufacturer,42(2002)237-244.

[16] S.F.Krar and A.F.Check, Electric Discharge Machining, in: technology of machine tools, Glencoe/McGraw Hill, New York, (1997)800

[17] M.J. Haddada and A. Fadaei Tehranib," Material removal rate (MRR) study in the cylindrical wire electrical discharge turning (CWEDT) process", journal of materials processing technology 199(2008) 369–378.

Application of SMED for Setup Reduction Time of Slitters (Steel Industry) in Indian Conditions

Shashank Srivastava[1], Rajendra Yadav[2], Rohan Narula[1]
[1]Department of Mechanical Engg., G.L.A.University, Mathura,U. P.
[2]Department of Mechanical Engg., Anand Engg. College, Agra,U. P.

***Abstract*:** The purpose of this paper is to show the application of Single Minute Exchange of Die (SMED) in a Steel industry. Shigeo Shingo developed SMED (single minute exchange of die) techniques for quick changeovers between products. By simplifying materials, machinery, processes and skills, changeover times could be reduced from hours to minutes. Quick changeovers meant products could be produced in small batches or even single units, with minimal disruption.

The paper represents the use of SMED to reduce the setup time of a Hot Rolled Slitter (HRS) and Cold Rolled Slitter (CRS) in Indian conditions.

The procedure adopted to achieve this is first to collect the time study data. The data is analyzed to obtain the standard time associated with each step/process and ultimately the HRS and CRS machine. The processes are then classified into external and internal activities. Some of the internal activities (related to arbor setting) are converted to the external activities as these can be performed outside the machine. This helps to reduce the setup time of HRS machine to some extent. Further by devising proper arrangement for cutter and separator rings, used in cutter arbor and separator arbor, the internal activities can further be improved.

With SMED the most important thing achieved is to decrease the production time. Four stages of SMED conduced, more external setups since internal setup steps were transformed into external and easier way of working because the needed tools can be found without any problems and the work area was organized better.

I. INTRODUCTION

SMED is the term used to represent the Single Minute Exchange of Die or setup time that can be counted in a single digit of minutes. SMED is often used interchangeably with "quick changeover". SMED and quick changeover are the practice of reducing the time it takes to change a line or machine from running one product to the next.

The successful implementation of SMED and quick changeover is the key to a competitive advantage for any manufacturer that produces, prepares, processes or packages a variety of products on a single machine, line or cell. SMED and quick changeover allows manufacturers to keep fewer inventories while supporting customer demand for products with even slight variations. SMED has a lot of hidden benefits that range from reducing WIP to faster ROI of capital equipment through better utilization.

To achieve the objective of setup reduction, a systematic study has been done to establish the standard time. Time study has been done to achieve it. The processes are broken down into smaller steps. The steps are classified as internal or external. A method has been devised to convert the external activities into internal activities. In this paper an attempt is made to implement SMED in steel industry to reduce the processing time of steel strips and coils. It ultimately reduces the lead time to dispatch the product to customers within a time limit.

I. ASSUMPTIONS OF TIME STUDY:

Rating
The rating is coded as given below:
1. A < 80%
2. B = 80%
3. C = 90%

4. D = 100%
5. E = 110%
6. F = 120%
7. G > 120%

Allowances Sheet

The Factors assumed in the allowances sheet are:

- Posture of the worker during working.
- Monotony of the work.
- Noise
- Temperature/ Humidity
- Ventilation

II.   SETUP TIMES

Setup time is "the time spent in preparation to do a job". In production area, setup has a more detailed definition as the time passed between the last good product and the first good product. Setup time consist of the preparation time and includes replacements, adjustments and attachments to and on a machine. In a big fraction of the setup time the machine doesn't run and doesn't produce. When the machine starts running it is still considered as setup time until the first good output is gained. During this ineffective time the outputs are scraps or require rework.

Traditional way of dealing with setups are using skilled workers for setup, using large lots in production, not having different type of products and combining different jobs with similar setup requirements but when the focus is on reducing setup time there are lots of benefits from it. Reducing setup time provides; better quality because the chance of making mistakes are less when the workload is less on worker while changing the dies, reduced costs because it allows to use small batches thus it reduces work-in-process (inventory level in process) and finished product inventory, better flexibility because small lot sizes are feasible and this makes daily demand changes possible.

There are two types of setup, internal and external.

1. Internal setup(IED) -- setup operations that can be performed only when the machine is stopped, such as mounting or removing dies

2. External setup (OED) – setup operations that can be completed while the machine is running, such as transporting dies to or from storage.

III.   BASIC STEPS IN SETUP REDUCTION:

- Preparation, checking up materials etc.
- Mounting, removing parts tools etc.
- Measurements, settings and calibration.
- Trial runs and adjustments

The greatest difficulty in a setup operation is adjusting the equipment correctly. A large portion of time associated with trial runs derives from these adjustments.

IV. IMPROVEMENT STUDY

For improvement, one must study the actual operation in great detail. It can be done in four ways.

A continuous production analysis
Work sampling study
Interviewing workers
Video recording the operation and studying it.

In the industry the areas identified for improvements are:

- Inventory Management
- Long lead time

The industry is unable to respond quickly to the frequent order change. To provide the finished product in time, proper coordination among the machines is required.

Basic machines used to process the raw material are

1.   Hot Rolled Slitter (HRS)
2.   Pickling
3.   Cold Rolled Mill (CRM)
4.   Cold Rolled Slitter (CRS)

HRS:

HRS is the abbreviation for Hot Rolled Slitters. HRS is used to reduce the thickness and gauge of raw coils for further operations. The thickness reduction is more in HRS then in CRS.

Pickling:

Pickling process is used for removal of corrosion from the coils.

CRM:

A 6 Hi mill is used for further thickness reduction of raw coils.

CRS:

CRS is the abbreviation for Cold Rolled Slitters. The operation of CRS is same as that of HRS with a difference that it is used for finishing operations. The gauge variations are more in CRS then in HRS.

Video recording of HRS and CRS has done to apply SMED on these machines.

IV. OBSERVATIONS AND ANALYSIS

Time study is performed on the machines to gather data.
Observed Time:
The time taken to perform an element or combination of elements obtained by means of direct observation.

Normal Time or Basic Time:

The time for carrying out an element of work at standard rating the observed time is converted into basic or normal

S.NO	Machines Processes	Standard Time (secs)
1.	CRS Arbor Setting	3967.6
2.	CRS running	655.77
3.	CRS Sleeve Setting	754.27
4.	CRM	1728.88
5.	Pickling	555.37
6.	HRS Arbor Setting	5394.69
7.	HRS Separator Setting	323.34
8.	HRS running	368.1
9.	CGL	1822.53
10.	GPCTL	1196.31
11.	CPCTL	917.03

time by multiplying it by rating factor and dividing it with standard rating.

$$\text{Basic time} = (\text{Observed Time} \times \text{Performance Rating}) \div \text{Standard Rating}$$

The representative time established from the observation data is the time, which an operator has taken while working at a certain pace. Standard performance is the rate of output which qualified worker will naturally achieve without over-exertion as an average over the working day or shift, provided that they know and adhere to the specified method and provided that they are motivated to apply themselves to their work.

This performance is denoted as 100 on the standard rating and performance scales

Allowances

It is shown that, during the Time study investigation which should be carried out before any job is timed, the energy expended by the worker in performing the operation should be reduced to a minimum through the development of improved methods and procedures, economic and effective method has been developed, and however, many jobs will still require the expenditure of human effort, and some allowance must therefore be made for recovery from fatigue and for relaxation. Allowance must also be made to enable a worker to attend to personal needs: and allowance (e.g. contingency allowances) may also be added to basic time

The determination of allowances is probably the most controversial part of work study. The difficulty experienced in preparing a universal accepted set of precise allowances

that can be applied to every working situation anywhere in the world is due to various reasons. The most important among them are:

Factors related to the individual.
Factors related to the nature of the work itself.
Factors related to the environment.
Standard time (ST)

Standard time is the time allowed to an operator to carry out specified task under specific conditions and at defined level of performance.

Standard Time = Basic Time + Allowances

Standard time of the various workstations is calculated from the data collected.

Standard time of the processes associated with the machines is shown in the table. From the table, it is inferred that a lot of time is wasted as setup time.

Total Standard time = 16929.59

CRS Arbor setting Standard time = 3967.6

HRS Arbor setting Standard time = 5394.69

23.4% of total standard time is related to CRS arbor setting and 31.8% of total standard time is related to HRS

The CRS operations consist of three parts. The first operation is arbor setting, in which rings are arranged to achieve the desired slitting. Second operation is running of CRS, in which the machine is operated practically to process the raw material. Third operation is sleeve setting operation, in which sleeves are fitted inside the coils for its proper removal from the turnstile stand. The amount of processing time associated with setup is shown in the fig.

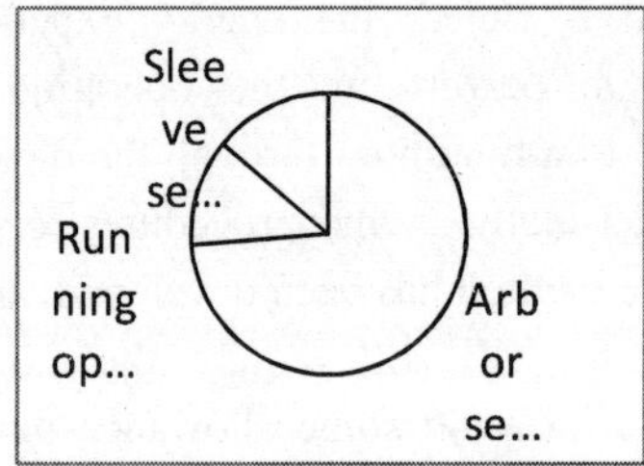

Fig.Standard time in percentage of HRS Machine.

From the pie chart it is inferred that 74% of the processing time of CRS machine is associated with the arbor setting operation. SMED is applied to CRS machine to reduce its setup time

To apply SMED in CRS machine following measures are taken:

1. A fixture is provided outside the machine to arrange the rings over the arbor.
2. The height of the fixture is equals to the shoulder height of the labour.
3. The arbor is moved to the fixture by overhead crane.
4. Extra arbor is provided to run the HRS machine while the new arbor setting is done.

The processing of raw material, after converting the internal operations related to arbor setting to external operations, becomes very smooth. 54 minutes is saved by applying SMED to the machine. The setup can further the reduced by applying SMED to the other processes of CRS.

The HRS operations consist of three parts. The first operation is arbor setting, in which rings are arranged to achieve the desired slitting. Second operation is separator setting, which is used to achieve proper loop of steel coils. . Third operation is running of HRS, in which the machine is operated practically to process the raw material. The amount of processing time associated with setup is shown in the fig.

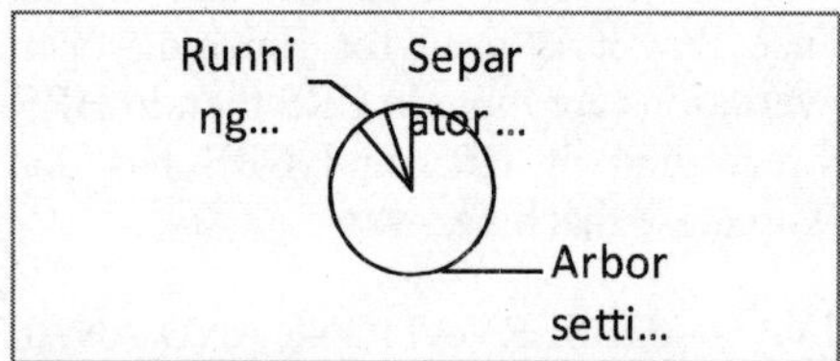

Fig. Standard time in percentage of CRS Machine

From the pie chart it is inferred that 89% of the processing time of HRS machine is associated with the arbor setting operation. SMED is applied to HRS machine to reduce its setup time

Similar measures are taken to achieve the setup reduction in HRS.87 minutes are saved by applying SMED to the machine. Further reduction in processing time is possible by applying SMED to separator setting. The separator setting is similar to that of arbor setting. The drag board is moved up for the arrangement of rings in the separator. The setup processes of separator have to be broken down into small elements. The steps of SMED are applied to reduce its setup.

Time to process the raw material is obtained by adding the standard time of HRS, pickling, CRM, CRS.

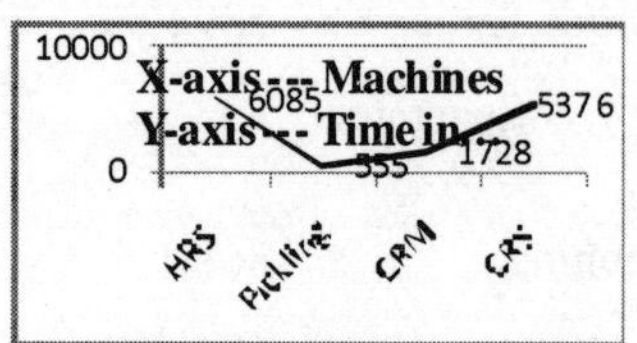

Graph.

Standard time to produce the product before SMED

The graph is showing the standard time of machines in seconds. It is inferred from the graph that maximum time is taken by HRS and CRS. This is the situation before SMED. The total time taken to process the raw material is 13744 seconds.

After applying SMED the standard time of machines is reduced and ultimately the processing time of raw material is reduced. This situation is shown by the graph . The total time to process the raw material is reduced to 4383 seconds.

V. RESULTS AND DISCUSSIONS

The graphs show that by SMED the setup, which is initially in hours, can be reduced to seconds. Moreover it would help in reduction of work in process inventory before CRS. SMED makes it possible to respond quickly to fluctuations in demand, and it creates the necessary conditions for lead-time reductions. Combining diversified, small lot production with SMED is the most effective way to achieve flexible production and maximum productivity. Most importantly SMED gives the chance to produce small batches thus the way of production will be similar to JIT. One machine can be used for many types of products with a little loss of time (Moxham, 2001). SMED reduces the non-productive time by streamlining and standardizing the operations for exchange tools, using simple techniques and easy applications. However the process doesn't give the specific actions to implement which can result in overlooking improvements. To overcome this, common statistical and industrial engineering tools can be integrated in the SMED approach to improve SMED implementation results.

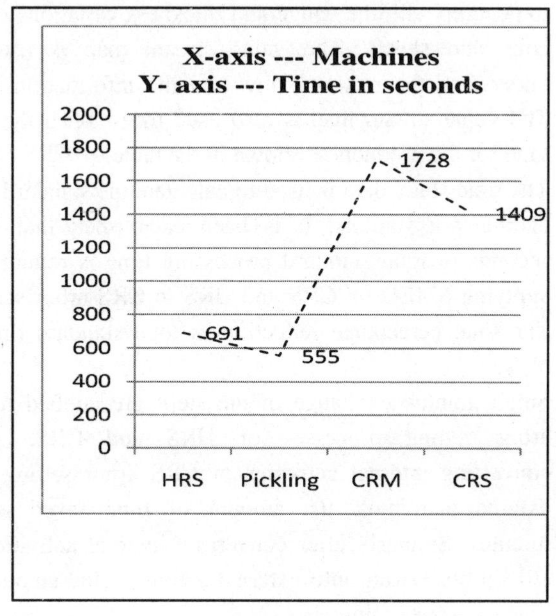

Graph

Standard time to produce the product after SMED

S. No.	Setup Steps of Cold Rolled Slitter Machine	Time (Sec.)	Before SMED Classification of Steps	After SMED Classification of Steps
1	The fitting tools are provided	Not defined	External	External
2	The tapes above the stripper are cut down	264.32	Internal	Internal
3	The housing is moved away and the check nut is removed	131.31	Internal	Internal
4	The ring from upper arbor is removed	265.22	Internal	External
5	The ring from lower arbor is removed	286.39	Internal	External
6	The spacer rings and cutter ring are fitted on upper arbor	1847.4	Internal	External
7	The stripper rings are fitted on upper arbor	415.67	Internal	External
8	Spacer rings and cutter rings are fitted on lower arbor	156.21	Internal	External
9	Stripper rings are fitted on lower arbor	285.26	Internal	External
10	Check nut is tightened	213.94	Internal	Internal
11	The gear is adjusted by gear and lever arrangements	101.88	Internal	Internal
12	Loading: The coil is moved near the roller cone by coil car	17.525	Internal	Internal
13	The roller cone is opened	6.684	Internal	Internal
14	The coil is aligned in the roller cone	11.609	Internal	Internal
15	Snubber roll is used for proper tightening	46.915	Internal	Internal
16	The coil is fed to the shearing table	55.55	Internal	Internal
17	The coil is moved to the separator arbor	38.428	Internal	Internal
18	The loop of the coil is adjusted	68.83	Internal	Internal
19	The coil is clamped to the recoiler	110.03	Internal	Internal
20	The tools are cleaned and put back to specified place	Not defined	External	External

VI. CONCLUSION

To sustain in market an industry has to fulfill the customer requirements in time. The medium scale process industry which is studied in the work has to respond quickly to the frequent change in orders. The industry has to cope up with the gauge and thickness variations of the coils and sheets. The value stream map is used to understand the material flow and the information flow. The value stream map is also used to establish the lead time for the products as shown in the table:

The time study data is used to calculate the standard time of the workstations. It is been established that 9361 seconds of total standard processing time is reduced by applying SMED on CRS and HRS to CRS arbor setting. The total percentage reduction in total standard time is 68%.

Single minute exchange of die steps are applied to the arbor setting processes of HRS and CRS. After converting internal activities of CRS arbor setting into external activities, the amount of time saved is 54 minutes. Similarly after converting internal activities of HRS arbor setting into external activities, the amount of time save is 87 minutes.

Thus it can be concluded with SMED, it is possible to reduce the setup times of machines. A proper understanding of the processes is the key to achieve it. Video taping of the machines can be done to get maximum details of the processes involved in it. Brainstorming sessions can be conducted to convert the internal activities into external activities. Creative ideas would help in further setup reduction.

VII. ACKNOWLEDGEMENT

The authors wish to express their thanks to
Mr. Fakruddin Ujjainwala, Professor, Industrial and Production Dept. SGSITS Indore (INDIA) for his valuable help in carrying out the experimental work

VIII. REFERENCES

[1]   Malkin,s (1989). Grinding technology.
Ellis Horwood Chichester, UK.

[2]   W.A.J. Chapman Fourth Edition,
Workshop Technology.1972.

[3]   Paul H. Blackie. Theory of Metal Cutting.

[4]   Alden,G.I.: Operation of Grinding Wheels in Machine Grinding, Transaction of the ASME,
Vol. 36, no.1446, p.451, 1914.

[5]   Bacher,W.R.,and M.E. Merchant: On the Basic Mechanics of the Grinding Process, Transaction of the ASME, VOl. 80, no.1, p.141,January, 1958.

[6]   Brecker, J.N. (1967). Ph.D dissertation, Carnegie-Mellon University.

[7]   ASTME, USA. Tool Engines handbook, second ed. McGraw Hill New YORk, 1959.

[8]   ASTME, USA Metal handbook, Vol 3: Machining. American Society for Metals, Ohio, 1967.

[9]   M.C.Shaw, Principles of Abrasive Processing Clarendom press, Oxford, 1996

[10] Backer, W.R. Marshall, E.R. and Shaw,
M.C. (1952) Trans. ASME 74.

[11]   A. Bhattacharyya, Metal Cutting,
Theory and Practice 1984.
Central Book Publishers Calcutta. India.

[12] Brecker, J.N. and Shaw, M.C. (1974).Ann. CIRP 23 (1),93

[13] Shaw,M.C. (1972) Metal cutting principles. Oxford. Clarendon Press.

[14] Malkin,s (1989). Grinding technology.
Ellis Horwood Chichester, UK.

[15]   S.Malkin, Grinding technology, Ellis-Horwood, chichester, 19

[16] Banerjee, J.K. and Hiller, M.J. (1969a) Tool Mfg Engr (ASTME), Feb., 59

[17] Brecker, J.N., Komanduri, R., and Shaw, M.C. (1973). Ann.CRIP 22 (2), 189.

[18] Browchuck,Roman." The inside scoop- Part 1".Grits & Grinds. Vol.60, no.21969,p.3

[19] Shaw, Milton C. "How to estimate grinding force and power?" Machining, USA, Vol. 74, No.7, March 1968 p.85

[20] Brecker, J.N. and Shaw, M.C. (1974). Ann.CRIP 2393

Casting of SiC Reinforced Metal Matrix Composites: A Review

Nishant Singh, Yogesh Dhote, Yashvir Singh
Hindustan College of Science & Technology, Farah, Mathura, U.P.
Email: nishant.singh78@gmail.com

Abstract--- One of the great challenges of producing cast metal matrix composites is the agglomeration tendency of the reinforcements. This would normally result in poor distribution of the particles, .igh porosity content, and low mechanical properties. One factor that, to date has restricted the widespread use of MMCs has been their relatively high cost. This is mostly related to the expensive processing techniques used currently to produce high quality composites. In this paper various method of casting of SiC reinforced metal matrix composites and their relative merits and demerits are discussed.
Keywords: Stir casting, Semi-Solid Processing, Al–SiC MMC, Composites etc.

I. INTRODUCTION

There are several fabrication techniques available to manufacture MMC materials. Depending on the choice of matrix and reinforcement material, the fabrication techniques can vary considerably. Fabrication methods can be divided into three types. These are solid phase processes, liquid phase process and semi-solid fabrication process. Solid state processes are generally used to obtain the best mechanical properties in MMCs, particularly in discontinuous MMCs. This is because segregation effects and inter-metallic phase formations are less for these processes, when compared with liquid state processes.

II. SOLID STATE METHODS

Powder Metallurgy----Powdered metal and discontinuous reinforcement are mixed and then bonded through a process of compaction, degassing, and thermo-mechanical treatment possibly via extrusion. Foil diffusion bonding: Layers of metal foil are sandwiched with long fibers, and then pressed through to form a matrix.

Liquid State Methods----
- Stir casting: Discontinuous reinforcement is stirred into molten metal, which is allowed to solidify.
- Squeeze casting: Molten metal is injected into a form with fibers preplaced inside it.
- Spray deposition: Molten metal is sprayed onto a continuous fiber substrate.
- Electroplating: A solution containing metal ions loaded with reinforcing particles is co-deposited forming a composite material.

Vapor Deposition Method----
- *Physical vapor deposition:* The fiber is passed through a thick cloud of vaporized metal, coating it.
- *Stir Casting----*Among the variety of manufacturing processes available for discontinuous metal matrix composites, stir casting is generally accepted as a particularly promising route, currently Practiced commercially. Its advantages lie in its simplicity and applicability to large quantity production. It is also attractive because, in principle, it allows a conventional metal processing route to be used, and hence minimizes the final cost of the product. This liquid metallurgy technique is the most economical of all the available routes for metal matrix composite production and allows very large sized components to be fabricated.

In preparing metal matrix composites by the stir casting method, there are several factors that need considerable attention, including,
- The difficulty of achieving a uniform distribution of the reinforcement material;

- Wetability between the two main substances;
- Porosity in the cast metal matrix composites; and
- Chemical reactions between the reinforcement material and the matrix alloy.

In order to achieve the optimum properties of the metal matrix composite, the distribution of the reinforcement material in the matrix alloy must be uniform, and the wetability or bonding between these substances should be optimized. The porosity levels need to be minimized, and chemical reactions between the reinforcement materials and the matrix alloy must be avoided.

L. Looney[1] reported experimental procedure for preparation of Al+SiC composite by stir casting according to this solidification synthesis of metal matrix composites involves producing a melt of the selected matrix material followed by the introduction of a reinforcement material into the melt, obtaining a suitable dispersion. The next step is the solidification of the melt containing suspended dispersion under selected conditions to obtain the desired distribution of the dispersed phase in the cast matrix. In this method, after the matrix material is melted, it is stirred vigorously to form a vortex at the surface of the melt, and the reinforcement material is then introduced at the side of the vortex. The stirring is continued for a few minutes before the slurry is cast. Harnby et al. [2] studied different designs of mechanical stirrers, as shown in Fig.1. Among them, the turbine stirrer is quite popular. During stir casting for the synthesis of composites, stirring helps in two ways: (a) transferring particles into the liquid metal, and (b) maintaining the particles in a state of suspension.

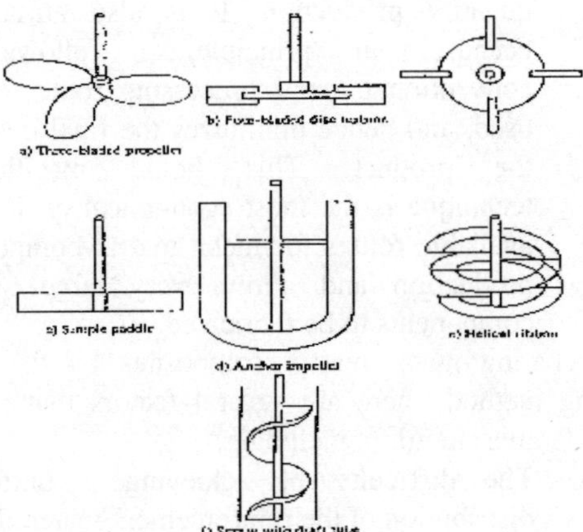

Fig.1 Mechanical stirrers for quick quench stir caster method

S. Naher,[3] reported the approach, the schematic drawing, of the stir caster is shown in Fig. 2.

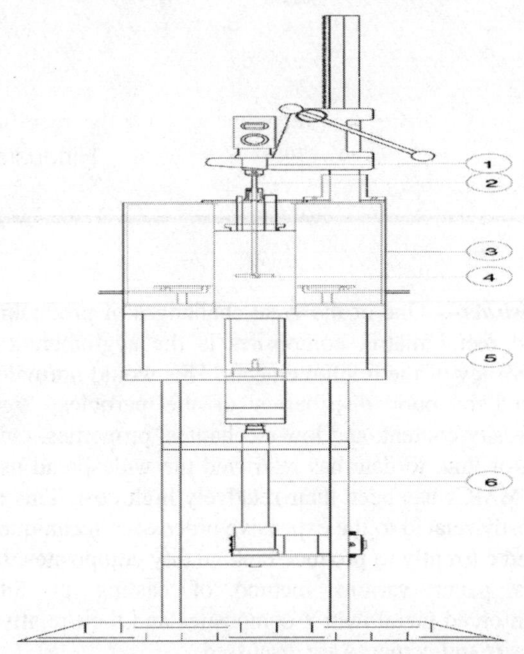

Fig.2. Schematic drawing of the Stir-Caster: (1) The Stirring Motor, (2) The Stirrer Rod, (3) Furnace, (4) Stirrer impeller, (5) Crucible, and (6) Aactuator

The stir-caster furnace was mounted on four legs a ceramic spacer was used to separate the crucible from the actuator. This arrangement allowed rapid extraction of the crucible from the furnace. The temperature within the stir-caster had to be precisely measured and accurately controlled (±1 ∘C), in order to control the fraction solid of the semi-solid alloy. Stainless steel material was chosen for the crucible, stirrer rod and impeller material.

The diameter of the crucible was 105 mm and the melt height was 65 mm. During experimental work, a four flat bladed 45° angled stirrer was chosen. The stirrer impeller diameter was 80 mm, blade height was 20mm and width was 2 mm. The stirrer was connected to a DC motor which was used to stir the molten matrix material. A lifting mechanism for the rotational drive unit and stirrer assembly was used to extract the stirrer from the melt before quenching of the melt and to facilitate the stirrer positioning, cleaning and replacement.

III. STIR-CASTER OPERATION

The stir caster operation was conducted according to the following sequence of events:

- Collection and preparation of the raw materials (A356 +SiC particles).
- Placing raw materials in the crucible under nitrogen gas into a furnace.

- Heating the crucible above the liquidus of A356 and allowing time to become completely liquid.
- During cooling stirring is started at the semi-solid condition then when the temperature was stabilized at the appropriate level the stirring was recommenced for the specified period and shear rate.
- The charge in the crucible was then quenched into water.
- MMC billets were produced.

The charge temperature was raised to 630 °C within 130 min at the start of each experiment. Stirring was then started and continued for the specified periods for the liquid state experiments. After the specified shearing periods the crucible was lowered and quenched. In the semi-solid experiments, the stirring was also started at 630 °C and continued for 5 min to promote the wettability. Stirring was then stopped and the temperature was lowered to 605 °C (0.30 fraction of solid), at a rate of 0.0011 °C /s.

IV. GRAVITY CASTING AND A NOVEL TWO-STEP MIXING METHOD

W. Zhou and Z. M. Xu [4] described Casting of SiC Reinforced Metal Matrix Composites by two step mixing method Two SiC particulate reinforced composites were produced. The matrix alloys of the composites were A356 and 6061 respectively. They are both aluminium alloys but differ greatly in amount of Si, For convenience in description, the composite with A356 as matrix and reinforced by 10.8 Vol % SiC particles is hereafter referred to as A35610.8% SiC, and the other composite as 6061-20%SiC. All the melting was carried out in a clay-graphite crucible in a resistance furnace. Scraps of alloy A356 or 6061 were preheated at 450°C for 3 to 4 hours before melting, and before mixing the SiC particles were preheated at 1100 °C for 1 to 3 hours to make their surfaces oxidized.

The furnace temperature was first raised above the liquidus to melt the alloy scraps completely and was then cooled down just below the liquidus to keep the slurry in a semi-solid state. At this stage the preheated SiC particles were added and mixed manually. Manual mixing was used because it was very difficult to mix using automatic device when the alloy was in a semi-solid state. After sufficient manual mixing was done, the composite slurry was re-heated to a fully liquid state, and then automatic mechanical mixing was carried out for about 20 minutes at an average stirring rate of 150-200 rpm. In the final mixing processes, the furnace temperature was controlled to be within 730 ±10 °C.

The pouring temperature was controlled to be around 720 °C. A preheated permanent steel mould with diameters in the range of 10 mm to 18 mm was used to prepare cast bars. The preheating temperature for the mould was either 50 °C for fast cooling or 350 °C for slower cooling.

V. ADVANTAGE OF TWO-STEP MIXING

When the SiC particles were added into the molten alloys, they were observed to be floating on the surface, though they have a larger specific density than the molten alloys. This was due to high surface tension and poor wetting between the particles and the melt. A mechanical force can usually be applied to overcome surface tension to improve wetability. Mechanical stirring could indeed mix the particles into the melt, but when stirring stopped, the particles tended to return to the surface. However, the fact that single particles also tended to return to the surface strongly indicates that the particles floated mainly because of the surface gas layers surrounding them. The gas layers might be the main factor for the poor wetability improve the particle distribution, the second mixing step is needed, i.e., to heat the slurry to a temperature above the liquidus and then to stir the melts using an automatic device. It was found that the two-step method resulted in a homogeneous distribution of particles.

VI. CENTRIFUGAL CASTING METHOD

T.P.D. Rajan [5] described Casting of SiC Reinforced Metal Matrix Composites by centrifugal method according to this; the conventional centrifugal casting method can be used for making functionally graded metal matrix composites. This process involves synthesis of MMC by stir casting method followed by centrifugal casting to form the gradient in microstructure due to centrifugal force. When particle-containing slurry is subjected to centrifugal force, two distinct particle enriched and depleted zones are formed. Depending on the density, the lighter particles segregate towards the axis of rotation, while the denser ones move away from axis of rotation. In the case of aluminium alloy, the particle enriched zone of heavier particles such as SiC, alumina and zircon is at the outer periphery, while that of lighter particles such as graphite and mica is at the inner periphery of horizontally cast centrifugal castings.

VII. MATERIALS AND METHODS

A schematic diagram of horizontal centrifugal casting equipment used for fabricating FGMMC is shown in Fig.3.

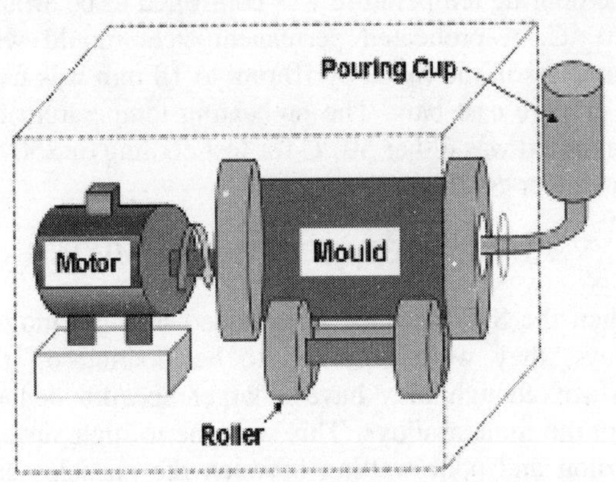

Fig.3. Schematic diagram of horizontal centrifugal casting equipment used for fabricating FGMMC

The two different matrix alloys used for synthesizing the FGMMC are 356 (Al-7.5Si-0.35Mg) cast and 2124 (Al-4.5Cu-1.6 Mg-0.25Zn-0.2Si) wrought aluminum alloys. Green variety SiC particles of 23 μm average particle size have been used as reinforcement. Initially, the Al(356)–15%SiC and Al(2124)–15%SiC composite melts are synthesized by liquid metal stir casting method and later shaped into hollow cylinder in a horizontal centrifugal casting machine. The composite melt at 750–760 °C is poured into a coated and preheated (250±10 °C) metal mould, which is rotated at 1100 rpm. Fig.3 and Fig.4 show the schematic diagram of horizontal centrifugal casting equipment used and the typical functionally graded cylindrical castings made respectively. The dimensions of the castings made (weighing 5 kg) are 380 mm length and 120 mm diameter with a wall thickness falling between 15 and 17 mm. Rings are first cut from the casting and later sliced to specimens of 20mm height and 40mm length for micro structural evaluation, heat treatment and hardness testing.

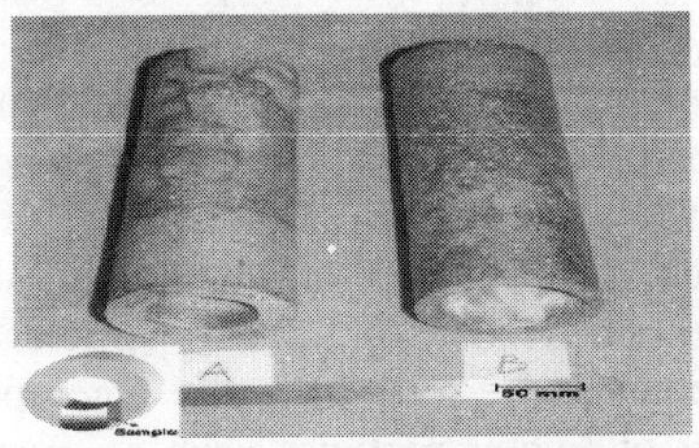

Fig.4. Typical functionally graded cylindrical castings made by centrifugal casting

The microstructures show the graded distribution of SiC particles in Al (356) alloy matrix. The outer periphery of the casting shows higher concentration of SiC particles than interior of the casting. By subjecting a homogenous composite melt of Al

(356)-15%SiC to centrifugal force, a maximum volume fraction of 45% has been obtained at the outer periphery leading to selective improvement in specific properties such as hardness and wear resistance. The hardness values measured from the outer to the inner periphery for as cast and heat treated Al(356)-SiC the outer edge hardness having a composition of Al(356)–44%SiC is 98 BHN compared to 58–60 BHN for the particle free zone. In Al (2124)–SiC FGMMC, the hardness near the outer surface is 115 BHN against 90 BHN for the particle free zone.

VIII. INJECTION METHOD

Sajjad Amirkhanlou, Behzad Niroumand [6] reported development of Al/SiC cast composites by injection of SiCp. The method involved injection of specially-made composite powders into the melt and was expected to result in gradual release of SiC particles in the liquid metal and overcome the agglomeration tendency of the reinforcement particles. Equal volumes of the pure aluminium and SiCp powders were milled for 50 h to achieve a mechanically interacting Al–SiCp composite powder ((Al–SiCp)cp). Schematic of the experimental set-up used in the production of the composites is shown in Fig. 6. Alloys were melted in a graphite crucible of 1.5 kg capacity and the temperature of the melt was raised to 700 °C. The melt was kept at this temperature for approximately 2 min while being stirred at 500 rpm and then injection of the reinforcements started. SiC particles were injected into the melt as untreated SiCp or as particulate composite powders. Pure argon (99.999%) was used as the carrier gas for injection of the reinforcements. After completion of the injection, the slurry was continuously cooled and stirred at an average cooling rate of 4.2 °C/min until reaching 650 °C (fully liquid, hence stir casting) or 607 °C (corresponding to 0.2 solid fraction according to Scheil equation, hence compo casting) and cast into a steel die placed below the furnace. In liquid metal stir casting, both injection of the particles and pouring of the resultant composite slurry were carried out in a fully liquid state. On the other hand, in the case of compo casting process, the reinforcement was added in a fully liquid state but the composite slurry was cast in semisolid state 1 wt. % magnesium was added to all composite specimens, either by direct addition to the melt or through the injected particulate composite powder, to increase the wetability between the matrix and the reinforcement.

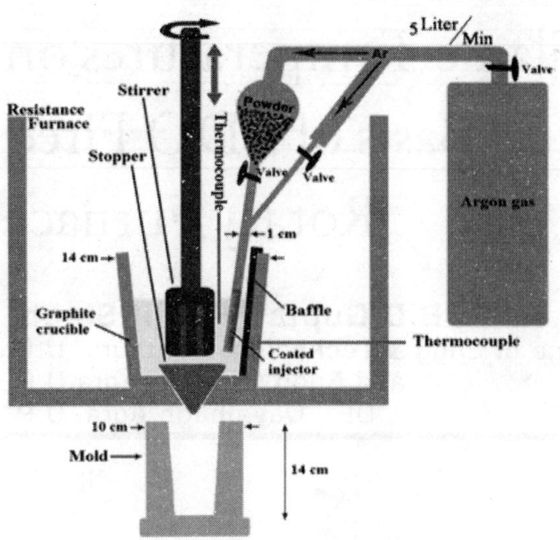

Fig.5. Schematic of the experimental set-up used in production of the composites

IX. RESULTS

Based on the results obtained under the experimental conditions used in the present work, the following conclusions can be drawn:

- Distribution of SiC particles in the matrix was greatly improved by injection of milled composite powders instead of untreated-SiC particles into the melt.
- Distribution of SiC particles were further improved by casting in a semisolid state (compo -casting) rather than in a fully liquid state (stir casting).
- The average size of SiC particles incorporated into the matrix was significantly reduced from about 8 to 3 lm by injecting the milled composite powders instead of untreated-SiC particles.
- The porosity content of the composites was considerably decreased when the reinforcements were injected as composite powders and the casting was performed in a semisolid state.

X. DISCUSSION AND CONCLUSION

A novel quick quenched stir-caster can be preferred for processing Al–SiC composites in liquid and semisolid state. Stirring the MMC slurry in semi-solid state, during the solidification process helps to incorporate ceramic particles into the alloy matrix. The quick quenched compo casting method was found to be successful to fabricate Al–SiC metal matrix composite.

It was found that the two-step method resulted in a homogeneous distribution of particles. Mechanical

stirring could indeed mix the particles into the melt, but when stirring stopped, the particles tended to return to the surface.

The gas layers might be the main factor for the poor wetability improve the particle distribution, because of this second mixing step is needed, i.e., to heat the slurry to a temperature above the liquidus and then to stir the melts using an automatic device. In centrifugal casting the outer periphery of the casting resulted in higher concentration of SiC particles than interior of the casting.

By subjecting a homogenous composite melt of Al-SiC to centrifugal force, a maximum volume fraction has been obtained at the outer periphery leading to selective improvement in specific properties such as hardness and wear resistance. Distribution of SiC particles in the matrix can be greatly improved by injection of milled composite powders instead of untreated-SiC particles into the melt.

The porosity content of the composites can be considerably decreased if the reinforcements will be injected as composite powders and the casting will performed in a semisolid state.

XI. REFERENCES

[1]   J. Hashim, L. Looney, M.S.J. Hashmi, "Metal Matrix Composites: Production by the Stir Casting Method".

[2]   S. Naher, D. Brabazon, L. Looney, "Development and Assessment of a New Quick Quench Stir Caster Design for the Production of Metal Matrix Composites".

[3]   W. Zhou, Z. M. Xu, "Casting of SiC Reinforced Metal Matrix Composites".

[4]   T.P.D. Rajan, R.M. Pillai, B.C. Pai, "Characterization of Centrifugal Cast Functionally Graded Aluminum-Silicon Carbide Metal Matrix Composites".

[5]   Sajjad Amirkhanlou, Behzad Niroumand, "Development of Al356/SiCp Cast Composites by Injection of SiCp containing Composite Powders".

[6]   D.M. Skibo, D.M. Schuster, L. Jolla, "Process for Preparation of Composite Materials Containing Nonmetallic Particles in a Metallic Matrix, and Composite Materials".

[7]   L.V. Ramanathan, "P.C.R. Nunes, Effect of Liquid Metal Processing Parameters on Microstructure and Properties of Alumina Reinforced".

[8]   N. Chawla, K. Chawla, "Metal Matrix Composites", USA: Springer Science-Business Media, 2005.

[9]   P. K. Rohatgi, R. Asthana and F. Yarandi, "Solidification of Metal Matrix Composites".

[10]   T.P.D Rajan, K. Narayan Prabhu, R. M.Pilla, B. C.Pai, "Solidification Characteristics of Discontinuous Aluminium Matrix Composites

Comparative Evaluation of Flame Temperatures on Experimental and Theoretical Basis of LDO Fired Rotary Furnace

R.K.Jain[1], B.D.Gupta[2], Ranjit Singh[3]
[1]BSA College of Engg & Technology, Mathura. U.P.
[2]Anand Engg College. Agra,U.P.
[3]DEI, Dayalbagh Agra, U.P.

Abstract- The flame temperature plays an important role in performance of LDO fired rotary furnace viz melting rate, specific fuel consumption, melting time, energy consumption, and emission levels. The authors have carried out experimental investigations on flame temperature of a 200 kg rotary furnace and its effect on other parameters. Further the mathematical computations of flame temperature based on different modes of heat transfer have been made and finally the comparison of experimental and theoretical results are being presented herein.

I. INTRODUCTION

The Rotary Furnace is very simple melting unit consisting mainly of a drum of required size having a cone on each side lined with refractory, fire bricks or ramming mortar generally having alumina as a constituent. This drum is placed on rollers so that they may be either locked or slowly rotated about their central axis. The rollers are driven by a small electric motor. At one end of the drum, a suitable burner is placed with appropriate blower system and combustion gases exit from other end. This drum or horizontal cylinder is flanked by two conical portions on both sides. One of the cones accommodate the burner whereas from the other cone hot flue gasses exit. Charging of the iron for melting is also done from this side. The cone on one side can accommodate different types of burners using the Light Diesel Oil (L.D.O.). The tap hole is located in the cylindrical wall halfway between the ends. This tap hole is used to take out the molten metal but it is kept closed during the melting of metal. A tap hole is made approximately in centre of the drum. The charging of material is done through the tap hole and other cone whereas the pouring is done through tap hole only.

A covered oil tank containing LDO is located at height of approx. 5-7 meters from burner end of the furnace which is connected to the burner through suitable diameter pipeline and control valves. A pump is installed to force the oil at desired pressure to the burner. There are a number of variables controllable to varying degrees which affect the quality and composition of the out-coming molten metal. These variables, such as, revolutions per minute of the drum, flame temperature, melting time, fuel consumption and melting rate play significant role in determining the molten metal's properties and should be controlled throughout the melting process. However, even an experienced operator may find it difficult to select the optimum input parameters which would yield ideal molten metal and often he may choose them by guessing which may not be effective and economical. Fig. 1.shows its layout

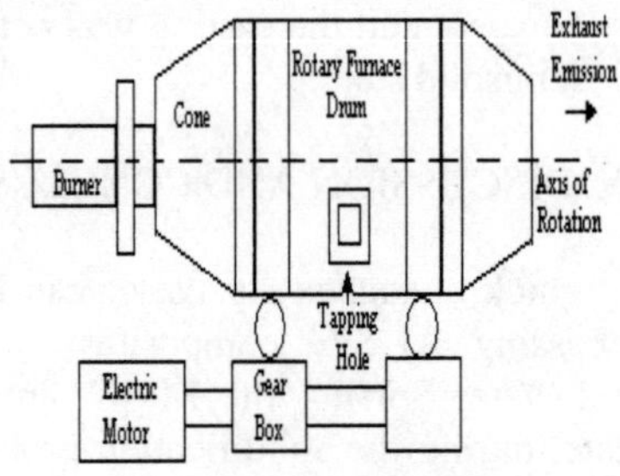

Fig. 1. Layout of Rotary Furnace

II. LITERATURE REVIEW

Baker EHW [1] explained the working of Rotary furnace. AFS [2] preferred to use Pulverized coal for Rotary furnace. Singh R, Jain RK [5] concluded that optimum RPM for rotary furnace is 1.4. Jain RK, Singh R [6] on basis of Experimental investigations confirmed that flame temperature depends upon preheated air and excess air. The optimum temperature of preheated air is 300 °C and excess air10 percent for optimum flame temperature within safe metallurgical limits. Jain RK, Singh R [8] Jain RK, Singh R [9] confirmed that based on the experimental investigations carried out on a 200 kg furnace in a foundry the optimum RPM of furnace rotation is 1 to give the maximum melting rate and minimum fuel consumption. Jain RK, Singh R, Gupta B.D. [11] presented an overview of energy consumption in ferrous foundry and stressed upon the need of an energy efficient furnace for foundries.Jain R.K, Singh R,[12].applied regression modeling and excel solver technique for mathematical modeling and optimization of critical parameters of rotary furnace viz. rpm, melting rate, specific fuel consumption etc. Mike Yeoman [10] has expressed deep concern on increasing energy consumption in ferrous foundries and stressed upon the need for its conservation.Jain RK Singh R, Gupta B.D.[13,14, 16] made a comparative study of energy consumption in melting by different furnaces used in foundries and concluded that for minimum emission level &energy consumption the optimum rpm for rotary furnace is 1 &further investigated the effect of flame temperature on performance of rotary furnace. Basu Navojit,.RoyP.K. [15] has evolved to plan and organize energy conservation activities in foundry sector. Baijayanath, Prosantopal Panigrahy K.C.[17]explained that most of the units are crippled with usage of rudimentary techniques the Indian foundry industry needs optimization of energy consumption. Singh Kamlesh Kumar[18]advocates the use of newer and cleaner technology for environmental and energy conservation. Arjunwadkar S.H, Pal Prosanto[19] stressed upon to use energy efficient melting techniques. Pandey G.N.,Singh Rajesh,Sinha A.K[20] emphasized upon To supply oxygen at 8kg/cm^2 pressure as it reduces melting time and emission levels. Basu Navojit,. RoyP.K.[22] explained the advantages of gaseous fuels to attain net adiabatic flame temperature as it is helpful for radiation heat transfer & leads to optimum energy input and lesser pollution. Yuonus ceingel[23-] have used differ rent formulas based on different modes of heat transfer for calculation of source (flame) temperatures.

MELTING OPERATION: The process The process of melting the charge is carried in following steps:

Preheating of oil and furnace- The oil is preheated up to 70 °C and forced at 1.5 kg/cm^2. to preheat the furnace and starting the combustion. A small volume of oil in form of stream jet, and primary air is forced inside the furnace. The droplets of oil come in contact with small volumes of primary air. At exit end of burner, initially drenched jute or warm cloth pieces are placed which are immediately ignited. As combustion proceeds, the volume of primary air is increased. When full ignition takes place, the secondary air at same pressure is started. The volume of primary and secondary air are controlled by valves to avoid the danger of backfire.

Charging After pre heating the furnace is charged.

Rotation- After sufficient pre heating and charging the furnace starts rotating.

Melting-The flame starts coming out of the exit end, which is initially yellowish in color. After approximately 1 hour, the color of flame changes to white indicating that metal has been thoroughly melted. The temperature of the molten metal is measured using pyrometer. If it is approximately 1250 to 1300°C, the rotation of furnace is stopped.

Tapping- the tape hole is slightly lowered and opened and metal is transferred into ladles, which are pre heated prior to transfer of molten metal to avoid heat losses.

Inoculation- The Ferro silicon and Ferro manganese are added in molten metal contained in the laddles.

Pouring - The ladles are then carried to moulds and pouring is completed. The experiments were conducted on self designed and developed furnace as shown in Figure 1

Experimental Investigations- The following experimental investigations were carried out on above self designed and developed 200 kg capacity rotary furnace installed at foundry shop, in Agra–(1)Exhaust gases and multipass counter flow heat exchangers (2) Effect of reducing excess air& using multipass counter flow heat exchanger on specific fuel consumption(3) Effect of reducing excess air& using compact heat exchanger on specific fuel consumption(4)Experimental Investigations (4)--Effect of oxygen enrichment, reducing volume , and increasing temperature of preheated air (by using compact flow heat exchanger) on fuel consumption.

Effect of combustion volume-- If the combustion volume is more than more fuel and time shall be required for reaching a certain temperature. Hence it is thought to optimize the combustion volume by reducing the amount of air and supplying oxygen externally. This was with 75% of theoretically required air and 6% oxygen additionally.

Temperature& specific fuel consumption-. No of experiments were conducted and effect of oxygen enrichment reducing volume, and using cross flow

Experimental Investigations (4)-Effect of oxygen enrichment, reducing combustion volume, and using compact flow heat exchanger on flame

heat exchanger) on flame temperature melting rate & specific fuel consumption is shown in table5—

Table.5—Effect of oxygen enrichment, using compact exchanger, reducing combustion volume, on flames temp. & Specific fuel consumption

S. N.	Preheat Air temp 0c	Flame temp 0c	Time heat Min.	Fuel /heat Lit.	Melting Rate kg/hr	Specif. fuel consumption Lit/kg	Oxygen Consum./ heat m^3	Air Consump /heat m^3
1	410	1710	33	56	363	0.280	39	459
2	418	1722	32	56	375	0.280	39	459
3	428	1730	32	55	375	0.280	38.5	451
4	449	1746	31.5	54	385	0.270	38	443
5	454	1752	31	53	387	0.265	37	434.5
6	458	1754	30.5	52	393.44	0.260	36.60	426.7
7	460	1755	30.5	52	393.44	0.260	36.50	426.5

I. FLAMETEMPERATURE CALCULATIONS ON THEORETICAL

BASIS—(Heat Transfer by Radiation Only)

Mainly the heat is transferred from flame to refractory by Radiation. The heat transfer by radiation is represented by the formula [23]—

$Q_{net} = \xi_s \times A_s \times \sigma \times (\xi_g.T_g^4 - \alpha_s.T_s^4)$

where, ξ_s = Emissivity of surface (Refractory)A_s = surface area of refractory mm^2

σ = Stefan Boltzman's constant = 5.78 x 10^{-8} watt/m^2 ° K^4

ξ_g = Emissivity of flame = 0.80= oxidizing flame

T_g = Temperature of flame

α_s = Absorptivity of refractory surface

T_s =Initial Temperature of refractory= 27 +273 = 300 K

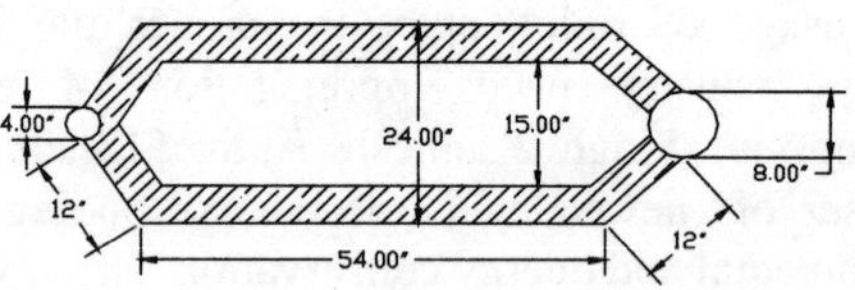

Fig 4 The dimensions of Rotary furnace
The rotary furnace consists of--

One horizontal cylinder of dimensions--
L= Length = 54 inches =54x25.4 mm =1371.6 mm

Outer diameter of furnace shell D(Before brick lining) = 24 inches = 24 x 25.4 mm = 609.6mm

Inner diameter of furnace Shell d (after brick lining) =15 inches = 15 x 25.4 mm = 381 mm

Surface area of cylinder=πd H=π d L=

22/7 x 381 x 1371.6 mm^2
=1642393.029mm^2 =1.642393.029 m^2

One cone at burner end- The frustum of cone at burner end has

Slant length L= 12 Inches = 12 x 25.4 mm = 304.8 mm,

Diameter d_2 at burner end = 4 inches = 101.6 mm,

Radius at burner end = r_2= 50.8 mm

Diameter at base = inner diameter of furnace shell = 381 mm

Radius at base = r_1 = 381/2 = 190.5 mm

The surface area of cone at burner end = π (r1—r2) x L   =22/7 x (190.5—50.8) x 304.8 mm^2
= 133824.617 mm^2-=0.133824617 **m^2**

One cone at heat exchanger end— The frustum of cone at heat exchanger end has

Slant length L = 12 inches = 12 x 25.4 mm = 304.8 mm

Diameter d_3 at exchanger end = 8 inches =203.2 mm,

Radius at exchanger end r_3 = 101.6 mm ,

Radius at base is same = 190.5 mm

The surface area of cone at exchanger end = π (r_1—r_3) x L.=
 22/7x (190.5-101.6) x304.8mm^2
 =85161.12 mm^2-=0.08516112m^2

 Total surface area=1+2+3= 1861378.766 mm^2= 1.861378766 m^2

Calculations

The theoretical flame temperature is being calculated on basis of fuel (LDO) consumed in experimental investigations as per table 6.5 (considering heat transfer by radiation only)--

(1) Observation no 1- Table no 5

The fuel (LDO) supplied to burneris= 56 liters

Qnet= Nett heat supplied= 56liter x 9.95192kwh/liter=557.3075kwh
= 557.307 x 10^3 watts

The refractory bricks are of 70% Al2O3 having Emissivity =0.31(at higher temperature of 2000 k)

and α = absorptivity = 0.65 [23]

The heat radiated **Q$_r$ = Q$_{net}$ = ξ_s x A$_s$ x σ x (ξ_g.T$_g^4$ – α_s.T$_s^4$)** [23]
$= \xi_s$ x A$_s$ x σ x (ξ_g.T$_g^4$ – α_s.T$_s^4$)
=**557.307** x 10^3 watts
= 0.31x1.86137x5.78 x 10 $^{-8}$ x (0.8 T$_g$.^{4} _.65 x 300 ^{4})
 = 3.3352117 x (0.8 T$_g$.^{4}_5.265 x 10^9)
=2.668169 T$_g^4$ -17.559889x10^9
or2.668169T$_g^4$=(**557.307**+0.17559889)x10^{11}
 or2.668169T$_g^4$=55.74825989x10^{12} T$_g^4$= 20.8938264 x10^{12}
T$_g$=2.137984x10^{30}K=2137.984^{0}K= **1864.984 °C**

(2) Observations no 2 of same table
The fuel supplied is same = 56 liters,
Q$_{nett}$ = 557.3075 x10^3 watts
=0.31x1.86137x5.78x10-^{8} x (0.8 T$_g$.^{4}._65 x 300 ^{4})
OrT$_g$ =2137.984^{0}K= 1864.984^{0}C

6.9.2-(3) Observations no 3
The fuel supplied is = 55liters,
Q$_{nett}$=547.3556x 10^{3}watts
=0.31x1.86137x5.78x 10 -^{8} x (0.8T$_g$.^{4} _.65 x 300 ^{4})
 T$_g^4$=(547.3556+.17559889)x10^{11}/ /2.66816
 T$_g$=2.1283786x10^{3}K=2128.3786k
=**1855.3786 °C**

(4) Observations no 4
The fuel supplied is = 54liters,
Q$_{nett}$.668169T$_g^4$=537.5792789 ,
T$_g^4$=20.147872x10^{12}
Tg = 2118.64^{0}k=1845.64^{0}c

 (5) Observations no 5
The fuel s supplied is = 53liters
Q=53liters =527.45176x10^{3}watts
45176+0.17559889) x10^{11}
T$_g$ = 2.108766x10^{3}K=1835.766^{0}C

6.9.2-(6&7) Observations no 6&7
The fuel supplied is = 52liters
Q=52x9.95192kwh= 517.49984x10^3 w
or 2.668169 T$_g^4$=517.67x10^{11}
T$_g^4$ =19.4018x10^{12}
Tg= 2098.750k = 1825.75^{0}c

II. COMPARISION OF ACTUAL AND THEORETICAL FLAME TEMPERATURES
(Heat Transfer By Radiation Only)

The comparison of actual and theoretical flame temperatures is shown in the following table 7-

SN	Fuel liters	Actual flame temperature	theoretical flame temp	percentage variation
1	56	1710	1864.98	8.310
2	56	1722	1864.98	7.666
3	55	1730	1855.37	6.757
4	54	1746	1845.64	5.398
5	53	1752	1835.76	4.562
6	52	1754	1825.75	3.929
7	52	1755	1825.75	3.875

Table 7 Comparison of actual and theoretical flame temperatures on basis of radiation heat transfer only

The graphical representation of comparison is shown in fig 5

The above flame temperature calculations are based on heat transfer by radiation alone. Although the major heat transfer takes place by radiation only but some of the heat transfer takes place by convection also when part of charge comes in contact with flame during rotation in initial stages.

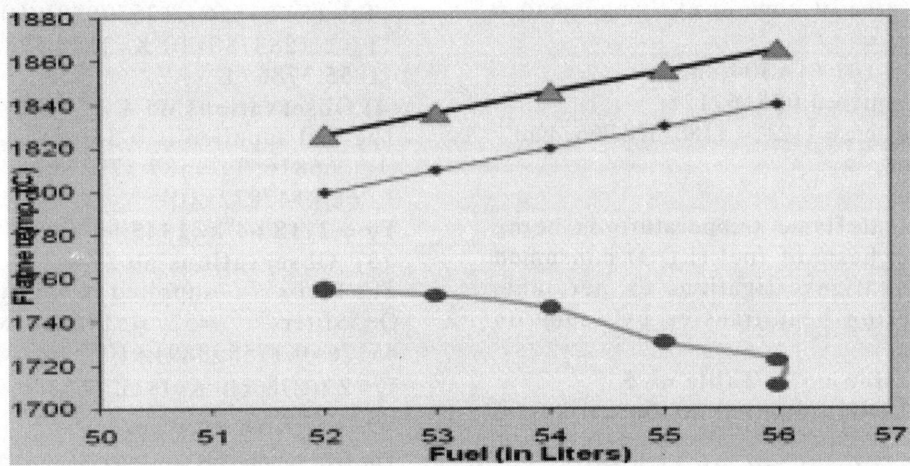

Fig5- the graphical representation of comparison of experimental and theoretic

III. RESULTS

Analysis and discussions (a) Experimental Investigations- The experimental investigations as per table 3 revealed that flame temperature can be significantly increased by reducing excess air and increasing preheated air temperature. The flame temperature was in range of 1510-1544 ^{0}C with 20% excess air, and preheated air temperature 240^{0}C using multipass counterflow heat exchanger whereas it increased to 1684-1707^{0}C with 10% excess air and preheated air temperature 407^{0}C using compact heat exchanger air. Further with oxygen enrichment of 6%, preheated air of 460^{0}c, and reducing air to 75% of theoretically required the flame temperature increased to range of 1710-1755^{0}c.

Flame temperature increased from 1510^{0}c to 1755^{0}c. The percentage increase is (1755-1510)* 100/1510 =16.25%

Fuel consumption reduced from 79 liters to 52 liters the percentage reduction is (79-52)*100/79 = 34.177%

The melting rate increased from 316 to 393.44 liters. The percentage increase is (393.44 - 316)*100/316 =24.50%

Specific Fuel consumption reduced from 0.3947 lit/kg to 0.260 lit/kg the percentage reduction is (0.3947 -- 0.260) *100/ 0.3947 =34.127%

The comparative results are summarized in following table 10

Sn	Increments	Percentage
1	(a)flame temperature (b) melting rate	16.225 % 24.50%
2	Reductions	Percentage
	(a) fuel consumption (b)specific fuel cons	34.177% 34.127 %

Table 10: The comparative results

The flame temperature plays an important role in performance of LDO fired rotary furnace .The higher flame temperature increases the rate of heat transfer which not only improves the performance of furnace but also the quality of castings produced is better.

(b) Mathematical computation the heat transfer inside the rotary furnace takes place mainly by radiation. Some heat transfer also takes place by convection in initial stages, when during rotation of furnace some charge comes in contact of flame. Considering heat transfer by radiation alone the average percentage variation between actual and mathematically computed flame temperature is 5.785 % as per table 5

IV. CONCLUSIONS

The flame temperature is an important parameter which is to be considered and controlled during operation. The long lazy luminous flame is generated due to oxygen enrichment which contains higher flame temperature and radiation heat transfer is increased whereas the short dull flame contains comparatively lower temperature and improves heat transfer by convection. The actual flame temperature is closer to mathematically computed flame temperature when both modes of heat transfer are considered .the average percentage variation is 4.434% which is fairly reasonable. Experimentally some heat is lot during combustion which mathematically cannot be accurately accounted for.

V. REFERENCES

[1] Baker EHW – "Rotary furnace" Modern Workshop Technology, part -1, Clever Hummer Press Ltd., London-2nd Edition.

[2] American Foundry men's Society: AFS 1979 "using Pulverized coal as fuel for Rotary furnace

[3] Jain R.K. Singh R etal "Economic justification of Coke less Cupola for pollution free castings in Indian environment with special relevance to Agra". Indian foundry journal Vol. 46, No.8 (2000) p.p.18-27.

[4] Singh R, Radhakrishna M, Patvardhan C, Rana G.P. "Rotary furnace – effect of rotational speed on rate of melting, fuel consumption, & pollution" Indian foundry journal, Vol. 52, No.2, 2006, p.p. 38-40.

[5] Singh R, Jain R.K. ,Patvardhan C, Kumar A, the effect of Rotational Speed, Air Preheating and excess air on performance of LDO Fired Rotary Furnace ,Trans.of 49th Indian Foundry Congress,2001, Pragati Maidan,New Delhi.

[6] Singh R, Jain R.K., Patvardhan C, Kumar A "Effect of air preheating & excess air on performance LDO fired Rotary furnace" Indian foundry journal, Vol. 46, No.11, 2006, p.p. 26-32.

[7] Tiwari S.N. "On status of Indian foundry Industry Indian foundry journal, No.8, 1998, p.p. 17-20.

[8] Jain R.K., Singh R, "Air pollution control in cast Iron foundries" proceedings of International workshop on electro magneto studies related to earth quake and Volcanoes, Nov. 20-22, 2006, Hotel Holiday Inn, Agra.

[9] Jain R.K., Singh R, "Optimization of Rotary furnace parameters" proceedings of International conference on Optimization Techniques in field of Engg. & tech Nov. 20-22, 2006, Hotel Holiday Inn, Agra

[10] Yeoman Mike, "Foundry Energy Consumption Survey report", "Foundry man", p 254-256, August 2000

[11] Jain R.K., Singh R,. Gupta B, "Energy considerations in Indian ferrous foundries'" Indian foundry journal, Vol. 54, No.8, 2008, p.p. 32-34

[12] Jain R.K., Singh R,. "Modeling and optimization of rotary furnace parameters using Regression & Numerical Techniques" 68th world Foundry Congress, Feb 7-10, 2008, Chennai

[13] Jain R.K., Gupta B,D. " "Mathematical modeling of critical parameters of rotary furnace"Thirteenth annual and first international conference of Gwalior academy of mathematical sciences, 10-13th January 2008 Agra

[14] Jain R.K., Gupta B,D. " Mathematical Computation of flame Temperature and its effect on performance of LDO fired Rotary Furnace" Thirteenth Annual and First International Conference of Gwalior Academy of Mathematical Sciences 10-13th January 2008, Agra

[15] Navojit Basu, P.K.Roy et al, "energy conservation in Indian foundries-planning and methods" Indian foundry Journal vol 53 No. 8 august2007, p.p. 31-40

[16] Jain R.K., Singh R,. 'Modeling and optimization of critical parameters of rotary furnace using computational techniques (Excel solver)" Indian foundry journal, Vol. 546, No.3, 2008, p.p. 28-34

[170 Baijya Nath, K. C.Panigrahi "energy conservation options among Indian foundries-a broad overview" Indian foundry Journal vol 53 No. 8 august2007, p.p. 27-30

[18] Kamlesh Kumar Singh" energy efficiency in foundry process and casting rejection control" Indian foundry Journal vol 53 No. 11 november 2007, p.p. 43-55

[19] S.H. Arjunwadkar, Prosanto Pal, et al "Energy savings and carbon credits- oppurtunities and challenges for Indian foundries" Indian foundry Journal vol 54No. 10 october 2008, p.p.33-37

[20] G.N.Pandey, Rajesh Singh &A.K.Sinha " Efficient energy measures in steel foundry"Indian foundry Journal vol 53No. 10 october 2007, p.p.49-51

[21] Singh R, Patvardhan C etal " Alternatives for Panacea to Ailing Foundries in Agra" Indian foundry Journal No. 8, 2003, p.p. 27-34.

[22] Navojit Basu, P.K.Roy et al "Use of gas in foundry sector" Indian foundry Journal vol 53No. 1 january2007, p.p.46-52

[23] Yunus A. Ceingel ',"Heat Transfer A Practical Approach" p468 By Yunus A.Ceingel, Published by Tata Mc Graw Hill,New Delhi

[24] Jain R.K., Gupta B.D. Singh R,. "effect of air preheating and excess air on flame temperature ,melting rate and fuel consumption of LDO fired rotary furnace" Indian foundry Journal vol 55No. 5 may2009, pp29-3

[25] Jain R.K., Gupta B.D. Singh R,. "effect of oxygen enrichment of preheated air on fuel (energy consumption) of LDO fired rotary furnace" Indian foundry Journal vol. 55No. 11 November 2009, pp29-3

Comparative Study of Different Electrode Materials in a Dielectric Medium (Castor Oil)

Sakendra Kumar and Abdul Aziz
Department of Mechanical Engineering, IIMT Engineering College, Meerut, U.P.

Abstract- **Electrical Discharge Machining (EDM) has been recognized as an efficient production method for precision machining of electrically conducting hardened materials. In this study, Experiment were performed to determine parameter effecting surface roughness (SR) along with structural analysis of surface with respect to material removal and electrode wear parameter. Experimental work conducted on AISI D204 die steel with copper, brass and stainless steel as tool electrode with Castrol oil as dielectric fluid. The data compiled during experimentation has been used to yield responses in respect of material removal rate (MRR), relative electrode wear (REW) and surface roughness (SR). Detailed analysis of structural features of machined surface was done by using optical microscope to understand the mode of heat effected zone (HAZ), which alternatively affects structure of machined workpiece and hence tool life. While investigating electric discharge machining (EDM) surface by microscopic views, it was observed that molten mass has been removed from surface as ligaments and sheets. In some cases, it is removed as chunks, which being in molten state struck to surface. All three specimens machined by different electrodes showed different microstructure of heat effected zones (HAZs).**

Keywords: Surface Roughness, Relative Electrode Wear, Electric Discharge Machining, AISI D204 (Die Steel Material) etc.

I. INTRODUCTION

Electric discharge machining (EDM) has widespread applications for manufacturing dies and tools to produce plastics mouldings die casting and sheet metal dies etc.Implementation of EDM process will awaken manufacturing engineers, product designers, tool engineer and metallurgical engineers about unique capabilities and benefits of this process.

In US, 4 fold increases in number of EDM machine installed in industry was observed during 1970-80. EDM can be used for machining of high precision of all type of conductive material (metals, alloys, graphite, ceramics etc.) of any hardness. In EDM process, material removal from workpiece is done by means of a series of electrical discharges. This paper presents work on machining by EDM for AISI D204 die steel. Maximum of MRR is an important indicator of the efficiency and cost effectiveness of the EDM process, however increasing MRR is not always desirable for all applications since this may scarify the surface integrity of the workpiece. A rough surface finish is the outcome of fast removal rates.

II. LITERATURE REVIEW

A significant amount of work has been focused on ways of yielding optimal EDM performance measures of high metal removal rate (MRR), low relative electrode wear (REW) and satisfactory surface roughness (SR). Soni (1994) has explained the migration of material elements between the electrode and work piece. The work piece used for this investigation is high carbon chromium die steel (T 215 Cr 12). They also studied the scanning electron microscope (SEM) investigation on changes in the chemical composition of resolidified layers of the tool and the work piece as well as debris. The changes in chemical composition often remain confined to within resolidified layer which was supported by others (George 1981, Gandadhar 1991). Abu Zeid (1996) investigated the role of voltage, pulse-off-time in the electro discharge machined AISI T1 high speed steel. He found that the MRR is not very sensitive to off-time changes at a low pulse-on-time corresponding to finish machining. Volumetric electrode wear has been found to be less with shorter off-times when finish machining but is independent of the off-time and direction of flushing. Lee and Lim (1988) investigated the surface transformation and damage in AISI O1, A2, D2, and D6 steels after EDM.

Experimental Set Up

A number of experiments were conducted to study the effects of the various machining parameters on EDM process. These studies have been undertaken to investigate the effects of current, pulse-on-time, duty factor on the MRR, hardness, and surface structure. All the experimental parameter shows in table.1 and experimental plan given in table2 .The AISI D204 die steel metal is machined with the copper, brass and stainless steel electrode tool. Castrol is the dielectric medium. 48 experiments are carried by varying EDM parameters such as current, pulse-on-time, and also duty factor while calculating the performance measures like MRR, REW, hardness and surface changes on each work piece. The specification of Die steel is AISI D204 D- stands for Die steel. The percentage of carbon in Die steel is 1.5 to1.75%. Other elements are added to get the mechanical and physical strength. The major alloying elements in Die steel are Chromium, Molybdenum, and Tungsten. Workpiece steel blocks were cut from the ingots. They were cut in to two pieces by tool and turret machine with dimensions of 69mm x 31mm x 31mm.

Experimental Work
Input Process Parameters

The electrical discharge machine (CHARMILLES D20 AND ISOCUT) with ISOPULSE 80 generator, servo-head and positive polarity for the electrode was used to conduct the experiments. Castrol Oil (SE-180) was used as dielectric and side flushing with a pressure of 1 kg/cm^2 which was maintained for the whole experiment. The sparking voltage was fixed at 200 V and discharge current was varied from 12 A to 32 A. In all, four values of discharge current, namely, 12 A, 16 A, 24 A, and 32 A were used for each of the three electrode materials. All the experiments were conducted with positive polarity of the electrode. Besides discharge current, two other variables related to the spark pulse wave are on-time and off-time which can be set independently. The electrode wear was calculated as the percentage of electrode material lost in the machining operation. The diameter of the machined cavity was taken at the tip and the overcut was calculated as half the amount by which this diameter exceeded the size of the electrode. Vertical taper in overcut was not considered due to lack of appropriate measuring instrument. Surface hardness and roughness readings were taken on the bottom surface of the machined cavity. Hardness was measured by

Rockwell Hardness Tester at three places for each cavity and average value was taken.

IV OUTPUT PARAMETERS

For each electrode material, the effect of variation in discharge current was studied on five output parameters, namely, material removal rate (MRR), tool wear rate (TWR), dimensional accuracy (lateral overcut), machined and surface hardness. The tool wear rate or relative electrode wear was calculated of electrode material lost in the machining operation. The diameter of the machined cavity was taken at the tip and the overcut was calculated as half the amount by which this diameter exceeded the size of the electrode. Vertical taper in overcut was not considered due to lack of appropriate measuring instrument. Surface hardness and roughness readings were taken on the bottom surface of the machined cavity. Hardness was measured by Rockwell Hardness Tester at three places for each cavity and average value was taken.

III RESULTS AND DISCUSSIONS

Line graphs have been drawn from the results obtained in the experiments. Each graph corresponds to one output parameter. The variation in discharge current has been uniformly taken along x- axis and the results of the three electrode materials in each dielectric medium have been displayed together in a single line graph to facilitate comparison.

Material Removal Rate

Graph No.1 showed the, the variation in material removal rate with different values of discharge currents and pulse on Time. In general, MRR increases with increase in discharge current for all the three electrode materials and with increase in pulse on Time decrease the value of MRR except copper electrode MRR increase from 50μs to 100μs then decreases for all current condition. Line graph shows lower MRR values when stainless steel is used as the electrode material for all conditions. However, if the comparison is made separately for each electrode, copper gives better MRR than steel and brass for discharge currents above 12 A. The maximum value of MRR reported by copper is 0.3595 gm/min and the maximum value of MRR by brass is 0.2976 gm/min at 32A discharge current and 100μs pulse on time. The value of MRR by copper is 9.43% higher than the corresponding value reported by Brass and 90.6% higher than stainless steel under the same machining conditions. This observation reaffirms the dependence of MRR on electrode material. Similarly, The maximum value of MRR reported by brass electrode is 0.4267 gm/min and the maximum value of MRR by copper is 0.2434 at 32 A discharge current and 50μs pulse on time. The value of MRR by brass is 27% higher than copper and 85.5% higher than steel. But overall, I found that Copper electrode give more material Removal Rate than both electrode in all the conditions of Current and Pulse On Time, except above condition. Hence, it may be concluded that the performance of copper for this parameter is best.

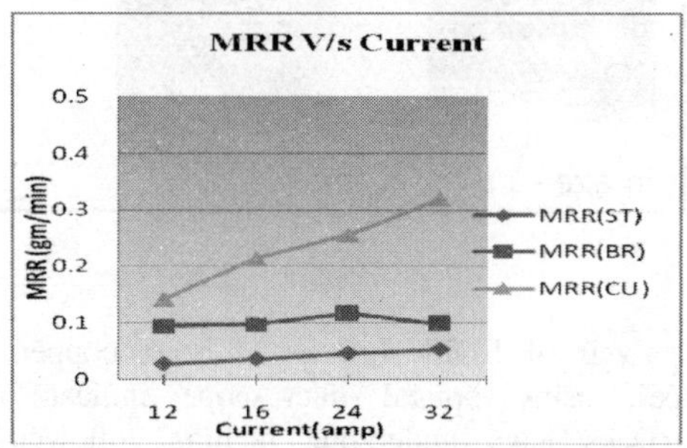

Graph no.1

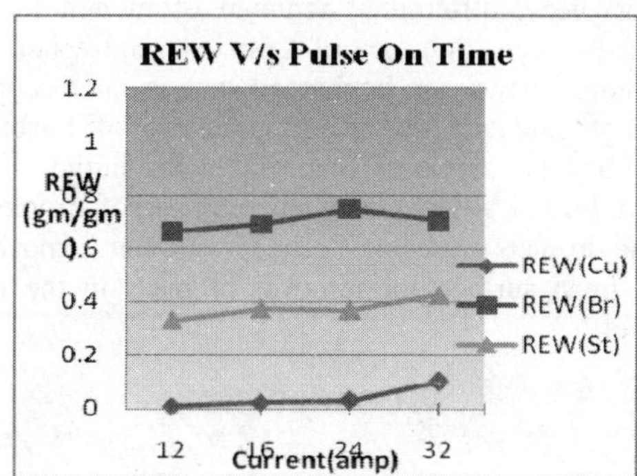

Graph no.2

Relative Electrode Wear

By Graph No.2 We can say that Copper has a much lower tool wear rate than brass and steel, it remains almost unaffected by increase in discharge current and increasing Pulse on Time. REW is higher Brass among all the electrode materials. Since, the machined surface is a replica of the electrode shape in EDM, electrode wear becomes a deciding factor in maintaining its accuracy and trueness. Any irregular deterioration of the geometric shape cannot be compensated even by overfeeding of the electrode. Minimum REW provided by copper is as low as 0.0044 gm/gm while that of steel is 0.2754 and brass is 0.5740 for same Current and Pulse on Time 16A, 200μs respectively. This observation reaffirms the dependence of REW on electrode material. Copper has a lower REW in all conditions of Current and Pulse On Time, it is conform by line graph. Hence, it may be stated that the performance of copper is better.

Machined Surface Hardness

The hardness of the workpiece material before machining was measured as 55 HRC. Surface hardness increases after EDM because rapid quenching of the machined surface takes place due to the flowing dielectric during pulse off time. It is observed from Graph no. 3 that in general, hardness increases after the EDM operation and there is a significant improvement in hardness when Brass is used. The use of brass results in maximum increase in hardness (from 55 HRC to 64 HRC at 9 A), whereas, the behaviour of copper and copper-chromium is almost identical. Other than that, each material shows random values of hardness which seems to be totally unrelated to the variation in discharge current and it may be concluded that there is no specific advantage offered by any of the electrode materials for this parameter.

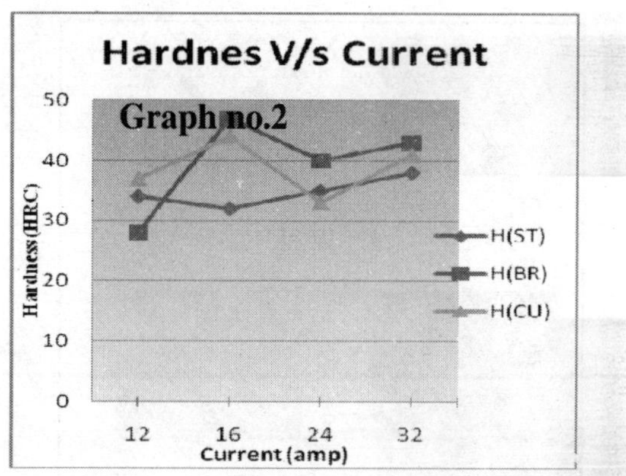

Graph no.3

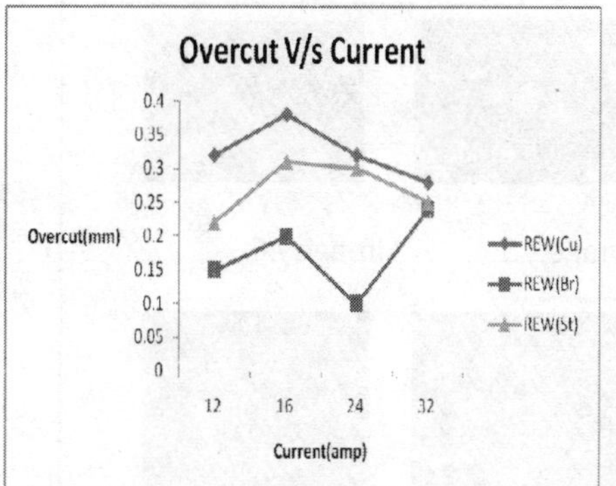

Graph no.4

Dimensional Accuracy (Lateral Overcut)

Overcut along vertical surfaces is unavoidable in an EDM operation and it takes a tapering form with increasing depth of cut. Though adequate compensation is given at the design stage, minimizing overcut is a long standing requirement of the die and mould industry. It is observed from

Graph no. 4 that overcut is lower (better dimensional accuracy) for Brass and lower for stainless steel. As the sparks at higher discharge currents carry higher energy, overcut also increases with discharge current16A. In this context, Brass shows two distinct advantages – first, it has the lowest overcut for all values of discharge current and second, the effect of increasing current on overcut is minimum for this material. The variation in overcut is only from 0.10 mm to 0.24 mm for Brass , from 0.22 mm to 0.31 mm for stainless steel and from 0.28 mm to 0.38 mm as the value of discharge current rises from 12 A to 32 A.

Surface structure study

Microscopic image (X=500) of AISI D204 after EDM machining) for different used electrode material

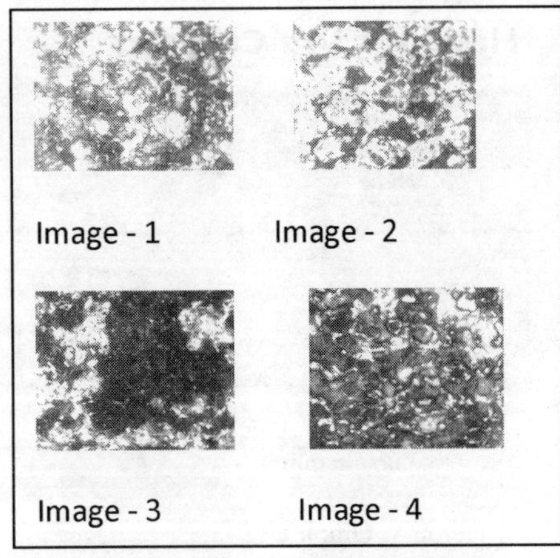

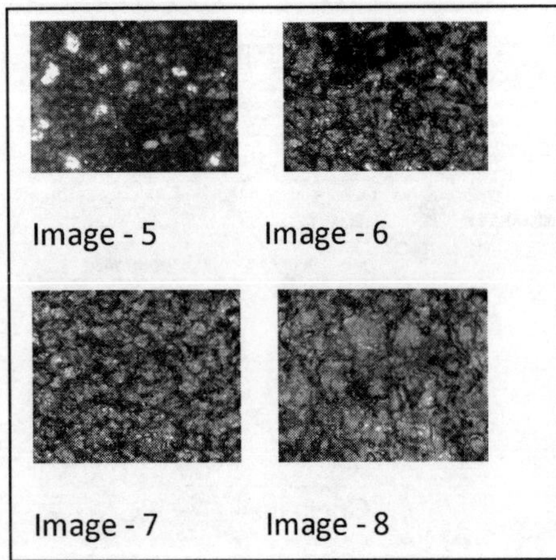

Fig. 2

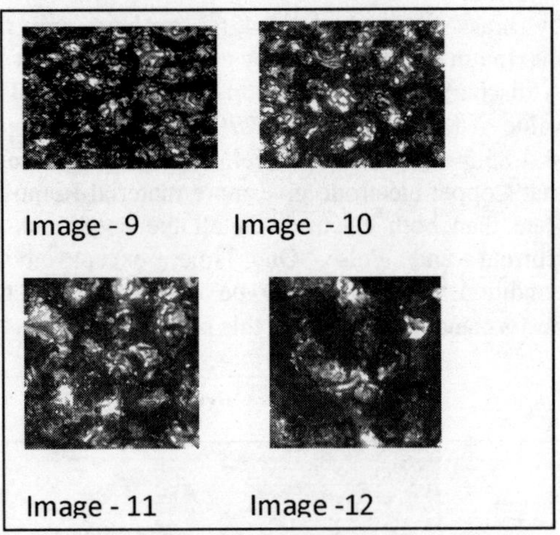

Fig. 3

Analysis of EDM surfaces of brass copper and steel, using optical microscope indicate that surfaces have undergone melting followed by washing out of molten substance. However, because of non uniform transfer of heat on workpiece, differential removal of molten metal can be seen. Presence of ligaments and spherical features, which are in charged state in microscopic image, indicate that , during course of flushing ,molten has come as ligament in the initial stage and further gets broken into number of droplets. The droplets are being flushed away, thus exposing of fresh surface for removal of mass in the die steel.

V. CONCLUSION

This experimental work has compared the performance of stainless steel, copper and brass as EDM electrode materials for machining of AISI D204 die steel in Castrol oil SE-180 as the dielectric media. The following conclusions may be drawn from this work:

- Copper clearly distinguishes itself as a superior electrode material which gives higher material removal rate, neglisble relative electrode wear good surface finish marginal lees than brass and least overcut (better dimensional accuracy) in all case of current and pulse on time. However, its impact on improving the hardness of the machined surface is only marginal.
- The relative electrode wear of copper is very low. It is especially important in EDM process because the machined surface is a mirror-image of the electrode shape and any unequal shape degeneration of the tool electrode may lead to the rejection of manufactured parts. Though copper is expensive than stainless steel and brass, the numerous advantages offered by it far outweigh the cost factor.

- Only one condition, MRR of copper is less and brass electrode at 32A current and 50μs pulse on time. But surface finish and hardness after machining show significant improvement, especially at higher values of discharge current. Dimensional accuracy (lateral overcut) remains almost same.
- From upper table, Copper clearly show better performance than other two electrodes. Very poor performance given by is stainless steel.
- During the experimentation, it is observed that a black layer forms on the tool electrode surface and machined surface when machining is carried out. This layer may be of carbon from the decomposition of the dielectric and some material may have come from the melted workpiece particles. This finding suggests that material deposition takes place during EDM process and it can be utilized for selective surface modification of the workpiece surface by selection of appropriate dielectric media and electrode tool material.

VI. REFERENCES

[1] Soni J.S, Chakraverti. G (1994), "Experimental investigation of migration of material during EDM of die Steel (T215Cr12)," pp.51-58.

[2] George Koshy, Philip P.K. and A Geddam (1981), "Hardening of surface layers using electrical discharge techniques," Proc. Conf. 11th AIMTDR, IIT, MADRAS, p. 315.

[3] Gandadhar A., Shunmugam M.S. and Philip P.K (1991), "Surface modification in electro discharge techniques with a powder compact tool electrode Wear", Vol. 143, No 1, p.45.

[4] Abu Zied O.A(1996), "The role of voltage pulse off-time in the electro discharge machined AISI T 1 high- speed steel," J. of Mat Process Technol. 61, pp.287-291.

[5] Lee L.C., L.C. Lim, Narayanan V. and Venkatesh V.C (1988), "Quantification of surface damage of tool steel after EDM," Vol. 28, No 4, pp. 359- 372.

[6] S T Jilani and P C Pandey (1984) 'Experimental investigations into the Performance of Water as Dielectric in EDM'. International Journal of Machine Tool Design and Research, vol 24, no 1, p 31.

[7] L.C. Lim, L.C. Lee, Y.S. Wong, H.H Lu (1991) "Solidification microstructure of electro discharged Machined surface of tool steels," Mater. Sci. Techno 7, pp. 239-248.

[8] Kruth J.P., Stevens L., Froyen L., Lauwers B (1995), "Study of the white layer of a surface machined by die- sinking electro discharge machining," Ann.CIRP, 44, 1, pp. 169-172.

[9] Ayers J.D., Moore K (1984), "Formation of metal carbide powder by spark machining of relative metals," Metal. Trans. A, 15A, pp.1172- 1127.

[10] Thompson P.F (1989), "Surface damage in electro discharge machining," Mater. Sci. Techno 5, pp.1153-1157.

[11] Kahng C.H. and Rajaurkar K.P., Michnigan Technological University, USA submitted by Prof.M.C. Shaw.

Diagnosing Major Policy Initiatives for Quality Assurance in Small Scale Industries through Directed Acyclic Graph (DAG)

Mukul Dev[1], S.K.Gaur[2], Abhay Benara[3] and Sunil Pasbola[4]
[1]Department of Mechanical Engineering, NIEM, Mathura, U.P.
[2]Department of Mechanical Engineering, Dayal bagh Educational Institute, Dayal bagh, Agra, U.P.
[3]Benara group of companies, Agra.
[4]RBS College Bichpuri, Agra, U.P.

*Abstract-*Small scale industries constitute an integral part of India's industrial structure. This sector covers a wide spectrum of industries categorized into village and cottage etc. It has been observed that a good number of industrial units especially those of small scale and medium scale category become sick after a sufficient lapse of time. The reason for such failures in industrial environment under consideration may be attributed to various determinative and cognitive parameters governing quality control, productivity etc. The spectra of small industrial slowdown loomed large over the Indian economy after factory output fell from 11.55 percent in 2006 to 8.3 percent in 2007. There has been a good deal of discussions in recent times on the urgent need for finding a solution to this problem. The present paper attempts to make to diagnose the major policy initiatives for quality assurance in small scale industries through directed acyclic graph (DAG).

Key words: Directed acyclic graph, Structural modeling

I. BACKGROUND AND MOTIVATION

Small scale industries constitute an integral part of India's industrial structure. This sector covers a wide spectrum of industries categorized into village and small cottage etc. It has been observed that a good number of industrial units especially those of small scale and medium scale category become sick after a sufficient lapse of time.

The reason for such failures in industrial environment under consideration may be attributed to various determinative and cognitive parameters governing quality control, productivity etc(1).

Table 1 shows average percentage growth in index for industrial productivity and signifies sharp fall up to 28 % in 2007-08 as compare to 2 % in 2005-06(6).

Table 1: Average percentage growth in Index for Industrial Productivity (IIP)

Year	Average Percentage Growth in IIP in April-March	Percentage change in percentage growth in IIP
2002-03	5.74	
2003-04	7.02	22.0 (rise approx.)
2004-05	8.35	18.0(rise approx.)
2005-06	8.15	2.0 (fall approx.)
2006-07	11.55	
2007-08	8.3	28.0(fall approx.)

According to information compiled by RBI from scheduled commercial banks, as on March 31, 2001, there were 2,52,947 sick/ weak units consisting of 2,49,630 units in the SSI sector and 3,317 units in the non-SSI sector. Among the 3,317 units, the private sector, public sector and joint/ co-operative sector accounted for 2,942 units, 255 units, and 106/14 units, respectively. The total number of sick SSI units has decreased from 3, 04,235 units to 2, 49,630 units but the number of sick/ weak units in the non-SSI sector has increased from 3,164 to 3,317. The total bank credit blocked in sick units has increased from Rs 23,656 crore (as on March 31, 2000) to Rs 25,775 crore (as on March 31, 2001). The small scale sector has Rs 4,506 crore (17.5 per cent) blocked in its units while the non-SSI sector has Rs. 21,270 crore (82.5 per cent). Bank credit blocked in the non-SSI sector in private, public and joint/co-operative units was Rs 17,705 crore, Rs 2,986 crore, and Rs 537 crore / Rs. 42 crore, respectively.

Since its inception in May 1987 till the end of September 2006, the Board of Industrial and Financial Reconstruction (BIFR) received 6,991 references under the Sick Industrial Companies (Special Provision) Act (SICA), 1985. These references included 296 references from CPSUs and State PSU (SPSUs). With 6,991 references received, 5,412 were registered under section 15 of SICA. 1,707 were dismissed as no maintainable under the Act. 760 rehabilitation schemes, including 12 by Appellate Authority of Industrial and Financial Reconstruction (AAIFR)/Supreme Court, were sanctioned and 1,303 companies were recommended to be wound up. 485 companies have been declared 'no longer sick' and were discharged from the purview of SICA on their net worth turning positive after the implementation of the schemes. 7.48 Among the 296 references for PSUs, 213 (91 CPSUs and 122 SPSUs) were registered up to September 30, 2006. Rehabilitation schemes were sanctioned for 28 CPSUs and 26 SPSUs. It was recommended that 29 CPSUs and 40 CPSUs be wound up. 9 CPSUs and 14 SPSUs were declared 'no longer sick'. As on March 31, 2002, 2003, 2004 and 2005 the gross disposal of cases was, 2,400, 2,867, 3,318 and 3,426, respectively. In the current year, the gross disposal of cases, as on September 30, was 4,115.

The occurrence of such declined in the state of small scale sector is the result of convergence of forces that can seldom be trace to a unique cause. In a system consisting of small scale environment, the problems that originate are mostly due to both sharp and precise taxonomic boundaries amongst various man-machine interactions or are solely dependent upon not sharply defined boundaries with ambiguous transition regions.

The failure to respond to proper functioning of an industry may be amalgamated effect of improper control of such parameters. Therefore, there exists a need to analyze these parameters hierarchically so that the sickness of the industrial units may be symptomatically treated and saved from ensuing disaster or collapse.

There has been a good deal of discussions in recent times on the urgent need for finding a solution to this problem.

The present paper attempts to made to diagnose the major policy initiatives for quality assurance in small scale industries through **directed acyclic graph (DAG).**

II. IDENTIFYING MAJOR POLICY INITIATIVES AND THEIR IMPACT STRUCTURE

Major policy initiatives responsible for avoiding sickness have been identified with the help of domain experts derived from industries, financial institutions and academia. They were also asked to identify symptoms of sickness. The identification process was carried out in a small workshop. A consensus was reached at 10 elements which were converged from initially 22 identified elements (7) through the process of nominal group technique. The lists of agreed identified elements are given in table 2.

Table 2. Elements of sickness

S. No.	Elements	Key Word.
1	Initial Finance	I.F.
2	Flow of Finance	F.F.
3	Availability of skilled labor	SKILLED LABOR
4	Product Quality	PRO. QUALITY
5	Credit facilities	CREDIT
6	Raw material Quality	R.M.QUALITY
7	Govt. policy in raw material	R.M.GOVT. POLICY
8	Bank credit	BANK CREDIT
9	Availability of Power	POWER
10	Govt. policy in general	POLICY

Developing impact structure:

ISM (impact structure methodology) was used to develop impact structure under the contextual relation affects for identified elements. The ISM (impact structure) process utilizes principle of graph theory and matrix theory to develop the structural model. It goes through the process of developing various model exchange isomorphisms from mental model to interpretive structural model through Self structural interaction matrix, reacheability matrix, Lower triangular matrix and minimum edge adjacent matrix. The Self structural interaction matrix so obtained is given in fig.2.1 followed by RM in fig. 2.2. Finally the ISM (impact structure) has been obtained as shown in fig. 2.4.

Table 2.1 Structural Self-Interaction Matrix

A	A	A	A	V	V	V	V	V	1 I.F.
A	A	A	A	A	A	A	A	2 F.F.	
O	V	O	O	O	O	O	3 S.L.		
A	O	O	O	O	O	6 P.Q.			
O	O	A	A	O	7 CREDIT				
A	O	O	A	9 R.M.Q.					
A	O	O	10 R.M.G.P.						
A	O	11 B.C.							
A	12 POWER								
13 POLICY									

Contextual relation "affects".

Table 2.2: Reachability Matrix

	1	2	3	4	5	6	7	8	9	10
1	1	1	1	1	1	1	0	0	0	0
2	0	1	0	0	0	0	0	0	0	0
3	0	1	1	0	0	0	0	0	1	0
6	0	1	0	1	0	0	0	0	0	0
7	0	1	0	0	1	0	0	0	0	0
9	0	1	0	0	0	1	0	0	0	0
10	1	1	0	0	1	1	1	0	0	0
11	1	1	0	0	1	0	0	1	0	0
12	1	1	0	0	0	0	0	0	1	0
13	1	1	0	1	0	1	1	1	1	1

Table 2.3 Minimum Edge Adjacent Matrix

	2	4	6	9	3	5	8	7	1	10
2	0	0	0	0	0	0	0	0	0	0
4	1	0	0	0	0	0	0	0	0	0
6	1	0	0	0	0	0	0	0	0	0
9	0	0	0	0	0	0	0	0	1	0
3	0	0	0	1	0	0	0	0	0	0
5	1	0	0	0	0	0	0	0	0	0
8	0	0	0	0	0	0	0	0	1	0
7	0	0	0	0	0	1	0	0	1	0
1	1	1	1	0	1	1	0	0	0	0
10	1	1	0	1	0	0	1	1	0	0

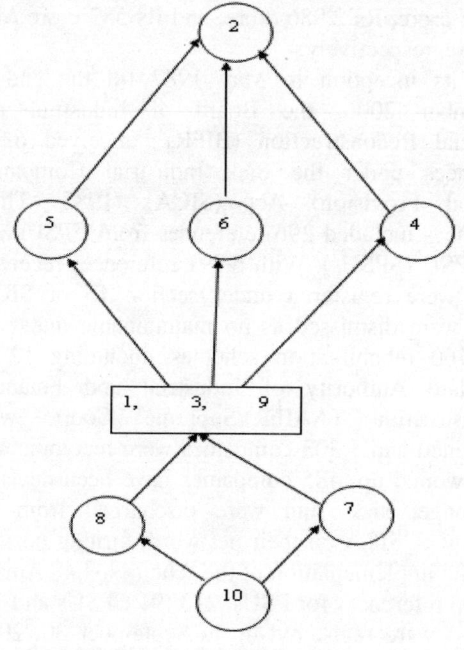

fig.2.4   ISM

Directed acyclic graph and related analysis(2,3,4,5):

Directed acyclic graph uses Bayesian inference for analysis. Wherein evidence or observations are used to update or to newly infer the probability that a hypothesis may be true. Bayesian inference uses aspects of the scientific method, which involves collecting evidence that is meant to be consistent or inconsistent with a given hypothesis. As evidence accumulates, the degree of belief in a hypothesis changes. Bayesian Networks (also knows as, Causal Probabilistic Networks, Causal Nets, Graphical Probability Networks, and Probabilistic Cause-Effect Models) are an emerging modeling approach of artificial intelligence (AI) research that aim to provide a decision-support framework for problems involving uncertainty, Complexity and probabilistic reasoning. A Bayesian Network is a graphical model that applies probability theory and graph theory to construct probabilistic inference and reasoning models. The nodes represent variables, events or evidence, whilst the arc between two nodes represents conditional dependences between the nodes.

A Bayesian Network is defined as a Directed Acyclic Graph (DAG) in which the arcs are unidirectional and feedback loops are not allowed in the graph. Because of this feature, it is easy to identify the parent-child relationship or the probability dependency between two nodes.

There are separate cross impact networks for different stages of ISM which are shown below one-by-one. The network inferences based on the theory of evidence and believe as explained in bayes theorem where belief can be updated and revised in the light of new evidence. The cycle consisting of elements 1, 3, and 9 in the ISM are considered as a single node (designated as A) for the purpose of analyzing the Bayesian network. With this modification in the structure, the ISM becomes a acyclic diagram with six distinct paths namely 10-7-A-5-2, 10-7-A-6-2, 10-7-A-4-2, 10-8-A-5-2, 10-8-A-6-2 and 10-8-A-4-2 (fig.3). The priori probabilities were obtained with the help of domain experts of a local industry in Agra manufacturing bearings and oil seals etc.limiting case for enhancing or inhibiting posterior probabilities of each elements of each path was obtained giving overall impact of probabilities for the ultimate extrimum element of the path (namely element 2 that is flow of finance). The analysis for these path is given below.

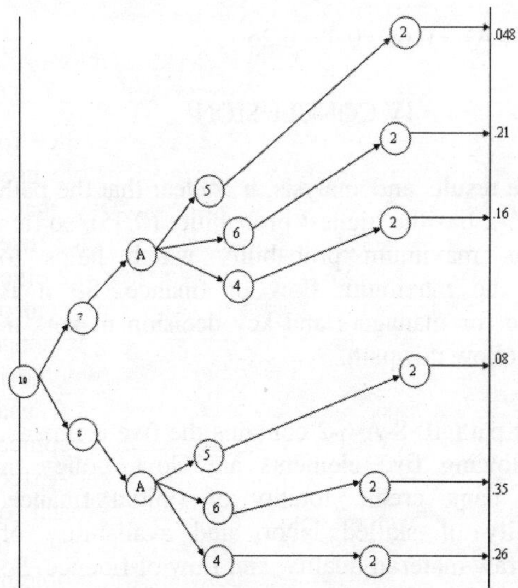

Fig. 3 Cross impact network

It has been assumed that impacted probability is a quadratic function of the impacted events such that

$$P(\,i\mid j\,) = P(\,i\,) + A_{ij}\,P(\,i\,)\,[\,1 - P(\,i\,)\,]$$

where $P(\,i\mid j\,)$ is the probability of event i occurring conditioned upon the knowledge that event j has occurred or will occurred. $P(\,i\,)$ is the probability of event i priori to event j. A_{ij} is an impact factor. If A_{ij} is 0 then event i and j have independence. If $A_{ij} > 0$ then event j enhances event i and if $A_{ij} < 0$ then event j inhibits i (8).

1. **for path 10-7-A-5-2 :**

$$P(7|10) = P(7) + A_{ij}\,P(7)\,[1 - P(7)\,]$$

Assume that case is enhancing, so $A_{ij} = 0.16$

$$P(7|10) = 0.38 + 0.16 * 0.38\,[1 - 0.38\,]$$

$$P(7|10) = 0.41$$

Now for limiting case :

$$P(i) <= P(i|j) <= P(i)/P(j)$$

$$0.38 <= 0.41 <= 0.63$$

So, $P(7|10) = 0.41$

Now , $P(A|10,7) = 0.41 * 0.60 = 0.24$

$$P(5|A) = P(5) + A_{ij}\,P(5)\,[1 - P(5)] \qquad \text{(enhancing)}$$

$$P(5|A) = 0.18 + 0.2 * 0.18\,[1 - 0.18]$$

$$P(5|A) = 0.2$$

Now for limiting case:

$$0.18 <= 0.2 <= 0.75$$

So, $P(5|A) = 0.2$

Now, $P(2|5,A) = 0.24 * 0.2 = .048$

2. **for path 10-7-A-6-2 :**

$$P(7|10) = 0.41$$

P(A|7,10) = 0.24

P(6|A) = P(6) + A_{ij} P(6) [1- P(6)]
(enhancing)

P(6|A) = 0.86 + 0.2 * 0.86 *0.14

P(6|A) = 0.88

Limiting case:

0.86 <= 0.88 <= 0.27

So, P(6|A) = 0.88

P(2|6) = 0.88 * 0.24 = 0.21

3.   for path 10-7-A-4-2 :

 P(4|A) = P(4) + A_{ij} P(4) [1- P(4)]
(inhibiting)

P(4|A) = 0.72 – 0.2 * 0.72 * 0.28

P(4|A) = 0.67

Limiting case:

1 + [P(i) – 1]/P(j) <= P(i|j) <= P(i)

-0.16 <= 0.67 <= 0.72

So, P(4|A) = 0.67

P(2|4,A) = 0.67 * 0.24 = 0.16

4.   for path 10-8-A-5-2 :

 P(8|10) = P(8) + A_{ij} P(8) [1- P(8)]
(enhancing)

P(8|10) = 0.64 + 0.2 * 0.64 * 0.36

P(8|10) = 0.68

Limiting case:

0.64 <= 0.68 <= 0.88

So, P(8|10) = 0.68

P(A|8,10) = 0.68 * 0.6 = 0.4

P(5|A) = P(5) + A_{ij} P(5) [1- P(5)]
(enhancing)

P(5|A) = 0.18 + 0.25 * 0.18 * 0.82

P(5|A) = 0.21

P(2|5,A) = 0.4 * 0.21 = 0.08

5.   for path 10-8-A-6-2 :

P(6|A) = P(6) + A_{ij} P(6) [1- P(6)]
(enhancing)

P(6|A) = 0.86 + 0.2 * 0.86 *0.14

P(6|A) = 0.88

P(2|6,A) = 0.88 * 0.4 = 0.35

6.   for path 10-8-A-4-2 :

P(4|A) = P(4) + A_{ij} P(4) [1- P(4)] (inhibiting)

P(4|A) = 0.72 – 0.25 * 0.72 * 0.28

P(4|A) = 0.66

P(2|4,A) = 0.66 *0.4 = 0.26

IV CONCLUSION

From the results and analysis, it is clear that the path 10-8-A-6-2 has the highest probability (0.35). so this path has maximum probability which helps to achieve the maximum flow of finance. So it is advisable for managers and key decision makers to strictly follow this path.

The best path 10-8-A-6-2 contains the five elements. The following five elements are Govt. policy in general, bank credit, totality of (initial finance, availability of skilled labor, and availability of power), raw material quality, and flow of finance. So managers should think to control these events to avoid flow of finance.

V. REFERENCE

[1] Gaur, S.K., Mishra D.S. Lal, A., Dominance of certain ergonomic parameters in an industrial environment by Fuzzy-set approach 1985-86.

[2] Gordon, Theodore, Hayward, "Initial Experiments with the Cross-Impact Matrix Method of Forecasting," Futures, Vol. 1, No. 2, 100-116, 1968. This was the original cross impact paper and introduced the method.

[3] Dalkey, Norman, "An Elementary Cross Impact Model," Technological Forecasting and Social Change, Vol.3, No.3, 341-351, 1972.

[4] Kane, Julius, "A Primer for a New Cross Impact Language- KSIM," Technological Forecasting and Social Change, Vol.4, No.2, 129-142, 1972.

[5] Duval, Fontela, and A.. Gabus,"Cross Impact: A Handbook of Concepts and Applications," Geneva: Battelle-Geneva, 1974.

[6] BIFR, Department of Economic Affairs, Ministry of Finance, 2006-07.

[7] Vasant Desai, "Problems and Prospects of small scale industries in India, 1983: pp. 10-11.

[8] dev, mukul, "cross impact analysis through Bayesian inference for dysfunction of small scale industries", unpublished m.tech desertation, department of mechanical engineering, dayal bagh educational institute, dayal bagh, agra, India, 2009.

Effect of Burner Shapes and Rotational Speed on the Performance of a Modified Rotary Furnace for Small Scale Ferrous Foundries Using Bio-fuels and their Blends

Ranjit Singh[1] and Purshottam Kumar[2]

[1]Faculty of Engineering
[2]Technical College, Dayalbagh Educational Institute, Dayalbagh, Agra

Abstract-Agra produces all grades of castings ranging from normal grey iron castings to graded and ductile castings. There are approximately 340 small and medium scale cast iron foundries in Agra. They manufacture general castings along with the graded and quality castings as required by large and renowned private and public sector undertakings and several units are export oriented exporting C.I. pipes and fittings. A number of small scale foundries produce general grey cast iron castings e.g. Cylinder block, flywheel, gearbox body etc. which are used in diesel engines, diesel pumps and generating industry. Thousands of small machining workshops/ industries have sprung up in Agra whose primary business is to procure castings from the foundries and after finishing and machining, to supply them to leading diesel engines manufacturers. Further, there exist hundreds of small scale foundries using crucible furnaces as melting technique. Unfortunately the gases emitted by these foundries are harmful pollutants. The major pollutants that are generated are suspended particulate matter (SPM), carbon monoxide (CO), carbon dioxide (CO_2) and sulphur dioxide (SO_2). A recent investigation carried out in foundries has shown that the emission levels are exceedingly high. In large number of foundries the average SPM level in the exhaust gasses is about 1500 mg/Nm^3 which is ten times the permissible limits. The gasses emitted by these foundries are pollutants that violate the Clean Air Legislation Act. Almost all the foundries used coke-fired cupolas for the melting. However, melting by coke- fired cupola does not obey the environmental regulations. The paper deals with the importance of a modified Rotary furnace for small scale ferrous foundries and the effect of various burner shapes and speed of rotation of the furnace has been studied on the performance of the furnace. A 200 kg. modified Rotary furnace when worked with a circular burner with number of holes on its periphery and a pre-heating arrangement of the charge using Jatropha and blends of Jatropha with LDO and rotated at 1.2 RPM gives a remarkable savings of about 38% in the fuel consumptions.

Key Words: Coke-less Cupola; S P M, Emission Level, Rotary Furnace, Bio-fuel, LDO.

I. INTRODUCTION

Agra produces all grades of castings ranging from normal grey iron castings to graded and ductile castings. There are approximately 340 small and medium scale cast iron foundries in Agra. They manufacture general castings along with the graded and quality castings as required by large and renowned private and public sector undertakings like ESCORTS, KIRLOSKAR, MARUTI, ABB, Punjab Tractors, and CIMMCO etc. The quality of castings produced is excellent and because of it there is a huge turnover. Beside this, several units are export oriented exporting C.I. pipes and fittings to Middle East and African countries. A few units are also exporting C.I. castings to Great Britain. A number of small scale foundries produce general grey cast iron castings e.g. Cylinder block, flywheel, gearbox body etc. which are used in Diesel engines, diesel pumps and generating industry. The diesel engines produced by these units are not only indigenously used in agricultural and domestic markets but are exported World wide also.

Thousands of small machining workshops/ industries have sprung up in Agra whose primary business is to procure castings from the foundries and after finishing and machining, to supply them to leading diesel engines manufacturers.

Further, there exist hundreds of small scale foundries using crucible furnaces as melting technique. Unfortunately the gases emitted by these foundries are harmful pollutants. The major pollutants that are generated are suspended particulate matter (SPM), carbon monoxide CO, carbon dioxide CO_2 and sulphur dioxide SO_2. A recent investigation carried out in foundries has shown that the emission levels are exceedingly high. In large number of foundries the average SPM level in the exhaust gasses is about 1500 mg/Nm^3 which is ten times the permissible limits. The gasses emitted by these foundries are pollutants that violate the Clean Air Legislation Act.

.The process being employed was melting iron in coke fired furnaces. A critical and in depth assessment of the above process revealed that the level of pollutants was almost three times the allowable limit. The efficiency of the process is poor and the environmental damage as a consequence of excessive pollution has manifested itself in tarnishing of the white marble of the Taj Mahal. Leading environmentalists raised a lot of hue and cry over this and Public Interest Litigation (PIL) was accepted by the honourable Supreme Court of India. Responding to one such PIL the honourable Supreme Court of India has ordered either summary closure of all foundries or their conversion to a very expensive gas fired set up. Most of the foundries are on there verge of closure. Half a million people are directly or indirectly affected. Millions of rupees worth of assets has been rendered futile. This paper aims at providing an alternative approach to melting, i.e., the rotary furnace and to establish a path breaking precedent for coexistence of industries and world heritage sights.(Ref.) Various experiments performed on the Rotary furnace fabricated and installed in the Institute showed the importance of using it (Ref). Here the furnace is used with different shapes of the nozzles and additional heating was provided to the charge by putting it on a mesh and placed it in the combustion chamber to recover the waste heat going to the atmosphere apart from the heat recovery by using a recuperator. The fuel used is Jatropha and blends of Jatropha with Light Diesel oil.

II. LITERATURE REVIEW

NML, Jamshedpur has tried to solve the problem of protection of Taj Mahal from SO_2 emission from Agra foundries. They have designed a lime scrubbing system and installed it in a foundry in Agra and studied the effect. Bandopadhaya, A. et al. [4] present an overview of pollution phenomenon in foundries of West Bengal and provide a set of strategies and necessary gas cleaning system required to bring down the level of pollution to the desired level. Ambardar, R. et al. [1] present an overview of pollution through metallurgical operations and analyse different pollutants emitted through the metallurgical operation. Bandopadhaya, A. et al. [5] emphasise energy,

environment and resource management in Indian mineral industries. Ranjit Singh, et al.[23] gives the economic Justification of Coke-less Cupola for Pollution Free Casting in Indian Environment with Special Reference to Agra. Pollution levels in foundries were measured by using a Entail Sack Monitoring kit and portable computerised onsite gas analyser. Panigrahi, S.C. [19] gives the scheme of cleaning of cupola gases and suggests the use of dry spark arrester, wet spark arrester, cyclones, cyclogel, bag filter, tower wet dust catcher with venturi pipe and waste water treatment etc. Maiti, B.R. [16] suggests detection and monitoring techniques on the basis of fundamental principles involved, mode of operation, accuracy of measurement etc. for environmental pollution from foundries Banerjee, S.N. [6] emphasises that besides gases and fumes, noise also poses serious problems. Karkharni, S.S. [14] Describes all the requirements of Central Pollution Control Board, Delhi for particulate matter, waste water discharge, ambient air quality standards and noise etc. She advocates installation of the air pollution control units developed by NEERI, NPC, MECON, NML etc. Chaudhary, S.P. [7] suggests different methods for efficient air systems like centrifugal or fan type blower, motor driven positive blower, wind belt, blast air piping and location of tuyers. Joshi, P. R. et al. [13] state that the Dr. Vardhrajan committee was set up in early seventies to study the environmental effects of Mathura oil refinery on Taj Mahal. The committee submitted its report that the effect of foundry pollution is more dominating on Taj Mahal than the pollution by Mathura Refinery. Landge, K.A., [15] recommends the use of high energy venturi scrabbling system. The scrubber contains automatically controlled variable orifice to meet the varying operating conditions. Ranjit Singh, et al.[24] present rotary furnace an efficient Answer to pollution, Proc. of National Conference on Energy and Environment. Parthasarathy, T.C. et al [20] recommend the use of ESP (Electro-Static Precipitators), bag filters, wet collectors and cyclones for the control of air pollution in the foundries. Tiwari, S.K. [26] stresses the need to install electric furnace to get rid of cupola emissions. The need to provide an un-interrupted supply of power to foundries at cheaper rates is also stressed. Sheshan, S. [25] also advocates the need of pollution control measures in foundries and suggests some measures for efficient control of pollution. Dastur, M. N. [9] advocates that major concern in foundries is the melting furnace emissions that make the environment foul. He suggests some other method to be developed which is technically feasible and economically viable.

The major pollutants emitted in various foundries units and their emission levels in Agra foundries are as follow:

MAJOR POLLUTANTS EMITTED IN VARIOUS FOUNDRY UNITS

CUPOLA EMISSION DATA - AGRA FOUNDRIES

Sl. No.	Unit	Cupola Inner Dia.	Emission Level	
			S P M (Mg/Nm^3)	SO$_2$ (Mg/Nm^3)
1	A	33"	1130	254
			2096	
2	B	33"	346	
3	C	32"	949	
4	D	33"	--	
5	E	34"	398	174
6	F	32"	309	255
7	G	26"	688	169
8	H	24"	--	--
9	I	34"	2529	239
10	J	34"	--	--

TYPICAL POLLUTANT CONCENTRATION IN CUPOLA EXIT GASES:

Location of Sampling   :   0.50 m inside and 0.25 m above the top of the door level of the Cupola

Foundry Code	Concentration of		
	Carbon monoxide		Carbon dioxide
	(g/Nm^3)	[%]	(%)
1.	50.43	[4.03]	5.7
2.	42.33	[3.39]	10.3
3.	124.93	[9.995]	10.6
4.	50.41	[4.03]	5.8
5.	61.21	[4.90]	8.5
6.	30.50	[2.44]	9.6
7.	28.96	[2.32]	5.3
8.	57.08	[4.57]	6.7

As per the above survey and the literature available, it is obvious that the coke fired cupola is not a suitable method for melting iron in the foundry. There have been considerable attempts in the foundry industry world-wide to tackle the problem of economically producing castings in a pollution free environment due to the increased awareness of the environmental protection and meeting the needs of the sustainable development. However, the conditions abroad are different from those applicable in Indian conditions. The solutions developed in the western countries are Electrical Induction Furnace, Crucible Furnace, Plasma Furnace, Laser Furnace, Rotary Furnace, Coke-less Cupola etc. The applicability of these technologies in Indian conditions has been studied by various researchers. Datta, S.K. et. al [12] advocate the installation of medium frequency core-less induction melting furnace. They prefer t   o   use

two induction furnaces connected to one power

A. Melting Units

Sl. No.	Section / Unit	Pollutants Emitted
1.	Cupola	Particulate, Metallic Oxides unburned hydrocarbons SO$_2$, CO$_2$
2.	Electric Arc Furnace	Particulate, SO$_2$, CO, oxides of Nitrogen, Cyanide, Fluoride.
3.	Electric Induction Furnace	Particulars, Carbon Zinc, Iron Oxides and Manganese Oxides

B. Non -Melting Units

Sl. No.	Section/Unit	Pollutants Emitted
4.	Moulding and Core making	Dust & CO
5.	Drying Oven Ladle Heating	Dust, CO & SO$_2$
6.	Sand Preparation	Dust
7.	Metal Pouring & mould cooling	Dust, CO
8.	Shake out	Dust
9.	Fettling	Dust
10.	Heat Treatment	CO & SO$_2$
11.	Pattern Shop	Saw Dust & wood chips

supply to optimise the power available. Mohammad Nazirudin, S. S. et al. [21] emphasise the installation of Induction Furnace for melting as they have lower emission values. Besides they suggest different pollution control equipment like general ventilation, canopy hood, box enclosure hood , side draft hood etc. Varshneya , K.K. et al. [35] point out a number of factors that account for the escalation in the energy required for melting, or even for making a casting in an induction furnace. Parmeshwaram, P.S. [25] gives an excellent review of various factors affecting the life of the induction furnace. Raizada, M. K. [27] suggests a water cooling   system for induction furnace. Rotary furnace has been in existence for a long time [4, 33]. Selby [33] discusses the advantages of the rotary furnace. However, not much literature is available on the design and use of the Rotary Furnaces.

III. BURNER DESIGN

The conventional burner having single circular hole is mounted to initiate the melting process. In this type of burner the fuel flows from the inner circle which is in large size.

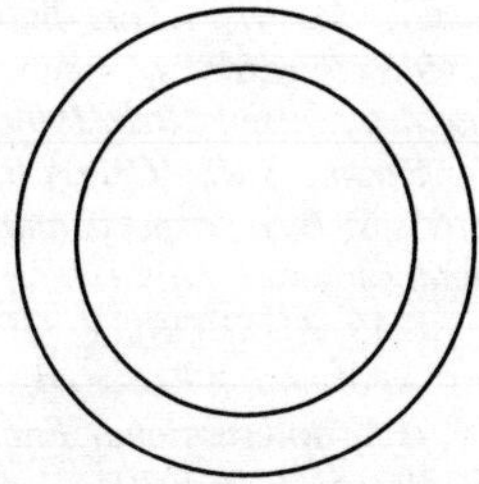

Fig. 1: Conventional Burner with one central large size nozzle

The shapes of the burner having a number of circular nozzles, which are used in melting process, are as follows-

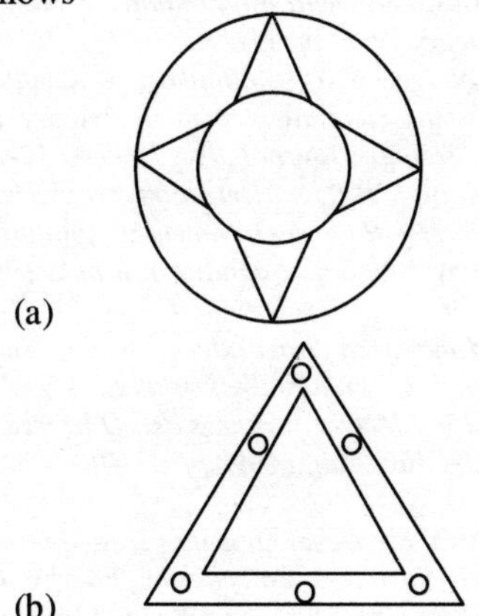

(a)

(b)

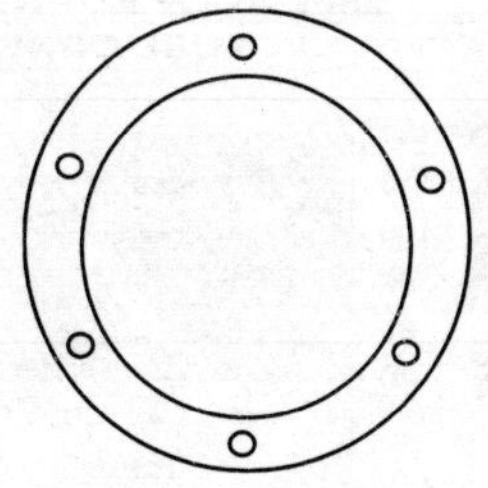

(c)

Fig. 2: Burner Shapes – (a) Star, (b) Annular, (c) Circular Ring

The above three types of burner shapes viz. Star Shape burner, Annular Ring (Triangular) shape burner and Circular Ring shape burner have been made to observe the effect of burner shapes on the melting process.

IV. RESULTS

BURNER SHAPE AND SPEED OF ROTATION

Experiments were carried out using Jatropha as fuel using the above three types of burners. The furnace was rotated at different speeds of rotation (RPM), 1.6 RPM, 1.4 RPM, 1.2 RPM, 1.0 RPM and 0.8 RPM and a number of heats were conducted.

Type / shape of burner	Fuel consumption (Litres)	Fuel consumption (Litres)	Fuel consumption (Litres)	Fuel consumption (Litres)	Fuel consumption (Litres)	Fuel consumption (Litres)
Rotational Speed →	2 RPM	1.6 RPM	1.4 RPM	1.2 RPM	1.0 RPM	0.8 RPM
Conventional	125	122	118	112	108	106
Conventional	120	118	115	111	106	105
Conventional	116	112	108	104	102	100
Conventional	**116**	**112**	**108**	**104**	**102**	**100**
Star	100	98	95	94	93	90
Star	96	94	93	93	91	88
Star	94	92	90	88	87	85
Star	**94**	**92**	**90**	**90**	**89**	**86**
Annular Ring	88	86	85	84	83	82
Annular Ring	86	85	83	82	82	81
Annular Ring	83	82	81	80	81	80
Annular Ring	**83**	**82**	**81**	**84**	**82**	**84**
Circular Ring	78	77	76	76	76	77
Circular Ring	77	76	76	76	75	76
Circular Ring	76	76	75	76	75	77
Circular Ring	**76**	**78**	**79**	**76**	**75**	**77**

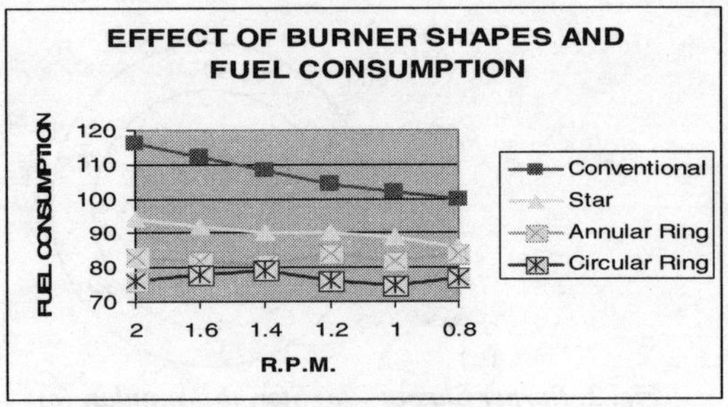

Fig.3: Fuel Consumption using different Burner shapes

V. CONCLUSIONS

By changing the conventional burner with Circular Ring type burner, the percentage saving in the fuel consumption is given by the following formula:

% saving in fuel consumption = Fuel saved per day x 100/ Fuel consumed per heat x number of heats per day

= 158 x 100 / 75 x 6 =35.1 %

Installation of Circular Ring burner in the rotary furnace saves 35.1 % of the fuel i.e. approximately one third of the fuel consumption, when compared with the Conventional burner.

The efficient thermal transfer of heat due to the slow rotation of the furnace. Also, the recovery of the heat from the waste gases is considerable which gives economical and reliable melting of iron in pollution free environment.

VI. REFERENCES

[1.] Ambardar, R., Chesti, A. R., "Pollution through metallurgical operations- an overview," Paper presented at the national seminar on pollution through metallurgical operations , Srinagar ,June 1987.

[2.] "American Foundry-men's Society" 1979 using Pulverised Coal as Fuel for Rotary Furnace.

[3.] BAKER, E. H. W, 1961, ROTARY FURNACE Modern Workshop Technology, Part 1, Cleaver Hume Press Ltd, LONDON, 2nd Edition.

[4.] Bandopadhaya, Amitava, Rao, Ramchandra, P. ,"Environmental pollution from cast iron foundries: The problems and strategies for their migration ." A Paper presented at workshop on , " Iron foundry energy scenario in raw material and technology." Organised by development commissioner for iron and steel Calcutta, May 5,1994.

[5.] Bandopadhaya, Amitava, Dutt, Ashimesh, and Gupta, K. N.," Proceedings of seminar on Indian mineral industries-energy environment and resource management," Bhuvaneshwar, Jan. 18-19 1996,pp 51-63.

[6.] Banerjee, S.N., "Clean air and control of pollution is a National requirement:, Indian Foundry Journal, Jan. 1995 , pp 30-34.

[7.] Chaudhary, S.P.," Efficient air system and control in modern cupola." Indian Foundry Journal, Aug.1997,pp 22-26.

[8.] Cupola melting." Indian Foundry Journal, Nov. 1995,pp 24-27.

[9.] Dastur, M. N. "The Indian foundry industry-Moving towards a new millennium" Indian Foundry Journal , Feb. 1999 vol. 45 ,no.2,pp 29-34.

[10.] Datta, S.K., Cele,A.B.," Induction melting practices: Basic aspects and recent trends"' Indian foundry journal, May 1995, pp 17-22.

[11.] GAS AUTHORITY OF INDIA Ltd "Natural Gas" information Received.

[12.] IEA "International Energy Authority" Natural Gas Prospects to 2010.

[13.] Joshi, Praveen R., and Rao Rama A. ," Energy conservation and air pollution control in Grey Cast Iron Foundries with emphasis on cupola melting, Indian Foundry journal, NOV. 1995,PP-24-27.

[14.] Karkhanis, S.S. ," pollution control and foundries," Souvenir 46th Indian Foundry congress, New Delhi, 1998, pp 11-17.

[15.] Landge, K.A., Kolhatkar, C.K. and Habbu, P.m.," Gas cleaning system in foundry industry." Indian Foundry Journal ,dec. 1995,pp 12-14.

[16.] Maiti, B.R.,' Detection and monitoring techniques for environmental pollution from foundries," Indian Foundry Journal, 1995,pp 25-29.

[17.] Mohammed Nazirudeen, S.S. , Sunder, M., Angelo, P.C. and Radhakrishnan, S.S. ,"Pollution control in Electric furnaces and Oil Fired Furnaces ,Foundry July /Aug., 1998,pp 37-40.

[18.] EUMANN F. "Development of Iron Melting Process in Shaft Furnaces Considering Energy usages " - GIESSEREI 79(1992) No.4 Pages 134-143.

[19.] Panigrahi, S.C., "Some aspects of pollution in foundries.", Indian Foundry Journal, June 1995,pp 1-9.

[20.] Parthsarthy, T.C. and Vijay Kumar, T.S., "Air pollution control in foundries" Indian Foundry Journal, Aug. 1998,pp 115-121.

[21.] Permeshwaram, P.S.," Pollution control in Electrical Induction furnace'" Indian Foundry Journal, May 1992,pp23-27.

[22.] aizada, R.K. ,"The role of induction furnaces in the development of castings" National Seminar on Pollution control m\easures,Calcutta,Aug,1993.

[23.] Ranjit Singh, et al., "Economic Justification of Coke-less Cupola for Pollution Free Casting in Indian Environment with Special Reference to Agra", Proc. of IPC-National Conference on Industrial pollution and control, Feb. 2000.

[24.] Ranjit Singh, et al., "Rotary Furnace an efficient Answer to pollution, Proc. of National Conference on Energy and Environment", Anand Engg. College, Agra, Dec. 2001.

[25.] eshan, S.,"Pollution in foundries-Need for pollution control in foundries. The Institution of Foundry-men, Coimbatore chapter, May 1997.

[26.] Tiwari, S.K., " On status of the Indian foundry Industry " , Indian Foundry Journal, Feb. 1998,pp 17-20.

[27.]Varshneya,K.K. AND Tank, C. BHARAT.," Energy conservation in induction furnace foundry operations" 46th Indian Foundry congress-,1998 , pp 335 - 341.

Effects of Process Parameters on EWR during Production through Electro Discharge Machining

Pankaj Upadhyay, S.K.Bajpai and Sarvesh Tiwari
Department of Mechanical Engineering, B.S.A College, of Engineering & Tech. Mathura U.P.

Abstract-**Electrical discharge machining (EDM) is a non-conventional process for manufacturing of complex or hard material parts that are difficult to machine by conventional machining processes. During EDM, the electrode shape is mirrored in the work piece. For this reason, the tool wear while machining a part in electric discharge machining (EDM) process, is of great concern in the researchers in recent years as the accuracy of tool wear directly affects the accuracy of the parts to be produced. The variation in different process parameters greatly affects the electrode wear rate (EWR).Present paper discuss the effects of different process parameters like ON time, OFF time, Sparking gap voltage, Servo speed on the electrode wear rate during the production through electro discharge machining process. Since it is impossible to study all the factors and determine their main effect (i.e., the individual effect) in a single experiment, Taguchi method is used for process optimization and identification of optimal combinations of factor for given responses.**

Key words: Electrode wear rate, On-time, Off- time, Servo speed.

I. INTRODUCTION

Electrical discharge machining is a non-conventional process for manufacturing complex material parts that are difficult to machine by conventional machining processes. During this process, the electrode shape is mirrored in the work piece. It is one of the most extensively used non-conventional material removal processes. It uses thermal energy to machine electrically conductive hard material parts regardless of their geometry. Many automotive and aerospace components, as well as mould and dies are manufactured using ED machining. During EDM process, there is no direct contact between the electrode and the work piece. This eliminates the mechanical stresses arising during machining. Electrical Discharge Machining is accomplished with a system comprising two major components: a machine tool and power supply. The machine tool holds a shaped electrode, which advances into work piece and produced a shaped cavity. The Power supply produces a high frequency series of electric spark discharges between the electrode and work piece which remove metal from work piece by thermal erosion or vaporization. A relatively soft graphite or metal electrode can easily machine hardened tool steels or tungsten carbide [1,2]. Several types of machines and industrial applications uses the EDM process for high precision machining of metals with die sinking and wire EDM being two major EDM variants.

Because of the nature of EDM process, optimization of the process parameters is required, in order to achieve the desirable performance specifications. In addition, the surface finish, dimensional accuracy and geometry of electrode as well as the material properties such as thermal conductivity and wear resistance affect EDM performance measures. Electro discharge machining has ability to machine any conductive material irrespective of their mechanical hardness. This ability of machining only conductive and semi-conductive materials is a disadvantage of the EDM process. The EDM process can process materials such as quenched steel and carbides which are mainly used for making cutting tools owing to their very high hardness and these materials are very difficult to machine using mechanical cutting processes.

II. DESIGN OF EXPERIMENTS

The design of an experiment includes following points.

- The set of treatments to be considered.

- The set of experimental units to be included.

- The procedures through which the treatment is to be assigned to the experimental units.

- The response variables to be observed

- The measurement process to be used.

III. TAGUCHI METHOD

Taguchi is the developer of Taguchi method [3]. We use Taguchi method, which is very effective to deal with responses influenced by multi-variables. This Method is a powerful Design of Experiments tool, which provides a simple, efficient and systematic approach to determine optimal machining parameters. Compared to the conventional approach to experimentation, this method reduces drastically the number of experiments that are required to model the response functions, wherein one variable is changed while the rest are held constant. The major disadvantage of this strategy is that at particular settings there may not be cutting at all due to the stochastic nature of EDM.

It is also impossible to study all the factors and determine their main effect (i.e., the individual effect) in a single experiment. Taguchi method is devised for process optimization and identification of optimal combinations of factor for given responses (4-9).

We will use different process parameters as shown in Table 1.

TABLE 1

Symbol	Machining parameters	Level 1	Level 2	Level 3
T-on	ON time	3	5	7
T-off	OFF time	2	4	6
SV	Sparking gap voltage	2	4	6
SS	Servo Speed	2	5	8

Orthogonal array

To select an appropriate orthogonal array for the experiments, the total degree of freedom needs to be computed. The degrees of freedom are defined as the number of comparisons between design parameters that need to be made to determine which level is better and specifically how much better it is. In the present study the interaction between cutting parameters is neglected. An L9 orthogonal array with four columns and nine rows is used. This array has 8 degree of freedom and it can be handle three level design parameters. Each cutting parameter is assigned to a column, nine cutting parameter combination being available. Therefore, only nine experiments are required to study the entire parameters space using the L9 orthogonal array. The experimental layout for the three cutting parameters using the L9 array is shown in TABLE 2

TABLE 2: Machining Parameters Level

Expt. no.	T-on	T-off	SV	SS
1	1	1	1	1
2	1	2	2	2
3	1	3	3	3
4	2	1	2	3
5	2	2	3	1
6	2	3	1	2
7	3	1	3	2
8	3	2	1	3
9	3	3	2	1

IV. EDM MACHINE AND MATERIALS

EDM machine with NC control in Z direction was used with negative polarity. Maximum machining current will be up to max 9-10 Amp. Brass tool electrode has been used, which is cylindrical in shape and having almost 1mm diameter. Maximum Cutting speed is 20mm/min.Work material is HCHC steel of size 25mm*20mm*15mm.The density of work piece is 7.6 gm/cc. Experiment are performed according to L9 orthogonal array, on work piece of weight 50 gm. A separate electrode has been used for each experiment. Time of machining is 02 minutes for all experiments. The EWR is calculated by weight difference of the electrodes and work piece after drilling. Electronic balance with 310 gm capacity with a precision of 0.001 gms has been used.

V. RESULTS AND DISCUSSIONS

The experimental results for EWR, based on L9 orthogonal array are shown in Table 3.

TABLE 3

Expt. No.	EWR(mm^3/min.)	Expt. No.	EWR(mm^3/min.)
1	2.9892	2	107416
3	080792	4	3.7992
5	0.22988	6	3.4982
7	1.2812	8	4.3982
9	2.9801		

In order to estimate the main effects and their differences, first the overall mean value of the observations for the experimental region is calculated. Similarly overall mean of EWR is also calculated and their values are given as:

Overall mean of EWR: 2.4139444 mm^3/min.

Effects of parameters on electrode wear rate (EWR)

The increase in the value of ON-time causes the increases in EWR. The plasma channel is believed to expand throughout the pulse duration however the EWR does not increase with the same rate because at the beginning of a discharge with a small diameter of the discharge channel, maximum heat density exists. As the channel diameter increases the heat admitting surface expands and the proportion of energy dissipated by radiation, convection and conduction rises markedly. (Refer Figure 1).

The electrode wear firstly decreases with the increase in Off-time and then increases as shown in fig1.With the diminishing of the pulse interval to the limit at which the discharges deteriorate into stationary arcs leads to the generation of socket discharges by favoring deposition of graphite protective film. This increases the resistance to electrical erosion of the brass tool electrode. Inversely an increase of pulse OFF-time pause interval determines both an increase diminishes the graphite film deposition and electrode wear increases .Thus the adequate conditions for the depositions of the graphite film and the diminishing of current at the beginning of discharge, causes the decreases in EWR. (Refer Figure 2).

The electrode wear rate decreases with increases in sparking gap voltage. The preset voltage determines the width of spark gap between the leading edge of the electrode and the work piece. High voltage setting increases the gap between the work piece and the tool electrode. This causes open circuit and hence does not contribute to any electrode wear rate. (Refer Figure 3).

The increases in servo speed increases EWR. Servo speed refers to the speed of the moving part of Electro Discharge machine on which tool is mounted. This greater servo speed causes more electrode wear i.e. EWR. However greater value of servo speed causes instability in the machining because tool comes down with a greater speed and cutting speed is slow which causes bending of tool electrode. (Refer Figure 4).

Analysis of variance (ANOVA) of EWR

We have used MINITAB statistical tool kit for the variance analysis. Results are shown in Table 4.The most significant parameter of the response function EWR is gap voltage followed by servo speed & the least significant parameter is T-OFF. Table 5 gives the optimum condition of the machining parameter and their level for electrode wear rate (EWR).

TABLE 4

Symbol	Parameter	DoF	Sum of squares	Mean square	F	Contribution %
T-ON	On-time	2	1.4474	.7759	2.84	8.806
SV	Gap voltage	2	12.9920	6.881	25.23	79.04
SS	Servo speed	2	1.4978	.7825	2.86	9.11
Error		2	.4982			3.031
Total		8	16.4359			100

TABLE 5: OPTIMUM CONDITION FOR EWR

Parameter	level	Level description(position)
T-ON	1	3
T-OFF	2	4
SV	3	6
SS	1	2

V. CONCLUSION

It is found that electrode wear rate decreases with increases in sparking gap voltage. The electrode wear rate firstly decreases with the increase in Off-time and then increases. Increase in the value of ON-time leads to the increases in electrode wear rate. The plasma channel is believed to expand throughout the pulse duration however the electrode wear rate does not increase with the same rate because at the beginning of a discharge with a small diameter of the discharge channel, maximum heat density exists. As the channel diameter increases the heat admitting surface expands and the proportion of energy dissipated by radiation, convection and conduction rises remarkably. The increases in servo speed increases electrode wear rate. However greater value of servo speed causes instability in the machining because tool comes down with a greater speed and cutting speed is slow which causes bending of tool electrode.

VI. REFERENCES

[1] Ho KH , Newman ST (2003) Int. J. Mach Manuf 43: 1287

[2] Brink D (2005) EDM : Principles of operation. http/www.edmtt.com, as on 14 June 2005.

[3] G. Taguchi, Introduction to quality Engineering, Asian Productivity Organization,Tokyo,1990

[4] L.C. Lee, L.C. Lim, V. Narayanan, V.C. Venkatesh, Quantification of surface damage of the tool steels after EDM, Int. J. Mach. Tools Manuf. 28 (4) (1988) 359-372.

[5] L.C. Lee, L.C. Lim, Y.S. Wong, H.H. Lu, Towards a better understanding of the surface features of electro-discharge machined too steels, J. Mater. Process, Techno. 24 (1990) 513-523

[6] T. Sato, T. Mizutani, K. Kawata, Electro-discharge machine for micro hole drilling, Natl. Techn. Rep. 31 (1985) 725-733.

[7] T. Masuzawa, C.L. Kuo, M. Fjino, Drilling of deep micro-holes by EDM, Ann. CIRP 38 (1989) 195-198.

[8] J.H. Zhang, T.C. Lee, W.S. Lau, Study on the electro-discharge machining.

[9] H. Lee, X.P. Li, Study of the effect of machining parameters on the machining characteristics in electrical discharge machining of tungsten carbide, J. Mater. Process. Technol. 115 (2001) 344-358.

Estimation of Optimal Cutting Parameters for Minimum Surface Roughness in Tangential Turn-Milling Process Using Differential Evolution Algorithm

R.S.S. Prasanth and K. Hans Raj

Dayalbagh Educational Institute, Dayalbagh, Agra, 282 005. INDIA.

Abstract—**Optimization techniques using evolutionary algorithms (EA) are becoming increasingly popular in engineering design and manufacturing processes due to affordable high end computational facilities. Surface roughness, a key indicator of quality of the surface is one of the most specified user demands in a machining process. This work, attempts to solve optimization problem for optimal cutting parameters (depth of cut, tool speed, work piece speed and feed rate) in order to obtain minimum surface roughness, in tangential turn milling operation, a relatively new manufacturing process in which both the tool and work piece are subjected to angular movement simultaneously. In this work we presents the Differential Evolutionary Algorithm (DEA) applied to find optimum cutting parameters in tangential turn milling process. This paper outlines the advantages of using DEA for highly nonlinear tangential turn-milling by comparing the results of with Genetic Algorithm and experimental investigations.**

I. INTRODUCTION

Generally, manufacturing of cylindrical work pieces is two phase process, i.e., first turning operation is performed and then the work pieces are ground for improved surface quality and to minimize the dimensional errors. Such process evidently slows down the production rate and increases its cost. The cost, time, and quality are the most important factors in production. As clarified, machining of cylindrical work piece with conventional turning and grinding processes has some disadvantages. To address such limitations, the process of turn milling was developed in the recent past. Turn-milling machine is a machine that is capable of both rotating-work piece operations (turning) and rotating-tool operations (namely milling and drilling). Generally these machines are based on lathes. The machine is typically recognizable as a horizontal or vertical lathe, with spindles for milling and drilling simply available at some or all of the tool positions. However, in the tangential turn-milling process, the cutting tool has a tangential contact with work piece and cutting process is carried out through the helix by side edge of cutting tool. The process of tangential turn milling is depicted in figure 1. During the process of simple turning, the cutting tool establishes contacts with the work piece at

only one point, resulting, heat and wear at the contact point between cutting tool and work piece. Consequentially tool life gets affected and surface roughness increases. Roughness is often a good predictor of quality and the performance indicator of a mechanical component since irregularities on its surface may form nucleation sites for cracks or corrosion. Although roughness is usually undesirable, it is difficult and expensive to control during manufacturing. Decreasing roughness of a surface will usually exponentially increase its manufacturing costs. However, in the tangential turn-milling process, contact time decreases and the tool life increases due to used many edges tools. In tangential turn-milling system, cylindrical surfaces are machined with these tools. Thus, using these tools some advantages have been gained, such as increased depth of cut [1].

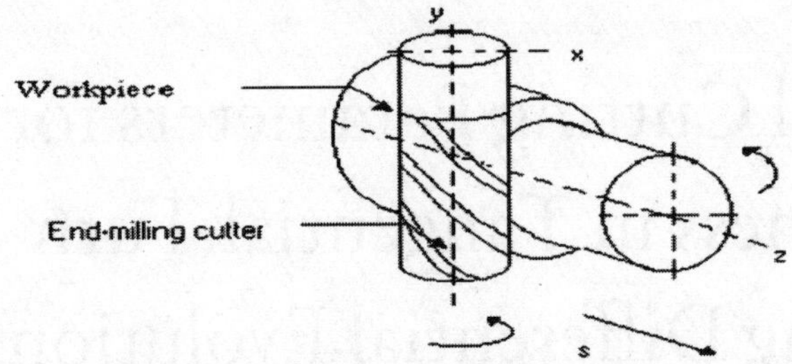

Fig 1. Tangential Turn-Milling Process

In the early 1960s, pioneering work on metal cutting using rotary tool technique was performed by Ramaswamy and Koeningsberg. They found that the tool life of rotary tools is twenty times more than that of stationary ones [2]. Venuviod and Barrow concluded that chip length ratio could be twice using rotary tools [3]. Armarego et al. the various types of rotary tool cutting operations and their performance advantage over conventional machining operations reviewed together with the simplified or fundamental operations used to study the cutting mechanics of these novel material removal operations [4]. Chen and Hoshi determined that cutting tool for machining the composite using rotary tools has high performance [5]. Joshi et al. performed a study of feasibility of rotary carbide tools in the intermittent machining of Al/ SiCp composites. They analyzed the influence of various factors and their interactions on the flank wear of rotary carbide tools during machining using Taguchi methods [6].

Lei and Liu developed a technique for the high-speed machining of titanium alloy (Ti–6Al–4V) with a new generation of driven rotary lathe tool. They designed and fabricated a new driving rotary tool based on the requirements of compact structure; sufficient stiffness and minimal edge run out and showed that, achieved in tool life was 60 times more than that of under certain conditions [7]. Kopac and Pocacnik theoretically and experimentally investigated the effect of the surface roughness in the process of turn milling and presented a comparison of the roughness between the centric and eccentric turn-milling with different process parameters [8].

Determination of optimum surface roughness at the production of the cylindrical work piece demands the estimation of cutting parameters. Estimation of optimum cutting parameters using the analytical methods during the experimental studies is very difficult. Using Taguchi method and ANOVA, W.H. Yang [9] designed optimization model for quality product. He found the optimal cutting parameters for turning operations by employing an orthogonal array,

calculating the signal-to-noise (S/N) ratio, and by analyzing the covariance of parameters.

Optimal cutting parameters such as depth of cur, feed rate, nose radius etc, were predicted for minimum average surface roughness in   orthogonal cutting process using differential evolution algorithm which outperformed a real coded genetic algorithm [10]. Tandon et al. generated a new evolutionary computation technique proposed and implemented to efficiently and robustly optimize multiple machining parameters simultaneously for the case of milling [11]. Choudhury and Mangrulkar investigated the process of orthogonal turn-milling for the machining of rotationally symmetrical work pieces. They determined that the surface finish of the machined surface improves with an increase in cutting speed and deteriorates with an increase in axial feed rate [12]. Bouzid investigated cutting parameter optimization to minimize production time in high speed turning and developed a method which consists of explaining the feed in relation to the roughness which depends on the cutting speed [13].

The paper is organized as follows. In section II methodology of differential evolution algorithm is explained. In the section III optimization problem is articulated. In the subsequent section results are presented, discussed and compared. Finally conclusions are drawn in last section.

II. METHODOLOGY

Differential Evolution Algorithm

In this section, Differential Evolution Algorithm, which has been applied for this work is briefly described. This computationally efficient evolutionary algorithm for Real-Parameter Optimization was originally developed by Storn and Price, in 1996, is a steady-state EA in which for every offspring a set of three parent solutions and an index parent are chosen [14]. The DEA is basically a population based stochastic search algorithm much like GA using the similar operators; crossover, mutation and selection. The main difference in constructing better solutions is that GA relies on crossover while DEA relies on mutation operator.

The main operation is based on the difference of randomly sampled pairs of solutions in the population. DEA uses mutation operation as a search mechanism and selection operation to direct the search towards the prospective regions of solution space. The DEA uses a non-uniform crossover that can take child vector parameters from one parent more often than it does from others. Employing the existing population members to construct trial vectors, the crossover operator efficiently shuffles information about successful combinations, enabling the search towards the better solution space. An optimization task consisting of D parameters can be represented by a D-dimensional vector. In DE, a population of NP solution vectors is randomly created at the start. This population is successfully improved by applying mutation, crossover and selection operators. The algorithm is outlined below:

Procedure DE
Input: Randomly initialized position and velocity of the particles: $xi_{(0)}$
Output: Position of the approximate global optima X*
Begin
    Initialize population;
    Evaluate fitness;
    For i=0 to max-iteration **do**
    Begin
    Create Difference-Offspring;
    Evaluate fitness;
        If an offspring is better than its parent
        Then offspring replacement in next generation;
        End If;
        End For;
End.
Operators used in the algorithm are:
Mutation
For each target vector x_i G, a mutant vector is produced by
v_i G +1 = x_i G +K (x_{r1} G − x_i G) + F (x_{r2}G − x_{r3} G)
Where i, r_1 r_2 r_3 $\in$ {1, 2 .....NP} are randomly chosen and must be different from each other and JF is the scaling factor which has an effect on the difference vector (x_{r2} G − x_{r3} G), K is the combination factor.

Crossover
The parent vector is mixed with the mutated vector to produce a trial vector u_{ji}G+1

$$u_{ji}G+1 = \quad v_{ji}; G+1 \text{ if (rnd}_j \quad \le CR) \text{ or } \quad j = rn_i,$$
$$q_{ji};G \quad \text{if (rnd}_j \quad > CR) \text{ and } \quad j \ne rn_i$$

where j= 1,2,.....D; r_j $\in$ [0, 1] is the random number; CR is crossover constant $\in$ [0, 1] and r_{ni} $\in$ (1,2,..... ,D) is the randomly chosen index.

Selection

All solutions in the population have the same chance of being selected as parents without dependence of their fitness value. The child produced after the mutation and crossover operations is evaluated. Then, the performance of the child vector and its parent is compared and the better one is selected. If the parent is still better, it is retained in the population. Figure 2 shows DEA process of determining a new proposal in detail: the difference between two population members (1, 2) is added to a third population member (3). The result (4) is subject to the crossover with the candidate for replacement (5) to obtain a proposal (6). The

proposal is evaluated and replaces the candidate if it is found to be better [15].

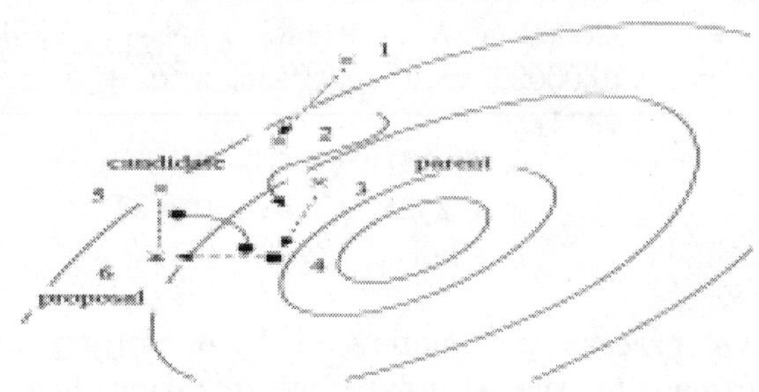

Fig. 2. Determining a new proposal

III. THE OPTIMIZATION PROBLEM

Engineering optimization problems are essentially mathematical models or functions of a process or a system (in this case average surface roughness R_a) that depends on many variables and that are required to meet one or more constraints. The effect over surface roughness of cutting parameters in the tangential turn-milling process is a new technique for machining of cylindrical work piece, which have been investigated [16] by Vedat Savas & Cetin Ozay. Investigation was performed with four independent (input) variables: tool speed (n), work piece speed (N), depth of cut (a) and feed (f). **Experiments on turn-milling were carried out on a CNC JHONFORD vertical milling machine in the accuracy of 0.001 mm. Cylindrical work piece of SAE 1050 steel of ∅40 mm diameter and 100 mm length were fixed between three jaw universal chucks and rolling centers on the machine bed using a special attachment. Rotation between 180 and 450 rpm were imparted to the job using four phase AC electrical motor of 0.75 kW power.**

Maximum possible tool speed in our condition was 2500 rev/ min, but at speeds higher than 800 rev/min tool wear was too high and therefore, tool speed level was varied between range 300 rev/min and 700 rev/min, work piece speeds N was varied between range 150 rev/min and 300 rev/min, depth of cut and feed were varied between range (f_{min}= 3.2 mm/min, f_{max}=20 mm/min, a_{min}= 0.1 mm and a_{max}= 1 mm). **Surface finish was measured using the MITUTOYO 211 Surf test instrument 0.01 ●m least count, which provided a numerical assessment of the surface roughness in terms of R_a values.** Using regression analysis and RSM method a second order polynomial [16] is evolved that fits the material removal process effectively. All the experiments were conducted with cooling. The effect of average surface roughness of cutting parameters was determined based on the aforesaid experimental investigations and is given by the following equation (1).

$$R_a = (0.000008 \times N^2 - 0.0082 \times N + 2.8734)$$
$$\times (0.00003 \times n^2 - 0.0135 \times n + 1.9924)$$
$$\times (0.0171 \times f + 0.4677)$$
$$\times (0.2525 \times a + 0.4087)$$
$$\times 5.3 \quad \ldots\ldots\ldots (1)$$

The process parameters of turn milling are defined in the standard optimization format that can be solved by a numerical optimization algorithm. An objective function to be minimized is necessary to define the standard optimization problem. For the process of tangential turn-milling, optimization problem can be mathematically articulated as follows:

$$\text{Min. } R_a \ (N, n, f, a)$$

Subject to:

$$300 \text{ rev/min} < N < 700 \text{ rev/min}$$
$$150 \text{ rev/min} < n < 300 \text{ rev/min}$$
$$3 \text{ mm/min} < f < 20 \text{ mm/min}$$
$$0.1 \text{ mm} < a < 1 \text{ mm}$$

Where N, n, f, a are function parameters 9cutting parameters) speed of the work piece, speed of the tool, feed rate, depth of cut respectively.

IV RESULTS

The results of the estimation of optimal cutting parameters to obtain minimum average surface roughness using differential evolution algorithm is presented here below. The results are also compared with the results of GA and with experimental investigations.

Cutting Parameters / Estimation Technique	GA [16]	DEA	Experimental Results [16]
Speed of Workpiece N	511.9	509.450596	530
Speed of tool n	234.9	224.5165349	224
Feed rate f	3.2	3.01385	3.2
Depth of cut a	0.1	0.1013	0.1
Objective Function R_a	0.456	0.436558187	0.42

The algorithm is executed on Laptop Machine with the following configuration: Intel Centrino Duo processor with 512 MB RAM, and 150 GB HDD. The details of program parameters are listed below.

Program parameters
DEA

Population size (NP)	= 24
Generations (G_{max})	= 350
Probability of cross over (CR)	= 0.90
Mutation Probability (F)	= 0.90
Dimension of the problem	= 4
Execution time	= 0.015 s

GA [15]

Population size (NP)	= 30
Generations (G_{max})	= 2000
Probability of cross over (CR)	= 1.0
Mutation Probability (F)	= 1.0
Dimension of the problem	=16

V. CONCLUSION

This paper outlines the DEA approach for optimization of cutting parameters tangential turn milling. This DEA approach is found to computationally efficient and easy to apply in order to obtain minimum average surface roughness values for a set of cutting parameters such as, work piece speed, depth of cut, feed rate, tool speed etc., subject to the operating constraints.

A good correlation is obtained between the value of surface roughness predicted by the DEA and surface roughness obtained from experimental measurements in process of tangential turn-milling. Based on the DEA estimations, it can be concluded that the range of percentage error in measurement of average surface roughness value is as small as 3.8%, which is comparatively quite low referring to GA based prediction whose percentage error in measurement of average surface roughness value is about 8.57%.

This work demonstrates that, DEA is computationally more efficient and quite appropriate to be applied for real world problems such as machining (tangential turn milling) as this would most surely lead us to experimentally verifiable results in continuous domain of optimization problems. This would be helpful for a manufacturing engineer to choose machining conditions for desired machining process for quality product which can simultaneously yield good paybacks. Integration of the proposed approach with an intelligent manufacturing system would certainly be a viable solution to deal with the trade off between the quality and the production cost in manufacturing industry.

VI. ACKNOWLEDGEMENTS

Authors gratefully acknowledge the inspiration and guidance of Most Revered Chairman, Advisory Committee on Education, Dayalbagh.

Authors also express their sincere thanks to Director, Dayalbagh Educational Institute, Dean Faculty of Engineering and the Principal, Technical College, for all their encouragement and support. Financial support of UGC R & D project is duly acknowledged.

VII. REFERENCES

[1] Schulz H, Lehmann T (1990) Krafte und antriebsleistungen beim orthogonalen drehfrasen (Forces and drive powers in orthogonal turn-milling). Werkstatt Betr 123(12):921–924.

[2] Ramaswamy N, Koeningsberger (1968) Experiments with Selfpropelled rotary cutting tools. Proc. of 9th IMTDR conference, Part 2, pp 945–959.

[3] Venuvinod PK, Barrow G (1972) Recent progess in machining with rotary tools, Proceeding of Fifty AIMTDR Conference, University of Roorkee, India, April 10–12, p 173–181.

[4] Armarego EJA, Karri V, Smith AJR (1994) Fundamental studies of driven and self-propelled rotary tool cutting process-I theoretical investigation. Int J Mach Tools Manuf 34(6):803– 815 Int J Adv Manuf Technol (2008) 37:335–340 339.

[5] Chen P, Hoshi T (1992) High performance machining of SiC whisker-reinforced aluminium composite by self-propelled rotary tools. Annals of CIRP 41:59–62.

[6] Joshi SS, Ramakrishnan N, Nagarwalla HE, Ramakrishnan P (1999) Wear of rotary carbide tools in machining of Al/SiCp composites. Wear 230:124–132.

[7] Lei S, Liu W (2002) High-speed machining of titanium alloys using the driven rotary tool. Int J Mach Tools Manuf 42:653–661 8. Kopac J, Pogacnik M (1997) Theory and practice of achieving quality surface in turn milling. Int J Mach Tools Manuf 37 (5):709–715.

[8] Kopac J, Pogacnik M (1997) Theory and practice of achievingquality surface in turn milling. Int J Mach Tools Manuf 37 (5):709–715.

[9] W.H. Yang, Y.S. Tarng, "Design optimization of cutting parameters for turning operations based on the Taguchi method", Journal of Materials Processing Technology 84 (1998), pp122–129.

[10] Prasanth R.S.S, K. Hans Raj K " Optimization of Orthogonal Cutting using differential evolution algorithm" International Conference Proceedings on Systemics, Cybernetics and Informatics" 9Jan – 2011), Hyderabad., pp 123- 127.

[11] Tandon V, El-Mounayri H, Kishawy H (2002) NC end milling optimization using evolutionary computation. Int J Mach Tools Manuf 42:595–605.

[12] Choudhury SK, Mangrulkar KS (2000) Investigation of orthogonal turn-milling for the machining of rotationally symmetrical work pieces. J Mater Process Technol 99:120–128.

[13] Bouzid W (2005) Cutting parameter optimization to minimize production time in high speed turning. J Mater Process Technol 161:388–395.

[14] Storn, R. and Price, K., Differential Evolution - a simple and efficient adaptive scheme for global optimization over continuous spaces, Technical Report TR-95-012, ICSI.

[15] Dervis KARABOGA, Selcuk OKDEM, "A Simple and Global Optimization Algorithm for Engineering Problems: Di_erential Evolution Algorithm" Turk J Elec Engin, VOL.12, NO.1 (2004), TUBITAK.

[16] Vedat Savas & Cetin Ozay, "The optimization of the surface roughness in the process of tangential turn-milling using genetic algorithm" Int J Adv Manuf Technol (2008) 37:335–340.

Experimental Analysis on MRR for Brass CZ131 in Electrochemical Machining

Amit Aherwar and Rohit Pandey

Department of Mechanical Engineering, Madhav Institute of Technology & Science, Gwalior.

Abstract-**Electrochemical machining (ECM) has tremendous potential because of its versatile applications and it is expected that it would be a promising, successful, and commercially viable machining process in the modern manufacturing industries. The analysis is based on the fundamental law of electrolysis. The principle of anodic dissolution of metal theory is the most accepted mathematical model for evaluating material removal from electrodes during electrochemical process. This paper shows an experimental study of feed rate and the effect of voltage on material removal rate for Brass CZ131 (Cu=62%, Pb=2%, Zn=rest) alloy. Sodium Chloride (NaCl) solution was taken as an electrolyte (250gm/lt) and inter electrode gap is 0.5mm. The said experimentation is carried out at Micromachining Cell MITS, Gwalior in the month of Nov 2010.**

Keywords: Electrochemical machining; Metal removal rate (MRR); Concentration; Inter-electrode Gap.

I.   INTRODUCTION

Electrochemical Machining developed in late 1950's has been accepted worldwide as a standard process in manufacturing and is capable of machining geometrically complex or hard material components, that are precise and difficult-to-machine such as heat treated tool steels, composites, super alloys, ceramics, carbides, heat resistant steels etc. being widely used in die and mold making industries, aerospace, aeronautics and nuclear industries [1]. Electrochemical machining is interesting because the removal of material is by an atom to an atom resulting in higher finish with stressed crack free surface and independent of the hardness of the work materials. Advance manufacturing technology has evolved thrust areas like automation of manufacture with high precession [2-4]. Now a days in ECM the main consideration is a mathematical model and a method for numerical modeling of two- dimensional electrolyte flow in the inter-electrode gap during ECM are presented. Programs for modeling electrolyte flow and observing the distribution of ECM parameters are designed. The modeling results are compared with experimental data on continuous ECM and ECM with a vibrating tool. Electrolyte properties (composition, concentration, pH value, temperature and concentration of foreign element) together with tool shape should be closely controlled because these are the important variables which determine the geometry of the machine component (anode profile).

My work is on parametric analysis on ECM in which selected variables are concentration and Inter-electrode gap (IEG). Concentration of electrolyte is important parameter for effect the MRR. At low concentration the MRR is low, because the ions were less in electrolyte and this shortage of ions reduce the MRR. At high concentration the MRR is also low, because the access of ions collides or impact to each other so reduce the chemical reaction and finally reduce the MRR. Material removal rate is directly related to the concentration of the electrolyte. Electrolyte used in ECM consists of basic salt dissolved in water flow at high velocity in the IEG serves multifarious function, viz dilutes the electrochemical reaction products and removes them out from the gap, dissipated heat at the faster rate and limits the concentration of ions at the electrodes surface to give higher machining rate. IEG is important parameters for affect the MRR. At high IEG the current flow is less because resistance increases the ions reduces kinetic energy and eroding action also decrease so finally the MRR of ECM is low. Since the first introduction of ECM in 1929 by

Gusseff, its industrial applications have been extended to electrochemical drilling, electrochemical deburring, electrochemical grinding and electrochemical polishing [5]. ECM was found particularly advantageous for high-strength alloys. For example, the semi-conductor industry frequently requires the machining of components of complex shape and high-strength alloys hence ECM is a major process candidate for semiconductor devices and thin metallic films [6–8].

ECM processes were also adopted in the aerospace and electronic industries for shaping and finishing operations of a variety of parts of the opening windows that are a few microns in diameter [7]. In 2001, J.W. Park, E-S Lee develops the electrochemical micro machining for air lubricated hydrodynamic bearings [9]. In 2004 J. KOZAK develops the thermal modes of pulse electrochemical machining (PECM) provides an economical and effective method for machining high strength, heat resistant materials into complex shapes such as turbine blades, die, molds and micro cavities. In 2005 S.K. Mukherjee, S. Kumar and P.K. Shrivastava give the effect of over voltage on MRR during electro-chemical machining. MRR in electrochemical machining is analyzed in context of over voltage and conductivity of the electrolyte solution [10]. In 2005 T. Paczkowski, L. Dabrawski give a mathematical model and a method for numerical modeling of two dimensional electrolyte flow in the IEG during ECM are programs fro modeling electrolyte flow and observing the distribution of ECM parameters are designed. The modeling results are compared with experimental data on continuous ECM and ECM with vibrating tool [11]. In 2006 Wansheng Zhao, Xiohai Lo, Zhenlong Wang done their study on micro electro chemical machining at micro to Meso-scale micro-ECM setup was developed to fabricate micro parts and explore the feasibility of micro-ECM at micro to meso-scale, including the design of high frequency micro-energy pulse power supply [12]. In 2007 S.K Mukherjee, S Kumar, P.K Shrivastava does the work on effect of electrolyte on the current-carrying process in ECM. D.S. Biligi, et al [13], developed the predicting radial over cut in deep holes drilled by shaped tube ECM. Ruszaj. A, Zybura, Skrabalak. M, investigate the electrochemical micromachining process supported by electrode ultrasonic vibration have been carried out in 2007. It has been proved that electrode ultrasonic vibration helps to decrease the surface roughness parameters R_a in comparison to the classical electrochemical process [14]. K. P. Rajurkar et al [15], discussed about the main advantages of the electrochemical machining (ECM) process, such as high material removal rates and smooth, damage-free machined surface, are often offset by the poor dimensional control .

II. PRINCIPLE OF ELECTROCHEMICAL MACHINING

The process of Electrochemical Machining is developed on the principle of Faraday's law and Ohm's law. In this process an electrolytic cell is formed by the anode (work piece) and the cathode (tool) in the midst of a flowing electrolyte. The metal is removed by the controlled dissolution of the anode according to the well known Faraday's law of electrolysis. When the electrodes are connected to about 20 V electric supply source, flow of current in the electrolyte is established due to positively charged ions being attracted towards the cathode and vice versa. Current density depends on the rate at which ions arrive at respective electrode which is proportional to the applied voltage, concentration of electrolyte, the gap between the electrodes and tool feed rate. Due to electrolysis process at the cathode, hydroxyl ions are released which combine with the metal ions of anode to form insoluble metal hydroxides. Thus the metal is mainly removed in the form of sludge's and precipitates by electrochemical and chemical reactions occurring in the electrolyte cell. In this way even hardest possible material can be given a complicated profile in a single machining operation.

Table 1: Typical electrolytes

Alloy	*Electrolyte*
Iron based	Chloride solution in water (mostly 20% NaCl)
Ni Based	HCL or mixture of brine and H_2SO_4
Ti Based	10% hydrofluoric acid + 10% HCL + 10% HNO_3
Co-Cr-W Based	NaCl
WC Based	Strong alkaline solution

III. ELECTROCHEMICAL AND CHEMICAL REACTION SCHEME

In aqueous solution of NaCl following reaction occurs.

$$NaCl \leftrightarrow Na^+ + Cl^-$$
$$H_2O \leftrightarrow H^+ (OH)^-$$

On passing the electric current through the solution positive ions moves towards cathode and negative ions moves towards anode. Each Na^+ ions gain an electron and is converted to Na. Hence Na^+ ions are reduced at the cathode by means of electrons.

Cathode reactions

Following reaction takes place at cathode2.

$$Na^+ + e^- \leftrightarrow Na$$
$$Na + H_2O \leftrightarrow Na\,(OH) + H^+$$
$$2H^+ + 2e^- \leftrightarrow H2\uparrow$$

It shows that only hydrogen gas will evolve at cathode and there will be no deposition.

Anode reaction

$$Fe \leftrightarrow Fe^{2+} + 2e$$
$$Fe^{2+} + 2Cl \leftrightarrow FeCl_2$$
$$Fe^{2+} + 2OH \leftrightarrow Fe\,(OH)_2$$
$$FeCl_2 + 2\,OH \leftrightarrow Fe(OH)_2 + 2\,Cl$$
$$2Cl^- \leftrightarrow Cl_2\,(g) + 2e^-$$
$$2FeCl_2 + Cl_2 \leftrightarrow 2FeCl_3$$
$$H^+ + Cl^- \leftrightarrow HCl$$
$$2Fe\,(OH)_2 + H_2O + O_2 \leftrightarrow 2Fe(OH)_3\downarrow$$
$$Fe(OH)_3 + 3HCl \leftrightarrow FeCl_3 + 3H_2O$$
$$FeCl_3 + 3\,NaOH \leftrightarrow Fe(OH)_3\downarrow + 3NaCl$$

IV. EXPERIMENTAL SETUP

In this paper we are going to discuss about the experimental work, which is consisting about experimental set up, selection of work piece material, and making of electrolytic solution. By taking all this information in account we are calculate the material removal rate. The set up consists of three major sub systems.

Machining Cell

This electro-mechanical assembly is a sturdy structure, associated with precision machined components, servo motorized vertical up/down movement of tool, an electrolyte dispensing arrangement, illuminated machining chamber with see through window, job fixing vice, job table lifting mechanism and sturdy stand. All the exposed components, parts have undergone proper material selection and coating/ plating for corrosion protection.

Figure 1: Experimental Setup of ECM

Technical Data

- Tool area - 30 mm^2.
- Cross head stroke - 150 mm.
- Job holder - 100 mm opening X 50 mm depth X 100 mm width.
- Tool feed motor - DC Servo type.

Control Panel

The power supply is a perfect integration of, high current electrical, power electronics and precision programmable microcontroller based technologies. Since the machine operates at very low voltage, there are no chances of any electrical shocks during operation.

Technical Data

- Electrical Out Put Rating - 0-300 Amps. DC at any voltage from 0 - 20 V.
- Efficiency - Better than 80% at partial & full load condition.
- Power Factor - Better than 85.
- Protections - Over load, Short circuit, single phasing.
- Operation Modes - Manual / Automatic.
- Operation Modes - Manual / Automatic.
- Tool Feed - 0.2 to 2 mm / min.
- Z Axis Control - Forward, reverse, auto forward/ reverse, through micro controller.
- Supply - 415 v +/- 10%, 3 Phase AC, 50 Hz.

Electrolyte Circulation

The electrolyte is pumped from a tank, lined by corrosion resistant coating with the help of corrosion resistant pump & is fed to the job. Spent electrolyte will return to the tank. The hydroxide sludge arising will settle at the bottom of the tank & can be easily drained out. Electrolyte supply shall be governed by flow control valve. Extra electrolyte flow is by- passed to the tank. Reservoir provides separate settling and siphoning compartments. All fittings are of corrosion resistant material or of SS., as necessary

- Initially Valve A, B, & C should be open
- All other valves should be closed
- Use By-Pass valve (D) if required
- And when machining is over close the valve A, D, & C and open the valve E, B, & F for cleaning purpose.

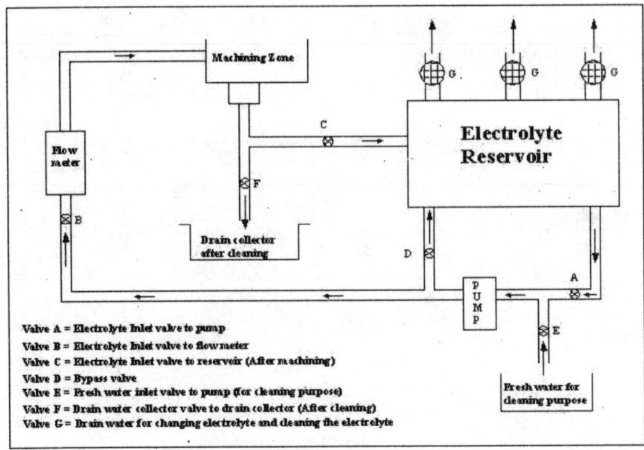

Figure 2: Electrolyte flow diagram

V. OPERATING PARAMETERS

Following are the some parameters which govern the ECM.

Feed Rate: A high feed rate results in higher metal removal rate. It decreases the equilibrium machining gap resulting in improvement of surface finish and tolerance control.

Voltage: Low voltage decreases the equilibrium-machining gap and results in better surface finish and tolerance control.

Nature of power supply and machining pulse: The nature of applied power supply may be of two types, such as DC (full wave rectified) and pulse DC. A full wave rectified DC supplies continuous voltage where the current efficiency depends much more on the current density. The efficiency decreases gradually when the current density is reduced, whereas in pulse voltage (duration of 1 ms and interval of 10 ms) the decrease is much more rapid. With decreasing current density the accuracy of the form of the work-piece improves.

Electrolyte type, concentration and flow: ECM electrolyte is generally classified into two categories; passivity electrolyte containing oxidizing anions e.g. sodium nitrate and sodium chlorate, etc. and non-passivity electrolyte containing relatively aggressive anions such as sodium chloride. Passivity electrolytes are known to give better machining precision. This is due to their ability to form oxide films and evolve oxygen in the stray current region. From review of past research, in most of the investigations researchers recommended NaClO3, NaNO3, and NaCl solution with different concentration for electrochemical machining (ECM). The pH value of the electrolyte solution is chosen to ensure good dissolution of the work-piece material during the ECM process without the tool being attacked.

Size, shape and material of the tool: The tool must match the required shape of the work piece depending on the material and the profile to be produced. Tool materials used in ECM must have good thermal and electrical conductivity; corrosion resistance must be highly machinable and should be stiff enough to withstand the electrolytic pressure without vibrating.

Table 2: Parameters

Work piece	Brass CZ131
Tool Material	Copper
Electrolyte	NaCl in water (Brine)
Voltage	12, 14, 15 Volts
Concentration of electrolyte	250gm/lit
Feed rate of servo mechanism	0.16mm/min, 0.19mm/min
IEG	0.5mm

Figure 3: Setting of work piece and tool

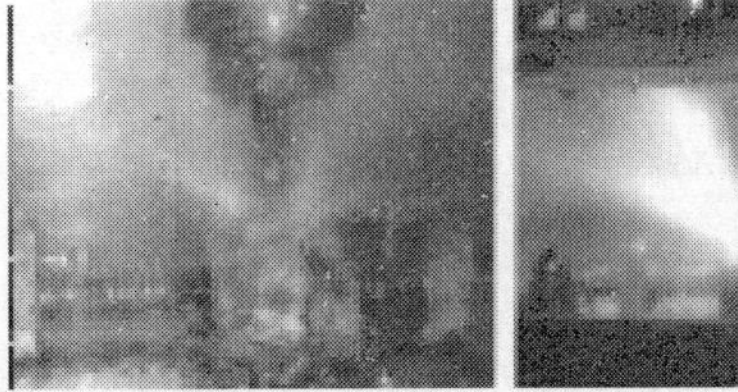
Figure 4: Electrochemical Machining process going.

Figure 5: work piece after machining

VI. RESULT AND DISCUSSION

Table 3: Observation Table

Run	Voltage (Volts)	Feed Rate (mm/min)	Time (min)	Initial wt. (gms.)	Final wt. (gms.)	MRR (gm/min)
1	12	0.16	12	141.2	131.6	0.80
2	12	0.19	12	131.6	120.6	0.91
3	14	0.16	12	118.3	108.1	0.85
4	14	0.19	12	108.1	96.8	0.94
5	15	0.16	12	95.2	84.7	0.87
6	15	0.19	12	84.7	73.0	0.97

VII. ANALYSIS OF MRR

Table 3 shows the general results for MRR, in all cutting. It shows that the material removal rate was influenced by feed rate. This result was expected because the material removal rate increases with feed rate. The electrochemical reactions did not produce the necessary and compatible effects with the increasing feed rate. According to the results, feed rate and voltage are control the material removal rate. As we can deduce from the graphs, the material removal rate increases with the feed rate as well as the voltage but it depends mainly on the tool feed rate.

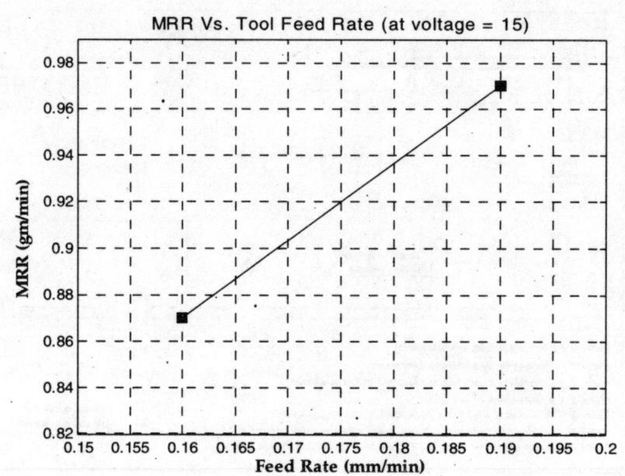

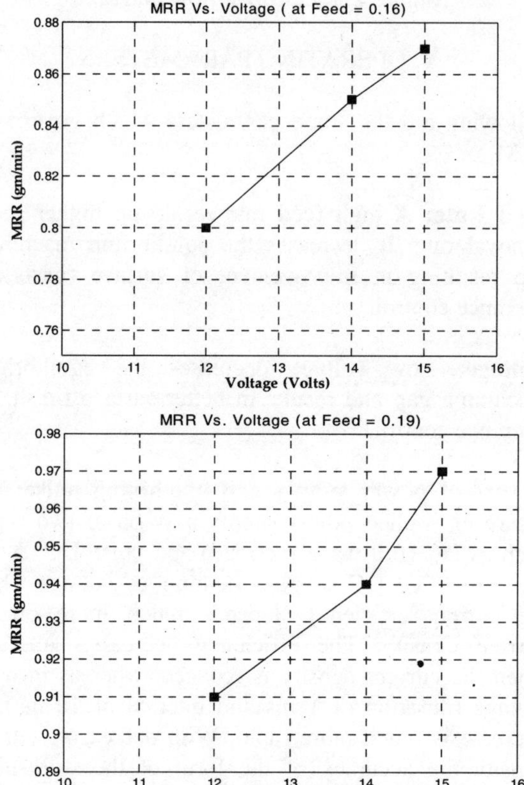

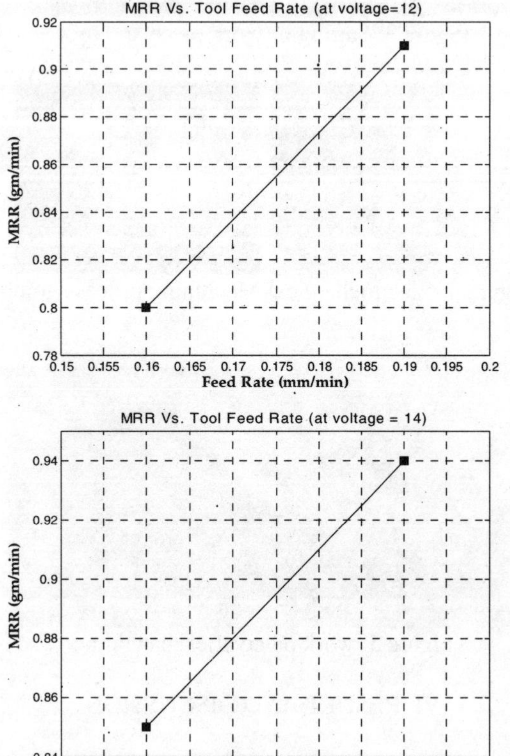

VIII. CONCLUSION

The experimentation work consists of study the influence of process parameters on machining criteria such as MRR. Some of the process parameters such as machining voltage and inter electrode gap (IEG) are successfully controlled with the help of unique setup available at MITS, Gwalior. The machining voltage and IEG was considered for the experimentation to study their influence on MRR. According to the results obtained in this work, the material removal rate increases with the increase of the feed rate and the applied voltage. IEG variable is maintained constant during the whole experimentation. The machining voltage 15V gives the maximum amount of MRR.

IX. ACKNOWLEDGEMENT

The authors would like to thanks Madhav Institute of Technology and Science, Gwalior, India for experimentation facility and support throughout the work.

X. REFERENCES

[1] E. Rumyantsev and A. Dovydev, "Electrochemical Machining of Metals", MIR Publication, Moscow, Russia (1989).

[2] Keown, Mc. P. A., "The Role of Precision Engineering in Manufacturing of the Future," Annals CIRP, Vol. 36, pp. 495-501 (1987).

[3] Ikawa, N. et al., "Ultra Precision Metal Cutting the Past, The Present and Future," Annals of CIRP, Vol. 40, pp. 551–554 (1991).

[4] Zhou, J. J., Zhai, X. B., Pang, G.B., Li, H. Y., Xu, W. J. and Guo, L. S., "Research on Pulse Electrochemical Finishing," Journal of Dallan University of Technology, Vol. 43, pp. 311_314 (2003).

[5] J. Wilson, Practice and theory of electrochemical machining (1971) 79-161.

[6] J.A. Mc Geough, Principle of Electrochemical Machining, Chapman & Hall, London, 1974.

[7] B. Bhattacharyya, B. Doloi, P.S. Sridhar, "Electrochemical micro-machining: New possibilities for micro-manufacturing", J. Mater. Process. Technol. 113 (2001) 301–305.

[8] O.L. Riggo, C.E. Locke, "Anodic Protection", Plenum Press, New York, 1981.

[9] J.W. Park, E.S. Lee, "Development of electrochemical micro machining for air lubricated hydrodynamic bearing", Journal Microsystems Technologies Publishers Springer Berlin/Heidelberg vol. 9, no.1-2/ November, 2002

[10] S.K. Mukherjee, S Kumar, and P.K. Shrivastava "Effect of electrolyte on the current-carrying process in ECM", in Proceeding of the Institution of Mechanical Engineers, Part C: Journal of Mechanical Engineering Science, vol. 221 no. 11, 2007

[11] T. Paczkowski, L. Dabrawski "Computer simulation of two-dimensional electrolyte flow in electrochemical machining" Journal Russian Journal of Electrochemistry Publishers MAIK Nauka by Springer Science + Business Media LLC, vol. 41, no. 1, 2005.

[12] Wansheng Zhao, Xiohai Lo, Zhenlong Wang "Study on micro ECM at Micro to Meso-Scale" in International Conference, vol no. 18-21, pp 325-329, 2006.

[13] D.S. Biligi, R. Kumar, V.K Jain and R. Shekhar "Predicting radial over cut in deep holes drilled by shaped tube ECM", International Journal of Advanced Manufacturing Technology, Springer London, vol 09, 2007

[14] A. Ruszaj, Zybura, M. Skrabalak. "Some aspects of the ECM process supportes by electrode ultrasonic vibrations optimization professional Engineering" publisher vol. 217, no.10, 2003

[15] T. Sekar, R. Marappan "Experimental investigations into the influencing parameters of electrochemical machining of AISI 202". Journal of Advanced Manufacturing Systems 2008; 7(2):337-43.

Parametric Analysis of Electrochemical Machining using Taguchi Method

[1]Mahesh Vishwakarma, [2]Tarun Soota, [3]Ashish Sharma

[1]College of Science & Engineering. Jhansi, U.P.
[2]BIET, Jhansi, U. P.
[3]MITS, Gwalior, M.P.

Abstract:- **The objective of present work is the parametric study of the machining of aluminum specimen using electro chemical machining (ECM) with a flat end copper electrode by using Taguchi methodology. The Taguchi method is used to formulate the experimental layout, to analyze the effect of each parameter on the machining characteristics and to predict the optimal choice for each ECM parameter such as feed rate, machining time, voltage and molar concentration of electrolyte. It is found that these parameters have a significant influence on machining characteristic such as material removal rate. In this work, experimental results are provided to verify this approach.**

Keywords: Electrochemical machining, Process parameters, Taguchi method.

I. INTRODUCTION

In the industrial and technological growth, the development of harder and difficult to machine materials, which find wide application in aerospace, nuclear engineering and other industries owing to their high strength to weight ratio.

Unconventional machining has grown out of the need to machine these exotic materials. The problems of high complexity in shape, size and higher demand for product accuracy and surface finish can be solved through un-conventional methods. Electrochemical machining is one of the most potential unconventional machining processes. This process may be considered as the reverse of electroplating with some modifications. Further it is based on the principle of electrolysis. Electrochemical machining (ECM) is the controlled removal of metal by anodic dissolution in an electrolyte cell in which the two electrodes used one is anode (work piece) and second is cathode (tool).

The electrolyte circulates by the pump through the inter electrodes gap, while direct current is passed through the cell, to dissolve metal from the work pieces. ECM is widely used in machining of jobs involving intricate shapes and to machine very hard or tough materials those are difficult or impossible to machine by conventional machining.

II. DESIGN OF EXPERIMENTS

Taguchi developed a method for experiment design to investigate how different parameters affect the mean and variance of a process performance characteristic. The experimental design proposed by Taguchi involves using orthogonal arrays to organize the parameters affecting the process and the levels at which they should be varies. Instead of having to test all possible combinations like the factorial design, the Taguchi method tests pairs of combinations. This allows for the collection of the necessary data to determine which factors most affect product quality with a minimum amount of experimentation, thus saving time and resources.

An Electrochemical machine, 0-300 amp and 0-20 volt DC supply is used for the machining. The specimens of aluminum are prepared with flat surfaces and better surface finish for machining. An aqueous solution of sodium chloride (NaCl) is used as electrolyte. Electrolyte flow rate is fixed as 5 liters per minute and an inter electrode gap of 0.5 mm is maintained for all the tests. The experimental layout for the machining parameters using the L_9 orthogonal array is used in this study. This array consists of four control parameters and three levels as shown in table 1 In Taguchi method, most all of the observed values are calculated based on 'the higher the better'.

S/N Ratio = - 10 log [$\sum$ (1/ y^2)/n]

To calculate the metal removal rate with the help of following equation,

MRR = $(m_i - m_f)$ / t.

Table 2.1 Factor Level

Levels	Feed mm/mn (A)	Machining time min (B)	Voltage V (C)	Molar concentration gm/lt (D)
1	0.16	10	10	250
2	0.19	15	12	300
3	0.21	20	15	350

Table 2.2 Observation Table

Run	A	B	C	D	Initial weight g	Final Weight g	Mrr g/min
1	1	1	1	1	171.6	166.0	0.59
2	1	2	2	2	166.0	155.4	0.71
3	1	3	3	3	155.4	140.2	0.76
4	2	1	2	3	179.6	173.6	0.60
5	2	2	3	1	173.6	162.8	0.72
6	2	3	1	2	162.8	146.8	0.80
7	3	1	3	2	175.8	169.4	0.64
8	3	2	1	3	169.4	158.9	0.70
9	3	3	2	1	158.9	142.5	0.82

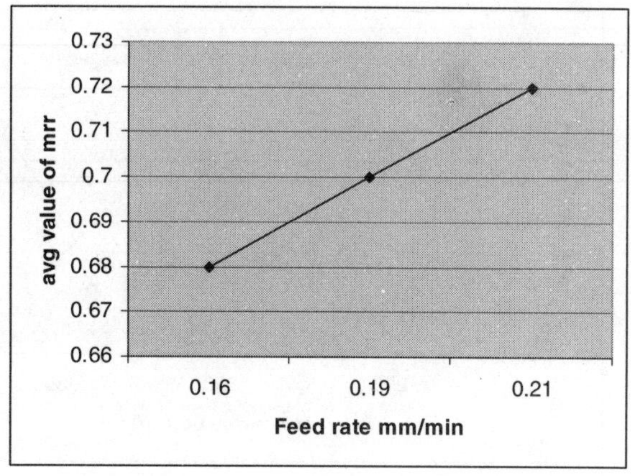

III. RESULTS AND DISCUSSIONS

The following three cavity is generated on three pieces(A,B,C) by using different time, feed rate, voltage but constant concentration of electrolyte which is 250g/l. similarly six reading at 300and 350g/l. in Fig 3.1

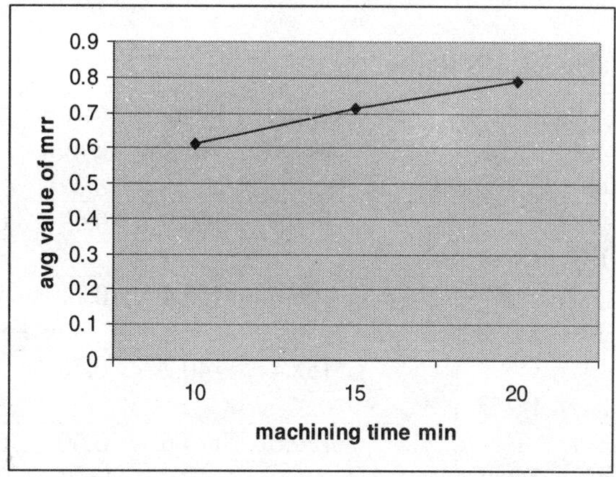

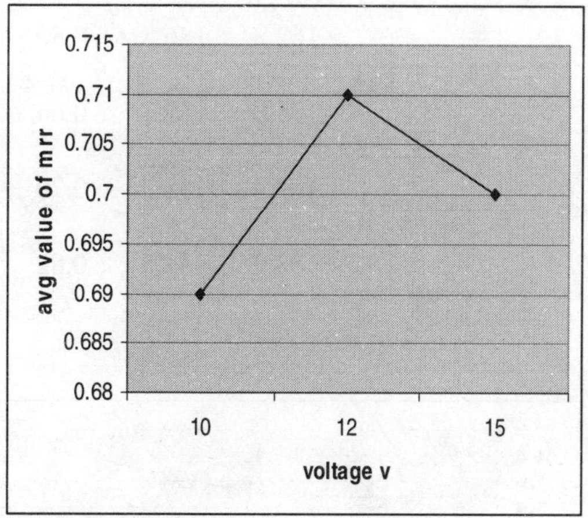

In order to estimate the effect of factor A (feed rate) on the average value of response variable, were summed together three observed response at level 1 of factor A (Feed rate). Then the sum was divided by 3 to obtain the average response at level of factor A (Feed rate). The average responses at level 2 and 3 were obtained in the similar manner. The estimated effects are presented graphically in Fig. 3.2

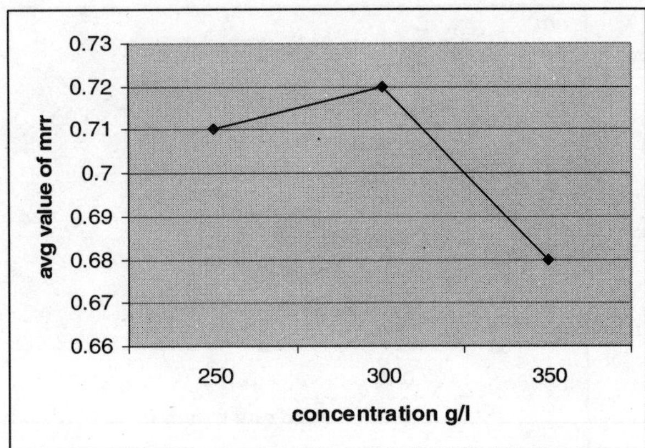

Fig 3.2 The graph between average value of mrr and process variables

Table 3.1 S/N Ratio

Run	A	B	C	D	Initial weight g	Final Weight g	Mrr g/min	SN
1	1	1	1	1	171.6	166.0	0.59	0.18
2	1	2	2	2	166.0	155.4	0.71	1.79
3	1	3	3	3	155.4	140.2	0.76	2.38
4	2	1	2	3	179.6	173.6	0.60	0.33
5	2	2	3	1	173.6	162.8	0.72	1.91
6	2	3	1	2	162.8	146.8	0.80	2.83
7	3	1	3	2	175.8	169.4	0.64	0.89
8	3	2	1	3	169.4	158.9	0.70	1.67
9	3	3	2	1	158.9	142.5	0.82	3.04

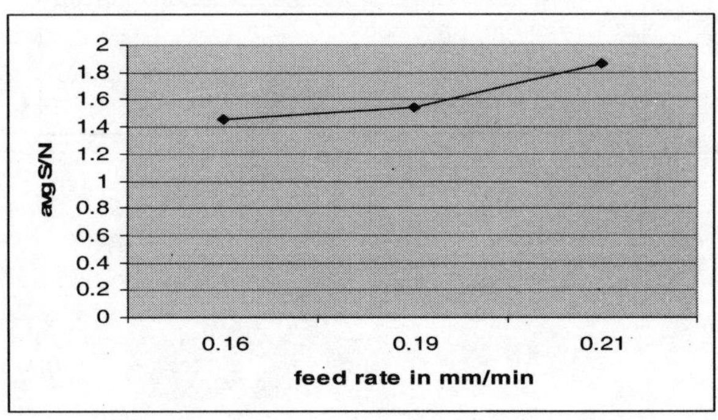

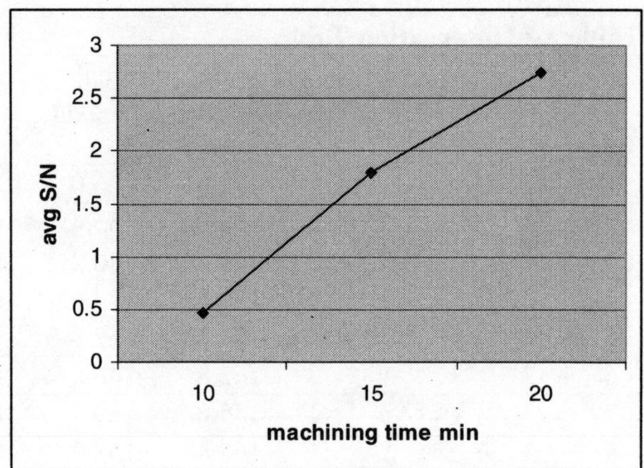

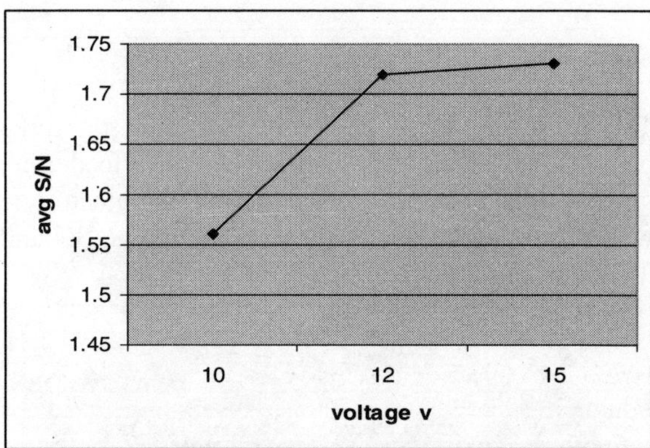

After finding the S/N ratio for the larger is better, we calculate the average response of A (Feed rate) at level 1 was obtained from the results of experiments 1,2,3 since level 1 of parameter A was used in these experiments. So, average response for this = S/N Ratio1 + S/N Ratio2 + S/N Ratio3 = 1.45. Similar calculations were performed for another factors and levels. The average responses for all parameters are given in Table 3.1 with overall mean S/N Ratio. The graph between average value of S/N ratio and process variables are shown in fig 3.3

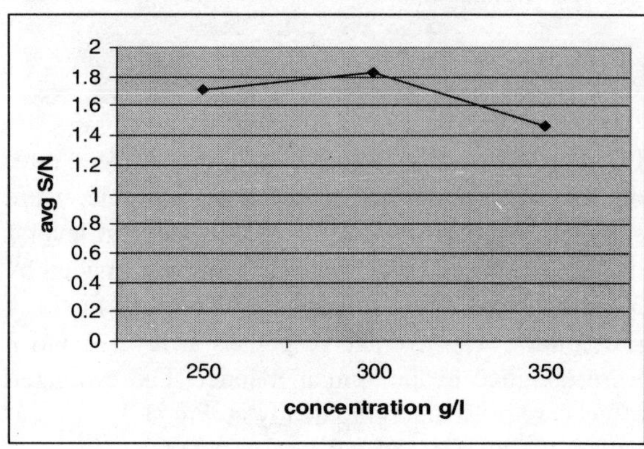

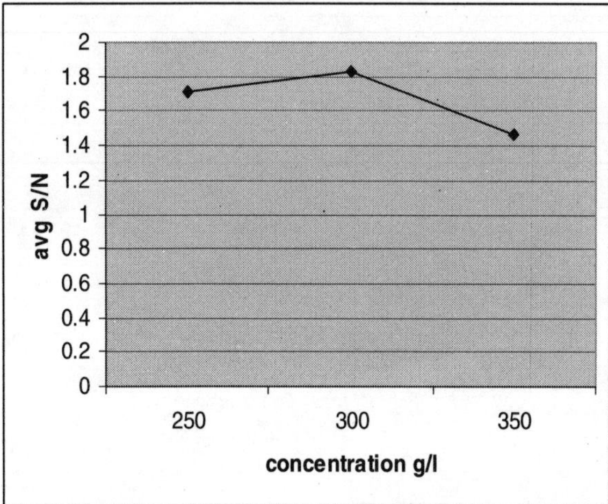

Fig 3.3 The graph between average value of S/N ratio and process variables
So the comparison between graph of average value of mrr and graph of S/N ratio there are no more changes to be found.

The delta matrix result shows that the machining time is higher the better and voltage is minimum effect on optimum value of material removal rate.

IV. CONCLUSION

According to Taguchi method the strongest parameter which is mostly affects the material rate is machining time and the lowest is the voltage. The machining time is directly effected on the material removal rate because the chemical reaction of electrolysis process is done during the machining process. The feed rate of electrode is responsible for the material removal and over cut on the work piece. The feed rate of electrode is directly proportional to max material removal and maximum over cut on the work piece. The concentration of electrolyte is directly related to material removal rate but highly concentration of electrolyte decrease the material removal rate due to excess movement of ion between two electrodes. Current density depended on the rate at which ions arrived at respective electrodes which was proportional to the applied voltage, concentration of electrolyte, gap between the electrodes, and tool feed rates. As the tool approached the work, the length of the conductive current path decreased and magnitude of current increased. This lessening of the gap and increase in current continued until the current was just sufficient to remove the metal at a rate corresponding to the rate of tool advance.

Table3.2 Overall mean S/N Ratio

Level	Average S/N Ratio by factor level				Overall mean S/N Ratio T
	A	B	C	D	
1	1.45	0.46	1.56	1.71	1.67
2	1.69	1.79	1.72	1.83	
3	1.86	2.75	1.73	1.46	

The effect of this factor is then calculated by determining the range:
Δ= Max-Min=1.86-1.45=0.41

Table 3.3 Result

Levels	A	B	C	D
1	1.45	0.46	1.56	1.71
2	1.69	1.79	1.72	1.83
3	1.86	2.75	1.73	1.46
Δ	0.41	2.29	0.17	0.37
Rank	2	1	4	3

V. REFERENCES

[1] S. K. Mukherjee, S. Kumar and P. K. Srivastava 'Effect of Over Voltage on Material Removal Rate During Electrochemical Machining',Tamkang Journal of Science and Engineering, Vol. 8, No 1, pp. 23_28 (2005).

[2] Sekar T., Marappan R. 'Improving Material Removal Rate of Electrochemical Machining AISI 1035 by Using Rotating Tool', web.tuke.sk/fvtpo/casopis/pdf08/2-str-60-63.pdf (2008).

[3] H. Hocheng, Y.H. Sun, S.C. Lin, P.S. Kao 'A material removal analysis of electrochemical machining using flat-end cathode', Journal of Materials Processing Technology Volume 140, Issues 1-3, 22 September 2003, Pages 264-268.

[4] Mr. S. S. Uttarwar , Dr. I. K. Chopade 'Effect Of Voltage Variation On MRR For Stainless Steel EN Series 58A (AISI 302B) In Electrochemical Machining: A Practical Approach', 2OHProceedings of the World Congress on Engineering 2009 Vol II WCE 2009, July 1 - 3, 2009, London, U.K.

[5] Jenny J. Sun E. Jennings Taylor 'Investigation of electrochemical parameters into an electrochemicalmachiningprocess',www.faradaytech nology.com/.../machining/namrc%20xxvi%20(1998). pdf Inc. 315 Huls DriveClayton, Ohio 45315-8983

[6] Thomas B¨ohlke, Ralf F¨orster 'Electrochemical Machining with Oscillating Tool Electrode;EstimationofMaximumPressure',www.itm. kit.edu/cm/download/Boehlke_Foerster_IJEM_2006. pdf (March 29, 2005).

[7] K. P. Rajurkar, B. Wei, L. Schnacker 'Monitoring and Control of Electrochemical Machining (ECM)', Journal of Engineering for Industry, Vol. 115/217 ,May 1993.

[8] S.K. Mukherjee, S. Kumar, P.K. Srivastava, Arbind Kumar 'Effect of valance on material removal rate in electrochemical machining of aluminum', journal of materials processing technology Volume 202, Issues 1-3, 20 June 2008, Pages 398-401 202

[9] R V Rao, P J Pawar, and R Shankar 'Multi-objective optimization of electrochemical machining process parameters using a particle swarm optimization algorithm', Journal of Engineering Manufacture Volume 222, Number 8 / 2008.

[10] Bhattacharyya, S.K. Sorkhel 'Investigation for controlled electrochemical machining through response

Surface methodology-based approach', Journal of Materials Processing Technology 86 (1999) 200–207.

Root Cause Failure Analysis of Gears

S. S. Chauhan

Department of Mechanical Engineering, IEC College of Engineering and Technology, Greater Noida

Abstract-**There are several defects and problems associated with the successful design of a new gear transmission.** *Knowledge* **and experience of several technical areas of engineering defects and problems associated with a new design are related to an inadequate evaluation of several areas such as, the lubrication and cooling requirements, complete static and dynamic load analysis, evaluation of materials and heat treatment and the latest manufacturing technology. In this paper some of the common problems of the gear design process are discussed with some recommendations made for avoiding these conditions.**

I.INTRODUCTION

Gear problems are a common occurrence in the gear industry and are often the results of improper design, a wrong selection of material for a given application or a lubrication system that was not adequate for the conditions encountered. Many times a noise or gear vibration condition will appear as an unexpected problem and may require considerable time and expense to correct.[1,2] There are quite a number of gear consultants who are kept busy with gear problems. Not all problems are design related but may be the result of some unavoidable or unexpected system condition. It is s always good practice with a new design to have experienced engineers or consultants conduct a design review of the system. Years of experience tends to alert one to some of the more common problems associated with the gear systems. The objective of this paper is to, point out a few of the more common problems encountered with gear systems and offer ways to avoid at least some of them.

Design Considerations [3, 4]

Sometimes problems are caused by failure to select the best type of gear for a given application. Particular gear type may not have been applied during gear selection. Several important parameters should be considered when selecting a gear type for a given system. For a parallel axis system, there is the possibility of using spur, helical, double helical, or other special types such as Wildhaber Novikov gears. Different contact ratios may also be considered for different applications. Generally spur gear designs are limited to pitch line velocities of 50 m/sec or lower. However there are applications where spur gears have been operated successfully at over 100 m/sec which shows that it can be done with the right choice of design parameters. When high dynamic loads, and air or oil trapping, can cause excessive wear or reduced life, helical gears are preferred for use at the higher speeds because there is a high total contact ratio which reduces the noise and dynamic loads. The air and oil can be pumped off the ends of helical gear teeth without producing high loads or excessive noise. Helical gears have thrust loads and overturning moments that must be accounted for. In some lightweight applications it is difficult to reduce the deflections caused by the overturning moment, to an acceptable level. Many helical gear designs use thrust rings and moment loads so that there are no external forces that require thrust bearings and heavy moment supporting structure. Space should be provided between the end of the teeth and thrust ring to allow the oil or air from the mesh to escape easily. The thrust ring should have a radius to increase the formation of an elastohydrodynamic (EHD) oil film by wedging action between the thrust ring and the ends of the gear teeth. These thrust rings when properly designed can take the thrust load Double helical gears are often used to eliminate thrust and moment loads of single helical gears. A double helical can cause noise and high dynamic loads if not properly manufactured. A double helical gear will shift axially to adjust for spacing error and helix error between the two sides of the gear. If these errors are very large the result will be noise and high dynamic loads. Because of this condition it is very difficult to manufacture a planetary gear system using double helical gears. As each planet gear tries to adjust for the sun gear and ring gear at the same time, severe vibration will be generated.

Accuracy selection criterion

Selection of the right gear accuracy for an application is very important as it affects the cost of manufacturing, the dynamic load and noise. The affect of accuracy on dynamic loads and noise are increased as the speed is increased. Reference 1 indicates that doubling the speed of a gear set increased the noise level by approximately 6 dB. Therefore a moderate to high speed gear set should have an AGMA tolerance class of 10 to 13 for low noise level and reduced dynamic loads. The highest gear accuracy is usually obtained by finish grinding on a good gear grinder. Good quality gears that are not too hard (less than RC 45) can be obtained by shaving. Hard gears (greater than RC 50) can also be finish cut to good accuracy by skiving with a carbide skiving cutter.

Gear Rim Thickness Considerations

Gears are often designed using empirical methods for some parameters such rim thickness. One design practice that has been used is to make the gear rim thickness equal to the tooth height. This practice has been very successful for gears that are fairly small in diameter compared to the tooth height. When the gear diameter is small the rim is very stiff and will have no low frequency vibration problems. However, when the diameter of the gear is large compared to the tooth height the rim becomes much more flexible and is subject to low frequency vibration modes. The increased flexibility may also cause an Increase in bending stress due to the gear tooth moment combined with the rim flexure. Constant ratio functions of r/t so that the rim thickness t increases proportionally to the gear radius thickness without causing a large increase in gear weight.

Gear Drawing and Specifications

There are many cases where a particular gear with a required tolerance and material condition has been requested from a gear manufacturer but the finished product was not the same as that required. This is primarily the result of inadequate gear drawings and specifications. It is not enough today to send a gear drawing with an AGMA class requested and a note that says carburize and harden. The drawing should include with the gear tolerance a profile chart. Drawing should also have a note listing inspection required and a list of material and heat treatment specifications which should be a part of the requirements. The material and heat treatment specification should be very specific as to the type of material , material certifications, heat treatment with requirement for certification and sample used for checking the method used. In addition to a good heat treatment specification, a good source should be selected that will do what is required in the specification and provide the required controls and record to verify the procedure. A material sample should be supplied with the finished gears with case depth and hardness verified on the sample and a report delivered with the gears giving a certification of all the required conditions.

Gear Vibration and Dynamic Loads

When a gear system is designed it should be evaluated for dynamic loads and resonant operating condition. There are many times when a gear system has failed because of high dynamic loads from the unexpected vibration of the systems. It is usually not enough to add a dynamic factor per AGMA 218.01(ref. 2) since these factors are general in nature and do not allow for resonant conditions that may exist. A vibration analysis should be used if available to predict vibration problems. There are several types of vibration conditions that may be encountered in a gear system. There may be shaft vibration, gear rim or web resonant vibration, shaft tonsorial vibration coupled with the gear teeth and housing vibrations.[5] The gear transmission should be evaluated for these vibration conditions before and after it is installed in the system. Testing of the gear system should be accomplished with adequate instrumentation to show any   vibration problem that might existing the operating range. If a gear system should happen to have a resonance near the operating speed a rapid failure of the gears could occur.

On gear rings that have vibration problems as shown in figure 1. These rings are designed to rub against the gear rim at a different frequency and thus it is often good practice to use vibration damper rings provide damping. The damping thus provided will considerably reduce the peak Stresses at the resonant conditions. Sometimes it may be necessary to change the weight of a gear to move its resonant vibration away from the gear operating speed.

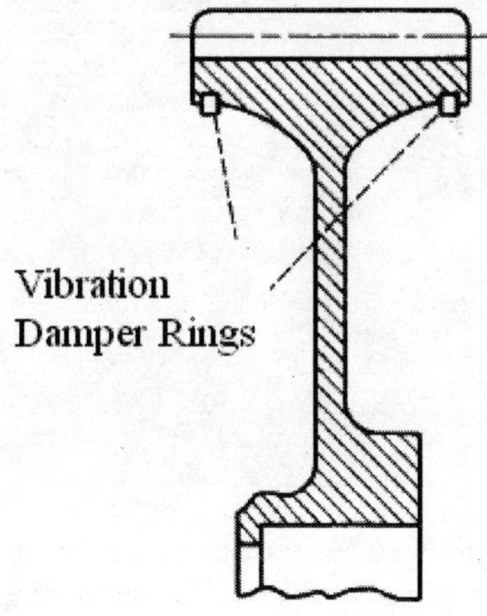

Figure 1 Gear with Vibration Damper Rings

Gear Life

When a gear set is designed is it normally required to have a life that will be equal or better than the other equipment in the system. The two areas of life of concern to the gear designer are the bending fatigue life and the surface fatigue life, assuming there is no significant wear. The bending fatigue life can be designed for essentially an infinite life by keeping the bending stress well below the stress that will give 10^7 cycles before failure, figure 2.

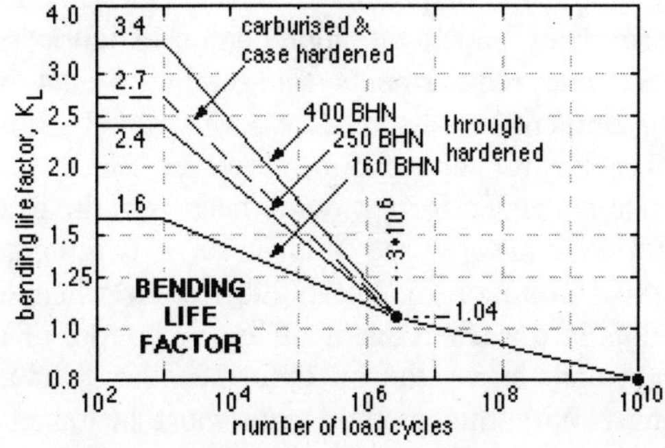

Figure 2 Bending Strength Life factor, K

Idler gears the stress should be further reduced by approximately 20 percent. In surface fatigue life there is no stress condition that will give infinite life. Therefore in applications where a large number of cycles are accumulated in a short period the stress must be reduced to account for this. At higher stress (life to approximately 10^6 cycles), test data reference 3 indicates that the life is inversely proportional to stress to the ninth power (fig. 3) at lower stress levels (above 10^7 cycles) but there are indications that the life in this region is inversely proportional to stress to the 20th power (ref. 2, fig. 4) that is very expensive to obtain because of the long hours involved in running tests on several specimens. For reverse bending application such as epicyclic gears and idler gears the stress should be further reduced by approximately 20 percent. In surface fatigue life there is no stress condition that will give infinite life. Therefore in applications where a large number of cycles are accumulated in a short period the stress must be reduced to account for this. At higher stress (life to approximately 10^6 cycles), test data reference 3 indicates that the life is inversely proportional to stress to the ninth power (fig. 3). at lower stress levels (above 10^7 cycles) but there are indications that the life in this region is inversely proportional to stress to the 20th power (ref. 2, fig. 4) that is very expensive to obtain because of the long hours involved in running tests on several specimens.

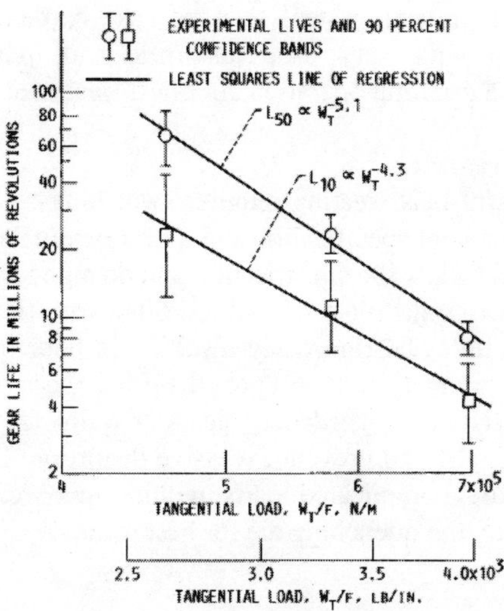

Figure 3 Load Life relationship for AISI 9310 steel spur gears speed 10000 RPM, Lubricant Super refined naphthenic mineral oil with additive

package.(Ref. Vallance & Doughtie, Designe of machine members, McGraw-Hill)

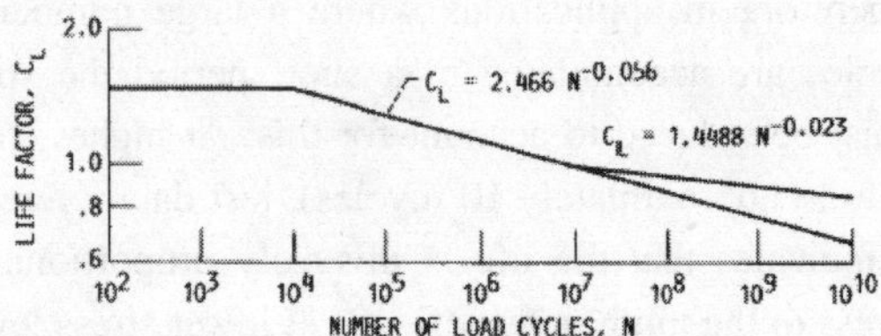

Figure 4 Pitting Resistance Life factor, C_L (Ref. Vallance & Doughtie, Designe of machine members, Mcgraw-Hill)

Materials

Many times a gear material is selected for an application which does not give the required surface durability or bending strength. Some steel materials that are adequate for small gears may not heat treat properly when used in a larger gear. Some lower alloy steels that will harden well in small sizes will remain soft with reduced strength when used for larger gears because it takes longer to quench. There is a wide variety of material available to the gear designer. Proper selection is required to meet the demand of low cost and reliability. There are many cases when plastic gears will perform very well because the loads and temperatures are acceptable. Soft steel gears are used for many applications but are generally much larger and weight more than a comparable case hardened gear. When weight size and reliability are important then a case hardened gear is usually the best choice. Most case hardened gears require finish grinding after heat treatment. However, nitriding steels can be case hardened by nitriding with very little distortion and may not require a grinding operation after heat treatment.

Heat Treatment

Successful heat treating requires two things, a good heat treatment specification and a heat treating source that will follow the specification and do a good job of heat treating the material. All too often specifications are not followed rigorously resulting in poor quality heat treatment and soft or distorted gear. Most carburized and hardened gears require a good quenching die to prevent excessive distortion. Larger and more complicated parts require more care in designing the quenching die for best results.

Figure 5 Damaged Esteem gear box (Courtesy Auto wings Automobile workshop, Greater Noida)

At corners and sharp points care must be taken in carburizing and nitriding gear teeth. Sometimes these areas will crack because of the surface compressive stress built up in the material from the carburizing or nitriding. A typical corner crack is shown in figure 5. This was eliminated by masking the ends and tips of the teeth to limit the amount of carbon near the corner of the tooth. The same type of thing can happen on narrow pointed gear teeth with nitriding. These type problems can be avoided by masking areas that do not required hardening such as the ends and of the teeth thus reducing the corner stresses.

II. LUBRICATION

Many gear system problems have been generated because of an inadequate appreciation of lubrication methods and requirements for gearing. Gears like rolling element bearing require a very small amount of lubricant for lubrication, however much more lubricant is needed for removing heat from the gears. Several methods are used to lubricate and cool gears. The most common is splash lubrication when one gear dips into a reservoir of oil in the bottom of the gearbox and throws the oil onto the other gear and bearings. With this method care must be taken to assure that an adequate flow of oil is fed to the bearings for good cooling. If this is not done bearing failure will occur because of insufficient cooling and not enough EHD oil film.[4] The most common method used to supply oil to the bearing with the splash lubrication method is to provide a channel to catch the oil splash and feed it to the bearings. When a gear operates at higher speed and loads the splash method of lubrication and cooling will generally not provide the cooling required for the gears. With inadequate cooling, early failure will occur. The failures may be due to scoring, wear, surface fatigue, or tooth breakage. The main cause for these failures sometimes is not recognized as over temperature of the gears from inadequate cooling. The first indication of excessive temperature of the gears may be superficial pitting or grey staining caused by a reduced EHD film thickness as the oil viscosity is reduced. In many applications the designer will use oil jet lubrication with the oil jet directed at the in to mesh position or at the out of mesh position. Neither of these methods provides the most effective cooling of the gears and could in many cases allow overheating of the gears. The into mesh method of oil jet lubrication will considerably reduce the efficiency of the gears since the oil going into mesh increases oil churning losses.

III.CONCLUSION

Many of the more common problems that occurring gear transmissions can be avoided in the design and manufacturing process. The gear type for the particular application should be selected and the accuracy requirements determined that will give the desired results with the minimum cost. For light weight designs rim thickness should be carefully evaluated for vibration problems and maximum bending stresses. The drawing and specifications should be specific and include the entire required information to assure delivery of the correct end product. Bending and surface fatigue design life should be adequate for the application. The gear material and heat treatment are very important to the life and cost for a given application and should be carefully determined. The lubrication system must be designed to provide good cooling and lubrication for the gears to prevent early failure.

IV. REFERENCES

[1]. Welbourn, D.B.: Gear Noise Spectra - A Rational Explanation. ASME Paper 77-DET-38, Sept. 1977.
[2]. AGMA Standard, Rating the Pitting Resistance and Bending Strength of Spur and Helical Involute Gear Teeth. AGMA 218.01, Dec. 1982.
[3]. Taylor, James I. (2000), The Gear Analysis Handbook (First ed.), Vibration Consultants,
[4]. Mobley, R. Keith (1999), Root Cause Failure Analysis (First ed.), Butterworth-Heinemann
[5]. Smith, J.D. (2003), Gear Noise and Vibration (Second ed.), Marcel Dekker

Study of Initial Die Design System for Diecasting Process

V. Kumar
[1]Department of Mechanical Engineering, Anand International College of Engineering,
Jaipur, Rajasthan (India)

Abstract-- **Diecasting is one of the die-based forming methods to manufacture large number of products with short time period and clean surface by high injection pressure of cast alloy. Initial or conceptual die design is composed of cast design, cavity layout design and design of gating system etc. In reality die design of diecasting has been performed by trial and error method, which causes monitory and time loss. This study presents a systematic approach for automation of initial die design process, especially layout design system using three-dimensional part product model. This system quantifies practical knowledge and experiences in die design as formulating procedure. By using this approach, it is possible for designers to make automatic and efficient design and it will result in reduction of required expenses and time. It is composed of cast design, cavity layout design, and gating system design stages. In addition, specific rules and equations for the system have been presented. The system has been successfully tested on several industrial components.**
Keywords: Diecasting, Die Design, Cast Design, Cavity Layout Design, Gating System Design.

I. INTRODUCTION

Diecasting is a fast and cost-effective manufacturing process for production of high volume, net-shaped, tight tolerance metal components. It has the benefit of increased quality and repeatability, often at lower costs compared to other processes. It is the method of rapidly producing metal components by injecting molten metal under high pressure into a permanent mould called a die. Molten metal solidifies rapidly (from milliseconds to a few seconds) to form a net-shaped component that requires no or little machining. The die-cast products are widely used in the automobile, aerospace, electronics and household appliance industries due to its high strength, low weight, excellent surface quality, and close dimensional tolerances.

The quality of the part produced is essentially determined by the die-casting die, in which the part is formed. A die-casting die consists of two mould halves known as *core* and *cavity* with a vertical parting surface when closed.

The quality of the part produced is essentially determined by the die-casting die, in which the part is formed. A die-casting die consists of two mould halves known as *core* and *cavity* with a vertical parting surface when closed.

The part of die which remains always stationary is called cavity half (or cover die) and the other half which is movable, is called core half (or ejector die). Two mould halves are assembled and poured with molten metal at high pressure. After solidification, these mould halves are separated and cast component is automatically ejected with the help of ejection mechanism. One or more side cores are usually employed if a part has undercut feature which is not accessible from core and cavity. Figure 1 shows die casting process with basic terminology.

Die design is a complex and time consuming process that requires knowledge and experience of die designer. The main tasks in computer aided design of a die casting die have been discussed by Fuh et al. [1] and shown in figure 2. It takes CAD file of the part as input and involves tasks such as Setting shrinkage, determining number of cavities and its layout, gating system design, die-base design, parting design and core/cavity creation, ejection design, cooling design, side core design, standard component design, etc. These die design tasks are generally divided into two broad categories i.e. initial (conceptual) design and detailed design. The initial design is composed of decisions made at the early stage of the die design, such as the selection of type of mould configuration, setting shrinkage, determining numbers of cavities and its layout, gating and runner system design, etc. The detailed design is composed of the core/cavity creation, the ejection system design, the cooling and venting component design, the assembly analysis and the final drafting.

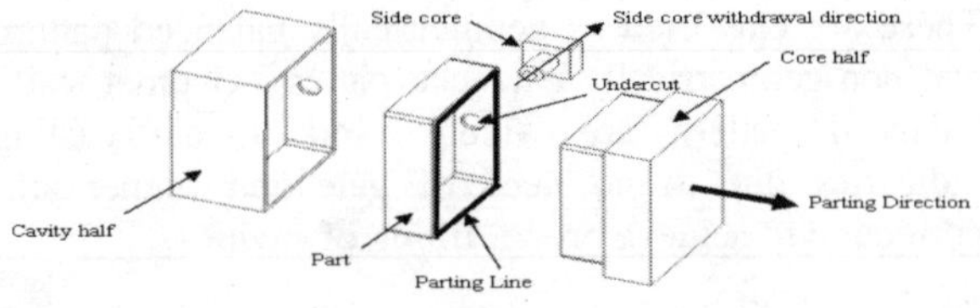

Fig. 1: Die-casting process terminology.

Most of the previous researches in die design of diecasting or injection moulding were focused on parting line, parting surface side only while little attention has been given to initial diecasting die design. A few numbers of researches on complete CAD/CAM systems for injection moulding and die casting has been carried out by researches [2-7]. Ye et al. [8] presented an algorithm for initial design of injection moulds. It calculates number of cavities and orients them in balanced layout pattern. Cast design module not considers draft. They also not consider the gating system design in their work. System for cavity layout design has been also developed [9-10]. Few systems also developed for gating system design [13-14]. But there is strong need for a system for initial die design process for diecasting.

This paper presents the methodology of initial die design system for diecasting process. The flow chart of cast design, layout design and gating system design of die casting die design process have been presented. The system is supported by rule base and knowledge base. It also use machine database to determine number of cavities.

II. METHODOLOGY OF DIE DESIGN SYSTEM

As shown in Figure 3, initial die design is mainly composed of cast design, cavity layout design and gating system design. At first, 3D part product model of the component is input and the design of the cast is begun. In cast design shrinkage factor and draft angle is provided on part model. In layout design first number of cavities has been calculated based on empirical relationship and then they are arranged in the specified pattern with usual clearances. Finally, gating system design has been carried out in which geometry of gate, runner and overflow are determined using rule base. Next sections describe these stages in details.

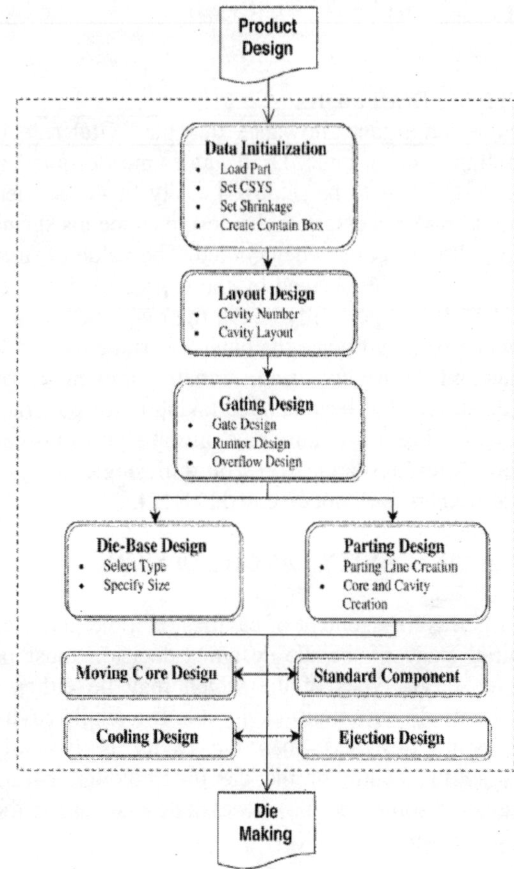

Fig. 2: CAD-based die casting dies design process [1]

III. CAST DESIGN

First of all, the cast must be designed because the dies can be generated from the cast in diecasting die design. The cast design consists of four parts; loading part model, applying material, application of shrinkage allowances, and draft application. In cast input, the part product model in commercial modeler is taken as input and orient in required viewing direction. The parting surface should be determined for detailed die design for diecasting. But the algorithm that determines the parting surface is not constructed, and it is supposed that user recognizes the location of parting surface in advance.

A. Setting shrinkage

In establishing the dimensions for cavities, an allowance must be added to the dimension specified for the part to be cast, to allow for shrinkage of the casting metal. The shrinkage allowances normally lies in the range of 0.005 to 0.018 inch per inch. In general, the calculation of shrinkage allowance at room temperature is given by the equation 1[14].

$$\Delta L = \beta\,(T - 20) - \alpha\,(t - 20)$$
$$(1)$$

B. Applying Draft angles

One more important allowance that play vital role in cast design is draft angle. Draft angles are designed to let the casted part to be removed easily from the dies, they are always necessary in die design as metals shrink or grip tightly to cores when cooled. The value of draft angles are normally defined by the degree of the taper of a sidewall and generally lies in the range of 0.5 to 2 for outside wall and twice the outside surface for inside wall depend upon alloy used. Finally draft angle on required surface has been applied taking parting surface as reference. For this feature recognition algorithms can be used. A few research works on draft angle analysis and addition are well documented [11-12].

IV. CAVITY LAYOUT DESIGN

Layout is the arrangement of cavities on a die plate. At the initial stage of the die design a decision must be made to select, types of die which may be either a single cavity die or a multi-cavity die. If a single cavity die is being designed, then the cavity is typically located in the centre of the die base, though gating requirements sometime may necessitate placing the die cavity off centre.

For multi-cavity dies arranging number of cavities is a complex task. To accomplish this task, there are essentially two fundamental types of cavity layout patterns are frequently used.

These are classified as geometrically balanced pattern and non-geometrically balanced pattern. Geometrically balanced pattern are widely used in die casting industries due to no need of gate and runner size correction to achieve proper filling of cavities.

- Geometrical Balanced patterns: Symmetric and circular type patterns fall under this category. These are also known as standard cavity layout pattern.
- Non-geometrical Balanced patterns: These types of patterns require gate and runner size correction for proper filling of cavities. They are also known as fishbone type pattern or series pattern and in-line pattern.

Figure 4 shows different types of cavity layout patterns used in diecasting process. Primarily, circular and symmetric patterns are frequently used in diecasting industries. The advantage associated with these patterns is equal flow length to all cavities without gate or runner size correction. Equal flow length ensures proper filling of all cavities. Use of circular pattern is limited to less number of cavities as compared to symmetric pattern. Number of cavities in circular pattern depends upon part geometry and part size. The most systematic arrangement of cavities is possible with symmetric pattern as it ensures proper balancing of die base. The disadvantage of symmetric pattern is large runner volume, much scrap and rapid cooling of melt as compared to circular pattern. Series pattern and in-line pattern are less preferred due to non equal flow length to all cavities although they can accommodate more number of cavities as compared to circular pattern. Due to non equal flow length of runner system computer filling analysis is needed to ensure proper filling of all cavities.

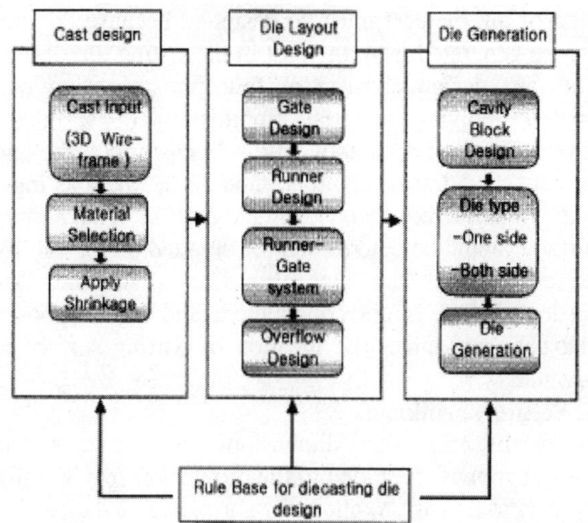

Fig. 3: Flow chart for initial die design system of die casting process

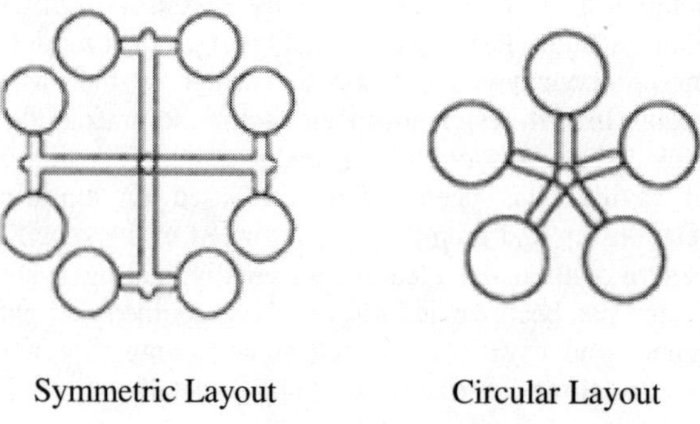

Symmetric Layout   Circular Layout

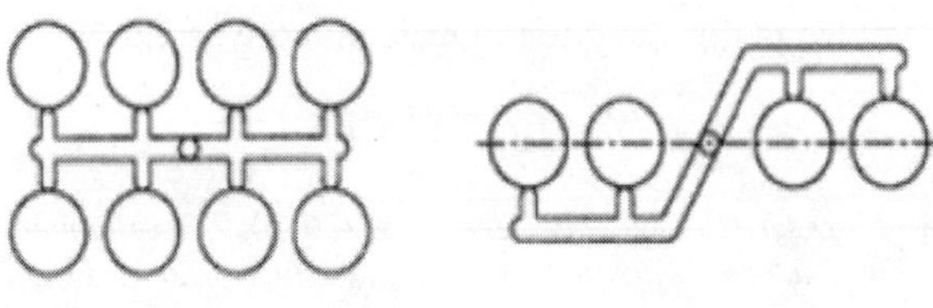

Fig. 4: Types of cavity layouts pattern.

To design the cavity layout of die-casting die, designer must first determine the number of cavities, select the layout pattern and arrange the orientation of individual cavities with necessary clearances. Each of these activities is dependent on a number of factors. For example, number of cavities depends primarily on the delivery date, the production cost, the machine parameters and part geometry features. Layout pattern depends on factors like number of cavities, size of die block and number and position of side-cores etc. Sufficient clearance between cavities should be maintained. Arrangement for orientation of the die cavities should take into account die clearances, which are required to accommodate gating system and, making a provision of side-cores etc.

Present work is an attempt to develop a system for design of cavity layout for a diecasting die. Proposed system has three stages to arrive at automatic cavity layout. The system depends upon the knowledge base of diecasting die design. It takes geometric information of part from CAD file and maintains a machine data base, while some other information is input by the user. First stage of the system determines the number of cavities considering technical, economical, time and geometrical limitations and presents feasible number of cavities as output. Second stage provides a possible solution for layout pattern, which is based on the number of cavities and uses a knowledgebase of layout patterns. Once the layout pattern has been selected, third stage of the system arranges the individual cavities by selecting a suitable mold base. Selection of mould base takes into account the clearances required to accommodate feeding system, side pulls and is based on the die design knowledgebase.

V. GATING SYSTEM DESIGN

The gating system is a series of passages through which the molten metal enters and fills the die cavity. It is composed of several gating elements i.e. gate, runner, overflow and shot sleeve (sprue). In the process of gating system design, the gate, runner and overflow are designed for constructing dies. In this system, the gating system design is divided into three parts; gate design, runner design and overflow design.

A. Gate design

Gate is the opening through which molten metal enters the cavity. Gate design needs cavity volume for determine gate dimensions. Once the mechanical properties of the cast are input and filling speed is known, the cross-sectional area of gate can be given by [14];

$$A_g = \frac{V}{v_g \times t_g} \tag{2}$$

Where,

A_g is gate area in m^2; V is cavity volume in m^3; v_g is gate velocity which depends on cast thickness in m/sec; t_g is filling time in sec.

If the gate having rectangular cross-sectional area then the width of gate w is given by [14];

$$w = \frac{A_g}{t} \tag{3}$$

Where, t is the gate thickness in mm and generally lies in the range of 0.5 to 3 mm.

B. Runner design

A runner is a channel located at the parting line to pass liquid metal from the sprue to the gate, and it may split into two or more as required to direct the liquid to various cavities. Generally, the runner is machined entirely in the ejector half and the cover half forms only the flat side of the runner. To get proper gate filling of the cavity, the cross-sectional area of runner must be larger than that of gate. A runner-to-gate area ratio of 1.15:1 to 1.5:1 is generally used. Oversize runners will increase metal losses and regrind costs. Finally shape of runner is selected and dimensions are calculated. One such shape i.e. trapezoidal is shown in figure 5 with terminology for example.

C. Overflow design

It is a chamber usually located at the opposite side of the gate to absorb undesirable non-metallic inclusion and exits air from casting cavity. They provide additional weight to casting and their weight is also added to cast weight in calculating total weight of shot. The volume of overflow is generally taken as 15% of the volume of casting. The volume of overflow is used to calculate dimensions of the overflow after selecting the geometry of overflow (the shape of cross-section, width and depth).

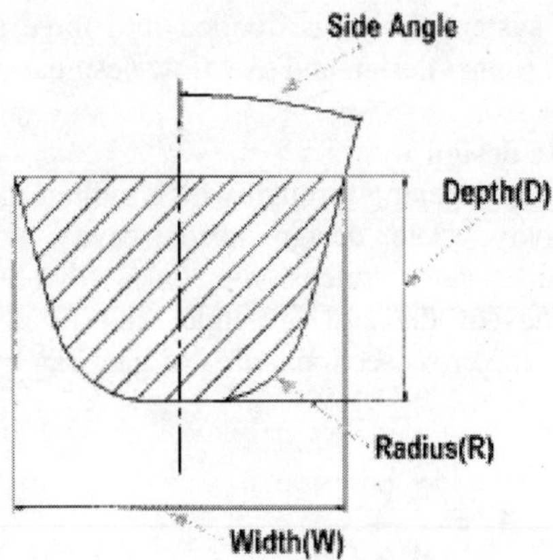

Fig. 5: Cross-sectional area of trapezoidal runner with terminology [6].

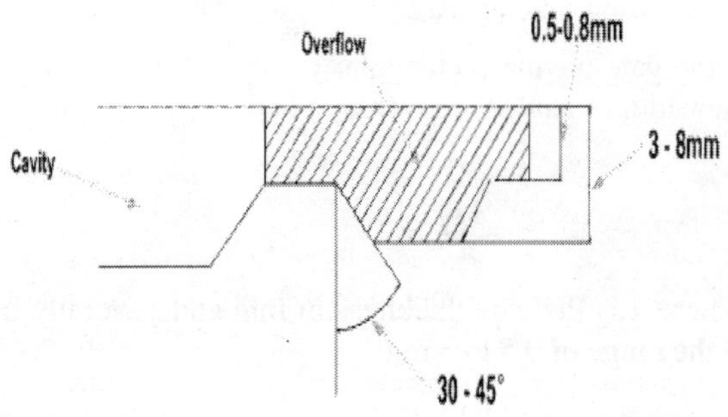

Fig. 6: Schematic diagram of general shape for overflow [6].

VI. CONCLUSIONS

The study presents the development of an automated CAD system for initial die design of diecasting. The primary conclusions of this study are as follows:

1. This study suggested a methodology for an easy and effective die design system so that the die designer can design the diecasting die, especially the cavity layout design system and gating design system.

2. A novice who may not have any experience of die design can perform die design even with only a little knowledge about diecasting. This system quantifies practical knowledge and experience for die design for diecasting as a formulated procedure of design.

In the proposed system it is assumed that the parting direction and parting surface is already known to the user. Identification of undercut features and their location in an automated manner would make the system more useful and increase the level of automation. Improvements in the system are being attempted in this direction to make it more useful

VII. REFERENCES

[1] Fuh, J.Y.H., Wu, S.H., Lee, K.S. (2002), "Development of a semi-automated die casting die design system", Proceedings of the Institution of Mechanical Engineers, Part B: Journal of Engineering Manufacture 216 (12), pp. 1575-1588.

[2] Zhang, W., Xiong, S. and Liu, B. (1997), "Study on a CAD/CAM system of diecasting", Journal of Materials Processing Technology, 63, pp. 707–711.

[3] Kruth, J. P. (1986), "Steps toward an integrated CAD/CAM system for mold design and manufacture: anisotropic shrinkage, component library and link to NC machining and EDM", Annals CIRP, 35.

[4] Chen Y. M. and Wei, C. L. (1997), "Computer-aided feature-based design for net shape manufacturing", Computer Integrated Manufacturing System, 10(2), pp. 147–164.

[5] Chan, W. M., Yan, L., Xiang, W., Cheok, B. T. (2003), "A 3D CAD knowledge-based assisted injection mould design system", Int J Adv Manuf Technol 22:387–395.

[6] Choi, J.C., Kwon, T.H., Park, J.H., Kim, J.H., Kim, C.H. (2002), "A study on development of a die design system for die casting", Int J Adv Manuf Technol 20:1-8.

[7] Woon, Y.K., Lee, K.S. (2004), "Development of a die design for die casting", Int J Adv Manuf Technol 23:399-411.

[8] Ye, X.G., Lee, K.S., Fuh, J.Y.H., Zhang Y.F., Nee, A.Y.C. (2001), "Automatic initial design of injection mould", Int. J Mater Prod Technol16 (6-7), pp.592-604.

[9] Hu, W., Masood, S. (2002), "An intelligent cavity layout design system for injection moulds" Int J of CAD/CAM, v.2, no.1, pp.69-75.

[10] Low, M.L.H., Lee, K.S. (2003), "A parametric controlled cavity layout design for a plastic injection mould", Int J Adv Manuf Technol 21, pp.807-819.

[11] Yan, Y. and Tan, S.T. (2004), "Adding draft angles on mechanical components containing constant radius blending surfaces", Computer-Aided Design, Vol. 7, pp. 565-580.

[12] Lee, S.H. and Lee, K. (1998), "An integrated CAD system for mold design in injection molding processes", Computer-aided design and manufacturing of dies and molds presented at the Winter Annual Meeting of ASME, pp.257-271.

[13] Wu, S.H., Fuh, J.Y.H., Lee, K.S. (2007), "Semi-automated parametric design of gating systems for die-casting dies", Computers and Industrial Engineering 53 (2), pp. 222-232.

[14] Kim, C. H. and Kwon, T.H. (2001), "A runner-gate design system for die casting dies", Materials and Manufacturing Processes, Volume 16, Issue 6, pp. 789 – 801.

Study of MRR in Drilling in EDM

Shyam Sunder Agarwal

Mechanical Engineering Department, Motilal Nehru National Institute of Technology, Allahabad (INDIA)-211004

Abstract- **The MRR while machining is a part in electric discharge machining (EDM) process is of great concern in the researchers in recent year as the accuracy of the parts to be produced. The change in MRR may occur due to variation in the machining parameters such as current, spark gap, pulse duration, voltage, flushing of dielectric fluid, mode of flushing of dielectrics, etc. Due to high precision and good surface quality that it can give Electo-discharge machining has been gaining popularity as a new alternative method to fabricate structures. The performance of the EDM process is evaluated in terms of MRR and the stability of the machining by considering various operating parameters of the EDM process, such as T-ON, T-OFF, Servo speed etc.**

Key words: Electrode wear, ON-time, Off-time, Servo Speed, Taguchi Method.

I. INTRODUCTION

Electrical discharge machining (EDM) is a non-conventional process for manufacturing of complex or hard material parts that are difficult to machine by conventional machining processes, During EDM, the electrodeshap is mirrored in the work piece. It is one of the most extensively used non-conventional material removal processes. It uses thermal energy to machine electrically conductive hard material parts regardless of their geometry. Many automotive and aerospace components, as well as mould and dies are manufactured using ED machining. During EDM; there is no direct contact between the electrode and the work piece. There upon EDM eliminates the mechanical stresses arising during machining. Electrical Discharge Machining is accomplished with a system comprising two major components: a machine tool and power supply. The machine tool hold a shaped electrode, which advances into work piece and produced a shaped cavity, the power supply produces a high frequency series of electric spark discharge between the electrode and work piece which remove metal form work piece by thermal erosion or vaporization.

A relatively soft graphite of metal electrode can easily machine hardened tool steels or tungsten carbide [1,2]. There are several types of machines and industrial applications that use the EDM process for high precision machining of metals with die sinking and wire EDM being two major EDM variants.

Generally depending upon the MRR, EDM can be characterized as: roughing. Semi-roughing and finishing. Because of the nature of EDM process, optimization of the process parameters is required, in order to achieve the desirable performance specification. In addition, the surface finish, dimensional accuracy and geometry of electrode as well as the material properties such as thermal conductivity and wear resistance effect EDM performance measures, Electro discharge machining has ability to machine any conductive material irrespective of their mechanical hardness. This ability of machining only conductive and semi-conductive materials is a disadvantage of the EDM process. The EDM process can process materials such as quenched steel and carbides which are mainly used for making cutting tools owing to their very high hardness and these materials are very difficult to machine using mechanical cutting processes.

II. DESIGN OF EXPERIMENTS

The design of an experiment includes following points.

- The set of treatments to be considered.
- The set of experimental units to be included.
- The procedures through which the treatment is to be assigned to the experimental units.
- The response variables to be observed.
- The measurement process to be used.

Taguchi Method

Taguchi is the developer of Taguchi method. [3] We use Taguchi method, which is very effective to deal with responses influenced by multi-variables. This Method is a powerful Design of Experiments tool, which provides a simple, efficient and systematic approach to determine optimal machining parameters. Compared to the conventional approach to experimentation, this method reduces drastically the number of experiments that are required to model the response functions, wherein one variable is changed while the rest are held constant. The major disadvantage of this strategy is that at particular settings there may not be cutting at all due to the stochastic nature of EDM.

It is also impossible to study all the factors and determine their main effect (i.e., the individual effect) in a single experiment, Taguchi method is devised for process optimization and identification of optimal combinations of factor for given responses (4-9).

We will use different process parameters.

Table 1

Symbol	Machining Parameters	Level 1	Level 2	Level 3
T-on	ON Time	3	5	7
T-off	OFF Time	2	4	6
Sv	Sparking gap voltage	2	4	6
SEN	Servo speed	2	5	6

Orthogonal Array

To select an appropriate orthogonal array for the experiments, the total degree of freedom needs to be computed. The degrees of freedom are defined as the number of comparisons between design parameters that need to be made to determine which level is better and specifically how much better it is. In the present study the interaction between cutting parameters is neglected. Therefore six degree of freedom owing to there being three machining parameters in the turning operations. In this study an L9 orthogonal array with four columns and nine rows was used. This array has 8 degree of freedom and it can be handle three level design parameters, each cutting parameter is assigned to a column, nine cutting parameter combination being available. Therefore, only nine experiments are required to study the entire parameters space using the L9 orthogonal array. The experimental layout for the three cutting parameters using the L9 array is shown in table.

Table 2

Experiment no.	T-on	T-off	Sv	SEN
1	1	1	1	1
2	1	2	2	2
3	1	3	3	3
4	2	1	3	3
5	2	2	1	1
6	2	3	2	2
7	3	1	2	2
8	3	2	3	3
9	3	3	1	1

III. EDM MACHINE & MATERIALS

EDM drilling machine with NC control in Z direction was used with negative polarity, Maximum machining current will be up to max 9-10 Amp. Brass tool electrode has been used, which is cylindrical in shape and having almost 1 mm diameter. Maximum Cutting speed is 16mm/min. Work material is HCHC steel of size 23mm*18mm*13mm. The density of work piece is 7.8 gm/cc. Experiment are performed according to L9 orthogonal array, on work piece of weight 50 gm. A separate electrode has been used fro each experiment. Time of machining is 02 minutes for all experiments. The MRR is calculated by weight difference of thee electrodes and work piece after drilling, Electronic balance with 300 gm capacity with a precision of .001 gm has been used.

IV. RESULT AND DISCUSSIONS

The experimental results for electrode wear rate and MRR, based on L9 orthogonal array is shown in following table.

Table 3

Experiment No.	MRR(mm^3/min)
1	6.3981
2	2.8871
3	1.4982
4	8.4521
5	0.76712
6	7.3982
7	2.3985
8	10.3212
9	3.9126

In order to estimate the main effects and their differences, first the overall mean value of the observations for the experimental region is calculated. Similarly overall mean of MRR is also calculated and their values are given in following table.

Table 4

	MRR(mm^3/min)
Overall Mean	4.8925688

MAIN EFFECTS OF PARAMETERS ON MRR (mm^3/min.)

Table 5

Parameters	Symbol	Level 1	Level 2	Level 3
ON time	T-on	3.5944	5.5391	5.5441
OFF time	T-off	5.7495	4.6590	4.2696
Sparking gap voltage	Sv	8.0391	5.0839	1.5546
Servo speed	SEN	3.692	4.2279	6.757

Main effect of ON-time on MRR

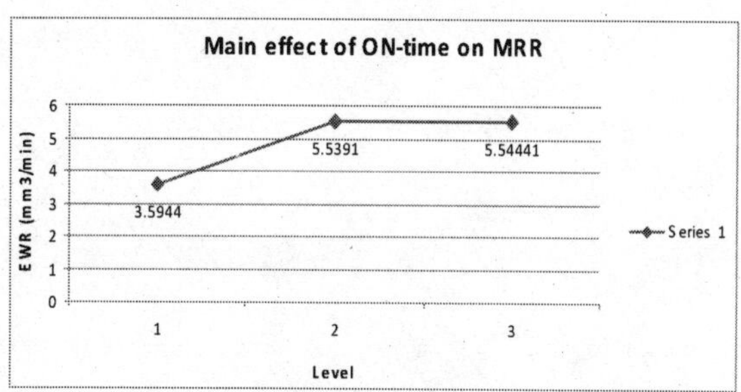

Figure 1

Main effect of OFF-time on MRR

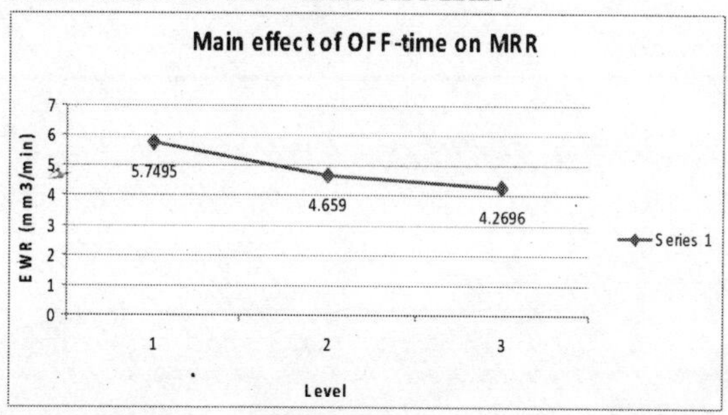

Figure 2

Main effect of sparking gap voltage on MRR

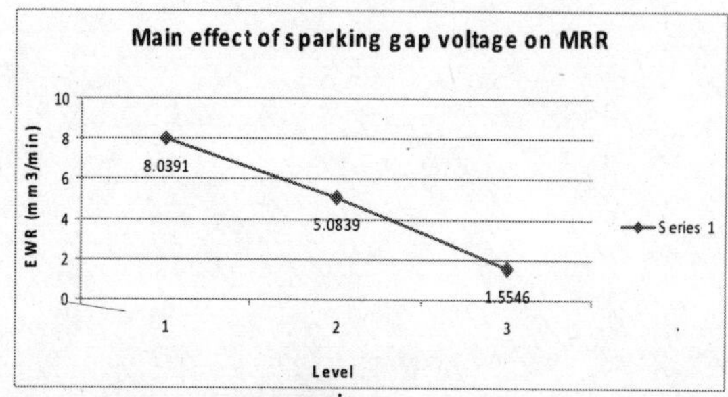

Figure 3

Main effect of servo speed on MRR

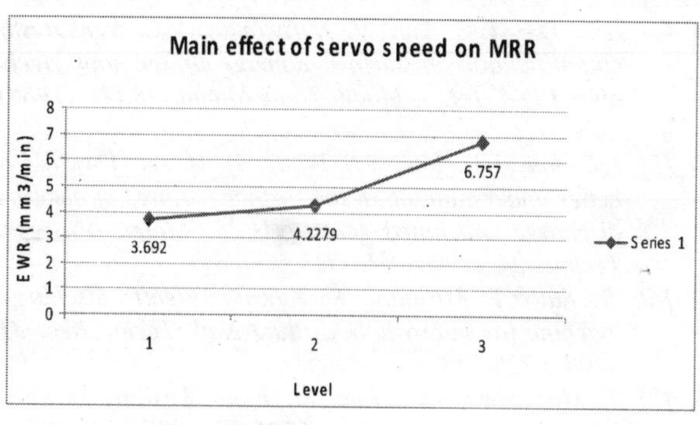

Figure 4

We have used MINITAB statistical took kit for the variance analysis. Result are shown in Table

ANOVA of MRR

Table 6

Paramet er	Do F	Sum of squares	Mean square	F	Contri bution %
On-time	2	7.295	3.598	1.9396	7.98
Gap voltage	2	64.398	31.720	17.099	70.45
Servo speed	2	15.987	8.116	4.375	17.49
	2	3.720	1.855		4.07
	8	91.40			100

Optimum condition for MRR

The Most significant parameter for MRR according to the % contribution given in table 6 is gap voltage followed by servo speed. Again the least significant parameter in OFF-time. The optimum condition of the machining parameter and their level for MRR is given in Table 7

Table 7

Parameter	level	Level description(position)
T-ON	3	7
T-OFF	1	2
Sv	1	2
SEN	3	8

V. CONCLUSION

The Individual effect of all the controllable parameters is predicted using Taguchi method.

VI. REFERENCES

[1] *Ho KH, Newman ST (2003) Int J Mach Manuf 43: 1287*

[2] *Brink D EDM: Principal of operation. http/www.edmtt.com, as on 14 June 2005.*

[3] *G. Taguchi, Introduction to quality Engineering, Asian productivity Organization, Tokyo, 1990*

[4] *L.C. Lee, L.C. Lim, V, Narayanan, V.C. Venkatesh, Quantification of surface damage of the tool steels after EDM, Int. J, Mach. Tools Manuf, 28 (4) ,(1988) ,359-372.*

[5] *L.C. Lee, L.C. Lim, Y.S. Wong, H, H. Lu, Towards a better understanding of the surface features of electro-discharge machined too steels, J, Mater, Process, Techno, 24, (1990), 513-523*

[6] *T. Sato, T. Mizutani, K. Kawata, Electro-discharge machine for micro hole drilling Natl. Techn. Rep. 31 91985) 725-733.*

[7] *T. Masuzawa, C.L. Kuo, M. Fjino, Drilling of deep micro-holes by EDM, Ann. CIRP 38 (1989), 195-198.*

[8] *J.H. Zhang, T.C. Lee, W.S. Lau, Study on the electro-discharge machining.*

[9] *S.H. Lee, X.P. Li, Study of the effect of machining parameters on the machining characteristics in electrical discharge machining of tungsten carvide, J. Mater. Process. Technol. 115 (2001), 344-358.*

Study of Wear Analysis of Poplar Wood and its Polyacrylonitrile Composites

Shalendra Kumar Pathak[1], Sushil Kumar Rai[1], Y. K. Tyagi[2] and Umesh Kumar Sharma[3]

[1]Department of Mechanical Engineering, Agra Public College of Technologyand Management, Agra, Uttar Pradesh, India.
[2]Department of Mechanical Engineering, Kumaun Engineering College, Almorah, Uttarakhand, India.
[3]Department of Applied science, ExcelInstitute of Management and Technology, Mathura, Uttar Pradesh, India.

Abstract - **In this study, analysis of wear test of Poplar wood and its acrylonitrile impregnated wood composites are investigated. Wood Polyacrylonitrile (PAN) composites from Poplar wood is synthesize. The synthetic process was carried out in presence of benzoyl peroxide (0.02 mol/l) catalyzed impregnation polymerization of acrylonotrile having concentrations in the range of 0M & 4.54M, were treated with properly shaped woods specimen in methanol medium at $(95\pm1)^{\circ}C$. Modification of the properties of wood polymer composites over untreated wood was evaluated in terms of wear index.**

Keywords: benzoyl peroxide, polyacrylonitrile, impregnation)

I. INTRODUCTION

The performance of wood as a construction material for outdoor applications deteriorates under accelerated weather environments due to fluctuation in weather and humidity for longed outdoor applications as well as decreasing the cost of wood and avoiding the need of frequent replacements in permanent and temporary constructions. A number of wood preservatives and new wood treatment processes have been developed during those wood treatment processes and are under continuous demands which can develop the modified wood materials with improved mechanical strength, thermo-oxidative stability, and resistance to bio-deterioration for their better outdoor applications.

The polymer loading of wood depends on the permeability of the wood species being treated. The void volume is approximately the same for sap wood and heart wood for each species. Because of this, it would be expected that the polymer would fill them to same extent [3].

In the past few decades a variety of commercially available vinyl monomers have been used for wood treatment to improve the mechanical and thermo-oxidative stability of low-grade woods [4, 5].
Advancement in the technology of thermoplastic impregnated wood composites have recently made great claims to replace quality woods with high grade wood polymer composites derived from low grade woods [1,6,7]. In many kinds of processing, wood has been subjected to treatment at elevated temperatures (e.g. drying), size stabilization, pulping, and production of particle and fiber boards.
Temperature affects the physical, structural, and chemical properties of wood. Several attempts have been made to establish the relationship between temperature and thermal stability of wood [8-12]. Reinforcement of several acrylic monomers like styrene, methylmethacrylate, and (chloropropyl)-2-propane phosphate has provided substantial thermal stabilities to various low grade woods.
Wear analysis is fundamentally the use of equations and models to evaluate wear behavior. However, for it to be effective, wear analysis must incorporate certain elements and considerations beyond the evaluation of equations. While wear behavior is complex, useful wear analysis often are not. Generally, the complexity of the analysis depends primarily on the engineering needs and secondarily on the wear situation.
As a consequence wear analysis is not limited to the evaluation of the effects of materials on wear

behavior. Wear analysis often enables the identification of nonmaterial solutions or nonmaterial elements in a solution to wear problems. For example, changes in or recommendations for contact geometry, roughnesses, tolerance, and so on are often the results of a wear analysis.

II. MATERIALS AND METHODS

This experiment was performed at the institute workshop and chemistry lab. The Wear test was done at CIRT, Pune.

Starting Materials
Acrylonitrite monomer was purchased from M/s- C. D. H. Chemicals India Pvt. Ltd., Mumbai. The monomer acrylonitrile was purified by extracting it with aqueous NaOH (10%) to remove inhibitor contents followed by repeated washings with distilled water. The fraction distilled at 82^0C was used for the impregnation polymerization reaction.

Preparation of Wood Specimens
First the wood specimens were prepared for their treatment as per IS: 1708-1960. The moisture content of wood was deduced according to ASTMD 1037-72a and was found to be 12.75%.

Preparation of Solution
The methanolic solution of acrylonitrile at concentration of 4.54M and methanolic solution of benzoyl peroxide at 0.02M have also been prepared.

Method of Treatment of Wood Specimens
The prepared wood specimens were placed in an airtight stainless steel chamber of the dimensions $20\times20\times30$ cm^3. The specimens were swelled in methanol (98%) for 5 hours. The solution of benzoyl peroxide (0.02M) and acrylonitrile (PAN) were added. The samples were then soaked in monomer solution for 12 hours at room temperature. The treated wood specimens were then wrapped in aluminum foil at $(95\pm1)^0$ C for 2 hours to induce the polymeric reaction.

Wear analysis of Wood and Impregnated Wood Composites
Specimens of untreated wood are dried to a moisture content of 7 or 8 percent. Specimens seasoned for 24 hours or longer in the conditioned atmosphere of the laboratory at 50% relative humidity and $(70 - 74)^0$ F temperatures. Wear resistance tests of natural and PAN Wood has been made on the Abraser.

Panels has selected for flatness, uniform thickness, and free from warp. Prepared the surface by sanding it smooth and free of indentations or other surface defects that might occur in the path of the abrading wheels.

A special extension nut, S-21 is available for holding material ¼ to ½ inch in thickness. This nut requires a 3/8 inch center hole in the specimen in place of the usual ¼ inch hole.

The section is the outline all the elements of a typical test procedure from analysis of the testing problem to final evaluation of results and mechanics of testing, the preparation and mounting of specimens, and the setup and operation of the Abraser is presumed.

The value of the Abraser in research and control programs depends to a considerable extent of the test problem, the service requirements and the desired wear characteristics of the material examined. Time and material may be saved by analyzing this problem and planning procedure before embarking the use of Polyacrylonitrite wood.

An air-conditioned test room is strongly recommended where reproducible precision results are required. Both heat and moisture affect the abrasion resistance of most materials, and particularly organic materials. Abrasion research projects are usually carried out in an atmosphere maintained at $(70-75)^0$ F temperature and 50% relative humidity. Without exception, samples which were to be tested should be seasoned in the test atmosphere for at least 24 hours. When tests were to be conducted.

Temperature and humidity conditions has specified kept constant in all tests.

Recording the Test
It was essential that record of every phase of test procedure be kept for purpose of comparison and in order that the test may be exactly duplicated at any future time.

Selecting the Method of Evaluating Test Results
Test results are expressed as a wear factor / numerical abrasion index of the test specimen.

The wear factor arrived at by one of the method of calculating results is not directly comparable. The method of calculating results should be expressed with the wear factor. The tests have been performed by Weight loss method.

This method of evaluation test results is recommended when the results are to be compared with those of similar materials having nearly the same specific gravity. The Taber wear index (rate of wear) is the loss in weight in milligrams per thousand cycles of abrasion for a test performed under a specific set of conditions. The lower the wear index, the better abrasion resistance quality of the material

100mg. x 1000 cycles 200 TABER WAER

INDEX

------------------------- = (Weight Loss Method)

500 cycles test

Wear Testing

Model 5130 Abrasers has used for wear tests as it has precision built test instruments designed to evaluate the resistance of surface to rubbing abrasion. Their field of application includes tests of solid materials, painted, lacquered, electroplated surfaces, and plastic coated materials, textile fabrics ranging from sheer silks to heavy upholstery, metals, leather, rubber and linoleum. In the hands of competent research technicians, both are capable of performing reproducible tests, accurate within the variations of quality inherent in the material itself.

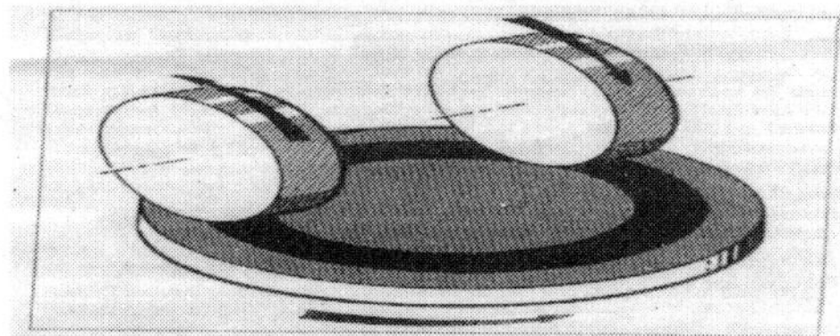

Fig. 1 Operational diagram

Characteristic rub-wear action of Abrasers is produced by contact of a test sample turning on a vertical axis, against the sliding rotation of two abrading wheels. The wheels are driven by the sample in opposite directions about a horizontal axis displaced tangentially from the axis of the sample. Figure diagrams the relative positions of sample and abrading wheels and the direction of their rotation.

One abrading wheel rubs the specimen outward toward the periphery and the other, inward toward the center. The resulting abrasion marks form a pattern of crossed arcs over an area approximately 30 square centimeters, Satisfactory for rating most materials.

An exclusive and important feature of the 5130 Abrasers is that the wheels traverse a complete circle on the specimen surface, revealing abrasion resistance at all angles relative to the weave or grain of the material.

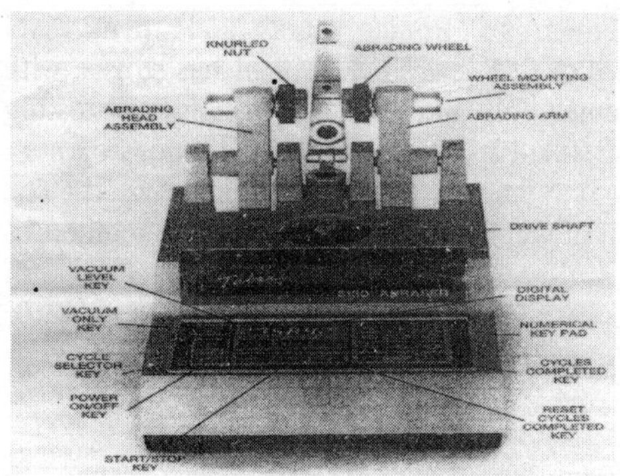

Fig. 2 Front-view of machine Abrasive with control panel

In practice, details of procedure for testing a particular material are established as specifications by the laboratory originating the test. These specifications may usually be obtained from the initiating laboratory. They should be followed exactly so that test results will be comparable in all cases.

In the event that no standard test has been established for a given material it was necessary to set up its own procedure. The following recommendations for specific materials are offered with the definite understanding that they do not apply to all variations of materials and that modifications may be required in performing a practical abrasion test.

In the wear test for specific materials it is assumed that, when possible, test should be conducted in an atmosphere in controlled humidity and $(70\text{-}74)^0$F temperature. The samples have been conditioned in the test atmosphere for at least 24 hours before testing.

The Vacuum control should be set high enough to remove abrading and abrasive particulars but not high enough to lift flexible specimens.

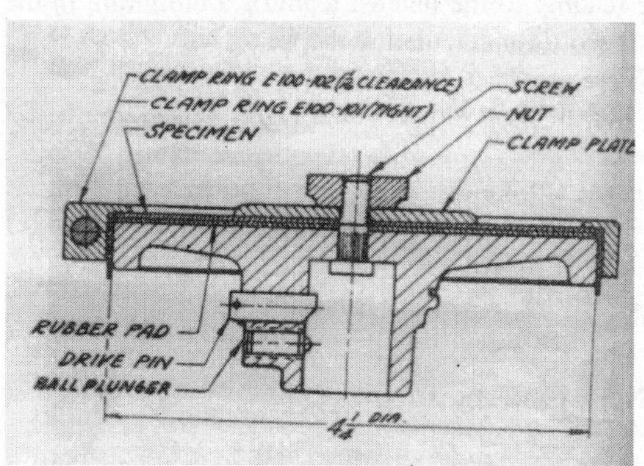

Fig. 3 Cross section of E-100-125 specimen holder with clamp ring

The specimen of Poplar wood and their PAN composites of 0M & 4.54M of 4 inches square with ¼ inch center hole and holded it in holder of testing machine E100 -125, tighten with S-21 extension nut . The abrasive wheels H - 18 calibrade used. The evaluation of wear is based on Weight loss method.

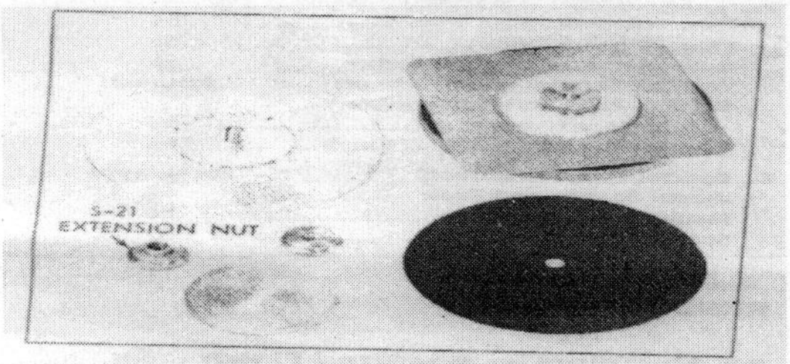

Fig. 4 Specimen holder E-100-125 with rubber pad with expansion nut

III. WEIGHING THE SPECIMEN

Before Testing

Immediately before these tested on the Abraser, the specimen has been weighted to the nearest tenth of a milligram in a sensitive laboratory balance and the weight entered in the record.

Accuracy of results depends on correct weighing of the specimen. A high precision analytical balance should be used and its accuracy periodically checked. Sufficient time should be taken to assure maximum accuracy in the weight determination.

After Testing

At the end of the abrading operation, clean the specimen thoroughly by brushing it free of any loose particles. Weight the specimen in the same balance and with the same care as before testing and entered the reading to the nearest tenth of a milligram in the test record for calculating the wear factor.

As per the wear index, (table 1) the wear strength of Poplar wood composites increases with the concentration of impregnation of polyacrylonitirite.

Table: 1
Wear index of Poplar wood

S.No.	Concentration	Increase concentration	Wear index
1	0M	0%	3560
2	4.54M	30%	3280

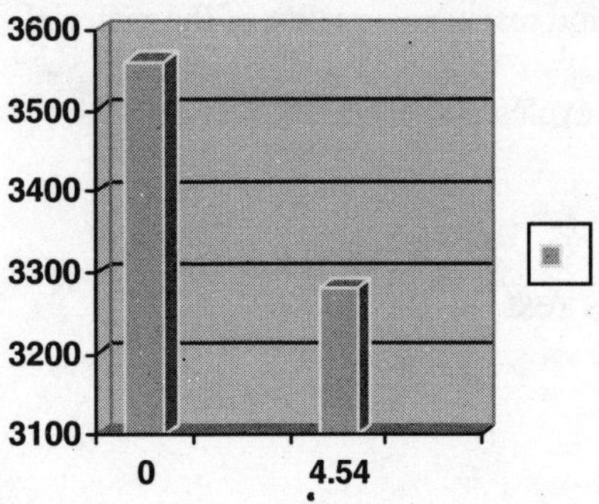

Fig 5. Wear Index of Poplar Wood with molar concentration

IV. CONCLUSION

It is concluded from the wear data presented in this paper that the wear stability of referenced wood was improved in comparison to untreated Poplar wood.

V. ACKNOWLEDGEMENTS

The authors are thankful to the Head of the Central Institute of Road transport, Pune, India for providing wear analysis data and are also thankful to Dr. J.C.Sharma and Dr Seema Garg for their continuous guidance.

VI. REFERENCES

[1] Schneider, M.S., Phillips, J.G., and Lande, S. 2000. "Physical and Mechanical Properties of Wood Polymer Composites". J. Forest Eng. 11: 83-89.

[2] Clemons, C. 2002. Forest Product Journal. 52:10.

[3] Balatinecz, J.J. and Woodhams, R.T. 1993. J. Forestry. 91:22-29.

[4] Dale Ellis, W. and O'Dell, J.L. 1999. J Appl Polym. Sci. 73: 2494-2505.

[5] Lingfi, M. and Yang, Y.Y. 1996. Zheziang Linxueynan Xebio. 13:104-108.

[6] Research Group of Wood Plastic Composites Research. 1997. China Wood Industry Application Prospect of Wood / Plastic Fiber Composites in the Passenger Car Industry. Institute of Wood Industry: Beijing, China. 11(5):22.

[7] Mapleston, P. 2001. "Additive Suppliers Turned their Eyes to Wood Plastic Composites". Modern Plastics. Aug: 52 (2001).

[8] Sanderman, W. and Augstin, H. 1963. Holz Roh-Werkst. 21:256-265, 305-315.

[9] Fengel, G. 1966. Holz-Roh-Werkst. 24:9-14, 98-109, 529-536.

[10] Kosik, M, Geratova, L., Rendos, F., and Domansky, R. 1968. Holzforch, Holzvenv. 20: 5-19.

[11] Kosik, M., Kozmal, F., Resiser, V., and Domansk y, R. 1968. Holzforch, Holzfrch, Holzusw. 20:11-15.

[12] Beall, F.C. and Eichkner, H.W. 1970. "Thermal Degradation of Wood Components. A Review of Literature". USDA, Forest Service Res. Paper. FPL 130.

[13] Oksmank, A. and Lindberg, H. 1995. Holzforschung. 49:243.

[14] Courturer, M.F., George, K., and Schneider, M.H. 1996. "Thermophysical Properties Wood Polymer Composites". Wood. Sci. Technol. 30(3):179-196; 125 (14). 1713145J.

[15] Ellis, D.W. and Sanadi, R.A. 1997. Proc 18th Riso Int. Symp. On Mat. Sci. 307:1673.

[16] Yap, G.S., Que, Y.T., Chio, L.H., and Chan, O.H.S. 1991. J Appl Polym Sci. 43:2057.

[17] Ibach, R.E. and Rowell, R.M. 2001. Holzforschung. 55:358.

[18] Joshi, T.K., Zaidi, M.G.H., Sah, P.L., and Alam, S. 2005. "Mechanical and Thermal Properties of Poplar Wood Polyacrylanitrile Composites". J. Polym. Int. 54: 198-201.

[19] Collins, E.A., Barc, J.M., Fred, W, Jr. 1973. Experiments in Polymer Science. Wiley Int. Science Publication: New York, NY. Chapter 9, pp. 216-262.

[20] Tyagi, Y.K., Zaidi, M.G.H., Singh Pratap, and Singh Dheer. 2006. BSME-ASME International conference on Thermal Engineering. Dhaka, Bangladesh.

[21] Pathak, S.K. 2010, Study of thermal properties of Babool Wood and its Polyacrylonitrite Composites, PJST 399-400.

To Analyze the Impact Strength of Weld Joint Through Electric Arc Welding Using Design of Experiment

[1]**Sandip Mishra**, [2]**Mahesh Vishwakarma**, [3]**Tarun Soota**

[1]SATI Vidisha

[2]CSE Jhansi

[3]BIET Jhansi

Abstract—**The objective of this paper is to obtain an setting of welding process parameter (joint angle of parent metals, electrode size, welding speed, cooling medium) resulting in an optimal value of impact strength which is welding on the structural steel with the flux coated electrode on electric arc welding machine. The effects of the selected process parameters have been accomplished using Taguchi method. After the welding the defects is to be measured with the help of microscope. The result indicate that the selected process parameters Joint angle, electrode size, welding speed and cooling rate has more influenced the weld strength and reduce the time and input cost of experiment.**

Keywords: Electric arc welding machine, Process parameters, Taguchi method.

I. INTRODUCTION

Welding is now a day's extensively used in the following fields; automobile industries, air craft machine frame, structural work, tanks, machine repair work, ship building, pipeline fabrication in thermal power plants and refineries, fabrication of metal structures. The electric arc welding is commonly used in every industry. In welding the strength of weld joint is very important characteristics. The various input welding parameters such as welding current, arc voltage, welding speed, electrode feed rate, electrode diameter and joint angle are most effected the output parameter such as deposition rate, weld bead geometry and strength of weld joint. All the welding processes are used with the aim of obtaining a welded joint with the desired weld bead parameters, excellent mechanical properties is with minimum distortion. After welding the cooling medium is also effected the mechanical properties of weld joint. In this paper the three medium of cooling is to be selected for improving the mechanical strength of weld joint such as natural air cooing, cooling in dry carbon powder medium and in brine solution.

II. DESIGN OF EXPERIMENTS

Taguchi method is a statistical tool which is to help of designing the experiment in less time and help to select the best parameter which is more affected the output in less no of experiment run. The experimental design proposed by Taguchi involves using orthogonal arrays to organize the parameters affecting the process and the levels at which they should be varies. Instead of having to test all possible combinations like the factorial design, the Taguchi method tests pairs of combinations. This allows for the collection of the necessary data to determine which factors most affect product quality with a minimum amount of experimentation, thus saving time and resources.

An Electric arc machine is used for the welding. The specimens a common structural steel is prepared with different V angle (30°, 45° and 60°). The dimension of work piece is 20mm*80mm*5mm flat surface. The E6011 types of electrodes are used for this experiment. The experimental layout for the welding parameters using the L_9 orthogonal array is used in this study. This array consists of four control parameters and three levels as shown in table 2.1 In Taguchi method, most all of the observed values are calculated based on 'the higher the better'.

S/N Ratio = - 10 log [$\sum (1/ y^2)/n$]

The Izod test is to be done for finding the impact strength of weld joint on the impact testing machine.

Table 2.1 Factor Level

Levels	Joint angle (A)	Electrode diameter mm (B)	Welding speed mm/sec (C)	Cooling medium (D)
1	30°	2.4	20	Air medium
2	45°	3.2	32	Carbon medium
3	60°	5.0	40	Brine medium

Table 2.2 Observation Table

Run	A	B	C	D	Impact Strength(J)	SN Ratio
1	1	1	1	1	155	48.57
2	1	2	2	2	105	45.19
3	1	3	3	3	94	44.23
4	2	1	2	3	120	46.35
5	2	2	3	1	155	48.57
6	2	3	1	2	167	49.22
7	3	1	3	2	135	47.37
8	3	2	1	3	105	45.19
9	3	3	2	1	82	43.04

Fig 2.1 Work pieces after welding

III. RESULTS AND DISCUSSIONS

The following Nine work pieces are to be welded according to the L9 orthogonal array. After finding the S/N ratio for the larger is better, we calculate the average response of A (Joint angle) at level 1 was obtained from the results of experiments 1,2,3 since level 1 of parameter A was used in these experiments. So, average response for this = S/N Ratio1, S/N Ratio2, S/N Ratio3, is 45.99. Similar calculations were performed for another factors and levels. The average responses for all parameters are given in Table 3.1 with overall mean S/N Ratio. The graph between average value of S/N ratio and process variables are shown in fig 3.1

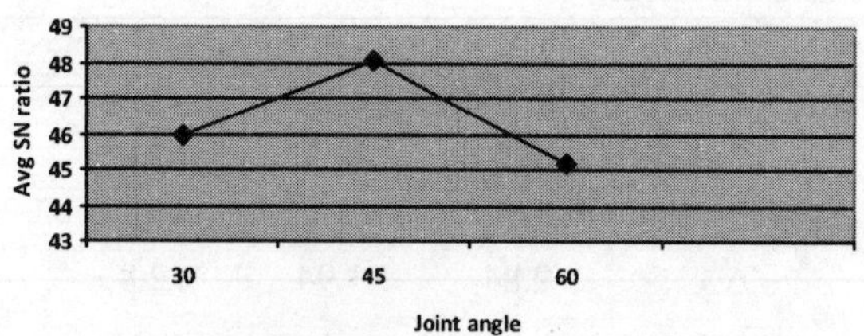

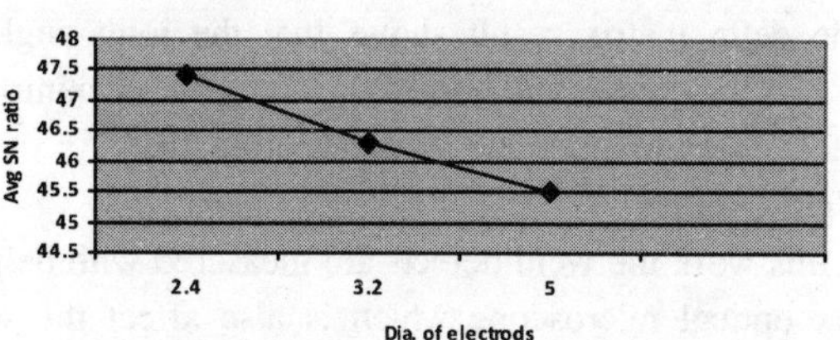

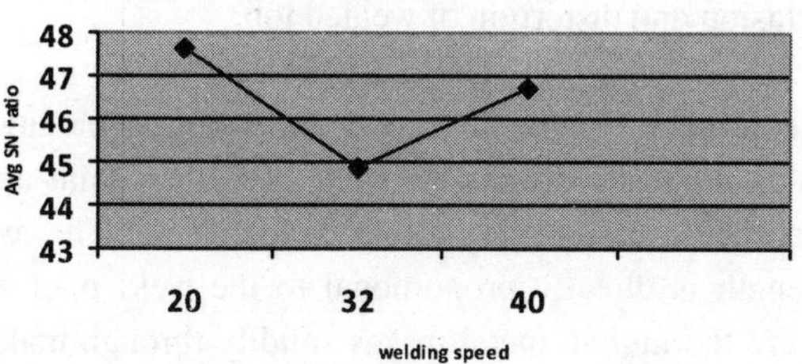

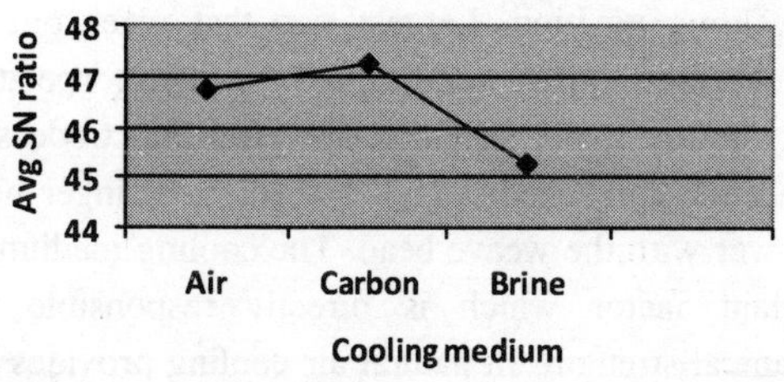

Fig 3.1 The graph between average value of S/N ratio and process variables.

Table 3.1 Overall mean S/N Ratio

Level	Average S/N Ratio by factor level				Overall mean S/N Ratio
	A	B	C	D	
1	45.99	47.43	47.66	46.72	46.41
2	48.04	46.31	44.86	47.26	
3	45.20	45.49	46.72	45.25	

The effect of this factor is then calculated by determining the range:

Δ= Max-Min=48.04-45.20=2.84

Table 3.2 Result

Levels	A	B	C	D
1	45.99	47.43	47.66	46.72
2	48.04	46.31	44.86	47.26
3	45.20	45.49	46.72	45.25
Δ	2.84	1.94	2.8	2.01
Rank	1	4	2	3

The delta matrix result shows that the joint angle is higher the better and electrode diameter is minimum effect on optimum value of impact strength.

Weld Quality

In this work the weld defects are measured with help of 15x optical microscope which is also affect the weld strength and weld quality. Normally cracks, slag inclusion, faulty weld size and profile and spatter are founds. The main defect is commonly found the slag inclusion and distortion of welded job.

IV. CONCLUSION

According to Taguchi method the strongest parameter which is mostly affects the weld strength is joint angle and the lowest is the electrode diameter. The weld strength is directly proportional to the weld pool area where the molten metal makes solidify through making a metallic bond. Electrode size is also a main parameter which depends on the joint thickness, edge preparation and welding position. Largest size that gives quality welds at high production rate. By welding speed is meant the arc travel speed. For a given electrode size and current, the speed is higher with the stringer bead and lower with the weave bead. The cooling medium is important factor which is directly responsible for mechanical strength. In natural air cooling provides the uniform good weld strength. In most cases after the welding the weld strength is improved by adding the carbon medium and brine solution is used for sudden quenching makes the weld joint harden. In this work the best parameters such as electrode diameter (45°), Electrode size (2.4mm), welding speed (40 mm/sec) and cooling medium (carbon powder medium) gives the better impact strength. For test confirmation the three tests is to be done at these levels.

V. REFERENCES

[1] P Kolhe, Prof C K Datta, Microstructure and Mechanical Properties of Multipass Submerged Arc Welding, IE(I) Journal Volume 89, October 2008

[2] Lenin N., Sivakumar M. and Vigneshkumar D Process Parameter Optimization in ARC Welding of Dissimilar Metals Thammasat Int. J. Sc. Tech., Vol. 15, No. 3, July-September 2010

[3] James Amanie N. A. Oguochaand S. Yannacopoulos Effect of Submerged Arc Welding Parameters on Microstructure of SA516 and A709 Steels Trends in Welding Research, Proceedings of the 8th International Conference, p 492-501 DOI: 10.1361/cp2008twr492

[4] Abdullah Al-Faruk, Md. Abdul Hasib, Naseem Ahmed, and Utpal Kumar Das Prediction of Weld Bead Geometry and Penetration in Electric Arc Welding using Artificial Neural Networks International Journal of Mechanical & Mechatronics Engineering IJMME-IJENS Vol: 10 No: 04

[5] S Kumanan, J Edwin Raja Dhas & Gowthaman Determination of Submerged arc welding process parameter using Taguchi method and Regression analysis, Indian Journal of Engineering & materials Vol.14 June 2007 pp, 177-183.

[6] Bipin Kumar Srivastava S.P. Tewari Jyoti Prakash A Review On Effect Of Arc Welding Parameters On

[7] Mechanical Behaviour Of Ferrous Metals/Alloys International Journal of Engineering Science and Technology Vol. 2(5), 2010, 1425-1432.

[8] Ayo Samuel Afolabi Effect of Electric Arc Welding Parameters on Corrosion Behaviour of Austenitic Stainless Steel in Chloride Medium AU J.T. 11(3): 171-180 (Jan. 2008)

Perspectives and Requirements for Manufacturing Flexibility

Ankur Das & Rahul Caprihan

Faculty of Engineering, Dayalbagh Educational Institute Dayalbagh, Agra

Abstract—**The paper elaborates on the differing viewpoints that researchers have about manufacturing flexibility together with the impact that various stimuli have on the flexibility construct. Pertinent factors that need to be integrated with manufacturing flexibility to render it useful in the modern day context are highlighted. The effect of information systems, global network infrastructures, and real-time information on manufacturing firms from a flexibility perspective is described. Some future research issues and challenges are also highlighted.**

1. INTRODUCTION

Although manufacturing system flexibility has been a buzzword in the academic research community for sometime now, it has only recently emerged as a ubiquitous concept in the present day manufacturing scenario. It has aptly been identified as the focus of the next competitive battle in the domain of manufacturing by Bolwijn and Kumpe (1990) and De Meyer *et al.* (1989). The recent proliferation of research articles is a pointer to the fact that flexibility has gained an unparalleled edge over many other areas in manufacturing research. Frazelle (1986) claims that flexibility is required in order to maintain competitiveness in a changing business environment, while Slack (1983) suggests the reasons to be the instability and unpredictability of the manufacturers' operational environment, the development in production technology, and the widening aims of production to progress beyond cost and productivity issues. With industries the world over competing fiercely for a chunk of the global market, and with the capacity to absorb fluctuations in demand economically and to develop and introduce new products quicker (Gaimon and Singhal, 1992) being seen as important competitive issues, we view flexibility as the panacea for a host of manufacturing related problems.

Past research (Eppink, 1978; Aaker and Mascarenhas, 1984; Adler, 1988; Gerwin, 1993; De Leeuw and Volberda, 1996) on flexibility acknowledges that it is not amenable to simple definition, since it is multidimensional (Suarez et al., 1995) and polymorphous (Evans, 1991). According to Upton (1994) the reason why flexibility is so difficult to define is that alternate definitions are often influenced by particular managerial situations or problems. Over the years, the applicability of flexibility in manufacturing systems has also changed. In contrast to the time when the concept evolved about two decades ago, the present day demands a more diverse definition. Today's environment is characterized by a much higher degree of volatility and variability (e.g. unpredictable demand, short product life cycles, high product variety and rapid technological change) than it was in the past. Despite two decades of research on flexibility, many important issues remain unresolved. For instance, how should flexibility be defined and measured? How can the economic benefits of flexibility be quantified? How should flexible resources be managed? What impact does flexibility have on operational performance? Questions such as these abound.

A significant factor in the flexibility achieved by flexible manufacturing systems (FMSs) is the role played by information systems in combining various complementary technologies. Information systems are the nerve center of most FMSs, a fact that is well accepted in the research and practicing community. Shaw (2000) observes that as manufacturing operations become increasingly global, the proper co-ordination between business and manufacturing units in the global value-adding chains needs special attention. Information systems can provide that co-ordination.

This paper is organized as follows. After a brief introduction in section one, section two reviews the evolution of flexibility from a mere concept to a now formidable competitive weapon in today's manufacturing arena. Section three surveys information based manufacturing research in FMSs. Specifically, the industrial applicability in relation to information based manufacturing is reviewed. The last section highlights some remaining challenges and offers scope for further research in this area.

II. EVOLUTION OF THE CONCEPT OF FLEXIBILITY

Manufacturing flexibility is a relatively new concept. There are diverse interpretations of flexibility in manufacturing, and a sizeable body of work now exists in providing workable definitions. Browne, Dubois, Rathmill, Sethi, and Stecke (1984) were one of the first to formally present a taxonomy and identified eight manufacturing flexibility types. This taxonomy laid the foundation for most subsequent research on manufacturing flexibility. In a very comprehensive review effort, Sethi and Sethi (1990) identified about 50 terms relating to 11 flexibility types. Recent activity has focused on developing valid measures for these flexibility types. To a lesser extent, empirical work has attempted to determine the "true" nature of manufacturing flexibility. These include in particular, the work of Dixon (1992) in the textile industry, Upton (1994, 1995) in the paper industry and Maffei_and Meredith (1995) in various other industries including machine tools, medical equipment, engines valves, and the injection molding machine industry.

Most studies on manufacturing flexibility provide implicitly or explicitly stated definitions of the manufacturing flexibility construct. Some representative definitions follow:-

- The ability of the production system to cope with the instability induced by the environment (Swamidass, 1988).
- The ability to respond effectively to changing circumstances (Gerwin, 1987; Gupta and Gupta, 1991).
- The ability to implement changes in the internal operating environment in a timely manner and at a reasonable cost in response to changes in market conditions (Watts, Hahn and Sohn, 1993).
- The ability to change or react with few penalties in time, effort, cost, or performance (Upton, 1994).

Various researchers have attempted to develop, to varying extents, some type of modeling framework on which to derive or classify flexibility terms. Zelenovic (1982) illustrates the interaction of the production system with its environment with the help of a cause and effect diagram. Brill and Mandelbaum (1989) develop a production framework for use in deriving measures of flexibility for a single machine or a group of machines. The applicability of this framework is limited because it consists of only a static view of the system.

Other researchers have attempted to derive flexibility dimensions based on a modeling framework, using existing definitions wherever possible. Correa (1994) reports 12 flexibility dimensions as proposed by seven different research efforts. These are *action* and *state* (Mandelbaum, 1978); *range* and *response* (Slack, 1987); *sensitivity*, *stability*, and *effort* (Gupta and Buzacott, 1989); *time* (Stecke and Raman, 1986); *organization* (Gerwin, 1987); and *range*, *switchability*, and *modifiability* (Dooner and De Silva, 1990). Interestingly, new dimensions are being introduced and older ones being continuously refined by researchers (Parker and Wirth, 1999, Golden and Powell, 2000 and Vokurka and O'Leary Kelly, 2000). Correa notes that these dimensions are not unique - many similarities exist. Therefore, the goal is to try to utilize a subset of these terms that do not overlap yet cover the different interpretations of flexibility.

The measurement of manufacturing flexibility has continued to pose a major challenge to researchers. The development of flexibility measures is extremely useful in order to exploit the benefits of a flexible system. By utilizing these measures, decision-makers have the opportunity to examine different systems at different flexibility levels. As manufacturing systems are operated and managed by people, it is necessary to record and utilize human knowledge and perceptions about flexibility in its measurements. This requirement is clearly documented in work of Gupta and Goyal (1990), Wharton and White (1988), and Gupta and Somers (1996). Numerous efforts have been reported which can be categorized by the aspect of flexibility they measure or by the approach used to determine flexibility. Kulatilaka (1988) and Hutchinson and Sinha (1989) concentrate on economic advantages of these measures, Mandelbaum and Buzacott (1990) highlight their effects on decision making, while Brill and Mandelbaum (1989), and Malek and Wolf (1991) describe the quantification of certain performance indices and operational characteristics of flexibility. From a methodological viewpoint, Yao (1985) and Kumar (1987) propose measures in the context of information theory, Kochikar and Narendran (1992) in graph theory, Chandra and Tombak (1992) in mathematical programming, and Barad & Sipper (1988) in Petri nets. In a recent research effort, Beach *et al.* (2000) present a survey based on a classification scheme originally developed by Kumar (1987), incorporating methods of measuring flexibility while simultaneously emphasizing the distinction between qualitative and quantitative measures. Their proposed taxonomy (mix, changeover, modification, volume, rerouting and material flexibility) is derived from the identification of the strategies used to accommodate specific uncertainties.

III. RECENT ADVANCES AND REQUIREMENTS

The recent years have seen a radical change in the market and competitive environment of various global and local firms. Consequently, the standards of flexibility established back in the 80's need to be modified in today's context. The acquisition of flexible technology as a direct response to changing markets is no longer the sole influence on manufacturing systems. Various other factors affect the performance of industries. Shaw (2000) describes an emerging manufacturing technology driven by information systems, the global network infrastructure, and new business models driven by the availability of real-time information. Information-based manufacturing focuses on using the correct information to assess what products to make, when to make them, and then making them the best possible. This becomes increasingly complex when a number of products, facilities, markets and companies are involved. More than just real time information is needed. Connectivity, the ability to coordinate and integrate, and an appropriate implementation strategy are all important. Information systems provide the infrastructure to carry out these objectives. Because of the need for an effective information infrastructure, the Internet has the potential to further enhance information-based manufacturing. Information-based manufacturing can be efficient only when the underlying supply-chain network is run efficiently, for the supply chain provides the infrastructure for directing all the activities from receiving raw material to the delivery of final products. Shaw (2000) illustrates how Web technology can help coordinate the supply-chain activities in manufacturing. He also illustrates the relationships between product types, supply-chain structures, information sharing, coordination, and the world-wide web.

Both globalization and innovation in information technology (IT) motivate the adoption of the network business model as the fundamental organizational structure of the new economy. According to Fulkerson (2000) the performance of network technology took a quantum leap in the early 1990s due to the convergence of three trends:

- digitalization of the telecommunications network,
- development of broadband transmission, and
- Increased performance of distributed computation that paralleled breakthroughs in microelectronics, operating systems, and programming languages.

These innovations enabled the development of fully interactive, computer-based, flexible processes of management, production, and distribution that involved simultaneous cooperation among internal divisions as well as external firms. The information technology advisory service The Gartner Group Inc. states that "IT is now the global language of business" (Pucciarelli et al., 1999).

IV. POSSIBLE AVENUES FOR FURTHER RESEARCH

It has been observed that the capacity to be flexible can be used *reactively* as an insurance policy to guard against shortening life cycles (Gaimon and Singhal, 1992), or *proactively* to gain competitive advantage (Zhao and Steier, 1993). It follows then, that in order to operate in a reactive as well as proactive environment, industries must adequately equip themselves with flexibility, for meeting market demands and to claim heavy stakes in competition. But empirical results indicate that employing an FMS in the industries did not 'magically' raise their efficiency or their profits. In fact in some instances, companies even indicated financial losses. It is obvious then that the acquisition of flexible technology does not necessarily guarantee success, and therefore to fully exploit the potential of manufacturing flexibility its acquisition must be elevated in the decision making hierarchy from the operational to the strategic level.

Although we possess an extremely powerful competitive weapon in the form of manufacturing flexibility, unfortunately we haven't been able to harness its complete potential. Many challenges remain which need to be addressed in order to render it useful. Specifically, we need answers to the following questions:

- Why are we still defining flexibility?
- Are the purported theories on manufacturing flexibility practically applicable?
- How do we relate flexibility to the return on investment?
- How do we relate the flexibility metrics / dimensions / measures / types / elements etc.?

Proceeding along these lines, several avenues open up for future research. We need to define the interrelationships between the various flexibility variables. For this purpose we need to understand the nature of the tradeoffs that may exist between the different dimensions of flexibility. A framework needs to be developed using the conceptual building blocks, each illustrating a particular aspect of manufacturing flexibility, which would provide a superstructure which could be viewed as a consolidation of contemporary manufacturing flexibility research, and hence a means of identifying research priorities.

The need of the day is not just the appreciation of the existence of a variety of flexibility types but also the existence of their inter-relationships and trade-offs. While referring to the required level of a particular flexibility type, the non-monetary costs of attaining this flexibility should also be comprehended. These costs could result in a decrease in other flexibility types, which in turn could adversely affect production, service or market objectives. Understanding the relationships among the flexibility types is paramount for developing an understanding of the managerial task required to manage enterprise flexibility.

Another important factor that needs to be addressed is the possible identification of relationships between flexibility vis-a-vis growth and financial performance. Fiegenbaum and Karnani (1991) found no direct relationship between volume flexibility and financial performance. Interestingly they found that for small size companies, output flexibility was positively related with performance. This multi-faceted behavior of flexibility can only be understood by a detailed evaluation of the relationships between flexibility and performance.

Still another important future research avenue is the development of a method for establishing a priori the appropriate flexibility type and development of flexibility measures that can span diverse industry groups. De Toni and Tonchia (1998) note that, notwithstanding the importance and constant interest raised by flexibility in academic and managerial circles, the measurement of flexibility is still in a stage of infancy.

In conclusion, we need to focus on the aforementioned research areas in order that the potential benefits of manufacturing flexibility be harnessed fully. In these demanding times of intense customer driven markets and turbulent manufacturing environments, new problems will continue to arise, and only a determined and focussed approach at resolving such issues will bear fruit.

V. REFERENCES

[1]   Aaker DA, Mascarenhas B. The need for strategic flexibility. Journal of Business Strategy, 1984; 5(2): 74-82.

[2]   Adler PS. Managing flexible automation. California Management Review, 1988; 30(3): 34-56.

[3]   Barad M, Sipper D. Flexibility in manufacturing systems: definitions and petri net modeling. International Journal of Production Research, 1988; 26(2): 237-248.

[4]   Beach R, Muhlemann AP, Price DHR, Paterson A, Sharp JA. A review of manufacturing flexibility. European Journal of Operational Research, 2000; 122: 41-57.

[5]   Bolwijn PT, Kumpe T. Manufacturing in the 1990's: productivity, flexibility and innovation. Long Range Planning, 1990; 23(4), 44-57.

[6]   Brill PH, Mandelbaum M. On measures of flexibility in manufacturing systems. International Journal of Production Research, 1989; 27(5): 747-756.

[7]   Browne J, Dubois D, Rathmill K, Sethi SP, Stecke KE, "Classification of Flexible Manufacturing Systems", The FMS Magazine, 114-117, 1984.

[8]   Chandra P, Tombak MM. Model for the evaluation of routing and machine flexibility. European Journal of Operations Research, 1992; 60: 156-165.

[9]   Correa HL, "Linking Flexibility, Uncertainty and Variability in Manufacturing Systems", Ashgate Publishing Co, Aldershot, England, 1994.

[10] De Leeuw A, Volberda HW. On the concept of flexibility: a dual control perspective. Omega, 1996; 24(2): 121-39.

[11] De Meyer A, Nakane J, Ferdows K. Flexibility: the next competitive battle-the manufacturing futures survey. Strategic Management Journal, 1989; 10, 135-144.

[12] De Toni A, Tonchia S. Manufacturing flexibility - a literature review. International Journal of Production Research, 1998; 36(6): 1587-1617.

[13] Dixon R. Measuring manufacturing flexibility: an empirical investigation. European Journal of Operational Research, 1992; 60: 131-143.

[14] Dooner M, De Silva A, "Conceptual Modeling to Evaluate the Flexibility Characteristics of Manufacturing Cell Design", In Proceedings of the 28th Matador Conference, UMIST, Manchester, United Kingdom, 1990.

[15] Eppink DJ. Planning for strategic flexibility. Long Range Planning,, 1978; 11: 9-15.

[16] Evans JS. Strategic flexibility for high technology maneuvers: a conceptual framework. Journal of Management Studies,; 1991; 28(1): 69-89.

[17] Fiegenbaum A, Karnani A. Output flexibility - a competitive advantage for small firms. Strategic Management Journal, 1991; 12(2): 101-114.

[18] Frazelle E. Flexibility: a strategic response in changing times. Industrial Engineering, 1986; 16-20.

[19] Fulkerson W. Information based manufacturing in the informational age. International Journal of Flexible Manufacturing Systems, 2000; 12(2): 131-143.

[20] Gaimon C, Singhal V. Flexibility and the choice of manufacturing facilities under short product life cycles. European Journal of Operations Research, 1992; 60: 211 – 223.

[21] Gerwin D. An agenda for research on the flexibility of manufacturing processes. International Journal of Operations Production Management, 1987; 7: 38-49.

[22] Gerwin D. Manufacturing flexibility: a strategic perspective. Management Science, 1993; 39(4): 395-410.

[23] Golden W, Powell P. Towards a definition of flexibility: in search of the holy grail?. Omega, 2000; 28: 373-384.

[24] Gupta D, Buzacott JA. A framework for understanding flexibility of manufacturing systems. Journal of Manufacturing Systems, 1989; 8(2): 89-97.

[25] Gupta Y, Goyal S. Flexibility of manufacturing systems: concepts and measurements. European Journal of Operations Research, 1990; 43: 119-135.

[26] Gupta Y, Gupta M. Flexibility and availability of flexible manufacturing systems: an information theory approach. Comput Ind, 1991; 17: 391-406.

[27] Gupta Y, Somers T. Business Strategy, Manufacturing flexibility and organizational performance relationships: a path analysis approach. Production Operations Management, 1996; 5: 204-233.

[28] Hutchinson GK, Sinha D. Qualification of the value of flexibility. Journal of Manufacturing Systems, 1989; 8(1): 47-56.

[29] Kochikar VP, Narendran TT. A framework for assessing the flexibility of manufacturing systems. International Journal of Production Research, 1992; 30(12): 2873-2895.

[30] Koste LL, Malhotra MK. A theoretical framework for analyzing the dimensions of manufacturing flexibility. Journal of Operations Management, 1999; 18: 75-93.

[31] Kulatilaka N. Valuing the flexibility of flexible manufacturing systems. IEEE Transactions on Engineering Management, 1988; 35: 250-257.

[32] Kumar V. Entropic measures of manufacturing flexibility. International Journal of Production Research, 1987; 25(7): 957-966.

[33] Maffei MJ, Meredith J. Infrastructure and flexible manufacturing technology: theory development. Journal of Operations Management, 1995; 13(4): 273-298.

[34] Malek LA, Wolf C. Evaluating flexibility of alternative fms design-a comparative measure. International Journal of Production Economics, 1991; 23(1): 3-10.

[35] Mandelbaum M, "Flexibility in Decision Making: An Exploration and Unification", Ph D thesis, Department of Industrial Engineering, University of Toronto, Canada, 1978.

[36] Mandelbaum M, Buzacott J. Flexibility and decision making. European Journal of Operations Research, 1990; 44: 17-27.

[37] Parker RP, Wirth A. manufacturing flexibility: measures and relationships. European Journal of Operations Research, 1999; 118: 429 – 449.

[38] Pucciarelli J, Claps C, Morello DT, Magee F. IT management scenario: navigating uncertainty. Gartner Group Inc. Gartner Analytics Service Report, R-08-6153, 1999.

[39] Sethi A, Sethi S. Flexibility in manufacturing, a survey. International Journal of Flexible Manufacturing Systems, 1990; 2: 289-328.

[40] Shaw MJ. Information - based manufacturing with the web. International Journal of Flexible Manufacturing Systems, 2000; 12(2): 115-129.

[41] Slack N. Flexibility as a manufacturing objective. International Journal of Operations and Production Management, 1983; 3(3): 5-13.

[42] Slack N. The Flexibility of manufacturing systems. International Journal of Operations and Production Management, 1987; 7(4): 35-45.

[43] Stecke KE, Raman N, "Production Flexibilities and Their Impact on Manufacturing Strategy", WP# 484, Graduate School of Business Administration, University of Michigan, 1986.

[44] Suarez FF, Cusumano MA, Fine, CH. An empirical study of flexibility in manufacturing. Sloan Management Review, 1995; 37(1): 25-32.

[45] Swamidass PM. Manufacturing flexibility, Monograph No 2. Operations Management Association, Waco, TX, 1988.

[46] Upton DM. The management of manufacturing flexibility. California Management Review, 1994; 36(2): 72-89.

[47] Upton DM. What really makes factories flexible? Harvard Business Review, 1995; 73(4): 74-84.

[48] Vokurka RJ, O'Leary-Kelly SW. A review of empirical research on manufacturing flexibility. Journal of Operations Management, 2000; 18: 485-501.

[49] Watts C, Hahn C, Sohn B. Manufacturing flexibility: concept and measurement. Operations Management Review, 1993; 9(4): 33-44.

[50] Wharton TJ, White EM. Flexibility and automation: patterns of evaluation. Operations Management Review, 1988; 6(3): 1-8,

[51] Yao DD. Material and information flows in flexible manufacturing systems. Material Flow, 1985; 3: 143-149,

[52] Zelenovic DM. Flexibility - a condition for effective production systems. International Journal of Production Research, 1982; 20(3): 319-337.

[53] Zhao B, Steier, F. Effective CIM implementation using socio-technical principles. Computer Integrated Manufacture, 1993; 27-29.

Quality Management and Lean Manufacturing Integrated System

[1]Lavlesh Kumar Sharma, [2]Ravi Mohan Saxena
[1]Student of M. Tech.,(C.I.M.), S.A.T.I., Vidisha (M.P.) India
[2]Lecturer in Department of Mechanical Engg., S.A.T.I., Vidisha (M.P.) India

Abstract— Now-a-days, in such a competitive environment, the level of customer expectations is increasing. The sustenance of an industry basically depends on the processes and applications. Many industries initially more focus on the ways to increase production volume yet quality all that matters; it is one of the most important criteria in ever increasing customer expectations. For keeping the cost low with quality is a great challenge for any kind of manufacturing or service industry. Lean manufacturing system (LMS) and Quality management system (QMS) i.e. ISO 9000 make the business more effective. This paper presents the ways to improve quality with having standard quality system and to reduce wastes through lean. In this paper there is a proactive integrated model.

Keywords— Quality Management (ISO 9000), Process Failure Mode and effect Analysis, Lean manufacturing, Value stream manufacturing (VSM), Kanban, Kaizen, TRIZ.

I. INTRODUCTION

ISO, the International Organization for Standardization, was founded in the year 1947 for developing technical standards. The International Organization for Standardization (ISO) created the Quality Management System (QMS) standards in 1987. They were the ISO 9000:1987 series of standards comprising ISO 9001:1987, ISO 9002:1987 and ISO 9003:1987; which were applicable in different types of industries, based on the type of activity or process: designing, production or service delivery. The standards are reviewed every few years by the ISO. The version in 1994 was called the ISO 9000:1994 series; consisting of the ISO 9001:1994, 9002:1994 and 9003:1994 versions. In 2005 ISO 22000, meant for food industries. The last major revision was in the year 2008 and the series was called ISO 9000:2000 series. The ISO 9002 and 9003 standards were integrated into one single certifiable standard: ISO 9001:2008. After December 2003, organizations holding ISO 9002 or 9003 standards had to complete a transition to the new standard. ISO released a minor revision, ISO 9001:2008 on 14 October 2008. It contains no new requirements. Many of the changes were to improve consistency in

Grammar, facilitating translation of the standard into other languages for use by over 950,000 certified organisations in the 175 countries (as at Dec 2007) that use the standard.

Womack, Roos, and Jones (1990) studied the implementation of lean manufacturing practices in the automotive industry. Other research efforts, such as the Lean Aerospace Initiative (2002), have looked at the unique challenges of implementing lean manufacturing practices in a specific type of industry. The larger research project for which this survey development effort was initiated had a similar focus-- understanding the implementation of lean manufacturing practices within the many kinds of manufacturing industrial sector. Lean manufacturing systems (LMS) come in existence in Japan after World War II. Japanese companies found to rebuild with limited resources such as man, material, machine, method, money. Japanese developed lower cost manufacturing practices by elimination of wastes *(muda, mura, muri & TIMWOOD)* and improve lead time.

An integration of Quality Management System (QMS) and Lean Manufacturing System (LMS) in the modern manufacturing industries is defined as the blending of process, application, and data with resources utilization Integration provides information, feedback and functionality of whole working environment. Quality is the main ingredient to customer satisfaction and Lean focuses on waste reduction and efficiency enhancement.

II. LITERATURE SURVEY

Briefly, it is called lean as it uses less, or the minimum, of everything required to produce a product or perform a service (Hayes and Pisano, 1994). David Holey (2000) mentions about ISO that for providing few concepts of paper, work, bureaucratic procedures, form non value added activities. These are codifying principles that have been established by businesses for generation. Liker and Wu (2000) defined it as a philosophy of manufacturing that focuses on delivering the highest quality product on time and at the lowest cost. Bill Graw (2001) states that now-a-days companies feel the need to right size, become learner, more agile, time based, faster, stronger, cheaper and better than their competitors. Incremental improvements, even gradual, continuous improvement to the existing order, won't bring most organizations to pre-eminent position in their areas of endeavour. Poppendieck (2002), Principles of lean thinking have been broadly accepted by many manufacturing operations and have been applied successfully across many disciplines. Javier Freire (2002) marked that an improvement methodology is proposed for the design process in construction projects. Based on concepts and principles of lean production, the methodology considers the design process as a set of three different models- conversion, flow, and value. According to the authors, four stages are necessary to produce improvements and changes   (1) Analysis /evaluation, (2) Changes implementation, (3) Feedback/control, (4) Standardization. Results of an application included an increase of 31% in the share of value adding activities, 44% reduction of unit errors in the products, up to 58% decrease of waiting times in the process, and an expansion of the utilization in the cycle times. Kristin Marshall (2002) had undertaken a study on Quality and Lean in which briefly discussed the backgrounds and contents of the ISO standards and lean initiatives. In his findings there are certain characteristics of both tools that would leave on believing that conflict would arise that could lead to discrepancies in quality audits or slowness in the implementation of lean driven change. One potential conflict of considerable is how the quick changes resulting from kaizen events in lean would exist with the traditional documentation and heavy requirements associated with ISO certification. Angappa (2003) asserts that productivity and quality are integral components of organization operational strategies. Productivity continues to play an important role both at macro and micro levels. At micro level, firms continue to use productivities as a performance measure to bench marks against the best in class companies with the objectives of identifying best practices. Quality management has become an important part of culture, particularly in new enterprises which are characterized by supply chain, e-commerce and virtual enterprise environments. Stanley (2003) state that companies can eliminate wide variation in quality program results by fusing strategic, tactical and operational elements of previously proposed quality disciplines. The synergies evolve to drive success. Worley (2004) defined it as the systematic removal of waste by all members of the organization from all areas of the value stream. Dennis (2005) in the International Journal of Productivity and Performance Management states that the viewing of Lean Implementation across the entire enterprise minimizes the possibility of overlooking opportunities for the further performance improvement. Lean principles can improve performance through quality on time delivery and customer satisfaction. ISO 2000 emphasizes that customer satisfaction is essential for any business. Working to recognizes quality management system can help to meet customer expectation. QMS provide a frame work for its process and activities. They can help a business to improve its efficiency by providing a best practice model for it to follow. Shao (2006) claims that to prosper in today's fiercely competitive global marketplace, enterprises especially small and medium sized ones, we must strive to provide products with shorter time to market, lower cost, higher quality, and better customer satisfaction. Quality management plays a vital role in achieving these goals. Terlaak (2006) in his article indicated that a 11 year panel of US manufacturing facilities was assigned to test whether certification with the ISO 9000 Quality management standard generate a competitive advantage. Results suggest that certified facilities grow faster after certification, and those operational improvements do not account for this growth. Results also indicate that the growth effect is greater when buyers have greater difficulty acquiring information about suppliers. Shah and Ward (2007), Lean manufacturing is most frequently associated with the elimination of seven important wastes to ameliorate the effects of variability in supply, processing time or demand. As an integrative concept, the adoption of lean manufacturing can be characterized by a collective set of key areas or factors. These key areas encompass a broad array of practices which are believed to be critical for its implementation. They are, *scheduling, inventory, material handling, equipment, work processes, quality, employees, layout, suppliers, customers, safety and ergonomics, product design, management and culture,* and *tools and techniques* (Wong et al., 2009).

III. TRADITIONAL APPROACH

The implementation of ISO may be done after the commencement of the production because the quality of product should be maintained as per our requirement during the whole production. Quality management is intended to investigate causes for problems; those may result in failure of the business due to lower customer satisfaction level. There are few findings those can reveal the lack of robustness in process measures. The findings are: i) Poor housekeeping, ii) Poor communication due to absence of functional team, iii) Imperfections in methodology, iv) less preventive and corrective approach, v) Unavailability of continuous improvement, vi) not clear parts identification in some cases etc.

Traditionally the resource utilization (People, Machine, Finance, material, process) during production is meant for bringing out only the desired production volume, sometimes with having undesirable wastes. Thus lean has changed the definition of resource utilization by eliminating wastes. This philosophy of reduction of wastes may be obtained by various techniques such as Kaizen, Kanban (Three bin systems), VSM, and TRIZ. Value stream mapping technique is one of the lean systems used to analyze the flow of material and currently required to bring a product or service to a consumer. Lean results an improvement in lead time.

IV. PROACTIVE APPROACH

System development:

A system will be developed to make the production standardize. Procedures shall be developed as per ISO 9000 in the basic frame work of systems in projects.QMS will deal with all the concerns of documents and preventive measures as Failure mode and effect analysis (FMEA).

Cross Functional Team Formation:

Cross functional team will be structured before the commencement of the production. This team will handle all the functions within and outside the department and communicate each team.

Integrated approach:

In order to maintain the quality as well as elimination of wastes in such a competitive market will be developed, procedures and supporting systems shall be developed for keeping lean concepts and quality.

In proactive approach VSM provides the value adding steps be drawn across the centre of the map and non-value adding steps be represented in vertical lines at right angles to the value stream by a cross functional team as Shigeo suggests. For elimination of wastes model provides the capability to recognize Muda, Mura, Muri; model also considers TIMWOOD seven wastes introduced by Toyota's chief Engineer Taiichi Ohno with Shigeo and Shingo as part of Toyota production System (TPS). For betterment and time reduction Kanban also accounts in this proactive approach; this scheduling system is a means to fine tuning and through the chain of customer-store processes controls the rate of production. TRIZ (Teoriya Resheniya Izobretatelskikh Zadatch) is also important concept for functional team in problem solving, some analysis and forecasting through SIT (Systematic Inventive Thinking), USIT (Unified Systematic Inventive Thinking) and algorithm of inventive problem solving. TRIZ process for creative problem solving may be shown as in figure 1.

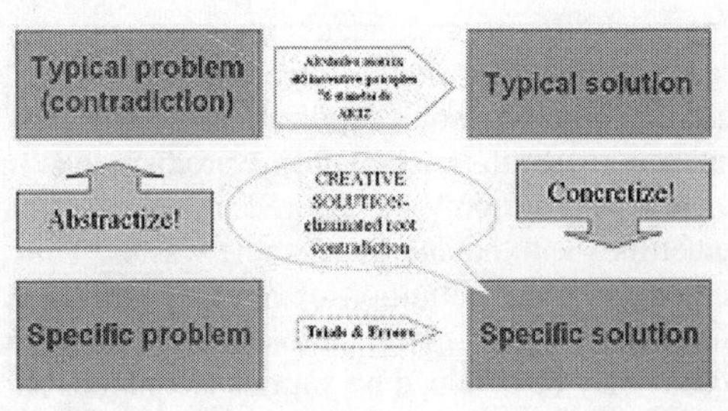

Figure 1.

For continuous improvement Kaizen is required which aims to eliminate waste by means of PDCA cycle with it five elements; Teamwork, Personal discipline, Quality circles, suggestions for improvement. The model identifies the linkage between lean and quality during PDCA cycle as depicted in figure2.

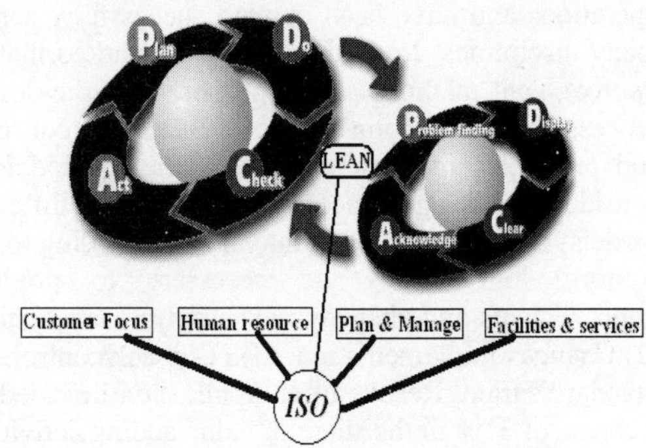

Figure 2

IV. QMS & LNS INTEGRATION SYSTEM

The organizational operations to run the business properly, a set of concepts are needed in the basic platforms of the managerial system; hence ISO is required to deal with. Initially every process starts with inputs followed by blending of inputs to outputs. Lean concepts require recognizing the target product, product family or service and breakdown the process into various stages for identifying opportunities for elimination or reduction of wastes with effective utilization of resources. The model identifies the possible linkage between two concepts as shown in figure3.

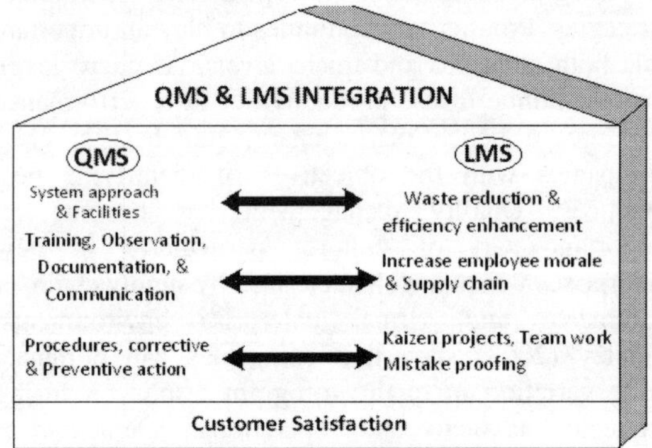

Figure 3

V. CONCLUSION

It has been observed that proactive approach of QMS & LMS model can assist the organization effectively in various manner like customer satisfaction, employee and organizational satisfaction, motivation, clear accountability, continuous improvement. The implantation of QMS helps to provide training and education to employee right from initial stages of production and services. It also helps to define responsibility and assigned role very clearly, and avoiding unnecessary confusions duplication of effort, and also focuses on well utilization of 5M. Simultaneously Lean puts the high morale of employees at all levels of system in terms of procedures, roles and responsibilities like 5S, mistake proofing. Integration of Quality Management and Lean Manufacturing enables to build up a high level of benefits include job satisfaction, control of rejections, elimination of wastes and customer satisfaction. Integration model envisages concerted efforts and eventually accelerates the yield of benefits.

VI. REFERENCES

[1]	*Learning to See: value-stream mapping to create value and eliminate muda by Mike Rother and John Shook.*

[2]	*Shongo, Shigeo A Revolution in Manufacturing: The SMED System, , Productivity Press, 1985, p5*

[3]	*Ohno, Taiichi (1988). Toyota Production System: Beyond Large-Scale Production. Productivity Press. Pp 25-29*

[4]	*Shingō, Shigeo (1989). A Study of the Toyota Production System from an Industrial Engineering Viewpoint. Productivity Press. pp. 30, 228*

[5]	*Hua, Z.; Yang, J., Coulibaly, S. and Zhang, B. (2006). "Integration TRIZ with problem-solving tools: a literature review from 1995 to 2006". International Journal of Business Innovation and Research 1 (1-2): 111–128.*

[6]	*S. Gajendra & Sampath Kumar, December 2009 International Journal of Manufacturing Science and Technology.*

[7]	*Altshuller, G. S.; Shapiro, R. B. (1956). "О Психологии изобретательского творчества (On the psychology of inventive creation)" (in Russian). Вопросы Психологии (The Psychological Issues) (6): 37–39.*

[8]	*Imai, Masaaki (1986). Kaizen: The Key to Japan's Competitive Success. New York, NY, US: Random House.*

[9]	*Rose, Kenneth H. (July, 2005). Project Quality Management: Why, What and How. Fort Lauderdale, Florida: J. Ross Publishing. p. 41*

[10]	*Paul H. Selden (December 1998). "Sales Process Engineering: An Emerging Quality Application". Quality Progress: 59–63*

[11]	*Cianfrani, Charles A.; West, John E. (2009). Cracking the Case of ISO 9001:2008 for Service: A Simple Guide to Implementing Quality Management to Service Organizations (2nd ed.). Milwaukee: American Society for Quality. pp. 5*

Current research trends in Electric Discharge machining: A Review

Rajeev Kumar

Assistant Professor Department of Mechanical Engineering Greater Noida Institute of Technology
Greater Noida

ABSTRACT

Electrical discharge machining (EDM) is a process for machining hard metals and forming deep complex shaped holes by arc erosion in all kinds of electro-conductive materials. In recent years, incessant research in material science has encouraged the engineering and development of advanced ceramic materials. Such materials satisfy the needs of high end applications in the areas of aerospace, automotive, defense, biological, surgical components and nuclear fields. Majority of work concentrates on improvement in process efficiency, optimization of process variables and process monitoring and control. This paper reviews the research trends in EDM on die sinking EDM, Rotating pin electrode(RPE), wire electric discharge machining (WEDM), dry EDM, rotary disc electrode discharge machining (RDE-EDM), micro-EDM ,EDM in water, EDM on ultrasonic vibration, EDM with powder additives and modeling technique in predicting EDM performances.

Keywords: EDM; Dry EDM, Modeling, Rotary pin electrode (RPE-EDM), Micro-EDM.

1. INTRODUCTION

Electrical Discharge Machining (EDM) is one such process which is widely used to machine electrically conductive materials. EDM is a thermo-electric process in which material removal takes place through the process of controlled spark generation. It is one of the most popular non-traditional machining processes being used today in the industry. EDM is commonly used in mould and die making industry and in manufacturing automotive, aerospace and surgical components. Since there is no mechanical contact between the tool and the workpiece, thin and fragile components can be machined without the risk of damage. This technique has been developed in the late 1940_s [1] where the process is based on removing small amount of material from a part by means of a series of repeated electrical discharge between tool called electrodes and the work piece in the presence of dielectric field [2]. A gap is maintained between tool & the work piece. The electrode is moved towards the work piece till the gap is small enough so that the applied voltage is so high to ionize the dielectric [3]. The duration of each spark is very short. The frequency of sparking may be as high as thousands of spark per second [4]. The area over which the spark is effective is also very small. However, temperature of the area under the spark is very high resulting in partly melting & partly vaporization of the work piece from localized area on both the electrodes, i.e. work piece & electrode. The material is removed in the form of craters which spreads over the entire surface of work piece. Finally, the cavity produced in the work piece is approximately the replica of the tool. In a paper published in 2010, Singh & Anand [5] have indicated

some of the variants of EDM: powder mixed EDM, machining advanced engineering materials, ultrasonic assisted EDM, WEDM, rotary EDM, dry EDM,micro-EDM performance and automation and control.

1.1 Overview of EDM

Electro discharge machining (EDM) is an electro-thermo non-traditional machining process, where electrical energy is used to generate electrical spark & material removal mainly occurs due to thermal energy of spark [6]. EDM is mainly used to machine difficult –to- machine materials and high strength temperature resistant (HSTR) alloys. EDM can be used to machine difficult geometries in small batches or even on job-shop basis. Work materials to be machined by EDM have to be electrically conductive.EDM doesn't make direct contact between the electrode & the work piece where it can eliminate chatter & vibration difficulties during machining [5]. The basic EDM can be traced as far back as 1770, when English chemist Joseph Priestly discovered the erosive effect of electrical discharge or sparks [6]. However it was only in 1943 at Moscow University where Boris & Lazarenko [7] exploited the destructive properties of electric discharge for constructive use. They were assigned by soviet Government to investigate the wear caused by sparking between tungsten electrical contacts, a problem which was particularly critical for maintenance of automotive engines during Second World War. Putting the electrodes in oil, researchers found that the sparks were more uniform & predictable than in air. They had then idea to reverse the phenomenon, & to use controlled sparking as an erosion method [8]. Lazarenko EDM system used resistance-capacitance type power supply, which was widely used EDM in 1950_s & latter served as the model for successive development in EDM [9]. In 1960_s, the development of the semi-conductor industry permitted considerable improvements in EDM machines [10].

There have been similar claims made at about the same time when three Americans employees came up with the notation of using electrical charges to remove broken taps & drills from hydraulic valves. During the following decades, efforts were principally made in generator design, process automatization, servo- control and robotics. Application of EDM in micro-machining became of interest during 1980_s [10]. After that period the world market of EDM started to increase strongly, & specific EDM research took place over basic EDM research [11]. Finally, in 1990_s, the new methods of EDM process control had arised using artificial neural networks (ANN), Fuzzy logic, Taguchi methodology, and response surface methodology (RSM) etc.

2. ELECTRICAL DISCHARGE MACHINING (EDM)

Basic fundamental of EDM process & process parameters and other material removal methods have been explained in this section.

2.1 Process

In EDM, a potential difference is applied between the tool and work piece. Both the tool & work piece are to be conductors of electricity. The tool & work material are immersed in a dielectric medium. Generally kerosene or demonized water is used as the dielectric medium. A gap is maintained between tool & the work piece. Depending upon the applied potential difference & the gap between the tool & work piece, an electric field is established. Generally the tool is connected to the negative terminal of the generator (G) & the work piece is connected to the positive terminal. As the electric field is established between the tool & the job, the free electrons on the tool are subjected to the electrostatic forces. If the bonding energy of the electrons is less, electrons would be emitted from the tool. Such emission of electrons is called as cold emission. The "cold emitted" electrons are then accelerated towards the job through the dielectric medium. As they gain velocity & energy, & start moving towards the job, there would be collision between the electrons & dielectric molecules. Such collision may result in ionization of dielectric molecules depending upon ionization energy of dielectric molecules & the energy of the electrons. Thus as electrons get accelerated, more positive ions & electrons would get generated due to collisions. This cyclic process would increase the concentration of electrons & ions ions in the dielectric medium between the tool and work piece at the spark gap. The concentration would be so high that the matter existing in that channel

could be characterized as "plasma". The electron resistance of such plasma channel would be very less. Thus all of a sudden, a large number of electrons will flow from the tool to the job & ions from the job to the tool. This is called avalanche motion of electrons. Such movement of electrons & ions can be visually seen as a spark. Thus electrical energy is dissipated as the thermal energy of the spark. The high speed of electrons then impinges on the job & ions on the tool. The kinetic energy of the electrons & ions on impact with the surface of the job (work piece) & tool respectively would be converted into thermal energy. Such intensed heat flux leads to extreme instantaneous confined rise in temperature which would be in excess of 10, 0000 C [12]. Such localized extreme rise in temperature leads to material removal. Material removal occurs due to instant vaporization of the material as well as due to melting. The molten metal is not removed completely but only partially. As the potential difference is withdrawn as shown in figure (1), the plasma channel is no longer sustained. As the plasma channel collapsed, it generates pressure or shock waves, which evacuates the molten material forming a crater of removed material around the site of spark. Thus to summarize, the material removal in EDM mainly occurs due to formation of shock waves as the plasma channel collapse owing to discontinuation of applied potential difference.

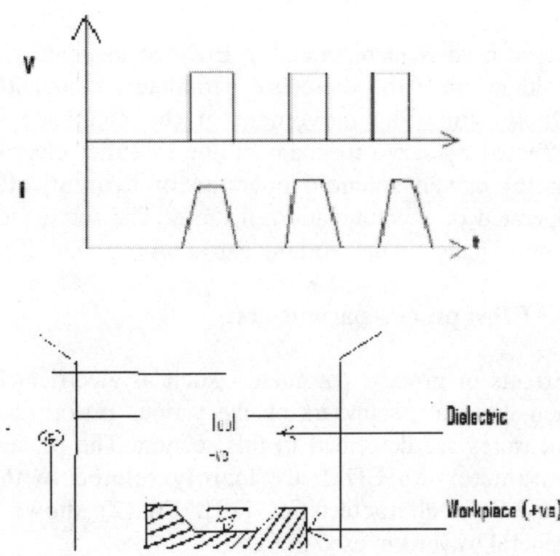

Figure (1): Schematic diagram of EDM

Generally the work piece is made positive & the tool negative. Hence, the electron strikes the job leading to crater formation due to high temperature & melting and material removal. Similarly, the positive ions impinge on the tool leading to the tool wear. In EDM, the generator is used to apply voltage pulses between the tool & the job. A constant voltage is not applied. Only sparking is desired in EDM rather than arcing. Arcing leads to the localized material removal at a particular point whereas sparks get distributed all over the tool surface leading to uniformly distributed material removal under the tool.

2.2 EDM Equipments

Fig. (2) Shows an EDM machine. EDM machine has the following major modules [13].

- Dielectric reservoir, pump & circulation System.
- Power generator & control unit
- Working tank with work holding device
- X-Y table accommodating the working table
- The tool holder
- The servo-system to feed the tool

2.2.1 Dielectric Unit:

The dielectric unit system consists of dielectric tank, a pumping unit & a filtering unit to handle types of dielectric. Dielectric used when considering the effect of alternating electric fields on the substance. Deionized water, kerosene, and mineral water are generally used as a dielectric fluid in EDM. A controlled electrical discharge occurs through a dielectric medium when sinker EDM is used. Dielectric is used as a quenching medium to cool & solidify the gaseous EDM debris resulting from the discharge [14].

2.2.2 Servo-system to feed the tool:

Servo head is incorporated in EDM to maintain gap voltage since the dielectric parameters constantly fluctuation. The movement of the electrode is affected by servo mechanism due to either electric motor driven solenoid operated or hydraulically operated or a combination of these. The servo feed control maintain the working gap at proper width [5].

2.3 EDM process parameters:

Effects of process parameters such as electrical & non-electrical parameters on the various performance measures are described in this section. The process parameters in EDM are mainly related to the waveform characteristics [12]. Fig.(2) shows a general waveform used in EDM.

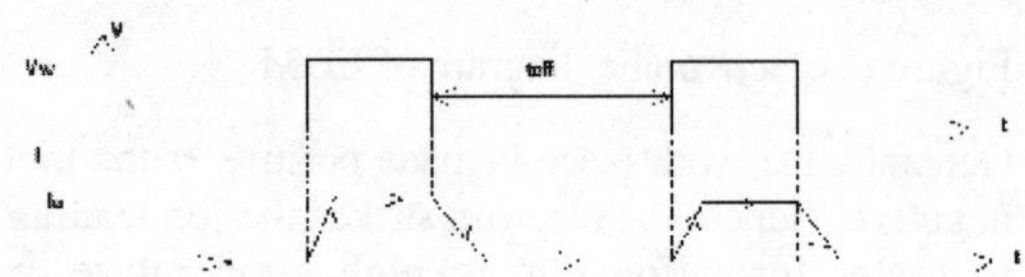

Fig. (2) Waveform used in EDM

The waveform is characterized by the

- The open circuit voltage-V_o
- The working voltage-V_w
- The maximum current- I_o
- The pulse on time – the duration for which the voltage pulse is applied
- The pulse-off time
- The gap between the work piece & tool- Spark gap (d)
- The polarity-straight-tool (-ve)
- The dielectric medium
- External flushing through the spark gap

2.4 Electrical parameters

This section discusses the process parameters of EDM. In this section focus has been given on the effects of process parameters such as electrical & non-electrical parameters on various performance measures.

2.4.1 Pulse Duration (t_{on})

Metal removal is directly proportional to the amount of energy applied during the pulse –on time $_{(ton)}$ [14]. The effect of the magnitude of the pulse-on duration on the surface texture of the specimen is more significant than the pulsed current [15]. During EDM, the primary parameters are pulsed current, pulsed voltage, pulse-on duration, and pulse-off duration. From the analysis carried out by Tung, Lee and Yur, in which the influence of these four parameters was Investigated for AISI 1045 carbon steel and H13 die-steel using the Taguchi method, it is known that pulsed current and pulse-on duration are the principal factors which influence surface roughness, thickness of the recast layer, and induced stress [15]. Experimentally It has been found that all that MRR can't be increased by increasing the Pulse on time, a suitable combination of peak current is also needed for increasing rate of removing unwanted material from the work piece. At constant current and constant duty factor, the MRR is decreased with increase in pulse on time [5].

2.4.2 Pulse Interval (t_{off})

It is defined as waiting time interval period during two pulses on time periods. Fig. (2), it is the duration of time in which no machining takes place and material becomes vaporize & removed from the machining zone [15]. The duration of time (µs) between the sparks (that is to say, on-time) . This time allows the molten material to solidify and to be wash out of the arc gap. This parameter is to affect the speed and the stability of the cut. Thus, if the off-time is too short, it will cause sparks to be unstable [16].

2.4.3 Electrode Gap (Spark gap)

The Arc gap is distance between the electrode and work piece during the process of EDM. It may be called as spark gap. Spark gap can be maintained by servo system. There are the two systems namely the hydraulic & the electromechanical are used to sense the average gap voltage. A suitable is required to be maintained to get good performance and gap stability. Gap width is not measured directly, but can be inferred from the average gap voltage [27].

2.4.4 Duty cycle (τ)

It is a percentage of the on-time relative to the total cycle time. This parameter is calculated by dividing the on-time by the total cycle time (on-time pulse off-time)

$$T = t_{on} / (t_{on} + t_{off})$$

2.4.5 Polarity

Polarity is defined as the type of potential given to the terminals.

It may be positive or negative connected to the tool electrode or work piece. Tool wear, processing speed, finish and stability of the EDM operation is affected by polarity. Experimentally it has been found that material removal rate (MRR) is more when the tool electrode is connected to the positive terminal than negative terminal. Positive polarity i.e., work piece '+ve' and tool '-ve' used during micro-EDM experimentation because tool wear is less in this case due to low sparking energy distribution the cathode, i.e., tool as compared to reverse polarity this helps in improving the micro-machining accuracy [18].

2.5 Non-Electrical parameters

Rotational movement of electrode, flushing of dielectric fluid and aspect ratio are some of the non-electrical parameters playing a significant role in delivering optimum performance measures. Various non-electrical parameters & their effects on performance measures have been discussed in this section.

2.5.1 Tool rotation

Tool electrode rotation is commonly used in small-hole EDM drilling operations. Tool rotation improves flushing and leads to a more uniform electrode wear. The effects of improved flushing are an increased MRR and lower Ra value. At the same time, process stability increases because tool rotation makes it easier to introduce fresh dielectric into discharge gap as the used up dielectric is thrown out due to the centrifugal force. Thus, even with low pulse off times and poor Flushing conditions good machining performance is obtained [19]. It has been found that the rotation speed plays a significant role in achieving the desired MRR by enhancing flushing. When speed is high, MRR increases irrelevant of feed rate and AR (aspect ratio) values.

2.5.2 Injection flushing

For efficient cutting & improved surface finish of machined surface, eroded particles must be removed from the gap by flushing. In addition to this, flushing also allows the fresh dielectric oil to be flowed into the gap & cools both the electrode & work piece. It has been found that the rotation speed plays a significant role in achieving the desired MRR by enhancing flushing [17].

The main characteristics which are required for dielectric used in EDM are high dielectric strength and quick recovery after break down [12]. Generally kerosene and deionized water is used as dielectric fluid in EDM. Tap water can't be used as it ionizes too early & thus breakdown due to presence of salts as an impurities occur [12]. Dielectric medium is also applied through the tool to achieve efficient removal of molten material. It was observed that the flushing has a considerable effect on MRR and TWR [19]. Further, the cooling rate of the tool increases with increase in the flushing pressure and hence reduced tool wear is observed [20].

2.5.3 Tool geometry

Tool geometry is concerned with the shape of the tool electrodes.ie. Square, rectangle, cylindrical, circular. etc.The ratio of length /diameter of any shaped feature of material. In case of rotating disk electrode the ratio becomes thickness/diameter. Murali et al [20] used graphite foil for straight grooving operation instead of pin shaped electrode. An aspect ratio of 2.3 was achieved by using FAST technique (Foil as tool electrode) which was improved to 8 by implementing GAME (Gravity assisted Micro EDM). Singh et al. [21] uses square and rectangular shaped electrodes having aspect ratio of 1.0 and 0.6 for machining 6061Al/Al2O3P composite .It concluded that shape of the electrode effects EWR. The tool having less aspect ratio gave higher value of EWR. Thus with increasing the size of electrode more good performance of ED Machining takes place.

2.5.4 Tool material

Electrode material should be such that it would not undergo much tool wear when it is impinged by positive ions. Thus the localized temperature rise has to be less by properly choosing its properties or even when temperature increases, there would be less melting. In addition to this, the tool should be easily workable as there is requirement of producing complex shapes in EDM [12]. Thus the basic characteristics of electrode materials are:

- High electrical conductivity
- High thermal conductivity
- Higher density
- High melting point
- Easy manufacturability
- Less expensive.

The following are the various electrode materials which are used commonly in the industry:

- Graphite
- Electrolytic oxygen free copper
- Tellurium copper-99% + 0.5% Tellurium
- Brass

Copper is comparatively a better electrode material as it gives better surface finish, low diameteral overcut, high MRR and fewer electrodes wear for En-31 work ma- trial, and aluminum is next to copper in performance, and may be preferred where surface finish is not the requirement. Of the four tested electrode materials, Cu and Al electrodes produce comparatively high surface roughness for the tested work material at high values of currents. Copper–tungsten electrode offers comparatively low values of surface roughness at high discharge currents giving good surface finish for tested work material [22].

2.5.5 Recast layer in EDM

Microscopic study of the machined components reveals three kinds of layers e.g. recast layer, heat affected zone (HAZ) and converted layer (fig. 3) [4]. If the molten from the work piece is not flushed out quickly, it will resolidify & harden due to cooling effect of dielectric & gets adhere to the machined surface. This thin layer is called recast layer. It is extremely hard & brittle. While machining the maraging steel with EDM process a recast layer was formed [23]. Due recast layer premature part failure shorter the life of part's fine surface finish.

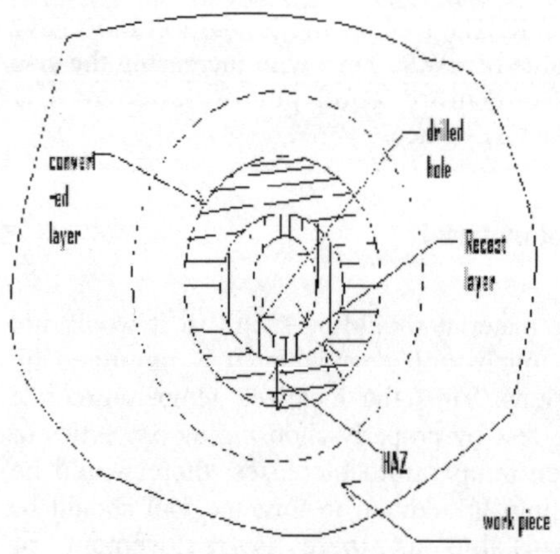

Fig. (3): Schematic diagram of three kinds of layers on an EDM'd component.

3.0 APPLICATION OF EDM

EDM can be employed to machine any material independent of their hardness, toughness; brittleness etc provided it has some minimum electrical conductivity. Hardened steel dies are manufactured by this process other than aerospace, automobile, tools, & machine tool components. It is also used for making through holes & miniature holes. It can be used for making dies for moulding, stamping, coining, forming, etc. it can also be employed for tiny holes (Say,-50 micron) and micro size slots.

4.0 RESEARCH AREAS OF EDM

In this section, the various research areas of EDM such as machining performance measures like material removal, tool wear and surface quality (SQ), the effects of process parameters including electrical and non-electrical variables, which are required to optimize the stochastic nature of the sparking process on the performance measures. Finally, research concerning the design and manufacture of electrodes is reported. Some of the research areas in EDM are shown in fig. (4).

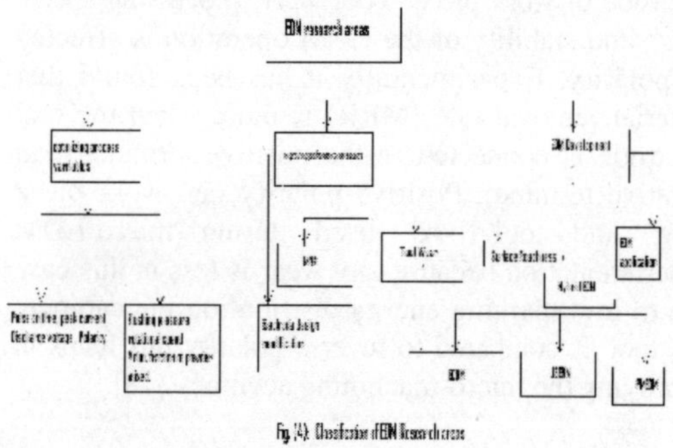

Fig. (4): Classification of EDM Research areas

5.0 TYPES OF EDM

5.1.1 Die sinking EDM

In die sinking process the tool electrode is the replica of the machined profile of the work material shown in fig (5.0). Die sinking process solves the problem of manufacturing accurate and complex-shaped electrodes of three-dimensional cavities. The highest material removal rates were achieved for die-sinking EDM of AISI P20 Tool Steel with negative graphite electrodes, much better than the results reached with copper tools [29]. Therefore graphite is more stable than copper when EDM as cathodes. A number of EDM machines can be used to machine smart materials or super alloys to manufacture aeronautical & aerospace parts. With the help of Wire-cut EDM (WEDM), Rotary disc electrode EDM (RDE-EDM) ,difficult to machine material, high strength and temperature resistance (HSTR) materials can be machined easily.

5.1.2 Wire EDM process

It was first introduced in late 1960_s. A special form of EDM that uses small diameter wire as electrode to cut a narrow kerf in work material. Wire diameter ranges from 0.076 to 0.3 mm. overcut is in the range of 0.02 to 0.05. Wire EDM Machining (also known as Spark EDM) is an electro thermal production process in which a thin single-strand metal wire (usually brass) in conjunction with de-ionized water (used to conduct electricity) allows the wire to cut through metal by the use of heat from electrical sparks. a thin single-strand metal wire, usually brass, is fed through the work piece, submerged in a tank of dielectric fluid, typically deionized water.

Wire-cut EDM is typically used to cut plates as thick as 300mm and to make punches, tools, and dies from hard metals that are difficult to machine with other methods [24]. Wire-cutting EDM is commonly used when low residual stresses are desired, because it does not require high cutting forces for removal of material. If the energy/power per pulse is relatively low (as in finishing operations), little change in the mechanical properties of a material is expected due to these low residual stresses, although material that hasn't been stress-relieved can distort in the machining process. Due to the inherent properties of the process, wire EDM can easily machine complex parts and precision components out of hard conductive materials. The WEDM process makes use of electrical energy generating a channel of plasma between the cathode and anode [25], and turns it into thermal energy at a temperature in the range of 8000–12,000 °C [26]. When the pulsating direct current power supply occurring between 20,000 and 30,000 Hz [27] is turned off, the plasma channel breaks down.

5.1.3 Micro EDM

When product manufacturing is in the range of 1 to 999 micron, then it is known as micro-manufacturing as per CIRP committee of physical & chemical processes [28-29]. In the very begning the term was used to present the miniaturization of electronic products & devices. However, with the present trend in the miniaturization of mechanical components, it is also being applied for the generation of microscopic mechanical components and devices [30].When machining micro-holes in carbide with copper tool electrode; positive-polarity machining must be used: this differs from the negative-polarity machining used commonly in normal EDM. Increasing the rotational speed of the tool electrode can improve the debris discharge and reduce the expansion of the micro-hole, but it has no great effect on the tool electrode wear rate.

A tool electrode with a notch can improve the debris discharge. At high rotation speed, it is appropriate to use greater width for the notch of a single-side notch electrode. Putting the electrode horizontally tends to cause concentration of the debris discharge, resulting in micro-holes of an elliptical shape. Furthermore, the mechanism can be improved by increasing the rotational speed [31]. Metal removal rate and tool-wear rate are found to increase monotically with the increase in peak current due to higher discharge energy at higher value of Ip .Also they (MRR and TWR) are found to increase when Ton increases from 1 to 10 jts but with further increase in Ton, MRR is observed to decrease. Flushing pressure and duty factor have no significant effect on both MRR and TWR. In addition to this Overcut of the machined micro-hole is mostly affected by the peak current and on-time and increased with increase in Ip and Ton. Also, like MRR, taperness of the machined hole is influenced [32]. Finally, through hole machining tests of Al2O3 ceramics, the difficulty of small hole machining has been identified, and the prerequisites of stable process have been given.

5.1.4 Dry EDM

Dry Electric Discharge machining (dry EDM) is a modification of the oil EDM process in which the liquid dielectric is replaced by a gaseous dielectric. High velocity gas flowing through the tool electrode into the inter-electrode gap substitutes the liquid dielectric. The flow of high velocity gas into the gap facilitates removal of debris and prevents excessive heating of the tool and work piece at the discharge spots. Providing rotation or planetary motion to the tool has been found to be essential for maintaining the stability of the dry EDM process. The dry EDM process schematic is shown in Figure 1.3. Tubular tools are used and as the tool rotates, high velocity gas is supplied through it into the discharge gap. Gas in the gap plays the role of the dielectric medium required for electric discharge. Also, continuous flow of fresh gas into the gap forces debris particles away from the gap. From experiments, it was found that discharge current, duty factor, air pressure and spindle speed are the significant factors which affect MRR and MRR increases with an increase in any of these factors. For MRR, most significant two-factor interaction effects are present among current and duty factor, current and spindle speed and pulse-on time and duty factor. Dry EDM uses gas as dielectric fluid, and high MRR can be obtained to cut high strength engineering materials with the presence of oxygen [33–34]. Later on Kunieda et al. [35] Confirm that by introducing oxygen gas into the discharge gap increases the material removal rate in water as a dielectric medium. Tao et al. [36] concluded that by combining oxygen gas with copper tool gives high MRR in dry EDM. Progress in Dry EDM research is shown in figure (5a &5b).

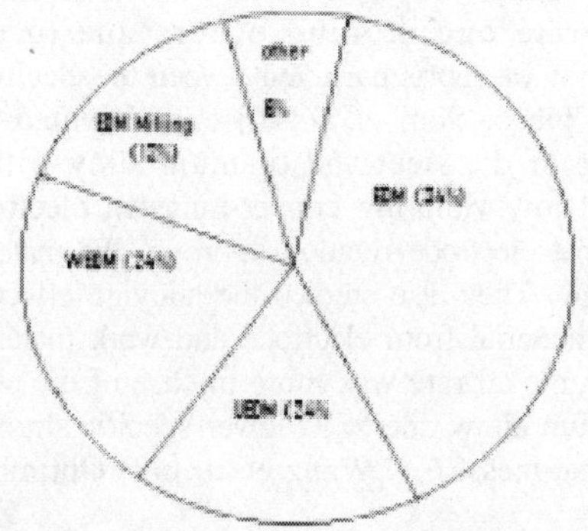

Fig. (5a): Research studies conducted in Dry EDM

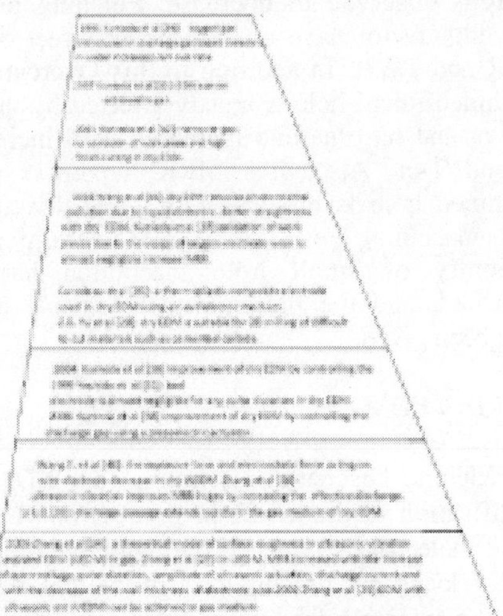

Fig. (5b) Progress in Dry EDM Research

5.1.5 Rotating Disc Electrode (RDE-EDM)-EDM

In recent years the study of micro electro mechanical systems (MEMS) has resulted in much research interest & manufacture of precision products such as micro engines, micro-pumps and micro robots have made successfully in industrial applications [37]. Rotary disc electrode electrical discharge machining is one of the variant process of electrical discharge machining process based on removing unwanted material in the form of debris from a part by means of a series of recurring electrical discharges (created by electric pulse generators in micro seconds)between a rotary tool called disc(thickness ranging from micron level to 1 mm) and the work material in the presence of a dielectric fluid(kerosene, distilled water). This fluid makes it possible to flush eroded particles from the gap.

This new application of RDE-EDM machining is achieved by locating the rotating disk electrode above or below the work piece to improve debris removal rate. Sato et al. [38] developed an electrical discharge machine for micro-hole boring. They claimed that electrode rotation served as an effective gap flushing technique, yielding better material removal.

The rotary electrode was placed above the work material. The difficulty of debris problem was encountered the research work i.e. lower metal removal rate and arching occurs due to the accumulation of debris particle between the electrode and work piece. Soni et al.[40] experimented the machining of die steel and titanium alloy with a rotary and low vibrating copper-tungsten electrode revealed that electrode rotation improves the material removal rate. They also studied the alloying effect of migrating material from electrode and work material. Material removal rate was more in case of die steel than titanium allow due to its lower specific strength and less hardness. C.C.Wang et al. [41] Optimizes

the blind-hole drilling of Al2O3/6061Al composite using rotary electro-discharging machining by using Taguchi methodology. Increase of rotational speed of the electrode or the injection flushing pressure of the dielectric fluid result in a higher MRR.

Mohan et al. [42] studied the effect of machining variables on MRR.EWR AND SR by blind hole drilling through cylindrical rotary electrode (copper) and compared its result from stationary electrode .MRR increases with increase in discharge current and for specific current it decreases with increases in pulse duration. MRR decreases with increase in vol. % of Sic but Increases with rotational speed of electrode which results in positive effect with MRR, TWR and better SR than stationary electrode.. Shih et al. [43] have reported about the electrical discharge grinding of AISI D2 tool steel work piece using a rotary disk copper electrode mounted on horizontal spindle. It was observed that lower electrode wear rate and higher material removal rate can be obtained in this process.

6.0 HYBRID EDM

6.0.1 Ultrasonic assisted EDM (USEDM)

The hybrid effect ultrasonic vibration of the electrode with EDM has come into the picture since mid 1980s.Ultrasonic is concerned with the vibratory wave of frequency above 16 Kc/s [44]. It is the combination of Electrical discharge machining that can improve material removal rate (MRR) as high as possible on account of decreasing surface finish. It can be used to machine high strength materials i.e. composites and super alloys. In USEDM process the abrasive slurry is replaced by suitable dielectric fluid (kerosene, distilled water,). The Ultrasonic vibration of tool or work piece in combination with electrical discharges produced by the electrode removes material effectively. Generally the tool vibrates in the direction normal to the work surface.

6.0.2 PMEDM

In spite of remarkable process capabilities, limitations such as low volumetric material removal & poor surface quality are associated with EDM. In order to enhance the process capabilities of EDM, abrasive powder mixed EDM (AEDM) process is used as one of the advance techniques [45]. In this method fine powder of a specific material is mixed into dielectric fluid of EDM to increase the process quality.

7. MODELLING

EDM process is influence by many input factors. Various techniques viz. dimensional analysis, artificial neural network (ANN) and thermal modeling are employed to predict the output of the process mainly the surface finish, tool wear and MRR. Neuro-Fuzzy model for MRR has been established & analyzed.

8.0 SUMMARY

In this paper, an overview of EDM process, EDM equipments, process parameters and applications are being discussed. This paper is essential for the development in the research to hybrid EDM process. The review of the research trends in EDM on rotary EDM, dry EDM, and EDM with powder additives, ultrasonic assisted EDM, WEDM, & micro EDM performances are presented.

9.0 REFERENCES

1. Singh. S, Maheshwari.S, Pandey P.C., Some investigation into the electric discharge machining of hardened tool steel using different electrode materials, Journal of Material Processing Technology 149 (2004), 272-277.

2. C.J. Luis, I. Puretas, G. Villa, material removal rate & electrode wear study on EDM of SiC, Journal of Material Processing Technology 164-165 (2005), 889-896.

3. B. Bojorquez, R.T. Marloth, O.S. Es-Said, Formation of a crater in the work piece on an electrical discharge machine, Engineering Failure analysis 9 (2002) pp 93-97.

4. Jain V.K (2008): Advance Manufacturing Processes, Allied Publication.

5. Pandey. A. Singh. S, Current research trends in variants of electrical discharge machining: A Review, Int. Journal of Engineering Science & Technology Vol. 2(6), 2010, pp 2172-2191.

6. S. Webzel, That first step in EDM, In: Machinery, 159,(4040), Findlay Publication Ltd., Kent UK, November 2001,pp 41

7. Anonymous, History & Development, in: The techniques & practice of Spark Erosion Machining. Spareatron, Gloucester, UK, pp 6

8. L. Houman, Total EDM, in: E.C, Jameson (Ed.), Electric Discharge Machining: Tooling, Methods & applications, Society of Manufacturing Engineers, Dearbern, Michigan,1983, pp5-19

9. A.L Livshit, Introduction in: Electro-erosion Machining of Metals, department of Scientific Research, Butterworth & Co., London, 1990, pp x.

10. Sato. T, Mizutani. T, Kuwata. K, (1985): Electrical discharge machine for micro hole drilling, National Technical Report, 31, pp 725-733.

11. Dauw, D.F,: Van. B (1995): On the evolution of EDM research, part 2: From fundamental research to applied research, Proceedings of the 1 1th. International Symposium for Electro Machining (ISEM-11), pp 133-142.

12. www.nptel.com, Module-9, Non-Conventional Machining, version-2 ME, IIT, Kharagpur.

13. Mishra P.K. (1997): Non-Conventional machining, The IEI, pp. 106-07

14. Pandey, P.C, Shan, H.S., (2003): Modern Machining Process, TMH, pp. 92-93.

15. www.openpdf.com

16. Kumar, Saurabh, M. Tech. Thesis, 2010, IIT Kanpur.

17. G. Karthikeyan, J. Ramkumar, S., Dhamodaran, S. Aravindan,: Micro electric discharge milling process performance: An experimental investigation, Int. J. of Machine tool & Manufacutre 50 (2010), pp. 718-728.

18. Pradhan, B.B , Magantam, A., Sarkar, B.R, Bhattacharya, B., Int. Journal of advance Manufacturing Technology (2009) 41, pp. 1094-1106.

19. P.K. Mishra, : Non-Conventional Machining, Narosa Publication, New Delhi,1997.

20. Murali, M.; Yeo, S.H. (2004): A novel sparks erosion technique for the fabrication of high aspect ratio micro-grooves, Microsystems Technologies, 10, pp. 628–632.

21. Singh, S.; Maheswari, S.; Pandey, P.C. (2007): Optimization of multiperformance characteristics in electrical discharge machining of Aluminium matrix composites (AMCs) using Taguchi DOE methodology, International Journal of Manufacturing Research, vol.2, No.2, pp.138-163.

22 Singh. S, Maheshwari.S, .S, Pandey, P.C. (2007): Journal of Manufacturing technology, 149 (2004), pp. 272-277

23. Krishna Mohan Rao, G.; Satyanarayana, S.; Praveen, M.(2008):Influence of machining parameters on electric discharge machining of maraging steels-An experimental investigation, Proceeding of the world congress on Eng., Vol. II.

24. M.Tech. Thesis, Dewangan Shailesh kumar, 2010, NIT Rourkela, india.

25. Shobert, E.I. (1983): What happens in EDM, Electrical Discharge Machining: Tooling, Methods and applications, Society of Manufacturing Engineers, Dearbern, Michigan, pp. 3–4.

26. Boothroyd, G.; Winston, A.K. (1989): Non-conventional machining processes, Fundamentals of Machining, Marcel Dekker, Inc, 491.

27. McGeough, J.A. (1988): Electro discharge machining, Advanced Methods of Machining, Chapman & Hall, London, 1988, pp. 130.

28. Masuzawa, T.; Tonshoff, H.K. (1997): Three-dimensional micromachining by machine tools", Annuals of the CIRP, 46, 2, pp. 621-628.

29. Crookall, J.R.; Heuvelman, C.J. (1971): Electro-discharge machining-the state of the art", Annual of the CIRP, 20, pp.113-120.

30. T. Sai, Study of effect of machining parameters on machining characteristics in EDM of WC, J. Material Processing Tech. 87 (1999), pp. 139-145

31. Chunhe Zhang, Hitoshi Ohmoni, Weill, Int. J. of Machine tool & manufacture, 40 (2000), pp. 661-674.

32. Yan, B.H.; Huang, F.Y.; Chow, H.M.; Tsai, J.Y. (1999): Micro-hole machining of carbide by electric discharge machining, Journal of Materials Processing Technology, 87, pp. 139–145.

33. Wansherg, Z.; Zhenlog, W.; shichun, D.; Guanxin, C.; Hongyu, W. (2002): Ultrasonic and electric discharge machining to deep and small hole

on titanium alloy, Journal of Materials Processing Technology vol.120, and pp.101–106.

34. Kaminski,P.C.; Capuano, M.N.(2003):Micro hole machining by conventional penetration electrical discharge machine, International Journal of Machine Tools & Manufacture, 43, pp. 1143–1 149.

35. Kunieda,M.; Furuoa,S. Taniguchi, N.(1991) :Improvement of EDM efficiency by supplying oxygen gas into gap, CIRP Annuals, Manufacturing Technology,40,pp.215-218.

36. Tao, J.; Shih, A.J.; Ni, J. (2008): Experimental study of the dry and near dry electrical discharge milling processes, Journal of Manufacturing Science and Engineering, 130.

37. H. M, Chow, B.H. Yan, F.Y Huang, Int. J. materials processing Technology. 91 (1999), pp. 161-166.

38. Sato, T.; Mizutani, T.; Yonemouchi, K.; Kawata, K.(1986): The development of an electro discharge machine for micro hole boring, Precision Engineering,8,pp. 163–168.

39. Soni, J.S.; Chakraverti.(1997):Performance evaluation of rotary EDM by experimental design technique, Defense Science journal, Vol. 47, No.1, pp. 65-73.

40. Sato,T.; Mizutani.; Yonemouchi,K.; Kawata,K.(1986): The development of an electro discharge machine for micro hole boring, Precision Engineering,8,pp. 163–168.

41. Yan, B.H.; Wang, C.C.; Liu, W.D.; Huang, F.Y. (1999): Experimental investigation of the blind-hole drilling of Al2O3/6061Al composite using rotary electro-discharging machining, International Journal of Advance Manufacturing Technology, 1999, 18, pp.372–379.

42. Mohan, B.; Rajadurai, A.; Satyanarayana, K.G. (2004): Electric discharge machining of Al–Sic metal matrix composites using rotary tube electrode, Journal of Materials Processing Technology, 153–154, pp.978–985.

43. Shih, H.R.; Shu, K.M. (2007): A study of electrical discharge grinding using a rotary disk electrode, International Journal of Advance Manufacturing technology.

44. Pandey, P.C.; Shan, H.S (2003): Modern Machining Process, Tata McGraw Hill publishing, pp.7.

45. Atkinson.; Ghoreishi, A.(2002): Comparative experimental study of machining characteristics in vibratory, rotary and vibro-rotary electro-discharge machining, International Journal of Material processing Technology, 120,pp.374-38

Recent Trends in Automotive Suspension

¹L.M. Jugulkar, ²Shankar Singh, ³Nitin V. Satpute
¹Assistant Professor, Rajarambapu Institute of Technology, Rajaramnagar, Sangli, Maharashtra.
² Associate Professor, Mechanical Engineering Department, (SLIET), Longowal, Distt. Sangrur, Punjab
³Research Scholar, Sant Longowal Institute of Engineering & Technology (SLIET), Longowal, Distt. Sangrur, Punjab

Abstract—**Function of a Shock Absorber is to isolate occupants from road induced vibrations. Vehicle undergoes vibrations due to uneven road profile, corner turning and braking. Hydraulic telescopic Shock Absorbers are used since inception of Automobiles, are very robust and is very reliable means of attenuating vibrations. With this type of Shock Absorber damping and spring properties do not change. Nowadays increased comfort requirement form the passenger has force automobile manufacturers to explore for Semi active and Active suspension, that changes it damping properties and uses an external active for attenuating road induced vibrations.
In this paper review has been taken from latest development in Automotive Suspension System.**

I. Introduction

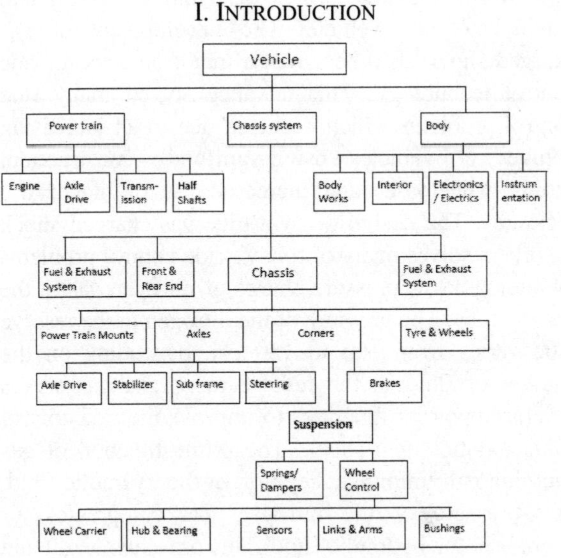

Fig: 1 Vehicle Components

The complete vehicle is traditionally divided into three groups as power train, chassis/suspension, and body. The power train contains elements which propel the vehicle, the body provides room for people, and the chassis and suspension allow the vehicle to ride, turn, and stop. Modern vehicles integrate the body and chassis into a structure known as unibody or monocoque. The chassis and suspension are main components of the automobile and together are made up of the wheels, wheel carriers, wheel bearings, brakes, wheel suspension, sub frames, springs (including stabilizers), dampers, steering gear, steering linkage, steering column and wheel, pedal cluster, motor mounts, drive shafts, differential.

In designing suspension of vehicle, increasing attention is being given to comfort of the passenger. Task of the suspension system is to isolate the vehicle from road induced vibrations. The conventional suspension consists of fluid damping and spring connected in parallel. For conventional passive dampers, properties of fluid damping and spring are constant.

A piston is attached to the end of the piston rod and works against hydraulic fluid in the pressure tube. As the suspension travels up and down, the hydraulic fluid is forced through tiny holes, called orifices, inside the piston. However, these orifices let only a small amount of fluid through the piston. This slows down the piston, which in turn slows down spring and suspension movement. The amount of resistance a shock absorber develops depends on the speed of the suspension and the number and size of the orifices in the piston. All modern shock absorbers are velocity sensitive hydraulic damping devices - meaning the faster the suspension moves, the more resistance the shock absorber provides. Because of this feature, shock absorbers adjust to road conditions. As a result, shock absorbers reduce the rate of:

Bounce

Roll or sway

Brake dive and Acceleration squat

Shock absorbers work on the principle of fluid displacement on both the compression and extension cycle. A typical car or light truck will have more resistance during its extension cycle then its compression cycle. The compression cycle controls the motion of a vehicle's unsprung weight, while extension controls the heavier sprung weight. During the compression stroke or downward movement, some fluid flows through the piston from chamber B to chamber A and some through the compression valve into the reserve tube. To control the flow, there are three valve stages each in the piston and in the compression valve. At the piston, oil flows through the oil ports, and at slow piston speeds, the first stage bleeds come into play and restrict the amount of oil flow. This allows a controlled flow of fluid from chamber B to chamber A. At faster piston speeds, the increase in fluid pressure below the piston in chamber B causes the discs to open up away from the valve seat.

At high speeds, the limit of the second stage discs phases into the third stage orifice restrictions. Compression control, then, is the force that results from a higher pressure present in chamber B, which acts on the bottom of the piston and the piston rod area. As the piston and rod move upward toward the top of the pressure tube, the volume of chamber A is reduced and thus is at a higher pressure than chamber B. Because of this higher pressure, fluid flows down through the piston's 3-stage extension valve into chamber B. However, the piston rod volume has been withdrawn from chamber B greatly increasing its volume. Thus the volume of fluid from chamber A is insufficient to fill chamber B. The pressure in the reserve tube is now greater than that in chamber B, forcing the compression intake valve to unseat. Fluid then flows from the reserve tube into chamber B, keeping the pressure tube full. Extension control is a force present as a result of the higher pressure in chamber A, acting on the topside of the piston area.

There are several shock absorber designs in use today

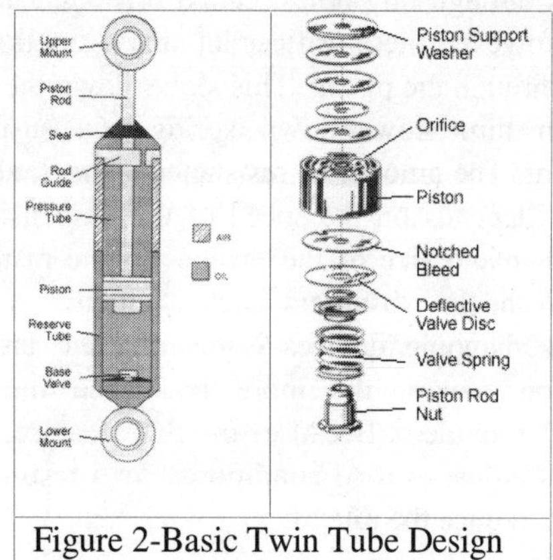

Figure 2-Basic Twin Tube Design

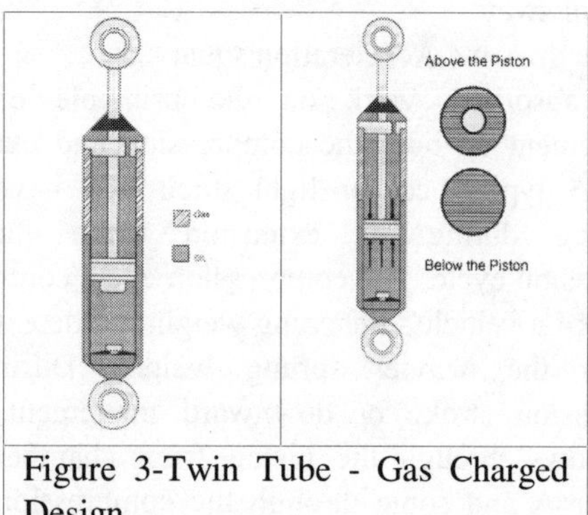

Figure 3-Twin Tube - Gas Charged Design

The twin tube design has an inner tube known as the working or pressure tube and an outer tube known as the reserve tube. The outer tube is used to store excess

hydraulic fluid. There are many types of shock absorber mounts used today. Most of these use rubber bushings between the shock absorber and the frame or suspension to reduce transmitted road noise and suspension vibration. The rubber bushings are flexible to allow movement during suspension travel. The upper mount of the shock absorber connects to the vehicle frame. Notice that the piston rod passes through a rod guide and a seal at the upper end of the pressure tube. The rod guide keeps the rod in line with the pressure tube and allows the piston to move freely inside. The seal keeps the hydraulic oil inside and contamination out. The base valve located at the bottom of the pressure tube is called a compression valve. It controls fluid movement during the compression cycle. Bore size is the diameter of the piston and the inside of the pressure tube. Generally, the larger the unit, the higher the potential control levels because of the larger piston displacement and pressure areas. The larger the piston area, the lower the internal operating pressure and temperatures. This provides higher damping capabilities. Ride engineers select valve values for a particular vehicle to achieve optimal ride characteristics of balance and stability under a wide variety of driving conditions. Their selection of valve springs and orifices control fluid flow within the unit, which determines the feel and handling of the vehicle. The development of gas charged shock absorbers was a major advance in ride control technology. This advance solved many ride control problems which occurred due to an increasing number of vehicles using uni-body construction, shorter wheelbases and increased use of higher tire pressures. The design of twin tube gas charged shock absorbers solves many of today's ride control problems by adding a low pressure charge of nitrogen gas in the reserve tube. The pressure of the nitrogen in the reserve tube varies from 100 to 150 psi, depending on the amount of fluid in the reserve tube. The gas serves several important functions to improve the ride control characteristics of a shock. The prime function of gas charging is to minimize aeration of the hydraulic fluid. The pressure of the nitrogen gas compresses air bubbles in the hydraulic fluid. This prevents the oil and air from mixing and creating foam. Foam affects performance because it can be compressed - fluid can not. With aeration reduced, the shock is able to react faster and more predictably, allowing for quicker response time and helping keep the tire firmly planted on the road surface. An additional benefit of gas charging is that it creates a mild boost in spring rate to the vehicle. This does not mean that a gas charged shock would raise the vehicle up to correct ride height if the springs were sagging. It does help reduce body roll, sway, brake dive, and acceleration squat. This mild boost in spring rate is also caused by the difference in the surface area above and below the piston. With greater surface area below the piston than above, more pressurized fluid is in contact with this surface. This is why a gas charged shock absorber will extend on its own. The final important function of the gas charge is to allow engineers greater flexibility in valving design. In the past such factors as damping and aeration forced compromises in design.

II. RECENT DEVELOPMENTS IN SHOCK ABSORBER

Cristiano Spelta et al. (2010) has described the controller does not consider mass supported by the vehicle. Variation of mass will affect performance of the shock absorber. While choosing damping level of the shock absorber in addition to the velocity, mass supported by the vehicle should be considered. [1]

In this the Semi active suspension system for Motorcycle is proposed. It works on input from single velocity sensor; variable damping is achieved by using an electro hydraulic valve. This valve will vary flow area between two chambers of the shock absorber, resulting in variable damping effect. Two current driven solenoid valves are used to vary damping factor. Control algorithms are decided so that they transmit very low acceleration to the occupants and also maintaining road contact of the tyre. Sky hook controller is used that will decide the optimum damping level based on relative velocity measurement. There are two damping levels as high or low, the controller will choose one of the level and suitable amount of current (300-1200 mAmp) will be send to the solenoid valve. This valve will decide extend of damping provided by the shock absorber. Various methods of control strategies are discussed with these highlights.

M. Zapateiro et al. (2009) This paper descries methodology to compute control voltage that has to be applied to the MR fluid damper to produce optimal damping force necessary to reduce vibrations. Current methodologies are based on Binghamand Bouc-Wen models, but it has limitations due to non linear model. Also present models give damping force created by specified voltage, but reverse is not possible.[2]

MR fluids create high force in compact size; they need low energy to operate. But their only drawback is that it has hysteresis force velocity loop of which it depends on many factors. Due to these non linear control methods needs to be applied.

In this paper, a neural network-back stepping controller for a class of semi active vehicle suspension systems equipped with MR dampers is proposed. Back stepping is a recursive design technique that consists in designing virtual controllers for each state of the system until the actual control input is reached. Each controller takes into account the previous one and must be asymptotically stable in the Lyapunov's sense. After the formulation of control law, a neural network is used as the inverse model of the MR damper, i.e., as the model that takes the desired damping force and calculates the damper control signal that generates the needed damping force.

Dahl model is used for MR modeling that uses control voltage & other variables affecting shape of control loop. Vibrations of the sprung mass, i.e. car body are reduced by reducing its angular velocity (i.e. velocity). Formula for control voltage is derived using back stepping approach. Later on neural networks can be trained to learn MR fluid dynamics in order to apply control voltage for optimum damping. The second test consists of a neural network which was designed to predict the control voltage given the piston displacement and velocity and the desired MR damper force. The output, i.e., the voltage, is fed back to the input.

Future work will investigate the performance of other controllers for this system based on other techniques such as the frequency-based QFT or mixed H2/HN control. Various models are presented in the paper for MR damper application. Work presented in the paper can be used for accurate modeling of MR damper performance.

W.L.Wang et al. (2009) propose use of low cost solenoid control valve for changing damping factor of the damper. The damper is simple in construction and is easy to control. For damping lateral vibrations two interchangeable dampers are installed between the bogeys. These two dampers are controlled by three high speed solenoid valves and two unloading valves. Combination of orifice and relief valve is obtained for getting desired damping. By changing position of three solenoid valves and two relief valves, flow passage and hence damping factor is varied. Skyhook control logic is used to decide control algorithm for this shock absorber. Totally six modes ranging from very soft to hard are obtained. This paper describes simple and low cost method of obtaining variable damping, rather than using costly and complicated method like MR dampers. In the proposed research work low cost solenoid valve can be used for changing cross sectional area of flow between two chambers, to vary damping factor of the shock absorber. This paper will be useful for designing such variable damper.[3]

M.J. Thoresson et al. (2009) Gradient based optimization method is used for optimizing shock absorber. Gradient based method is least time consuming and all the design variables are efficiently considered. Design factors are damper scale factor and static gas volume. Objective function is written in the form of comfort and safety criteria. For this acceleration at sprung mass and tyre displacement are considered. Optimization with four design variable sis carried out in ADAMS software. The results of two design variable optimization using MATLAB software are also discussed. Result shows that combined optimization is compromise between ride and handling. The optimization procedure discussed in this paper may be useful in performing optimization of the proposed semi active shock absorber.[4]

Yanqing Liu et al (2008). This paper proposes use of Voigot element to achieve variable spring stiffness. Spring stiffness can be varied up to three times its initial value. This uses using variable damper along with springs connected in series and parallel. Also if variable damper is connected in parallel, the system with variable stiffness as well as variable damping can be achieved. Formulas for equivalent stiffness are written by assuming single degree of freedom system. Since changing stiffness will change natural frequency o the system, significant change in performance of the system can be obtained using the system with variable stiffness. Control laws are written to decide level of stiffness required at the particular instant. In the paper change in damping properties is achieved using MR damper. But any other means of changing the damping factor can also be used. The paper describes system that changes its stiffness using vaiable MR damper connected in series and in parallel with spring. This method can be extended for using in automotive shock absorbers for changing equivalent stiffness of the supporting spring. For changing damping factor, methods other than using MR fluid can also be used. [5]

Semi active damping only expends a small amount of energy to change system parameters, such as damping and stiffness. They are simple to implement. Spring with variable stiffness is not fully explored by the researchers. By changing damping c2, equivalent stiffness of sub damper changes and so equivalent stiffness k'. Results are obtained for various settings of the various elements. Peak displacement is found to be significantly reduced with optimum settings of c1 & c2.

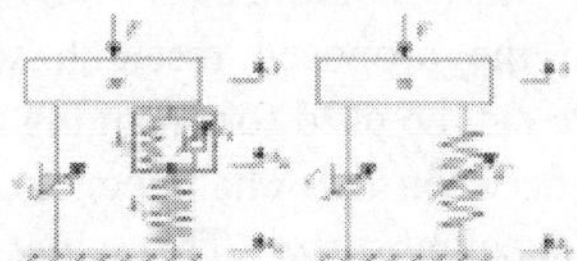

Fig 4:Mechanical configuration of variable stiffness and damping: (a) original model and (b) equivalent model (after Yanqing Liu et al., 2008)

Three distinct settings for damper are chosen based on soft, low & high. This system offers excellent control over stiffness and stiffness can be varied up to three times the given value. It is very easy & quick to control with MR dampers. But this paper does not give its application for use in automobile suspension design. This research paper will be useful for designing a system with variable stiffness for use in the proposed shock absorber.

Ping Yang et al (2008). Describes new type of shock absorber consisting of rubber ball, viscous fluid is explained. This absorber is used for delicate electronics items. It has non linear characteristics. Energy is dissipated through forcing of oil by piston on rubber air balls.

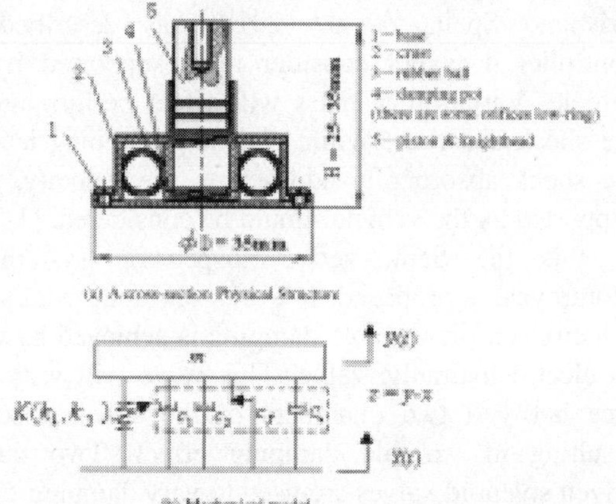

Fig 5:Physical structure and mechanical model of the prototype shock absorber(after Ping Yang et al 2008)

The shock absorber performance can be distributed in three bands as, low transmission, resonance transmission & again low transmissibility isolation band. The components considered to compute damping force are: - throttling damping force, laminar flow force, inertia damping force, structural damping force in the components & friction force. Based on test & analytical data, there is some light difference between the two. Mathematical model constructed is not very accurate. Factors affecting performance of damper are:- oil viscosity, number of orifices in the damper, frequency(force increase with increase in frequency & vice versa). The analysis can be used as basis for design since performance varies as per frequency, etc. also mathematical model is not very accurate. Non linear damping effect of rubber ball is discussed in the paper. [7]

C. Lauwerys J et al (2004) given presently available control methods for active shock absorbers are complex, & hence difficult to employ. They have either very simple mathematical model (than actual system), or the parameters are so many that they are difficult to implement. To avoid complexity of mathematical models, a model-free control structure and design approach is developed. This approach is based on physical principles of semi-active shock absorbers and cars in general, but does not require a model of its dynamics. Therefore, it is applicable to any semi-active or active suspension system and any type of car.

The control structure incorporates many physically interpretable parameters. The tuning of these parameters is based on the principles of passive shock-absorber tuning. Skyhook method is used to calculate desired damping force, and then control module will decide the required current flow through circuit. Accelerometers at four corners of vehicle are used to; compute velocities, which are used to calculate vehicle roll, pitch & bounce. These three velocities are used to compute proper damping factor in the above three directions. Non linear skyhook model is used. This controller will change damping intensity based on feedback from this algorithm. [8]

Choon Tae Lee et al (2006) studied displacement sensitive shock absorber has variable flow passage between compression & rebound chamber. For lesser displacement of suspension flow area is greater & as suspension displacement increases flow area reduces accordingly. It is achieved by using displacement sensitive orifice at the cylinder wall. Such a DSSA improves ride comfort on the paved road driving conditions because of low damping force caused by small piston stroke. Displacement-sensitive orifices can be divided into three zones such as the soft, transient and hard zone. The flow continuity equation is used that takes in to account piston velocity, coeff. Of discharge, pressure, discharge rate areas of all the orifice & valves at various locations. Based on this equation damping force is calculated, this force is same as of the experimental readings taken. Detailed mathematical model constructed in this paper gives exact value of damping force developed by the shock absorber. This model can be used to determine response of the shock absorber for various operating parameters. The developed mathematical model does not take in to account loss of laminar flow and temperature effect. Also it is not applicable to piston with disc valves.[9]

T. Pranoto et al (2005) have developed two types of dampers are discussed as rotary type & with two degree of freedom. For applications like bed in ambulance or any other, in order to avoid motions, the damping force has to be large in usual usage, but it has to be small when the shock occurs. i.e. at start damping should be less & there should be deflection, but later damping should be more.[10]

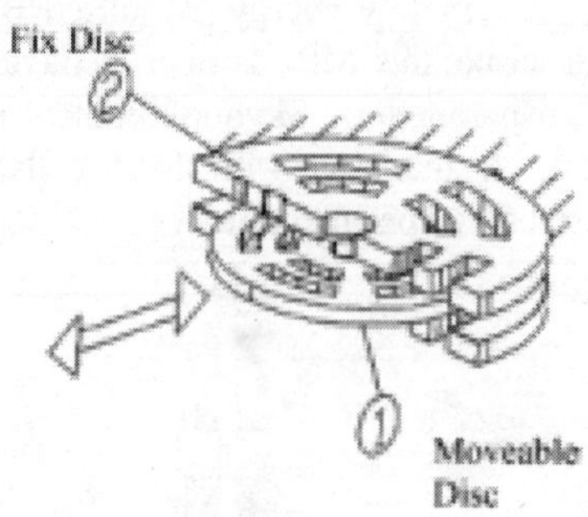

Fig 6 : The disc of 2DOF-type.(after T. Pranoto et al,2005)

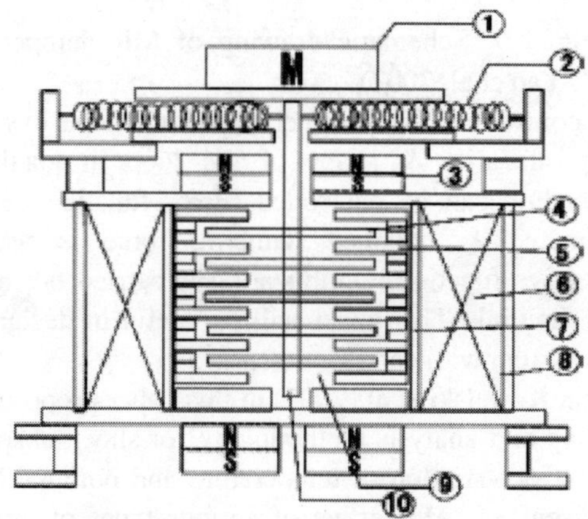

Fig 7 : 2DOF-type Damper. Notes: (1)mass; (2) spring; (3) permanent magnet; (4) inner disc; (5) outer disc; (6) coil; (7) inner pipe; (8) outer pipe; (9) MRF; (10) shaft.(after T. Pranoto et al, 2005)

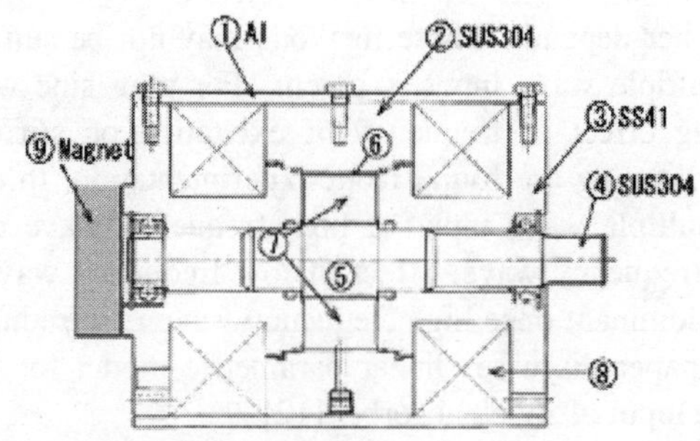

Fig 8:Rotary-type damper.(1) outer Cylinder,(2) inner Housing, (3) outer housing, (4) shaft, (6) grooves to fix outer disc, (7)MRF, (8) coils,(9) permanent magnet. (after T. Pranoto et al, 2005)

Rotary type gives better efficiency. Shock is improved significantly. MR fluid damper can be made in rotary or reciprocating configuration. The work discusses various configurations their effectiveness. Formulas presented in the work can be useful for damper modeling.

G.Z. Yao et al,(2002) MR fluid offer much higher change in viscosity than ER fluid. It takes place very quickly, needs very low energy. Modified Bouc–Wen model to describe the MR damper behavior, which consists of 16 parameters. Skyhook control method is adopted here. A machine with load cells, velocity sensors is used for experimentation.

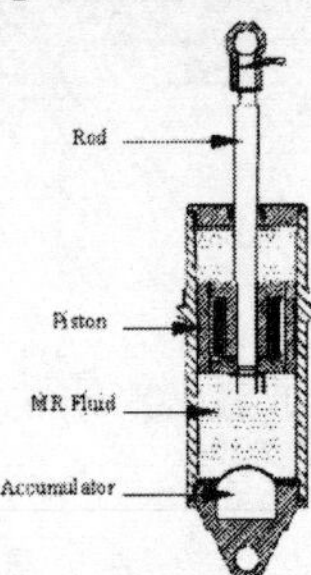

Fig 9 :Schematic drawing of MR damper.(after G.Z. Yao et al, 2002)

By controlling the parameters in the shape hysteresis loop, linearity & shape of the loop in loading & unloading can be controlled. Error function between estimated & obtained damping force is used as objective function. Optimization is carried out in Mat lab simulink. This paper will be useful in designing a sub system with variable stiffness.[11]

Darin Kowalski et al,(2001) in this paper proposes new testing and analysis methodology for shock absorbers. Effect of variation of temperature and nominal length is discussed. Also effect of various types of inputs is discussed. Stepped sine wave and sine on sine wave type input is discussed. A linear model is constructed base on stiffness and damping factor. The parameters are then found using data fitting methods. In earlier research on shock absorber researchers have found 14 to 84 parameters on which function of the shock absorber depends. These methods may not be suitable if multiple wave input is given. For pure sine wave testing effect of frequency of excitation on stiffness and damping are found from experimentation. In case of multiple wave input of high frequency wave over low frequency wave, effect of low frequency wave is less dominant once high frequency wave is introduced. The paper discusses linear parametric model for sine wave input of shock absorber.[12]

J. A. Tamboli et al,(2001) describes In the earlier literatures, power spectral density (PSD) is treated as white noise & velocity of vehicle as constant to calculate Root mean square acceleration (RMSA). But in actual practice, PSD varies exponentially & velocity is not same. The output based on the above assumption Root mean square acceleration (RMSA) varies significantly from actual conditions. RMSA for highway road conditions is calculated for actual road input and vehicle velocity. Using FFT power PSD is calculated at rear wheel. Effect of vehicle velocity on PSD is studied. For this a new variable in form of ratio of amplitude at rear & front suspension is introduced

based on a time lag between two suspension displacements. Optimum values of front & rear suspension are obtained based on criteria of minimum RMSA. Half car model is constructed for the analysis. q0i (t) =q(t + time lag), in this manner displacement at front and rear wheel are interrelated. The time lag depends on velocity of the vehicle. Factor corresponding to the time lag is included in the analysis. Method of transfer function is used to find RMSA. Again by using method of power spectral density, values of RMSA are found and compared with the earlier method. Optimization is done for minimizing objective function corresponding to acceleration of the sprung mass. Humans are most sensitive to vibrations of the frequency range of 4-8 Hz. Therefore the optimization is done for frequency of 5 Hz. Non linear optimization techniques are used for optimizing damping parameters for reduction in RMSA subjected to some boundary conditions .[14]

III. CONCLUSIONS

Even though conventional hydraulic telescopic Shock Absorber provides very robust and reliable solution for Automobile Shock Absorber, it has some limitations since its damping and stiffness properties can not be changed.

Ideally damping properties and spring rate should be varied depending on vibration velocity. For lower velocities if spring rate and damping are kept low lesser acceleration is transmitted to the occupants, increasing comfort, whereas at higher velocities damping and spring rate needs to be increased to ensure that the tyres are not getting lifted off the ground. Cost of Semiactive suspension is very high as compared to that of passive suspension.

In case of semiactive suspension damping properties can be varied by using MR fluid damper or by varying flow area between compression and rebound chamber. Magnetic circuits can be used in both the cases, but inductance of the magnetic circuits has to be kept at minimum to avoid phase lag between the signals.

In using the semiactive suspension, damping rate is kept lower for lower velocities and vice versa. No major attempt has been made till date to change stiffness of the suspension. Change in stiffness changes natural frequency of the suspension, hence effect of change in stiffness is affecting the performance of the Shock Absorber.

IV. REFERENCES

[1] Cristiano Spelta , Sergio M. Savaresi , Luca Fabbri, Experimental analysis of a motorcycle semi-active rear

suspension, *Control Engineering Practice 18 (2010) 1239–1250*

[2] M. Zapateiro, N. Luo , H.R. Karimi, J. Vehi, Vibration control of a class of semi active suspension system using neural network and back stepping techniques, journal of Mechanical Systems and Signal Processing 23 (2009) 1946–1953

[3] W.L.Wang, G.X. Xu, Fluid formulae for damping changeability conceptual design of railway semi-active hydraulic dampers, International Journal of Non-Linear Mechanics 44 (2009) 809-819

[4] M.J. Thoresson, P.E. Uys, P.S. Els 1, J.A. Snyman, Efficient optimization of a vehicle suspension system, using a gradient-based approximation method, Part 2: Optimization results, Mathematical and Computer Modeling 50 (2009) 1437-1447

[5] Yanqing Liu, Hiroshi Matsuhisa, Hideo Utsuno, Semi-active vibration isolation system with variable stiffness and damping control, Journal of Sound and Vibration 313 (2008) 16–28

[6] Variable Damping & spring stiffness for semi active damping Science direc.com, Journal of sound & vibrations (2008)

[7] Ping Yang, Ninbo Liao, Jianbo Yang, Design test and modeling evaluation approach of a novel Si-oil shock absorber for protection of electronic equipment in moving vehicles, Mechanism and Machine Theory 43 (2008) 18–32

[8] C. Lauwerys J. Swevers P. Sas , Model-free control design for a semi-active suspension of a passenger car, Proceedings of ISMA2004, 75-86

[9] Choon Tae Lee, Byung Young Moon, Simulation and experimental validation of vehicle dynamic characteristics for displacement-sensitive shock absorber using fluid flow modeling, Mechanical Systems and Signal Processing 20 (2006) 373–388

[10] T. Pranoto, K. Nagaya, Development on 2DOF-type and Rotary-type shock absorber damper using MRF and their efficiencies, Journal of Materials Processing Technology 161 (2005) 146–150

[11] G.Z. Yao, F.F. Yap, G. Chen, W.H. Li, S.H. Yeo, MR damper and its application for semi-active control of vehicle suspension system, Mechatronics 12 (2002) 963–973

[12] Darin Kowalski, Mohan D. Rao, Jason Blough, Scott Gruenberg, Dave Griffiths, The Effects of Different Input Excitation on the Dynamic Characterization of an Automotive Shock Absorber, Society of Automotive Engineers, Inc. 2001

[13] N. Chandra Shekhar, H. Hatwal, A. K. Mallik, Performance of non-linear isolators and absorbers to shock excitations, Journal of Sound and Vibration 227, (1999), 293-307

[14] J. A. Tamboli, S. G. Joshi, Optimum Design of a Passive Suspension System of a Vehicle Subjected to Actual Random Road Excitations, Journal of Sound and Vibration (1999),193-205

[15] John C. Dixon, The Shock Absorber Handbook(2007), Second Edition, Publisher: John Wiley and Sons, England

[16] Emanuele Guglielmino, Tudor Sireteanu, Charles W. Stammers, Gheorghe Ghita, Marius Giuclea, Semi-active Suspension Control, 2008 Springer-Verlag London Limited

Experimental Investigation into the Electrical Discharge Machining of Nickel Based Superalloy (Inconel- 600)

Ravish Arora, Shankar Singh

Department of Mechanical Engineering, Sant Longowal Institute of Engineering and Technology, Longowal District Sangrur, Punjab

Abstract- **Advanced materials such as nickel-based Super alloys belong to the group of hard-to-machine materials that has gained widespread attention in the defence, aerospace, energy and medical industries. They are supposed to exhibit outstanding mechanical strength and creep resistance at high temperatures as well as corrosion and oxidation resistance. Present manufacturing industries are facing challenges from machining such advanced and difficult-to-machine materials, the stringent design requirements & machining costs. Literature identifies that Nickel based alloy suffers poor machinability during conventional-type machining.**

The present study attempts to find optimal parameter settings for EDM of Inconel 600 superalloy. The work specimens were machined by electrolytic copper tool electrodes having circular and square geometry/shape by changing the aspect ratio. The performance measures such as metal removal rate (MRR-g/min and overcut (OC-mm) has been chosen as observed performance criteria. In addition, four process parameters namely one noise factor viz. Tool/electrode geometry and three control factors viz. pulse current (Ip-A) , Pulse On time (T_{on}-μs) and duty cycle (ζ-%) have been considered for analysis . The experimental runs were performed on the basis of L18 orthogonal array ($2^1 \times 3^3$). The effect of EDM process parameters and their interaction were discussed and the optimal settings have been suggested.

Key Words: Advanced materials, Super alloys, EDM, DoE, Optimization, MRR, OC.

I. INTRODUCTION

EDM is one of the most important technologies among non-conventional machining processes. It can diminish mechanical stresses, distortions and vibration problems which are generated during machining conventional processes. EDM process uses electrical discharge spark companying with an extremely high temperature to carry out material removal. It means that EDM is basically an electro thermal process which transforms electric energy into thermal energy. So, the parameters like current, voltage, spark gap, pulse duration etc. are important for study of machining process. A pulse discharge occurs in a small gap between the work piece and the electrode which removes the unwanted material from the parent metal through melting and vaporising. The electrode and the work piece must have electrical conductivity in order to generate the spark. Nickel base super alloys are extensively used in high temperature applications such as gas turbines, electric power generation equipment, nuclear reactors and high temperature chemical vessels [1]. Tsai observed that material removal rate (MRR) is increased with the increase of the current keeping same current rising slope. Also, the MRR increases with the increase of the current rising slope keeping the same current [2]. Puertas observed that increase in pulse time produces a decrease in material removal rate. In the case of boron carbide, MRR decreased with pulse time down to a minimum value, after which it tends to increase. On the other hand, it has been observed that material removal rate tends to increase when duty cycle is increased for all the cases [3]. Hewidy also observed that VMRR increases with the increase in peak current value and the water pressure but after certain limit the increase of peak current leads to decrease of VMRR. He also found that wear ratio increases with the increase of peak current and also surface roughness increases with the peak current and decreases with the increase of duty factor and wire tension[1].

The present work highlights the development of optimal machining parameters like noise factor i.e., shape/ geometry of tool electrode, and control factor like pulse current (I_p), duty cycle (τ) and pulse on (T_{on})

time for electric discharge machining of superalloy, Inconel-600 on MRR and overcut (OC). This work has been established based on the Taguchi's Design of Experiment approach. Taguchi technique is basically intended as a guide and reference for industrial practitioners like managers, engineers and scientists involved in product or process experimentation and development. Taguchi technique used two elevators for optimisation of process i.e., Taguchi quality loss function and Signal to Noise Ratio. Taguchi defined the loss function as a quantity proportional to square deviation from the nominal quality characteristics. A high value of S/N implies that the signal is much higher than the random effects of noise. A process consistent with highest S/N always yields optimum quality with minimum variation. Generally S/N ratio is evaluated as Lower the better, Nominal the best and Higher the better. Lower the Better for performance characteristics, whose values are preferred when low, like surface roughness etc. Nominal the Best is used for performance characteristics, whose values are preferred to be nominal and Higher the better is used for performance characteristics, whose values are preferred to be high like MRR etc. In the current experimentation the for MRR Higher the better and for over cut lower the better S/N ratio has been used. For the purpose of experimentation L_{18} orthogonal array has been used.

II. TOOLS AND EQUIPMENT USED

This section of paper describes about the tools and equipment used in the experimentation.

Figure 2.1 EDM machine used for the experimentation

The EDM

The machine used for the purpose of experimentation was ELECTRAPULS PS50. The dielectric fluid used in the experimentation was EDM oil DNR Spark SPO-A. The electrode used was electrolytic copper electrode of 12 mm diameter, circular and 11 mm square electrode.

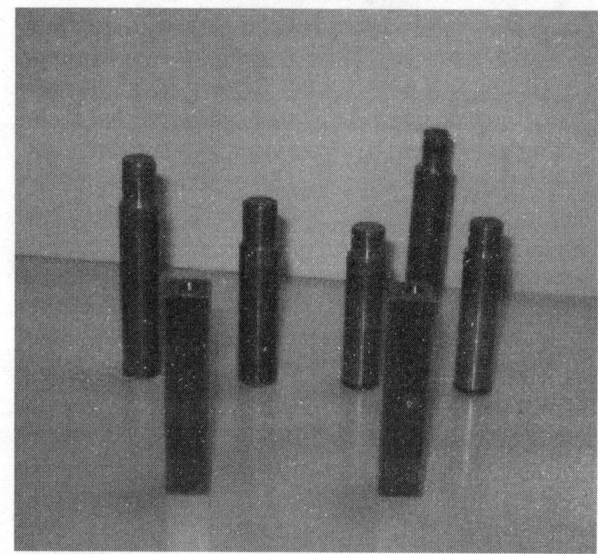

Figure 2.2 Electrodes used in the experimentation.

Instruments for MRR measurement and Over Cut

MRR has been measured by measuring the weight of the work piece before and after experimentation. The machine used for measurement of MRR was Mettler Toledo having least count 0.00001gm. The cylindrical work pieces used in experiment were 20 mm in diameter of Inconel- 600. Over cut was measured using optical microscope.

III. THE METHOD OF EXPERIMENTATION

The experimentation was carried out on ELECTRAPULS PS50 machine. For the purpose of experimentation specially designed fixture was used because Inconel -600 is non-magnetic and the machine could hold only magnetic work material. The fixture was made from mild steel and was having the arrangement for holding the workpiece in correct position desired for machining. The workpiece was kept on the worktable of EDM machine and the machining parameters were set as per L_{18} array.

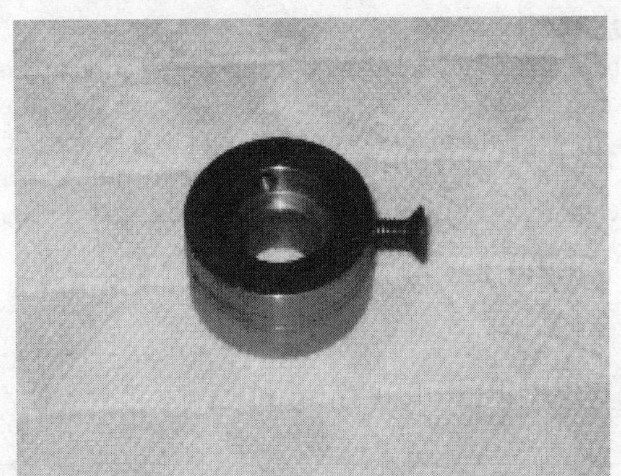

Figure 2.3 Fixture used for experimentation

The machined workpiece was weighed before and after experiment. The overcut was measured with the help of optical microscope.

IV. RESULTS AND DISCUSSIONS

MRR and overcut were observed in the experimentation. For the purpose of experimentation the Taguchi's Design of Experiment approach was selected. The orthogonal array used is L_{18} and is mixed level design. Control factors used were fixed at three levels. Controls factors used were peak current, pulse on time and duty factor, whereas the noise factor viz. shape/ geometry was set at two levels. Peak current (I_p) was having values of 8A, 10A and 12A respectively. Pulse on time (T_{on}) was set at 100µs, 150 µs, 200 µs respectively. Similarly duty cycle (τ) was set at 33%, 50% and 66.66% respectively. For noise factor viz., tool shape/ geometry square shape and circular shape of tool electrode was used. The degree of freedom for a factor is calculated as number of levels of the factor minus one. Therefore degree of freedom for peak current (B), pulse on time (D) and duty cycle (C) is two and tool shape/ geometry (A) is two. The degree of freedom for interaction (A×B) between tool shape/ geometry and peak current is multiplication of individual factor's degree of freedom. Hence degree of freedom for the interaction is two. The total degree of freedom for the experimentation is nine. The array to be selected for the experimentation should have higher degree of freedom than the

calculated one. The allocation of degree of freedom is shown below in the table 4.1. Other factors like flushing rate etc. were kept constant during experimentation.

Table 4.1 DOF allocated to various factor combinations

Interaction	Units	DOF
Tool shape/ geometry (A)	-	1
Peak current (B)	Ampere	2
Duty cycle (C)	%	2
Pulse on time (D)	µs	2
Interaction (A×B)	-	2
Total	-	9

The response variables selected for this study is material removal rate (MRR) and the overcut. The material removal rate (MRR) in the form of an equation is defined as follows

$$MRR = \frac{Volume-of-material-removed}{time\_of\_machining}$$

The overcut is defined as the increase in size of the cavity produced due to machining than tool size. The material removal rate (MRR) and overcut values observed were recorded in the table 4.2.

Table 4.2 Values of various response variables

Tool Shape	I_p	τ	T_{on}	Mean MRR (g/min)	Mean OC (mm)	S/N Ratio(MRR)	S/N Ratio(OC)
R	8	33.33	100	0.134	0.04	17.4579	27.96
R	8	50	150	0.222	0.04	13.0729	27.96
R	8	66.66	200	0.237	0.0457	12.505	26.8
R	10	33.33	100	0.21	0.015	13.5556	36.48
R	10	50	150	0.318	0.0375	9.95146	28.52
R	10	66.66	200	0.36	0.0432	8.87395	27.29
R	12	33.33	150	0.434	0.0397	7.25021	28.02
R	12	50	200	0.515	0.0315	5.76386	30.03
R	12	66.66	100	0.414	0.0307	7.65999	30.26
S	8	33.33	200	0.273	0.0837	11.2767	21.55
S	8	50	100	0.176	0.0756	15.0897	22.43
S	8	66.66	150	0.245	0.0722	12.2167	22.83
S	10	33.33	150	0.325	0.0717	9.76233	22.89
S	10	50	200	0.426	0.0705	7.41181	23.04
S	10	66.66	100	0.313	0.0787	10.0891	22.08
S	12	33.33	200	0.627	0.0742	4.05465	22.59
S	12	50	100	0.431	0.0837	7.31045	21.55
S	12	66.66	150	0.588	0.8225	4.61245	1.697

The main effects plot for S/N ratio for MRR against tool geometry, peak current, duty cycle and duty cycle has been shown in graph 4.1. The main effects plot show that with square tool having same area as that of round tool, S/N ratio for MRR is higher. With increase in peak current S/N ratio for MRR increases. With increase in duty cycle it has been observed there is increase in S/N of MRR but rate of increase of S/N is very small. Further with increase in pulse on time S/N ratio for MRR increases [4]. Thus for all the machining parameters decided, with increase in their values, increase in S/N ratio for MRR is observed. Further the interaction plot for current and tool geometry has been shown in graph 4.2.

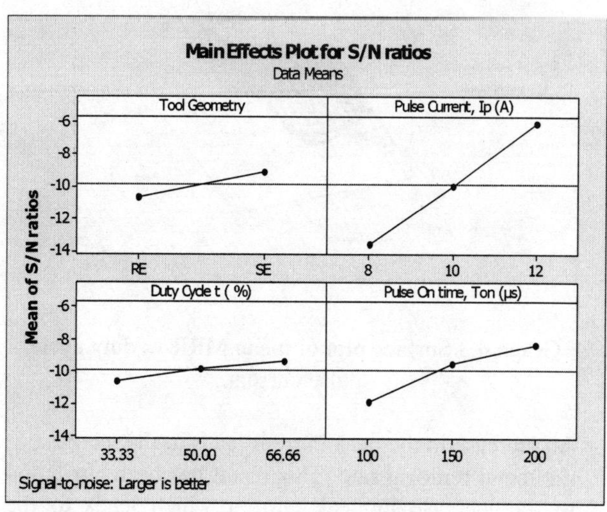

Graph 4.1: Main effects plot for S/N ratio for MRR

Graph 4.2, interaction plot shows that there is no interaction between tool geometry and pulse current.

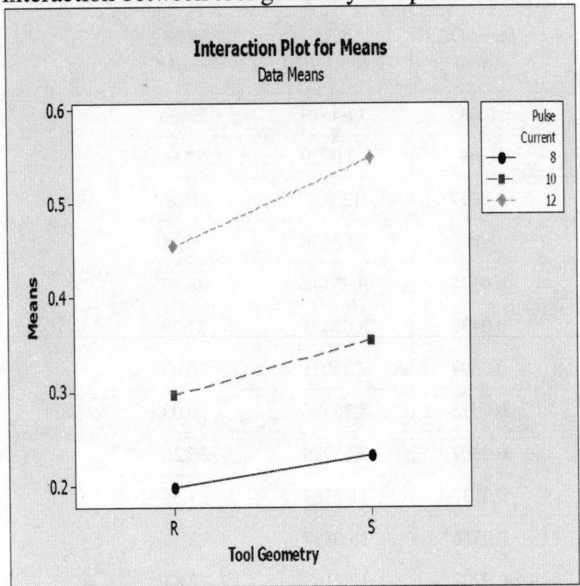

Graph 4.2 Interaction plot for current and tool geometry.

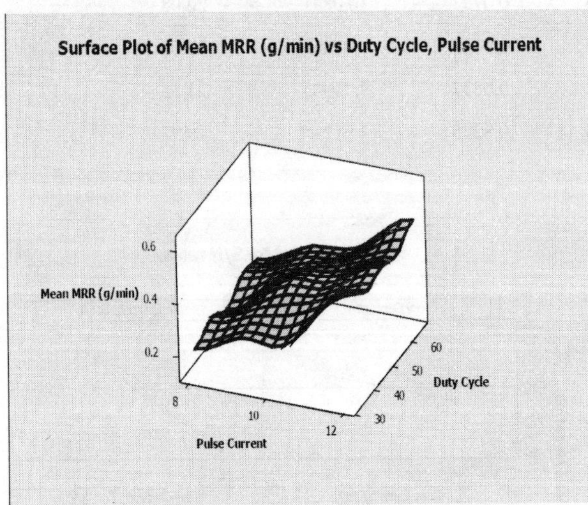

Graph 4.3 Surface plot of mean MRR vs duty cycle, pulse current.

An increase in the peak current leads to the increase of the metal removal rate. This result has been attributed to the increase in peak current which leads to the increase in the rate of the heat energy and hence in the rate of melting and evaporation [1,5,6]

Graph 4.3, Surface plot for mean MRR v/s duty cycle and pulse current shows that MRR is maximum at maximum values of duty cycle and peak current. The graphs for main effects, interaction plot and surface plot has been plotted using minitab 15.

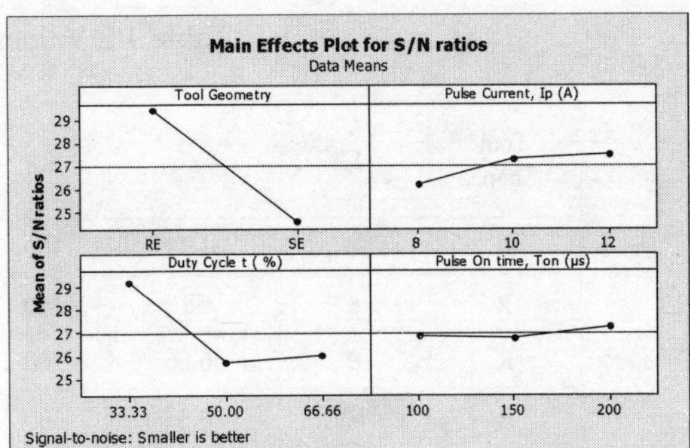

Graph 4.4 Main effects plot for S/N of OC

Graph 4.4, main effects plot for OC shows that S/N for over cut is minimum for square shape of tool. Over cut is minimum at current rating of 8A, duty cycle 50% and pulse on time of 150μs.

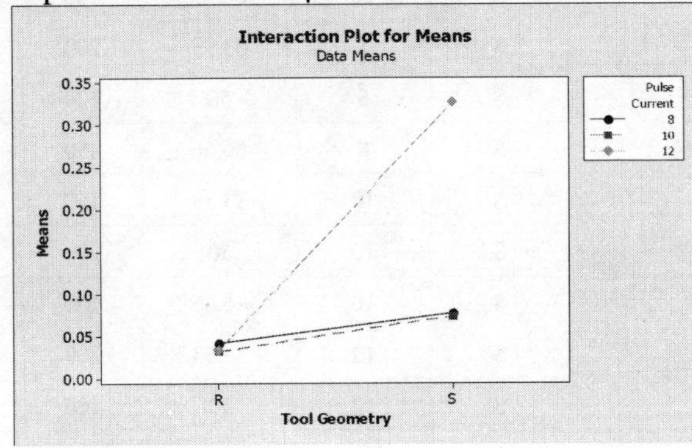

Graph 4.5 Interaction plot of over cut.

Graph 4.5, the interaction plot for over cut shows that over cut shows that there is almost nop interaction or small interaction for round shape tool geometry.

V. CONCLUSION

S/N ratio for material removal rate is maximum when pulse current, duty cycle and pulse on time are maximum. Also it is maximum when square geometry of tool is used. It can also be observed from the data obtained from experimentation that MRR increases with increases with pulse current, duty cycle, and pulse on time. MRR is also high for square tool geometry. The cause for high MRR and high over cut at high current rating is that at high at higher peak current value, the energy released per spark increases leading to higher MRR and higher overcut values. For higher pulse on time, the time for spark increases which upto a certain extent helps in increasing the MRR but after that MRR decreases. Duty cycle is basically ratio of pulse on time to sum of pulse on time and pulse off time. Increase in duty cycle increases the energy delivered per spark and hence MRR increases. The square geometry of tool causes higher MRR. The reason for this may be that for square geometry tool periphery is more as compared as circular geometry tool. Because of this longer periphery area available for machining on sides increases leading to higher MRR.

VI. REFERENCES

[1] M. S. Hewidy, T. A. El -Taweel, M. F. El - Safty (2005), Modelling the machining parameters of wire electrical discharge machining of Inconel 601 using RSM, Journal of Materials Processing Technology, 169,328-336.

[2] Y. Y. Tsai, C. T. Lu, Influence of current impulse on machining characteristics in EDM , Journal of Mechanical Science and Technology, 21,2007,1617-1621

[3] I. Puertas, C. J, Luis, A study on electrical discharge machining of conductive ceramics, Journal of Materials Processing Technology 153-154, 2004,1033-1038.

[4] S. H. Tomadi, M. A. Hassan, Z. Hamedon, IAENG R. Daud, A. G. Khalid, Analysis of the influence of EDM parameters on surface quality material removal rate and electrode wear of tungsten carbide, Proceedings of the International MultiConference of Engineers and computer scientists ,Vol II, March 18-20,2009, Hong Kong.

[5] M. Ramulu, G. Paul, J. Patel, EDM Surface effects on the fatique strength of a 15 vol% SiCp/Al metal matrix composite material, Composite Structures 54, 2001, 79-86.

[6] J. L. Lin, C. L. Lin, The use of the orthogonal array with grey relational analysis to optimize the electrical discharge machining process with multiple performance characteristics, International Journal of Machine Tools and Manufacture 42, 2002,237-244.

Electro less Copper Plating of Silicon Carbide for Cu/SiC$_p$ Metal Matrix Composite

[1]Rajesh Choudhary, [2]Harmesh Kumar Kansal, [3]Shankar Singh
[1]MIMIT Malout, India
[2]Punjab University, Punjab, India
[3]Sant Longowal Institute of Engineering & Technology, Punjab, India

Abstract- **Metal matrix composites are being used in many industrial applications because of their unique and outstanding physical and mechanical characteristics. Powder metallurgy processes are used to produce reinforced MMCs. The main problem in forming Cu/SiC$_P$ composites is the formation of reaction by product i.e.Cu$_3$Si which is formed by the reaction of solid state SiC and copper at 1173 K, which leads to poor thermal and electrical properties. [1]. The possible way of avoiding the Si attack by Cu is by coating SiC particles with copper or some other metal. This paper deals with the electroless plating of SiC particulates with copper. The coated particles will be used for producing copper matrix composites. We have covered three different grain sizes of particles. We have applied the nonelectrolytic method of deposition for coating. Before coating the surface needs to be catalysed. We have analysed the coated surfaces of different grains by means of the scanning electron microscopy.**

Keywords: Electroless copper coating, Copper matrix composite

1. INTRODUCTION

The research in fabricating copper based composites including Cu-Al$_2$O$_3$, Cu-Zr-Al$_2$O$_3$, Cu-TiO$_2$, Cu-Si$_3$N$_4$ and Cu-SiC.is being extensively explored. Method of fabricating these composites include casting, co-precipitation, internal oxidation and powder metallurgy is selected as reinforcement for the metal matrix composite. Owing to these characteristics, SiC has been applied for service at high temperatures under corrosive conditions and in areas where wear must be prevented [8]. Grinding function can be performed with Cu/SiC$_P$ electrode in addition to thermal erosion in electrical discharge grinding. process. The authors have suggested the procedure to apply coating on SiC particulatesfor the fabrication of copper silicon carbide composite. The combination of both materials provides special features (strength, thermal conductivity), which are suitable for electrical discharge machining applications. These features can be ensured, if there is a suitable contact between the matrix and the reinforcement. The production of the composite has many difficulties, for instance suitable bonding between SiC and copper, inhomogenity of SiC particles in the matrix and the The silicon carbide is characterized by its hardness, excellent high temperature creep, high thermal conductivity and excellent corrosion resistance and hence control of porosity. The possible solution of these problems is the coating of the SiC grains with suitable material to avoid interfacial reactions at interface. The most recommended plating material for SiC is copper.

II. Experimentation

Material used

Particulate material and items used for the electrolessplating is given in table 2.

S.No.	Item	Make	Purity
1.	120 mesh SiC	TedPella,USA	99%
2.	220 mesh SiC	CDH	98%
3.	400 mesh SiC	CDH	98%
4.	Potassium sodium tatrtrate	Rankem	99%
5.	Copper sulphate	Rankem	99%
6.	Formaldehyde	Rankem	99%

Table 1. Material used in experimentation

Copper plating by electroless process

The coating of SiC particles with copper can be done by electroless method. In electroless process we need metal salt solution and a reducing agent. But before starting plating, the surface of SiC particles has to cleaned, sensitized and activated. For optimal bonding between SiC particles and copper, following three steps were performed.

(a)   Surface cleaning: Surface cleaning was done by immersing SiC powder in acetone for half an hour with continuous stirring. The acetone dissolves dust particles and impurities, which can remove by decanting off the solution. This process was repeated unless until clear acetone was obtained after decanting off. Then Sic powder was immersed in distilled water again for half an hour with stirring. After removing water the sample was heated at 250°C in an air drying oven. Dried SiC particles were ground to obtain fine particles.

(b)   Sensitization and Activation: It can be achieved by using catalysts. Transition metal salts like $PdCl_2$, $SnCl_2$ have been found to be excellent sensitizer and activator [9]. Catalytic property of such salts is due to their ability to show variable oxidation state. They first, provide surface for adsorption of the substrate by entering into chemical reaction (chemisorptions) and then weaken the bonds of the substrate. SiC particles powder obtained from step 1 was placed in solution containing acidified $SnCl_2$ for one hour. Then SiC powder thus obtained was placed in acidified solution of $PdCl_2$. The composition of these solutions is given in table no.2

Table 2. Composition of the solution for SiC sensitization and activation

Chemical	Amount per 250ml	Temperature(°C)	Time(min)
$SnCl_2$	2g		
HCl	8 ml	20°C	60
$PdCl_2$	0.1g		
HCl	2ml		

(c)   Copper Plating: For electrolessplating cupric salt i.e. copper sulphate has been used by Shu et.al [9]. A number of common reducing agents have been suggested for use in electroless copper baths, like formaldehyde, sugars, hypophosphyte [10]. Formaldehyde was used as reducing agent for this process. It shows its reducing behaviour only at basic pH i.e.pH>12 but copper salts are insoluble at pH>4, so if formaldehyde was to be used as reducing agent the use of some chelating agent was must(ref) and for that potassium sodium tartrate was used for this purpose. SiC powder was first placed in CuSO4 solution with continuous stirring for half an hour. SiC powder was then recovered from above solution and then placed in potassium sodium tartrate solution with stirring for half an hour. Then two solutions were mixed together and stirred continuously. To this solution 40% formaldehyde was added. In order to achieve high pH (pH=13) NaOH was also added. The amount and concentration of the various solutions is given in table no.3

For Cu (II), the relevant half-cell reaction for electroless deposition is:

$$Cu^{2+} + 2e^- \Leftrightarrow Cu^0$$

The copper ions get reduced and deposited on the surface of the SiC particles.

Table 3. Composition of the solution for SiC sensitization and activation

Chemical	Amount per 250ml	Temperature(°C)	Time(min)
Copper Sulfate($CuSO4.5H_2O$)	2g		30
Potassium Sodium Tartrate($KNaC_4H_4O_6$)	5g	20°C	30
40%Formaldehyde(HCHO)	10ml		30(after mixing above two solutions)
Sodium Hydroxide(NaOH)	4g		

III. RESULTS AND DISCUSSION

Three different sizes of SiC powder have been used for coating by this method i.e. 120, 220 and 400 mesh. The coating procedure was same for all the three sizes. It has been observed that the reaction was most vigorous in case of SiC particulate having size 120 mesh and least vigorous in case of 400 mesh size. Fig.1 shows morphology of bared SiC powders of three selected sizes at magnification of 500X. All the three sizes of SiC were examined by SEM after electroplating.It was observed that, if coating time increases, the numbers of the particles and the plating rate also increases as observed by [11].

The coating of finer particle 400 grain size was difficult. If the size of particles decreases, the filtration

and the cleaning of the grains becomes difficult as observed by Tomolya et.al.. Moreover activation of such particles is not uniform because particles are under cohesive forces. SEM images of 400 grain size SiC particles are shown in Fig. 3 at a magnification of 500X and 5000X. It is clear from these micrographs that the coating is not as uniform as in 120 grain size. SiC 220 mesh size shows sufficient coating but coating is not very uniform because of smaller size. The light colored copper layer on the particles can be seen unequivocally. It was observed from SEM of all the three plated SiC powders that all sizes were successfully coated with copper to more and less extent.

IV. CONCLUSION

$SnCl_2$ is good sensitizer and $PdCl_2$ is a suitable activator which activates the surface of silicon carbide particulate to accelerate the plating process. Formaldehyde was observed to be very good reducing agent which released copper at considerable rate which then got deposited on already activated surface of the SiC particulate. The potassium sodium tartrate helped in solubilising copper sulfate at higher pH by making complex with copper sulphate which otherwise remain insoluble at higher pH.

Smaller size of particles introduces difficulties in cleaning filtration and activation because of cohesion resulting into non-uniform coating.

For uniform plating proper contact between particle and reducing copper for sufficient time is required, but due to smaller size contact time get reduced and occasional plating leads to fin shaped plating of the particles

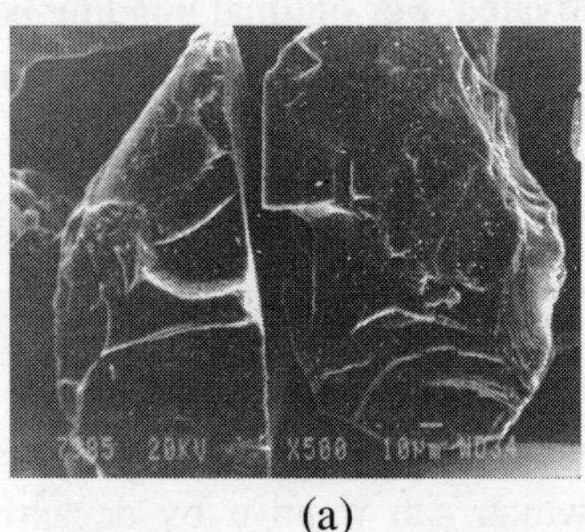

(a)

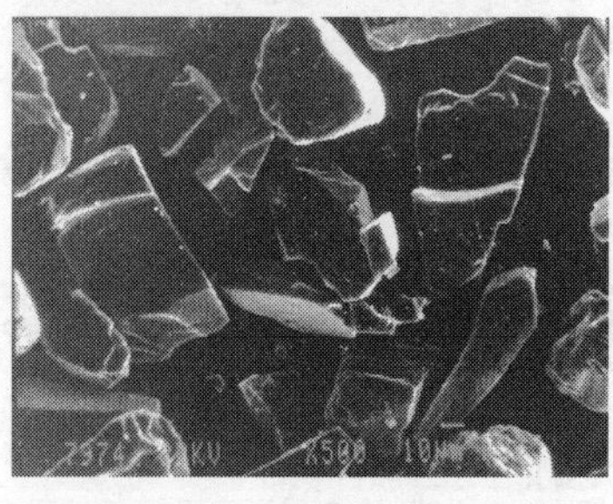

(b)

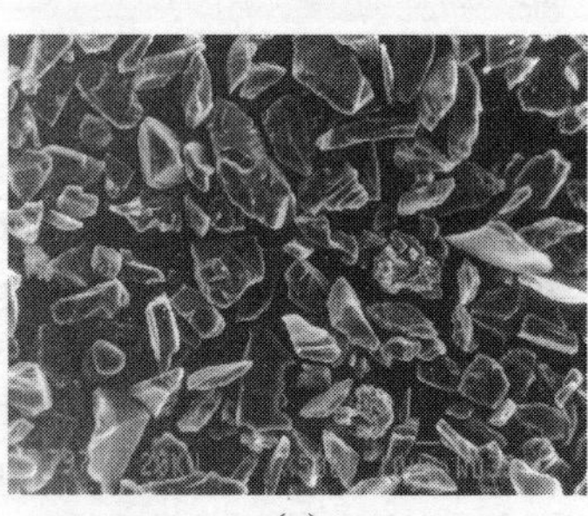

(c)

Fig. 1 The morphology of as received SiC powders a) 120 mesh b) 220 mesh c) 400 mesh at 500X

(a)

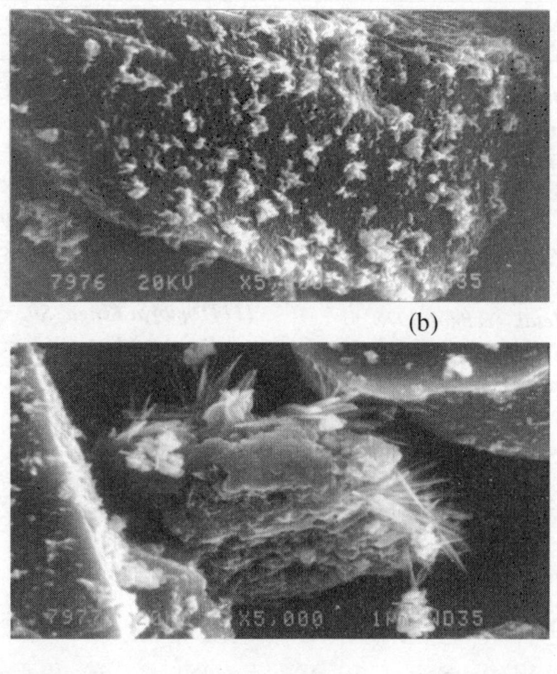

(b)

(c)

Fig. 2. SEM images of coated SiC particles grains a) 220 mesh size at 500X b) 220 mesh size at 5000X after plating

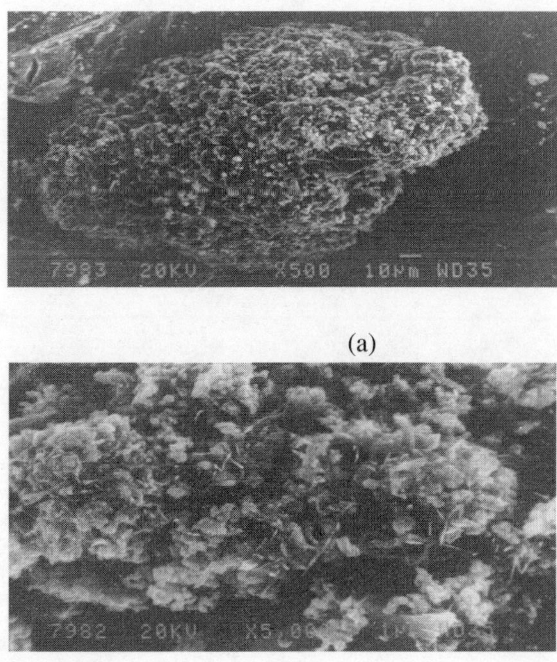

(a)

(b)

Fig. 3 SEM images of coated SiC particles grains a) 120 mesh size at 500X b) 120 mesh size at 5000X after plating

V. REFERENCES

[1] *Mingyuan Gu, Zhi Mei, Yanping Jin, Zhenga Wu: Srtucture and amorphization of the oxide on the silicon carbide surface in an SiC_p/Al composite, Scripta Materialia, Vol. 40., pp. 985-991, 1999.*

[2] *Zwisky, K.M.; Grant, N.J. Metal Prog. 1961, August, 209.*

[3] *Denisenko, E.T.; Polushko, A.; Filatova, N.A. Porosh Kovaya Metall. 1971, 10, 49.*

[4] *Surappa, M.K.; Rohatgi, P.K. Metal. Technol. 1978, 5 (10), 358.*

[5] *Grimes, J.H.; Scott, K.T.B. Powder Metall. 1968 11 (22), 213.*

[6] *Rogers J. A.; Miles, D.E.; Hopkins, B.E. Powder Metal. 1973, 16, 66.*

[7] *Kobayashi, K.F.; Tachibana, N.; Shingu, P.H. J. Mater. Sci. 1990, 25, 3149.*

[8] *Occhionero Mark, Richard Adams, and Kevin Fennessy: A new substrate for electronics packaging: aluminum-siliconcarbide (AlSiC) composites, 1999, http://www.alsic.com/papers/cpseuro992.pdf*

[9] *Shu K. M., Tu G. C., Fabrication and characterization of Cu-SiCp composites for electrical discharge machining applications, Material & Manufacturing Processes, Vol. 20, 483-502, 2006.*

[10] *Cheryl A. Deckert, Shipley Company, Inc.: Electroless Copper Plating, Surface Engineering, Volume 5, ASM Handbook, 1994, pp. 311-322*

[11] *Tomolya Kinga: SiC szemcsék bevonása rézzel, Miskolci Egyetem, Kutatási beszámoló IV., 2003. május 5., pp. 7-26*

Surface Roughness Optimization in Turning Operation based on Taguchi Design of Experiment

Amit Agarwal, Vikas Agarwal Pankaj Gupta, Pramod kumar, Ashish

Department of Mechanical Engg, F.E.T,.R.B.S College, Bichpuri Agra, U.P.

Abstract- **Turning operation is a very important material removal process in modern industry, In this paper Taguchi Method is applied to find Optimum process parameters while machining of EN9(Medium carbon steel) material. A L9 array, signal-to noise ratio and analysis of variance (ANOVA) are applied to study performance characteristics of machining parameters (Cutting speed, feed rate and depth of cut) with consideration of surface finish. Results obtained by Taguchi Method match closely with ANOVA and feed rate are most influencing parameter. Mathematical regression equation is formulated for estimating predicted value of surface roughness and results are verified experimentally**

Keywords: Taguchi design, Turning operation, Surface roughness, Profilometer.

I. INTRODUCTION

The increase of consumer instinct for quality metal cutting related products which have more precise tolerances and better product surface roughness has lead the metal cutting industry to continuously improve quality control of metal cutting processes. Due to the need for different parameters in a wide variety of machining operations, a large number of newly developed surface roughness parameters were developed. Several characteristics contribute to the evaluation of surface quality which is defined by the American National Standards Institute (ANSI).

Surface Roughness

Surface roughness is the measure of the finer surface irregularities in the surface texture. These are the result of the manufacturing process employed to create the surface. The ability of a manufacturing operation to produce a specific surface roughness depends on many factors. In Turning, the final surface depends on cutting speed (m/min); Tool feed rate (mm/rev.) and depth of cut (mm), the amount and type of lubrication at the point of cutting, and the mechanical properties of the piece being machined. A small change in any of the above factors can have a significant effect on the surface produced. Surface finish could be specified in many different parameters viz. Roughness average (R_a), Root-mean-square (rms) roughness (R_q), Maximum peak-to-valley roughness height $(R_y$ or $R_{max})$. In this study Average roughness was measured with a Federal Pocketsurf Profilometer, at three different locations which are shown in Fig. 4. Average Surface Roughness Height is calculated by the relation;

$$Ra = \frac{1}{L}\int_0^L |y(x)|dx \quad (1)$$

Taguchi Approach

The mechanism behind the formation of surface roughness is very dynamic, complicated and process dependent and thus it is very difficult to calculate its value through theoretical analysis. Therefore, trial and error approach is used to set up lathe machine cutting conditions to achieve the desired surface roughness. Obviously, the trial and error method is neither effected nor efficient and the achievement of the desirable value is a repetitive, costly and can be very time consuming. The dynamic nature and wide range of turning parameters usage of turning operations in practice have raised a need for selecting a systematic approach namely, Taguchi method. Taguchi suggested the use of Orthogonal Arrays (OAs), in the experiment implementation stage, investigated and predicted the response value (R_a) which might be affected during the product manufacturing phase.

II. EXPERIMENTAL SETUP AND DATA COLLECTION

External longitudinal turning was performed in a centre lathe of excellent operational condition. Cylindrical bars of EN9 steel of 105 mm in length and of 40 mm in diameter were processed by a carbide tool, which was kept practically sharp during the experiment and cutting fluid was used.

The cutting parameters and their levels are listed in Table I. To elaborate the plan of experiments the method of Taguchi was used for the 3 factors considered (cutting speed, feed rate and depth of cut) at three levels, respectively.

The array chosen was L9 for the main effects of the factors and the interactions (Table II).

Table I Process Parameters & their levels

S.No.	Process Parameter Design	Process Parameter	Levels		
			Low	Medium	High
1.	A	Cutting Speed (m/min.)	80	115	150
2.	B	Feed Rate (mm/rev.)	0.05	0.08	0.10
3.	C	Depth Of Cut (mm.)	0.10	0.25	0.40

Table II The basic Taguchi L9 Orthogonal array

Trial No	Column Factors		
	1	2	3
1	1	1	1
2	1	2	2
3	1	3	3
4	2	1	3
5	2	2	1
6	2	3	2
7	3	1	2
8	3	2	3
9	3	3	1

Average roughness was measured with a Federal Pocket surf profilometer at three different locations (1, 2, 3) as per the isometric view Fig.1, the average roughness height readings are as given below in Table III and Table IV

Table III Design of Experiment or Matrix Experiment Table

S. No.	Cutting Parameter			Average Surface Roughness (µm)	S/N Ratio
	Cutting Speed 'A' (m/min.)	Feed 'B' (mm/rev.)	Depth of Cut 'C' (mm)		
1	80	0.05	0.10	4.26	-12.67
2	80	0.08	0.25	5.16	-14.28
3	80	0.10	0.40	14.87	-23.45
4	115	0.05	0.40	5.84	-15.34
5	115	0.08	0.10	7.21	-17.17
6	115	0.10	0.25	12.52	-21.96
7	150	0.05	0.25	3.17	-10.09
8	150	0.08	0.40	6.93	-16.82
9	150	0.10	0.10	15.33	-23.71

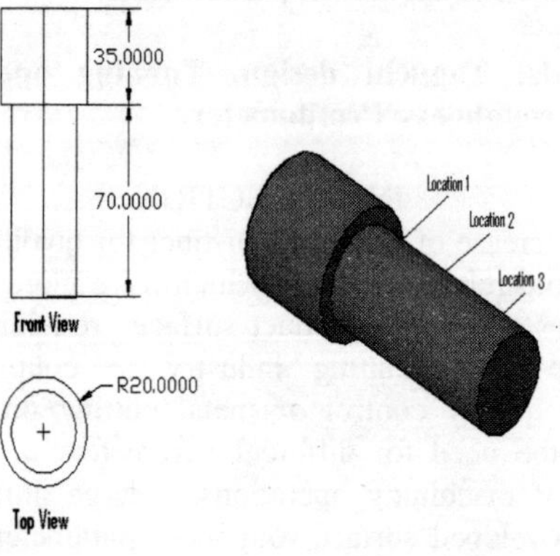

Figure 1 Isometric view of work piece with specified locations

Table IV Average Surface Roughness Measurement at Specified Locations

EX. No.	SURFACE ROUGHNESS MEASUREMENT AT LOCATIONS (µm.)			AVERAGE SURFACE ROUGHNESS µm.
	1	2	3	
1	3.48	4.36	4.94	4.26
2	4.98	5.67	4.84	5.16
3	14.32	14.93	15.36	14.87
4	5.79	6.20	5.53	5.84
5	7.54	7.42	6.68	7.21
6	11.98	12.39	13.19	12.52
7	3.62	2.63	3.26	3.17
8	6.58	7.37	6.84	6.93
9	15.72	14.69	15.58	15.33

III. EXPERIMENT RESULTS AND ANALYSIS

As per the Taguchi parameter design an important attribute is NOISE or uncontrollable factors for the analysis. Here one of the noise factor considered is the measurement location of the work piece and was measured at three locations. The description of various location considered in this project is shown in an isometric view of the work piece (Fig. 1).

Signal to Noise ratio or (S/N ratio)

The response variable considered in this study, surface roughness, is of smaller the better kind. Therefore, signal to noise ratio is defined by

S/N RATIO

$$(\eta) = -10 \log 10 \frac{1}{n} \sum_{i=1}^{n} Y_i^2$$

where, Yi are the individual measurement of surface roughness at each location.

Analysis of Means and Response Graphs for Mean

Analysis of Means

The analysis of each controllable factor is studied and the main effect of the same is obtained in Table V. Mean effect of each factor at individual level i.e. at 1, 2, 3 levels is equal to the average of surface roughness of all the experiments with the factor at individual level.

The term Delta represents the maximum change in the value of surface roughness between the three levels. The larger value of Delta represents the larger influence on the surface roughness. Hence the rank of each factor due to its influence on the surface roughness is found.

Table V Average (Mean) Effect Response Table for the Raw Data

Levels	Controllable Factors		
	A	B	C
	Cutting Speed (m/min.)	Feed Rate (mm/rev.)	Depth of Cut (mm.)
1	*8.10	*4.42	8.93
2	8.52	6.43	*6.95
3	8.48	14.24	9.21
Max.-Min. (Delta)	0.42	9.82	2.26
Rank	3	1	2

From the Table V the value of Delta (Maximum – Minimum) corresponding to the factor (B) Feed rate represents the largest influence on surface roughness followed by factor (C) Depth of Cut and finally factor (A) Cutting Speed.

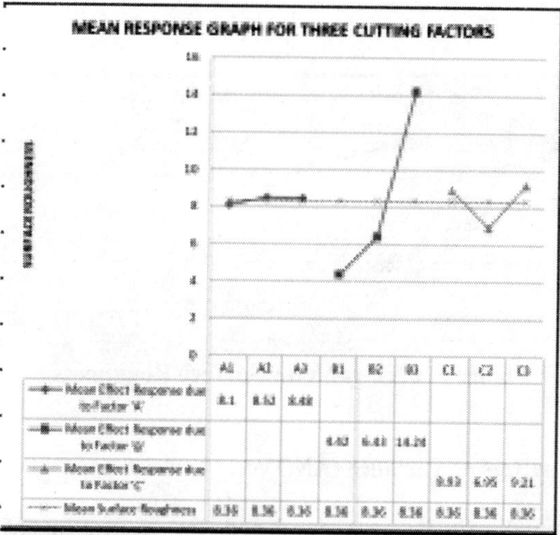

Figure 2 Mean response graph for three cutting factors

Fig. 2 shows the values obtained from the response table are plotted to visualize the effect of the three factors at three levels.

Analysis of Signal to Noise (S/N) Ratio

The objective of using the S/N ratio as a performance measurement is to develop products and processes insensitive to noise factors. The S/N ratio indicates the degree of the predictable performance of a product or process in the presence of noise factors. Process parameter settings with the highest S/N ratio always yield the optimum quality with minimum variance. Thus the level that has a higher value determines the optimum level of each factor. Table 6 represents the Signal to Noise ratio based on Design of experiments.

Table VI: Average Effect Response Table for S/N ratio

Levels	Controllable Factors		
	A Cutting Speed (m/min.)	B Feed Rate (mm/rev.)	C Depth of Cut (mm)
1	**-16.80**	**-12.70**	-17.85
2	-18.16	-16.09	**-15.44**
3	-16.87	23.04	-18.54
Max.-Min	1.36	10.34	3.1
Rank	3	1	2

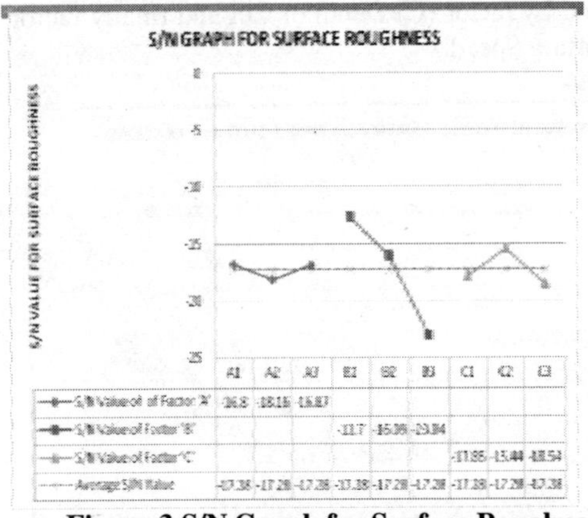

Figure 3 S/N Graph for Surface Roughness

Analysis of Variance (ANOVA)

As experiment was developed for assessment of surface roughness which was influenced by cutting speed, feed rate, and depth of cut. The Table 4 illustrates the responses of surface roughness. Analyses were done with ANOVA results for the identifying factors which are affecting the performances.

Table VII. ANOVA Table for S/N RATIO

S. No.	Factor	DOF	Sum of square	Mean square	F-Value	Contri bution
1	Speed	2	3.51	1.75	3.12	1.87%
2	Feed	2	166.74	83.37	148.87	89.05%
3	Depth of Cut	2	15.87	7.93	14.16	8.47%
4	Error	2	1.12	0.56		0.6%
5	Total	8	187.24			

Figure 4 Contribution of the various parameters affecting surface roughness of EN9.

The planning layout of the confirmation experiments while considering the noise factor total of two experiments were conducted and surface roughness measurement at various locations is tabulated in Table VIII.

Regression Analysis by MATLAB

From the factor level combination, to validate the optimum cutting conditions A1, B1, C2, the result is predicted by regression modeling as,

$$S = 382.53 \,(\text{Cutting Speed})^{0.0413}(\text{Feed Rate})^{1.5876}(\text{Depth of cut})^{0.0096}$$
$$S = 382.53 \,(80)^{0.0413}\,(0.05)^{1.5876}\,(0.25)^{0.0096}$$
$$= 3.89\,\mu m.$$

Table VIII Final Confirmation Experiment

		A1-B1- C2				
	Exp. No.	R1	R2	R3	R_a (μin.)	S/N Ratio
N1	1	2.62	3.12	2.99	2.93	-9.30
N2	2	2.90	2.43	3.04	2.79	-8.95

Table IX Comparison of the Confirmation Experiment for Surface Roughness

	Initial Cutting Parameter	Optimal Cutting Parameters		
		Prediction (Taguchi Method)	Prediction (Regression Modelling)	Experiment
Level	A3B1C2	A1B1C2	A1B1C2	A1B1C2
Surface Roughness (µm.)	3.17	2.75	3.89	2.86
S/N Ratio (dB)	-10.09	-10.38	N.A	-9.12

IV. CONCLUSIONS

Experiments conducted and subsequently analysis performed by using the Taguchi Method and Regression Modeling. The optimum characteristics for the least surface roughness in turning operation are identified and the confirmation experiments are conducted, then the results obtained are compared with the above said optimization methods and are discussed as follows:

- From the (Table V) Average (Mean) Effect Response Table for the Raw Data, the factor (B) Feed rate represents the largest influence on surface roughness followed by factor (C) Depth of Cut, and finally factor (A) Cutting Speed which is further verified with the (Table VI) Average Effect Response Table for S/N Ratio.

- From Table VII, it is apparent that the F- ratio values of factor B (Feed rate) and factor C (Depth of cut) are all greater than $F_{0.10, 2, 2}$ =9.00. While factor A (Cutting Speed) is proved to be a least significant machining factor affecting surface roughness.

- The result obtained from the confirmation experiments reveals that the Taguchi Design of Method has provided the best prediction for the response variable (surface roughness) i.e. 2.75 µm. very close to 2.84 µm. (actual result).

In regard to S/N ratio the results are found to be -10.38 dB in compare to -9.12 dB.

- The paper has discussed an application of the Taguchi method for optimizing the cutting factors in turning operation and indicated that the Taguchi design of experiment is an effective way of determining the optimal cutting factors for surface roughness. The surface roughness achievement from the confirmation runs under the optimal cutting factors identified through Taguchi design of experiment that is accomplished with a relatively small number of experimental runs.

V. ACKNOWLEDGEMENT

The authors are indebted to Dr. B.S. Kushwah, Associated Professor & Head, Department of Mechanical Engineering, F.E.T.R.B.S College, Bichpuri, and Agra for his guidance and facilitating the study.

VI. REFERENCES

[1] Kackar N. Raghu, (1985), "Off-line Quality Control Parameter Design & Taguchi Method" Journal of Quality Technology, Vol. 17, No. 4, pp. 176-188.

[2] Kackar N. Raghu, (1986), "Taguchi's Quality philosophy analysis and commentary", Quality Progress, Dec., pp. 21-28.

[3] Phadke, S.M., (1989), "Quality Engineering Using Robust Design," Prentice Hall, Englewood Cliffs, New jersey.

[4] Barua, P.K. (1997), "Investigation and optimization of V-Process parameters affecting the quality of A1-7% Si Alloy casting using Taguchi Technique". Ph.D. Thesis, University of Roorkee, April.

[5] W.H.Yang, Y.S.Tarng, (1998), "Design Optimization of cutting parameters for turning operations based on the Taguchi Method". "Journal of Materials Processing Technology ", Vol.84, pp.122-129.

[6] Lou Mike S., Chen, Joseph C. Li, Caleb M., (1998), "Surface Roughness Prediction Technique for CNC End–Milling." Journal of Industrial Technology, Vol. 15, No. 1. November, pp.1-6..

[7] Lou, M. S. & Chen, J. C., (1999), "In process surface roughness recognition system in end-milling operations", Journal of Advanced Manufacturing Technology, Vol. 15, pp. 200-209.

[8] Lambda Research , Diffraction Notes,(2000), " Efficiently Optimizing Manufacturing Processes Using Iterative Taguchi Analysis", No.25, Winter, www.lambda-research.com

[9] Yang, J. and Chen, J., (2001), "A Systematic Approach for Identifying Optimum Surface Roughness Performance in End Milling Operations". Journal of Industrial Technology, Vol. 17, No.2. November.

[10] Nicolo Belavendram, (2001), "Quality by Design-Taguchi Technique for Industrial Experimentation, Prentice Hall, Great Britain".

[11]   *Hazim El-Mounayri, Zakir Dugla and Haiyan Deng, (2002) , "Predicting Surface Roughness in End Milling using Innovative Technique from EC(Evolutionary Computation)", ASME-International Mechanical Engineering Congress & Exposition; Symposium on Process Planning and Process Optimization.*

[12]   *Ganter W.A., (2004), "An observation of Taguchi method", Quality and Reliability Intl. Journal Vol. 15, pp. 3-5.*

[13]   *Phillip J Ross, (2005), "Taguchi Techniques for Quality Engineering" Tata Mc-Graw Hill 2nd edition.*

[14]   *E.Daniel Kerby, (2006), "A Parameter study in a turning Operation using the Taguchi Method", The Technology Interface/ Fall, Iowa State University pp. 1-14.*

[15]   *Sang-Heon Lim, Choon-Man Lee and Won Jee Chung, (2006), " A study on the optimal cutting condition of a high speed feeding type laser cutting machine by using Taguchi method", Intl. Journal Of Precision engineering and manufacturing, Vol. 7, No.1, January, pp. 18-23.*

[16]   *Neelesh K. Jain, V.K. Jain and Kalyanmoy Deb, (2007), "Optimization of process parameters of mechanical type advanced machining processes using genetic algorithms" Intl. Journal of Machine Tools and Manufacture, Vol. 47, Issue 6, May, pp. 900-919.*

[17]   *S.Thamizhmanii, S. Saparudin, S. Hasan, (2007), " Analysis of surface roughness by turning process using Taguchi method", Journal of Achievements in Materials and Manufacturing Engineering (JAMME), Vol.20, Issue 1-2, pp. 503-506.*

[18]   *Hari Singh, Pradeep Kumar (2007), "Mathematical models of tool life and surface roughness for turning operation through response surface methodology" Journal Of Scientific & Industrial Research, Vol.66, March , pp. 220-226.*

[19]   *M. Sanjari, A. Karimi Taheri and M. R. Movahedi (2008), "An optimization method for radial forging process using ANN and Taguchi method" Intl. Journal of Advanced Manufacturing Technology, Vol. 40, February, pp. 776-784.*

[20]   *F.C. Tsai, B.H. Yan, C.Y. Kuan and F.Y. Huang , (2008), "A Taguchi and experimental investigation into the optimal processing conditions for the abrasive jet polishing of SKD61 mild steel". Intl. Journal of Machine Tools and Manufacture, Volume 48, Issues 7-8, June, pp. 932-945.*

[21]   *Hang Rae Kim, Kang Yong Lee (2009), "Application of Taguchi method to determine hybrid welding conditions of Aluminum alloy", Journal of Scientific & Industrial Research, Vol.68, April, pp. 296-300.*

Study and Comparative Analysis of Unconventional Manufacturing Processes

Dheeraj Nanda , Himanshu Ranjan Verma , Ashutosh Singh

Department of Mechanical Engineering Anand Engineering College Keetham Agra

*Abstract-*This paper entitled to provide a review of machining of materials in the context of applications and comparison to Non conventional manufacturing & machining processes. The important concepts and background of the processes are introduced briefly. After which, the applications of these processes are discussed as applied to relevant themes in machining processes. A brief overview of research work on experimental and theoretical studies on various process monitoring and control is considered. Selection of a particular process is explained with the help of tables and graphs.

I. Introduction

Given the importance of machining for most industries, there exists a wealth of research focusing on improving the performance of the machining process. Manufacturing is process of making of goods & articles & providing services to meet the human needs. Since beginning of the human development, people have evolved tools and energy sources to power these tools to meet the requirements for making the life easier and enjoyable. In the early stage, tools were made of stone for the item being made. When iron tools were invented, desirable metals and more sophisticated articles could be produced. In twentieth century products were made from the most durable and consequently, the most un machinable materials. In an effort to meet the manufacturing challenges created by these materials, tools have now evolved to include materials such as alloy steel, carbide, diamond and ceramics and to produce complex shapes. Initially, tools were powered by muscles; either human or animal.

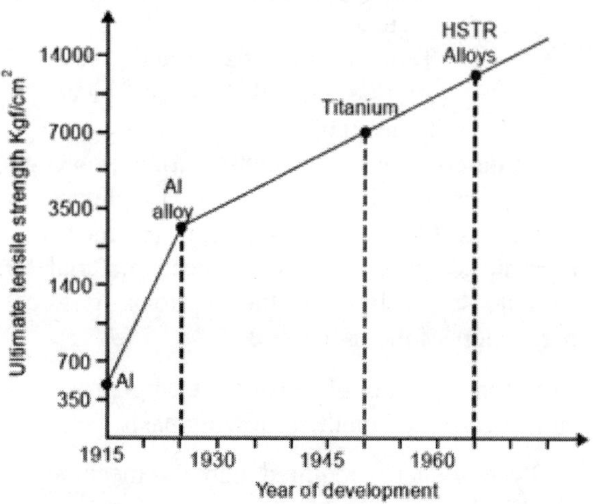

Fig. 1   *Trend of increase of material strength*

In the figure 1, Merchant had displayed the gradual increase in strength of material with year wise development of material in aerospace industry. By the Non-traditional manufacturing processes material removal can be done precisely by various ways such as electro-chemical reaction, high temperature plasma and high-velocity jets of liquids and abrasive.

During the last 80 years, over 20 different non-traditional manufacturing processes have been invented and successfully implemented into production processes.

II. CLASSIFICATION OF UNCONVENTIONAL MANUFACTURING PROCESSES

Various mechanical properties such as Hardness, Toughness and brittleness etc basically restrict the formation of the component of desired shape and dimensions. In this era, with the development of industrialization there is a need of new material and the processes to meet the requirements efficiently.

There are various situations where traditional machining processes are not economical such as

- Materials of high strength and hardness.
- Too flexible work piece to support the cutting forces.
- Complex shape and close tolerance.
- High surface finish and dimensional accuracy.
- Temperature rise and residual stresses in the work piece is undesirable.

The non-conventional manufacturing processes are not affected by hardness, toughness or brittleness of material and can produce any intricate shape on any work piece material by suitable control over the various physical parameters of the processes.

The non-conventional manufacturing processes can be classified on the following basis

(*i*) Type of energy required, namely, mechanical, electrical, chemical etc.

(*ii*) Basic mechanism involved in the processes, like erosion, ionic dissolution, vaporization etc.

(*iii*) Source of immediate energy required for material removal, namely, hydrostat

ic pressure, high current density, high voltage, ionized material etc.

(*iv*) Medium for transfer of energies, like high velocity particles, electrolyte, electron, hot gases, etc.

On the basis of above requirements, the various processes may be classified as shown in table 2.

III. COMPARATIVE ANALYSES OF UNCONVENTIONAL MANUFACTURING PROCESSES

A comparative analysis of the various unconventional manufacturing processes should be made so that a guide-line may be drawn to find the suitability of application of different processes.

A particular manufacturing process found suitable under the given conditions may not be equally efficient under other conditions. Therefore, a careful selection of the process for a given manufacturing problem is essential. The analysis has been made from the point of view of:

- Physical parameters involved in the processes
- Capability of machining different shapes of work material
- Applicability of different processes to various types of material, *e.g.* metals, alloys and non-metals
- Operational characteristics of

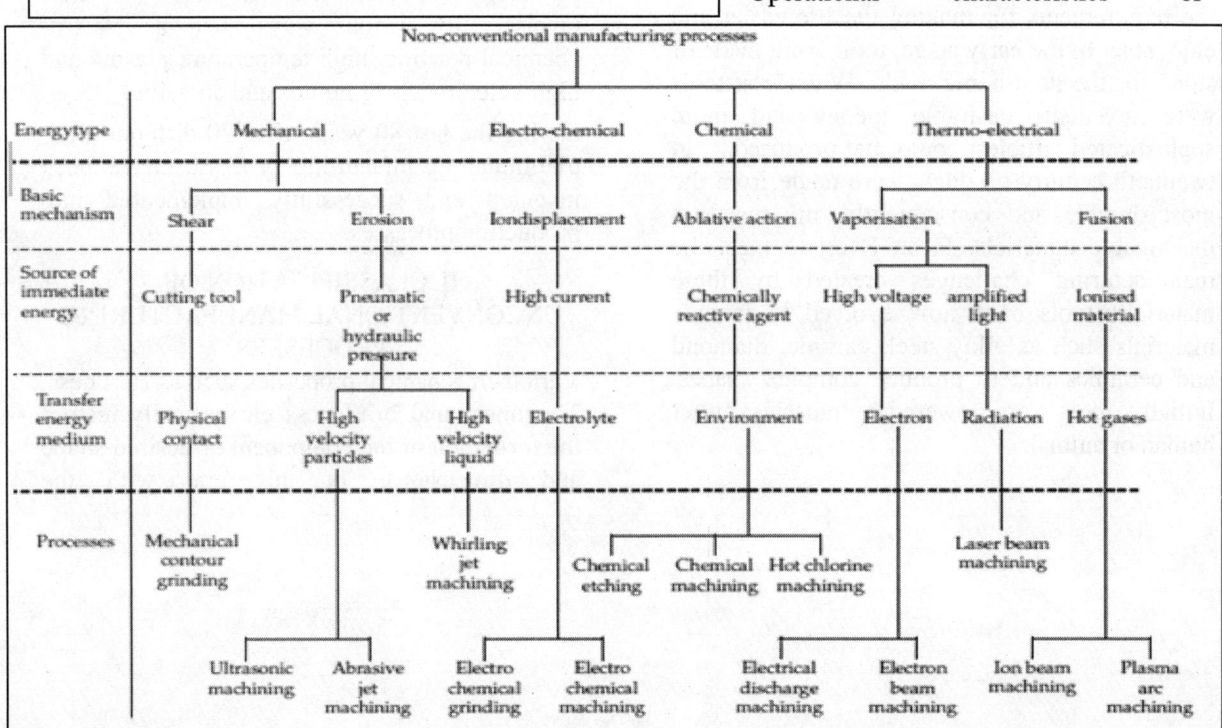

Table-1 Classification of Non Conventional processes

manufacturing
- Economics involved in the various processes.

A- PHYSICAL PARAMETERS

The physical parameters are basically physical properties. These parameters have a direct impact on the metal removal as well as on the energy consumed in different processes. (Table 2)

Thus, the capital cost involved in the processes (EBM, ECM etc.) lying above the mean line is high   whereas for the processes below that

line (*e.g.*, EDM, PAM, MCG) is comparatively low

B- CAPABILITY TO SHAPE APPLICATION

The capability of different processes can be analyzed on the basis of various machining operation point of view such as micro-drilling, drilling, cavity sinking, pocketing (shallow and deep), contouring a surface, through cutting (shallow   and   deep)   etc.

TABLE 2 Physical Parameters of the Non-conventional Processes

Parameters	USM	AJM	ECM	CHM	EDM	EBM	LBM	PAM
Potential (V)	220	220	10	—	45	150000	4500	100
Current (Amp)	12 (A.C.)	1.0	10000 (D.C.)	—	50 (Pulsed D.C.)	0.001 (Pulsed D.C.)	2 (Average 200 Peak)	500 (D.C.)
Power (W)	2400	220	100000	—	2700	150	—	50000
Gap (m.m.)	0.25	0.75	0.20	—	0.025	100	150	7.5
Medium	Abrasive in water	Abrasive in gas	Electrolyte	Liquid chemical	Liquid dielectric	Vaccum	Air	Argon or hydrogen

From the comparative study of the effect of metal removal rate on the power consumed by various non-conventional machining processes shown in graph.   fig. 2

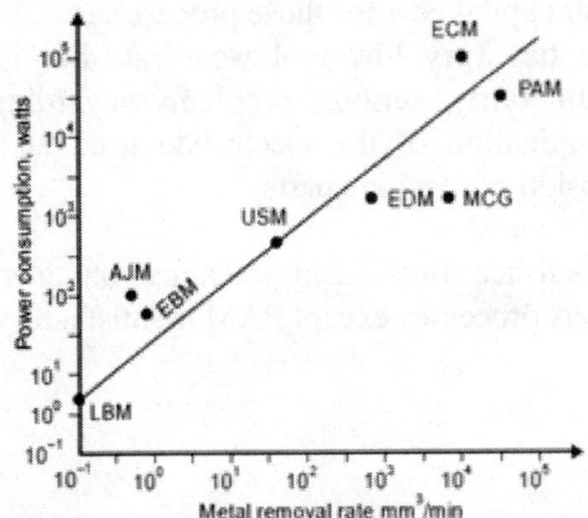

Fig . 2 shows effect metal removal rate on power consumption

It is found that some of the processes (*e.g.* EBM, ECM) above the mean power consumption line consume a greater amount of power than the

processes (*e.g.* EDM, PAM, ECG) below the mean power consumption line.

For micro-drilling operation, the only process which has good capability to microdrill is laser beam machining while for drilling shapes having slenderness ratio, $l/D < 20$, the process USM, ECM and EDM will be most suitable.

EDM and ECM processes have good capability to make pocketing operation (shallow or deep). For surface contouring operation, ECM process is most suitable but other processes except EDM have no application for contouring operation.

C-   APPLICABILITY TO MATERIALS

Materials applications of the various machining methods are summarized in the table 3 and table 4. For the machining of electrically non-conducting materials, both ECM and EDM are unsuitable, whereas the mechanical methods can achieve the desired results

TABLE 3. Capability of Shape Application

ii. Tolerance maintained

Process	Holes				Trough cavities		Surfacing		Trough cutting	
	Precision small holes		Standard		Precision standard		Double contouring	Surface of revolution	Shallow	deep
	Dia < .025 mm	Dia > .025 mm	Length < 20 mm	Length > 20 mm						
USM	—	—	good	poor	good	good	poor	—	poor	—
AJM	—	—	fair	poor	poor	fair	—	—	good	—
ECM	—	—	good	good	fair	good	good	fair	good	good
CHM	fair	fair	—	—	poor	fair	—	—	good	—
EDM	—	—	good	fair	good	good	fair	—	poor	—
LBM	good	good	fair	poor	poor	poor	—	—	good	fair
PAM	—	—	fair	—	poor	poor	—	poor	good	good

TABLE-4 Material Application

Metals Alloys					
Process	Aluminium	Steel	Super alloy	Titanium	Refractory material
USM	Poor	Fair	Poor	Fair	Good
AJM	Fair	Fair	Good	Fair	Good
ECM	Fair	Good	Good	Fair	Fair
CHM	Good	Good	Fair	Fair	Poor
EDM	Fair	Good	Good	Good	Good
EBM	Fair	Fair	Fair	Fair	Good
LBM	Fair	Fair	Fair	Fair	Poor
PAM	Good	Good	Good	Fair	Poor

USM is suitable for machining of refractory type of material while AJM are for super alloys and refractory materials.

TABLE-5 Material Application

	Non-Metals		
Process	Ceramics	Plastic	Glass
USM	Good	Fair	Good
AJM	Good	Fair	Good
ECM	—	—	—
CHM	Poor	Poor	Fair
EDM	—	—	—
EBM	Good	Fair	Fair
LBM	Good	Fair	Fair
PAM	—	Poor	—

D- MACHINING CHARACTERISTICS

The machining characteristics of different non-conventional processes can be analyzed with respect to :
i. Metal removal rate
iii. Surface finish obtained
iv. Depth of surface damage
v. Power required for machining

The process capabilities of non-conventional manufacturing processes have been compared in table 6.

The metal removal rates by ECM and PAM are respectively one-fourth and 1.25 times that of conventional whereas others are only small fractions of it.

Power requirement of ECM and PAM is also very high when compared with other non-conventional machining processes. This involves higher capital cost for those processes.

ECM has very low tool wear rate but it has certain fairly serious problems regarding th contamination of the electrolyte used and the corrosion of machine parts

The surface finish and tolerance obtained by various processes except PAM is satisfactory.

Table-6 Process Capabilities

Process	MRR (mm³/min)	Tolerance (μ)	Surface (μ) CLA	Depth of surface damage (μ)	Power (watts)
USM	300	7.5	0.2-0.5	25	2400
AJM	0.8	50	0.5-1.2	25	250
ECM	15000	50	0.1-2.5	5.0	100000
CHM	15	50	0.5-2.5	50	—
EDM	800	15	0.2-12	125	2700
EBM	1.6	25	0.5-2.5	250	150 (average) 2000 (peak)
LBM	0.1	25	0.5-1.2	125	2 (average)
PAM	75000	125	Rough	500	50000
Conventional machining	50000	50	0.5-5.0	25	3000

The metal removal efficiency is very high for EBM and LBM than for other processes.

IV. CONCLUSION

In conclusion, the suitability of application of any of the processes is dependent upon various factors such as capability of machining, applicability, operational characteristics and economics involved in the various processes. These must be considered all or some of them before applying nonconventional processes.

V. REFERENCES

[1]. Boothroyd, G. and Knight, W.A (1989) *Fundamentals of Machining and Machine Tools*, 2nd edition, New York: Marcel Dekker Inc.

[2]. Cook, N.H. and Nayak, P.N. (1969) 'Development of improved cutting tool materials', *Technical Report of AFML-TR-69-185*, US Air Force Materials Laboratory.

[3]. Ezugwu, E.O. (2005) 'Key improvements in the machining of difficult-to-cut aerospace superalloys', *International Journal of Machine Tools and Manufacture*

E- ECONOMICS OF THE PROCESSES

The economics of the various processes are analyzed on the basis of following factor and given in Table 7.

(*a*) Capital cost

(*b*) Tooling cost

(c) Consumed power cost

(d) Metal removal rate efficiency

(*e*) Tool wear.

Table-7 Economics of Processes

Process	Capital cost	Tooling cost	Power consumption cost	Material removal rate efficiency	Tool wear
USM	L	L	L	H	M
AJM	VL	L	L	H	L
ECM	VH	M	M	L	VL
CHM	M	L	H*	M	VL
EDM	M	H	L	H	H
EBM	H	L	L	VH	VL
LBM	L	L	VL	VH	VL
PAM	VL	L	VL	VL	VL
MCG	L	L	L	VL	L

The capital cost of ECM is very high when compared with traditional mechanical contour grinding and other non-conventional machining processes whereas capital costs for AJM and PAM are comparatively low.

EDM has got higher tooling cost than other machining processes. Power consumption is very low for PAM and LBM processes whereas it is greater in case of ECM.

Optimization & Analysis of Welding Operation parameters for Achieving the Highest Weld - Strength by using Taguchi Design of Experiment

[1]A.K.Sehgal, [2]Anuj Agarwal, [2]Meenu
[1]Department of Mechanical Engineering, K.P.Engineering College, Agra
[2]Department of Mechanical Engineering, N.I.T, Kurkshetra, Haryana

Abstract- **The purpose of this study is to efficiently determine the optimum welding operation parameters for achieving the highest breaking strength in the range of parameters. In order to meet the purpose in terms of both efficiency and effectiveness, Taguchi parameter design methodology is studied. This may include selection of parameters, utilizing an L_9 orthogonal array, conducting experimental runs, data analysis, regression modeling, determining the optimum combination and finally verification experimentally. Hardness of the material to be welded, electrode diameter and edge preparation are suggested as controlled parameters for the analysis. Taguchi suggested the use of Orthogonal Arrays (OA[s]), in the experiment implementation stage, to investigate and predict noise factors which might affect the quality of a given product during the product manufacturing phase. Through OA experiment analysis, the quality influencing factors of a product can then be identified, controlled, and hence compensated during early product design stage.**

1. INTRODUCTION

Welding is a fabrication or sculptural process that joins materials, usually metals or thermoplastics, by causing coalescence. This is often done by melting the work pieces and adding a filler material to form a pool of molten material (the weld pool) that cools to become a strong joint, with pressure sometimes used in conjunction with heat, or by itself, to produce the weld[1]. The same consumable finds application for the welding of high hardness quenched and tempered steels to meet the service requirements in the construction of combat vehicles[2]. However, use of stainless steel filler for a non stainless steel base metal must be avoided as ASS fillers are considerably more expensive. In the recent years, the developments of low hydrogen ferritic (LHF) steel consumables that contain no hygroscopic compounds are utilized for welding quenched and tempered steels [4, 6].

Quenched and tempered steel welds must be of good quality especially when used for construction of combat vehicles in military applications[13]. In order to achieve the highest weld strength in terms of both efficiency and effectiveness, this study will utilize theTaguchi parameter design methodology. This includesselection of parameters, utilizing an orthogonal array,conducting experimentalruns, data analysis, determining the optimum combination and verification. As suggested in the introduction, hardness, electrode diameter, edge preparations are included as controlled parameters in this study. The Taguchi quality strategies discussed are derived from several experiment techniques used in the product design stage to implement the quality concepts into the product. Taguchi suggested the use of Orthogonal Arrays (OA), in the experiment implementation stage, to investigate and predict noise factors which might affect the quality of a given product during the product manufacturing phase. Through OA experiment analysis, the quality influencing factors of a product can then be identified, controlled, and hence compensated during early product design stage. Considering that the literature suggested that the feed rate has a much higher effect on breaking strength than the other two parameters, it was determined that a robust but efficient experiment would include more attention than the other factors.

II. EXPERIMENTAL PROCEDURE

Test Set-up and Experimentation

Selection of Ranges & Levels of Process Variables

The experimental setup is constructed for the material and the various factors and their levels are chosen, which are dependent on the following properties of the material under welding:

- Structure of the material
- Hardness of the material
- Tensile Strength of the material

Hardness

The hardness depends upon the type of material being used. The original material is taken as medium (M), and the other two were prepared by establishing heat treatment with the standard procedure and are designated as soft (S) & hard (H) respectively. Selection of ranges for hardness is as follows:

S	=	279 BHN
M	=	294 BHN
H	=	326 BHN

Diameter of Electrode

The second parameter is electrode diameter and it has three levels. The three levels are low, medium and high. Considering the constant parameter:

Type of Welding – Electric Arc Welding

Skill of worker – Same Worker

For the experimentation, from literature review the range of welding electrode diameter is kept at three levels that are given below: (Fig. 1)

Fig. 1 Welding Rods

L = Lower Level = 3.70 mm.
M = Medium Level = 4.70 mm.
H = High Level = 5.70 mm.

Weld Design (Edge Preparation)

Weld design are of different types but we use here three levels called three weld designs such as V, J and Bevel type. V type of edge is one in which both the edge is at 30 degree angle. In Bevel type one edge is straight and the other is at 45 deg. Angle. In J type one edge is straight and the other is in J- shape. Selections of weld designs are as follows: (Fig. 2.1, Fig. 2.2 & Fig. 2.3).

L = V
M = Bevel
H = J

Fig. 2.1 V -joint   Fig. 2.2 J-joint   Fig. 2.3 Bevel joint

Experimental Layout

The process parameter and their levels have been decided on the basis of past experience and literature. The experimental setting of the work piece & the welding process is as shown in Fig. 3. The experiments are to be conducted in three levels with four controllable factors. These parameters and their levels are tabulated in Table I.

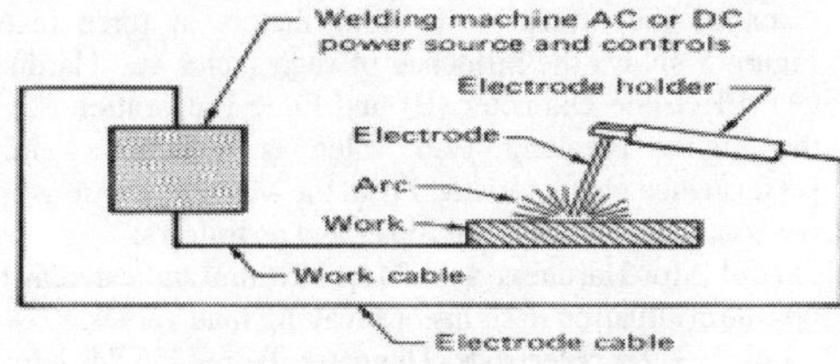

Fig. 3 Experimental Setting of the Work Piece & the Welding Process

Fig. 4 Universal Testing Machine (UTM)
Courtesy: PPDC, Foundry Nagar, Industrial Site, Agra

TABLE I
Process Parameters and their Levels

SN	Exp. No.	Hardness	Electrode Diameter (mm)	Edge Preparation	Breaking Load (N/mm^2) I	Breaking Load (N/mm^2) (II)	Mean Breaking Load (I + II)/2 (N/mm^2)
1	1	S	D1=3.70	V	197.51	206.69	202.10
2	2	S	D2=4.70	BEVEL	220.65	220.65	220.65
3	3	S	D3=5.70	J	298.89	290.44	294.66
4	4	M	D1=3.70	J	240.97	230.08	235.52
5	5	M	D2=4.70	V	325.10	391.83	358.46
6	6	M	D3=5.70	BEVEL	420.85	438.24	429.55
7	7	H	D1=3.70	BEVEL	228.97	228.97	228.97
8	8	H	D2=4.70	J	345.55	273.30	309.42
9	9	H	D3=5.70	V	329.75	356.32	343.03

Specification of work material

Low carbon mild steel
Breaking strength= 535.22N/mm^2
Hardness (BHN) = 292
Weight = 500 gm

Chemical specification

Carbon(C) = 0.06%
Silicon (Si) = 0.09%

TABLE II

S.No	Process Parameter Design	Process Parameter	Levels		
			Low	Medium	High
1.	A	Hardness	S	M	H
2.	B	Electrode Diameter(mm)	D1=(3.70)	D2=(4.70)	D3=(5.70)
3.	C	Weld Design	V	Bevel	J

Consolidated Design of Experiment Table

Manganese (Mn) = 0.37%
Phosphorus (P) = 0.063%
Sulphur (S) = 0.065%

Dimension of work material

Length = 300 mm
Width = 25 mm
Thickness = 10 mm

Measurement of Breaking Strength

The tensile test on the universal test machine is carried out by gripping the two ends of the specimen and applying an increasing pull on the specimen till it fractures. During the test, the tensile load as well as elongation of the previously marked gauge length in the specimen is measured with the help of load dial of the machine and extensometer respectively.

III. TAGUCHI ANALYSIS

Calculation for Degree of Freedom

Total Degree of freedom (D.O.F): Each of the three factors is to be studied at three levels; therefore each factor has D.O.F. of 2

Since, D.O.F. = number of levels -1.

The degree of freedom for the interaction is computed by multiplying the D.O.F. of each of interacting factors. Thus;

D.O.F. for A is 2
D.O.F. for B is 2
D.O.F. for C is 2

Therefore, the total D.O.F. or,

$$(D.O.F._T) = 6$$

Selection of Orthogonal Array

The OA (Orthogonal Array) to be selected must satisfy the following conditions:

Since, D.O.F. of O.A. selected $\geq$ D.O.F. required.

The experiment under consideration has 6 D.O.F. and therefore requires an O.A with 9 or more D.O.F., Hence an O.A. with at least 9 experiments is to be selected to estimate the effect of each factor and the desired interaction. Taguchi has tabulated 9 basic orthogonal arrays which are called standard orthogonal arrays. An arrays name indicates the number of rows, number of columns and number of levels in each of the columns. As per Taguchi philosophy; usually it is expensive to conduct various experiments therefore, the optimum possible OA must be selected. Here, the selected O.A. is L_9 from the book of standard orthogonal Arrays (Table II). The factors & their chosen levels are listed in Table I. The starting levels before conduction the matrix experiments for the 3 factors, identified by hardness, electrode diameter and weld design (edge preparation).

Condensed Experimental Results

With reference to Table I and Table II; the results are collected for the design the experiment or matrix experiment (Table III). As discussed earlier in this project, the controllable factors taken are Hardness (A), Electrode Diameter (B) and Weld Design (C). Since they affect weld strength in welding operations, and since these factors are controllable in the welding process, they are considered as controllable factors. Further, as per the Taguchi parameter design an important attribute is uncontrollable factors for the analysis. The gauge length while testing in UTM is selected as uncontrollable factor.

Signal to Noise ratio or (S/N) ratio

The response variable considered in this study, breaking load is of larger the better kind. Therefore, signal to noise ratio is defined by considering the reciprocal of the quality characteristic. The objective function is to be maximized:

$$S/N \text{ Ratio} = -10 \, Log_{10} \, 1/n * \Sigma_i \, (1/ \, Y_i^2)$$

TABLE III

Mean breaking load & S/N ratio summary sheet

Symbol	Controllable factors	Level I	Level II	Level III
A	Hardness	239.13	**341.17**	293.80
B	Electrode Diameter	222.19	296.17	**355.74**
C	Edge Preparation	**301.19**	293.056	279.86

Average of S/N Ratio = 48.99 dB

Where, Y_i are the individual measurement of breaking load at each location.

Analysis of Means and Response Graphs for Mean

Analysis of Means

The analysis of each controllable factor is studied and the main effect of the same is obtained in Table IV. Main effect of each factor at individual level i.e. at 1, 2, 3 levels is equal to the mean of breaking load of all the experiments with the factor at individual level.

For example;

(a)   The main effect of the hardness on breaking load at various level is calculated as follows:

(b)

$$S = (202.10 + 220.65 + 294.66)/3$$
$$S = 239.13 \text{ N/mm}^2.$$

Similarly, $M = 341.17$ N/mm²
$$H = 293.80 \text{ N/mm}^2$$

(c)   The main effect of the electrode diameter at various levels are calculated as:

$$D1 = (202.1 + 235.52 + 228.97)/3$$
$$D1 = 222.19 \text{ N/mm}^2$$
Similarly: $D2 = 296.17$ N/mm²
$$D3 = 355.74 \text{ N/mm}^2$$

(d)   The main effect of the edge preparation at various levels are calculated as:

$$V = (202.1 + 358.46 + 343.03)/3$$
$$V = 301.190 \text{ N/mm}^2$$
Similarly: $B = 293.056$ N/mm²,
$$J = 279.860 \text{ N/mm}^2$$

TABLE IV

Factors Affect Table for Mean Breaking Load

From the Table IV the value in bold shows the larger the better criteria as proposed by Taguchi method.

Exp. No.	Breaking Load (N/mm²) I	Breaking Load (N/mm²) (II)	Mean Breaking Load (I + II)/2 (N/mm²)	S/N ratio (dB)
1	197.51	206.69	202.10	46.11
2	220.65	220.65	220.65	46.87
3	298.89	290.44	294.66	49.37
4	240.97	230.08	235.52	47.44
5	325.10	391.83	358.46	50.97
6	420.85	438.24	429.55	52.65
7	228.97	228.97	228.97	47.20
8	345.55	273.30	309.42	49.65
9	329.75	356.32	343.03	50.69

Response Graphs for mea

The values obtained from the response table are plotted to visualize the effect of the four factors at three levels. Figure 5 shows the influence of each factor viz. Hardness (A), Electrode Diameter (B) and Edge Preparation (C) as the higher breaking load value is a better welding performance characteristic. From the Mean response graph, observational findings are illustrated as follows:

1. Level 2 for Hardness $A_2 = 341.17$ N/mm² indicated as the optimum situation in terms of breaking load values.
2. Level 3 for Electrode Diameter $B_3 = 355.74$ N/mm² indicated as the optimum situation in terms of breaking load values.
3. Level 1 for Edge Preparation $C_1 = 301.19$ N/mm² indicated as the optimum situation in terms of breaking load values.

Fig. 5: Mean response graph for three welding factors

3.6 *Analysis Of Signal To Noise Ratio & Response Graphs For Breaking Load*

Analysis of Signal to Noise (S/N) Ratio

The objective of using the S/N ratio as a performance measurement is to develop products and processes insensitive to noise factors. The S/N ratio indicates the degree of the predictable performance of a product or process in the presence of noise factors. Process parameter settings with the highest S/N ratio always yield the optimum quality with minimum variance. Thus the level that has a higher value determines the optimum level of each factor. From the Table V, the value in bold shows the larger the better criteria as proposed by Taguchi method.

TABLE V
Factor Effect Table of S/N Value for Breaking Load

Symbol	Controllable factors	Level I	Level II	Level III
A	Hardness	47.45	**50.68**	49.18
B	Electrode Diameter	46.91	49.16	**50.90**
C	Edge Preparation	**49.25**	48.90	48.82

The values obtained from the response table are plotted to visualize the effect of the three parameters. Fig. 6 shows the influence of each factor viz. hardness (A), electrode diameter (B), edge preparation (C) on the breaking load and as the higher breaking load value is a better welding performance characteristic. From the S/N response graph observational findings are illustrated as follows:

(a)   Level 2 for hardness A_2 = 50.68 dB indicated as the optimum situation in terms of S/N values.

(b) Level 3 for electrode diameter B_3 = 50.90 dB indicated as the optimum situation in terms of S/N values.

(c) Level 1 for edge preparation C_1 = 49.25 dB indicated as the optimum situation in terms of S/N values.

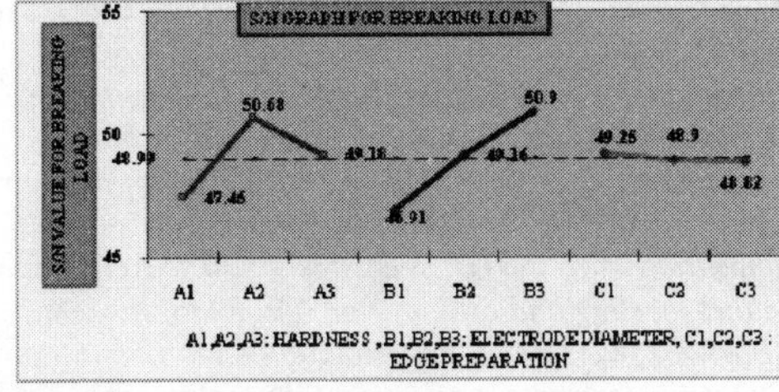

Fig. 6: S/N Graph for Breaking Load

Analysis Of Variance (ANOVA)

Analogy with Fourier analysis [5] of the power of an electrical signal and ANOVA is displayed in Fig. 6. The experiments are arranged along the horizontal axis like time. The overall mean is plotted as a straight line like a DC component. The effect of each factor is displayed as a harmonic. The level of factor A for experiments 1, 2, 3 is A1. So the height of the wave for A is plotted as m_{A1} for these experiments. Similarly the height of the wave for the experiments 4, 5, 6 is m_{A2} and the height of the wave for the experiments 7, 8, 9 is m_{A3} .The waves of other factor are also plotted similarly. By virtue of this analogy, the observed S/N ratio for any experiment is equal to the sum of the height of the overall mean and the deviation from mean caused by the levels of the three factors. By referring to the waves of the different factors (Fig. 6) it is clear that factors A, B, C are in the decreasing order of importance. Orthogonal decomposition of the observed S/N ratio can be described as the sum of the overall mean, Effect of factor A, Effect of factor B, Effect of factor C. In order to determine the significant cutting parameters affecting the quality characteristics (Ra), ANOVA analysis of raw data is used and the result is shown in Table VI.

(i) Grand total sum of squares = $\sum\limits_{i=1}^{n} \eta\iota^{2} = 21643.28\ dB^2$

(ii) Overall mean of S/N ratio = m = 48.99 dB

(iii) Sum of square due to mean = 9 * $(48.99)^2$ = 21600.18 dB^2

(iv) Sum of square due to factor A = $3(m_{A1}-m)^2 + 3(m_{A2}-m)^2 + 3(m_{A3}-m)^2$ = 15.79 dB^2

(v) Sum of square due to factor B = $3(m_{B1}-m)^2 + 3(m_{B2}-m)^2 + 3(m_{B3}-m)^2$ = 24.01 dB^2

(vi) Sum of square due to factor C = $3(m_{C1}-m)^2 + 3(m_{C2}-m)^2 + 3(m_{C3}-m)^2$ = 0.3138 dB^2

(vii) Total sum of squares = $^{n}\Sigma_{i=1}(\eta_i - m)^2$ = 39.5419 dB^2

From Table VI, it is apparent that the F- ratio values of factor A (Hardness), factor B (Electrode Diameter), and factor C (Edge Preparation) are all greater than $F_{0.05,\ 2,\ 45}$ =3.2 and thus proven to be significant.

TABLE VI
ANOVA Table for S/N RATIO

Source of Variation	Degree of Freedom	Sum of Square	Mean Square	F- Value
Hardness	2	15.79	7.895	13.64
Electrode Diameter	2	24.01	12.005	20.844
Edge Preparation	2	0.3138	0.1569	59.715
Error	45	60.4691	1.3437	-
Total	53	100.582	_	_

Confirmation Experiment

To validate the optimum welding conditions (A2, B3, C1) that were suggested by the experiments (refer 3.5.2), so that it must corresponds with the predicted value by the Taguchi methodology the planning layout of the confirmation experiments, while considering noise factors (measured locations); the total of four experiments were conducted and breaking strength measurement at various pieces have been taken and is tabulated in Table VII.

TABLE VII
Final Confirmation Experiments
Level Combination (A2 - B3 - C1) as suggested by Taguchi

Calculation for Actual Result
After the confirmation experimentations have been conducted, the results for the surface roughness found to be,
Total Average of breaking load values = (303.61+ 305.18) / 2
= 304.4 N/mm²
Mean S/N ratio = 49.17 dB.

Comparison of Results
The results obtained from the confirmation experiments are hereby compared by the predicted result of Taguchi Design experiment.

Actual Result
Mean of Response value (breaking load) = 304.4 N/mm²
S/N ratio = 49.17 dB

Predicted Result (By Taguchi Method)
Predicted mean Breaking load = 291.36 N/mm²
Predicted S/N ratio = 48.99 dB
Variation % = + {(304.4-291.36) / 20.03} × 100 = 4.476 %

IV. CONCLUSIONS AND FUTURE WORK

The outcome of the calculations and formulation for the optimization by the methods i.e. Prediction by Taguchi Method. By using the optimum factor – level combination suggested by Taguchi methodology the experiments are conducted and the results are summarized in the Table VIII.

TABLE VIII
Comparison of the Confirmation Experiment for Welding Joint Strength

Parameters	Prediction (Taguchi Method)	Actual Experimental Values
Level	A2B3C1	A2B3C1
Breaking strength (N/mm²)	291.36	304.40
S/N Ratio (dB)	48.99	49.17

Experiments conducted and subsequently analysis performed by using the Taguchi Method. The optimum characteristics for the high breaking strength in welding operation are identified and the confirmation experiments are conducted, then the results obtained are compared with the above said optimization methods and are discussed as follows:

1. From the Average (Mean) Effect Response Table for the Raw Data, the factor (C) Edge Preparation represents the largest influence on weld strength followed by factor (B) Electrode Diameter, factor (A) Hardness
2. It is apparent that the F- ratio values of factor A (Hardness), Factor B (Electrode Diameter), and factor C (Edge Preparation) are all have effect.
3. From response graphs for mean and S/N ratio, observational findings are illustrated as following:

(a) Level 2 for Hardness A₂ = medium indicated as the optimum situation in terms of mean value.
(b) Level 3 for Electrode Diameter B₃ = 5.70 mm, indicated as the optimum situation in terms of mean value.(c) Level 1 for Edge Preparation C₁ = V type, indicated as the optimum situation in terms of mean value.
4. The result obtained from the confirmation experiments reveals that the Taguchi Design of Method has provided the

Exp. No.	Breaking Strength (N/mm²)	S/N ratio (dB)
1	303.61	49.64
2	305.18	49.69
Average	304.4	49.67

good prediction for the response variable (breaking strength) i.e.291.36 N/mm² and is close to 304.4 N/mm² (actual result). In regard to S/N ratio the results are found to be 49.17dB in compare to 48.99 dB.
5. This Research work has discussed an application of the Taguchi method for optimizing the welding factors in Welding operation and indicated that the Taguchi design of experiment is an effective way of determining the optimal welding factors for breaking strength. The welding strength achievement from the confirmation runs under the optimal welding factors identified through Taguchi design of experiment that is accomplished with a relatively small number of experimental runs.

However the research work can be extended as future scope by taking various other factors and level combinations which affects the welding strength like skill of the worker, heating before welding, quality of electrode and speed of the welding operation etc. with other noise factors.

V. References

[1]   *Kackar N. Raghu, (1985), "Off-line Quality Control Parameter Design & Taguchi Method" Journal of Quality Technology, Vol. 17, No. 4, pp. 176-188.*
[2]   Kackar N. Raghu, (1986), "Taguchi's Quality philosophy analysis and commentary", Quality Progress, Dec., pp. 21-28.
[3]   Mitchell, J.P., (June 1987), "Reliability of Isolated Clearance Defects on Printed Circuit Boards." Proceedings of Printed Circuit World Convention IV, Tokyo, Japan, pp.50.1-50.16.
[4]   Deb P, Challenger K D, Clark DR. Transmission Electron Microscopy Characterizations of Preheated and Non-Preheated Shielded Metal Arc
Weldments of HY-80 Steel [J]. Materials Science and Engineering, 1988, 77: 155.
[5]   Phadke, S.M., (1989), "Quality Engineering Using Robust Design," Prentice Hall, Englewood Cliffs, New jersey.
[6]   Ade F. Ballistic Qualification of Armour Steel Weldments [J].Welding Journal, 1991, 70(1): 53.
[7]   Stanley, D.O. and Unal, R., (1992), "Application of Taguchi methods to dual mixture ration propulsion system optimization for SSTO Vehicles". 30ᵗʰ Aero Space Sciences meeting and exhibit, Jan. 6-9, Reno, Nevada.
[8] Barua, P.K. (1997), "Investigation and optimization of V-Process parameters affecting the quality of Al-7% Si Alloy casting using Taguchi Technique". Ph.D. Thesis, University of Roorkee, April.
[9] Cary H B. Modern Welding Technology [M]. 3rd ed. London: Prentice-Hall Inc, 1999

[10]   *Lambda Research , Diffraction Notes,(2000), " Efficiently Optimizing Manufacturing Processes Using Iterative Taguchi Analysis", No.25, Winter, www.lambda-research.com*

[11]   *Nicolo Belavendram, (2001), "Quality by Design-Taguchi Technique for Industrial Experimentation, Prentice Hall, Great Britain".*

[12]   *Ganter W.A., (2004), "An observation of Taguchi method", Quality and Reliability Intl. Journal Vol. 15, pp. 3-5.*

[13]   *Phillip J Ross, (2005), "Taguchi Techniques for Quality Engineering" Tata Mc-Graw Hill 2nd edition.*

[14]   *M. Sanjari, A. Karimi Taheri and M. R. Movahedi (2008), "An optimization method for radial forging process using ANN and Taguchi method" Intl. Journal of Advanced Manufacturing Technology, Vol. 40, February, pp. 776-784.*

[15]   *F.C. Tsai, B.H. Yan, C.Y. Kuan and F.Y. Huang , (2008), "A Taguchi and experimental investigation into the optimal processing conditions for the abrasive jet polishing of SKD61 mild steel". Intl. Journal of Machine Tools and Manufacture, Volume 48, Issues 7-8, June, pp. 932-945.*

Comparative Performance Analysis of a Diesel Engine Using JOME, KOME and MOME Blends

Vivek Kumar[1], L.M. Das[2], Karu Ragupathy [1]
[1]Amity School of Engineering & Technology, Amity University, Noida
[2]Centre for Energy Studies, Indian Institute of Technology Delhi, New Delhi

Abstract-Fossil fuel resources are depleting due to increasing use and causing environmental degradation. Therefore, renewable, carbon neutral alternative fuels are necessary for environmental and economic sustainability. Biodiesel derived from vegetable oils and animal fats is potential renewable, alternative fuel. Biodiesel comprises mono alkyl esters of long chain fatty acids. Among the various methods of production of biodiesel such as direct blending, microemulsification, pyrolysis, and transesterification; transesterification is widely accepted technique. In transesterification, oil is reacted with monohydric alcohol in presence of a catalyst. In the present paper, experimental investigation is carried out to examine performance and emissions of different blends of JOME, KOME and MOME. Results indicated that Jatropha have closer performance to diesel. Karanja has equally good thermal efficiency with the blends except when it is with B100. Mahua oil ester is having the lower efficiency with 50% load. The SFC for all the biodiesel fuel blends has higher value at higher loads due to its lower calorific value than diesel. SFC is lower for Jatropha blends compared to other two biodiesel blends.

Keywords: Biodiesel, Transesterification, JOME, KOME, MOME, etc.

I. INTRODUCTION

India imports almost 35% of the crude petroleum and at present rate of consumption its oil reserves will last for the next 25 – 30 years. The country is facing a dilemma because it is not possible to afford to import more petroleum oil, nor it is possible to discard petroleum-fuelled prime movers. Viable alternative fuels from renewable resources that are environment friendly are being developed due to twin crises which the world is currently confronted: 1. Fast Depletion of Fossil Fuel Resources due to increased use. 2. Environmental Degradation due to emissions. Research is directed to explore plant based fuels and plant oils and fats as fuels.

Fuels derived from the renewable biological resources to be used in diesel engine are known as Biodiesel. Biodiesel is a clean burning, non-toxic, renewable fuel that can be manufactured from vegetable oils, animal fats, or recycled restaurant oils. It is safe, biodegradable, and reduces serious air pollutants such as particulate, carbon monoxide, hydrocarbons, and air toxins. Blends of 20% biodiesel with 80% petroleum diesel (B20) can generally be used in diesel engines without any modifications in the engine. Biodiesel can also be used in its pure form (B100), but it may require certain engine modifications to avoid maintenance and performance problems. With the increase of biodiesel concentration in blends, density, kinematic viscosity, cetane number and flash point of blends increase [1]. Studies of many International Researchers show that biodiesel has excellent lubricity. Even 1% of biodiesel blend can provide adequate lubricity as compared to mineral diesel.

With Mahua based biodiesel the brake specific fuel consumption (BSFC) and exhaust gas temperature increases, whereas brake thermal efficiency (BTE) decreases with increase in the proportion of mahua oil biodiesel in the blends at all compression ratios (18:1 to 20:1) and injection timings (35-45° before TDC). The reverse trend for these parameters were observed with increase in the CR and advancement IT, engine performance becomes at par with diesel. Pure mahua biodiesel could also be used in the engine without affecting its performance [2]. B20 of PME, JME and NME have closer performance to diesel and B100 has lower brake thermal efficiency mainly due to its high viscosity. PME gives better performance compared to JME and NME. Smoke, HC, CO emissions at different loads are reduced for biodiesel blends [3].

II. BIODIESEL PRODUCTION PROCESS OPTIMIZATION

The use of heterogeneous catalyst is suggested for transesterification of high free fatty acid content non-edible oils to produce biodiesel [4]. The quantity of catalyst, amount of methanol, reaction temperature and reaction time are the main factors affecting the production of methyl esters. The optimal conditions for production of methyl esters from edible oils are: the reaction time of 90 min at 50 °C, 180 ml of methanol 1000 ml of oil and 1.5 wt % of NaOH catalyst. For non-edible based methyl ester's production, the requirement of amount of methanol is changed to 210 ml / 1000 ml of oil. Addition of excess catalyst causes saponification that leads to reduction in ester yield. Biodiesel production process is incomplete when the methanol amount is less than the optimal value. Higher reaction temperature increases the rate of transesterification and reduces the reaction time. Excess reaction time favors reverse reaction of transesterification[5]. The biodiesel production from feed stocks with high FFA is extremely difficult using alkaline catalyst because alkaline catalyst reacts with FFAs to form soap that prevents the separation of the glycerin and ester. A two step transesterification for Jatropha and Karanja and a three stage transesterification for Polanga oil are developed to convert the high FFA oils to their esters. The conversion efficiency is strongly affected by the amount of alcohol. The volumetric ratio 11:1, 11.5:1 and 12:1 of alcohol favors the alkaline catalyzed transesterification process within 2h for the formation of JOME, KOME and POME, which is sufficient to give 93%, 91% and 85% yield of ester respectively [6].

The optimum conditions for methanolysis of Karanja oil was 1 % KOH as catalyst, alcohol to oil molar ratio 6:1, reaction temperature 65 °C, rate of mixing 360 rpm for a period of 3 h. The reaction was incomplete with a low rate of stirring (180 rpm) [5,7].

Economics

- The cost analysis indicates that the cost of oil is major component (about 75-80 %) of total cost of biodiesel.
- Use of lower cost feed stocks would have tremendous impact on biodiesel economics.
- Another approach which leads to reduction in operating cost is improvement in technology.

III. METHODOLOGY

In view of the escalating energy crisis, environmental degradation and the potential of biodiesel being adopted as an alternative fuel for diesel engines, the present study aims at comparative evaluation of performance characteristics of compression ignition engine using different blends of biodiesel of different feedstocks.

Biodiesel Preparation

Biodiesel is a mono alkyl ester (methyl) of long chain fatty acids derived from vegetable oils and animal fats. Biodiesel can be used in any C.I. Engine without the need of modification.

The important factors to be considered for Process include:

- Must be able to process high free fatty containing oils/ feed stocks.
- Must be able to process raw both expelled and refined oil.
- Process should be environment friendly; almost zero effluents.
- Able to produce marketable byproducts glycerin, fatty acids, soap if any.
- Must be able to produce fuel grade esters; Biodiesel produced should meet the standard specifications.

- The process should be adaptable over a large range of production capacities.

Vegetable oils have larger molecules with straight chain and complex structure (Triglycerides): upto four times larger than typical diesel molecules. High viscosity of vegetable oils due their large molecular mass (30 – 200 cSt) as compared to mineral diesel (4 cSt) leads to poor fuel atomization, Inefficient mixing with air contributes to incomplete combustion which leads to more deposit formation, carbonization of injector tips. Vegetable oils have about 10 % lower heating value than mineral diesel and Low volatility.

Problems Encountered With Vegetable Oil

- Clogging of Fuel Lines
- Carbonization of injector tips
- Deposit on Cylinder Walls
- Poor Ignition and combustion due to improper atomization

Fuel formulation techniques adopted to overcome problem of high viscosity

- Pyrolysis or thermal cracking
- Blending
- Emulsification
- Esterification

Transesterification was eventually successful in bringing about the viscosity close to conventional diesel.

Transesterification

Transesterification, also called alcoholysis, is the process of using an alcohol (methanol) in the presence of a catalyst, such as potassium hydroxide, to chemically break the molecule of the raw renewable oil into methyl esters of the renewable oil with glycerol as a by-product.

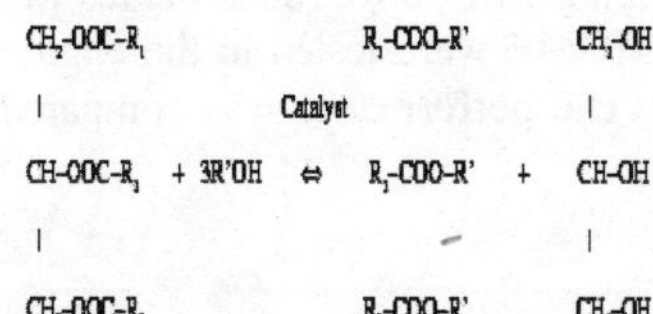

Fig.1. Chemical equation for transesterification

This process is suitable for feed stocks having FFA content 1% to about 20%. The oil is pretreated before transesterification at moderate temperature (60-80°C) in presence of a base catalyst. FFA is also converted to biodiesel resulting in higher yield of biodiesel. For high FFA feedstock – acid catalyzed esterification followed by base catalyzed transesterification is used or FFA can be removed first and the purified oil is transesterified. Different catalysts e.g. NaOH, KOH, Non alkaline catalysts, acids, metal complexes and bio catalysts etc. can be used. Anhydrous ethanol, isopropanol or butanol can be substituted for methanol. Alcohols other than methanol may require additional process steps and quality control. Basic transesterification is carried out at atmospheric pressure and temperature around 60°-70°C. Some technologies use higher temperatures and elevated pressure, typically in super critical range of methanol.

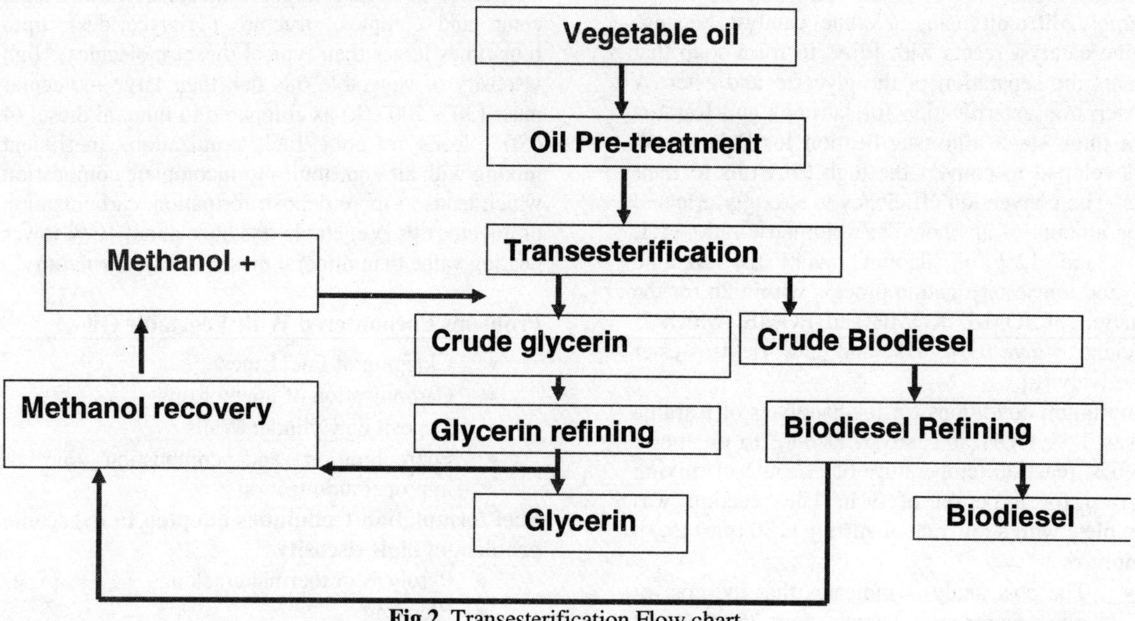

Fig.2. Transesterification Flow chart

Table 1. Specifications of the test engine

IV. EXPERIMENTAL SETUP

Tests were conducted on a single cylinder four-stroke, naturally aspirated, constant speed compression ignition engine. Schematic diagram of experimental setup is shown in Figure 3. The engine was tested at a rated speed of 1500 rpm under steady state conditions. The specifications of the test engine are shown below in Table 1. All engine performance tests were conducted in the Engines and Unconventional Fuel Lab at the Centre for Energy Studies, Indian Institute of Technology, Delhi.

Performance of the engine with conventional diesel fuel was used as the basis for comparison. Diesel, Biodiesel ranging from 5 to 100% blends of JOME, KOME and MOME were tested in the engine for its performance. The performance was compared in the next session

Engine	Kirloskar naturally aspirated DI diesel (DAF 8)
Rated power	5.9 kw @ 1500 rpm
Cylinder bore	95 mm
Stroke length	110 mm
Compression ratio	17.5
No. of cylinders	One
Stroke type	Four

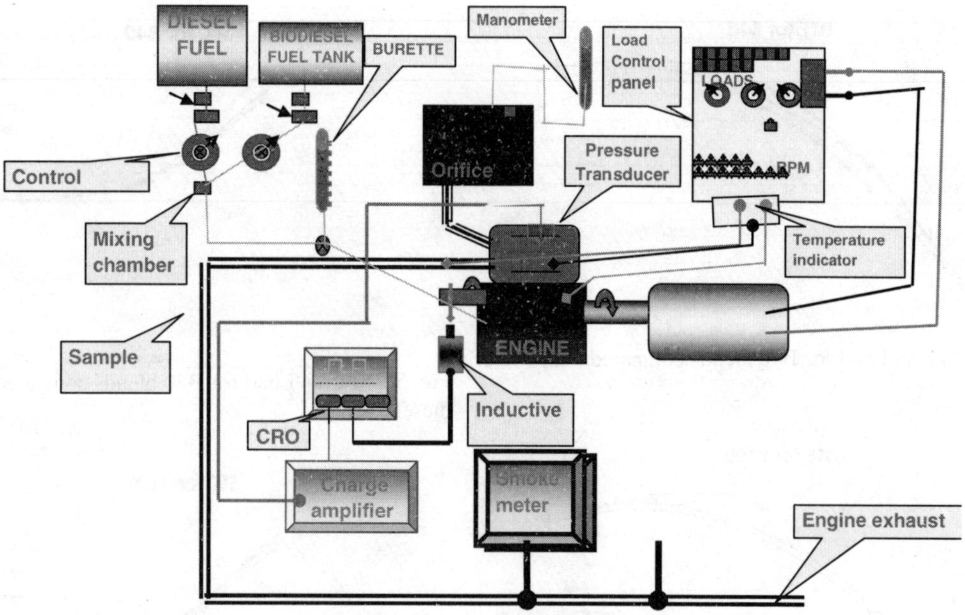

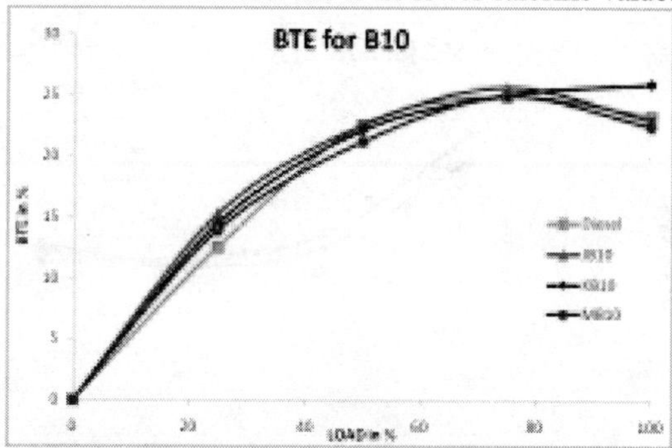

Fig.3. Experimental setup

V. RESULTS AND DISCUSSIONS

Experimental investigations were carried out for different blends of Jatropha (JOME), Karanja(KOME) and Mahua(MOME) methyl esters (Biodiesel) and performance was evaluated and compared with mineral diesel. All the tests were conducted on compression ignition engine without any engine modifications. The performance of the engine with diesel, blends of biodiesel and diesel, and neat biodiesel are presented and discussed below.

Fig.4a to 4d shows the variation of brake thermal efficiency (BTE) with load. In the part load operation, B20 blend of MOME shows the minimum efficiency. The low efficiency may be due to low volatility, slightly higher viscosity and higher density of the methyl ester of Mahua oil, which affects mixture formation of the fuel and thus leads to slow combustion. A slight lower efficiency for esters was reported in general due to the lower heating value of the esters than with diesel.

Fig.5a to 5d shows the variation of brake specific fuel consumption (BSFC) with load for different diesel–biodiesel blends and neat diesel fuel. As the load increases, BSFC decreases for all fuels. At 100% load, the BSFC increases for the entire diesel - Biodiesel blends compared to diesel.

Fig.6a to 6d shows the variation of brake mean effective pressure (BMEP) with load for different diesel–biodiesel blends and neat diesel fuel. As the load increases, BMEP increases for all blends. With lower blends (B10 and B20), BMEP is higher for JOME at higher load and it is lower than diesel for higher blends. The BMEP for other fuel blends of KOME and MOME is lower than the diesel for all loads.

The performance of JOME is very good and it is having higher BTE with lower SFC except for B40 blend. At lower loads the SFC is less for all the biodiesel blends with diesel and at higher loads BTE and SFC are almost similar to diesel.

At 100% load the SFC of biodiesel blends are higher than that of diesel. In general, KOME is having higher SFC than other blends due to its lower calorific value.

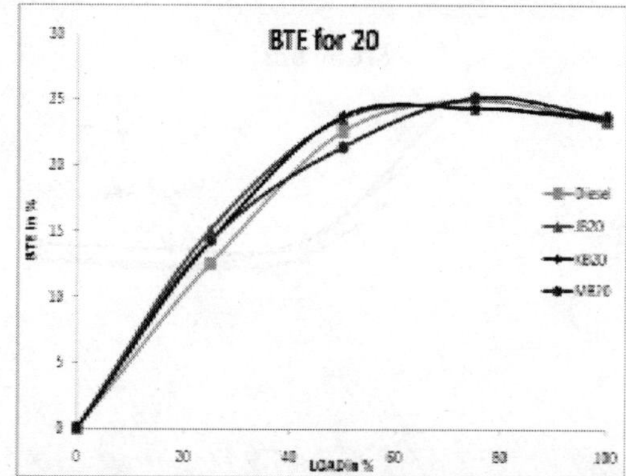

Fig. 4a. BTE vs Load for B10 blends compared with diesel

Fig. 4b. BTE vs Load for B20 blends compared with diesel

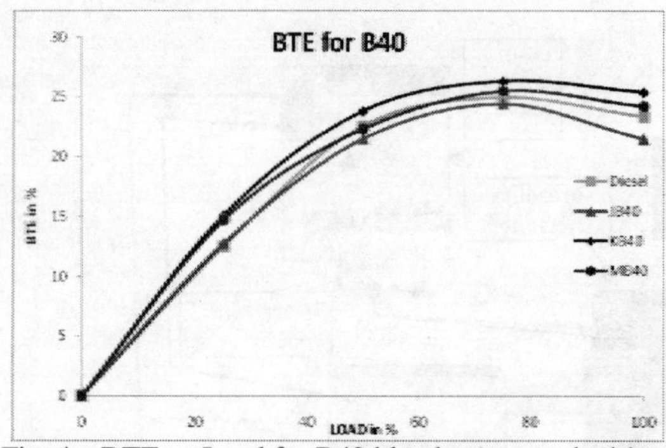

Fig. 4c. BTE vs Load for B40 blends compared with diesel

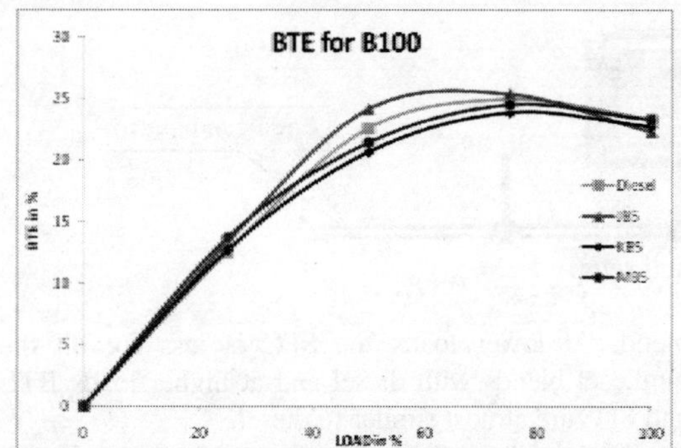

Fig. 4d. BTE vs Load for B100 blends compared with diesel

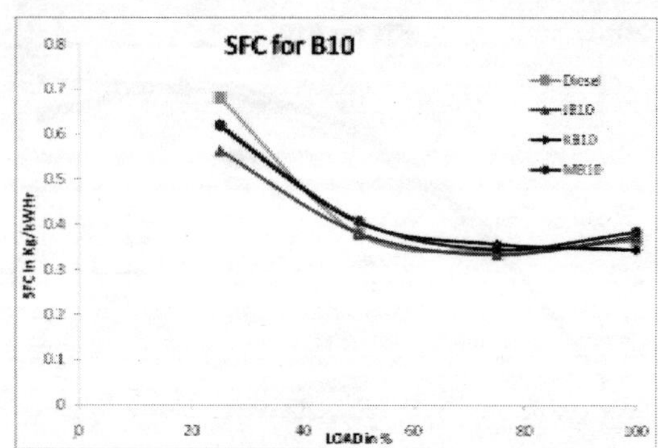

Fig. 5a. SFC vs Load for B10 blends compared with diesel

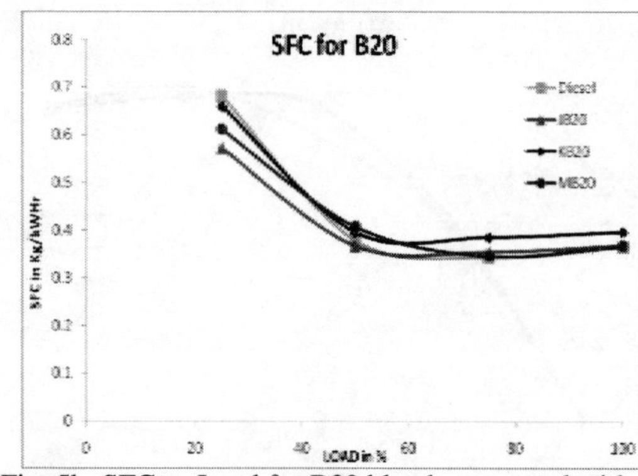

Fig. 5b. SFC vs Load for B20 blends compared with diesel

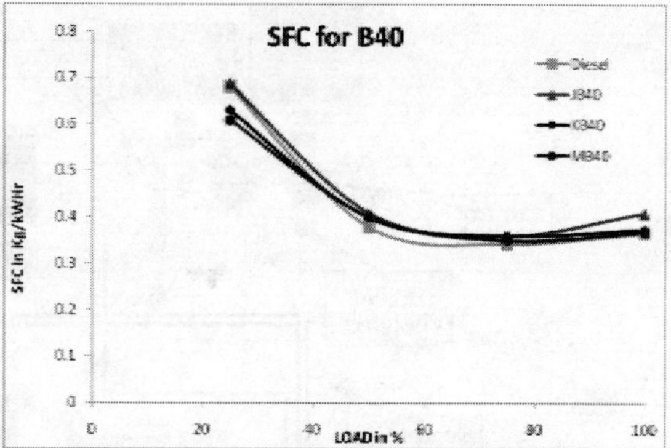

Fig. 5c. SFC vs Load for B40 blends compared with diesel

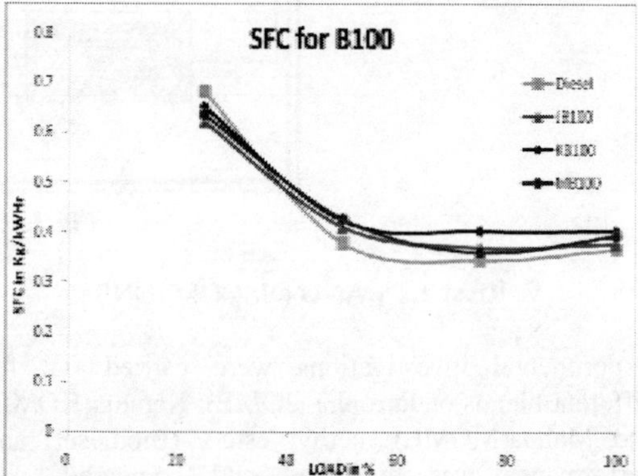

Fig. 5d. SFC vs Load for B100 blends compared with diesel

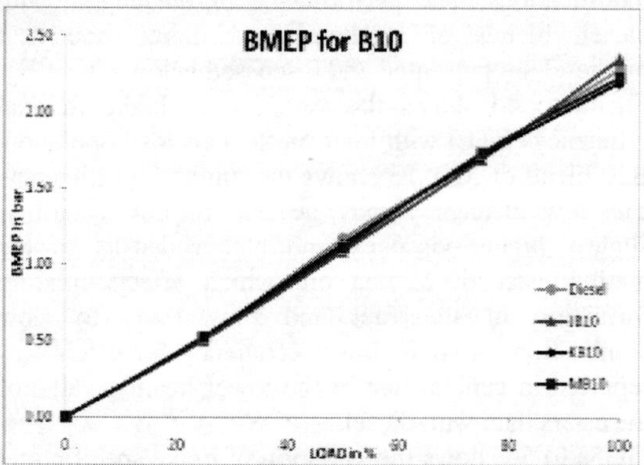

Fig. 6a. BMEP vs Load for B10 blends compared with diesel

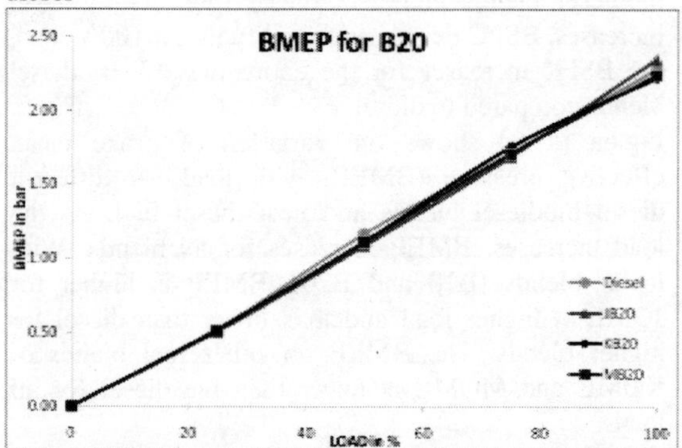

Fig. 6b. BMEP vs Load for B20 blends compared with diesel

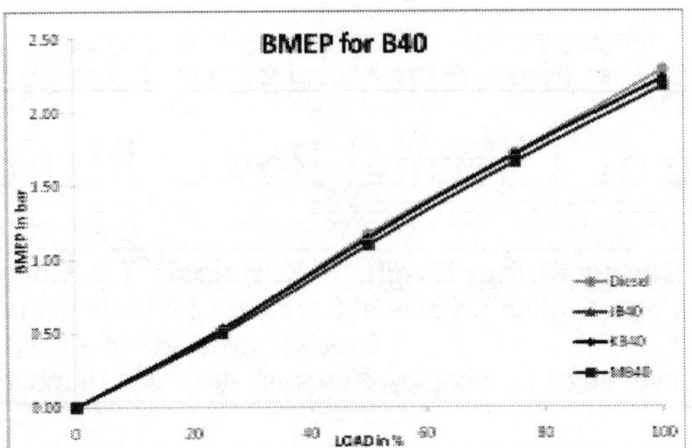

Fig. 6c. BMEP vs Load for B40 blends compared with diesel

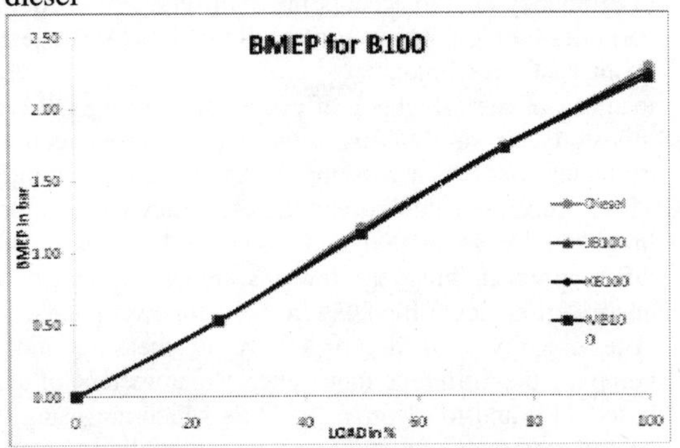

Fig. 6d. BMEP vs Load for B100 blends compared with diesel

VI. CONCLUSIONS

As the first step for the appropriate fuel development, the preparation of the corresponding biodiesel (methyl esters) of Jetropha, Karanja and Mahua oil was done. The properties of the methyl esters (biodiesel) were experimentally observed to be close to those of diesel fuel. The effects of biodiesel addition to diesel fuel on performance characteristics of a single cylinder constant speed compression ignition engine fuelled with diesel and biodiesel blended fuels have been investigated, and compared with the baseline diesel fuel. The main conclusions of this study are:

- The existing diesel engine performs satisfactorily on biodiesel fuel without any significant engine hardware modification
- Engine performance with biodiesel does not differ greatly from that of diesel fuel. A little power loss, combined with an increase in fuel consumption, is often encountered due to the lower calorific value of the biodiesel blends.
- The performance of JOME is very good with higher BTE and lower SFC except with B40 blend.
- At higher load BTE and SFC are almost similar to diesel. At 100% load the SFC of biodiesel blends are higher than that of diesel.
- At 100% load the SFC of biodiesel blends are higher than that of diesel. In general, KOME is having higher SFC than other blends due to its lower calorific value.
- With lower blends (B10 and B20), BMEP is higher for JOME at higher load and it is lower than diesel

for higher blends. The BMEP for other fuel blends of KOME and MOME is lower than the diesel for all loads.

- In view of the petroleum fuel shortage, biodiesel can certainly be considered as a potential alternative fuel.

VII. REFERENCES

[1] Sahoo P.K., Das L.M., Babu M.K.G., Arora P, Singh V.P. "Comparative evaluation of performance and emission characteristics of Jatropha, Karanja and Polanga based biodiesel as fuel in tractor engine" Fuel 88 (2009) 1698-1707.

[2] Raheman H., Ghadge S.V., "Performance of diesel engine with biodiesel at varying compression ratio and ignition timing". Fuel 87 (2008) 2659-2666.

[3] Rao T. Venkateswara, Rao G. Prabhakar and Reddy K. Hema Chandra, "Experimental investigation of Pongamia, JAtropha and Neem Methyl Esters as Biodiesel on C. I. Engine". JJMIE, Volume 2, Number 2, Jun 2008, page 117 – 122.

[4] Sinha Shailendra, Agarwal Avinash Kumar, "Biodiesel development from Rice bran oil: Transeterification process optimization and fuel charcterisation". Energy conversion & management 49 (2008) 1248 – 1257.

[5] Naik S.N. Vidhya Sagar D, Meher L.C, "Technical Aspects of Biodiesel Production by Transesterification- A review", Renewable & Sustainable Energy Review, 2004, 29: 1-21

[6] Eevera T., Rajendran K., "Biodiesel production process optimization and charcterisation to assess the suitability of the product for varied environmental conditions". Renewable Energy 34 (2009) 762 – 765.

[7] Sahoo P.K., Das L.M. "Process optimization for biodiesel production from Jatropha, Karanja and Polanga oils". Fuel 88 (2009) 1588-1594.

[8] Report of the committee on development of biofuels. Planning Commission, Govt. of India; 2003.

[9] Meher LC, Vidya S, Dharmagadda S, Naik SN. Optimization of alkali-catalyzed transesterification of Pongamia pinnata oil for production of biodiesel. Bioresource Technol 2006;97(12):1392–7.

[10] Agarwal AK, Das LM. Biodiesel development and characterization for use as a fuel in compression ignition engines. Trans ASME 2006;123:440–7.

[11] Kumar C, Babu MKG, Das LM. Experimental investigations on a Karanja oil methyl ester fueled DI diesel engine. SAE; 2006-01-0238. 2006.

[12] DK Bora, M Polly, V Sanduja, LM Das. Performance evaluation and emission characteristics of a diesel engine using Mahua oil methyl ester. SAE; 2004-28-0034. 2004.

[13] Schwabm A.W, M.O. Bag by, and B. Freedman, "Preparation & Properties of Diesel Fuels from Vegetable Oils", Fuel, Vol.66, pp, 1372, 1987

[14] B.K. Barnwal, M.P. Sharma, "Prospects of Biodiesel Production from vegetable oils in India", Renewable & Sustainable Energy Review, 2005, 9:363-378.

[15] Chandan Kumar, M.K.G, Babu & L.M. Das, "Experimental Investigation on a Karanja oil Methyl Ester Fueled DI Diesel Engine", SAE 2006-01-0238, 2006.

[16] Baiju.B, Naik.M.K, Das.L.M. "A Comparative Evaluation of Compression Ignition Engine Characteristics using methyl and ethyl esters of karanja oil" , International journal of renewable energy May 2009, Elsevier publications

Measuring Energy Efficiency Using DEA: The Case of Thermal Power Plant

[1]Sudhir Kumar Singh, [2]V.K. Bajpai, [2]T.K Garg
[1]Department of Mechanical Engineering, Skyline Institute of Engineering &Technology,
Greater Noida, 201306 India.
[2]Department of Mechanical Engineering, National Institute of Technology Kurukeshetra, 136119 India.

Abstract- **In recent years the Indian economy has achieved a rapid growth rate. However, one of the challenges facing the economy is the large energy supply that would be needed to sustain this growth path. Hence attaining energy efficiency is crucial for the economy. This paper examines the 34 Thermal power plants (TPP) and utilizes Data Envelopment Analysis (DEA) to measure the energy efficiency of the TPP. The result reveals that two plants are using the resources efficiently and one plant is rated as least efficient under constant return to scale assumption. Also the slack analysis indicates the causes of inefficiency in TPP. The results of such studies would help policy makers and top management towards improving the performance levels of coal based TPP in India.**

Key words: Decision making unit; Data Envelopment analysis, Thermal power plants.

I. INTRODUCTION

In every five year plan, energy resources consumption in Indian coal based Thermal Power Plants (TPP) is growing rapidly. Energy efficiency measurement is an important issue in TPP from energy resources consumption point of view. The ability to distinguish between efficient and inefficient TPP is important for government and plant operators so that they can take the corrective action in inefficient TPP.

The reason for selecting the coal based TPP for the study is based on the installed power generation capacity criteria. The installed power generation capacity under utilities has increased from 1362 MW in 1947 to 155859.23 MW in Nov 2009. Similarly electricity generation in utilities in the country has grown up from 4.1 billion kWh in 1947 to the level of 699.191 billion kWh in 2007-2008.Reference [12] shows that thermal power plants have the largest share of installed capacity generation as well as in electricity generation followed by hydro and nuclear power plants. Coal based TPPs are the leading providers in installed capacity (53.3% of the total installed capacity in 2009) and electricity generation (68% of the total generation of electricity in 2007-2008). If we see the sector wise installed power generating capacity than state sector is the leader followed by central sector and private sector.

The usual measure of efficiency in TPP often relies on a single indictor like electricity generated per energy input [3].This ratio takes only the heating value of fuels into account and neglecting the other variables such as force outage and auxiliary power used etc. This usual measure of efficiency fails to account the multi-input, multi-output variables problem in power plant. Reference [4] suggested an alternative technique
to measure the efficiency of power plant using DEA. In this study, we apply DEA to measure the cross sectional efficiency of TPP across the different regions of India. There are some studies about the efficiency measurement in electricity generation and also in other fields using DEA approach but very few researchers evaluated the energy efficiency using DEA in thermal power plants.

The objectives of the paper are to measure and to compare the different input energy parameters of coal based TPP and to identify the TPPs which are using the energy resources optimally and sub-optimally in relation to other plants. In this paper, we have also computed the super efficiency scores of the efficient TPP for complete ranking. In addition, we also made an effort to identify the major causes for inefficient TPP.

This paper is structured as follows: a brief introduction is provided in the section I followed by the section II review of literature. Section III Data description, while in section IV presents the methodology used in this paper. In section V the results and discussion are presented. Finally, the main conclusions are summarized in section VI.

II. LITERATURE REVIEW

There are some studies on efficiency measurement in electricity generation and in other fields also using DEA approach are – Reference [3] evaluated the power generation efficiency of major thermal power plants in Taiwan during 2004-2006 using DEA and also conducted stability test. According to their results, all power stations achieved acceptable overall efficiencies and the combined cycle power plants were the most efficient among all plants. Reference [5] analyzed the energy use and efficiency in Turkish manufacturing sector SME'S using DEA approach. In their study, they concentrated on the main energy consumption components as inputs and leaving other inputs such as raw material, manpower etc from the study. The results indicate that there are significant potentials to save energy in the companies that are the inefficient users. Reference [6] examined the efficiency of the 26 state owned electric utilities in India using DEA. The results indicate the performance of several SOEUs is sub optimal; there is significant potential to reduce number of employees from several utilities. It was also found that the bigger utilities display greater in efficiencies and have distinct scale inefficiencies. Reference [7] has examined the efficiency of 56 coal based thermal power generation in India during 1994-1995 to 2001-2002 using the stochastic frontier analysis function methodology.

The results indicate that highest efficiency attained is 96% by Dhanu plant in west region. Therefore inefficient power plant should adopt the best practices of efficient plants, such as Dhanu in west region in this study. Reference [8] measure the technical efficiency of China's thermal power generation based on cross- sectional data for 1995 and 1996.

Their results indicate that municipalities and provinces along the eastern cost of china and those with rich supplies of coal achieved the highest levels of technical efficiency. The presence of labor in many regions indicates the labor redundancy was a serious problem. Result of second stage regression analysis shows that fuel efficiency and the capacity factor significantly affect the technical efficiency.

III. DATA DESCRIPTION

The study is limited to coal based TPPs because they are the key player in electricity generation in India. Our cross sectional study (2007-2008) covers 34 TPPs from the different regions of India out of 78 plants because of non-availability of secondary oil data. Detailed information about the selected input and output variables of TPP can be obtained from the annual report of electricity generation published by the CEA [13]. We considered the plant wise data of TPPs in our study on an annual basis. For meaningful DEA analysis, it is important that the sample size of the DMUs should be greater than either product of number of inputs and outputs or 2 to 3 times the sum of inputs and outputs [1].

To construct a DEA model for evaluating the energy efficiency, the input variables need to be selected based on the consideration of energy consumption component only while only output variable is the net electricity produced. The energy consumption components are broadly classified as auxiliary power consumption, coal consumption and secondary oil consumption in coal based TPP as input variables.

Coal is the primary fuel for boiler, Oil is the secondary fuel for the boiler which is used at the time of starting, tripping of plant and for standby proposes. Auxiliary power is used to operate the electric equipment within the plant. The variables chosen for the study are also in line with the choice of variables as identified by [5].

IV.METHODOLGY

In this paper, DEA methodology is employed to measure the energy efficiency scores of homogeneous decision making unit (DMU). DEA technique is non-parametric and does not require the functional form of the production process. The technique of frontier analysis has been described by [9], but mathematical formulation was provided by[10] which known as Charnes, Cooper and Rhohde (CCR) model.

Measures of efficiency are based on whether the inputs or outputs are controllable variables. A DMU can be made efficient either by reducing the input values and without disturbing the output values (input oriented) or by increasing the output values with the same input values (output oriented). In this paper, input oriented CCR model is adopted because the different input energy variables are to be minimized without affecting the generation of electricity. We describe the appropriate input oriented DEA models that are used in the study in each subsection given below:

A CCR Efficiency Model

The measurement of relative efficiency introduced by CCR in the form of ratio of outputs to inputs of the $DMU_j = DMU_x$ to be evaluated relative to the ratios of all. CCR construction is the reduction of the multiple outputs to multiple inputs (for each DMU) to that of a single virtual output and virtual input. That is a function of multiplier for a particular DMU the ratio of this single virtual output to single virtual input. Suppose that there are N DMUs $(j = 1,2 \dots \dots N)$,each producing s outputs from m inputs.

If DMU_k uses the multi-input $x_{ik} = (x_{1k}, x_{2k}, \dots, x_{mk})$ to produce the multi output $y_{rk} = (y_{1k}, y_{2k} \dots, y_{sk})$ Than

$$\text{Efficiency of unit } k = \frac{\sum_{r=1}^{s} u_r y_{rk}}{\sum_{i=1}^{m} v_i x_{ik}} \tag{1}$$

Where u_r, v_i are referred to as the DEA-weight on the rth output and i^{th} input respectively. $y_{rk} = $ the amount of output r produced by DMU k and $x_{ik} = $ the amount of input i consumed by DMU k. k = Decision making unit being evaluated in the sample.

In mathematical programming parlance, this ratio, which is to be maximized forms the objective function for the particular DMU to be evaluated, subject to the constraints reflect that the efficiency of all units must be less than or equal to one.

$$\text{Max } h_k = \frac{\sum_{r=1}^{s} u_r y_{rk}}{\sum_{i=1}^{m} v_i x_{ik}} \tag{2}$$

$$\text{Subject to } \frac{\sum_{r=1}^{s} u_r y_{rk}}{\sum_{i=1}^{m} v_i x_{ik}} \leq 1$$

$$u_r, v_j \geq 0 \; \forall \; i \text{ and } r$$

Where $h_k = $ Energy efficiency score of kth DMU relative to other DMUs.

Full version of CCR model, just replace the $u_r, v_j \geq 0$ with $u_r, v_j \geq \varepsilon > 0$ to define the decision variables of the DEA programs to be strictly positive [1].

It is difficult to solve the model (2) because of its fractional objective function. This non- linear ratio model can be converted to linear programming problem (LPP) by setting the numerator or denominator of the ratio equal to1.Details of the linear form of CCR model may be found in Ramanathan [1].

$$\text{Max } h_k - \sum_{r=1}^{s} u_r y_{rk} \tag{3}$$

$$\text{Subject to} \sum_{j=1}^{n} v_j x_{ik} = 1$$

$$\sum_{r=1}^{s} u_r y_{rk} - \sum_{i=1}^{m} v_j x_{ik} \leq 0$$

$$u_r, v_j \geq \varepsilon > 0$$

Where ε is a non- Archimedean constant. The basic theory of LP States that every another closely related linear problem called dual. It is possible to write the dual of any LP problem using certain rules. The dual has the benefit of solving the LPP more efficiently than the primal problem [12].

Because the number of constraints are less and equal to number of inputs and outputs considered where as primal problem has the number of constraints equal to the number of DMUs. Mathematically, the number of constraints $z + m + n + 1$ in primal reduces to $m + n$ in the dual [6].The dual is given below [2], [5]:

$$\text{Min } \theta_k - \varepsilon[\textstyle\sum_{i=1}^{m} s_{ik}^- + \sum_{r=1}^{s} s_{rk}^+] \tag{4}$$

$$\text{Subject to} \sum_{j=1}^{N} \lambda_j y_{rj} = s_{rk}^+ - y_{rk}$$

$$\sum_{j=1}^{N} \lambda_j x_j = \theta_k x_{ik} - s_{ik}^-$$

$$\lambda_j \geq 0 . \, s_{ik}^-, \, s_{rk}^+ \geq 0 \quad \forall \; i \text{ and } r, \theta_k \text{ free}$$

Where θ_k = Energy efficiency score of DMU k in relation to others under CCR assumption and λ_j = the dual coefficient to DMUs. s_{rk}^+ = Slack variables for output constraint and s_{ik}^- = Slack variables for input constraint. y_{rj} = the amount of output r produced by DMU j and x_{ij} = the amount of input i consumed by DMU j.

CCR implies model (4) as envelopment model because this model measures the relative efficiency with reference to a PPS boundary which envelops the input and output value of observed DMUs [12]. Optimal solution of CCR envelopment model yields three possibilities-

1 If $\theta_k = 1$ and $s_{rk}^+ = 0 . \, s_{ik}^- = 0$ then DMUs are strongly efficient.

2 If $\theta_k = 1$ and one slack value is positive than DMU is weakly efficient.

3 If $\theta_k < 1$ and some slack values are positive then the DMU is inefficient.

B. Super Efficiency Model

Efficient TPP can not be ranked because efficiency score to all DMUs are assigned to one. In this study, super efficiency DEA model is employed to further discriminate the level of efficient TPP as obtained from the result of CCR DEA model. Reference [11] proposed a model with slight modification to standard form of DEA model (Which is known as super efficiency model).The super efficiency modification to CCR model excludes $y_k \lambda_k$ and $x_k \lambda_k$ from the L.H.S while retaining its inputs and outputs in the R.H.S of the constraints (i.e. excludes the DMU j_k which is under evaluation from the input- output constraints.).Thus CCR model is modified as follows:

$$\text{Min } \varepsilon_k - \varepsilon[\textstyle\sum_{i=1}^{n} s_{ik}^- + \sum_{r=1}^{s} s_{rk}^+] \tag{5}$$

Subject to

$$\sum_{j=1}^{N} \lambda_j y_{rj} = s_{rk}^+ - y_{rk}$$

$$\sum_{j=1}^{N} \lambda_j x_j = \theta_k x_{ik} - s_{ik}^-$$

$$j \neq j_k , \, \lambda_j , s_{ik}^- . \, s_{rk}^+ \geq 0 \quad \forall \; j , i \text{ and } r, \theta_k \text{ free}$$

Super efficiency score of efficient TPP obtained by this model will always have greater than 1. Thus efficient TPP and Chanderpur showed best performance. Of which Budge-Budge is top ranked with the highest score i.e

can be ranked in order of their measured levels of efficiency. Energy efficiency score of deficient TPP remains identical as obtained by the standard DEA model. Superefficient firms have greater room for reducing its output without becoming inefficient.

The CRS model is solved to obtain the energy efficiency score of TPP with the help of DEAP software. The AP model is solved to obtain the super efficiency score of a TPP with the help of excel solver software which is an inbuilt function of Microsoft Excel.

V. RESULT AND DISCUSSION

In this paper, the results of the CCR Model (4) show that energy efficiency score varies across TPP from 55% to 100% relatively. The results of the CCR model are in the Table-1. The overall efficiency had a mean score of 82% for all TPPs and majority of the TPPs lie above the mean score. Based on the optimized results of CCR Model (4), it is observed that Budge –Budge plant and Chandrapur plant under regulatory of private and Mahagenco respectively are relatively efficient. The overall energy efficiency score of aforesaid TPP is equal to 1. The inefficiency score of the remaining 32 power plants indicate that they should reduce their existing input levels by the same proportions of input mix to produce the current level of generation. For example NR-Kota has the efficiency score 84% and indicates that 16% of its current inputs level should be reduced to become energy efficient. Fig.1 shows the energy efficiency score of TPP.

Peer TPP is revealed in Table 1 Column 4 have significant practical value in case of inefficient Power plants. The efficient Peer TPP is the TPP correspond to positive λ at the optimal solution. The maximum frequency of TPP as a peer is likely to be a better role model for inefficient power plants because such a peer is genuinely a well performing TPP. Budge -Budge plant operating practices and environment matches more closely to those of other less efficient TPP. Efficient TPP such as Chandrapur plant is an efficient peer for few TPP to be very productive but operating in an environment, which are very dissimilar to the rest of the TPP.

In addition, we analyzed slack variables to get improvement directions for the inefficient power plants, as shown in Table 1. The input slack value indicates the excessive amount of input used by inefficient TPP. Thus inefficient TPPs can become efficient, if energy consumption by an amount equal to the slack values present in the respective energy variables. For example E.R. Bokaro requires decrease of 1061.1 thousands liter in its secondary oil, 703 tons in coal and 95.13 MU in auxiliary consumption.

The super efficiency model results are reported in the second last column of Table 1.Only two DMUs, namely Budge-Budge

Budge-Budge plant had an outstanding performance compared to other plants.

In addition, result of the analysis shows that Faridabad is the least performer and Harduaganj is the second least performer in the set. In the present work, we also identify the causes for excessive use of resources in TPP after discussing with the experts of this area. The reasons for inefficiency may be managerial, scale size, factors beyond management control for example age of plant, lack of proper maintenance of main components like boiler etc.

Table – 1
Energy Efficiency Score, Reference Set and Slack in Inputs

S.No.	Region-TPS	CRS Efficiency (%)	Peer Units	Auxiliary Power (MU)	Oil Cons. (In'000Lt)	Coal Cons. (In '000 T)	Super Efficiency Score	Ranking
1	ER-Budge Budge	100	34	0	0	0	1.167	1
2	WR-Chandrapur	100	21	0	0	0	1.071	2
3	SR-Mettur	96	12,21,34	21.84	121.04	180.72	0.550	3
4	WR-Dhanu	96	21,34	16.08	27.26	262.95	0.839	4
5	ER-Bandel-	95	34	14.92	3686.2	60.89	0.819	5
6	NR-GHTP	95	12,21,34	13.37	34.04	101.17	0.954	6
7	SR-Tuticorin	99	12,21,34	3.51	139.4	34.17	0.931	7
8	NR- Ropar	93	21,34	56.79	8431.4	471.73	0.844	8
9	SR-Raichur	89	12,21,34	101.9	852.21	864.5	0.882	9
10	SR-North Chennai	90	12,32,34	40.89	312.43	296.49	0.618	10
11	WR-Ukai-	90	12,21,34	46.17	1143.6	416.85	0.571	11
12	NR-Suratgarh	88	34	110.1	748.79	793.74	0.835	12
13	WR-Nasik	87	12,21,34	73.42	8629.4	631.59	0.896	13
14	WR-Korba-West	85	12,21,34	81.11	5249.1	682.29	0.828	14
15	NR-Kota	84	12,21,34	122.6	629.9	939.49	0.792	15
16	NR-Pariccha	83	21,34	37.22	27552	339.41	0.853	16
17	ER-I.B. Valley	82	12,21,34	51.18	340.11	477.1	0.869	17
18	NR-Panipat	84	12,32,34	154.3	1797.3	1114.4	0.766	18
19	ER-Chandrapura	80	34	52.78	4370.5	335.98	0.770	19
20	NR-GNDTP	82	32,34	59.89	750.58	390.58	0.753	20
21	WR-Gandhi Nagar	83	12,32,34	95.91	2605.3	720.22	1.071	21
22	WR-Sikka Rep	79	34	34.26	4577.3	225.03	0.956	22
23	SR-Rayal Seema	79	12,21,34	84.13	323.57	669.13	0.792	23
24	WR-Bhusawal	77	12,32,34	73.91	25561	621.66	0.888	24
25	WR-.Koradi	79	21,34	132.6	3976.7	1140.5	0.702	25
26	ER-Bokaro	77	21,34	95.31	1061.1	703	0.951	26
27	WR-Parli	73	21,34	104.3	7820.4	1173.3	0.960	27
28	ER-Santaldih	72	34	62.37	12836	347.23	0.901	28
29	SR-Ennore	70	21,34	67	278.93	640.34	0.802	29
30	NR-Rajghat	70	34	46.49	2683.7	222.98	0.767	30
31	NR –Panki	62	34	57.18	3296.5	376.25	0.820	31
32	NR-I.P.Station	60	34	57.95	9710	394.71	0.954	32
33	NR- Harduganj	57	34	53.96	6432.8	322.47	0.721	33
34	NR-Faridabad	55	21,34	45.17	5974.4	386.4	1.167	34
	Average Efficiency	82						

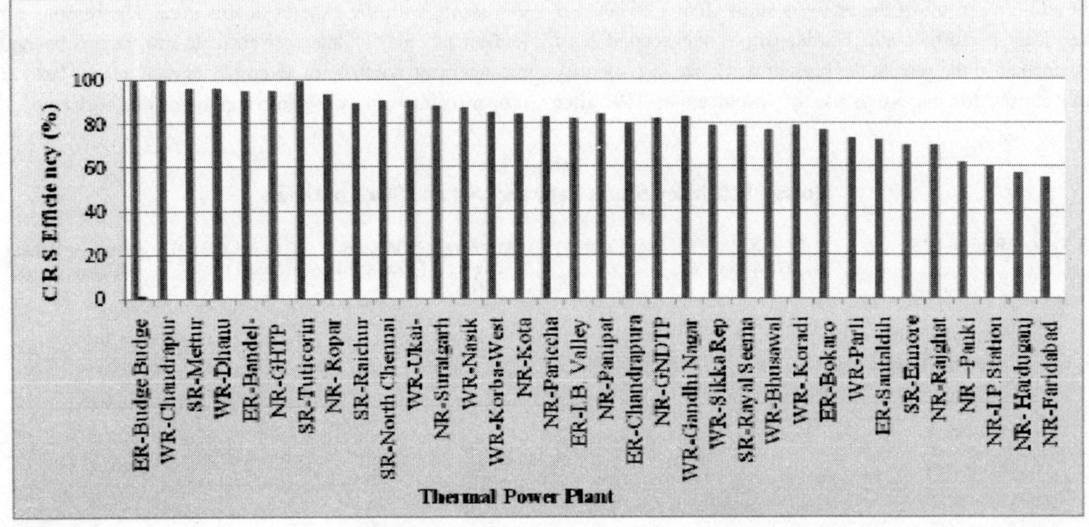

Fig.1 Efficiency score of thermal power plant

VI. CONCLUSION

This paper evaluates energy efficiency score of 34 TPPs owned by central, state and private sector across the country using the DEA technique. The CRS analysis shows that Budge-Budge and Chandrapur plant are using the resources efficiently and Faridabad plant is rated as least efficient. Out of these efficient plants, Budge-Budge plant had an outstanding performance compared to other plants. Inefficient plants can become efficient if they give proper attention on the causes identified in this study. To improve the performance of inefficient TPP, a number of good operating practices of efficient energy users can be considered. The private sector plants perform significantly better than state sector. Maximum higher and lower efficiency score are observed in Southern and Northern region plants respectively. It is suggested that, CEA should plan for renovation and modernization of plant after considering such analysis. This DEA analysis highlights huge savings in energy consumption in inefficient TPP. The results of such studies would help policy makers and top management towards improving the performance levels of coal based TPP in India. The approach of this study could be applied for other manufacturing sectors.

VII. REFERENCES

[1] R. Ramanathan, "An introduction to data envelopment analysis", Sage publication, 2003.

[2] Emmanuel Thanassoulis, "Introduction to the theory and application of data envelopment analysis",Norwell Kluwer Academic publishers, 2001.

[3] C.H. Liu and Sue J. Lin, "Evaluation of thermal power plant operational performance in Taiwan by data envelopment analysis," Energy Policy, vol. 38, pp.1049-1059, 2010.

[4] Boaz Golany,Yaakov Roll and David Rybak, "Measuring Efficiency of power plants in Israel by data envelopment analysis", IEEE Transactions on engineering management,vol. 41, pp.291-301,1994.

[5] Semih Onut , Selin Soner, "Analysis of energy use and efficiency in Turkish manufacturing sector SMEs. Energy Conversion Management", vol.48, pp.384-394, 2007.

[6] Tripta Thakur, S.G. Deshmukh and S.C. Kaushik, "Efficiency evaluation of the state owned electric utilities in India", Energy Policy, vol.34, pp.2788-2804, 2006.

[7] Shnmugam,K.R and Kulshreshtha,P, "Efficiency analysis of coal –based thermal power generation in India during post-reform era", Int. J. of Global energy Issues,vol.23,pp.15-28,2005.

[8] Pun-Lee Lam and Alice Shiu, "A data envelopment analysis of the efficiency of China's thermal power generation", Utility Policy, vol.10, pp.75-83, 2001.

[9] M.J.Farrell,"The measurement of productive efficiency"Journal of royal statistical society,vol.120,pp7-24,1957.

[10] A Charnes, W.W. Cooper and E. Rhodes, "Measuring the efficiency of decision making units", European Journal of Operational Research,vol 2,pp.429-444,1978.

[11] Andersen P., Petersen, N.C.A., "Procedure for ranking efficient units in Data Envelopment Analysis" Management Science, vol. 39 (10), pp.1261–1264, 1993.

[12] Sudhir Kumar Singh,V.K Bajpai and Seema Sharma, "Energy efficiency evaluation of thermal power plants in India using data envelopment analysis", Proceedings of the International conference on advances renewable energy, Bhopal, India, 2010.

[13] CEA,"Review of performance of Thermal power stations Central electric authority, Ministry of power, India. 2008.

Operation and Development of Small Hydropower Plant

Gaurav Dwivedi, Sachin Mishra, S.K.Singal
Alternate Hydro Energy Center, Indian Institute of Technology Roorkee, Roorkee.

Abstract: — **Energy is an essential ingredient of socio-economic development and economic growth. Electricity is a basic part of nature and it is one of the most widely used forms of energy. Small hydropower is one of the most cost-effective and environmentally benign energy technologies for production of electrical power worldwide. In India, the available potential of small hydro is at about 15,600 MW Hydropower is based on a simple process, taking advantage of the kinetic energy freed by falling water. In practice, this process is applied in many different ways depending on the electrical services sought and the specific site conditions. Accordingly, there is a wide variety of hydroelectric projects, each providing different types of services and generating environmental and social impacts of different nature and magnitude. This paper focuses on development of small hydropower plant and their operation.**

I. INTRODUCTION

Energy is the basic requirement for economic development of a country. The growing demand for energy has resulted in a dependency on fossil fuels. The price of fossil fuels and security of energy supply has been a concern in India for many decades. India has a large hydro potential in the range of medium and bigger size projects. After independence in 1947, large hydroelectric projects were executed, some of them are still under construction and some are planned for the future. The inherent drawbacks associated with large hydro are large gestation periods, submergence of large areas along with vegetation, and resettlement of the people. Political and environmental implications have made planners look for alternatives to large hydro. Electricity from small hydro is probably the oldest, yet reliable renewable energy source. In India, the potential for small hydro has been estimated to be 15,000 MW. Already, 674 small hydropower (SHP) plants with total installed capacity of about 2430 MW are in operation and 188 projects with another 483 MW are under various stages of implementation [6]. This shows that a good potential of large as well as small hydropower is available for development.

Small Hydro-Power Plants (SHPPs) have found special importance due to their relatively low administrative and executive costs, and a short construction time compared to large power plants. These SHPPs are in the "run-of-river" category because their generated capacity is based on the deviated water flow of river runoff and consists of a diversion dam, conveyance of water system, head pond, forebay, penstock, power house, and tailrace structure of the body of the SHPP as well as other electrical and mechanical equipment. The deviated flow of a river reaches the forebay after running in a path to the diverted point, and then enters into the SHPP structure via penstock pipes. Daily regulation of the water volume in the head pond is used to get maximum power from the SHPP during peak hours. The amount of energy generated during different daily hours and/or different seasons of the year are the most important issues worthy of study in the run-off river SHPP studies. In other words, calculating the optimal installation capacity (optimal designed flow) is one of the most important factors in planning SHPPs.

II. CLASSIFICATION OF SHP

There is no worldwide consensus on definitions regarding SHP, mainly because different development policies in different countries. Based on installed capacity of hydropower projects, classification of hydropower varies differently in various countries. Classification of SHP in India is as below:

- Pico 5 kW & below
- Micro 100 kW & below
- Mini 2000 kW & below
- Small 25000 kW & below
- Medium 100,000 kW & below
- Large above 100,000 kW

III. SHP TECHNOLOGY

SHP plant generates electricity or mechanical power by converting the power available in flowing water of rivers, canals and streams. The objective of a hydropower scheme is to convert the potential energy of a mass of water flowing in a stream with a certain fall, called head, into electric energy at the lower end of the scheme, where the powerhouse is located.The power of the scheme is proportional to the flow and to the head. A well designed SHP system can blend in with its surroundings and have minimal negative environmental impacts. SHP schemes are mainly run-of-river with little or no reservoir impoundment. For run-of-river, a portion of rivers water is diverted to a water conveyance, that delivers the water to a turbine. The moving water rotates the turbine, which spins a shaft. The motion of the shaft can be used for mechanical processes such as pumping water or it can be used to power an alternator or generator to generate electricity.

Component of SHP

Small hydropower is not simply a reduced version of a large hydro plant. Specific equipments are necessary to meet fundamental requirements with regard to simplicity, high-energy output and maximum reliability. Figure 1a and 1b shows the major components of a SHP scheme as detailed below.

Weir and Intake

A SHP scheme extracts water from the river in a reliable and controllable way. A weir is used to maintain the water level and ensure a constant supply to the intake. The following are required for an intake:

- The desired flow must be diverted,
- The peak flow of the river must be able to pass through the weir and intake without causing damage to them,
- lesser possible maintenance and repairs,
- It must prevent large quantities of loose material from entering the canal,
- It must have the possibility to more piled up sediments.

Power channel

The channel conveys water from the intake to the forebay tank. The length of the channel depends on the topographical conditions. In one case a long canal combined with a short penstock or in other cases a combination of short canal with long penstocks may be there. The channels are lined with cement, clay or polythene sheet to reduce friction and prevent leakages. Size and shape of a channel is a compromise between cost and reduced head. The following are incorporated in a channel:

- Settling basin – these are basins which allow particles and sediments, coming from the river flow, to be settled on the basin floor. These deposits are periodically flushed.
- Spillways – these divert excess flow at certain points along the channel.

I. Forebay tank

The forebay tank forms the connection between the channel and the penstock. The main purpose is to

provide immediate water supplyto the turbine for power generation.

IV. PENSTOCK

Penstock is a pipe, which conveys water under pressure from the forebay tank to the turbine. In front of the penstock, a trashrack (Figure 1b) is installed to prevent large particles from entering the penstock. .Pipes are generally made and supplied in standard lengths and have to be joined together at site. There are several ways to join the pipe; flanged, spigot and socket, mechanical and welded. Expansion joints are used to compensate for maximum possible change in length under temperature variation..Penstock pipes can either be buried or surface mounted. This depends on nature of terrain and environment considerations. Buried pipelines should be 0.75m below the surface so that vehicles do not damage it. However, one disadvantage can be, that if, leaks occur in the pipes it would be difficult to detect and rectify. When pipes run above ground, anchors or thrust blocks will be needed to counteract the forces which can cause undesired pipeline movement. The pressure rating of the penstock is critical because the pipe wall must be thick enough to withstand the maximum water pressure. This pressure depends on the head; the higher the head the more will be the pressure.

IV. POWERHOUSE AND TAILRACE

Powerhouse is a building that contains the turbine generator and the control units. The tailrace is a channel that allows the water to flow back to the stream after it has passed through the turbine. (Figure 1b)

V. TURBINES

A turbine unit consists of a runner connected to a shaft that converts the potential energy in falling water into mechanical or shaft power. The turbine is connected either directly to the generator or is connected by means of gears or belts and pulleys, depending on the speed required for the generator. The choice of the turbines depends mainly on the head and the design flow for the SHP installation. All turbines have power-speed characteristics. They will perform most efficiently at a particular speed, head and flow combination. There are mainly two types of turbines; impulse and reaction depending upon head, discharge and capacity.[2,3]

VI. DRIVESYSTEMS

The drive system transmits power from the turbine shaft to the generator shaft. It also has the function of changing the rotational speed from the one shaft to the other when the turbine speed is different to the required speed of the generator. The following can be considered for SHP drive system:

- Direct drive
- Flat belt and pulley
- V or wedge belt and pulleys
- Chain and sprocket
- Gearbox

VII. GENERATORS

These convert the mechanical (rotational) energy produced by the turbine to electrical energy. The basic principle of generator is that voltage is induced in a coil of wire when the coil is moved in a magnetic field. There are two types of generators; synchronous and asynchronous. Synchronous generators run at steady speed and are standard in electrical power generation. Asynchronous generators are known as induction generators. These are appropriate for smaller systems and have the advantage of being rugged and cheaper than synchronous generators. The generator speed depends on the number of poles. Electrical power can be generated in either AC or direct current (DC). AC can be connected directly to household appliances and AC is much more economical for transmitting power to homes. DC can be used in two ways, either directly as DC or converted to AC through the use of an inverter. The main advantage of DC is ease of battery storage.

VIII. CONTROLLERS

SHP systems with lead acid batteries require protection from overcharge and overdischarge. Overcharge controllers redirect the power to an auxiliary or shunt load when the batteries reaches a certain level. (Figure 2). This protects the generator from overspeed and overvoltage conditions. Overdischarge control involves disconnecting the load from the batteries when the voltage drops below a certain level.

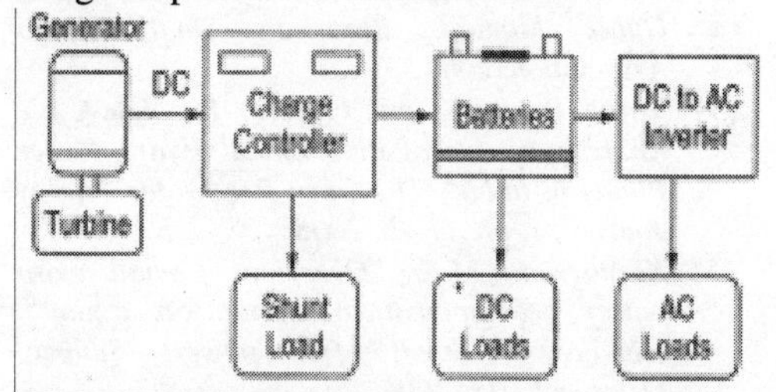

Figure 2: Electrical block diagram of a battery-based small hydro system.

Over the last two decades, electronic load controllers (ELCs) have been developed that have increased the simplicity and reliability of modern SHP system. An ELC is a solid-state electronic device designed to regulate output power of SHP systems. Maintaining a near-constant load on the turbine generates stable voltage and frequency. The controller compensates for variation in the main load by automatically varying the amount of power dissipated in a resistive load, generally known as the ballast or dump load, in order to keep the total load on the generator and turbine constant. Water heaters are generally used as ballast loads. an elcs constantly senses and regulates the generated frequency. The frequency is directly proportional to the speed of the turbine. The major benefit of elcs is that they have no moving parts, are reliable and virtually maintenance free.

X. TRANSMISSION / DISTRIBUTION NETWORK

The size and type of electric conductor cables required depends on the amount of electrical power to be transmitted and the length of the power line. For most SHP systems, power lines would be single phase, however sometimes three phase power lines are used.

I. Advantages of hydropower

A. Economics

The major advantage of hydroelectricity is elimination of the cost of fuel. The cost of operating a hydroelectric plant is nearly immune to increases in the cost of fossil fuels such as oil, natural gas or coal, and no imports are needed. Hydroelectric plants also tend to have longer economic lives than fuel-fired generation, with some plants now in service which were built 50 to 100 years ago. Operating labor cost is also usually low, as plants can be automated and have few personnel on site during normal operation. Where a dam serves multiple purposes, a hydroelectric plant may be added with relatively low construction cost, providing a useful revenue stream to offset the costs of dam operation.

B. co$_2$ Emissions

Since hydroelectric plants do not burn fossil fuels, they do not directly produce carbon dioxide. While some carbon dioxide is produced during manufacture and construction of the project components, this is a tiny fraction of the operating emissions of equivalent fossil-fuel electricity generation. According to this project, hydroelectricity produces the least amount of greenhouse gases and externality of any energy source. Coming in second place was wind, third was nuclear energy, and fourth was solar photovoltaic. The extremely positive greenhouse gas impact of hydroelectricity is found especially in temperate climates.

C. Other uses of the reservoir

Reservoirs created by hydroelectric schemes often provide facilities for water sports, and become tourist attractions themselves. In some countries, aquaculture in reservoirs is common. Multi-use dams installed for irrigation support agriculture with a relatively constant water supply. Large hydro dams can control floods, which would otherwise affect people living downstream of the project.

XI. BENEFITS FROM CDM SHP PROJECTS

The basic principle of the CDM is that both developed and developing countries benefit from participating, because synergies between global carbon abatement goals and local sustainable development goals are exploited. From the developing country perspective, the benefits arise both from the increased investment flows and from the requirement that these investments should advance host country sustainable development goals. More specifically, the cdm may contribute to several developing countries sustainable development objectives, including following: [4, 5]

- Increased energy efficiency and conservation.
- Transfer of technologies and financial resources.
- Local environmental benefits, e.g. cleaner air and water.
- Local environmental side benefits, such as health benefits from reduced local air pollution.
- Poverty alleviation and equity considerations through income and employment generation.
- Sustainable energy generation.
- Private and public sector capacity development.

In addition to these benefits, CDM projects may have a number of additional side benefits (or indirect benefits) on other national development objectives such as rural development, energy access, capacity building, education, and health.[6]

XII. COMPARISON WITH OTHER SOURCES OF POWER GENERATION

Hydroelectricity eliminates the flue gas emissions from fossil fuel combustion, including pollutants such as sulfur dioxide, nitric oxide, carbon monoxide, dust, and mercury in the coal. Hydroelectricity also avoids the hazards of coal mining and the indirect health effects of coal emissions. Compared to nuclear power, hydroelectricity generates no nuclear waste, has none of the dangers associated with uranium mining, nor nuclear leaks. Unlike uranium, hydroelectricity is also a renewable energy source. Compared to wind farms, hydroelectricity power plants have a more predictable load factor. If the project has a storage reservoir, it can be dispatched to generate power when needed. Hydroelectric plants can be easily regulated to follow variations in power demand.Unlike fossil-fuelled combustion turbines; construction of a hydroelectric plant requires a long lead-time for site studies, hydrological studies, and environmental impact assessment. Hydrological data up to 50 years or more is usually required to determine the best sites and operating regimes for a large hydroelectric plant. Unlike plants operated by fuel, such as fossil or nuclear energy, the number of sites that can be economically developed for hydroelectric production is limited; in many areas the most cost effective sites have already been exploited. New hydro sites tend to be far from population centers and require extensive transmission lines. Hydroelectric generation depends on rainfall in the watershed, and may be significantly reduced in years of low rainfall or snowmelt. Long-term energy yield may be affected by climate change. Utilities that primarily use hydroelectric power may spend additional capital to build extra capacity to ensure sufficient power is available in low water years

XIII. CONCLUSION

Small hydropower is a proven technology. SHP provides reliable, cheaper and clean energy to remote communities and islands that are not connected to the grid. SHP system is simple to operate and maintain. Its lifespan is measured in decades. These are always more cost-effective than any other form of renewable energy. The maximum electrical power that can be harnessed from SHP system depends on the head, the efficiency of generator, and turbine.

XIV. REFERENCES

[1] Montes, G.M., Lopez, M.M.S., Gamez, M.C.R., and Ondina, A.M., "An overview of renewable energy in Spain. The small hydropower case", J. Renewable and Sustainable Energy Reviews, 9, pp. 521–534, 2005.

[2] Bruno, G.S., Fried, L., "Focus on Small Hydro", renewable energy focus November/December 2008.

[3] "How to develop Small Hydropower plant", Guide Manual, European Small Hydro Association, 2004.

[4] Varun, Bhat, I.K., and Prakash, R., "Life Cycle Analysis of Run-of River Small Hydro Power Plants in India", The Open Renewable Energy Journal, 1, pp. 11-16, 2008.

[5] Kesharwani, M.K., "Overview of small hydro power development in Himalayan region", Himalayan Small Hydropower Summit, October 12-13, 2006.

[6] www.mnes.nic.in (Assecced on 22/01/2011)

[7] Sadden, B., "Hydropower Development in Southern and Southeastern Asia", IEEE Power Engineering Review, pp. 5 – 9, 2002.

Simulation of a Combined Cycle (Power and Refrigeration) Operated by a Solar Heat Source

[1]Bijendra Singh, [2]K.P. Tyagi and [1]Arvind Kumar

[1]MIT, Bulandshahr
[2]KIET, Ghaziabad

Abstract

Exergy thermodynamics is employed to analyze a binary ammonia water mixture thermodynamic cycle that produces both power and refrigeration. The analysis includes exergy destruction for each component in the cycle as well as the first law and exergy efficiencies of the cycle. The optimum operating conditions are established by maximizing the cycle exergy efficiency for the case of a solar heat source. Performance of the cycle over a range of heat source temperatures of 320–460°K was investigated. It is found that increasing the heat source temperature does not necessarily produce higher exergy efficiency, as is the case for first law efficiency. The largest exergy destruction occurs in the absorber, while little exergy destruction takes place in the boiler.

Nomenclature

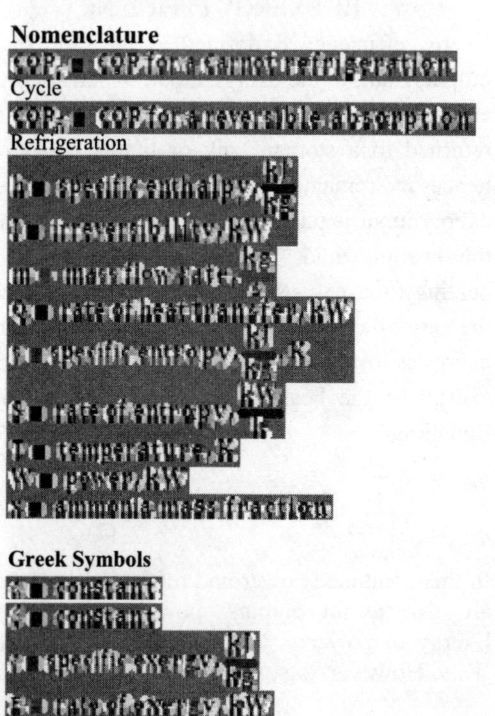

COP_c = COP for a Carnot refrigeration Cycle

COP_a = COP for a reversible absorption Refrigeration

h = specific enthalpy, $\frac{kJ}{kg}$

I = irreversibility, kW

m = mass flow rate, $\frac{kg}{s}$

Q = rate of heat transfer, kW

s = specific entropy, $\frac{kJ}{kg \cdot K}$

S = rate of entropy, $\frac{kW}{K}$

T = temperature, K

W = power, kW

x = ammonia mass fraction

Greek Symbols

α = constant

ζ = constant

ϕ = specific exergy, $\frac{kJ}{kg}$

ψ = rate of exergy, kW

Δ = change

η = efficiency

Subscripts

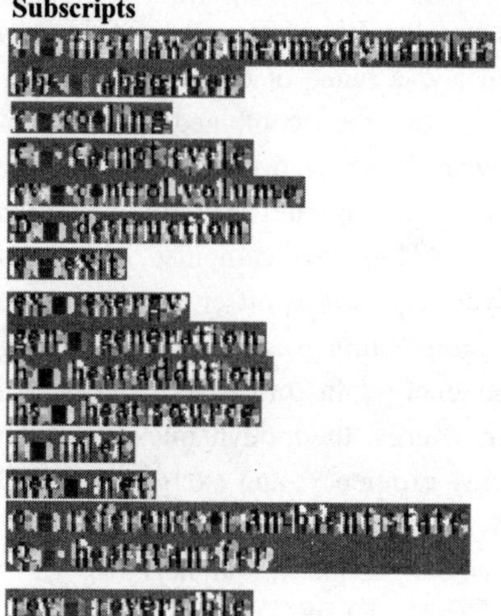

o = first law of thermodynamics

ab = absorber

c = cooling

cc = Carnot cycle

cv = control volume

D = destruction

e = exit

ex = exergy

gen = generation

h = heat addition

hs = heat source

i = inlet

net = net

o = difference from ambient state

Q = heat transfer

rev = reversible

Introduction

Binary mixture thermodynamic cycles are receiving more attention from researchers in recent years. Binary mixtures boil over a range of temperatures; as boiling progresses the temperature of the mixture increases due to the change in its composition. The change in boiling temperature leads to a good match with a sensible heat source (1-3). The increased effectiveness in the heat transfer during the heat addition process reduces the cycle irreversibility and improves the cycle performance. A novel cycle that employs ammonia water binary mixture as a working fluid was proposed by Goswami. As a binary mixture cycle, it has a good match between the heating fluid and the working fluid during the heat addition process. In addition, the Goswami cycle can produce both work and refrigeration, and it can utilize flat plate solar collectors with a potential reduction in the capital costs of solar thermal power by as much as 50% and applications in buildings that require power and air conditioning. A schematic of the cycle is shown in Fig. 1, in which an ammonia water vapor mixture with over 99% ammonia mass fraction expands in the turbine to a temperature below the ambient temperature. An absorption process replaces the conventional heat rejection and condensation process in this cycle. A simulation program was developed by for the combined power and refrigeration cycle, the program uses material and energy balances as well as the thermodynamic properties of the binary ammonia water mixture. The simulation program computes the state conditions at different locations of the cycle and the energy transfer for each component in the cycle, including work and heat transfer. Parametric study of the cycle was carried out and a range of operating conditions was suggested for the combined power and refrigeration cycle. Optimization of the cycle was investigated and some optimum operating conditions were examined by Lu. The combined power and refrigeration cycle can utilize different heat sources including low temperature waste heat, solar energy and geothermal energy. In this paper, the cycle is analyzed using exergy thermodynamics. First law efficiency, exergy efficiency, and exergy destruction in the cycle are examined over a range of heat source temperatures that correspond to solar heat sources.

II. EXERGY

Exergy analysis is used as a tool to enhance the understanding of thermodynamic processes and improve their performance. Exergy analysis reveals the irreversibility in the cycle and shows the possibilities where improvements in efficiency could be made.

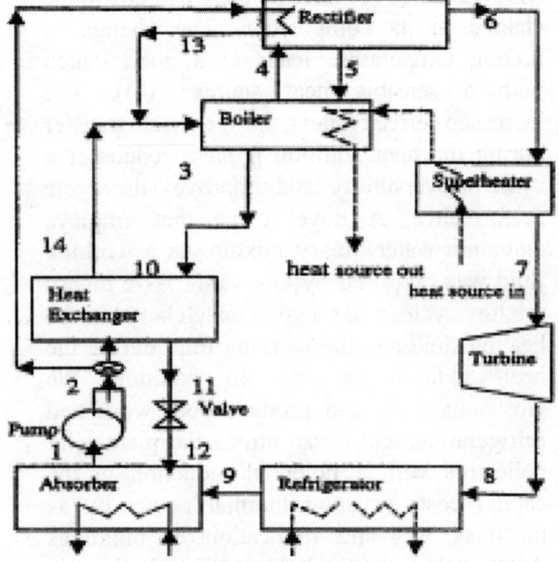

Figure 1.Schematic of the combined power and refrigeration

Exergy is defined as the maximum amount of reversible work a substance can do during the process of reaching equilibrium with its surroundings. If surrounding temperature T_0 is taken as a reference temperature, then exergy per unit mass of a stream, ε is given as

$$\varepsilon = (h - h_o) - T_o(s - s_0) \qquad (1)$$

Where h is enthalpy, s is entropy, and the subscript o refers to reference state. For a mixture, the exergy is given in terms of exergy of pure components evaluated at component partial pressure and mixture temperature. Szargut suggested that for a binary mixture, exergy could be given in terms of enthalpy, entropy and composition of mixture as follows,

$$\varepsilon = (h - T_o s) + \alpha + \beta X \qquad (2)$$

Where x is the mass fraction of one component in the mixture, and a and b are two constants whose values are set arbitrarily such that exergy in the cycle is always positive. Actually the constants a and b are functions only of pure components at the reference state and their values do not change in the cycle. It can be shown using material and exergy balances that in calculating the exergy destruction in the cycle for any control volume, the constants a and b vanish and, therefore, have no effect on the value of exergy destruction in the cycle. In our calculation, a is set as 50, b is set as 250, the reference temperature T_0 is 290°K, and the reference pressure is 0.1013 MPa. Ammonia water mixture properties are calculated based on the method developed by Xu and Goswami

III. EXERGY EFFICIENCY

Exergy efficiency is defined as the ratio of exergy output, Eout, to the exergy input. When solar thermal energy is used as a heat source the heating fluid is returned to a storage tank or to the collectors and hence its remaining exergy is not lost. Thus the exergy input is taken here as the change in exergy of the heating fluid, ΔE_{hs} as shown in Eq(3). If the heating fluid exhausts to the environment such as in the case of a gas turbine or a geothermal resource, its exergy is lost, so the efficiency is based on the initial exergy of the heating fluid, , Ehs-in, as given in Equation(4)

$$\eta_{ex} = \frac{\sum E_{out}}{\sum \Delta E_{hs}} \qquad (3)$$

$$\eta_{ex} = \frac{\sum E_{out}}{\sum E_{hs-in}} \qquad (4)$$

In this combined power and refrigeration cycle, there are two useful outputs: power and refrigeration. Exergy of power is just the net power of the cycle *Wnet*. However, the exergy EQ_c of a refrigeration load

Qc is given by Zsargut as,

$$E_{QC} = \frac{Q_C}{COP_C} \qquad (5)$$

Where COP_C is the coefficient of performance for a Carnot refrigeration cycle and is given as,

$$COP_C = \frac{T_C}{T_O - T_C} \qquad (6)$$

Where T_C is the cold reservoir temperature and T_O is the ambient temperature. Therefore, the exergy efficiency, η_{ex} of the combined power and refrigeration cycle may be given as,

$$\eta_{ex} = \left(\frac{W_{net} + Q_C/COP_C}{\Delta E_{hs}} \right) \qquad (7)$$

IV. EXERGY DESTRUCTION AND IRREVERSIBILITIES

Exergy destruction is equal to the Irreversibilities as given by Guoy-Stodola equation. Exergy destruction ED is calculated by rearranging the exergy balance equation for a control volume at steady state in the following form.

$$E_D = \sum m_i \varepsilon_I - \sum m_e \varepsilon_e - W_{cv} + \sum \left(1 - \frac{T_O}{T} \right) Q \qquad (8)$$

Where W_{cv} is the work of control volume, m is the mass flow rate, E_D is exergy destruction within the control volume, Q is heat transfer with surroundings or other fluids, and subscripts i and e are used for inlet and exit, respectively. Average temperature is used whenever temperature is not constant. Irreversibility is calculated from entropy generation and ambient temperature as,

$$I = T_O S_{gen} \qquad (9)$$

Where I is irreversibility and S_{gen} is the entropy generation

V. REVERSIBLE COMBINED POWER AND REFRIGERATION CYCLE

Figure 2 shows a simple thermodynamic representation of the combined cycle. At steady state, an energy balance for the cycle is given as,

$$Q_h + Q_C = Q_O + W_{net} \qquad (10)$$

where Q_h is the heat addition from the heat source at average temperature T_h , Q_c is the refrigeration at average temperature T_c , Q_o is the heat rejection to the ambient heat sink at T_o , and Wnet is the net power output of the cycle. The second law for a reversible cycle, where entropy generation is zero, is given as,

$$Q_h/T_h + Q_C/T_c - Q_o/T_o = 0 \qquad (11)$$

By substituting Q_o from Eq. (10) and rearranging, the following equation is obtained:

$$Q_h \eta_c + \frac{Q_C}{COP_C} - W_{net} \qquad (12)$$

Where the Carnot efficiency η_c is given as,

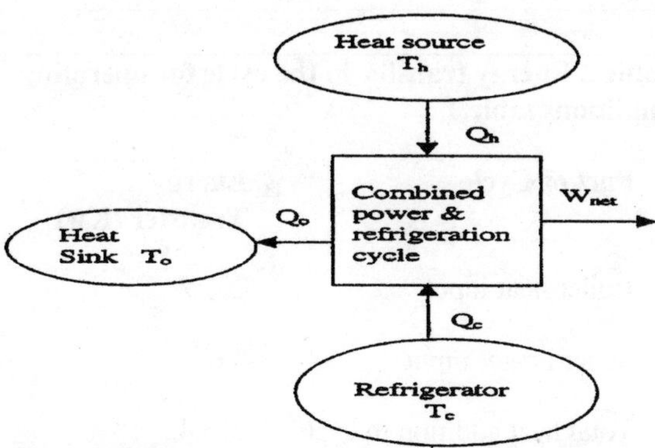

Figure 2 Thermodynamic representation of the combined power

Table 1 Optimized operating conditions for the combined cycle, heat source at 350°K

State point	T (K)	P MPa	Enthalpy (kJ/kg)	Entropy (kJ/kg.K)	Exergy (kJ/kg)	Concentration (kg. ammonia/kg solution)	Flow rate m/m1
1	295.0	0.52	113.1	4.190	44.7	0.652	1.000
2	295.1	1.71	-111.7	4.190	46.3	0.652	1.000
3	325.1	1.71	34.8	4.780	60.9	0.652	1.000
4	340.8	1.71	1367.4	1.85	402.3	0.502	0.161
5	300.6	1.71	80.9	0.767	105.1	0.738	0.083
6	300.6	1.71	1209.8	4.283	307.9	0.907	0.157
7	308.2	1.71	1302.8	4.351	400.3	0.907	0.157
8	301.0	1.71	1267.2	4.351	294.7	0.901	0.157
9	305.0	0.52	1065.3	4.356	293.1	0.907	0.157
10	340.8	1.71	80.1	0.891	31.8	0.306	0.843
11	300.1	1.71	107.7	0.249	17.1	0.306	0.843
12	300.3	0.52	-107.7	0.254	15.7	0.306	0.843
13	325.6	1.71	29.3	0.645	55.4	0.653	0.002
14	303.9	1.71	78.2	0.790	61.4	0.653	0.018

$$\eta_c = \left(\frac{T_h - T_c}{T_h} \right) \qquad (13)$$

And the COP_C for the refrigeration cycle operating between To and Tc is given by Eq.(6). $Q_h \eta_c$ is the exergy of the heat source, which represents the upper limit on the output exergy from the combined cycle. For the case of all refrigeration, and no net power, Eq. (12) simplifies to Eq. (14) which is the Coefficient of Performance of the reversible absorption refrigeration cycle, COP_R

$$\frac{Q_c}{Q_h} = \frac{T_C(T_h - T_o)}{T_h(T_o - T_h)} \qquad (14)$$

Equation (12) gives the exergy balance for a reversible cycle, while the exergy balance for an irreversible cycle includes the exergy destruction term E_D, as shown below,

$$Q_h \eta_c + \frac{Q_C}{COP_C} - W_{net} = E_D \qquad (15)$$

The exergy for refrigeration as given in the exergy balance Eq. (12) is the same as given earlier in Eq. (5)

Table 2 Energy transfer in the cycle for operating conditions table 1

Part of Cycle	Energy Transfer (Kw)
Boiler heat input	231.8
Super heater input	3.6
Total heat addition to cycle	235.4
Solution heat exchange	174.2
Rectifier heat rejection	11.6
Absorber heat rejection	221.6
Refrigeration output	9.1
Turbine power output	24.5
Pump power input	1.5
Cycle net power	23.0

Turbine is low at 3.3, thus small amount of power is produced from the cycle at the above operating conditions. Only 16% of the basic solution flows through the turbine and the refrigerator producing power and refrigeration, while the rest of the solution returns to the absorber. Exergy in the cycle is highest at the inlet of the turbine, where it has the highest potential of doing work. Exergy is lowest where the weak ammonia solution returns to the absorber. At the entrance of the refrigerator the working fluid still has a considerable amount of exergy, though the temperature is below the ambient. The exergy values in Table 1 are relative ones and, absolute exergy values can be calculated by adding the chemical exergy at the dead state to the above given physical exergy,

VI ENERGY ANALYSIS.

Table 2 shows the energy transfer for different parts in the cycle for the conditions of Table 1, and assuming 1 kg/s as the flow rate of basic solution. In this simulation, it is assumed that the turbine and the pump are both isentropic, throttling process in the valve is a constant enthalpy one, and pressure losses

are negligible in the cycle. The first law efficiency for the combined cycle is calculated as given in Eq.(16),

$$\eta_I = \left(\frac{W_{net} + Q_C}{Q_h}\right) \qquad (16)$$

The first law efficiency includes both useful outputs: the power W_{net} and the refrigeration Q_c. Heat input includes the heat added to the cycle in both the boiler and the super heater. The efficiency of the optimized cycle at the conditions shown in Table 1 is 13.6%. A Carnot heat engine operating between 350°K and 290°K would have a thermal efficiency of 17.2%. The combined power and refrigeration cycle has efficiency, which is 79.1% of the Carnot efficiency at the optimum operating conditions in Table 1. A Lorenz cycle, which is a reversible cycle that accounts for a

Table 3 Exergy destruction and Irreversibilities, heat source

Part of Cycle	Irreversibilities .kW.	Exergy Destruction .kW.	% of Total
Boiler	1.7	1.7	11.8
Superheater	0.1	0.1	0.7
Solution heat exchanger	3.4	3.4	23.6
Rectifier Absorber	1.0	1.0	6.9
Refrigerator	0.1	0.1	0.7
Turbine	0.0	0.0	0.0
Pump	0.0	0.0	0.0
Mixing at boiler inlet	0.0	0.1	0.7
Pressure valve	1.2	1.2	8.3
Total cycle	14.4	14.4	100

Sensible heat source as opposed to a constant temperature heat source of the Carnot cycle is more appropriate for comparison with this cycle. The efficiency of a Lorenz cycle operating between the same conditions would be 15.9%.

VII EXERGY DESTRUCTION.

Table 3 shows the Irreversibilities and exergy destruction in the cycle. Exergy destruction is calculated using Eq.(8) and Irreversibilities by using Eq.(9). As expected by the Guoy-Stodola equation, exergy destruction and Irreversibilities are the same. As explained earlier, the turbine and the pump are assumed isentropic. Irreversibilities in the refrigerator are close to zero since the heat transfer in the refrigerator is relatively small and occurs at temperatures close to the ambient. The absorber has the highest Irreversibilities at 47.2% of the total cycle Irreversibilities, next is the solution heat exchanger at 23.6%, followed by the boiler at 11.8% and then the pressure reducing valve. The absorber has the highest Irreversibilities since it involves two highly irreversible processes of condensation and mixing of ammonia and water components. The solution heat exchanger involves the process of heat exchange between the two streams of strong and weak ammonia solutions with a large temperature difference. The boiler involves the boiling process, which is irreversible exergy efficiency for the cycle, as defined in Eq.(7) and at the operating conditions of Table 1, is 61.8%. Hence 39.2% of the heat source exergy change is being lost in the cycle, these losses are in the form of exergy destruction, and exergy losses from the absorber during the heat rejection process to the environment. The total exergy destruction is also calculated using the exergy balance Eq. (15) which gives a total exergy destruction of 14.4, the same as in Table 3. Average temperatures are used in calculating the Carnot efficiency and COP in Eq.(15)

Effect of Heat Source Temperature.

The optimum performance of the combined cycle was examined over a heat source inlet temperature range of 320–460°K. This low and medium temperature range can be obtained from flat plate collectors or medium temperature concentrators. Figure 3 shows the effect of heat source temperature on the performance of the cycle, including net power and refrigeration capacity as a fraction of the heat addition, first law efficiency and exergy efficiency. The refrigeration, as a fraction of the heat addition Qc/Qh, changes little as the heat source temperature increases. As temperature of the heat source approaches the ambient temperature, refrigeration approaches zero. The highest refrigeration fraction is around a source temperature of 390°K. The refrigeration fraction decreases above this source temperature and becomes zero around 480°K. The net power, as a fraction of heat addition $Wnet/Qh$, increases as the heat source temperature increases. Since the turbine output power is related mainly to the pressure ratio across the turbine, net cycle power curve can be explained in relation to the pressure ratio in Fig. 4, which shows a continuous increase with the heat source temperature. The first law efficiency curve, which is the sum of the refrigeration and power curves, shows a corresponding behavior to the power curve up to a maximum value of 23.6% at 440°K. After the maximum point, the efficiency starts decreasing slowly in a similar behavior to the refrigeration curve. The exergy efficiency shows a maximum value of 65.2% at 380°K. The sharp increase in the exergy efficiency between 320 and 380°K is due to the increase of outputs, the net power and refrigeration, as shown in the same figure.

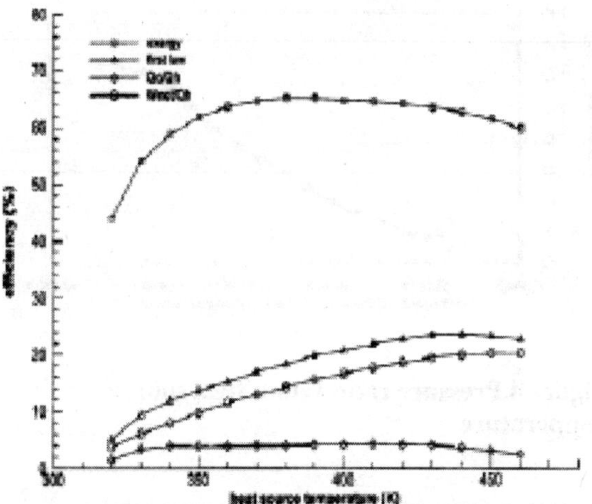

Figure 3.Efficiency versus heat source temperature

Refrigeration decreases above 400°K while the net power continues to increase slowly, the net result being that the exergy curve shows a maximum. Figure 5 shows the refrigeration to net power ratio versus the heat source temperature. The maximum refrigeration to net power ratio occurs at 330°K. Increasing the heat source temperature above 330°K would favor the production of net power rather than refrigeration. Figure 6 shows normalized exergy destruction in the cycle as a function of the heat source temperature. Total exergy destruction in the cycle increases with an increase in the heat source temperature. It can be seen in Fig. 6 that exergy destruction in both absorber, and heat exchanger change little as the source temperature increases. Super heater has almost zero exergy destruction becauseof its small heat load. The boiler exergy destruction is much lower than the absorber. Exergy destruction in theRectifier increases as the heating load in the rectifier increases as depicted in Fig. 7, and becomes the largest above 420°K. Therefore, to improve the design of the cycle, the rectifier may be eliminated above a heat source temperature of 420°K, especially if low temperature (<273°K) refrigeration is not required. From an exergy efficiency point of view, if the heat source is between 320 and 460°K, then the best operating heat source temperature is around 380°K, since it gives the maximum exergy efficiency. A solar heat source using a flat plate collector at 360°K (87°C) gives an exergy efficiency of 63.7%, with refrigeration as 3.7% of the heat addition and net power as 11.5% of the heat addition to the combined cycle.

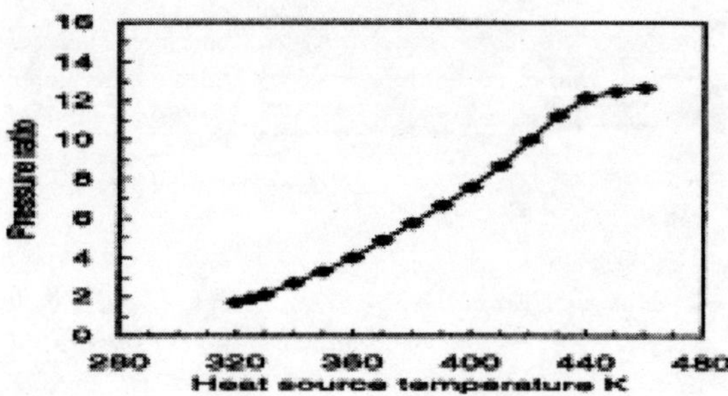

Figure 4.Pressure ratio versus heat source temperature

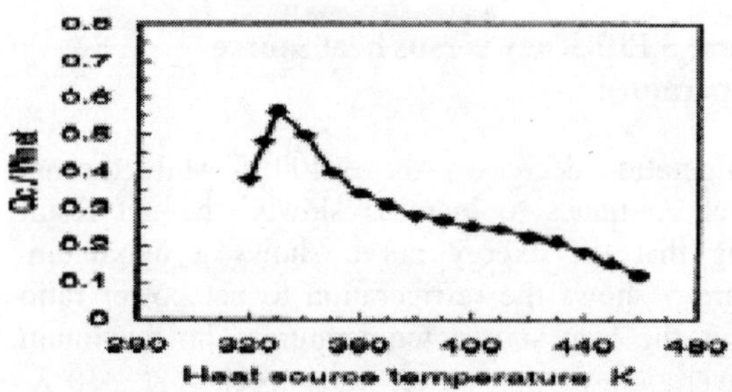

Figure 5. Refrigeration to net power ratio versus heat source

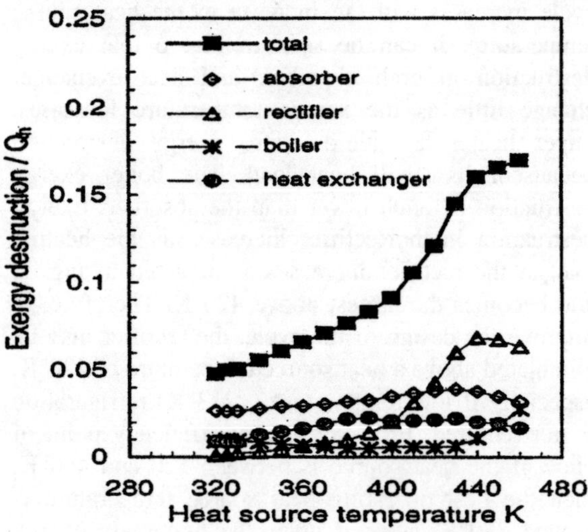

Figure 6.Normalized exergy destruction versus heat source

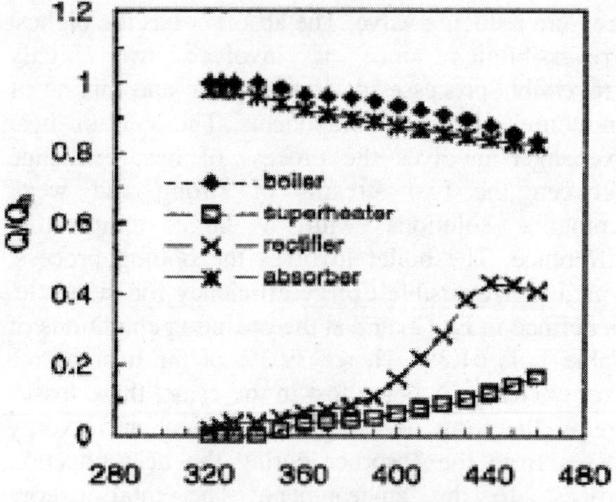

Figure 7.Normalized heat transfer versus heat source

VIII CONCLUSIONS

The combined power and refrigeration cycle, operating under optimum conditions, can produce different fractions of power and refrigeration depending on its heat source temperature. Exergy analysis can be used to improve the cycle performance. Total exergy destruction in the combined power and refrigeration cycle increases as the heat source temperature increases. The absorber has the highest exergy destruction for a heat source temperature of 320 to 400°K, while above 400°K the rectifier would have the highest contribution to the exergy destruction in the cycle. The boiler has moderate contribution to the cycle exergy destruction. This leads us to a conclusion that it may be better to eliminate the rectifier for source temperatures above 400°K, especially if low temperature refrigeration below 273°K is not required. The results also lead us to a study of a better absorber.The exergy efficiency has a maximum point over the investigated range of 320–460°K heat source temperatures. Increasing the heat source inlet temperature does not necessarily lead to higher exergy efficiency as it is the case for Carnot thermal efficiency and the first law efficiency. Heat from flat plate solar collectors ~temperature of 90°C or less can drive the combined power and refrigeration cycle, and produce power and refrigeration at the same time. The cycle has a good thermal efficiency reaching 25.3% at 430°K, which constitutes 78% of the Carnot engine efficiency operating between the same upper and lower boundary conditions.

VIII REFERENCES

[1] *Kalina, A. I., 1984, "Combined Cycle System With Novel Bottoming Cycle," ASME J. Eng. Gas Turbines Power, **106**, 737–742.*

[2] *Kanlina, A. I., and Leibowitz, H. M., 1987, "Applying Kalina Technology to a Bottoming Cycle for Utility Combined Cycles," ASME paper 87-GT-35*

[3] *Ibrahim, O. M., and Klein, S. A., 1996, "Absorption Power Cycles,"*

[4] *Goswami, D. Y., and Xu, F., 1999, "Analysis of a New Thermodynamic Cycle for Combined Power and Cooling Using Low and Medium Temperature Solar Collectors," ASME J. Sol. Energy Eng., **121**, 91–97.*

[5] *Xu, F., 1997, "Analysis of a Novel Combined Thermal Power and Cooling Cycle Using Ammonia-Water Mixture as aWorking Fluid," Ph.D. thesis, Univ. of Florida.*

[6] *Lu, S., 2001, "Combined Power/Cooling Cycle," Ph.D. thesis, Univ. of Florida.*

[7] *Mcllvried, H. G., Ramezan, M., Enick, R. M., and Venkatasubramanian, S., 1998, "Exergy and Pinch Analysis of Advanced Ammonia-Water Coal Fired Power Cycle," Proc. of ASME Advanced Energy Systems, **38**, 197–203.*

[8] *Szargut, J., Morris, D. R., and Steward, F. R., 1988, Exergy Analysis of Thermal, Chemical, and Metallurgical Processes, Hemisphere Publishing Corp*

[9] *Kotas, T. J., Mayhew, Y. R., and Raichura, R. C., 1995, "Nomenclature forExergy Analysis," Proc. IMechE., **209**, 275–279.*

Study of Thermal Properties of Babool Wood and its Polyacrylonitrite Composites

Mohd Arif[1], Shalendra Kumar Pathak[2], Lokesh Upadhyay[3] and Vikas Sharma[1]
[1]Department of Mechanical Engineering, Anand Engineering College,
Keetham Agra, Uttar Pradesh, India.
[2]Department of Mechanical Engineering, Agra Public College of Technology & Management,
Artoni, Agra, Uttar Pradesh, India.
[3]Department of Mechanical Engineering, Sachdeva Institute of Technology Farah- Mathura,
Uttar Pradesh, India.

Abstract -In this study some thermal properties of Babool wood and its acrylonitrite impregnated wood composites were investigated. Polyacrylonitrite (6.81 mole/l) as impregnated into Babool wood through benzoyl peroxide (0.02 mol/l) to initiate the polymerization process forming free radicals in methanol medium at 75±1°C. Modification of the thermal properties over untreated wood was evaluated in terms of differential thermogravimetry-thermogravimetric-differential thermal analysis (DTG-TG-DTA) in air. Resistance of wood against thermo-oxidation was improved with impregnation of polyacrylonitrite (PNA). Impregnation of polyacrylonitrite into Babool wood was confirmed through scanning electron microscopy.

Keywords: Benzoyl peroxide, polyacrylonitrite impregnation, thermo-oxidative stability.

I. INTRODUCTION

The performance of wood as a construction material for outdoor applications deteriorates under accelerated weather environments due to fluctuation in weather and humidity for longed outdoor applications as well as decreasing the cost of wood and avoiding the need of frequent replacements in permanent and temporary constructions.

A number of wood preservatives and new wood treatment processes have been developed during those wood treatment processes are under continuous demand which can develops the modify wood materials with improved mechanical strength, thermo-oxidative stability and resistance towards bio-deterioration for their better outdoor applications.

The polymer loading of wood depends on the permeability of the wood species being treated. Because the void volume is approximately the same for sap wood and heart wood for each species. It would be expected that the polymer would fill them to same extent [3]. In the past few decades a variety of commercially available vinyl monomers has been used for wood treatment to improve the mechanical and thermo-oxidative stability of low-grade woods [4, 5]. Advancement in the technology of thermoplastic impregnated wood composites have recently made great claims to replace quality woods with high grade wood polymer composites derived from low grade woods [1,6,7]. In many kind of processing wood has been subjected to treatment at elevated temperatures e.g. drying, size stabilization, pulping, production of particle and fiber boards. As temperature affect the physical, structural and chemical properties of wood. Several attempts have been made to establish relation between temperature and thermal stability of wood [8-12]. Reinforcement of several acrylic monomers like styrene, methylmethacrylate, and (chloropropyl)-2-propane phosphate has provided substantial thermal stabilities to various low grade woods. Recently dynamic mechanical thermal analysis has been recognized as a useful thermo analytical method of detecting relations polymers and composite molecules and the temperature is scanned over a range from sub ambient to above the material glass transition. This analysis is more sensitive than other thermo analytical methods [13-15]. Thermo-oxidative stability of wood polymer composites (WPCs) from the tropical wood Geonggang (Cratoxylum arborescence) and methyl-methacrylate, methylmethacrylate acrylonitrile and styrene acrylonitrile combination was investigated through thermo gravimetric analysis and differential scanning calorimetry. The thermal data indicate that wood has been thermally modified [2, 16, and 17]. Polymerization of methylmethacrylate into Babool wood has also been reported and the composites indicated excellent moisture resistance and thermo-oxidative stability [18]. In the present research work efforts have been made to develop such polyacrylonitrite impregnated composites in methanol medium having improved thermal stabilities for their commercial exploitation for desirable purposes.

II. MATERIALS & METHODS

Starting materials

Acrylonitrite monomer was purchased from M/s-C. D. H. Chemicals India Pvt. Ltd. Mumbai. The monomer acrylonitrite was purified by extracting it with aqueous NaOH (10%) to remove inhibitor contents followed by repeated washings with distilled water. The fraction distilled at 82°C was used for the impregnation polymerization reaction.

Preparation of wood specimens

First the wood specimens were prepared for their treatment as per IS: 1708-1960. The moisture content of wood was deduced according to ASTMD 1037-72a and was found to be 12.75%.

Preparation of solution

The methanolic solution of acrylonitrite at concentration of 2.27M and methanolic solution of benzoyl per oxide at 0.02M have also been prepared.

Method of treatment of wood specimens

The prepared wood specimens were placed in an airtight chamber of stainless steel of dimensions $20\times20\times30cm^3$. The specimens were swelled in methanol (98%) for 5 hrs. The solution of benzoyl peroxide (0.02M) and acrylonitrite (PAN) were added. The samples were then soaked in monomer solution for 12 hours at room temperature. The treated wood specimens were then wrapped in aluminum foil at 95±1°C for 2hours to induce the polymeric action reaction. Impregnation of polyacrylonitrite into Babool wood was confirmed through scanning electron microscopy.

Characterization of wood and impregnated wood composites

Perkin Elmer (Pyris Diamond) thermal analyzer model STA-78 has been employed to study differential thermogravimetry-Thermogravimetry-differential thermal analysis (DTG-TG-DTA) of untreated wood and its PAN impregnated wood composite in the atmosphere of static air at a hating rate of 10°C / minute up to 550°C using alumina as reference. The sample size was taken 10mg. The crystallization temperature (T_c) and oxidation temperature (T_{ox}) have been deduced from DTA curve [19], whereas the maximum decomposition temperature (T_{max}) and final decomposition temperature (T_f) were measured from DTG. TG scans were exploited to evaluate the range for various decomposition stages electron micrographs of woods and their PAN reinforced wood composites were scanned on LEO-435 SEM. The morphologies of wood and its PAN reinforced composites were studied in view to get a clear understanding about the affinity of PAN with wood.

III. RESULTS AND DISCUSSION

The various thermo analytical data of above said wood and its PAN reinforced wood composites as deduced from DTG-TG-DTA in air have been summarized (Table 1). Comparison of scanning electron micrographs with impregnation of polyacrylonitrite into babool wood lumens was not uniform [17]. TG data has been used to study the weight loss in wood and related composites at various temperature range 0-550°C. Graph-1 of untreated Babool wood TG profiles indicates that thermo-oxidative decomposition of wood was started at 209°C with 11.6% weight loss. The further weight loss in wood was recorded in the temperatures range 255 – 314°C with 20%of weight loss and the weight loss further intensified to 86.45%% up to 383°C. The first & second DTA endotherms have represented the crystallization temperature (T_c) at 310°C and oxidation temperature (T_{ox}) at 411°C. Similarly the maximum (T_{max}) and final (T_f) decomposition temperature were recorded at 303°C & 397°C respectively from DTG endotherms. Thermo-oxidative decomposition of PAN impregnated wood composites referring Graph-2 started at 211°C with 9.5% weight loss. From 292 to 365°C a rapid wt. loss of 66.8% in wood polymer composite was recorded which was further intensified to 94% at 457°C. Polyacrylonitrile (PAN) impregnated composites have shown all such thermal parameters more improved than untreated wood.

Table 1: DTG – TG – DTA Properties of Babool Wood Acrylonitrile Impregnated   Wood Composites.

S. No.	Concentration	moles/ liter	Sample (mg)	Decomposition ranges Th(°C)			DTA (°C)		DTG (°C)	
				I (%)	II (%)	III (%)	T_C	T_{OX}	T_{max}	T_f
1	0%	0	10	228-292	292-335	355-454	354	435	330	358
2	45%	6.81	10	208-325 5.77	325-363 75	363-399 99.72	328	395	307	373

T_c :   Crystallization temperature
T_{ox}:   Oxidation temperature
T_{max}:   Maximum decomposition temperature
T_f: Final decomposition temperature

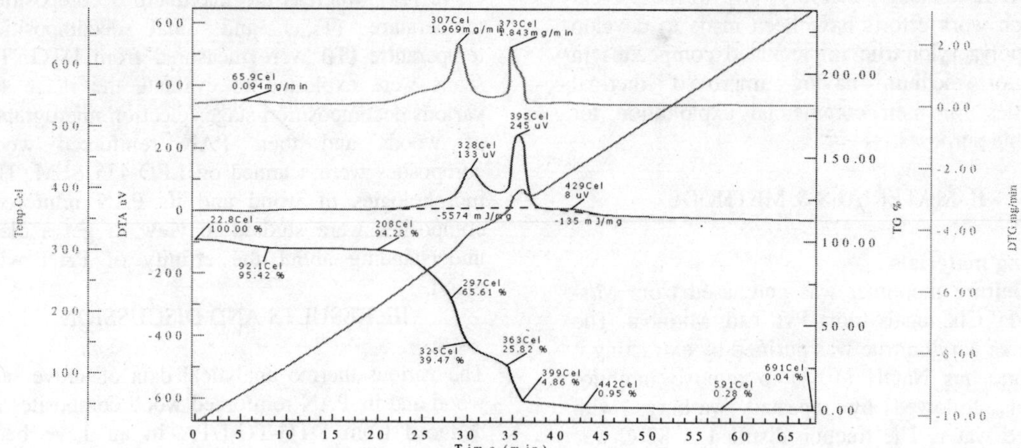

Figure 1 DTG–TG–DTA Curve for Untreated Babool Wood.

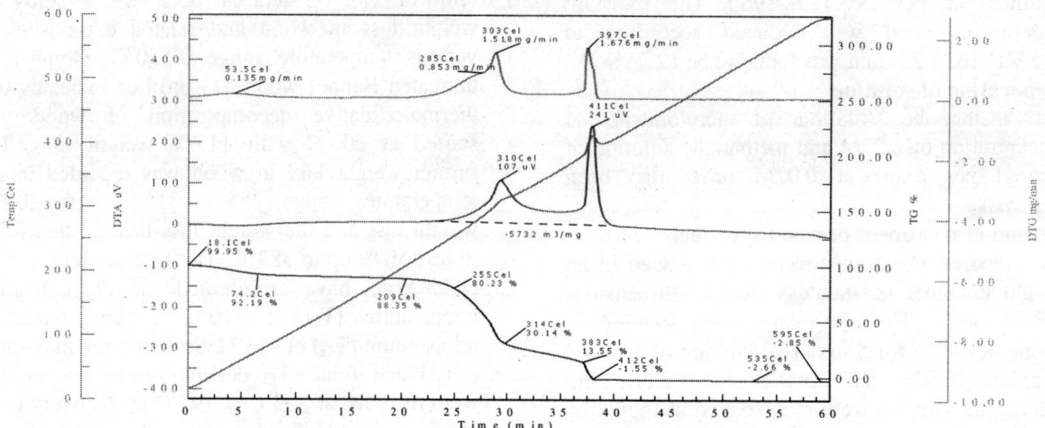

Figure 2 DTG–TG–DTA Curve for Polyacrylonitrite Affinity of 6.81M Concentration Treated Babool Wood

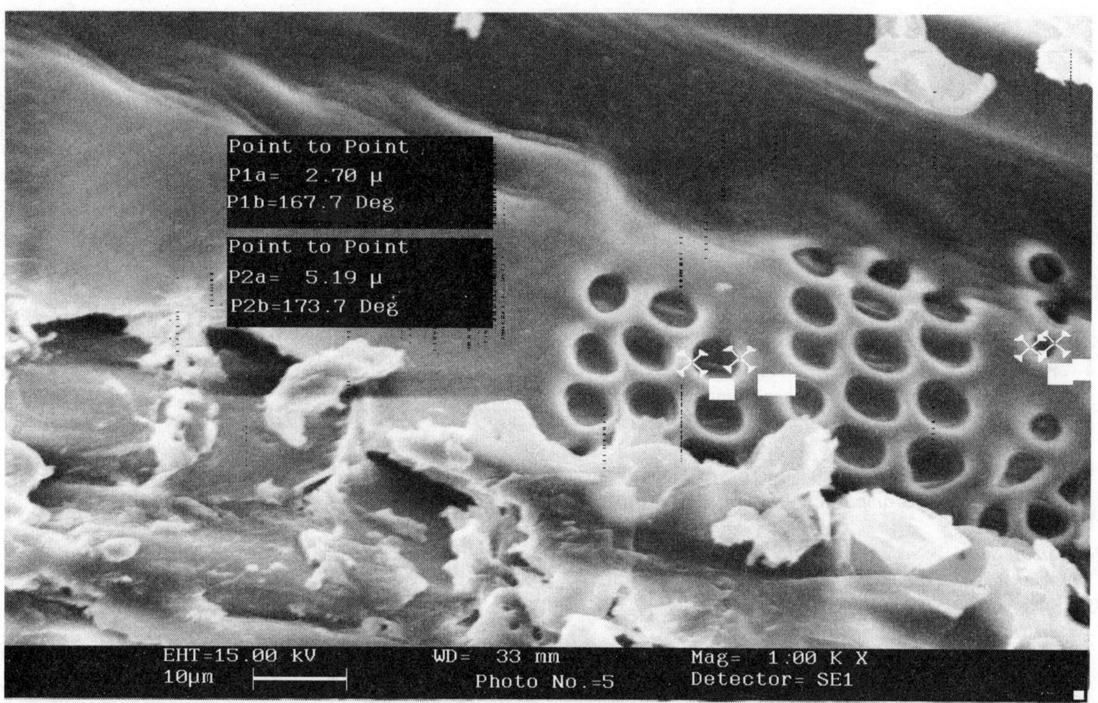

Figure 3 Microscopic View of Untreated Babool Wood.

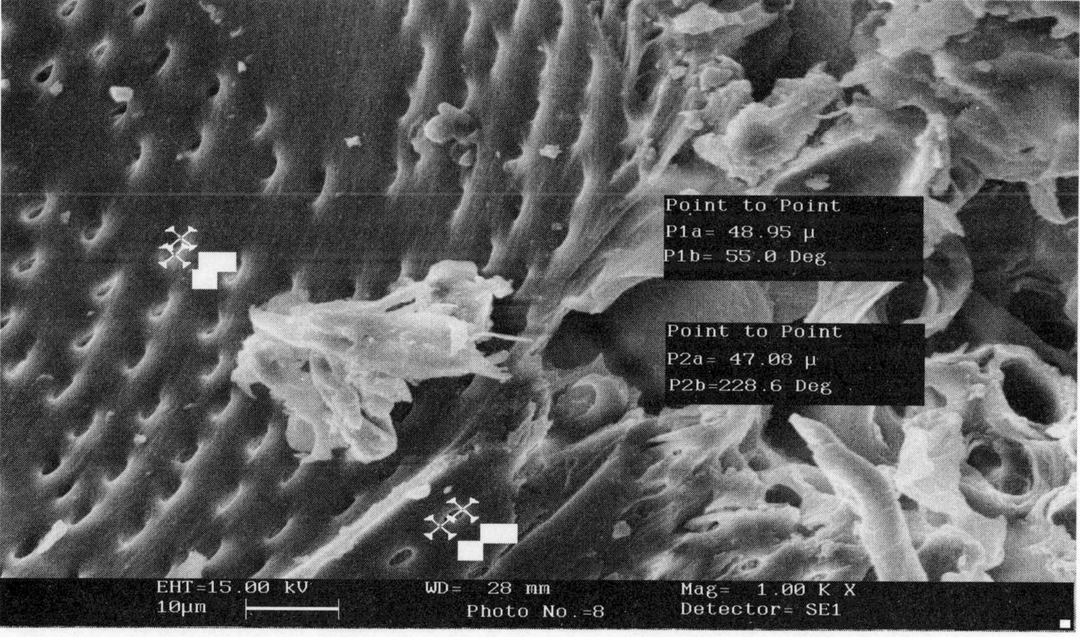

Figure 4 Microscopic View of Polyacrylonitrite Affinity of 6.81M Concentration Treated Babool Wood

IV. CONCLUSION

It is concluded from these thermal data that the thermal stability of PAN reinforced composites was improved in comparison to untreated Babool wood.

V. ACKNOWLEDGEMENT

The authors are thankful to the Head, Institute Instrument Centre, Indian Institute of Technology, Roorkee, India for providing thermal analysis data and scanning electron micrographs and also thankful to Dr. S. K. Mishra for his continuous guidelines.

VI. REFERENCES

[1] *Schneider M.S; Phillips, J.G. and Lande, S. 2000. Physical and mechanical properties of wood polymer composites, J. Forest Eng 11, 83-89.*

[2] *Clemons C, 2002, Forest Product Journal 52 : 10*

[3] *Balatinecz, JJ and Woodhams, R.T., 1993, J. Forestry, Vol. 91, pp. 22-29.*

[4] *Dale Ellis W and O'Dell JL, 1999. J Appl Polym. Sci. 73: 2494-2505.*

[5] *Lingfi M and Yang YY, 1996 Zheziang Linxueynan Xebio 13, 104-108.*

[6] *Research group of wood plastic composites research, China Wood Industry Application Prospect of Wood / Plastic Fiber Composites in the Passenger Car Industry, Institute of Wood Industry, Beijing 11(5) : 22(1997)*

[7] *Mapleston P, Additive Suppliers Turned their Eyes to Wood Plastic Composites, Modern Plastics Aug: 52 (2001).*

[8] *Sanderman W & Augstin, H. 1963. Holz Roh-Werkst, Vol. 21pp, 256-265, 305-315.*

[9] *Fengel, G. 1966. Holz-Roh-Werkst. Vol. 24. pp. 9-14, 98-109, 529-536.*

[10] *Kosik, M, Geratova, L, Rendos, F. and Domansky, R, 1968, Holzforch, Holzvenv Vol. 20, pp. 5-19.*

[11] *Kosik, M, Kozmal, F., Resiser, V. and Domansky, R, 1968. Holzforch, Holzfrch. Holzusw, Vol, 20, pp. 11-15.*

[12] *Beall, F.C and Eichkner, H.W. 1970. Thermal Degradation of Wood Components. A Review of Literature USDA. For Service Res. Paper. FPL 130.*

[13] *Oksmank and Lindberg H, Holzforschung 49; 243, (1995).*

[14] *Courturer, M. F, George, K and Schneider, M.H 1996. Thermophysical Properties Wood Polymer Composites, Wood. Sci. Technol. Vol. 30(3), pp. 179-196; 125 (14). 1713145J.*

[15] *Ellis DW and Sanadi RA, Proc 18th Riso Int. Symp. On Mat. Sci. 307; 1673 (1997)*

[16] *Yap GS, Que YT, Chio LH and Chan OHS, J Appl Polym Sci. 43; 2057 (1991)*

[17] *Ibach RE and Rowell RM, Holzforschung 55 : 358 (2001)*

[18] *Joshi TK, Zaidi MGH, Sah PL & Alam S, 2005, Mechanical & Thermal Properties of Poplar Wood Polyacrylanitrile Composites, J. Polym. Int. 54 : 198-201 (2005)*

[19] *Collins, Edward, A., Barc, Jan. Mayer, Fred W., Bill Jr. 1973. Experiments in Polymer Science, Wiley Int. Science Publication, Chapter 9, pp. 216-262.*

[20] *ASME Tyagi Yogesh kumar, Zaidi MGH, Singh Pratap and Singh Dheer, 2006. BSME International conference on Thermal Engineering, Dhaka, Bangladesh.*

Performance Analysis of Domestic Refrigerator Using Different Refrigerant

[1]**Amitesh Paul**, [2]**Rahul Sharma**

[1]Department of Mechanical Engineering, Shri Satya Sai Institute Of Science and Technology,
Sehore, M.P., India,

[2]Department of Mechanical Engineering, Swami Vivekanand College of Technology, Indore, M.P., India,

Abstract-The fundamental reason for having a refrigerator is to keep food cold. Cold temperatures help food stay fresh longer. The basic idea behind refrigeration is to slow down the activity of **bacteria** (which all food contains) so that it takes longer for the bacteria to spoil the food. For example, bacteria will spoil milk in two or three hours if the milk is left out on the kitchen counter at room temperature. However, by reducing the temperature of the milk, it will stay fresh for a week or two – the cold temperature inside the refrigerator decreases the activity of the bacteria that much. By freezing the milk you can stop the bacteria altogether and the milk can last for months (until effects like freezer burn begin to spoil the milk in non-bacterial ways). Refrigeration and freezing are two of the most common forms of food preservation used today.

The present work involves the analysis of domestic refrigerator using different refrigerant. In this analysis different refrigerants are used in a test rig to check the performance of a domestic refrigerator under different load conditions. The refrigerants used are R-12, R-134a and Hydrocarbon Blend.

Key words: Refrigerating effect, Coefficient of Performance, Evaporator Temperature, Condenser Temperature, R-12, R-134a and Hydrocarbon Blend.

I. INTRODUCTION

The thermodynamics of the vapor compression cycle can be analyzed on a temperature versus entropy. At point 1 in the diagram, the circulating refrigerant enters the compressor as a saturated vapor. From point 1 to point 2, the vapor is isentropic ally compressed (i.e., compressed at constant entropy) and exits the compressor as a superheated vapor.

From point 2 to point 3, the superheated vapor travels though part of the condenser, which removes the superheat by cooling the vapor. Between point 3 and point 4, the vapor travels through the remainder of the condenser and is condensed into a saturated liquid. The condensation process occurs at essentially constant pressure.

Between points 4 and 5, the saturated liquid refrigerant passes through the expansion valve and undergoes an abrupt decrease of pressure. That process results in the adiabatic flash evaporation and auto-refrigeration of a portion of the liquid (typically, less than half of the liquid flashes). The adiabatic flash evaporation process is isenthalpic (i.e., occurs at constant enthalpy).

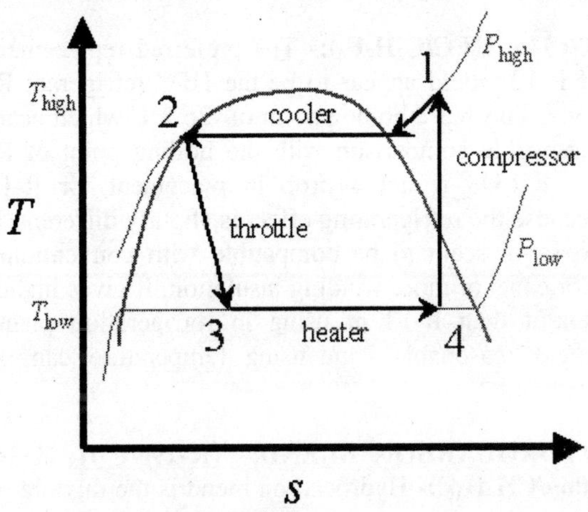

Between points 5 and 1, the cold and partially vaporized refrigerant travels through the coil or tubes in the evaporator where it is totally vaporized by the warm air (from the space being refrigerated) that a fan circulates across the coil or tubes in the evaporator. The evaporator operates at essentially constant pressure. The resulting saturated refrigerant vapor returns to the compressor inlet at point 1 to complete the thermodynamic cycle.

Many systems still use HCFC refrigerants, which contribute to depletion of the Earth's ozone layer. In countries adhering to the Montreal Protocol, HCFCs are due to be phased out and are largely being replaced by ozone-friendly HFCs. However, systems using HFC refrigerants ten to be slightly less efficient than systems using HCFCs. HFCs also have an extremely large global warming potential (GWP) because they remain in the atmosphere for many year and trap heat more effectively than carbon dioxide. With disruption of the status quo already a certainly, alternative non-halo alkane refrigerants are gaining popularity. In particular, once-abandoned refrigerants such as hydrocarbons (HCs, such as butane) and CO_2 are coming back into broader use. For example, Coca-Cola's vending machines at the World Cup 2006 in Germany used refrigeration utilizing CO_2.

II. PROPERTY OF DIFFERENT REFRIGERANT

R-12(FREON/CCl_2F_2):-This is the most widely used and most popular refrigerant. It is commonly used for all refrigeration purposes. It is colourless and odourless liquid. It is non-toxic, non-flammable, non-explosive and non-corrosive. It condenses at atmospheric pressure. This refrigerant is commercially available in different cylinder size. Generally 0.7 kg of refrigerant is required in refrigeration system per cubic meter of air conditioned space.

R-134a (HFC/$C_2H_2F_4$):- The preferred replacement of R-12 would appear to be the HFC refrigerant R-134a. This has a boiling point of -26.2°C which bears reasonable comparison with the boiling point of R-12. R-134a is not a drop in placement for R-12 because the refrigerating effect is slightly different. It does not seem to be compatible with conventional lubricants or more winding insulation. It gives higher benefit than R-12 in using in refrigeration plants where reasonable condensing temperature can be specified.

HYDROCARBON BLEND: (R-290/C_3H_8 & R-600a/CH_4H_{10}):- Hydrocarbon blend is the mixture of two hydrocarbons one is R-290 (propane) 50% and other is R-600a (isobutane) 50%. This is the best replacement for R-12 and R-134a. Hydrocarbons are naturally occurring substances that are obtained from the refineries after distillation. They have been used as refrigerants for many decades, mostly in very large industrial plant but also in small low temperature systems. This is the most appropriate blend of propane and isobutane as only minimum changes to the compressor and refrigeration system is needed.

Calculation for COP Using the Refrigerant Chart:

1. Take observation for temperature and pressure from test rig using different refrigerant
2. Using refrigerant chart (p-h curve) for R-12, R-134a and Hydrocarbon Blend obtain values of enthalpies for different load.

TABLE FOR R-134a

PARAMETER	NO LOAD	LOAD I	LOAD II	LOAD III	LOAD IV
T_1 (°C)	-1.2	-0.7	-0.2	0.1	0.6
T_2 (°C)	68.2	70.3	72.4	75.0	77.2
T_3 (°C)	54.2	55.1	56.2	56.9	57.2
T_4 (°C)	-11.2	-10.3	-9.4	-7.9	-6.4
h_1 (KJ/kg)	397.98	398.27	398.56	398.74	399.03
h_2 (KJ/kg)	428.57	428.92	429.15	429.24	429.28
h_{f3} (KJ/kg)	277.56	278.04	281.36	282.49	283.2
h_4 (KJ/kg)	185.21	186.39	189.22	189.54	191.5
V (volt)	232	232	232	232	232
I (ampere)	0.95	0.95	0.95	0.95	0.95
P_d (bar)	12.7	12.7	12.9	12.9	12.9
P_s (bar)	0.39	0.45	0.45	0.45	0.45

Where,

T_1= Suction temp. of compressor inlet.

T_2= Discharge temp. of compressor outlet.

T_3= Discharge temp. of condenser outlet.

T_4= Evaporator Temperature.

h = Enthalpy (J/kg)

Ps = Suction Pressure of Compressor.

Pd = Discharge Pressure of Compressor.

LOAD	P	m_r	R.E	WD	COP
NO	187.34	6.12	736.97	187.2	3.93
I	187.34	6.11	734.60	187.3	3.92
II	187.34	6.13	718.44	187.5	3.83
III	187.34	6.14	713.78	187.2	3.81
IV	187.34	6.19	717.00	187.3	3.82
V	187.34	6.24	714.77	187.3	3.81

3. Determination of mass flow rate for different loads.
4. Calculation of Refrigerating effect, work done and COP for different loads.
5. Plotting of graphs for:
   a. COP Vs Power.
   b. R.E. Vs Evaporator Temperature.
   c. R.E. Vs Condenser Temperature.
   d. Mass flow Rate Vs Evaporator Temperature.
   e. Mass Flow Rate Vs Condenser Temperature.

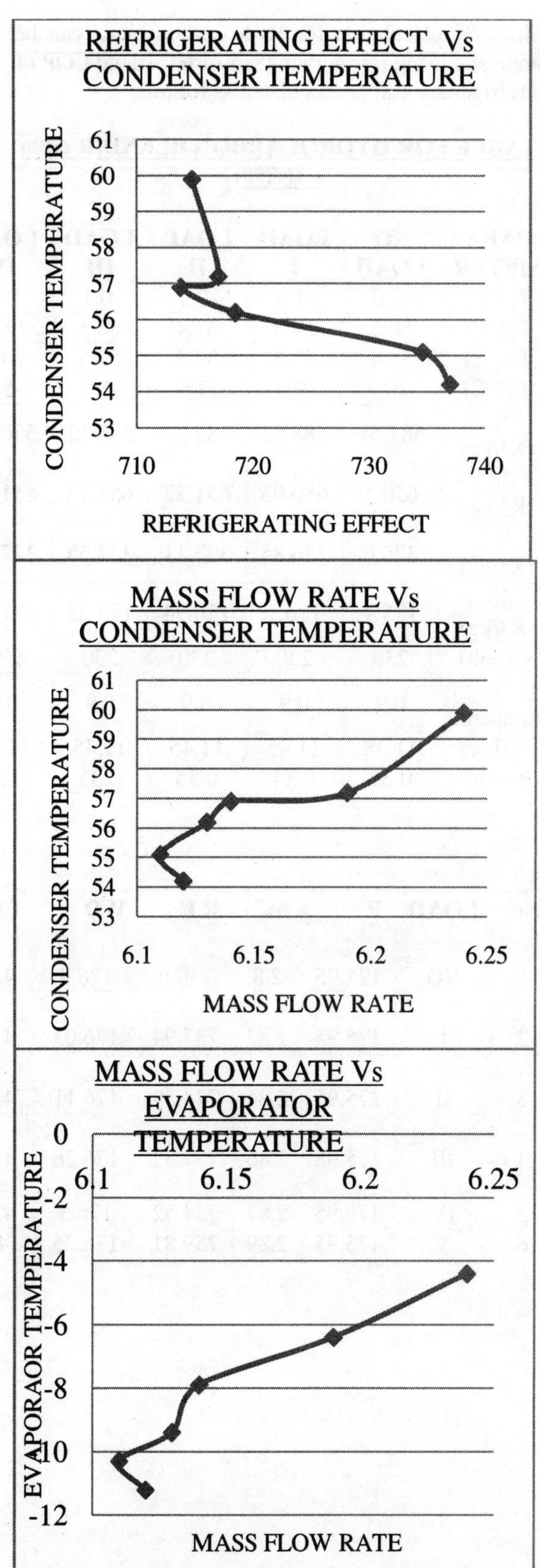

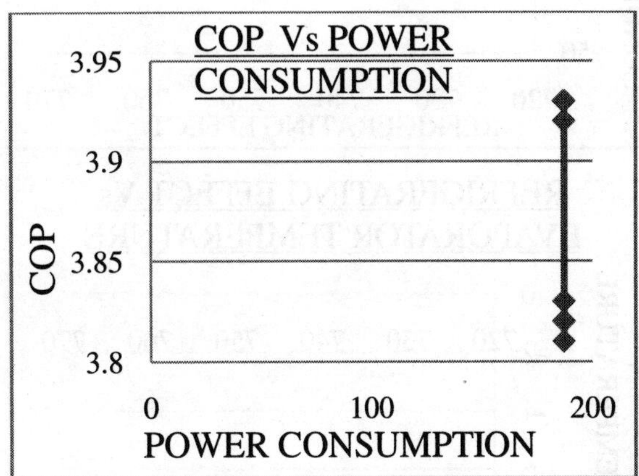

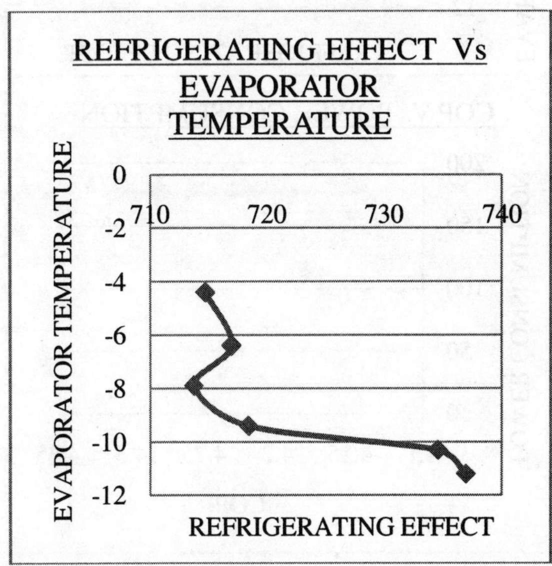

Similarly calculation for Hydrocarbon blend can be done and using refrigerant chart of HC Blend COP of a refrigerator can be calculated as follows:

TABLE FOR HYDROCARBON BLEND(R-600a + R-290)

PARA METER	NO LOAD	LOAD I	LOAD II	LOAD III	LOAD IV
T_1 (°C)	-2.0	-1.6	-0.9	-0.1	0.6
T_2 (°C)	50.5	57.3	63.4	66.5	68.2
T_3 (°C)	50.4	51.3	52.5	53.6	54.3
T_4 (°C)	-10.9	-9.8	-7.6	-6.3	-5.1
h_1 (KJ/kg)	587.54	588.53	589.22	590.12	590.51
h_2 (KJ/kg)	650.5	650.95	651.23	651.73	651.94
h_{f3} (KJ/kg)	326.6	326.85	327.11	327.35	327.61
h_4 (KJ/kg)	175.4	176.1	176.94	177.23	177.84
V (volt)	230	230	230	230	230
I (ampere)	0.9	0.9	0.9	0.9	0.9
P_d (bar)	11.38	11.45	11.45	11.45	11.45
P_s (bar)	0.31	0.34	0.35	0.35	0.35

S No.	LOAD	P	m_r	R.E	WD	COP
1	NO	175.95	2.8	730.6	176.3	4.14
2	I	175.95	2.82	737.94	176.03	4.19
3	II	175.95	2.84	744.39	176.11	4.22
4	III	175.95	2.86	751.52	176.26	4.26
5	IV	175.95	2.87	754.52	176.3	4.28
6	V	175.95	2.89	759.81	176.75	4.3

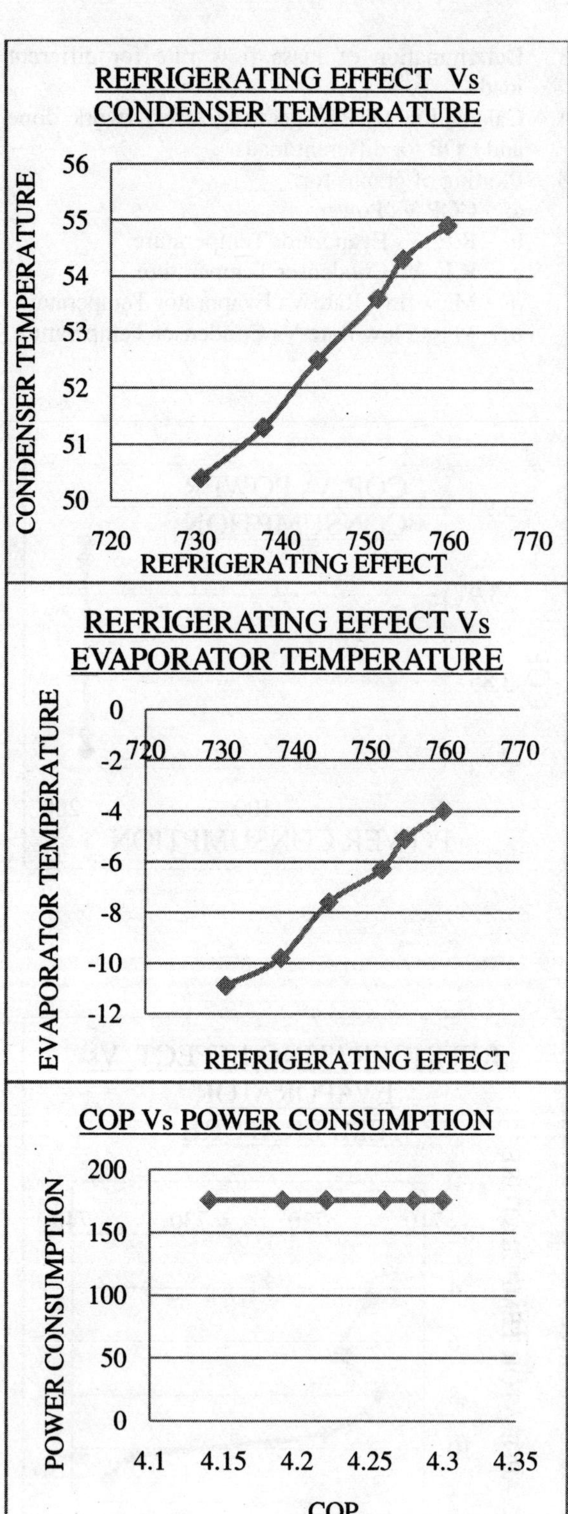

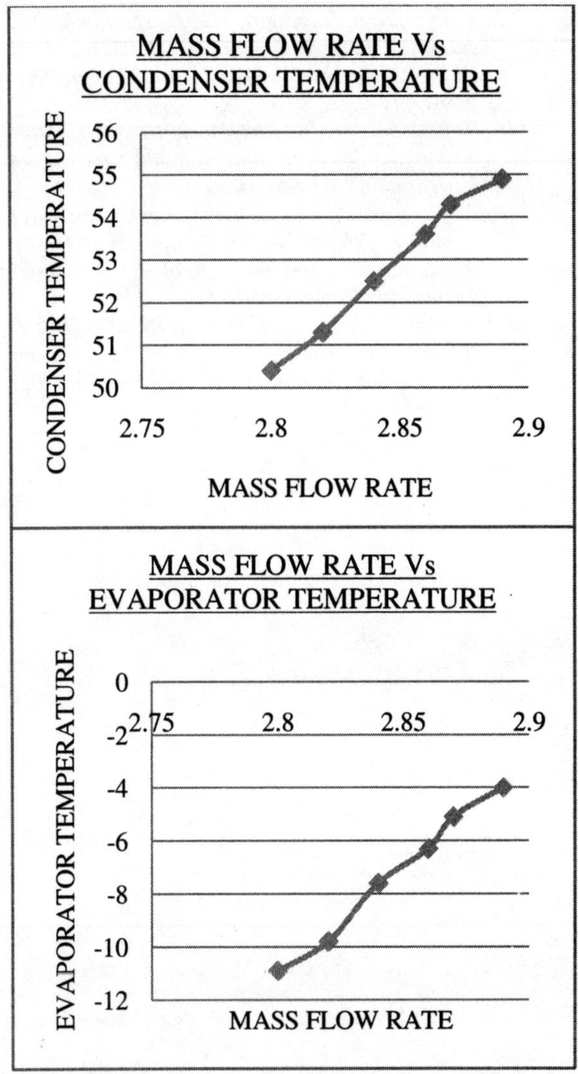

III. RESULT AND DISCUSSION

In the above analysis it is found that the COP of the refrigerator using hydrocarbon blend is higher than that of the COP of the refrigerator using R-134a.Thus the performance of the refrigerator is increased if we use hydrocarbon blend as a refrigerant.

The hydrocarbon blend compressor must have an increased volumetric displacement, about 65% to 70% greater than the R 12 model, to have similar levels of refrigerating capacity. The enthalpy difference of R 600a is significantly greater than that of R 12, as can be observed in section B. Thus, a lower mass flow rate is required to obtain the same refrigerating capacity. The monitored temperature at the discharge of the compressor operating with hydrocarbon blend is lower than that of the R 12 model. As can be observed, the volume flow with hydrocarbon blend is around 1.3% lower than that of R 12, i.e., in principle no alterations are required in the capillary tube of refrigeration systems when

hydrocarbon blend is used as a substitute for R 12. There are many other advantages of hydrocarbon blend over R-12 and R-134a those are given below which increases the performance of any refrigerator.

Parameter	R-12	R-134a	Hydrocarbon Blend
Evaporating Pressure, bar	0.8	0.6	0.65
Condensing Pressure, bar	11.9	12.9	11.1
Pressure Ratio	7.2	8.7	7.3
COP compared to R-12	-	lower	higher
Volumetric Capacity	-	similar	similar
Noise Level	-	similar	similar
Discharge Temperature	77	72	63
ODP	0.9	0	<0.01
GWP	3.1	0.27	0
Atmospheric Life, years	102	14.6	1

IV. REFERENCES

[1] Khurmi R S, Refrigeration and Air-Conditioning, S.Chand publication.

[2] Arora C P, Refrigeration and Air-Conditioning, Tata McGraw Hill Pub.

[3] Ballaney P L, Thermal Engineering, Khanna Publication.

[4] General Manual of Domestic Refrigerator Test Rig.

[5] Ji J, Chow TT, Pei G, Dong J, He W. Domestic air-conditioner and integrated water heater for subtropical climate. Applied Thermal Engineering. 2003; 23(5):581-592.

[6] Inan C, Gonul T, Tanes MY. X-ray investigation of a domestic refrigerator. Observations at 25°C ambient temperature. Int. J Refrig. 2003;26(2):205-213.

[7] Suzuki S. Noise control of domestic facilities. 1998; Int. J. Jpn. Soc. Precis. Eng. 1998;32(3):66-170

[8] Jung D, Kim CB, Song K, Park B. Testing of propane/isobutane mixture in domestic refrigerators.Int. J Refrig.. 2000;23(7):517-527. Hoffman, J. S., 1987, assessing the Risks of Trace Gases that can modify the Stratosphere, Office of Air and Radiation, U.S. Environmental Protection Agency, Washington DC.

[9] IPCC, 1994, Radiative Forcing of Climate Change, The 1994 report of the scientific assessment working group of IPCC, Summary for

policy makers, *Intergovernmental Panel on Climate Change*, 28 p.

[10] Lohbeck, W., 1995, editor *Hydrocarbons and other progressive answers to refrigeration, Proceedings of the International CFC and Halon Alternatives Conference, 23–25th October, Washington DC*, published by Green peace, Hamburg.

[11] Molina, M. J. and Rowland, F. S., 1974, *Stratospheric sink for chlorofluoro methanes: chlorine atom catalyzed destruction of ozone*, Nature, Vol. 249, June 28, pp. 808–812.

[12] www.google.com

[13] www.freeengineering.com

[14] www.wikipedia.com

[15] B. Donald, B. Nagengast (1994). *Heat and Cold mastering the great indoors*. ASHRAE, Inc. 1791 Tullie Circle NE Atlanta, GA 30329.ISBN1-883413-17-6.

[16] R. Radermacher, K. Kim, *Domestic refrigerator: recent development*, International journal of refrigeration 19(1996) 61-69.

[17] Y. S. Lee, C. C. Su, *Experimental studies of isobutene (R600a) as refrigerant in domestic refrigeration system*. Applied Thermal Engineering 22 (2002) 507-519.

[18] S.J.Sekhar, D.M.Lal, *HFC134a/HC600a/HC290 mixture a retrofit for CFC12 system*, International journal of refrigeration 28(2005) 735-743.

Production of Biofuel Oil from Algae in Agra Region

[1]Sunil Bhadauria, [2]Vishwajeet Singh, [2]A.K.Singh, [3]Vishnu Singh Sikarwar
[1]Sachdeva Institute of Technology, Mathura, U.P.
[2]Department, R.B.S .College, Agra, U.P.
[3]Singhania University, Jhunjhunu, Rajasthan.

Abstract—**Biodiesel is a clean burning fuel that is produced from renewable resources like algae, Jatropha etc. It contains no petroleum, but can be blended at any level with petroleum diesel to create a biodiesel blend. Many algae are exceedingly rich in oil which may exceed to 80%. The biodiesel production from algae also has the beneficial by-product of reducing carbon and NOx emissions from power plants, if the algae are grown using exhausts from the power plants. Microalgae, like higher plants comprise of proteins, carbohydrates, fats and nucleic acids. They produce storage lipids in the form of triacylglycerols (TAGs). The biodiesel, can be synthesized from TAGs via a simple transesterification reaction in the presence of acid or base and methanol. The samples of algae were collected from Keetham lake (Soor Sarowar Pakshi vihar). The oil was extracted using hexane solvent extraction method. On the day of collection following algal species were identified:*Spirogyra aequinoctialis, S. crassa, Chlorella sp., Ulothrix sp., Hydrodictyon sp., Chlamydomonas sp., Cladophora sp., Anabena azollae, Oscillatoria sp., Oscillatoria prolifica, Scenedesmus perforatus, Vaucheria sp.* diatoms and some unidentified algal species. The biodiesel from algae is not any different from biodiesel produced from vegetable/plant oils. The difference is however in the yield of oil, and hence biodiesel. Besides, the production of algae is sustainable and they do not require as much area and time as crop.**

I. INTRODUCTION

An important problem for our environment lies with the production and use of fuels. The natural sources will not be sufficient anymore for our gigantic consumption. Additionally the production process creates an enormous emission of carbon dioxide. When the diesel is burnt in the engines, more carbon dioxide will be released into our atmosphere. The concentration gets too high and the greenhouse effect is worsened. There is urgent need of alternate fuel that is ecofriendly.

Biodiesel is a clean burning fuel that is produced from renewable resources like algae, Jatropha etc. It contains no petroleum, but can be blended at any level with petroleum diesel to create a biodiesel blend. Biodiesel is already used in several countries. In Germany for example cars drive on a mixture of 60% of conventional diesel and 40% biodiesel. The experiences so far are highly positive. As yet biodiesel is too coarse for most of the diesel engines to run on exclusively. But even if 40% of all diesel would be replaced by biodiesel, a great part of our goal would be achieved. Even the US's first commercial jet flight (Boeing 737-800) completed 2 hours test flight with one engine powered by a 50-50 blend of regular petroleum based jet fuel and a synthetic alternative made from Jatropha and algae.

Generally the biodiesel is made from rapeseed. However it is impractical and not lucrative to produce biodiesel from rapeseed on a large scale because the cultivation of it takes in too much space. That is why we want to use algae as material for our biodiesel. Algae produce relatively much more oil than rapeseed. Many algae are exceedingly rich in oil (Chisti, 2007; Banerjee et al. 2002) which can be converted to biodiesel using new technology. Oil content of some algae exceeds 80% of the dry weight of algal biomass (Chisti, 2007; Banerjee et al. 2002). Algae for our fuels will make us less dependant of the fossil fuels. Also the production process of diesel from algae is much better for the nature. The algae capture more than 40% of emitted CO_2 (on sunny days, up to 80%) along with over 80% of NO_x emissions. Algae reduce NO_x day and night, regardless of weather or lighting condition. The process is essentially an effect of the surface configuration of the algae cell walls. Even dead algae can provide significant NO_x reduction, upto 70% (Sheehan et al. 1998). The algae reuse all carbon dioxide that is emitted during the process for more oil production. The exhausts of biodiesel are cleaner than that of "fossil diesel" as well. Thus the biodiesel is a promising candidate for sparing the fossil fuels and reducing the current immense pollution.

Biodiesel has advantages over conventional diesel fuel in that

- Higher yield and hence – hopefully – lower cost
- Algae can grow practically in every place where there is enough sunshine
- The biodiesel production from algae also has the beneficial by-product of reducing carbon and NOx emissions from power plants, if the algae are grown using exhausts from the power plants.
- The properties of biodiesel blend are fabulous, the biopart of the blend has a lower freezing point. It does not freeze at high altitude temperature, delivers the same or more power to the engines and is lighter as well.

it is renewable, biodegradable, and produces less SO_x and particulate emissions when burned.

Microalgae, like higher plants comprise of proteins, carbohydrates, fats and nucleic acids. They produce storage lipids in the form of triacylglycerols (TAGs). Although TAGs could be used to produce of a wide variety of chemicals, on the production of fatty acid methyl esters (FAMEs), which can be used as a substitute for fossil-derived diesel fuel. This fuel, known as biodiesel, can be synthesized from TAGs via a simple transesterification reaction in the presence of acid or base and methanol.

II. MATERIALS AND METHODS

The samples of algae were collected from Keetham lake (Soor Sarowar Pakshi vihar) about 21 km from Agra situated at Mathura road. One of the areas of Pakshi Vihar was isolated region comprising still water among stony habitat. The habitat was quite suitable for growth of algae. The algae were collected in sterilized sample poly-bags from different regions of water to obtain complete diversity. These algae were taken to laboratory and grow in lab conditions in Beneck's broth. The algae genera and species were identified with the help of light microscope using suitable literature {Abdul – Majeed (1935), Fritsch (1948), Smith (1950), Desikachary (1959), Prescott (1951)}.

The diversity of algal species again observed after 15 days. After the duration of one month the oil was extracted by hexane solvent method.

CO_2 supply – this is an essential requirement of algae production. CO2 will be supplied from the fermentation procedure produced as byproduct of petha waste (*Saccharomyces cerevisiae* decomposes petha waste into carbon dioxide and alcohol. This process is

called fermentation. The carbon dioxide that is released here will be used later in the process for the cultivation of the algae).

Light, nutrients and pH will be monitored from time to time in lab.

III. HEXANE SOLVENT METHOD

Algal oil can be extracted using chemical hexane, which is relatively inexpensive. Hexane solvent extraction will be used in isolation or it will be used along with the oil press/expeller method. After the oil has been extracted using an expeller, the remaining pulp will be mixed with cyclo-hexane to extract the remaining oil content. The oil dissolves in the cyclohexane, and the pulp will be filtered out from the solution. The oil and cyclohexane will be separated by means of distillation. These two stages (cold press & hexane solvent) together will be able to derive more than 95% of the total oil present in the algae.

IV. RESULTS AND DISCUSSION

On the day of collection following algal species were identified:

Spirogyra aequinoctialis, S. crassa, Chlorella sp., Ulothrix sp., Hydrodictyon sp., Chlamydomonas sp. of **Chlorophyceae**, *Cladophora sp., Anabena azollae, Oscillatoria sp., Oscillatoria prolifica, Scenedesmus perforatus, Vaucheria sp.* of **Cyanophyceae,** diatoms of **Bacillariophyceae** and some unidentified algal species. The members of Chlorophyceae and Cyanophyceae were abundant at different regions of water.

After 15 days of incubation the diversity changed toward dominance of two species – *Oscillatoria sp.* and *Chlamydomonas sp.*

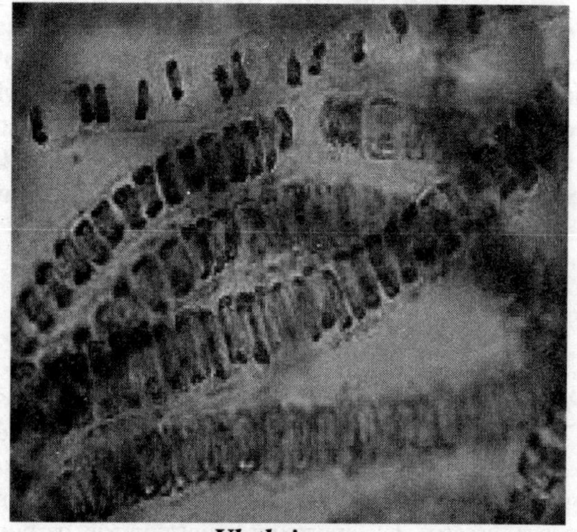

Ulothrix sp.,

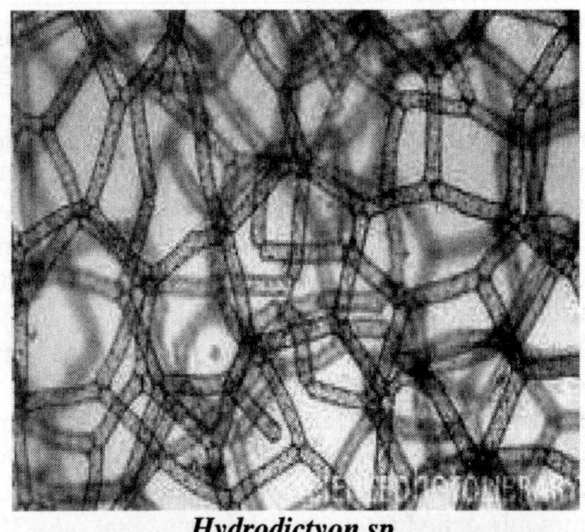

Hydrodictyon sp

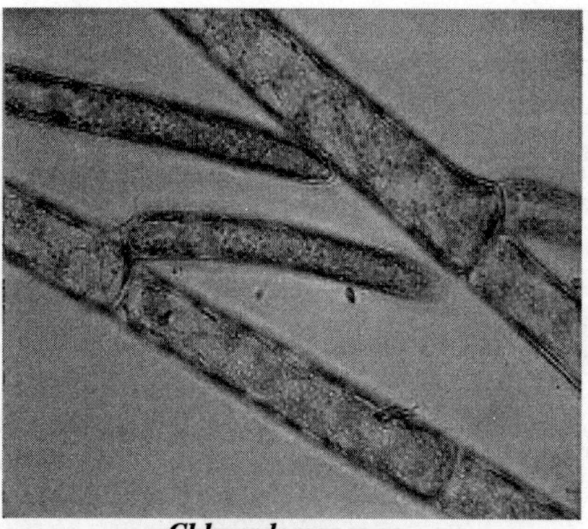

Chlamydomonas sp.

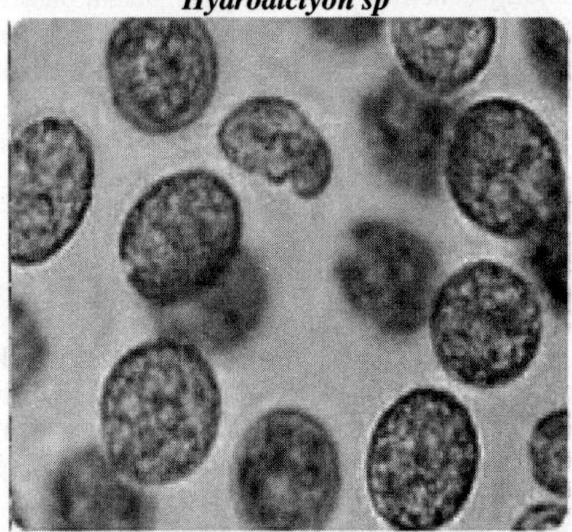

Cladophora sp

Oscillatorla sp

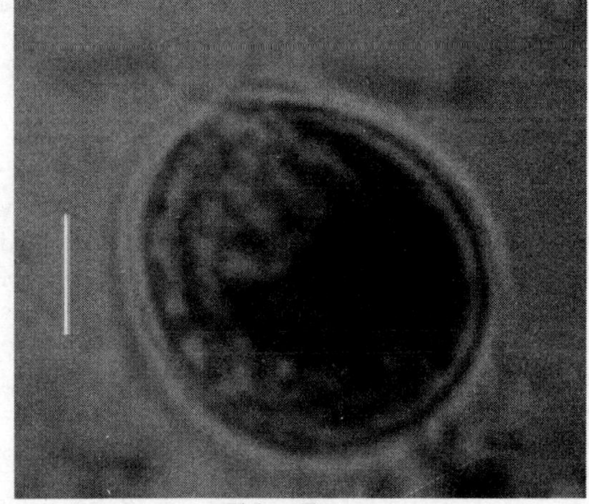

Chlorella sp.,

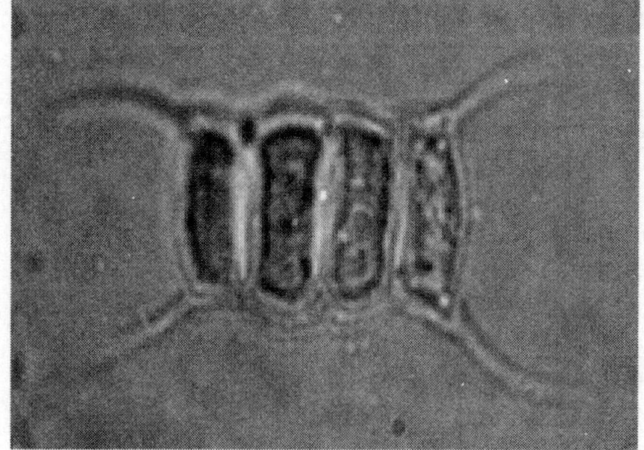

Scenedesmus perforates

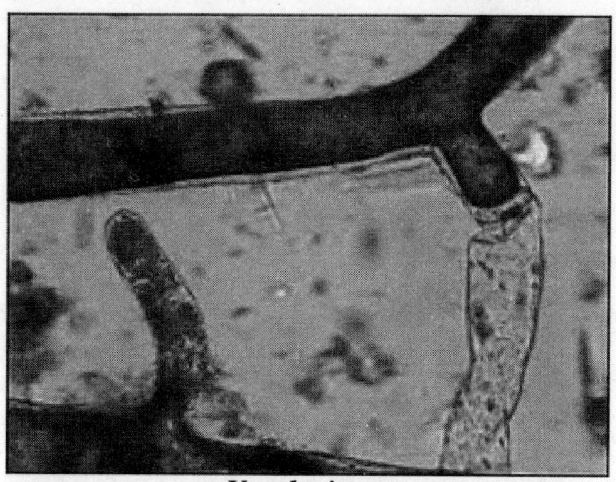

Vaucheria sp

Microalgae contain lipids and fatty acids as membrane components, storage products, metabolites and sources of energy. Algal strains, diatoms, and cyanobacteria (catagorised collectively as "Microalgae") have been found to contain proportionally high levels of lipids (over 30%). Lipid accumulation in algae typically occurs during periods of environmental stress, including growth under nutrient-deficient conditions.

The following species listed are currently being studied for their suitability as a mass-oil producing crop, across various locations worldwide.

Neochloris oleoabundans, Scenedesmus dimorphus are microalga belonging in the class Chlorophyceae, *Euglena gracilis,Phaeodactylum tricornutum* (diatom),*Pleurochrysis carterae, Prymnesium parvum, Tetraselmis chui, Tetraselmis suecica, Isochrysis galbana*. Algal strains such as *Botryococcus braunii* can produce long chain hydrocarbons representing 86% of its dry weight. The other favoured algae is Bacilliarophyceae (diatom algae).

Since algae need for their growth sunlight, carbon-di-oxide and water, they can cultivated in open ponds. Fresh and brackish water from lakes, rivers and aquifers can be used. Growth media are generally inexpensive (Chisti, 2007). However, the unassisted growth in open ponds is slow, owing to the lower concentration of carbon-di-oxide; where carbon-di-oxide concentrations are increased artificially, higher growth rates can be achieved in open ponds as well. Alternatively, algae could be grown in closed structures called photobioreactors, where the environment is better controlled than in open ponds.

Photobioreactors

Microalgae can be grown on a large scale in photobioreactors (Camacho et al. 1999; Molina et al. 1999; Chisti 2006). A photobioreactor is an equipment that is used to harvest algae. Photobioreactors can be set up to be continually harvested, or by harvesting a batch at a time (like polyethlyene bag cultivation). A batch photobioreactor is set up with nutrients and algal seed, and allowed to grow until the batch is harvested. A continuous photobioreactor is harvested either continuously, as daily, or more frequently. Many different designs of photobioreactors have been developed, but a tubular photobioreactor seems to be most satisfactory for producing algal biomass on the scale needed for biofuel production.

Difference between biodiesel from algae and biodiesel from other plant/vegetable oils

The biodiesel from algae is not any different from biodiesel produced from vegetable/plant oils. All biodiesel essentially are produced using triglycerides (commonly called fats) from the plant/algal oils. The difference is however in the yield of oil, and hence biodiesel. According to some estimates, the yield of oil from algae is over 25 times more than a source of biodiesel of palm oil and 300 times better than soy. Crop derived biodiesel are unsustainable and they can not displace petroleum-derived transport fuels. Microalgal biodiesel has the potential to be able to completely displace petroleum-derived transport fuels without adversely impacting supplies of food and other agricultural products. It is further demonstrated that microalgal biodiesel is a better alternative than bioethanol from sugarcane, which is currently the most widely used transport biofuel (Gray et al. 2006).

The biomass residue that remains after extraction of oil could be used partly as high protein animal feed and possible as a source of small amount of other high value microalgal products (Molina et al. 1999; Grvrilescu and Chisti 2005; Chisti 2006).

V. CONCLUSION

Petroleum is made form the conversion of oily substance under conditions of high pressure and temperature. It is exhausting continuously because of growing demand day by day. One best alternative has been suggested as oil from algae – which is easy to produce, better fuel quality, biodegradable and ecofriendly. Besides, algae decrease pollution by consuming carbon-di-oxide and nitrogen oxides during its cultivation and growth.

VI. REFERENCES

[1] Abdul –Majeed, (1935). *The Fresh water algae of the Punjab. Part I, Bacillariophyta (Diatomae). The University of Punjab, Lahore Publication.*

[2] Banerjee, A., Sharma R., Chisti Y. and Banerjee, U. C. (2002). *Botruococcus braunii: a renewable source of hydrocarbons and other chemicals. Crit. Rev. Biotechnol. 22 (3): 245-279.*

[3] Benemann, J.R., Van Olst J.C., Massingill, M.J., Weissman J.C. and Brune D.E.

The Controlled Eutrophication Process: Using Microalgae for CO2 Utilization and Agricultural Fertilizer Recycling. Pp 1-6. http://www.unh.edu/p2/biodiesel/pdf/algae_salton_sea.pdf.

[4] *Camacho, F.G., Sanchez Miron A., Gomez, A.C., Grima, E.M. and Chisti Y. (1999). Comparative evaluation of compact photobioreactors for large-scale monoculture of microalgae. J. Biotechnol. 70: 249-270.*

[5] *Chisti, Y. (2007) Biodiesel from microalgae. Biotechnol. Adv. 25: 294-306.*

[6] *Chisti Y. (2006).Microalgae as sustainable cell factories. Environ. Eng. Manag. J. 5:261-274.*

[7] *Desikacharya, T. V. (1959). Cyanophyta. Pub. ICAR Publ. New Delhi. 686.*

[8] *Fritsch, F. E. (1935). Structure and Reproduction of the algae , vols I & II. Pub. Cambridge University Press, Cambridge.*

[9] *Gavrilescu, M. and Chisti, Y. (2005). Biotechnology – a sustainable alternative for chemical industry. Biotechnol. Adv. 23: 471-499.*

[10] *Gray, K. A., Zhao, L. and Emptage M. (2006). Bioethanol. Curr. Opin. Chem. Biol. 10(2): 141-146.*

[11] *Molina, E., Fernandez, F.G.A., Camacho F.G. and Chisti Y.(1999). Photobioreactors: light regime, mass transfer, and scaleup. J. Biotechnol. 70(1-3) :231-247.*

[12] *Prescott, G.W. (1951). Algae of the Western Great lakes area. Cranbrook Instt. Of Science, Bloom field Hills, Michigan. Bulletin No. 30, 946.*

[13] *Sheehan, J., Dunahay, T., Benemann, J. and Roessler, P.(1998) A Look Back at the U.S. Department of Energy's Aquatic Species Program— Biodiesel from Algae pp. 328.*

[14] *Smith, G. M. (1950). Fresh water algae of the United States. IInd ed. New York.*

Screening and Characterization of Chlorella for its Potential to Produce Biodiesel

[1]Ankita Shrivastava, [2]Brajesh Singh Kushwah, [3]Sunil Bhadauria, [4]Sadhana Jadon, [5]S.A.Mallick, [1]Vishwajeet Singh, [4]Seema Bhadauria

[1]Institute of Research and Innovative Studies (IRIS EDU SOLUTIONS), Agra.
[2]Department of Mechanical Engineering, R.B.S. College, Agra.
[3]Department of Mechanical Engg., Sachdeva Institute of Technology, Mathura.
[4]Department of Botany, R.B.S. College, Agra.
[5]Sai Nath College, Agra.

Abstract—Microalgae, like higher plants, produce storage Lipids in the form of triacyglycerols (TAGs). Comparatively algae produce more oil than any other oilseeds which are currently in use. Many microalgal species can be induced to accumulate substantial quantities of lipids, often greater than 60% of their biomass. Algae include a diverse group of microorganisms and occur in a variety of natural habitats, including terrestrial habitats such as soil and aquatic habitats ranging from freshwater and brackish waters to marine and hyper-saline environments. Biofuel from algae may be produced from microalgae, macroalgae, i.e. seaweed, or cyanobacteria. Cultivation of microalgae and cyanobacteria can be done through so-called photoautotrophic methods in open or closed ponds or through heterotrophic methods. Photoautotrophic refers to the fact that in these processes algae need light to grow and generate new biomass. In a heterotrophic process, algae are grown without light and feed on carbon sources, for example sugars, in order to create new biomass. Biomass is one of the better sources of energy. Among biomass, algae (macro and microalgae) usually have a higher photosynthetic efficiency than other biomass .Biodiesel (monoalkyl esters) is one of such alternative fuel, which is obtained by the transesterification of triglyceride oil with monohydric alcohols. It has been well-reported that biodiesel obtained from algal species. Biodiesel is a nontoxic and biodegradable alternative fuel that is obtained from renewable sources. Petroleum diesel combustion is a major source of greenhouse gas (GHG). Apart from these emissions, petroleum diesel is also major source of other air contaminants including NOx, SOx, CO, particulate matter and volatile organic compounds. The purpose of this study was to select and characterize *Chlorella* algal species which tolerate high light intensities~ temperature variations and accumulate lipids. Samples have been collected from Kheetham lake, Agra. Samples were screened through a multi-step process and have been examined for growth requirements. Approximate cellular composition of these species was determined. The burning of an enormous amount of fossil fuel has increased the CO_2 level in the atmosphere, causing global warming. Biomass has been focused on as an alternative energy source, since it is a renewable resource and it fixes CO_2 in the atmosphere through photosynthesis. If biomass is grown in a sustained way, its combustion has no impact on the CO_2 balance in the atmosphere, because the CO_2 emitted by the burning of biomass is offset by the CO_2 fixed by photosynthesis.

I. INTRODUCTION

Microalgal species are capable of producing biomass yields containing high percentages of oils (Aaronson et. al., 1980). Microalgal systems can use law value natural resources, such as saline water and arid lands, therefore offering the potential for large biomass energy contributions without competition for prime agricultural or forest land. The growth of microalgal biomass and lipid harvesting represent potential for harvesting solar energy products. The microalgal energy technology offers following advantages:

Algal lipids are in general high in polyunsaturated fatty acids (Miller,1962), and are comparable to that of plants: linseed, cotton-seed oils (Nurris, 1983). Microalgae have the ability to survive in extreme environments. Cells of many algal species have negligible ash content.

One of the methods by which the energy storage capacity of photosynthesis can be maximized is by controlling the metabolism of the organism. Tuning the metabolism of algae can lead to enhanced production of energy-rich compounds such as fatty acids and glycerol. A single algal species may show remarkable variation in its metabolism, according to the conditions to which it is exposed, such as carbon dioxide supply light intensity, temperature, nutrient concentrations, and salinity (Holm-Hansen, et. al.,1959). Changes in the supply or consumption of metabolites may have considerable effects on metabolic patterns. The accumulation of energy storage compounds in algae, such as fats and oils, can be induced by manipulating the environmental conditions under which the algae are grown (Shifrin and Chisholm, 1981).

Nutrient deficiencies generally lead to a decrease in protein and photosynthetic pigments and an increase in energy-rich products such as carbohydrates and lipids (Healey, 1973). Nitrogen starvation in particular can lead to remarkable changes in algal cell composition (Fogg, 1959). Opute (1974) demonstraned that lipids accumulate in the diatom Nitzschiapalea, under N-deficient conditions.

Only biodiesel and bioethanol are produced on an industrial scale. They are the petroleum replacement for internal combustion engines, and are derived from food crops such as sugarcane, sugar beet, maize (corn), sorghum and wheat, although other forms of biomass can be used, and may be preferable [Van der Laaka, et al, 2007]. The most significant concern is the inefficiency and sustainability of these first generation biofuels. In contrast, the second generation biofuels are derived from non-food feedstock. They are extracted from microalgae and other microbial sources, ligno-cellulosic biomass, rice straw and bio-ethers, and are a better option for addressing the food and energy security and environmental concerns. Microalgae, use a photosynthetic process similar to higher plants and can complete anentire growing cycle every few days. In fact, the biomass doubling time for microalgae

during exponential growth can be as short as 3.5h [Chisti,2007]. Some microalgae grow heterotrophically on organic carbon source. However, heterotrophic production is not efficient as using photosynthetic microalgae [Chisti, 2007], because the renewable organic carbon source required is ultimately produced by photosynthetic crop plants.

Microalgae are veritable miniature biochemical factories, and appear more photosynthetically efficient than terrestrial plants [Pirt,1986] and are efficient CO_2 fixers [Brown, ,1993]. The ability of algae to fix CO_2 has been proposed as a method of removing CO_2 from flue gases from power plants, and thus can be used to reduce emission of GHG. Many algae are exceedingly rich in oil, which can be converted to biodiesel. The oil content of some microalgae exceeds 80% of dry weight of algae biomass [Chisti, 2007, Banerjee, et al,2002].

The net annual harvest of algal biomass cultivated in subtropical areas can be as high as 40 tons ha-1 (dry matter), even higher if CO_2 is supplied [Klass,1998]. It is possible to produce about 100 g m-2 d-1 of algal dry matter in simple cultivation systems [Patil, et al,2005]. In theory, high oil content algae could produce almost 100 times of soybean per unit area of land [Kong, 2007]. The calculations made by Chisti [Chisti, 2008]clearly demonstrate the strong scenario for microalgal biofuels. The use of algae as energy crops has potential, due to their easy adaptability to growth conditions, the possibility of growing either in fresh- or marine waters and avoiding the use of land. Furthermore, two thirds of earth's surface is covered with water, thus algae would truly be renewable option of great potential for global energy needs. This paper aims to analyze and promote integration approaches for sustainable microalgal biodiesel production, with emphasis on hydrothermal technology for direct liquefaction of algal biomass with no need to dry the feedstock.

Chlorella pyrenoidosa (Chlorophyceae), produces high concentrations of lipids, from 28% to 70% dry weight (Fogg, 1959) when grown in N-starved cultures. Fat production is also stimulated by light. Spoehr and Milner (1948), showed that N-deficient *Chlorella pyrenoidosa* attains a greater lipid content at high light intensity than at low light intensity. it is evident that hydrocarbon production by algae is occurring in natural systems and can be maximized by the variety of environmental conditions (Shifrin and Chisholm, 1981). In order to select promising algal species as potential producers of oils for energy technology, these growth conditions need to be identified. In this project, the most promising lipid forming species collected in and were screened and characterized with respect to their growth properties and oil-liquid productivity.

The specific objectives of the research are:

To collect algal samples from the Keetham lake.

To isolate oleaginous algal species capable of growth under high temperature and light intensity.

The specific research objectives stated above are fundamental to the acquisition of the following benefits and goals:

Isolation of oleaginous algal species. Defining the conditions which increase lipid production in oleaginous microalgae, e.g., temperature, light, salinity, and nutrients.

II. MATERIALS AND METHODS

Sites of Sampling

The region of Agra, Keetham is characterized by freshwater resources, e.g., lakes, rivers and ponds. The soil has a high percentage of carbonate rocks. The temperature ranges from 2-18°C in January and 21-35°C in July. The waters of the Agra Keetham are mixed and consists a wide variety of living organisms that can tolerate a wide range of salinities.

Field Area

Field trips have been conducted to: rivers, lakes, ponds, streams, and swamps under environmental stress . Fresh water and marine algal samples were collected from dry and wet inhabitants. Temperature, pH, and salinity of samples were recorded. Salinity was measured with a conductance meter. Collected samples were enriched with nitrate and phosphate media to maintain the dominant species. They were protected from temperature changes during the trips, by being kept in a cooler.

Algal Sample Processing

Screening Procedure

It involves many steps for selection of oleaginous algal species tolerant of high temperature and light intensity.

Growth Conditions Culture Room: A small room (3m D x 2.45m W x 2.14m H), was designed specifically for this project. It has been provided with shelves, which have been illuminated with cool white light fluorescent tubes. Light intensity varied from

400 to 500 foot-candles on the shelves. Intermittent illumination was used for culturing (14 h:10 h light dark cycle).

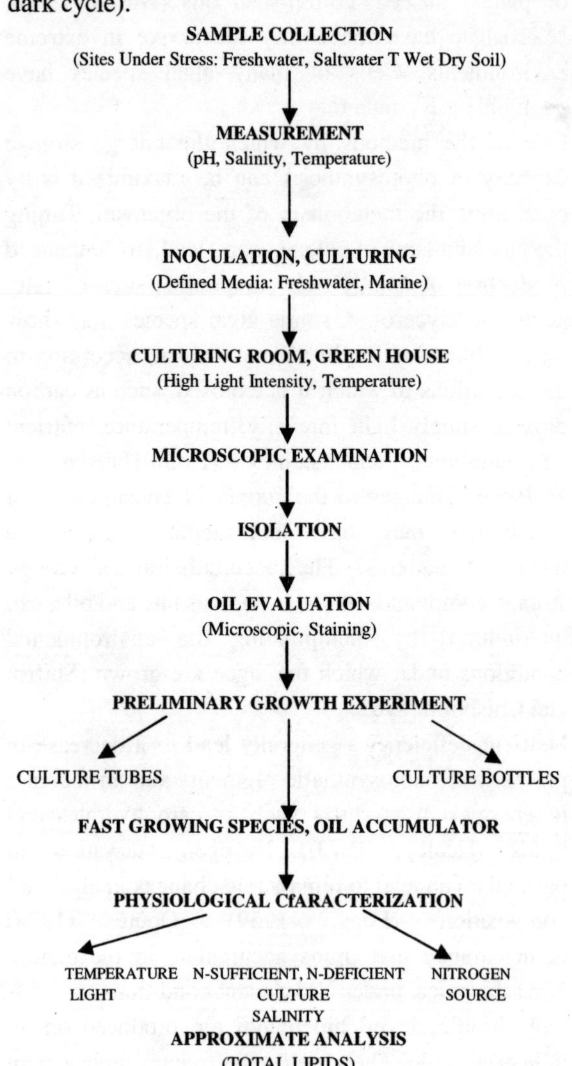

Figure1. Screening Procedure of Oleaginous Algal Species

Growth Media

Growth media for fresh water and marine algal species were used for isolation. Defined media were supplemented with artificial sea salts at different strengths. Among the principal basic media are: Isolation Algal species have been isolated by streaking enriched agar plates or by micropipetting under a light microscope. Species were chosen on the basis of rapidity of growth and on dominance.

Qualitative Evaluation of Lipids

Oil accumulation was identified by microscopic examination. For these algal species, Lien's (1984) staining technique for lipids has been used.

Bold's Basal Medium

Stock Solutions in 400ml Distilled water

- NaNO$_3$ 10.0 g
- MgSO$_4$.7H$_2$0 3.0 g
- NaCl 1.0 g
- K$_2$HPO$_4$ 3.0 g
- KH$_2$PO$_4$ 7.0 g
- CaCl$_2$.2H$_2$0 1.0 g
- Trace Element Solutions in 1 litre Distilled water
  ZnSO$_4$.7H$_2$0 8.82 g
  MnCl$_2$.4H$_2$0 1.44 g
  MoO$_3$ 0.71 g
  CuSO$_4$.5H$_2$0 1.57 g
  CoNO$_3$.6H$_2$O 0.49 g
  (Autoclave to disssolve)
- H$_3$BO$_3$ 11.42 g
- EDTA – KOH Solution
- EDTA 50.0 g
- KOH 31.0 g
- FeSO$_4$.$_7$H$_2$O 4.98 g
- H$_2$SO$_4$ (conc) 1.0 ml
- Final Medium (made up to 1 litre with Distilled water)
- Stock Solutions 1-6 10.0ml each
- Trace Element Solutions 7-10 1.0ml each
- Soil Extract 250ml

Species Identification

Isolates were identified in our laboratory.

Culture Test Tubes: A multi-wrist shaker was modified to simultaneously shake 150 culture test tubes. Isolates were inoculated in 3 ml growth media and shaken at 29-30°C. Growth was evaluated visually

Culture Bottles: Small culture bottles (60 ml capacity), containing 45 ml growth medium, have been inoculated with algal strains, and aerated, using a small air pump. Growth of the cultures was evaluated visually.

Growth Measurement

Temperature, Salinity, Light: Selected species were evaluated for temperature, salinity and light tolerances on a gradient plate. Temperature could be adjusted in range from 10°C to 40°C. Illumination was provided by eight cool white fluorescent tubes (40 W) with a 14h : 10h light, dark cycle. Different light intensities were obtained by varying the distance between the cultures and light source. Selected algal species were cultured in small bottles (60 ml capacity) containing 20 ml growth medium enriched with sea salt or sodium chloride. Culture bottles in triplicate were inoculated from stock cultures in the exponential phase and were shaken once a day. The temperature and salinity combinations on the gradient plate were evaluated by determining cell concentrations when cultures reached the stationary phase.

Nitrogen Source: Selected algal strains were grown in culture flasks of 100 ml capacity containing 50 ml basic medium, and enriched with different concentrations of urea or nitrate. The treatments were run in triplicate and inoculated with exponentially dividing cells. The culturing flasks were placed on the shelves in the culture room and were shaken by hand twice a day. Yields of all treatments were measured by cell counting.

Batch Culturing: Experimental amounts of algal cells were grown in one liter bottles, containing 800 ml sterile growth medium, by inoculating them with 50 ml of pre adapted rapidly growing culture in a 125 ml Erlenmeyer flask. The batch cultures were aerated with air mixed with 3% CO$_2$ at the rate of 150 ml/min. Cultures were illuminated continuously by placing them in front of six cool white fluorescent lamps (32 W). Light intensity, measured at the surface of culture bottles with a light meter, was 600 feet candle. The cultures were grown in a water bath kept at 29-30°C by the use of a heater thermostat combination.

III. RESULTS AND DISCUSSION

Screening the samples for oleaginous algal strains, the growth of the isolates in preliminary experiments was evaluated and represented by signs (+). The preliminary growth experiments were designed to test the growth of the isolates in liquid media. Most of the isolated strains were uni algal or mixed species, but not axenic. No significant Bacterial growth was observed during the growth experiments. The following strains were selected for characterization on the basis of fast growth rate and microscopic identification of oil accumulation in the cells:

Chlorophyceae (Unicellular Green):

Chlorella sp. MB-31

Table-1.Algal Species Isolated From Field Trips; Water Characteristics; Preliminary Growth and Evaluation

Species	Temp.	pH	Salinity ppt.	Growth Condition	Preliminary Growth
Chlorella sp	28°C	7.2	0	FRESH WATER	++
Chlorella sp	23⁰C	7.6	12	SALT WATER	+
Chlorella sp	22°C	6.5	0	FRESH WATER	+
Chlorella sp	30°C	7.2	0	FRESH WATER	+++

Temperature, Light, Salinity Requirements: The growth parameters of Chlorella species were examined under different combinations of light intensity, temperature and salinity, in order to define the optimal growth conditions as well as high and low limits. The gradient block was used for this experiment. The maximum densities reached by each species was determined by evaluating cell counts in the stationary phase. Chlorella species grew in 0.0 ppt. as well as 15 ppt. salinities. Higher yield was favored by 800 ft-C and 35°C. Lower temperatures (15°C, 20°C) arrested completely the growth of the alga at all salinities.

Growth conditions for Chlorella sp. MB-3l are:

Temperature: 25°C - 35°C;

Light intensity: 400 ft-C - 800 ft-C;

Salinity: 0.0 ppt. - 32 ppt.

It is evident that Chlorella sp. has the ability to tolerate wide salinities from 15 ppt. to 32 ppt. and temperatures from 20°C to 35°C.

In this experiment, all the selected species were grown in duplicate batch cultures. The cultures were maintained under similar temperatures and light intensities. One batch of Chlorella species was analyzed in the exponential growth phase (5 days old), when the cells were actively dividing and sufficient nitrogen was available in the medium. The second batch was analyzed in the stationary phase (14 days old), when the cells ceased dividing and the medium had become N-depleted. Data for approximate cellular compositions were expressed on the basis of organic weight and represented in Table 2. Chlorella, in freshwater or saline medium, did not show a clear difference in composition. Nevertheless, N-starved cells contained more lipids 28.6% (freshwater) and 32.4% (saline) than N-sufficient cells which contained 15.3% and 26.5% respectively. Proteins decreased on the expense of carbohydrates which increased relatively.

Table 2.Approximate Cellular Composition of Selected Algal Species

Species	Cell Size (uM³)	Growth Rate	Growth Conditions	% Organic Wt		
				Protein	Carbohydrate	Fat
Chlorella sp	2-3	0.92	FW, NE	49.2	11.3	15.4
			FW, ND	24.4	27.2	28.5
			SW, NE	22.5	26.7	26.6
			SW, ND	17.3	27.6	32.8

Direct Liquefaction of Algae for Biodiesel Production

The microalgal biomass has relatively high water content (80-90%) and this is major bottleneck for usage in energy supply. As most other biomass, the high water content and inferior heat content makes the microalgal biomass difficult to be used for heat and power generation. Thus necessitating pre-treatments to reduce water content and increase the energy density. As consequence the energy cost increases and makes the alternative less economically attractive [Klass,1998; Braun,1996. Agarwal, 2007;Demirbas and Balat,2006].

IV. REFERENCES

[1]   Aaronson, S. 1973. Effect of incubation on the macromolecular and lipid content of the phytoflagellate O~hromonas danica. J. Phycol. 9: 111-113.

[2]   Miller, J. D. A. 1962. Fats and steroids. In: R. A. Lewin (ed.) Physiology and Biochemistry of Algae, Academic Press, New York, p. 357-370.

[3]   Holm-Hansen, 0., Nishida, K., Mosses, V., and Calvin, M. 1950. Effects of mineral salts on short-term incorporation of carbon dioxide in Chorella. J. expo Bot. 10: 109-124.

[4]   Nurris, N. 1983. Three-, six-, and nine-carbon ozonolysis products from cotton seed oil and crude Chlorella lipids. JAOCS, 60: 806-811.

[5]   Shifrin, N. S., and Chisholm, S. W. 1981. Phytoplankton lipids:interspecific differences and effects of nitrate, silicate and light-darkcycles. J. Phycol. 17: 374-384.

[6]   Healey, F. P. 1973. The inorganic nutrition of algae from an ecological viewpoint. eRe Critical Rev. Microbial. 3: 69-113.

[7]   Fogg, G. E. 1953. The metabolism of Algae. London: Methuen & Co.

[8]   Opute , F. I. Ann. Bot. 38: 1974. Studies on fat accumulation in Nitzschia palea Kutz. 889-902.

[9]   Van der Laaka, W.W.M.; Raven, R.P.J.M.; Verbong, G.P.J. Strategic Niche Management for Biofuels: Analysing Past Experiments for Developing New Biofuel Policies, Energy Policy 2007, 35, 3213–3225.

[10]   Chisti, Y. Biodiesel from Microalgae. Biotechnol. Adv. 2007, 25, 294-306.

[11]   Pirt, S.J. The Thermodynamic Efficiency (Quantum Demand) and Dynamics of Photosynthetic Growth. New Phytol. 1986, 102, 3-37.

[12]   Brown, L.M.; Zeiler, B.G. Aquatic Biomass and Carbon Dioxide Trapping. Energy Convers. Manage 1993, 34, 1005-1013.

[13]   Banerjee, A.; Sharma, R.; Chisti, Y.; Banerjee, U.C. Botryococus Braunii: A Renewable Source of Hydrocarbons and Other Chemicals. Crit. Rev. Biotechnol. 2002, 22, 245-279.

[14]   Klass. D.L. Biomass for Renewable Energy, Fuels, and Chemicals; Academic Press: San Diego, USA, 1998; pp. 651. Patil, V.; Reitan, K.I.; Knudsen, G.; Mortensen, L.; Kallqvist, T.; Olsen, E.; Vogt, G.; Gislerød, H.R. Microalgae as Source of Polyunsaturated Fatty Acids for Aquaculture. Curr. Topics Plant Biol. 2005, 6, 57-65.

[15]   Kong, Q.; Yu, F.; Chen, P.; Ruan. R. High Oil Content Microalgae Selection for Biodiesel Production. Proceedings of 2007 ASABE Annual International Meeting, Minneapolis, Minnesota, USA, June 17-20; American Society of Agricultural and Biological Engineers: St. Joseph, Michigan, USA, 2007; 077034.

[16]   Chisti, Y. Biodiesel from Microalgae Beats Bioethanol. Trends Biotechnol. 2008, 26, 126-131.

[17]   Braun, A.R. Reuse and Fixation of CO2 in Chemistry, Algal Biomass and Fuel Substitutions in the Traffic Sector. Energy Convers Manage 1996, 37, 1229-1234.

[18]   Agarwal, A.K. Biofuels (Alcohols and Biodiesel) Applications as Fuels for Internal Combustion Engines. Prog. Energ. Combust. 2007, 33, 233-271.

[19]   Demirbas, M.F.; Balat, M. Recent Advances on the Production and Utilization Trends of Biofuels: A Global Perspective. Energy Convers. Manage 2006, 47, 2371-2381.

Performance Evaluation of Eco-Friendly Transesterified non-edible oil in C.I. Engine

Ashok Yadav[1], Onkar Singh[2], Naveen Kumar[3]
[1]Department of Mechanical Engineering,Sachdeva Institute of Technology,Farah, Mathura
[2]Department of Mechanical Engineering,Harcourt Butler Technological Institute, Kanpur
[3]Department of Mechanical Engineering, Delhi Technological University, Delhi

I. INTRODUCTION

Under the depleting conditions of petroleum products a search for an alternative fuel for compression ignition engines is in progress. Vegetable oils are good alternatives to diesel oil since they are renewable and can be produced in rural areas where there is critical need for modern forms of energy. This was stated with remarkable foresight by none less than the inventor of the diesel engine, Dr. Rudolf Diesel, more than hundred years ago, "The diesel engine can be fed with vegetable oils and would help considerably in the development of the countries which will use it. This may appear a futuristic dream but I can predict with great conviction that this way of using a diesel engine may in future be of great importance",[4].

Now, the world faces the problem of energy crisis due to depletion of resources and increased environmental problems. Under the depleting conditions of petroleum products a search for an alternative fuel for compression ignition engines is in progress. For developing countries, fuels of bio-origin, such as alcohol, vegetable oils, biomass, biogas, Synthetic fuels, etc. are becoming important. Such fuels can be used directly, while others need some sort of modification before they are used as substitute of conventional fuels. Statistics shows that India has imported about 82 Million Tones of crude oil which is 70% of its requirement during year 2003-04, causing a heavy burden on foreign exchange and the country's energy dependence of oil is about 35%[5,6]. The known world wide reserves of petroleum are 100 billion barrels and these petroleum reserves are predicted to be consumed in about 40 years [5,6]. So the availability of petroleum is uncertain in the future.

II. ALTERNATIVE DIESEL FUEL

Alternative fuels should be easily available, environment friendly and techno-economically competitive. One of such fuels is triglycerides (vegetable oils/animal fats) and their derivatives. Vegetable oils, being renewable, are widely available from a variety of sources and have low sulphur contents close to zero, and hence cause less environmental damage (lower greenhouse effect) than diesel. Besides, vegetable oils and their derivatives are produced widely in the country for food and other purposes. Since there is ever increasing demand for edible vegetable oils for food purposes, we can think only of non-edible oils for this field. Karanja tree (*Pongamia Pinnate*) is one of the best substitute since extraction of oil from its seeds is 30% of seeds weight. Its oil has high scope of its conversion into biodiesel.

Vegetable oils (Triglyceride) as diesel fuels

The use of vegetable oils, such as palm, soya bean, sunflower, peanut, and olive oil, as alternative fuels for diesel engines dates back almost nine decades, but due to the rapid decline in crude oil reserves, it is again being promoted in many countries. Depending upon the climate and soil conditions, different countries are looking for different types of vegetable oils as substitutes for diesel fuels. For example, soya bean oil in the US, rapeseed and sunflower oils in Europe, palm oil in South-east Asia (mainly Malaysia and Indonesia) and coconut oil in the Philippines are being considered. The following table shows the production oil seeds in India, [7,8].

Table 1
Production of oilseeds in 2002–2003 in India

Oilseed	Production (million tones) World India		Total oil availabili ty (million tons)	% Recove ry	Oil cost (Rs. per ton)
Soya bean	123.2	4.30	0.63	17	4300
Cottonse ed	34.3	4.60	0.39	11	3200
Groundn ut	19.3	4.60	0.73	40	6200
Sunflowe r	25.2	1.32	0.46	35	5360
Rapeseed	34.7	4.30	1.37	33	5167
Sesame	2.5	0.62	-	-	6800
Palm kernels	4.8	-	-	-	-
Copra	4.9	0.65	0.42	65	3035
Linseed	2.6	0.20	0.09	43	-
Castor	1.3	0.51	0.21	42	-
Niger	0.8	0.08	0.02	30	-
Rice bran	-	-	0.60	15	2000
Total	253.6	21.18	4.92	-	

Table 2

Properties of Vegetable oils

Vegetable oil	Kinematic vis-cosity at 38 °C (mm²/s)	Cetane No. (°C)	Heating value (MJ/kg)	Cloud point (°C)	Pour point (°C)	Flash point (°C)	Density (kg/l)
Corn	34.9	37.6	39.5	-1.1	-40.0	277	0.9095
Cottonseed	33.5	41.8	39.5	1.7	-15.0	234	0.9148
Crambe	53.6	44.6	40.5	10.0	-12.2	274	0.9048
Linseed	27.2	34.6	39.3	1.7	-15.0	241	0.9236
Peanut	39.6	41.8	39.8	12.8	-6.7	271	0.9026
Rapeseed	37.0	37.6	39.7	-3.9	-31.7	246	0.9115
Safflower	31.3	41.3	39.5	18.3	-6.7	260	0.9144
Sesame	35.5	40.2	39.3	-3.9	-9.4	260	0.9133
Soya bean	32.6	37.9	39.6	-3.9	-12.2	254	0.9138
Sunflower	33.9	37.1	39.6	7.2	-15.0	274	0.9161
Palm	39.6	42.0	–	31.0	–	267	0.9180
Babassu	30.3	38.0	–	20.0	–	150	0.9460
Diesel	3.06	50	43.8	-	-16	76	0.855

Use of vegetable oils as diesel fuel

It has been found that these neat vegetable oils can be used as diesel fuels in conventional diesel engines, but this leads to a number of problems related to the type and grade of oil and local climatic conditions. The injection, atomization and combustion characteristics of vegetable oils in diesel engines are significantly different from those of diesel. The high viscosity of vegetable oils interferes with the injection process and leads to poor fuel atomization. The inefficient mixing of oil with air contributes to incomplete combustion, leading to heavy smoke emission, and the high flash point attributes to lower volatility characteristics. These disadvantages, coupled with the reactivity of unsaturated vegetable oils, do not allow the engine to operate trouble free for longer period of time. These problems can be solved, if the vegetable oils are chemically modified to biodiesel, which is similar in characteristics to diesel. Table2, [8, 14] shows the properties of vegetable oils in relevance of its use in C.I. engine as fuel.

Derivatives of vegetable oils as diesel fuels

Considerable efforts have been made to develop vegetable oil derivatives that approximate the properties and performance of the hydrocarbon-based diesel fuels. The problems with substituting triglycerides for diesel fuels are mostly associated with their high viscosities, low volatilities and polyunsaturated character. These can be changed in at least four ways: pyrolysis; micro emulsification; dilution; and transesterification.

Pyrolysis

Pyrolysis refers to a chemical change caused by the application of thermal energy in the presence of air or nitrogen sparge. Many investigators have studied the pyrolysis of triglycerides to obtain products suitable for diesel engines. The liquid fraction of the thermally decomposed vegetable oil is likely to approach diesel fuels.

Microemulsification

Micro emulsions are isotropic, clear, or translucent thermodynamically stable dispersions of oil, water, surfactant, and often a small amphiphilic molecule, called cosurfactant [10, 11]. Microemulsions because of their alcohol content have lower volumetric heating values than diesel fuels, but the alcohols have high latent heat of vaporization and tend to cool the combustion chamber, which would reduce nozzle coking. A microemulsion of methanol with vegetable oils can perform nearly as well as diesel fuels.

Dilution

Dilution of vegetable oils can be accomplished with such materials as diesel fuels, a solvent or ethanol. The dilution of sunflower oil with diesel fuels in the ratio of 1:3 by volume has been studied and engine tests were carried out by Ziejewski et al. [12]. The viscosity of this blend was 4.88 cSt at 40°C. They concluded that the blend could not be recommended for long-term use in the direct injection diesel engines because of severe injector nozzle coking and sticking. A comparable blend with high oleic safflower oil was also tested and it gave satisfactory results, but its use in the long term is not applicable as it leads to thickening of lubricant.

Transesterification

Transesterification [9, 13], also called alcoholysis, is the displacement of alcohol from an ester by another alcohol in a process similar to hydrolysis, except than an alcohol is used instead of water. This process has been widely used to reduce the viscosity of triglycerides. The transesterification reaction is represented by the general equation:

$$RCOOR\,'+ R''OH \xleftrightarrow{catalyst} RCOOR\,'' + R'OH \quad (1)$$

Ester   Alcohol   Ester   Alcohol

If methanol is used in the above reaction, it is termed methanolysis. The reaction of triglyceride with methanol is represented by the general equation:

```
        H                                    H
        |                                    |
    H-C-OOR                              H-C-OH    ROOCH3
        |                                    |        +
                              CATALYST
    H-C-OOR'    + 3 CH3OH   <----------->  H-C-OH      +
R'OOCH3  (2)
        |                                    |        +
    H-C-OOR"                              H-C-OH
R"OOCH3
        |                                    |
        H                                    H      (Fatty acid
   (Triglyceride)     (Methanol)         (Glycerol)    methyl
esters)
```

The fatty acid methyl esters (known as biodiesel) are attractive as alternative diesel fuels.

III. EXPERIMENTAL SETUP

The objectives of the present work is to evaluate the feasibility of using oil and biodiesel made from the seeds of Jatropha trees in different proportions with fossil diesel by evaluating the performance and exhaust emission characteristics of a diesel engine. It is also aimed to find out the optimal concentration of oil blend and biodiesel blend, based on the performance characteristics and also to observe whether there is any specific combustion related problem of oil and biodiesel fuel with unmodified C.I engine.

To achieve the above-mentioned goal, biodiesel made from Jatropha is prepared and subjected to emission and performance tests on the unmodified diesel engine running at 1500 rpm. The testing was done in Center of Advanced Studies & research in Automotive Engineering, Mechanical Engineering Department, Delhi College of Engineering, Delhi. Engine performance and exhaust emission data were recorded and relevant parameters like BMEP, BSFC, BSEC and thermal efficiency etc were calculated. Based on these parameters, various curves were drawn and compared to base line diesel curve in order to assess the performance of the engine with different biodiesel blends.

For fuel characterization of Jatropha methyl ester, the samples were subjected to several property tests in accordance with standard testing procedures. These tests included higher calorific value, specific gravity, kinematic viscosity at 40°C, and flash points. Results are tabulated

Engine System

The engine selected for emission and performance testing best represents the engine system most common among the rural section & is widely used in agricultural sector of India. It is a single cylinder, four stroke, direct injection, and water-cooled natural aspirated vertical diesel engine. Kirloskar India Limited has manufactured the engine, & it develops 5.2 KW power output at the rated speed of 1500 rpm.

Table 4: Engine Specifications	
No. of Cylinders	1
No. of strokes	4
Fuel	H.S. diesel
Rated power	5.2 KW @ 1500 rpm
Cylinder diameter	87.5 mm
Stroke length	110 mm
Compression Ratio	17.5 : 1
Orifice diameter	20 mm
Inlet Valve Opens	4.5° Before TDC
Inlet Valve Closes	35.5°After BDC
Fuel Injection	23° Before TDC
Exhaust Valve opens	35.5° Before BDC
Exhaust Valve Closes	4.5° After TDC

Test Procedure

S. No	Fuel sample	Flash Point (°C)	Calorific Value KJ/Kg	Specific cgravity (gm/cc)	Kinematic Viscosity (40⁰C)(cSt)
1.	H.S.D.	68	42450	0.832	4.623
2.	J.M.E.	169	37250	0.879	6.848

The engine is started at no load by pressing the inlet with decompression valve lever and it is released suddenly when the engine is hand cranked at sufficient speed.Feed controlled is adjusted to obtain the engine rated speed and it is allowed to run for about 30 minutes till the steady state conditions reached. To assess the present condition of the engine, a constant speed test with diesel as a fuel is carried out and base line data is generated. Test results with all other fuels are compared with base line data to evaluate the performance of the engine. With the help of fuel measuring device and stopwatch, the time elapsed for consumption of 20cc of fuel is measured. RPM, power output (in terms of weight applied and spring scale reading) and smoke density are also measured. The engine is loaded gradually keeping the rated speed within variation of permissible range and the observations of various parameter are recorded. Short-term performance test are carried out on the engine with various blend of biodiesel and diesel.

IV. RESULTS AND DISCUSSIONS

The variations in the values of Thermal Efficiency with respect to BMEP are shown in figure 1. Among the all blends of Biodiesel, JME10 has the highest value of thermal Efficiency for a given BMEP. The lead diversifies as the Brake Mean Effective Pressure increases. High Thermal Efficiency can be interpreted as the cause of complete combustion, hence meager requirement of fuel for the same power output. The reduction in the mass flow rate of various blends due to better Combustion is counter acted by the reduction in the Calorific Values of the same. Hence, a drastic difference is not observed in the values of thermal efficiencies. After, JME10 the next higher thermal efficiency is of JME30. JME5 and JME20. It can be observed from the graph that at low load there is small increase in thermal efficiency as soon the load is further increased; there is sudden rise in thermal efficiency till certain point, as the load is increased further thermal efficiency decreases. It is also seen that diesel has least value of thermal efficiency in comparison to all other biodiesel blends.

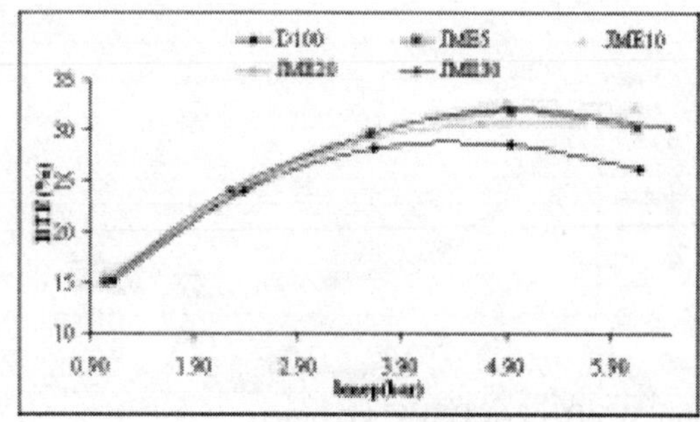

Figure 1: effect of BTE vs. bmep

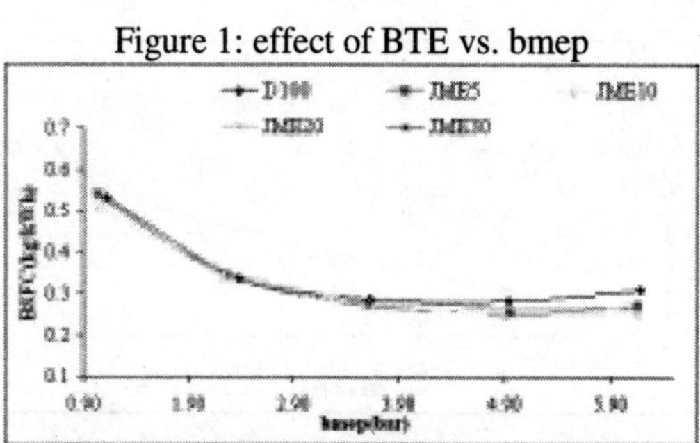

Figure 2: effect of BSFC vs. bmep

Figure 2 suggests that at low engine loads i.e. at low BMEP, Diesel with 10 % biodiesel blend has the least BSFC, followed by JME5.This is because of better combustion of fuel which may be probably of oxygen content & high cetane rating of biodiesel. The difference however narrows as the load progresses up to a certain limit and then it starts diversifying. As the load increases, the BSFC starts decreasing for all the fuels, which implies that at higher loads compression ignition engines run more efficiently than at part loads. At the full load condition the fuel sample JME10 has the least value of BSFC.

BSFC is not a very reliable parameter to compare the two fuels as the calorific values and specific gravity of the blends follow different trends. Hence, brake specific energy consumption is more reliable parameter for comparison. The graph presents the variation of BSEC w.r.t. BMEP for various blends & neat diesel. Brake Specific Energy Consumption is a function of BSEC & Calorific Value. As it can be observed from the graph JME10 has the lowest BSEC closely followed by JME20, JME30 & JME5. All the blends have their BSEC lower than neat diesel. At lower load minimum BSEC of JME10 is due to better combustion of fuel & higher cetane no. of fuel.

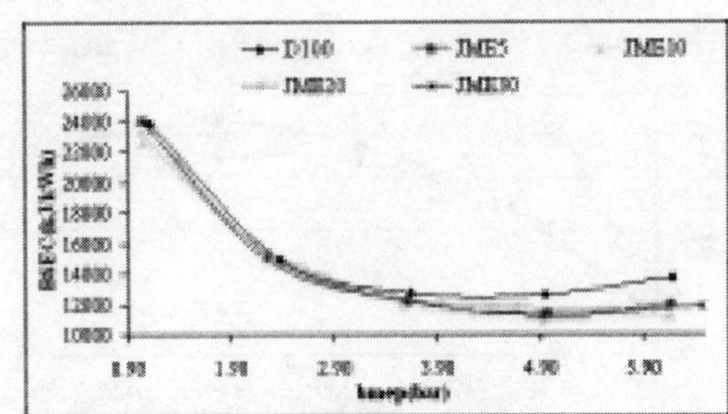

Figure 3: effect of BSEC vs. bmep

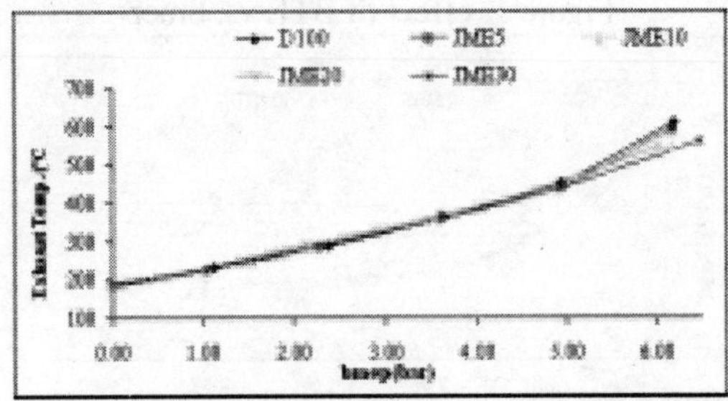

Figure 4: effect of Exhaust temp. vs. bmep

The variation of exhaust temperature with load for different blends of Jatropha biodiesel is compared with diesel fuel in figure 4. For JME5-JME30 the exhaust temperature measured varied between 183°C and 560°C as compared to 185°C and 610°C for diesel indicating very low variation in exhaust temperature. This could be due to nearly the same quantity of fuel being consumed per hour for both diesel and biodiesel blends in each load setting of the engine.

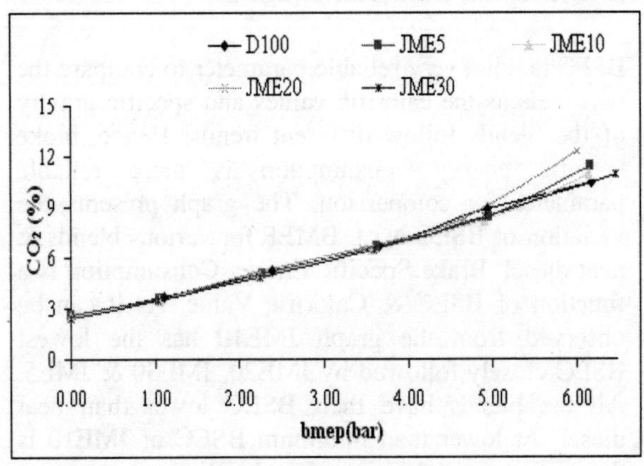

Figure 5: effect of CO_2 vs. bmep

The figure 5 shows that amount of CO_2 in the exhaust increases as the load increases on the engine. This is due to complete combustion as the load increases.

The trend and percent values of CO_2 emission was same for all blends and diesel as the load increases. The explanation for the same values could be lower specific fuel consumption than diesel. But at maximum load there is a variation in the amount of CO_2 emission for different fuel samples.

The variation of CO produced by running the diesel engine using JME5 to JME 30 is compared with diesel in the above figure 6. There is a drastic decrease in CO production in combustion when compared with diesel fuel. This is due to the presence of oxygen in the biodiesl. This oxygen helps in complete combustion of the fuel and emission of CO_2 only. At very high loads there is increase in the CO emission.

The variation of HC emissions in combustion process in the engine is shown in the figure 7. At very low load there is a variation of HC emissions from different blends but the values are very low compared to diesel fuel. As the load increases HC emission becomes same for all biodiesel blends with little increase in its value. Again the lesser values of this emission are the result of complete combustion due to presence of oxygen in biodiesel.

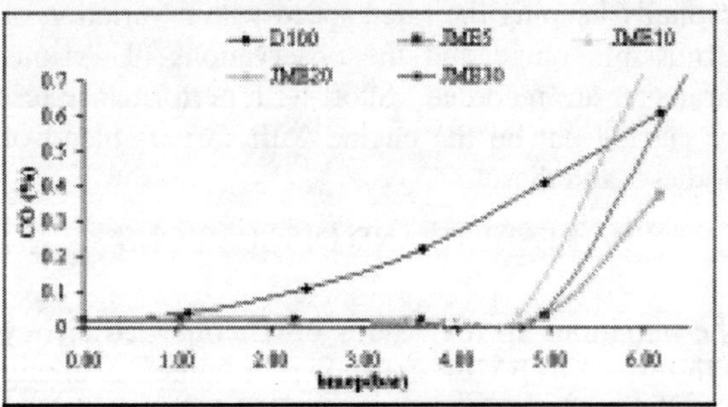

Figure 6: effect of CO vs. bmep

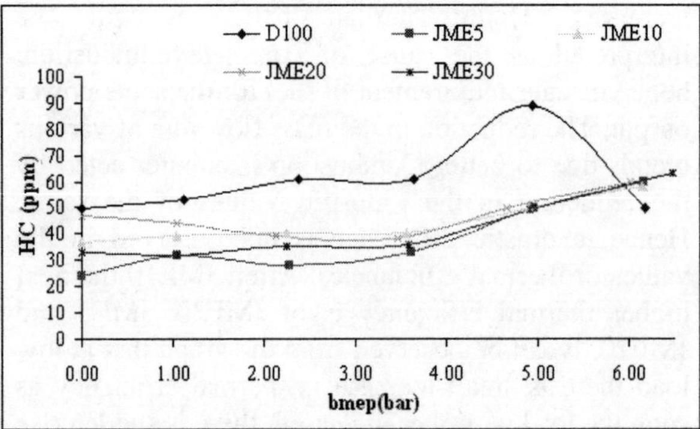

Figure 7: effect of HC vs. bmep

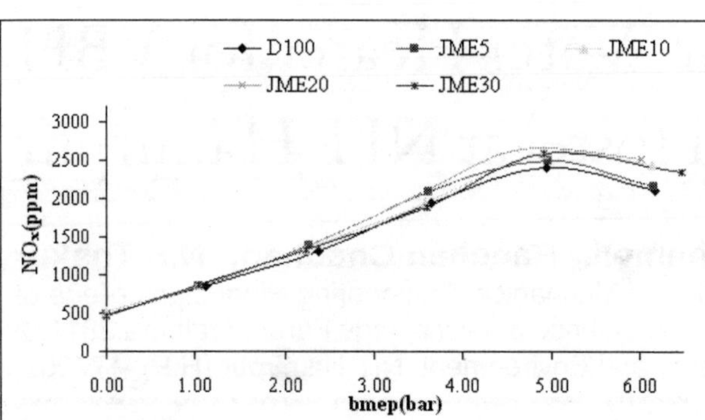

Figure 8: effect of NO_x vs. bmep

The variation of NOx emission is shown here in figure 8. There is very little difference in the values for various blends and diesel. In the general curve of NOx production, the amount increases at higher loads. This is attributed to increase in suction of atmospheric air for complete combustion at high loads.

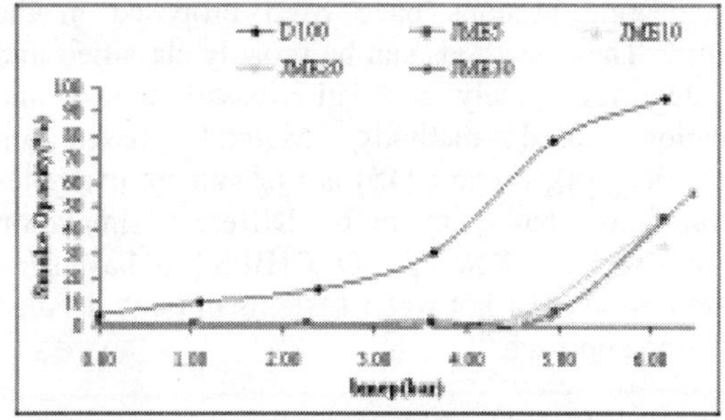

Figure 9: effect of Smoke opacity vs. bmep

It can be observed that the smoke density has reduced with the use of biodiesel in compression ignition engine in comparison to the diesel, it can be seen that at low load, the smoke density is some what lower for all blends as the load on the engine increases the smoke density increases. It may be noted that smoke density is maximum for HSD and minimum for JME10 at throughout the engine operation. It emphasizes the fact that at 10 % biodiesel blend complete combustion will be there.

V. CONCLUSION

In view of environmental considerations, the derivative of triglyceride, biodiesel is considered 'carbon neutral' because all the carbon dioxide released during consumption had been sequestered from the atmosphere for the growth of vegetable oil crops. In comparision to diesel fuel, a little amount of power loss happened with vegetable oil fuel operations. Particulate emissions of biodiesel fuels were higher than that of diesel fuel, but on the other hand, NOx emissions were less. Jatropha oil Methyl Esters gave performance and emission characteristics closer to the diesel fuel. So, they are more acceptable substitutes for diesel fuel.

VI. REFERENCES

[1] Gill W. Paul, James Smith and Eugene Ziuryas., " Fundamentals of IC Engines".

[2] Fergusen, "Internal Combustion Engines", Mc Graw Hill.

[3] Ganesan V., "Internal Combustion Engines", TMH, 1994.

[4] Ayham Demirbas, "Biodiesel from vegetable oils via transesterification in supercritical methanol", Energy Conversion and Management, Vol. 33, 2002.

[5] K.A. Subramanian, S.K. Singal, Mukesh Saxena, Sudhir Singhal, "Utilization of liquid biofuels in automotive diesel engines: An Indian perspective", Biomass and Bioenergy 29 (2005) 67-72.

[6] Orhan AS, Dulger Z, Kahraman N, Veziroglu TN, "Internal combustion engines fuelled by natural gas–hydrogen mixtures", International Journal of Hydrogen Energy 2004;29:1527–39.

[7] B.K.Barnwal, M.P. Sharma, "Prospect of biodiesel production from vegetable oils in India", Renewable and Sustainable Energy Reviews 9 (2005) 363-378.

[8] Goering CE, Schwab AW, Daugherty MJ, Pryde EH, Heakin AJ. "Fuel Properties of eleven vegetable oils", Trans. ASAE 1982; 25: 1472-83.

[9] Anjana Srivastava, Ram Prasad, "Triglycerides-based diesel fuels", Renewable and Sustainable Energy Reviews, 4(2000) 111-133.

[10] Bagby MO. Vegetable oils for diesel fuel: opportunities for development. International Winter Meeting of the ASAE, Hyatt Regency Chicago, 15-18 December, 1987.

[11] Schwab AW, Bagby MO, Freedman B. Preparation and properties of diesel fuels from vegetable oils. Fuel 1987; 66:1372-8.

[12] Ziejewski M, Kaufman KR, Pratt GL. Vegetable oil as diesel fuel. Seminar II, Northern Regional Research Center, Peoria, Illinois, 19-20 October, 1983.

[13] Otera J. Transesterification. Chem Rev 1993; 93(4):1449-70.

[14] SEA News Circular. The Solvent Extractors Association of India, vol. X; 1996.

Optimization of Solar Water Heater for VBH Hostel at NIT Hamirpur

[1]Manoj kumar ,[2]Sunil Chamoli, [2]Ranchan Chauhan, [1]N.S Thakur

[1]Department of Mechanical Engineering Hindustan college of science & Technology Farah, Mathura 281122

[2]Centre for Excellence in Energy and Environment, NIT Hamirpur (H.P) -177005,

*Abstract-***A solar water heating system is designed for VBH hostel at NIT Hamirpur with some assumed collector parameters. This study involves utilization of solar energy through the use of the solar collectors combined with the heat exchanger to keep the tank at the optimum temperature. Investigation has been carried out on the thermal and economic aspects of solar water heater to arrive at the optimal size of solar system. Plots have been prepared to represent the optimum solar collector area. A design procedure has been proposed to arrive at the optimum value of solar collector area.**

Key Words: Solar hot water system, Economic analysis, Optimization

I. INTRODUCTION

Although solar energy is the most important renewable energy source it has not yet become widely commercial even in nations with high solar potential such as India. There are limited applications and most of them are inefficient both in terms of energy use and economical benefits. The economical feasibility of a solar energy system is mainly determined by its initial cost and long term efficiency. The cost of the conventional energy replaced by solar means is, of course, another important parameter. Therefore, in the use of solar energy systems careful consideration is vital to find out the system capacity for optimum useful energy collection at the installation site. Thermosyphon-type flat plate collectors have been used in India since 1970, and at present about 50% of the installed systems are still of this type. However, the installations are mostly by trial and error or say a thump rule. There are quite a large number of different manufacturers producing collectors with varying types and performances. Although the performance of solar water heating systems largely depends on the design storage tank, a flow mixer to mix the water from the mains at times when the water from the tank is above the desired load temperature, an electric heater to heat reverse flow at time of low or no radiation and the necessary piping.

parameters [1–10], such as hot water consumption rate [1, 11, 12] and climatic data [1, 11–13], systems are installed without prior determination of the optimum system size and capacity thus proper design of solar water heating system is
important to assure maximum benefit to the user, especially for a large system. Designing a solar hot water system involves appropriate sizing of different components based on predicted solar insolation and hot water demand. A number of design methods for solar water heating systems have been proposed in the literature. These methods can be broadly classified into two categories, namely, correlation based methods and simulation based methods. Methods based on utilizability [14], F chart [15] are prominent examples of correlation based methods. Different simulation programs such as TRNSYS, SOLCHIPS [16] have been used to design solar hot water systems through detailed simulation approach.

II OBJECTIVES

It is well known that the ambient temperature during winter at Hamirpur is very less thus using solar energy for water heating is a viable option. However there is a need to determine optimum size of the solar collector system to suit the given load requirement. . In the present work, it is proposed to investigate the economic aspects of solar water heating system to arrive at the optimal size of solar system. The following are major objectives of present study
1. Thermal analysis of solar water heating system.
2. Economic analysis of the system.
3. Optimization of solar collector area.

III SYSTEM DESCRIPTION

A diagram of the solar water heating Thermosyphon system is shown in Fig. 1 its real diagram is shown in fig. 2. It consist of a flat plate collector, a vertical the water in the tank when the solar contribution from the collector is insufficient, a check valve to prevent

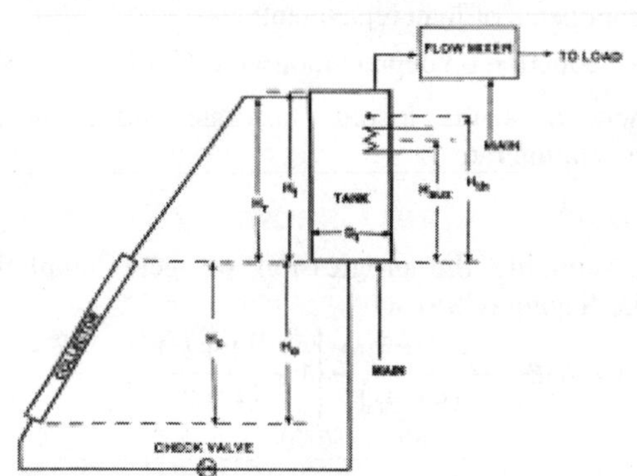

Fig. 1: Shows the schematic diagram of the system
Fig. 2: Shows the actual installed solar water heater

IV THERMAL ANALYSIS OF THE SYSTEM

4.1 Load Calculation

If V_L is the required amount of water and T_h is the desired hot water temperature then energy required Q_{Load} is expressed as

$$Q_{Load} = C_P \rho V_L (T_h - T_c) \qquad (1)$$

Where C_P is the heat capacitance of water (C_P=4200J/kg^{0}C),ρ its density (ρ=1kg/L), T_c is the cold mains water temperature Q_{Load} is prorated by the number of days the system is used per week.

Total Load for a month

$$Q_{\text{Total Load}} = \sum_{k=1}^{k=12} Q_{Load}(k) \qquad (2)$$

Where k is the month number from January to December.

4.2 Useful energy gain from solar collector

Useful energy gain from solar collector can be calculated as shown under:

For a flat plate collector the useful energy gain in a day is given by the equation

$$E_{Gain} = A_c F_R \sum_{i=1}^{i=24}[Hs(i)(\tau\alpha) - U_L(Tc - Ta(i))] \qquad (3)$$

Where T_c is the inlet fluid temperature (^{o}C), F_R is the collector heat removal factor, A_c is the collector area of absorbing surface (m^2), H_s is the solar intensity on the plane of absorber (W m^{-2} a $\tau\alpha$ h transmittivity absorptivity product of collector absorber

(Using: E_{Gain} =0 when $E_{Gain} < 0$)

The value of ambient temperature and Insolation is taken from metrological data of Hamirpur.

The heat gain for a year is given by

$$E_{\text{Total Gain}} = \sum_{k=1}^{k=12} E_{Gain}(k) \qquad (4)$$

Where k is the month number from January to December.

V ECONOMIC ANALYSIS OF THE SYSTEM

For evaluating the economic viability of a solar thermal energy system installed for heating water, the savings which will accrue annually and on a long term basis, have been calculated.

5.1 Solar Fraction

Solar fraction is defined as a fraction of the load supplied by solar energy. Monthly Solar Fraction is written as

$$f = \frac{Energy\ collected\ in\ a\ month}{Total\ load\ in\ a\ month}$$

(Using: f=1 when f >1.)

Annual solar fraction is defined as

$$f' = \frac{\sum_{n=1}^{n=12} Q_{load} f}{\sum_{n=1}^{n=12} Q_{load}} \qquad (5)$$

Where n is the month number

5.2 Cumulative Solar Savings

The Cumulative solar savings over a period of n years is obtained by summing the present worth of the annual solar saving and considering the initial down payment made at the time of installation of the solar system.

Considering a solar energy system

1. Assume that the system requires a total investment C of which a fraction f_l is taken as loan. The interest rate on loan is d_l and the loan is to be paid back in equal installments over a period of n years.

   Total Investment C= Unit cost of collector x Area of collector, i.e. Total Investment C= C_A x A_c

2. Let the annual energy load to be met be $E_{\text{Total Load}}$ and assume that the solar system supplies a fraction f' of this load. This would result in an annual saving of (f' $E_{\text{Total Load}}$) of conventional energy. Assume that the cost of this energy is C_f per unit of energy and that it increases at the rate of i_f every year.

3. The solar system requires maintenance cost M per year and this will increase at the rate of i_m every year.

4. The tax deductions are allowed both on the interest component of the annual loan repayment installment as well as on depreciation of the system. The depreciation is assumed to be at a uniform rate r_d per year. The income tax rate is r_t.

In any year j

Fuel savings= $C_f(1 + i_f)^{j-1} f' E_{Total\ Load}$ \qquad (6)

Where C_f is the cost of conventional energy in Rs/MJ, i_f is the rate of increase every year

Annual repayment on loan= $\dfrac{d_l f_l C}{[1-\frac{1}{(1+d_l)^n}]}$ if $j\leq$

Where d_l is the interest rate on loan, f_l is the fraction taken as loan and n is the CSS calculation period

Maintenance charges= $(1+ i_m)^{j-1}$ M (7)

Where i_m is the rate of increase in maintenance cost and M is the annual maintenance cost in Rs

Tax deduction on the interest

Thus annual solar savings in the year j=

$$C_f(1 + i_f)^{j-1} f' E_{Total\,Load} - \frac{d_l f_l C}{[1-\frac{1}{(1+d_l)^n}]} - (1+ i_m)^{j-1} M$$

$$+[1 - \frac{(1+d_l)^{j-1}-1}{(1+d_l)^{n}-1}]r_t d_l f_l C + r_t\, r_d\, C$$

(10)

The cumulative solar saving over a period of n years is obtained by summing up the present worth of the annual solar savings and considering the initial down payment.

Thus Cumulative Solar Saving (CSS) =

$$-(1 - f_l)C + C_f f' E_{Total\,Load} \sum_{j=1}^{j=n} \frac{(1+i_f)^{j-1}}{(1+d)^j} -$$
$$\frac{d_l f_l C}{[1-\frac{1}{(1+d_l)^n}]}\sum_{j=1}^{j=n} \frac{1}{(1+d)^j} - M \sum_{j=1}^{j=n} \frac{(1+i_m)^{j-1}}{(1+d)^j} +$$
$$r_t d_l f_l C \sum_{j=1}^{j=n} \frac{1}{(1+d)^j}\left[1 - \frac{(1+d_l)^{j-1}-1}{(1+d_l)^{n}-1}\right] +$$
$$r_t r_d C \sum_{j=1}^{r_d} \frac{1}{(1+d)^j}$$

(11)

Where d is the market discount rate and n_l is the payback period of loan   .

For the above equation; $n \geq$ ₁, $n \geq$ r_d), $d\neq$ f, $d\neq$ m & $d\neq$ ₁

A programme in Matlab is designed for the above calculations.

VI RANGE OF PARAMETERS

Table 1 indicates the range of operating parameters

	Parameter	Value
1.	C_A (Rs/m^2)	1000 to 50000
2.	C_f (Rs/MJ)	0.1 to 10
3.	D	0.05 to .25

VII RESULTS AND DISCUSSION

The results obtained from the mathematical simulation of the solar heated biogas plant are discussed in detail to understand the effect of parameters on the performance of solar collector. Figs.3,4 & 5 show the variation of CSS as a function of solar collector area for different values of Unit Collector Cost (C_c), Unit Fuel Cost (C_f), and Market Discount Rate (d), respectively.

Fig. 3 shows the variation of CSS as a function of solar collector area for different values of unit cost of

$$= [1 - \frac{(1+d_l)^{j-1}-1}{(1+d_l)^{n}-1}]r_t d_l f_l C \quad \text{if } j\leq \tag{8}$$

Components of loan repayment:

Tax deduction on depreciation= $r_t\, r_d\, C$ if $j\leq \dfrac{1}{r_d}$ (9)

Where r_t is the income tax rate and r_d is the depreciation rate

On summing the progression we get Cumulative Solar Saving (CSS) =

$$-(1 - f_l)C + \frac{C_f f' E_{Total\,Load}}{(d - i_f)}\left[1 - \frac{(1 + i_f)^n}{(1 + d)^n}\right]$$
$$-\frac{d_l f_l C}{[1 - \frac{1}{(1+d_l)^{n_l}}]}\frac{1}{d}\left[1 - \frac{1}{(1 + d)^{n_l}}\right]$$
$$-\frac{M}{(d - i_m)}\left[1 - \left[\frac{1 + i_m}{1 + d}\right]^n\right]$$
$$+ r_t d_l f_l C \left[\left[\frac{(1 + d_l)^{n_l} - 1}{(1 + d)^{n_l} - 1}\right]\right.$$
$$-\frac{1}{(1+d_l)^{n_l} - 1}\frac{1}{(d - d_l)}\left[1\right.$$
$$\left.\left.- \left[\frac{1 + d_l}{1 + d}\right]^{n_l}\right]\right] + \frac{r_t r_d C}{d}\left[1\right.$$
$$\left.-\frac{1}{(1 + d)^{\frac{1}{r_d}}}\right]$$

(12)

collector. It can be observed that CSS increases on decreasing the cost of solar collector and the optimum value of collector area decreases as the collector cost increases.

Fig. 4 shows the variation of CSS as a function of solar collector area for different values of unit fuel cost. It reveals that with the hike in fuel cost, CSS also increases as the fuel saving increases. The optimum collector area also increases with the increase in conventional fuel cost.

Fig 5 shows the variation of CSS as a function of solar collector area for different values of market discount rates. It can be seen that the maximum value of CSS is obtained at market discount rate of 5%. For all the values of market discount rates, CSS attains maxima at the same value of collector area. It can be concluded that for fixed values of unit collector cost and fuel cost, the optimum value of collector area is not affected by the discount rate.

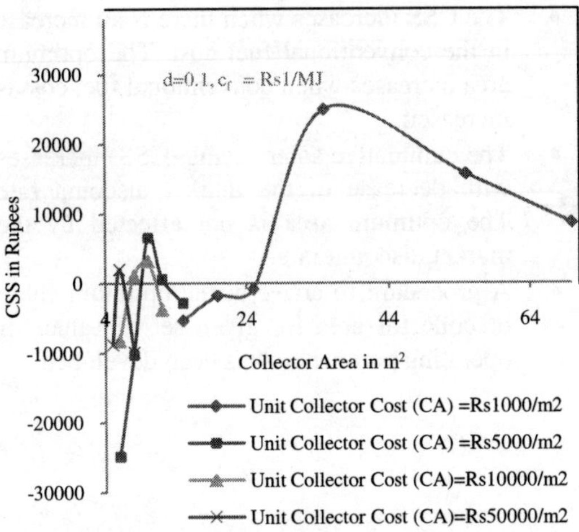

Fig 3 Effect of solar collector area on CSS for different Values of unit collector cost.

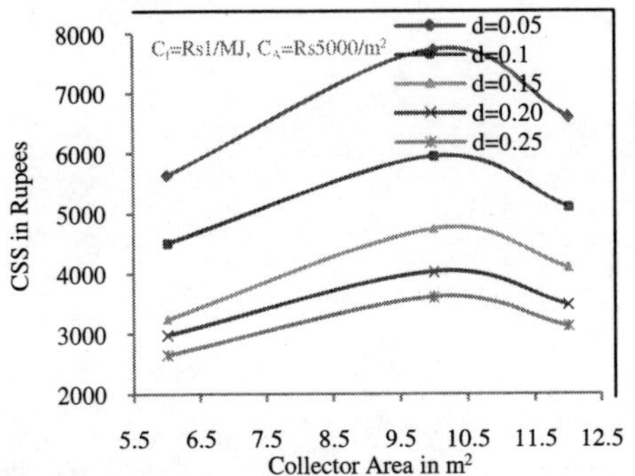

Fig. 5 Effect of solar collector area on CSS for different Values of market discount rate.

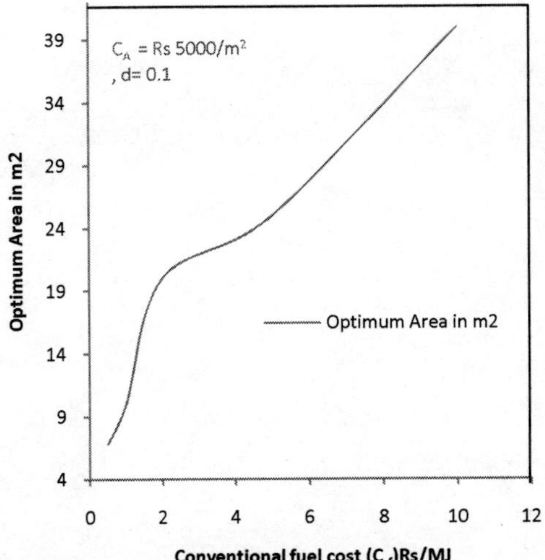

Fig. 7 Optimum value of collector area for different value of conventional fuel cost.

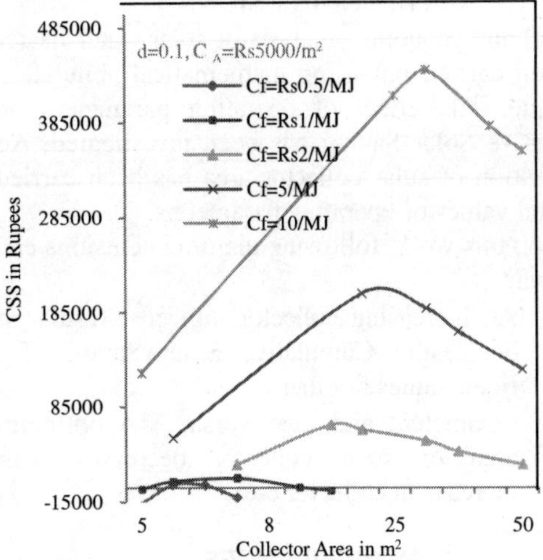

Fig 4 Effect of solar collector area on CSS for different Values of conventional fuel cost.

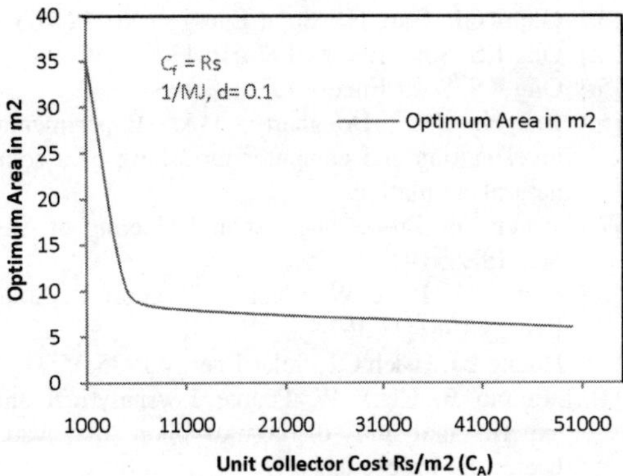

Fig. 6 Optimum values of collector area for different Values of unit collector

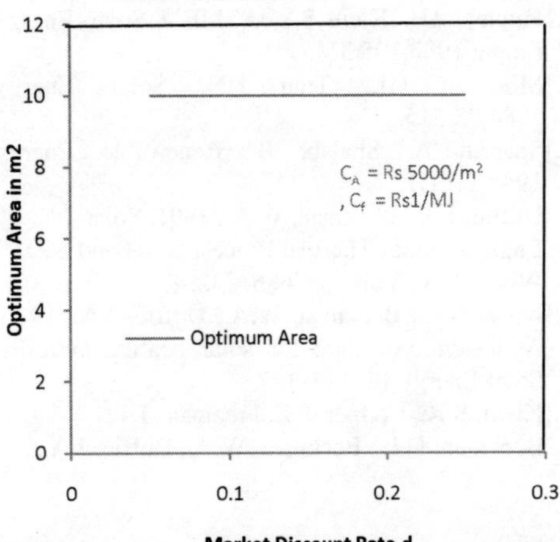

Fig. 8 Optimum value of collector area for different value

A designer can make use of graphs to arrive at the optimal value of collector area for different sets of operating parameters for a given plant capacity. For instance, at C_A= Rs 5000/m^2, C_f= Rs 1/MJ and d=0.1, one can determine the optimum collector area as 10 m^2.

VIII CONCLUSIONS

Thermal and economic analysis of solar water heater has been carried out using mathematical simulation technique. The effect of operating parameters on Cumulative Solar Saving has been investigated. An optimization of solar collector area has been carried for given values of operating parameters.

Based on this work, following major conclusions can be drawn:

- An increasing collector unit cost results in decreasing Cumulative Solar Saving for fixed values of other
- parameters and vice versa. The optimum area of solar collector decreases with increase in collector cost.

- The CSS increases when there is an increase in the conventional fuel cost. The optimum area increases when conventional fuel cost is increased.
- The cumulative solar saving (CSS) increases with decrease of the market discount rate The optimum area is not affected by the market discount rate.
- A procedure to arrive at the optimum value of collector area for given set of values of operating parameters has been developed.

IX REFERENCES

[1.] Morrison GL, Braun JE. Solar Energy 1985;34:389.

[2.] Close DJ. Solar Energy 1962;6:33.

[3.] Gupta GL, Garg HP. Solar Energy 1968;12:163.

[4.] Ong KS. Solar Energy 1974;16:137.

[5.] Ong KS. Solar Energy 1976;18:183.

[6.] Baughn JW, Dougherty DA. Experimental investigation and computer modelling of a solar natural circulating

[7.] system. In: Proceedings Annual Meeting of Am. Sec. ISES, 1977;1:425.

[8.] Mertol A, Place W, Webster T, Grief R. Solar Energy 1981;27:367.

[9.] Huang BJ, Hsieh CT. Solar Energy 1985;35:31.

[10.]Nimmo B, Clark W, Pearce J. Analytical and experimental study of thermosyphon solar water heater. In: Proceedings

[11.]Annual Meeting of Am. Sec. ISES, 1977;4:30.

[12.]Hobson PO, Norton B. ASME J Solar Energy Engng 1988;110:282.

[13.]Fanney AH, Klein SA. ASME J Solar Energy Engng 1983;105:311.

[14.]Morrison GL, Tran HN. Solar Energy 1984;33:515.

[15.]Shariah A, Shalabi B. Renewable Energy 1997;11:351.

[16.]Duffie, J.A., Beckman, W.A., 1991. Solar Engineering of Thermal Processes, second ed. Wiley, New York, pp. 686–732.

[17.]Klein, S.A., Beckman, W.A., Duffie, J.A., 1976. A design procedure for solar heating systems. Solar Energy 18, 113–127.

[18.]Klein, S.A., Cooper, P.I., Freeman, T.L., Beekman, D.L., Beckman, W.A., Duffie, J.A.,

Automatic Railway Gate Opening

Nidhi Gupta, Bhawna Bisht

Department of Mechanical Engineering, Madhav Institute of Technology & Science, Gwalior, M. P.

Abstract-In this paper we are concerned of providing an automatic railway gate control at unmanned level crossing replacing the gate operated by gate keepers and also the semi automatically operated gates it deal with two things. Firstly , it deal with the reduction of time for which the gate is being kept closed .And secondly, to provide safety to the road users by reducing the accidents that usually occur due to carelessness of road users and at times errors made by the gatekeeper. By employing the automatic railway gate control at the level crossing the arrival of train is detected by the sensor placed on either side of the gate at about 2-3 km from the level crossing. Once the arrival is sensed, the sensed signal is sent to the main controller

I.   INTRODUCTION

Now a days, India is the country which having worlds largest railway network .Over hundred of railways running on track every days. As we know that it is surely impossible to stop the running train at instant in some critical situation or emergency arises. Therefore at the place of traffic density there is a server need to install a railway gate in view of protection purpose and every gate there must be an attendant to operate and maintain it. In view of that, if we calculate the place of railways crossing and such places where it would to be install and overall expenditure. The graph arises and arises but, our country is a progressive country. It has already enough economical problems which are ever been unsolved so, to avoid all these things some sort of automatic and independent system come in picture .Now a days automatic system occupies each and every sector of Applications as it is reliable, accurate and no need to pay high attentions.

II. WORKING PRINCIPLE

We are providing an automatic railway gate control at unmanned level crossing replacing the gates operated by gate keepers and also the semi automatically operated gate. It deals with two things .firstly, it deal with the reduction of time for which the gate is being closed and secondly to provide safety to the road users by reducing the accident that usually occur due to carelessness of road users and at time errors made by the gate keeper. The arrival of train is detected by the sensor placed on either side of the gate at about 2-3 km from the level crossing. Once the arrival is sensed, the sensed signal is sent to the main controller. Subsequently, buzzer indication and light signal on either side are provided to the road users indicating the closure of gate, motor is activated and the gate is closed. The departure of train is detected by sensors placed at about 1 km from the gate. The signal about the departure is sent to the main controller, which in turn operate the motor and opens the gate. Thus, the time for which the gate is closed is less compared to the manually operated since the gate is closed depending upon the telephone. Let, R1 sensor on the track, placed at reference point about 2 km from the gate to detect the train arrival on either direction both the UP and DOWN   tracks of RAILWAYS. Initially when power is switched on, IC555 is reset by power on reset, and the entire circuit is in the idle state. LED indicates the power status. In the event of vibrations, IC is clocked by the pulses from the piezoceramic element connected to its clock pin. Immediately after clocking, the output would go high and NPN transistor would conduct thus, in turn, energizes relay. Zener diode at the clock input of IC is used to protect against high voltage input. The relay contacts

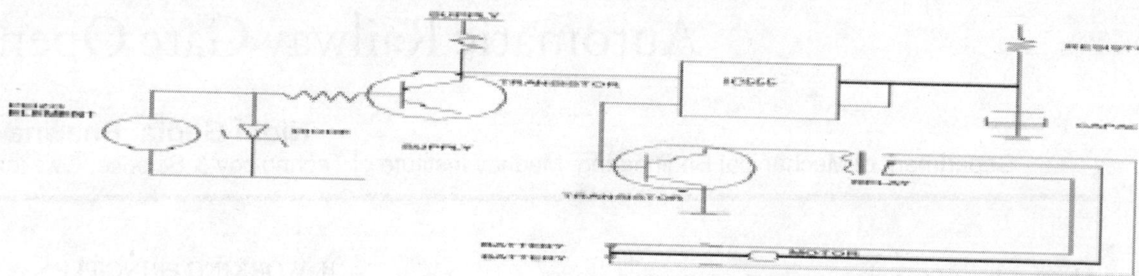

Fig:1

III. REFERENCE POINT SETTINGS DEPEND ON THE FOLLOWING FACTORS.

- Minimum breaking distance
- Speed allowed on that track
- Road traffic density

So as to clear the traffic before actually the gate is closed on the arrival of train, vibration sensor sense the signal, the detector will sense vibration caused by this. The circuit works off a 9V battery or 9V regulator power supply and uses a piezoceramic element as the vibration detector. can be used to switch any alarm device to indicate vibration detection. The same output is used to run the gear motor (BI-DIRECTIONAL)   .The movement of gear motor can be controlled by the use of limiting switch (ELECTREOMECHANICAL SWITCH) S1 andS2. These switches are placed at both the end of railway crossing gate. The movement of gate on either side i.e. opening or closing of gate result in activation of either switch S1 and S2 these switch have been arranged such that during closing or opening of gate one or the other switch is eliminate from the circuit result in stopping of the gear motor movement.
Let, S1   switch remained pressed when gate is open . When vibration is sensed gear motor start its movement as a result S1 switch is released and as soon as gate is closed switch S2 is pressed. Pressing off switch S2 results in stopping of gear motor movement. The gate remained closed for the specific time duration. When the train passes off the track, the output of the circuit would go down, thereafter the gas reopens. When the gate open up switch S2 is released while S1 is pressed. As soon as switch S1 is pressed gear motor again stop the movement

IV. EXPERIMENTAL SETUP

Vibration Sensor: It uses piezoelectric effect to detect the vibration in the rails due to arrival or departure of train and direction of vibration indicate the rival or departure. This could sense the train position roughly at 2000 to 2500 m away. Piezoelectric vibration sensors used for detecting from various vibration sources are generally classified in to two larger type, resonant type and non resonant type. Vibration is typically measured using analog vibration sensing elements, such as analog accelerometer, positioned on machinery at strategic locations.

Gear motor: Gear motor work in a dc motor, having one electromagnetic coil to produce movement. Due to electromagnetic induction coil is energized; the motor rotates a few degrees. Repeating the sequence because the motor to move a few more degrees called the step angle .Repeating the sequence causes the motor to move a few more degree and so on, resulting in a constant rotation of the motor shaft.

Advantage of gear motor:
Gear motor can be operated from very low speeds of 0.01rpm to high speeds of the order of 60 rpm
Drift free.
Stopping and reversing.
Higher torque per package size
Holding torque at standstill
Bi-directional operation
No extra feedback components Rapid response to starting, required

Application:
- Metal punching
- Welding
- Robotics

Piezoceramic Element
It is a small piece that conducts electricity. They are found in microwaves, engines and even computers and cell phones. When a manufacture makes a piezoceramic element they create a crystal structure using metal particles. The most commonly used choices are titanium, zirconium, lead and barium. These are then mixed with a binder and fired into a specific shape or size. Once they are shaped, electrical current is introduced to make the part polarized in a specific way. Even when the current is removed, the part will stay polarized permanently, making these ideal for moving electricity throughout a system.

4.4 Electrical limit switch: An electrical limit switch includes a shaft which is selectively rotated by a switch actuating rocker arm. A pinion cooperates with a symmetrically biased rack for effecting a centered rest pinion for the shaft and rocker arm , and a cam is affixed to the shaft to selectively connect the disconnect electrical contact acting through a plunger linkage. In accordance with other aspect of the present invention, additional; plunger actuating embodiments are provided to effect contact switching responsive to associated applied mechanical displacement.

Capacitor

A capacitor is a passive electronic component consisting of a pair of conductors is separated by a dielectric. When a potential difference exists across the conductor, an electric field is present in the dielectric. This field stores energy and produces a mechanical force between the conductor .The effects is greatest when there is a narrow separation between large areas of conductor, hence capacitor conductors are often called plates. Capacitors are widely used in electronic circuit to block the flow of direct current while allowing alternating current to pass, to filter out interference, to smooth the output of power suppliers, and for many other purposes. They are used in resonant circuits in radio frequency equipment to select particular frequencies from a signal with many frequencies.

Resistor

A resistor is a two terminal electrode component that produce a voltage across its terminal that is proportional to the electric current passing through it in accordance With ohm's law: V=IR Resistors are elements of electrical network and electronic circuit and are ubiquitotious in most electronic equipment. Practical resistor can be made of various compounds and films, as well as resistance wire. The primary Characteristic of a resistor is the resistance, the tolerance, maximum working voltage and the power rating. Other characteristic include temperature coefficient, noise and inductance. Resistors can be integrated in to hybrid and printed circuits, as well as integrated circuits

Zenor Diode

Zenor diode is used to maintain a fixed voltage. They are designed to breakdown in a reliable and non destructive ways to so that they can be used in reverse to maintain a fixed voltage across their terminal.

Light Emitting Diode

LED emits light when an electric current passes through them. LED can be damaged by heat when soldering, but the risk is small unless you are very slow. No special precaution are needed for soldering most LEDs

Relays

A relay is an electrically operated switch. Current flowing through the coil of the relay creates a magnetic field which attracts a lever and changes the switch contact. The coil current can be on or off so relays have two switch position and most have double through switch contact relay allow one circuit to switch a second circuit which can be completely separated from the first. For ex a low voltage battery circuit can be use a relay to switch a 230 volt AC mains circuit. There is no electrical connection inside the relay between the two circuits; the link is magnetic and mechanical.

Advantage of relay

Relay can switch AC and DC, transistor can only switch DC.
Relay can switch higher voltage than standard transistors
Relay are often a better choice for switching larger currents (>5A)

Disadvantage of relays

Relays are bulkier than transistors for switching small current. Relay use more power due to current flow through their coil.

(NPN): NPN is one of the two types of bipolar transistor, in which the letter N and P refers to the majority changes carrier inside the different region of the transistor. NPN transistor consists of a layer of P-doped semiconductor between two N-doped layers. A small current entering the base in common emitter mode is amplified in there collector output. In other terms, an NPN transistor is on when its base is pulled high relative to the emitter

V. RESULT AND DISCUSSION

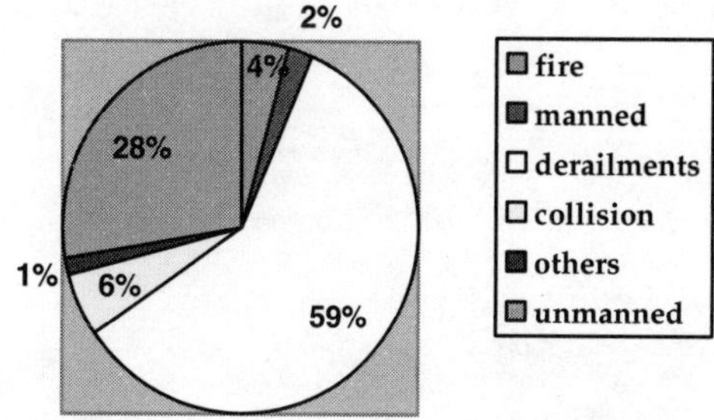

Fig 2: Details of accident on Indian railway in past year (2004-2005)

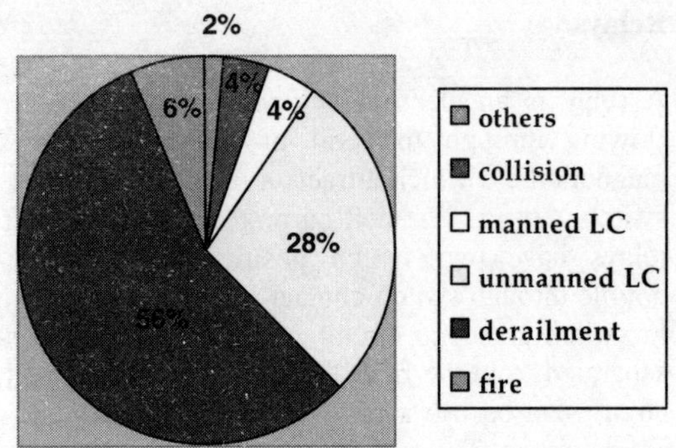

Fig 3: Detail of accident on Indian railway in past year (2005-2006)

Accident Cases

coupled light engines dashed with auto rickshaw at gate no. 78 at 14/2-3 of Hyderabad division of south railway on 10-01-03 (killed 9,injured 2)

2.   5027 up dashed with tractor on gate at km 429/10-11 between Bhatparrani and Bhatni station of BSB division of nearly on 30-03-03 (killed 11, injured 8)

321 UP (tata Nagpur passenger) dashed with dumper at unmanned level crossing gate no. 452 between Durg and Murlipur station on 28-08-01 ,the train driver loss his life

5RN passenger train dashed with one tractor trolley on unmanned level crossing at km

1115/3 -4(58 killed, 4 injured)

On 30 -05-06 Kapil vastu express dashed with a tractor tailor at un manned level crossing near Kuian village of Balrampur district (up),where 3 people died

VI. CONCLUSION

In the advent of modernization in Indian Railways, the bottleneck in safety like accidents on unmanned level crossing is to be dealt with more in a scientific manner and rational thinking for minimizing accidents to utmost with modern equipment and more awareness to the road users

VII. FUTURE SCOPE

This project is developed in order to help the INDIAN RAILWAYS in making its present working system a better one, by eliminating some of the loopholes existing in it

Based on the response and report obtained as a result of significant development in the working system of INDIANRAILWAYS, this project can be further extended to meet the demands according to the situation

This can be further implemented to have control room to regulate the working of the system. Thus become the user friendliness

VIII. REFERENCE

[1] *Australian transport safety Buereo, Monograph 10, level crossing accidents, ASTB January 2002.ISN 1444-3503*

[2] *RAILWAY SAFETY (2002): GI/RT 7011 Provision, risk assessment and review of level crossing*

[3] *Case study made in north Coimbatore railway control room"*

[4] *Railway safety (2003) railway group safety plan*

[5] *Kenneth.J.AyalaThe89C51   Microcontroller Architecture programming and Applications, Pen ram International.*

[6] *-D.Roychoudary   and   Sail   JainL.I.C,   New   Age International.*

[7] *Principles of Electronics by V.K.MEHTA.*

[8] *-Communication Systems by Simon Hawkins.*

[9] *Electrical Technology â€œ vol. 2- B.L. Theraja*

Biomass As A Non-Conventional Energy Resource

[1]Nitin Wahi, [1]A.K. Bhatia, [1]Anjana Goel, [2]Seema Bhadauria
[1]Dept. of Biotechnology, GLA University, Mathura, U.P.
[2]Dept. of Botany, RBS College, Agra,U.P.

Abstract: Each and every process in the world is just based on the energy exchange in this universe. Energy is the key input that drives our life cycle in every manner. Primarily, it can be said as a gift to mankind from nature. With growing population, increasing industrialization in developing countries, the global demand for energy is expected to increase by leaps and bounce. The primary and present energy source has been fossil fuels, however the over exploitation of fossil fuels reserves and large scale environment degradation caused by their widespread use, particularly global warming, urban air pollution, acid rains, health hazards compel humans, to harness the non-conventional, renewable and environment friendly energy resource as vital global energy supply for the betterment of life on the earth. The non conventional energy resources includes: Solar energy, Wind energy, Geothermal energy, Energy from Biomass. Although there has been number of non conventional energy resources, among them some are most important but least explored that is from the biomass. It has been revealed that biofuels which provided only 1.8% of the total world transport fuel in 2008, in 2010 provides approx 15% of the total energy supply. With the recent advances and breakthrough in biotechnology; the new energy revolution is on the verge of creating potent energy source by utilizing the potential of plants and microorganisms.

I. INTRODUCTION

Life on earth is driven by energy.[14]

"Energy can neither be created nor be destroyed but can only be transformed from one form to another"(Law of conservation of energy).
Autotrophs (plants) take it from solar radiation and heterotrophs (animals) take it from autotrophs. Energy captured slowly by photosynthesis is stored up, and has been converted into denser reservoirs of energy in the form of fossil fuels during the course of Earth`s history.[14]

So, humans have always explored the nature since ancient times to find suitable energy source, which have led to the discovery of fossil fuels. Fossil fuels includes coal, natural gas, petroleum, etc.[2] With increasing population, industrialization and improvement in the living standards of human beings there has been a phenomenal increase in the demand for energy globally.[1]

II. INDIAN ENERGY NEED

India ranks second in the world in terms of population, while sixth in position in terms of total energy consumption. With increasing population of the country the energy requirement of the country has increased day after day.

More than 70% of the primary energy needs are being met through imports from other countries in the form of crude oil[2]. Since independence India has increased its power generation capacity from 1362 MW to approx 112,058 MW.[3] It seems to be impressive in terms of numbers but with increasing population it is unsatisfactory. Approx 400 million Indians have no access to electricity, the per capital consumption of 600 Kwh is one of the minimum globally. The ministry of power has started a program targeting supply of power called as 'Power on demand' by 2012.[3] Demand as per 16th Electric Power Survey requires an additional 100,000MW uptill 2012 through non conventional energy source. It will require an additional Rs. 8,00,000 crores.[1]

Although government has made commissions; bills; projects; acts for meeting the energy need of the country, such as:

- Electricity regulatory commission, 1991.
- Environmental audits for power projects, 1992.
- Energy conservation bill, 2000.
- Renewable energy promotion bill, 2005.

The basic problem in India is not is not of framing laws or acts but of implementing them at the ground level. For electricity, the problem is not the one of distribution but of provision to all.[5] Many people try or attempt to steel electric power, similarly others switches over their vehicles from petrol to LPG, without any sort of recommendations from experts thus risking their lives with that of others even.

In India not only the government has to enforce impressive laws for the development of renewable energy but also has to spread awareness among people about the overexploitations of natural sources.

III. NEED FOR NON CONVENTIONAL ENERGY SOURCE

Fossil fuels supply almost the entire energy consumed today. These are easy to exploit; provides cheaper energy source and technically easy to be refined.[1]

There are several limitations with these fossil fuels:

- There are only 200 years for which the reserves of coal will last.[1]
- The raw material for nuclear power will last only for 50 years.[2]
- With increasing use of conventional energy sources the problem of global warming is becoming bad to worse. According to the 2007, fourth Assessment Report by the Intergovernmental Panel on Climate change (IPCC) global surface temperature has increased by $0.74 \pm 0.18°c$ during the 20th Century.[2]
- The burning of fossil fuels have lead to the deposition of sulfur dioxide, nitrous oxide in the environment generation the danger of acid rains.
- In the words of Mahatma Gandhi: ''Earth has everything for everybody need but not for everybody greed'', the need for sustainable development is the greatest today and if humans will continue to exploit nature for themselves like today then they will also perish themselves as the mighty dinosaur's were in the past.

IV. NON CONVENTIONAL OR RENEWABLE ENERGY SOURCE

To meet the need of the present generation without compromising with the need of the future generation one must focus on the sustainable development.[4] The present world attention is being focused on natural, clean and renewable source of energy. These includes among them:

- SOLAR ENERGY.
- WIND ENERGY.
- GEOTHERMAL ENERGY.
- ENERGY FROM BIOMASS.

SOLAR ENERGY

The solar energy is perhaps the most easily available renewable source of energy present in nature. Solar energy is the energy that is released by the Sun as electromagnetic radiations. The Earth receives 174 petawatts (PW) of incoming solar radiation at the upper atmosphere.[8] Approximately 30% is reflected back to space while the rest is absorbed by clouds, oceans and land masses. This energy reaching the earth's atmosphere consists of about 8% UV radiation, 46% visible light and 46% infrared radiations.

The solar energy can be utilized in two forms:

- Solar Electricity.
- Solar Heating.

Solar electricity refers to the production of electric power by entrapping solar energy using solar cells.

Solar heating is to capture sun`s energy for cooking foodstuffs and heating buildings. With the development in the field of semiconductors now much more effective and cheap solar cooker`s are available. Solar cooker`s can be grouped into three broad categories: box cookers, panel cookers and reflector cookers.[9] The simplest solar cooker is a basic box which consists of an insulated container with a transparent lid. It can be used effectively with partially overcast skies and will typically reach temperatures of 90–150 °.[9]

India is a vast country with an area of approx 3.2 million sq.km. Most part of the country receives 250-300 days of sunlight so has a tremendous potential.

- 140 MW solar thermal/naphtha hybrid power plant with 35 MW solar tough component is to be constructed in Rajasthan.[5]
- Grid interactive solar photovoltaic power projects aggregating to 2490 KW have been installed.[6]

Wind Energy

The origin of wind energy is due to sunlight.[1] When the sun rays falls over the land and the oceans the land air gets heated up to a greater extent than the air above the oceans thus creating a pressure difference leading to the generation of wind. The kinetic energy stored in the wind can be directly utilized in the generation of electricity with the help of wind mails or turbines.

Turbines generally require a wind in the range of 20 Km/hr and there are relatively only few places which have significant prevailing winds.[6] Its potential is believed to provide 5 times energy than current global energy consumption.

- India has the 5th largest wind power installed capacity of 3595MW in the world[1]
- Estimated gross wind potential in India is 45,000 MW.[6]

Geothermal Energy:

Geothermal energy is a very clean source of power. It comes directly from the interior heat of the earth which converts water to steam directly and which moves out from earth surface.

This form of the energy is site specific and can only be applicable to a particular site on the earth.[1]

It can be used in two ways:

- Geothermal heating.
- The geothermal energy can be used in the heating purpose of houses buildings.

- Geothermal electricity.
- The geothermal energy can be used in the generation of electricity with the help of turbines.

Energy from Biomass

Biomass is the biological material derived from living, or recently living organisms, such as wood, waste and alcoholic fuels.[10]

Biomass is the most efficient and effective renewable energy source available in the world.[1]

Biomass is the oldest energy source used by humans. Biomass is the organic matter that composes the tissues of plants and animals. Until the industrial revolution prompted a shift to fossil fuels in the mid 18th century, it was the world's dominant fuel source.

Biomass can be burned for heating and cooking, and even generating electricity.

Previously biomass energy was generated from the burning of wood, burning animal manure (dung), peat (partially decomposed plant and animal tissues), or converted biomass such as charcoal. With the recent advance in science and technology biomass can also be converted into a liquid biofuel such as ethanol or methanol or CNG, etc.[10]

One example of biomass energy in developed nations is the burning of municipal solid waste. In Delhi,

several plants have been constructed to burn urban biomass waste and use the energy to generate CNG.

Green Chemistry

Green chemistry, also called sustainable chemistry, is a philosophy of chemical research and engineering that encourages the design of products and processes that minimize the use and generation of hazardous substances. Green chemistry seems to focus on industrial applications, and sustainable development. With a lots and lots of attention being paid over green house effect, environmental pollution the need for developing new technology has emerged and one of them is the biofuels.

GENERATION OF BIOFUELS

Generations of biofuels refers to the major technological and strategic change that fundamentally changed the way in which the biofuels are produced.

The term biofuel covers solid biomass, liquid fuels and various biogases.

The fuels that are easiest to burn cleanly are typically liquids and gases. Thus liquids (and gases that can be stored in liquid form) meet the requirements of being both portable and clean burning.

Biofuels provided 1.8% of the world's transport fuel in 2008. Investment into biofuels production capacity exceeded $4 billion worldwide in 2007 and till the end of 2011 biofuels will be providing 15% of the world transport fuel.

FIRST GENERATION BIOFUEL

First generation biofuels' are biofuels made from sugar, starch, vegetable oil, or animal fats using classical technology.[11] The basic feedstock for the production of first generation biofuels are often seeds or grains such as sunflower seeds, which are pressed to yield vegetable oil that can be used in biodiesel, or wheat, corn which yields starch that is fermented into bioethanol.[12]

These feedstock could instead enter the animal or human food chain, and as the global population has risen their use in producing biofuels has been critesed for diverting food away from the human food chain, leading to food shortage and price rises.[12]

Second Generation Biofuels

The second generation biofuels laid there focus or implementation on non- food crops.

Non food crops include waste biomass, the stalks of wheat, corn, wood, and biomass crops[12]. Some second generation (2G) biofuels use biomass to liquid technology i.e. converting biomass into liquid fuels. Many second generation biofuels are under development.

Cellulosic ethanol production uses non-food crops or inedible waste products and does not divert food away from the animal or human food chain. Lignocellulose is the "woody" structural material of plants[13]. This feedstock is abundant and diverse, and in some cases it is in a significant disposal problem.

Third Generation Of Biofuels

Algae fuel, are also called as third generation biofuel. Algae are low-input, high-yield feedstocks to produce biofuels. Based on laboratory experiments, it is claimed that algae can produce up to 30 times more energy per acre

than land crops such as soybeans, but these yields have yet to be produced commercially[10].

One of the greatest advantage of many biofuels over most other fuel types is that they are biodegradable, and so relatively harmless to the environment if spilled. The United States Department of Energy estimates that if algae fuel replaced all the petroleum fuel in the United States, it would require only 15,000 square miles[10].

Fourth Generation Of Biofuels

Fourth generation biofuels are the biofuels produced by pathways other than those being utilized for the production of first, second or third generation biofuels.

fourth generation technology pathways include: pyrolysis, gasification, upgrading, solar-to-fuel, and genetic manipulation of organisms to secrete hydrocarbons.

Hydrocarbon plants or petroleum plants are plants which produce terpenoids as secondary metabolites that can be converted to gasoline-like fuels. Latex producing members of the Euphorbiaceae plant family members have been studied for their potential energy uses.

Scientist's are also working and experimenting with the help of recombinant DNA technology tools on oil producing plants and microorganism's for the production of genetically modified organism with enhanced potential for the production of biofuels.

With the recent advances and breakthrough in biotechnology; the new energy revolution is sure to arrive, as many of the world leading scientists have claimed of creating potential energy source in the form of genetically modified plants and microorganism

V. REFFERENCES

1. Chapter 1: Development of Energy; Science Classified>> Energy alternatives. Website: www.Scienceclarified.com/Scitech/Energy-Alternatives/ The development of energy.html.

2. Non- conventional energy resources; K.S.Sidhu Director/research Punjab state electricity board, PEC Campus, Chandigarh.

3. Overview of power sector in India 2005- India Core.com.

4. Kadambini Sharma "Renewable Energy: The way to Sustinable Development; Electrical India", Jul.2002 42 (14) 20-21 .

5. B. Siddarth Baliga, "Renewable Energy Sources", Electrical India: Dec 2004 44 (12) 150-152.

6. E.C. Thomas, "Renewable Energy in India", Electrical India: Dec 2004 44 (12), 150-152.

7. Renewable Energy Technologies in Asia: A regional research and assimilation program by Diwaker Basnet.

8. Agrafiotis, C.; Roeb, M.; Konstandopoulos, A.G.; Nalbandian, L.; Zaspalis, V.T.; Sattler, C.; Stobbe, P.; Steele, A.M. "Solar water splitting for hydrogen production with monolithic reactors". Solar Energy 79 (4): 409–421. doi:10.1016/j.solener.2005.02.026

9. Anderson, Lorraine; Palkovic, Rick. Cooking with Sunshine (The Complete Guide to Solar Cuisine with 150 Easy Sun-Cooked Recipes). Marlowe & Company. ISBN 156924300X.(1994)

10. "Towards sustainable production of biofuels from microalgae" by Vishwanath Patil, Khanh-Quang Tran, and Hans Ragnar Giselrod.

11. "High yield bio-oil production from fast pyrolysis by metabolic controlling of Chlorella protothecoides" by Xiaolong Miao and Qingyu Wu.

12. Demirbas, A. (2009). "Political, economic and environmental impacts of biofuels: A review". Applied Energy 86: S108–S117. doi:10.1016/j.apenergy.2009.04.036. edit

13. Andrew Bounds "OECD warns against biofuels subsidies". Financial Times. (2007-09-10) http://www.ft.com/cms/s/0/e780d216-5fd5-11dc-b0fe-0000779fd2ac.html. Retrieved 2008-03-07.

14. "Earth and human evolution", by David Price. From Population and Environment: A Journal of Interdisciplinary Studies: March 1995.Human Science press,Inc16(4),301-19

Green Manufacturing

Sanjeev Gaur & Ravindra Kumar Malviya
Department of Mechanical Engineering, IVSIT, Mathura-281004

Abstract-The term green manufacturing is recently being practiced for meeting the demand of nature in making the environment clean and green. This term is used to describe those manufacturing practices which are environment-friendly during any part of the manufacturing process. These manufacturing methods support and sustain a renewable way of producing products and/or services that are entirely harmless to the environment. It emphasizes the use of processes that do not pollute the environment or harm the consumers or employees. Green manufacturing includes various manufacturing issues like recycling, conservation, waste management, water supply, environmental protection, pollution control and other related issues. Green manufacturing minimizes waste and pollution. These aims can be achieved through the right production and process design. Actually green manufacturing is more of a philosophy rather than an adopted process or standard.

Since today people have become so conscious about global warming and the ramifications of pollutive industries that the manufacturers have started seeking practical implementable solutions to sustain green manufacturing practices. The aim of green manufacturing is to support future generations by attaining sustainability by preserving natural resources. This paper will include a brief introduction to green manufacturing; its benefits from environmental and technological point of view, the way it may be implemented and the challenges that it might face.

I. INTRODUCTION

Green manufacturing is a method for manufacturing that minimizes waste and pollution. These goals are often achieved through product and process design. The center for Green Manufacturing at the *University of Alabama* defines the purpose of green manufacturing as:

"To prevent pollution and save energy through the discoveryand development of new knowledge that reduces and/or eliminates the use or generation of

hazardous substances in the design, manufacture, and application of chemical products orprocesses."

In *Bridge to a Sustainable Future (April 1995),* the Clinton White House defined an environmental technology as a technology that

- reduces human and ecological risks,
- enhances cost effectiveness,
- improves process efficiency, and
- creates products and processes that are
- environmentally beneficial or benign."

Areas of applications of green manufacturing:
Green Manufacturing addresses various manufacturing issues like recycling, conservation, waste management, Lean manufacturing attention paid to waste generated along the way, energy reduction in streamlined logistics– Materials reuse, Green plastics (biodegradable) Product design, use of recycled materials, design for service, disassembly etc.

II. RECYCLING

Green living is a concept which deals with the necessity of protecting the environment which is very important for all of us nowadays as it influences the way in which we all live and we should all influence it positively so as for us to be able to have a good quality life Recycling is also important to be mentioned when discussing about green living as it is very important and should also be of great interest to us. We should all be aware of the necessity for good recycling to be done by all of us so as for the environment not to be damaged anymore by throwing things away without taking in consideration the effect that these actions can have. Recycling is very important from this point of view and a fundamental aspect related to green living the concept and we should all become aware of it and we should all take it in account if we want to be able to improve the quality of life on earth so as to improve our health and the way in which we can all enjoy living surrounded by a clean environment. When discussing about recycling as an important aspect in the green living concept it should also be mentioned the fact that there have been created many latest recycling technologies which enable us to use recycled types of fiber. Two examples are represented by the no waste technology and the carpet fiber type of recycling technology. Recycling can be easily understood by a diagram which shows the recycling process clearly and properly.

The first has been created in the idea of offering global solutions for a proper waste disposal management. It relates to hygienic type of products which should be held within a Know waste type of plant which should be able to sterilize these products and in the end to separate each of the components which might be reused in order to be produced recycled fiber in various forms. The latter deals with a technology related to the use of recycled types of fibers such as the nylon. Both of these latest technologies in recycling, an important aspects related to the green living concept are important to be analyzed as certain interesting and useful ideas can be considered from them. Recycling is vital in a world in which green living should be a concept followed by all of us if we want to improve the quality of life on earth and in this way the quality of our own lives.

III. ENERGY CONSERVATION

Energy conservation refers to efforts made to reduce energy consumption. Energy conservation can be achieved through increased efficient energy use, in conjunction with decreased energy consumption and/or reduced consumption from conventional energy sources.

Energy conservation can result in increased financial capital, environmental quality, national security, personal security, and human comfort. Individuals and organizations that are direct consumers of energy choose to conserve energy to reduce energy users can increase energy use efficiency to maximize profit.costs and promote economic security. Industrial and commercial

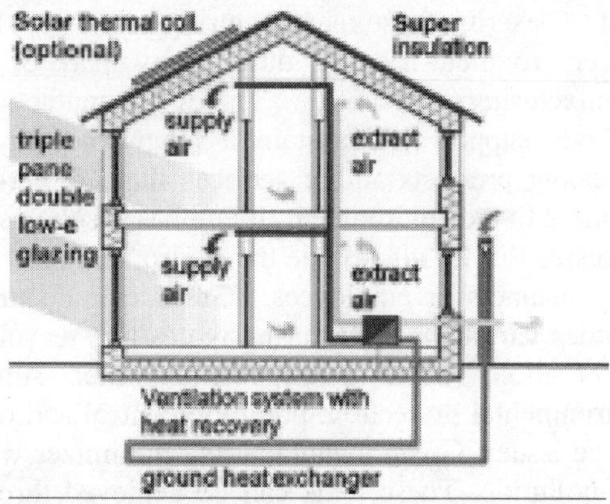

Low-energy building techniques and technologies in an Energy conserving building.

Electrical energy conservation is an important element of energy policy. Energy conservation reduces the energy consumption and energy demand per capita and thus offsets some of the growth in energy supply needed to keep up with population growth. This reduces the rise in energy costs, and can reduce the need for new power plants, and energy imports. The reduced energy demand can provide more flexibility in choosing the most preferred methods of energy production.

IV. WASTE MANAGEMENT

Waste management is the collection, transport, processing, recycling or disposal, and monitoring of waste materials. The term usually relates to materials produced by human activity, and is generally undertaken to reduce their effect on health, the environment or aesthetics. Waste management is also carried out to recover resources from it. Waste management can involve solid, liquid, gaseous or radioactive substances, with different methods and fields of expertise for each.

Waste management practices differ for developed and developing nations, for urban and rural areas, and for residential and industrial producers. Management for non-hazardous waste residential and institutional waste in metropolitan areas is usually the responsibility of local government authorities, while management for non-hazardous commercial and industrial waste is usually the responsibility of the generator.

V. REFERENCE

[1] Green Consumption and green manufacturing by David DorrnfieldThe Vivid Edge;Green Living Recycling
[2] Life Cycle Environmental Assessment of Municipal Solid Waste to Energy Technologies
[3] Life Cycle Assessment (LCA) of Municipal Solid Waste   Management in the State of Kuwait
[4] Lumina Technologies, Analysis of energy consumption in a San Francisco Bay Area research office complex, for (confidential) owner, Santa Rosa, Ca. May 17, 1996

Green Manufacturing Emerging Vistas of Mechanical Engineering in 21st Century

Ami Chand Sharma, Prashant Saraswat

Department Of Mechanical Engineering IVS Engineering College of Technology & Management, Mathura

Abstract -Green manufacturing is a method for manufacturing that minimizes waste and pollution. These goals are often achieved through product and process design. Green manufacturing is a method that support and sustain a renewable way of producing products and/or services that do no harm to you or the environment.

Basic Terms

Here are some terms to help you better understand the idea of green manufacturing:
Product design – includes the definition of the product architecture and the design, production, and testing of a system for production.
Three life cycle approaches to product design:
Design for reuse – refers to designing products so that they can be used in later generations of products.
Design for disassembly – a method for developing products so that they can be easily taken apart.
Design for remanufacture – a method for developing products so that the parts can be used in other products.

Object
The purpose is to support future generations by attaining sustainability by the means of preserving natural resources.
Industries in the US spend over $100 billion/year waste treatment, control, and disposal. 1996 Dupont spent $ 1 billion for environmental compliance (research budget $ 1 billion; chemical sales of $18 billion)

Strategy

"Environmentally friendly manufacturing will become one of industry's greatest strategic challenges, not only from an engineering perspective, but from a business and marketing perspective as well." A stable economy that uses energy, the environment and human health. and resources efficiently. Social and political systems that lead to a just society

Functions

- **Manufacturing**
- **Environmental Protection**
- **Optimal Use of Resources.**

Concepts such as Lean and Green, Corporate Social Responsibility, Waste to Profit, Design for Environment, Zero Waste and Environmental Sustainability through mind-opening sessions and intensive workshops. The program's goals are to familiarize delegates with new approaches to reducing costs and increasing profits, while at the same time conserving energy and resources. Delegates can network with key industry contacts and gain the tools and ideas that will benefit the environment and their companies' bottom lines. The basic strategy is
• Creating a Sustainable Facility through Energy Efficiency Design
• Energy Efficiency through Lean—Government and Industry Lessons in Collaboration
• Best Practices in Being Green, Productive, and Profitable

• Increasing Sustainability of Lean Six Sigma—Environmental Wastes Audit
• Design for Sustainability—Integrating Life-Cycle Thinking

1) Rethink product and process technology
2) Explore the market potential
3) Supply goods and services
4) Extend producer responsibility
5) First page example
6) Reduce energy consumption
7) Integrate promotion
8) Incorporate goals
9) Promote development
10) Integrate environmental costs
11) Do not Use natural raw ingredients, including organics hazardous chemical conditioners, chemical anti-bacterial or chemical preservatives, use botanically-derived, colors, spices etc. Cold manufacturing process for soap Metalworking fluids (emulsions of oil, water and stabilizing agents)

Examples

❖ The need for fossil fuels has lead to discoveries of different methods of manufacturing that replace renewable resources. Examples
- Compact fluorescent light bulbs,
- Software distribution via the Internet,
- Hybrid electric vehicles
- More efficient refrigeration
❖ European directorate for the environment said it would eliminate brominates flame   retardants used in electronics Siemens could not find a replacement
Sony Corp. produced a viable alternative
❖ New clean air emissions limits for automobiles
❖ Honda and Toyota produced new engines to meet and exceed requirements

Manufacturing is when you take raw materials and combine the raw materials together to form a finished product. Most people when they think of manufacturing they think about having to order all of the numerous parts that are going to be needed to put together the finished product that is going to be for sale. For example, computers require a variety of little parts to be put together to sell the finished product. With manufacturing, you are probably aware of the various forms of manufacturing that can be used, but you might not be familiar with green manufacturing.

When people think of manufacturing they do not think of green manufacturing because the thought of manufacturing is large companies, putting numerous emissions into the air to manufacture their product. For example, chemical processing plants and gas refineries are types of manufacturing plants, which nobody thinks about them using green methods of manufacturing. The problem is that more and more people are becoming concerned with the environment, so more companies have to change their manufacturing methods to help protect the environment.

Disposable products

Detergents, newspapers and other disposable items can be designed to decompose, in the presence of air, water and common soil organisms. The current challenge in this area is to design such items in attractive colors, at costs as low as competing items. Since most such items end up in landfills, protected from air and water, the utility of such disposable products is debated.

Eco fashion and home accessories

Creative designers and artists are perhaps the most inventive when it comes to up cycling or creating new products from old waste. A growing number of designers up cycle waste materials such as car window glass and recycled ceramics, textile off cuts from upholstery companies, and even decommissioned fire hose to make belts and bags. Whilst accessories may seem trivial when pitted against green scientific breakthroughs; the ability of fashion and retail to influence and inspire consumer behavior should not be underestimated. Eco design may also use bi-products of industry, reducing the amount of waste being dumped in landfill, or may harness new sustainable materials or production techniques e.g. fabric made from recycled PET plastic bottles or bamboo textiles.

Energy Sector

Solar energy and Alternative energy

Sustainable technology in the energy sector is based on utilizing renewable sources of energy such as solar, wind, hydro, bio energy, geothermal, and hydrogen. Wind energy is the world's fastest growing energy source; it has been in use for centuries in Europe and more recently in the United States and other nations. Wind energy is captured through the use of wind turbines that generate and transfer electricity for utilities, homeowners and remote villages. Solar power can be harnessed through photovoltaic, concentrating solar, or solar hot water and is also a rapidly growing energy source.

The availability, potential, and feasibility of primary renewable energy resources must be analyzed early in the planning process as part of a comprehensive energy plan. The plan must justify energy demand and supply and assess the actual costs and benefits to the local, regional, and global environments. Responsible energy use is fundamental to sustainable development and a sustainable future. Energy management must balance justifiable energy demand with appropriate energy supply. The process couples energy awareness, energy conservation, and energy efficiency with the use of primary renewable energy resources.

Water Sector

Main articles: Reclaimed water, Rainwater harvesting, and Storm water harvesting

Sustainable water technologies have become an important industry segment with several companies now providing important and scalable solutions to supply water in a sustainable manner.

Beyond the use of certain technologies, Sustainable Design in Water Management also consists very importantly in correct implementation of concepts. Among one of these principal concepts is the fact normally in developed countries 100% of water destined for consumption that is not necessarily for drinking purposes, is of potable water quality. This concept of differentiating qualities of water for different purposes has been called "fit-for-purpose". This more rational use of water achieves several economies, that are not only related to water itself, but also the consumption of energy, as to achieve water of drinking quality can be extremely energy intensive for several reasons.

Sustainable technologies

Sustainable technologies use less energy, fewer limited resources, do not deplete natural resources, do not directly or indirectly pollute the environment, and can be reused or recycled at the end of their useful life. There is a significant overlap with appropriate technology, which emphasizes the suitability of technology to the context, in particular considering the needs of people in developing countries. However, the most appropriate technology may not be the most sustainable one; and a sustainable technology may have high cost or maintenance requirements that make it unsuitable as an "appropriate technology," as that term is commonly used.

Encouraging sustainability

- Training meeting with factory workers in a stainless steel eco-design company from Rio de Janeiro - Brazil.
- The use of sustainable technologies may be encouraged through means such as reducing the capacity of the electrical cable supplying a home, such as Australia's Crystal Waters Village. In some cases the electricity supplier charges a higher rate for the energy used when the capacity

METHODS FOR WASTE REDUCTION

Waste reduction seems to be a very simple concept, but there are many different ways to consider it and to start implementing processes that will ensure it works. Waste can be reduced in almost everything we do in our personal lives, but are also a very important consideration for manufacturing companies and everything they do. Here is some more information about waste reduction in manufacturing.

The basic belief behind waste reduction for most business is, sadly, less than benevolent and is mostly a consideration for the bottom line of the company. While recent pressure to become green may have increased the desire of companies to waste less, they also want to decrease waste to ensure greater profit margins. If something is sold at the same price but less material, labor and effort is needed to produce it, then this is seen as a gain for the company. While waste minimization for a company often requires an investment of capital and time, it is almost always paid for with increased efficiency and more goodwill towards the company as well.

There are many different processes that can be used to reduce the amount of waste used in any organization or process. One of the most obvious ones is called resource optimization. This involves using raw materials more efficiently to decrease costs and the impact on the environment. One interesting example is the use of fabric and the way that a garment is designed using computers to make the most out of every piece of fabric that goes onto the production line. This is done by having a computer program arrange the various pieces of a pattern in such a way that it gets the maximum amount of use from a small piece of fabric as possible.

Another aspect of resource optimization is the reuse of scrap material. This is only useful in some cases, but the most common use is that of scrap metal. The properties of most metals allow it to be melted and used again for the same purpose. This process can be challenging to implement and can also be costly, but should always be weighed against the alternative of purchasing more material. The decision to reduce the amount of material is almost always a conscious decision and there are numbers behind the idea to back up the actions.

Improving quality control measures can also ensure that less material is wasted and ultimately the end consumer will be more satisfied as well. Process monitoring can quickly alert someone if there is a fault or defect that will compromise the quality of the product or cause it to fail prematurely. Many high tech processes have these kinds of measures in place and they are especially important in the manufacture of consumable goods as well.

Other parts of the process, like product design, can also be improved resulting in lower costs. This may require a significant amount of time to be invested in the process and management of the system before it is actually implemented, but it will help give returns on the investment as the process is carried out. The design of the product and the process that will be used to produce it is very important in making the most of the manufacturing process.

Improving the durability and quality of products can also reduce the amount of waste. Many consumers believe that this method is rarely if ever used because most companies are only interested in the short term profits that will come from producing materials to sell to the public. But, making a product more durable and making it last longer will use less material over time and require that people purchase less of the product or a similar product in the long run. It is rare to find companies with this view, but they are out there and many more are joining the cause to create less waste.

Indian Role-

India aims to be Green manufacturing hub

This is the brand that will give India Inc a pole position in the global manufacturing sweepstakes. The government plans to market Indian manufacturing as Green to make it stand apart from Chinese or even say US products, which are typically energy-intensive. There is sound economics behind the tag.

Top government officials say Indian manufacturers in most sectors are far more energy efficient than their counterparts abroad. This means in addition to labour intensive character, products manufactured here are more environmentally compatible. For instance, in cement industry, against the global lowest energy intensity of 65 kilowatt per tonne, the domestic manufacturers are at a comfortably close 69. The same picture is evolving in other manufacturing sectors like steel.

According to Ajay Shankar, secretary, department of industrial policy and promotion, India is ideally suited to develop Green as the USP of its manufacturing profile. "Our best is already there, the strategy is to make the followers catch up with them", he said.

Measurements of overall energy intensity of GDP in India also point to the same trend. According to the government's own estimates, energy intensity here is the same as in OECD countries, when GDP is calculated in terms of the purchasing power parity (PPP). Industry chambers, like FICCI and CII also aver that the move has started vigorously.

"Our manufacturing is at par with the best in the world and far better than the developing world," said Arun Kumar, associate director, KPMG.

Why Green Manufacturing

No sector of the economy comes close to the manufacturing sector in generating vast volumes of waste. And, with the world's eyes more focused on corporate environmental responsibility than ever, scrutiny of manufacturing practices has reached an all-time high. North American businesses are in a particularly vulnerable position with new, stricter regulations coming into force, pressure from all-sides to act now, tight budgets, sharp rises in energy prices, and an all-around lack of relevant and reliable information on how to green the manufacturing process.

The Green Manufacturing Summit is the first event to address all of these crucial current concerns. At no other event will you get a total picture of what the current climate is, how it will change over the next few years as new environmental legislation is enacted and supply chain pressure increases, and how to position yourself as a green manufacturer, while finding the ROI in environmental practices.

As the process that contributes most to the carbon footprint of the majority of companies, manufacturing holds the key to greening the supply chain in a strategic and profitable way, and the goal of the Green Manufacturing Summit is to show manufacturers how to go green using real case studies by leading companies who are working towards these goals and know what the key competencies and potential pitfalls are. The best way to learn is by doing – but the best way to ensure success is by seeing what others have done and learning from their successes and mistakes. This gathering of industry experts is where you will determine what's right for your business, and find out once and for all where the ROI in green manufacturing is for you. The event is end-user led, so the focus is on the concerns of manufacturers and retailers, and what the industry leaders can teach you. Key topics include:

- How to **achieve energy efficiency in factories and warehouses**
- What the **new demands of customers** will be, and where they'll speak with their wallets
- How to **predict and plan for upcoming environmental legislation**
- How **a green agenda can help you cope with increased global competition and growing supply chain pressure**
- How to **go green while meeting the bottom line, and where to find your green manufacturing ROI**

- The truth behind manufacturing **carbon footprint measurement and management – what to measure, how to measure it, and how to report it**
- **PLUS** Recycling and waste management, cost effective implementation of green manufacturing technologies, Design for Environment, water sustainability, focus on facilities, and sustainable packaging

Over the course of the conference you will meet expert speakers and your fellow attendees, discuss the specific challenges you face in small, focused groups, hear the key case studies that will change the way you think of 'greening' from a fad to a business necessity, and come away with a concrete plan for how to implement green practices into your operation in a strategic and profitable way.

You'll also have a chance to meet with experts on the latest technologies and software for managing and reducing your environmental footprint in our focused green supply chain exhibition zone. This is your one-stop shop for the latest information and advice on going green the smart way, and your chance to finally get clear on what's going on in the industry, how you can be a part of it, and how it can make your operation more efficient and profitable. Customer-facing businesses face a unique challenge when it comes to green manufacturing – and the Green Manufacturing Summit is the place where that challenge gets translated into opportunity. Don't miss out – places will sell out – book your pass today and join us at this crucial industry gathering.

Future of Green Manufacturing

The demand for sustainability is growing stronger every day, and manufacturers are looking for ways to comply in order to reduce their impact on the environment and increase profitability.

Many are discovering that lean practices—which seek to eliminate all forms of waste—offer effective tools for sustainable manufacturing. To help practitioners better understand how green efforts can create growth and improve competitiveness, the Society of Manufacturing Engineers (SME) is presenting a Lean to Green Manufacturing Concept. According to Society of Manufacturing Engineers (SME),"The leading similarity between the benefits of lean and the benefits of green is waste."

This concept offers proven business practices to dramatically reduce the overall impact of manufacturing on the environment."

The features of SME presentations, workshops, and case studies of successful methods attendees can immediately implement to reduce environmental concerns and enact changes in their manufacturing operations.

The Lean to Green Manufacturing Conference also includes a panel discussion and case histories on the Environmental Protection Agency's E3 initiatives. The **E3: Economy Energy and Environment** program, designed as a model for collaboration among manufacturers, utilities, municipal governments, and federal agencies, is already saving manufacturers millions of dollars annually.

Attendees can also discover how colleges and universities are providing the next generation of manufacturing professionals with both classroom and real-life learning experiences in manufacturing sustainability, energy efficiencies, and the elimination and recycling of waste.

Additionally, attendees can choose to go on the "Green Crawl Tour" and visit area facilities that have successfully implemented lean principles and green strategies.

Challenges

Range of coordinated actions
Trade and environment policies (mutually supportive)

Eliminating environmental harmful subsidies
Promoting the transfer of technologies and financial resources
Efficient operation of markets
Achieving greater international cooperation

Hurdles

Businesses have a responsibility of influence
Prices of raw material and subsidized energy are essential
Lack of availability and information High-profile leadership
Training programs need more support
Accessibility to loans from the government
National cleaner production centers need to be established

Summary

Cleaner production is a preventive strategy that aims at promoting the use and the development of cleaner, processes, products, and services.

A key to more sustainable development is long-term structural changes in the way our economies work.

The measure of success of this conference of 'green' manufacturing is the extent to which the participants successfully reduce their products' and companies' environmental footprints upon their return to their workplaces

Harmonisation of SAARC Standard

Syamal Datta
Anand Engineering College, Agra

<u>Background and necessity</u>

SAARC(South Asian Association for Regional Cooperation) was established during 1983 with a proclaimed charter for increased cooperation among south Asian nations. Among its objectives, notable was to promote welfare of people of south Asia and improvement of their quality of life , to accelerate economic growth and to promote active collaboration and mutual assistance in scientific and technical field. To fulfill these objectives, apart from social and cultural aspect, growth friendly environment for trade and commerce is the need of the day.

We are all aware of the famous saying that standardisation pays but in quest of prosperity in globalised atmosphere of trade and commerce ,only adherence to standardisation is not adequate assurance of quality for the purpose of export or acceptance by a third country. To sustain more than 20% growth in export would become increasingly difficult.

As the developed economies have stopped growing or growing at a snail's pace, the emerging markets of developing nations offer challenging opportunity not only for traditional items of goods and services but for high & medium technology product also. Among all economic zones , SAARC is the least organised and developed to function as free trade regeon.

SAARC is the most trade hungry regeon housing one sixth of the humanity immersed in abject poverty and their economic condition can be drastically changed by increased free trade. Though use of standard remains a voluntary activity but necessary legislation and political will on the part of state, industry, trade & as a whole society can boost not only trade but competitiveness, innovation and acceptance of innovation by the market.

<u>Why harmonisation</u>

For seamless movement of goods and services, essential requirement is to remove or minimize all tariff or non tariff barriers and harmonisation of standard which means compatibility of standards of different nationality or organisation and transposing these national standards to a unified, harmonisied SAARC standard achieved through negotiation and consultation by authorised standard organisations of respective countries.

Presumption of conformity is the essence of harmonised standards within the region of cooperation and it can be conferred in legal and regulatory terms only to those national standards which has been transposed to a harmonised standard.

In short, aim of harmonized standard are the followings

- One standard
- One certification
- One testing

In the Preamble of SARSO (South Asia Regional Standard Organisation) ,established at Dhaka ,the aim has been stated as to develop harmonised standard for the region to fecilitate intra regional trade and to have access to global market. In addition to breaking technical barrier for free flow of goods and services in the region, developing new standards for the product of regional/sub regional interest and conformity assessment with the international standards would be primary function of SARSO.

<u>Stake holders</u>

Though standardization and harmonization is a voluntary process of developing specification in consultation with all interested parties i.e. Industry, consumers, trade union, environmental NGOs , public authorities, educational institutes, R&D laboratories etc, and carried out by independent statuary bodies, need for regulation and legislation are increasingly being felt necessary. The other important activity area of SARSO would be training of technical man power and establishment of accredited laboratories.

<u>Bottlenecks</u>

In comparison to other economic cooperation regions like EU , NAFTA or ASEAN. , SARSO would encounter more difficulty during its infancy because of the followings

(a) Lack of political will on the part some member countries

(b) No fully developed standardization organization

(c) Fear of domination by more technically and economically developed neighbor.

(d) Protection to domestic industry

(e) Lack of trained manpower

(f) Lack of sufficient accredited testing laboratories

India's dominance in terms of sheer size of economy is also source of discomfort for other major economies like Pakistan and Bangladesh. That is why no FTA(Free Trade Agreement) has yet been signed by Pakistan and Bangladesh with India though similar agreements have been signed by Srilanka, Nepal, Bhutan and Maldive. The proposed time frame of 0% import duty in stages among SAARC nations and realization of SAFTA was by the year 2020,looks a distance dream.

Another hurdle in harmonisation is the time taken in formulation of standard, BIS has taken steps to reduce the time from average 44 months to 24/12 months but the same are unlikely to be achieved by other national standardisation bodies due to inherent inadequecies.

Progress made

The example of harmonisation of standard for food items is relevent here.In spite of initial hiccups it has made significant progress in terms of distribution of responsibilities taking in view of history, geography, user base and expertise available. General distribution was as follows

- Grains such as wheat, rice, pulses & sugar----- India
- Meat and meat products------------------------------- Pakistan
- Fish and fish products------------------------------- Bangladesh
- Fats and oils------------------------------------ Nepal
- Spices and herbs-------------------------------- Srilanka

All these countries have existing standards on these general group of items and their expertise in finding a common platform in the region and then finding assurance of conformity with international standard was appropriate division of responsibility.Considerable deviation from international standards was not neither envisaged nor encountered as all national standards were formulated in conformity with CODEX for greater acceptability in export market.

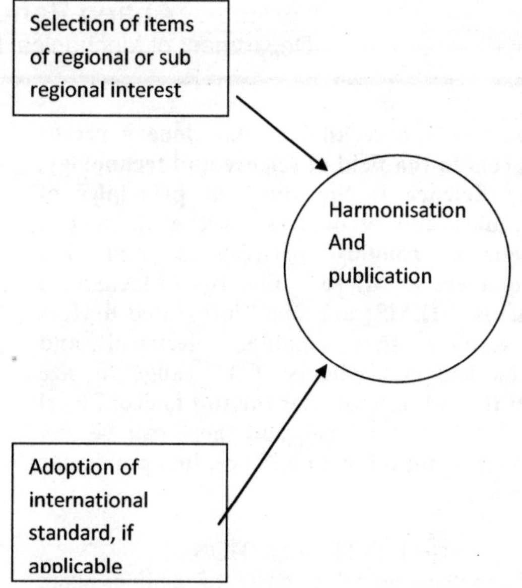

Direct benifit

Most notable example of benefit of harmonization was seen during cold war period, immediately after the second world war when western world was divided into NATO and Warshaw pact countries, the need arose to develop harmonized NATO and GOST standards as same war equipments were manufactured in different countries using components available from different countries to be used anywhere in the region. Greater success of harmonization in soviet block resulted in mass production of high tech equipment at a much cheaper cost. The crescendo of harmony was reached when soviet union used an aircraft engine to power their most famous tank T-34 which had the distinction of being largest number tank produced in the world.

What next

Technical committee formed to prepare draft standards are dominated by large industries. Inadequate representation from medium and small scale industries tilts heavily in favour of large industries. R&D and Quality Assurance persons must have more representation and say in these matters.

Nanotechnology
A Recent Applications in MEMS

Anand Pandey, Nikhil Sharma and Shailendra Singh

Department of Mechanical Engineering, Anand Engineering College-Agra, U.P.

Abstract — **Nanotechnology has done a recent progress in the field of science and technology. Nano Science is the study of principles of molecules and structures with at least one dimension roughly between 1 and 100 nanometers. Micro Electro Mechanical Systems (MEMS) are small integrated devices or systems that combine electrical and mechanical components. They range in size from the sub micrometer (or sub micron) level to the millimeter level, and there can be any number, from a few to millions, in a particular system.**

I. INTRODUCTION

Nanotubes are tiny tubes of carbon about 10.000 times thinner than a human hair. These consist of rolled up sheet of monolayer or multilayer carbon atoms bonded together in a hexagon. There are two main types of nanotubes: single walled nanotubes and multilayer walled nanotubes.

Nanotubes are composed entirely of sp2 bonds, similar to those of graphite .This bonding structure, stronger than the sp3 bonds found in diamond, provides the molecules with their unique strength .Nanotubes naturally align themselves into ropes held together by vander waals forces. Under high pressure,nanotubes can merge together ,trading some sp2 bonds for sp3 bonds, giving possibility for producing strong, unlimited length wires through high pressure nanotubes linking.

Carbon nanotubes are cylindrical carbon molecules with novel properties that make them potentially useful in a wide variety of applications. They exhibit strength and unique electrical properties, and are efficient conductors of heat.

Nanotubes can be made into metals or semiconductors depending on how these are rolled up.Roll the sheet of carbon atoms the way a cigarette is rolled, with edges of carbon atoms touching along their length and a nanotubes is formed that acts like a tiny metal wire conducting electricity.

If the tube is wound like a paper straw, a miniature semiconductors is formed that could replace silicon transistors-the building blocks of chips. The unique properties of these nanotubes, including their strength, electrical properties and conducting capabilities, make them useful in electronics and mechanical applications.

II. FUNCTIONING OF NANOTUBES

Conductive carbon nanotubes have been used for several years in brushes for commercial electronics motors. They replace traditional carbon black ,which is mostly impure spherical carbon fullerence.The nanotubes improve electrical and thermal conductivity because they stretch through the plastic matrix of the brush. This permits the carbon filler to be reduced from 30% down to 3.6%,so that more matrixes are present in the brush.Nnaotubes composite motor brushes are better –lubricated(from the matrix),cooler-running(both from better lubricationand superior thermal conductivity),less britlle ,stronger and more accurately moldable.

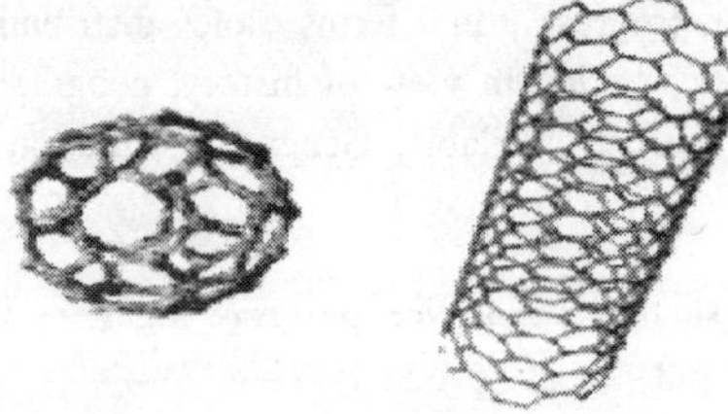

Fig.1.Structure of carbon Nanotubes

One use for nanotubes that has already been developed is as extremely fine electron guns, which could be used as miniature cathode ray tubes in thin high brightness low energy low weight displays .Instead of a single electron gun, as in a traditional cathode ray tube display ,here there is a separate electron gun for each pixel in the display. This type of display would consist of a group of many tiny CRT's each providing the electrons to hits the phosphor of one pixel, instead of having one giant CRT whose electrons are aimed using electric using and magnetic fields. These displays are known as field emission displays (FEDs).

Nanotubes have been shown to be superconducting at low temperature .Nanotubes conduct electricity better than copper, which makes them a contender for replacing the delicate wires that connect components together inside computer chips. Not only that, they can carry heat far more efficiently than diamond-one of the best heat conductors around. if the processor chips are given a nanotubes coatings, billions of semiconductors devices could be packed into a tiny space with little risk of them burning up.

III. FABRICATION TECHNOLOGIES OF MEMS

The three characteristic features of MEMS fabrication technologies are miniaturization, multiplicity, and microelectronics. Miniaturization enables the production of compact, quick-response devices. Multiplicity refers to the batch fabrication inherent in semiconductor processing, which allows thousands or millions of components to be easily and concurrently fabricated. Microelectronics provides the intelligence to MEMS and allows the monolithic merger of sensors, actuators, and logic to build closed-loop feedback components and systems. The successful miniaturization and multiplicity of traditional electronics systems would not have been possible without IC fabrication technology. Therefore, IC fabrication technology, or micro fabrication, has so far been the primary enabling technology for the development of MEMS. Micro fabrication provides a powerful tool for batch processing and miniaturization of mechanical systems into a dimensional domain not accessible by conventional (machining) techniques. Furthermore, micro fabrication provides an opportunity for integration of mechanical systems with electronics to develop high-performance closed-loop-controlled MEMS.

The plastic body of Renault car contains carbon nanotubes.This makes the plastic panels so conductive that they can be earthned while the car is sprayed with paint droplets charged up to 20,000 volts. The droplets seek ground instead of floating away, making spray painting more efficient and less polluting.

Nanotubes are incorporated into carry cases and trays used to transport chips and hard drives. Carry boxes made from nanotubes –spiked plastics carry away any charge before it build up.Due to its good thermal properties, CNTs can also be used to dissipate heat from tiny computer chips.

Some of the most exciting application comes from piezoelectric plastic materials that produce a viltage when pressed. Researchers in USA have discovered that polyvinylidene fluoride (PVDF), a polymer widely used in ultrasound sensors, become three times as sensitive to pressure when nanotubes are sprinkled in. Addition of just one nanotube for every 8000 strands of PVDF is enough to produce such super sensitivity. These improvements come about because nanotubes keep the polymer in a stable piezoelectric state.

IV.APPLICATIONS OF MEMS IN NANOTECHNOLOGY

The present and near-future applications for NPs administered to human body can be grouped Under the following categories:

- MEMS pressure micro sensors typically have a flexible diaphragm that deforms in the presence of a pressure difference.
- The application of MEMS technology to accelerometers is a relatively new development.
- Imaging Agents
- Artificial Blood (respirocytes)
- Detoxifying Agents (nanobullets)
- Electromagnetic sensors
- Biosensors
- Protein engineering
- Nano bricks and building blocks
- Very tiny motors, pumps, gyroscopes, and accelerometers
- Supercomputer in your palm
- Tiny bio- and chemical-sensors

V. FUTURE OF MEMS

Three basic Microsystems technology processes we have seen, bulk micromachining, sacrificial surface micromachining, and micromolding/LIGA, employs a different set of capital and intellectual resources. MEMS manufacturing firms must choose which specific Microsystems manufacturing techniques to invest in [14]. Future MEMS applications will be driven by processes enabling greater functionality through higher levels of electronic-mechanical integration and greater numbers of mechanical components working alone or together to enable a complex action. Future MEMS products will demand higher levels of electrical-mechanical integration and more intimate interaction with the physical world. The high up-front investment costs for large-volume commercialization of MEMS will likely limit the initial involvement to larger companies in the IC industry.

VI. CONCLUSION

MEM S have already invaded our daily life and have used for large number applications. They have been of great interest, both from a fundamental point of view and for future applications. The most eye-catching features of these structures are their electronics, mechanical, optical, thermal and chemical characteristics, which open a way to future applications. They will have important role in nanotechnology engineering.

II. REFERENCES

[1] Micromachine Devices, European Study Sees MEMS Market at   More Than $34 billion by 02, May 1997, pp.213-225.

[2] Micromachine Devices,2,1996

[3] M. Mehregany and S. Roy, Introduction to MEMS, 2000, Microengineering Aerospace Systems, El Segundo, CA, Aerospace Press, AIAA, Inc., 1999.

Self Healing Materials

Rahul Jain, Sachin Singh

Department of Mechanical Engineering, FET Agra College, Agra

Abstract - Self-healing materials, where does it come from? Indeed, this is what everyone saw at least several times. Wounds or skin cuts heal after some time. So, it has been natural to try to create materials possessing such a wonderful property. One can list thousands of possible applications for such materials in variety of different fields. Increased reliability and lifetime can be critical in medicine, space missions, traffic, military, construction, and so on. In principle, it is possible to "heal" (recover) different properties. For now self-healing means mostly recovery of mechanical properties. The others properties will definitely be addressed in the future. As the idea of self-healing materials is far from being novel, recent promising results are the result of fruitful fusion of physics, chemistry, and biology, which is called nanoscience or nanotechnology. Composite materials capable of decreasing the amount of their mechanical degradation have attracted a lot of attention and interest these days. Recent news from Nissan about successful commercial release of "self-healing" car painting has heated public interest in such type of materials. Common names for these materials are self–healing, self–repairing, autonomic–healing, autonomic-repairing materials. Because all these names mean the same thing in nature, we will use just one, self-healing. As usual, these names are used for quite a broad variety of materials with very different healing/repair mechanisms. Here we briefly overview this variety of materials and the mechanisms of healing, possible applications, general technical challenges.
The ability of materials to self-heal from mechanical and thermally induced damage is explored in this paper and has significance in the field of fracture and fatigue. The history and evolution of several self-repair systems is examined including nano-beam healing elements, passive self-healing, autonomic self-healing and ballistic self-repair. Self-healing mechanisms utilized in the design of these unusual materials draw much information from the related field of polymer–polymer interfaces and crack healing.
The relationship of material damage to material healing is examined in a manner to provide an understanding of the kinetics and damage reversal processes necessary to impart self-healing characteristics. In self-healing systems, there are transitions from hard-to-soft matter in ballistic impact and solvent bonding and conversely, soft-to-hard matter transitions in high rate yielding materials and shear-thickening fluids. These transitions are examined in terms of a new theory of the glass transition and yielding, viz., the twinkling fractal theory of the hard-to-soft matter transition. Success in the design of self-healing materials has important consequences for material safety, product performance and enhanced fatigue lifetime.

Keywords: Self healing, Micro Capsules.

I. INTRODUCTION

Self-healing materials are polymers, metals, ceramics and their composites that when damaged through thermal, mechanical, ballistic or other means have the ability to heal and restore the material to its original set of properties. Only few materials intrinsically possess this ability. This is a very valuable characteristic to design into a material since it effectively expands the lifetime use of the product and has desirable economic and human safety attributes. In self-healing systems, there are transitions from hard-to-soft matter in ballistic impact and solvent bonding and conversely, soft-to-hard matter transitions in high rate yielding materials and shear-thickening fluids used in liquid armor. The biological analogy of self-healing materials would be the modification of living tissue and organisms to promote immortality, and many would agree that partial success in the form of expanded lifetime would be acceptable. Self-healing materials are materials designed to recover strength from low-level damage done to the material over the course of its service lifetime. The self-healing technique is particularly useful when applied to composite materials, since composites have low damage detectability and is susceptible to sudden and brittle failure. This study is aimed at two self-healing methods that had been implemented into composite materials with self-healing capabilities, that is: (1) using embedded hollow glass fibres (HGF) storing epoxy resin and hardener, and (2) using microencapsulated epoxy resin with 2-methylimidazole/$CuBr_2$ as the hardener.

Types of self-healing materials and the healing mechanisms

Although all types of these materials have their own self-healing mechanism, we start from describing some common features. Virtually all materials with long degradation time deteriorate through development of microcracks (fatigue). A sharp apex of each crack works as a knife cutting the materials with ease. This results in larger cracks, and consequently, mechanical degradation. Example of such material would be plastics used for construction, artificial bones, dental cement, etc. To heal such materials, one needs to seal those microcracks before their further growing. The other type of degradation and the healing mechanism is important for materials that can degrade sufficiently fast. Example of such materials can be various coatings, armor, all surfaces that can suffer sudden impact or collision with a projectile. In such a case, not only cracks, but even holes should be sealed and healed. Definitely there are materials of dual purposes, which would degrade through both of the above mechanisms. To classify self-healing materials, one can consider four different classes: plastics/polymers, paints/coatings, metals, and ceramics/concrete. We will discuss each of these classes below.

Plastics/Polymers

Polymers/plastics are attractive from mechanical and chemical points of view. Many plastic materials are strong and resistant to breaking. However, once fractured, the material

deteriorates irreversibly. Even under normal wearing, plastics used to develop small cracks that also grow irreversibly. This leads to degradation of their mechanical properties and decreasing life time of such materials. This is where self-healing is needed the most.

The working principle of self-healing mechanism is based on having small capsules filled with healing glue. The capsules can be made of another polymer or can be made of glass. These capsules are mixed within the polymer body. When microcracks are developed in the polymer body, these also rapture the capsules. The glue leaks in the cracks and heals them before cracks can get any bigger. Specifically, a healing agent was stored in polymeric capsules (hundreds of microns in diameter) and mixed with the polymeric body. The glue activator (needed to rigidify the glue inside the cracks) was also added to the polymer body. When a crack propagates it ruptures the capsules, glue leaks out into the crack, seals, and rigidifies. This repairs the crack, and to some extent recovers mechanical integrity of the polymer.

Another approach is based on using hollow fibers instead of microcapsules. Hollow fibers based on glass tubes are filled with either resin or hardener, which are released into the damaged area when the fibers are fractured. When the resin and the hardener are mixed in the crack plane, the resin hardens, repairing the crack. In contrast to the microcapsules, the fibers used

should be small, because otherwise multiple healing events will not be favored. No catalyst is needed, because of the use of a two-component hardener resin system. Similar approach is used at ESA's European Space Technology Research Centre (ESTEC) in the Netherlands and Ian Bond, University of Bristol, UK.

It is worth noting that thermoplastic materials demonstrate interesting natural healing property. Being heated, they can recover their mechanical integrity and properties. This can be

used to fix some impact damage even autonomically. For example, after collision with such a plastic, there can be a dent/hole/scratch. However, as a part of the collision energy transfers into heat. So the area of the damage can be melted and heal itself. By manipulating thermally reversible Diels-Alder reactions, a transparent polymer material with self-repairing functionality at ~120°C is developed.

Paint

Apart from cosmetic reason, paint is typically serves to protect surfaces. Self-healing protection coating for cars from Nissan is one of such examples. In principle, the mechanism of healing here can be similar to the described previously. However, main cause of wearing of paint coating is due to scratches, abrasion, and mechanical damage (collisions). It implies a specific restriction to a possible healing mechanism. Specifically, recover of mechanical recovery is not as important as recovery of protective property. This means, for example, that the healing agent can seal or inhibit corrosion of the surface underneath the crack rather than seal the crack itself. To fix scratches cosmetically, and up to some extend protect coated surface, a rather viscous polymer can be used instead of glue.

Metals

Metals being superior materials in many respects, suffer from cracks, dents and corrosion. Presently, the issue of corrosion is addressed by various coating. Self-healing of metals is not as developed as that for plastics. So far this activity was mainly computational, and focused on modeling of a possible design of such metals.

Electroconductivity of metals can be used in self-healing of both metals and ceramics. New methods involving electric-field induced colloidal aggregation are being explored. In this approach the self-healing part is primary an insulation layer between conductive parts, electrodes inserted in some conductive solution. If a crack develops inside such insulation layer, it is being sealed by non-conductive colloidal nanopartciles suspended in the solution. This technology was tested with a system that contains two concentric cylinders (the inner one

has a thin layer of insulating ceramic coating), with an electrical field applied to the between the conductive cylinders. Solution containing colloidal dispersion of either polystyrene or silica particles was placed between the two cylinders. When a defect occurs in the insulating coating, metal is exposed and creates high current density at the damaged site. This leads to fluid flow though the crack, causing colloidal particles to coagulate around the defect, and consequently, seal it.

Ceramics/Concrete

There are at least three different directions in autonomic healing of structural materials. The first one is the "classical" use of healing capsules. The second one is inhibiting corrosion of inner reinforcement frame (like the frame in concrete). Combination of both showed promises. An encapsulated healing compound was added to concrete. Both corrosion mitigation (using a time-release corrosion inhibitor) and crack sealing studies have demonstrated these materials to have the potential for increasing the life of reinforced concrete structures.

The other interesting approach to self-healing concrete was found in. It is suggested to use chalk as a part of concrete materials that have direct contact with water. If a crack appears the water the material is standing in gets inside. While for modern concrete that leads to irreversible deterioration, in the chalk concrete, the water dissolves the chalk in the mortar. That suspension of chalk penetrates into the cracks and settles there calcifying, sealing the crack. This approach is rather promising because chalk is relatively cheap.

II . SAMPLE PREPARATION

Hollow Glass Fibre Samples

HGFs hand-drawn from borosilicate tubes were embedded in an E-glass/epoxy composite. This 2.5 mm thick laminate was made by wet hand lay-up, and allowed to post-cure. This laminate was machined to 12 mm wide (90° direction) and 100 mm long specimens using a diamond-tipped blade, and was cleaned using an ultrasonic bath followed by a compressed air gun. Coloured resin and hardener were drawn into the fibres using a vacuum, with an end result as shown in Figure 1. Not all fibres were filled due to possible clogging or improperly drawn fibres. The resin used was Uroxsys ECS Epoxy Resin (green) along with ECS Winter Hardener (red). The long fibres with hardener were sealed using hot glue, and a polythene film was used to seal the resin-filled middle section.

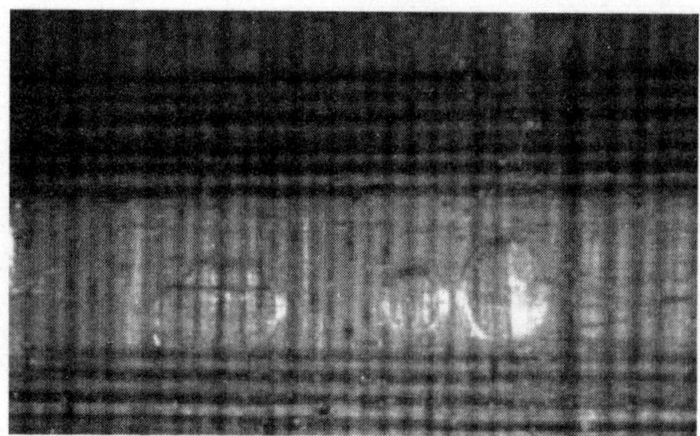

Figure 1. Optical micrograph of filled HGF specimen

Microcapsule Samples

The microcapsules were manufactured by means of applying melamine-formaldehyde to encapsulate epoxy and its hardener mercaptan, which is pentaerythritol tetrakis (3-mercaptopropionate).

A resin by Adhesive Technologies, ADR246 resin with ADH145 hardener, was used to cast two panels – one with regular resin and hardener, and another with microcapsules mixed in, which has approximately 6-7% of microcapsules by weight. Addition of E-glass fibres through the panel allowed easy cutting as well as allowed the double torsion crack test to be done on it.

Each specimen size was cut by a band-saw to Approximately 150 mm long, 60 mm wide and 6 mm thick. Both panels were post-cured after casting. Following this, a stress concentration was made by milling a 1 mm deep groove in the middle of each specimen, along the longitudinal direction. Then, a 25 mm deep notch was cut into one end of the specimen, overlapping the groove. This allowed a good starting point for crack propagation during testing.

III. APPLICATIONS

Applications of self-healing materials are expected to be very broad. In the future it will have a massive impact on virtually all industries, from the automotive industry to the energy sector. It will be able to extend product lifetimes, to increase safety, to reduce maintenance cost. The major applications being developed today are in automobile, building/construction, and aerospace industries. Because nowadays the self-healing materials are in their baby stage of development, research interest and funding focus mostly on the development of the materials rather than their applications. Below we overview the existing and probable applications in which the self-healing is expected to be the most valuable. It makes sense to divide all applications on low and high cost sensitivity. Low cost sensitive applications In average, these applications will be developed first because self-healing can be attained through a fast development cycle, using rather cost consuming mechanisms and materials. Medical dental/ artificial body replacements Nowadays an artificial bone replacement can last up to 10-15 years. Development of good biocompatible self-healing composite materials may extend this time. Another application will be in dentistry. In making artificial teeth and tooth filling materials, self-healing would benefit their functional lifetime. All such material would be in big demand virtually independent of price.

Aero/Space

Extending lifetime of a satellite in orbit around Earth, say twice, would approximately decrease the cost of the mission two times. Furthermore the increase of a spacecraft lifetime will result in longer time missions to the destinations far away in the Solar System, and maybe beyond. Having satellites made of lighter self-healing polymer materials instead of metal, which is relatively heavy, is a very cost-effective solution. A single space carrier will be able to deliver multiple satellites. And finally, safety of air- and spacecrafts can be improved by using self healing components.

Military

Having armor, body protection that could heal itself even during the battle will be beneficial for the Army. Air force and Navy can additionally benefit from fast self disappearing holes in the skin of a jet or ship. A prototype of such material already exists. Dupont's Surlyn show good properties to heal after ballistic damage

High cost sensitive applications

Car painting

Cost here is one of the main issues. Self-healing should definitely be cheaper than just repainting. One of the first commercial self-healing materials, "Scratch Guard Coat" was released by Nisan in December of 2005. According to the press release, Scratch Guard Coat contains a newly developed high elastic resin that helps prevent scratches from affecting the inner layers of a car's painted surface. With Scratch Guard Coat a car's scratched surface will return to its original state anywhere from one day to a week, depending on temperature and the depth of the scratch. Moreover, the paint is hydrophobic. While the composition and healing principle has not been resealed, the healing mechanism is clearly within the mechanisms described above. Presumably it was possible to create a similar paint a while ago. The real state-of-the-art of the Nissan paint is its fairly low cost and long lifetime.

Civil Construction

Tones of these materials are required. Self-healing capsules might solve some problems. However, this action is unlikely to be within the range of reasonable cost. So far a reasonable solution was using the chalk. Calcium for self-healing concrete is cheap. Self-healing coatings on structural steel components in, for example, bridges can be very popular. Again, here the healing mechanism is not in recovery mechanics of the coating but rather in protection against rust. This helps sustaining mechanical integrity of the coated steel constructions. The working mechanism of the self-healing coating is the release of healing/inhibiting corrosion compounds when microcapsules containing these compounds are abraded.

IV. REFERENCES

[1]   Pang J. W. C., Bond I. P.: 'Bleeding composites'— damage detection and self-repair using a biomimetic approach. Composites, Part A: Applied Science and Manufacturing, 36, 183–188 (2005)

[2]   Williams G., Trask R. S., Bond I. P.: A self-healing carbon fibre reinforced polymer for aerospace applications. Composites, Part A: Applied Science and Manufacturing, 38, 1525–1532 (2007).

[3]   Nissan Motor Co., L. Nissan develops world's first clear paint that repairs scratches oncar surfaces. 2005 [Available from: http://www.nissanglobal.com/EN/NEWS/2005/_STORY/051202-01-e.html.

[4]   Beyer, M. and U. Keil, from abrasion resistant to self healing. Kunststoffe-Plast Europe, 2003. 93(6): p. 88-+.

[5]   Dry, C., Two views on self-healing polymers. Chemical & Engineering News, 2001. 79(11): p. 8-8.

[6]   Fab, R., K.H. Kochem, and K. Muller-Nagel, New BOPP capacitor film for metallisation with improved, performance at higher temperatures. Electronics Information & Planning, 2000. 28(3): p. 103-109.

[7]   Gobin, P.F., et al., Smart materials group at the national institute of applied sciencerecent data and trends. Materials Transactions, 2004. 45(2): p. 166-172.

[8]   Lee, J.Y., G.A. Buxton, and A.C. Balazs, Using nanoparticles to create self-healingcomposites. Journal of Chemical Physics, 2004. 121(11): p. 5531-5540.

[9]   Vuillaume, P.Y., A.M. Jonas, and A. Laschewsky, Ordered polyelectrolyte"multilayers".

[10] Photo-cross-linking of hybrid films containing an unsaturated and hydrophobized poly(diallylammonium) salt and exfoliated clay. Macromolecules, 2002. 35(13): p. 5004-5012.

[11] White, S., et al., Two views on self-healing polymers - Reply. Chemical & Engineering News, 2001. 79(11): p. 8-8.

[12] White, S.R., et al., Autonomic healing of polymer composites. Nature, 2001. 409(6822):p. 794-797.

[13]   Brown, E.N., N.R. Sottos, and S.R. White, Fracture testing of a self-healing polymer composite. Experimental Mechanics, 2002. 42(4): p. 372-379.

[14]   11. Brown, E.N., S.R. White, and N.R. Sottos, Microcapsule induced toughening in a selfhealing polymer composite. Journal of Materials Science, 2004. 39(5): p. 1703-1710.

[15]   Kessler, M.R., N.R. Sottos, and S.R. White, Self-healing structural composite materials. Composites Part a-Applied Science and Manufacturing, 2003. 34(8): p. 743-753.

[16]   Kessler, M.R. and S.R. White, Self-activated healing of delamination damage in woven composites. Composites Part a-Applied Science and Manufacturing, 2001. 32(5): p.683-699.

[17]   White, S.R., et al., Autonomic healing of polymer composites (vol 409, pg 794, 2001). Nature, 2002. 415(6873): p. 817-817.

[18]   Sokolov, I., Use of self-assembled, electrospun glass tubes and capsules, coated glass capsules and tubes for self-healing materials. 2006, Clarkson University, Patent Pending.

[19]   Bleay, S.M., et al., A smart repair system for polymer matrix composites. Composites Part a-Applied Science and Manufacturing, 2001. 32(12): p. 1767-1776.

[20]   Blaiszik B. J., Sottos N. R., White S. R.: Nanocapsules for self-healing materials. Composites Science and Technology, 68, 978-986 (2008).

[21]   Pang J. W. C., Bond I. P.: A hollow fibre reinforced polymer composite encompassing self-healing and enhanced damage visibility. Composites Science and Technology, 65, 1791–1799 (2005)

[21]   Yin T., Rong M. Z., Zhang M. Q., Yang G. C.: Self-healing epoxy composites- Preparation and effect of the healant consisting of microencapsulated epoxy and latent curing agent. Composites Science and Technology, 67, 201–212 (2007).

[22]   Pang J. W. C., Bond I. P.: 'Bleeding composites'—damage detection and self-repair using a biomimetic approach. Composites, Part A: Applied Science and Manufacturing, 36, 183–188 (2005)

[23]   Brown E.N., White S.R., Sottos N. R.: Retardation and repair of fatigue cracks in a microcapsule toughened epoxy composite — Part II: In situ self-healing. Composites Science and Technology, 65, 2474–2480 (2005).

[24]   Richard P. Wool, Advance Article on the web
10th January 2008
DOI: 10.1039/b711716g

[26]   Processing, Microstructure and Performance of Materials IOP Publishing IOP Conf. Series: Materials Science and Engineering 4 (2009) 012017
DOI:10.1088/1757-899X/4/1/012017

[27]   Igor Sokolov, Department of Physics,   Department of Chemistry, Center for Advanced Material Processing, Clarkson
University, Potsdam, NY 13699, USA

Use of Smart Materials for Society

Kaushal Mahesh Chandra, Jyoti Vimal, Sharma Prateek

Department of Mechanical Engineering, Madhav Institute of Technology & Science, Gwalior-474005

Abstract — **Smart materials are materials that have one or more properties that can be significantly changed in a controlled fashion by external stimuli such as stress, temperature, moisture, pH, electric, magnetic fields**

Smart materials are widely used in therapy for chronic wounds as living tissue, structural and medical applications, health monitoring, heating and cooling systems, smart dyes & inks, smart paints and coatings, self cleaning materials etc., that is, using in today's human-beings life. Smart materials and structures, intelligent structures and Biomimetics are a new rapidly growing interdisciplinary technology embracing the fields of materials and structures, sensors and actuator systems and information processing and control. Smart materials & system is divided into two classes. Type one materials undergo change in one or more of their properties (chemical, electrical, magnetic and mechanical or thermal) in direct response to a change in external stimuli in the surrounding environment. A type two smart material transforms energy from one form to another as like photovoltaic, thermoelectric, piezoelectric etc. Shape Memory Alloy (SMA) & Smart hybrid composites are also good example of smart materials for the 21st centuary.

The purpose is to better define the smart materials and their usefulness in every aspect of life as current status for society.

Keywords : Smart materials, Biomimetics, Piezoelectric materials, Shape Memory Alloy.

I. INTRODUCTION

Until relatively recent times, most periods of technological development have been linked to change in the use of materials (eg the stone, bronze and iron ages). Single material can't respond to today's life requirement alone. Then a material felt which can be capable of controllable response to their environment in every aspect of life. Smart materials are materials that have one or more properties that can be significantly changed in a controlled fashion by external stimuli such as stress, temperature, moisture, pH, electric, magnetic fields [1]. Smart materials can be used as "Actuators", "Sensors", both or in some cases as self -sensing actuators. Smart materials present a role not only new and exciting design in civil engineering as like as Smart bridge, smart houses etc. but also revolutionized in Automobile, Space, Aircraft, Health monitoring system, Domestic appliances, Orthopaedic implants etc. Basically, there is no standard definition for smart materials, and the term smart material is generally defined as a material that can change one or more of its properties in response to an external stimulus.

HISTORY OF SMART MATERIALS

In the glamorization of Smart materials, we often forget the legacy from which Smart materials sprouted seemingly so recently and suddenly. Texts from as early as 300BC were the first document the size of alchemy. Metallurgy was by then a well developed technology. It used to be practiced by Greeks and Egyptians, but many Philosophers were concerned that this empirical practice was not governed by a satisfactory scientific theory [2].

In space, the first foreseen application of smart materials was onboard satellites in the middle of the 1980 [3]. However, it can be difficult to present various Smart materials due to great range of performance and availability.

II. DUMB MATERIALS

Most familiar engineering materials and structures until recently have been 'dumb'. They have been preprocessed and/or designed to offer only a limited set of responses to external stimuli. Such responses are usually non-optimal for any single set of conditions, but optimised to best fulfill the range of scenarios to which a material or structure may be exposed.

Advanced composites such as glass and carbon fibre reinforced plastics, which are often thought to be the most flexible engineering materials since their properties (including strength and stiffness) can be tailored to suit the requirements of their applications, can only be tailored to a single combination of properties.

III. BIOMIMETICS

Dumb materials and structures contrast sharply with the natural world where animals and plants have the clear ability to adapt to their environment in real time. The field of biomimetics, which looks at the extraction of engineering design concepts from biological materials and structures, has much to teach us on the design of future manmade materials. The process of balance is a truly smart or intelligent response, allowing, in engineering terms, a flexible structure to adapt its form in real time to minimise the effects of an external force.

The natural world is full of similar properties including the ability of plants to adapt their shape in real time limping reflex to heat and pain. The materials and structures involved in natural systems have the capability to sense their environment, process this data, and respond.

IV. SMART MATERIALS

Science and technology have made amazing developments in the design of electronics and machinery using standard materials, which do not have particularly special properties (i.e. steel, aluminum, gold). Imagine the range of possibilities, which exist for special materials that have properties, can manipulate. Some such materials have the ability to change shape or size simply by adding a little bit of heat, or to change from a liquid to a solid almost instantly when near a magnet, these materials are called smart materials.

Smart materials have one or more properties that can be dramatically altered. Most everyday materials have physical properties, which cannot be significantly altered, for example if oil is heated it will become a little thinner, whereas a smart material with variable viscosity may turn from a fluid which flows easily to a solid. A variety of smart materials already exists, and is being researched extensively. These include piezoelectric materials, magneto-rheostatic materials, electro-rheostatic materials, and shape memory alloys. Some everyday items are already incorporating smart materials such as coffeepots, cars, the International Space Station, eyeglasses and the number of applications for them is growing steadily.

Each individual type of smart material has a different property which can be significantly altered, such as viscosity, volume, and conductivity. The property that can be altered influences what types of applications the smart material can be used for.

V. PIEZOELECTRIC MATERIALS

These materials possess the property by which they develop electricity when mechanically deformed (called direct affect). They also have reverse effect i.e. experiences a dimensional change when an electric voltage is applied to them. Piezoelectric materials are widely used as a sensor in different environments. Piezoelectric ceramics are generally manufactured from components of lead zirconate/ lead titanate and exhibit greater sensitivity relative to ferroelectric ceramics of other compositions.

Piezoelectric materials have two unique properties which are interrelated. When a piezoelectric material is deformed, it gives off a small but measurable electrical discharge. Alternately, when an electrical current is passed through a piezoelectric material it experiences a significant increase in size. Piezoelectric materials are most widely used as sensors in different environments. They are often used to measure fluid compositions, fluid density, fluid viscosity, or the force of an impact. An example of a piezoelectric material in everyday life is the airbag sensor in car. The material senses the force of an impact on the car and sends and electric charge deploying the airbag.

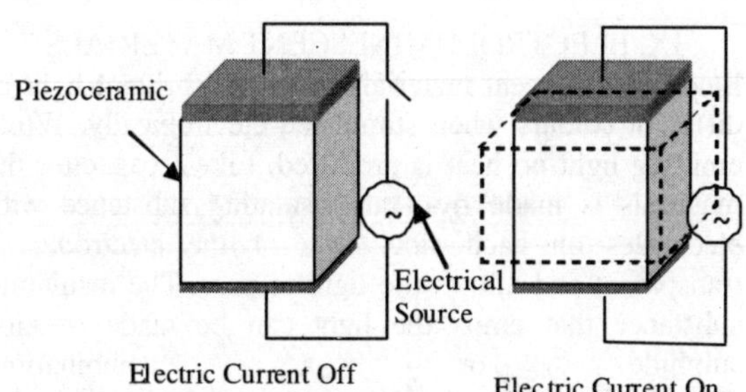

Fig. 1 An illustration of the Piezoelectric Effect.

VI. MAGNETOSTRICTIVE MATERIALS

Magnetosriction is a property available in magnetic material such as Iron, Cobalt and Nickel and also in rear earth materials like Lanthanum and Terbium upon the application of magnetic field. It is also known as piezomagnetic materials. As these materials are magnetized it changes it shapes. This coupling between magnetic and mechanical energies gives the transduction capability in such material and can be used in both actuation and sensing devices. $CoFe_2O_4$ magnetostrictive material structure is very useful for making magneto-electro-elastic sandwich composite plate.

VII. PHOTOCHROMIC MATERIALS

Photochromic materials change reversibly colours with changes in light intensity. Usually, these are colourless in a dark place, and when sun light or ultraviolet radiation is applied molecular structure of the material changes and it exhibits colours. When the relevant light source is removed the colours disappears. Changes from one colour to another colour are possible by mixing photochromic colours with base colours. They are used in paints, inks, and mixed to mould or casting materials for different applications.

VIII. THERMOCHROMIC MATERIALS

Thermochromic materials change reversibly colour with changes in temperature. They can be made as semi-conductor compounds from liquid crystals or using metal compounds. The change in colour happens at a determined temperature which can be varied doping the material. They are used to make paints, inks or are mixed to moulding or casting materials for different applications.

Fig. 2

IX. ELECTROLUMINESCENT MATERIALS

Electroluminescent materials produce a brillant light of different colours when stimulated electronically. While emitting light no heat is produced. Like a capacitor the materials is made from an insulating substance with electrodes on each side. One of the electrodes is transparent and allows the light to pass. The insulating substance that emits the light can be made of zinc sulphide or a combination. They can be used for making light stripes for decorating buildings, industrial and public vehicles safety precautions.

Fig. 3

X. FLUORESCENT MATERIALS

Fluorescent materials produce visible or invisible light as a result of incident light of a shorter wavelength i.e. X-rays, UV-rays, etc. The effect ceases as soon as the source of excitement is removed. Fluorescent pigments in daylight have a white or light colour, whereas under excitation by UV radiation they irradiate an intensive fluorescent colour.

They can be used for paints, inks or mixed to moulding or casting materials for different applications.

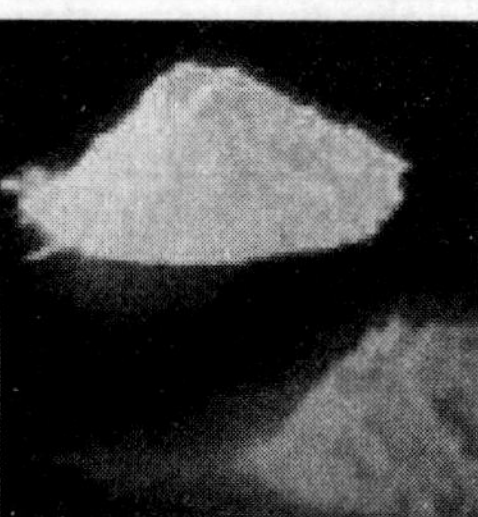
Fig.4

XI. PHOSPHORESCENT MATERIALS

Phosphorescent materials are known as afterglow materials. Phosphorescent materials produce visible or invisible light as a result of incident light of a shorter wavelength. These are detectable only after the source of the excitement has been removed. Afterglow effect pigments are polycrystalline inorganic zinc sulphide or alkaline earth sulphides, and can be used in paints, inks or mixed to moulding or casting materials for different applications.

Fig. 5

XII. POLYMER GELS

Polymer gels consist of a cross-linked polymer network inflated with a solvent such as water. They have the ability to reversibly swell or shrink (up to 1000 times in volume) due to small changes in their environment such as pH, temperature, electric field. Micro sized gel fibres contract in milliseconds, while thick polymers layers require minutes to hours or days react.

Fig. 6

XIII. SHAPE MEMORY ALLOYS

It is also called Smart metal/Memory alloy/Muscle wire/Smart alloy.SMA is a metal/alloy which can remember and return to a specific shape after considerable deformation. It is widely used as the sensing, actuator and damping instruments [4].

A Type of SMAs
- Shape memory polymers,
- Ferromagnetic SMAs.

B Products
- Super elastic glasses,
- Coffeepot thermostat, and
- Electrical connector, etc.

XIV. ADVANTAGES OF SMART (OR ACTIVE) MATERIALS

- High energy density (compared to pneumatic and hydraulic actuators),
- Excellent bandwidth,
- Simplified packaging, and
- Novel functions such as the huge volume change as a function of temperature exhibited by smart gels.

XV. APPLICATIONS OF SMART MATERIALS

There are many possibilities for such materials and structures in the man made world. Engineering structures could operate at the very limit of their performance envelopes and to their structural limits without fear of exceeding either. These structures could also give maintenance engineers a full report on performance history, as well as the location of defects, whilst having the ability to counteract unwanted or potentially dangerous conditions such as excessive vibration, and effect self repair.

A SMART MATERIALS IN AEROSPACE
Some materials and structures can be termed sensual devices. These are structures that can sense their environment and generate data for use in health and usage monitoring systems. To date the most well established application of health and usage monitoring systems are in the field of aerospace, in areas such as aircraft checking.

B SMART MATERIALS IN CIVIL ENGINEERING APPLICATIONS
They could be used in the monitoring of civil engineering structures to assess durability. Monitoring of the current and long term behaviour of a bridge would lead to enhanced safety during its life since it would provide early warning of structural problems at a stage where minor repairs would enhance durability, and when used in conjunction with structural rehabilitation could be used to safety monitor the structure beyond its original design life. This would influence the life costs of such structures by reducing upfront construction costs and by extending the safe life of the structure.

C OTHER EXAMPLES OF APPLICATIONS
- Fast response valves,
- High-power-density hydraulic pumps,
- Active bearings for reduction of machinery noise,
- Footwear,
- Sports equipment,
- Precision machining,
- Vibration and acoustic sensors, and
- Dampers, etc.

XVI. CONCLUSION
The advances and improvements in smart materials allow them to provide to a diverse set of applications, especially in the defense, aerospace, healthcare, electronics, and semiconductor industries. Although very few of these applications are at present commercially viable, their potential for future acceptance is undeniable. With future advances, smart materials are also likely to be useful for fabricating important appliances devices.
For faster development and use of innovative smart materials, base technologies such as robotics, micro-fluidics, and micromachining need to be brought on balance with the production requirements of the materials.

FUTURE
The development of true smart materials at the atomic scale is still some way off, although the enabling technologies are under development. These require novel aspects of nanotechnology and the newly developing science of shape chemistry. The concept of engineering materials and structures which respond to their environment, including their human owners, is a somewhat alien concept. It is therefore not only important that the technological and financial implications of these materials and structures are addressed, but also issues associated with public understanding and acceptance.

XVII. REFERENCES
[1]. John F. Mc CABE et. al "Smart materials in dentistry-future prospects", Dental Materials journal 2009; 28 (1), pp.37-48.

[2]. D. Michelle Addingto,Da Daniel L. Schodek , A text Book "Smart Materials in architecture& design".

[3]. Eric M. Flint, JSrg Melcher, Holger Hanselka, "The 'Promise' Of Smart Materials For Small.

[4]. Dazhi Yang,"Shape memory alloy and smart hybrid composites-advanced materials for the 21st Century", Materials and Designs 2000;21,pp.503-505.

[5]. Cliff Friend "Smart materials: the emerging technology" abstract from Materials World, vol. 4,pp. 18-18, 1996.

[6]. Thomas Nissen (computer Graphics), Intelligent Materials-Smart materials workshop.

[7]. Jianguo Wang, Linfeng Chen, Shisheng Fang "State Vector approach to analysis of multilayered magneto-electro-elastic plates" International Journal of Solids and Structures 40 (2003) 1669-1680.

[8]. Rajesh K. Bhangale, N. Ganesan "Static analysis of simply supported functionally graded and layered magneto-electro-elastic plates" International Journal of Solids and Structures 43 (2006) 3230-3253.

Eco Industrial Park : A Foundation for the Development of Indian Industries

Dilip Johari and Dhairya Pratap Singh

Department of Mechanical Engineering, B.S.A.College of Engineering and Technology, Mathura

Abstract : **An Eco-Industrial Park involves a network of firms and organizations, working together to improve their Environmental and Economic performance. This is a community of manufacturing and service businesses seeking enhanced environmental and economic performance through collaboration in managing environmental and resource issues including energy, water, and materials. By working together, the community of businesses seeks a collective benefit that is greater than the sum of individual benefits each company would realize if it optimized its individual performance only. This paper considers how Eco-Industrial Parks relate to the generic concept of Industrial Ecology and examines their viability. In particular, this paper looks at the future role of Eco-Industrial Parks in developing and a newly industrialized country. Industrial Ecosystem is a system in which the consumption of energy and materials is optimized, waste generation is minimized and the effluents of one process serve as the raw material for another process. The goal of an EIP is to improve the economic performance of the participating companies while minimizing their environmental impact.**

Keywords : Industrial Ecology, Industrial Ecosystem, Eco-Industrial Park, Eco-Industrial Development;

I. INTRODUCTION

Eco-Industrial Park is based on the concept of low energy and raw material consumption, and environment friendly. The eco-industry is a kind of industrial system that is managed by the ecology. In this kind of eco- industry park, there are some factories and plants that are connected with each other like a food chain, which means the waste from one industry is the raw material for another industry, so that these industries form a closed ecosphere. The eco-industry park can efficiently decrease the pollution and increase the material utilization. On the other hand, the companies do not need to spend much money on the transportation for the material because the factories are in same reason. Eco-Industrial Parks represent a promising strategy to promote sustainable industrial development and implement industrial ecology concepts. They also provide a new model for local economic development. The benefits Eco-Industrial Parks provide may serve as incentives for companies to improve their environmental performance in terms of management of materials, energy and waste. The potential they offer in terms of local development is already encouraging communities to invest in concepts incorporating this approach to industrial development. Some planners and researchers of EIPs have used the team "industrial ecosystem" to describe the type of symbiotic relationships that develop amongst participating firms. At its core, an EIP is very simple. It strives simultaneously to increase business success while reducing pollution and waste. Rooted in the emerging discipline of industrial ecology, an EIP mirrors natural systems. As single organisms can be viewed alone or in a larger ecology, single enterprises can organize themselves in more complex business ecologies. When we refer to EIPs it is far more than a share plot of land. By moving to higher levels of interdependent organization quantum level improvements can be realized in resiliency, flexibility and resource conservation. This pays off for the business and the environment.

II. MISSION OF EIP DEVELOPMENT

- Provide a national development model that promotes business, people, economy, and natural and cultural resources
- Create family-wage jobs and training opportunities;
- Protect and enhance natural and cultural resources, demonstrate conservation and efficient resource use, develop and use industrial ecology principles;
- Support private businesses and industrial development, and revitalize the local economy;
- Develop the next generation of industrial facilities which combine profit, resource, efficiency, industrial ecology, and pollution prevention;

III. COMPONENTS OF EIP DESIGN

EIPs have a rich menu of design options, including site design, park infrastructure, individual facilities, and shared support services. The following highlights some major strategies an EIP design team can draw upon in planning a park.

Natural Systems

An industrial park can fit into its natural setting in a way that minimizes environmental impacts while cutting operating costs. The Herman Miller design plant in Phoenix illustrates the use of native plant reforestation and the creation of wetlands to minimize landscape maintenance, purify storm water run-off, and provide climate protection for the building. At another level, plant design, landscaping, and design choices in materials, infrastructure, and building equipment, can reduce a park's contributions to global climate change and its consumption of non-renewable resources.

Energy

More efficient use of energy is a major strategy for cutting costs and reducing burdens on the environment. In EIPs, companies seek greater efficiency in individual building, lighting, and equipment design. Examples include flows of steam or heated water from one plant to another (energy cascading), or steam connections from firms to provide heating for homes in the area. Finally, in many regions, the park infrastructure can use renewable energy sources such as wind and solar energy.

Material Flows

In an eco-park, companies perceive wastes as lost opportunities that ideally are potential products to be re-used internally or marketed to someone else. Individually, and as a community, they work to optimize use of all materials and to minimize the use of toxic materials. The park infrastructure may include the means for moving by-products from one plant to another, warehousing by-products for shipment to external customers, and common toxic waste processing facilities.

Water Flows

In individual plants, designers specify high efficiency building and process equipment. Processed water from one plant may be re-used by another (water cascading), passing through a pre-treatment plant as needed. The park infrastructure may include mains for several grades of water (depending on the needs of the companies) and provisions for collecting and using storm water runoff.

IV. A SUCCESSFUL APPROACH TO EIP DEVELOPMENT

The literature study revealed that the successful development of an EIP would require the active participation from a number of stakeholders:

- Public sector stakeholders from local, regional and national government agencies;
- Representatives of local companies and potential future tenants in the EIP;
- Leaders in the industrial and financial community;
- Labour representatives;
- Educational institutions;
- Practitioners with the full complement of capabilities needed in the project: architecture, engineering, ecology, environmental management, and education and training;

V. STRATEGIES FOR DESIGNING AN EIP

Several basic strategies are fundamental to developing an EIP or industrial ecosystem. Individually, each adds value; together they form a whole greater than the sum of its parts.

Integration into Natural Systems

Minimize local environmental impacts by integrating the EIP into the local landscape, hydrologic setting, and ecosystems;

Minimize contributions to global environmental impacts, i. e. greenhouse gas emissions.

Energy Systems

Maximize energy efficiency through facility design or rehabilitation, co-generation, energy cascading, and other means;

- Achieve higher efficiency through inter-plant energy flows;
- Use renewable sources extensively.

Materials Flows and Waste Management for the Whole Site

Emphasize pollution prevention, especially with toxic substances;

Ensure maximum re-use and recycling of materials among EIP businesses;

Reduce toxic materials risks through integrated site-level waste treatment;

Link the EIP to companies in the surrounding region as consumers and generators of usable by-products via resource exchanges and recycling networks.

Water

Design water flows to conserve resources and reduce pollution through strategies similar to those described for energy and materials.

Effective EIP Management

In addition to standard park service, recruitment, and maintenance functions, park management

- Maintains the mix of companies needed to best use each others' by-products as companies change;
- Supports improvement in environmental performance for individual companies and the park as a whole;
- Operates a site-wide information system that supports inter-company communications, informs members of local environmental conditions, and provides feedback on EIP performance.

VI. THE FUTURE OF EIP

There are some trends that are predictable about the future: concern for the environment will grow; resources will become scarcer and less reliable; demands for quality and cost competitiveness will increase. These challenges can be overcome with a new paradigm of industrial organization that asks each community, each company, and each employee how they work together to use resources more wisely. The development of EIPs can create good jobs, strong companies, sustainable communities, and improved regional environments.

Emerging Projects of EIP

- Trenton, New Jersey
- The Duwamish Coalition/Seattle, Washington
- Plattsburgh Air Force Base/ Plattsburgh, New York
- Burnside Industrial Park/Dartmouth, Nova Scotia

VII. BENEFITS OF EIP

Eco-industrial Parks based on industrial ecology have following benefits:

- Waste products from one industry provide the input for another, reducing the input cost;
- Waste now has an economic value, increasing profits;
- The creation of a larger and more varied economic base;
- The potential for the job creation from firm formation;
- Reduced emission means less need separate industrial and residential land uses and consequently reduced movement cost between two;

VIII. CONCLUSION

From these points, one can identify those which one should keep in mind when developing an EIP.

First and foremost, one should assure active company/industry participation in the planning stages of the project.

Second, the costs of EIP planning should not be solely carried by the government. Companies should also be financially committed to the planning phases. This will also enhance company commitment in the realization phases of the project.

Furthermore, the initial focus of the EIP project should not be on the establishment of physical energy, water, and material waste exchanges but on the establishment of utility sharing projects. The project should initially be focused on such projects because these projects, compared to the physical waste exchange projects, require relatively small economic investments while at the same time they offer a possibility for a reasonable economic and environmental benefit.

Finally, when the project is well established—that is when companies are fully aware of the benefits that are to be gained—the development can move along to the more company-specific and economically challenging projects, although the projects should always render an economic as well as environmental benefits.

IX. REFERENCES

[1] Lowe, Ernest (1997): "The Eco-Industrial Park: A Foundation for Sustainable Communities?"

[2] R.R.Heeres, W.J.V. Vermeulen, F.B. de Walle, "Eco-industrial park initiatives in the USA and the etherlands: first lessons " Journal of Cleaner Production 12 (2004) 985–995

[3] Ernest A Lowe, (2001), Eco-Industrial Park Hand book for Asian developing countries.

[4] Ray Cote, Whither Industrial Ecology; Transforming Industrial Park, Faculty of Management Dalhousie University.

[5] Braden R. Allenby & Deanna J. Richards, (1994), The Greening of Industrial Ecosystem Advisory Committee on Industrial Ecology and Environmentally Preferable Technology, National Academy of Engineering, ISBN: 0-309-58580-5

[6] Cosgriff Dunn B, Steinemann A. Industrial ecology for sustainable communities. Journal of Environmental Planning and Management 1998; 41(16):661–72.

[7] Frosch RA. Industrial ecology; adapting technology for a sustainable world. Environment 1995; 37(10):16–24, 34–7.

[8] Heeres R. Eco-industrial park development moving towards a sustainable society. Graduation Paper. Environmental Science Department, Utrecht University, 1999.

[9] Gertler N. Industrial Ecosystems: Developing Sustainable Industrial Structures. Master Thesis. Massachusetts Institute of Technology, 1996.

[10] Deppe M, Leatherwood T, Lowitt P, Warner N. A planner's overview of eco-industrial development.

American Planning Association Annual Conference. 2000.

[11] Lau SM. Eco-industrial park development: manufacturing changes. Minnesota: The Green Institute; 1998.

[12] Cohen-Rosenthal, Ed,1998 "Eco-Industrial Development: New Frontiers for Organizational Success", Proceedings Fifth International Conference on Environmentally Conscious Design and Manufacturing

[13] Chertow, Marian. 2000. "Industrial Symbiosis: A Review." Annual Review of Energy and the Environment Vol. 25.

[14] Fleig, Anja-Kathrin. 2000. Eco-Industrial Parks as a Strategy towards Industrial Ecology in Developing and Newly Industrialised Countries.

[15] Hauff, Michael & Wilderer, Martin (1999): "Industrial Estates - The Relevance of Industrial Ecology for Industrial Estates", Green Business Opportunities, Vol. 5 No. 2

[16] Koenig, Andreas. 2000. Development of Eco-Industrial Estates in Thailand, Project Development and Appraisal, June to December 2000. Overheads for Roundtable Meeting November 27. GTZ, Bangkok.

[17] Martin, Shiela. 1996. Eco-Industrial Parks: A Case Study and Analysis of Economic, Environmental, Technical, and Regulatory Issues. Research Triangle Institute.

[18] Research Triangle Institute (1994): "Eco-industrial parks; a case study and analysis of economic, environmental, technical and regulatory issues." Final report for the U.S. Environmental Protection Agency.

[19] Cohen-Rosenthal, Edward & McGalliard, Thomas (1998): "Eco-Industrial Development: The case of the United States",

[20] Esty, Daniel C. & Porter, Michael E. (1998): "Industrial Ecology and Competitiveness – Strategic Implications for the Firm.", Journal of Industrial Ecology, Vol. II No. 1.

[21] Garner, Andy and Keoleian, Gregory A. (1995): "Industrial Ecology: An Introduction" National Pollution Prevention Centre for Higher Education, University of Michigan

Design of Electromagnetic Suspension System-A Review

Yashvir Singh, Nishant Kumar Singh and **Anand Poras**

Deptt. of Mechanical Engineering, Hindustan College of Science and Technology, Mathura, U. P.

Abstract-This paper suggested that the main drawback of a mechanical suspension system is that there is a mechanical inertia of every spring. Due to this inertia force the shocks experienced by the wheels by the irregularities transferred to the passenger carriage. This disability of the mechanical suspension system can be easily removed by an electromagnetic spring, which serves as a spring of variable spring constant. An electromagnetic suspension system is equivalent to a mechanical spring with variable spring constant, for the light shocks this electromagnetic spring acts as a spring with very low spring constant and for the heavy shocks it will act similar to the spring capable to bear the shock. The main advantage of the electromagnetic suspension system is that spring can achieve any value of the spring by varying the current in the electromagnetic coil. This serves better as the inertia of the spring is a function of the load applied to the spring. Therefore, the inertia force interrupts the action of the spring in a least amount.

Keywords: Inertia, Electromagnetic suspension system (EMS), shocks.

Nomenclature

F = Magnetic force between two coils
B_1 = Magnetic field of the first electromagnet
B_2 = Magnetic field of the first electromagnet
r = Distance of the fix magnet and floating magnet
F_{FPM1} = Force due to fix permanent magnet-1 under shock
F_{FPM2} = Force due to fix permanent magnet-2 under shock
F_{EM1} = Force due to electromagnet-1 under shock
F_{EM2} = Force due to electromagnet-2 under shock
B_{EM1} = Magnetic field due to electromagnet-1
B_{EM2} = Magnetic field due to electromagnet-2

I. INTRODUCTION

Suspension is the term given to the system of springs, shock absorbers and linkages that connects a vehicle to its wheels. Suspension systems can not only contribute to the car's handling and braking for good active safety and driving pleasure, but also keep vehicle occupants comfortable and reasonably well isolated from road noise, bumps, and vibrations. The suspension also protects the vehicle itself and any cargo or luggage from damage and wear. The ride quality of a vehicle is significantly influenced by its suspension system, the road surface roughness, and the speed of vehicle. A vehicle designer can do little to improve road surface roughness, so designing a good suspension system with good vibration performance under different road conditions become s a prevailing philosophy in the automobile industry. Passive suspension systems use conventional dampers to absorb vibration energy, the dampers and stiffness coefficients are constant. The active suspension system use extra power to provide a response-dependent damper, which is capable of producing an improved ride comfort. Over the years, both passive and active suspension systems have been proposed to optimize a vehicle's ride quality.

In this paper author try to design an electromagnetic suspension system which can provide ride comfort to the consumers.

II. CONSTRUCTION

The basic parts of EMS shown in fig.1.

Figure 1
Assembly view of EMSS

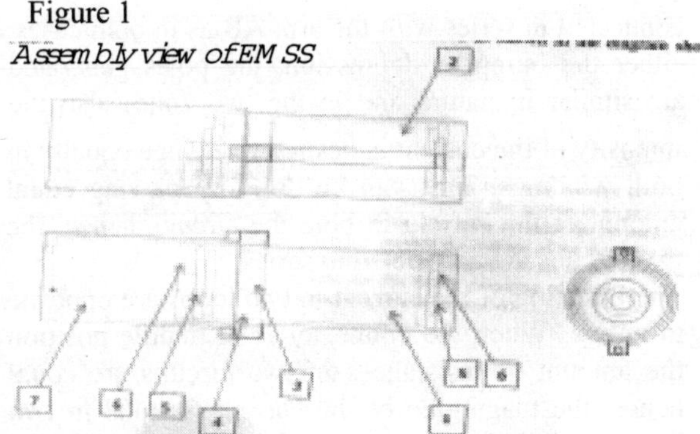

Inner Cylinder

It is a cylindrical part having two parallel grooves on peripheral surface as shown in fig.3. The main purpose of the grooves is to provide the passage for the sliding of the floating assembly arms. The inner cylinder can be made up of any diamagnetic materials like plastic to prevent the leakage of the magnetic flux to the outer environment; this ensures the isolation of the magnetic fields to the surrounding ferrous materials. The inner cylinder is inserted into the outer cylinder with a slide fit. But for the purpose of the electromagnetic assembly the inner cylinder is kept fix and the outer cylinder reciprocates over it. The constrain is provided for the placement of the fix permanent magnet.

Figure 2
Design of EMSS

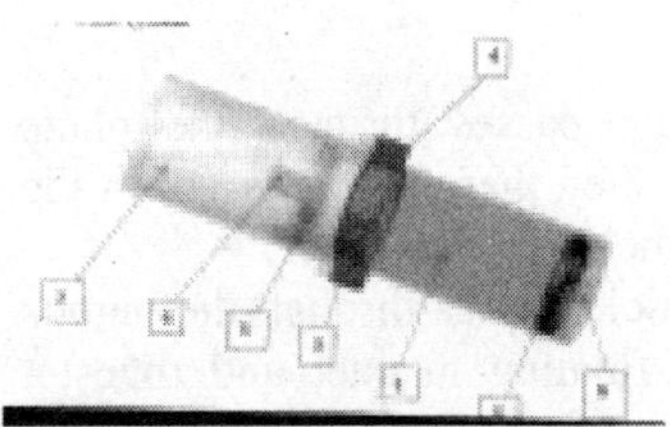

Figure 3
Inner Cylinder

Outer Cylinder

The outer cylinder contains inner cylinder with sliding fit. The inner cylinder can slide inside the outer cylinder; the flange of the outer cylinder is connected to the arms of the floating magnet assembly so that it can float in the grooves of the inner cylinder with the floating magnet.

Figure 4
Outer Cylinder

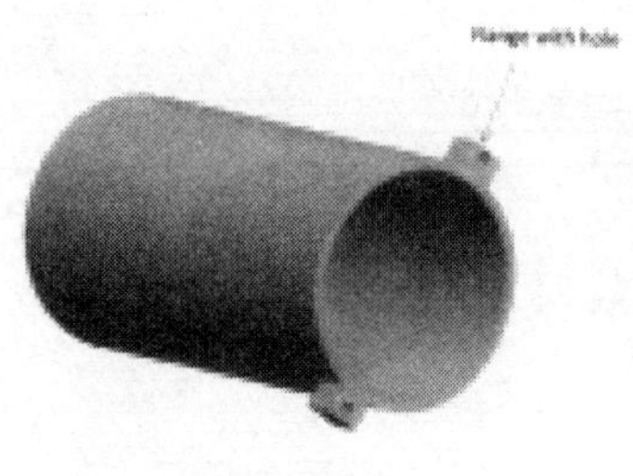

Floating magnet

The floating magnet is sandwiched in a part made up of plastic, called as floating magnet holder, which covers the floating magnet completely.

Figure 5
Floating magnet

Floating magnet Holder

This part has two arms, both in a straight line. hese arms are inserted between grooves of the inner cylinder the arms are connected to the flange of the outer cylinder which enables it to move with the floating magnet assembly.

Figure 6
Floating magnet holder

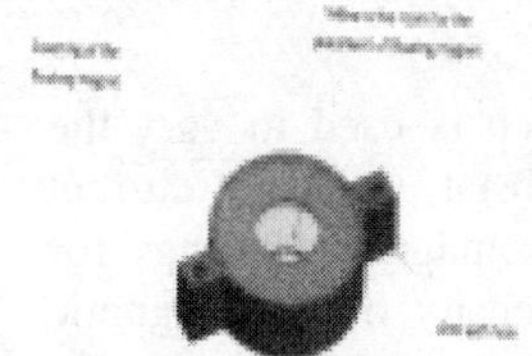

Fixed Permanent Magnet

It is placed between constrain and the electromagnet. The main purpose of this magnet is to generate the balancing force for the balance of the weight of the vehicle at rest. The balancing force is generated by mutual interaction of the two pairs of floating a d fix permanent magnet. One pair generates repulsive force and other pair generates attractive force. The sums of these forces balance the weight of the v hicle when it is kept in rest.

Figure 7
Fixed permanent magnet assembled view

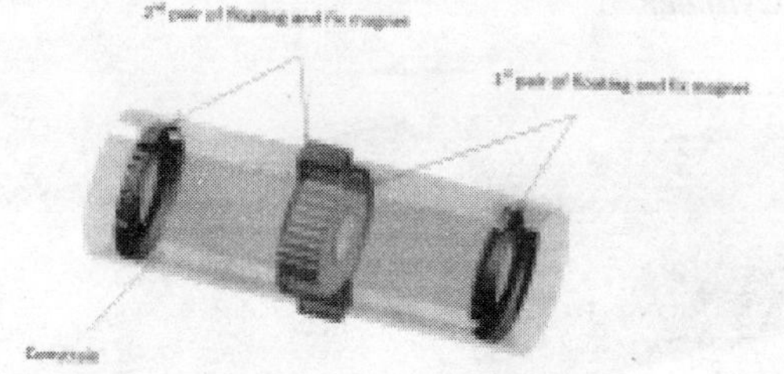

Electromagnet

There are two poles electromagnet as shown in fig.8. There is a copper winding wrapped on mild steel core which acts as electromagnetic field conductor. The magnetic field is generated by passing the current in the copper winding wrapped around it. The intensity of magnetic field depends upon intensity of current.

Figure 8
Electromagnet assembly

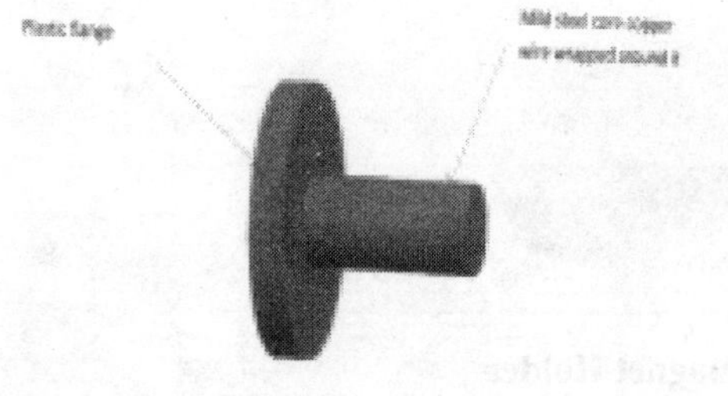

Connector

It is part which connects the floating magnet holder, flange of outer cylinder and rheostat slide together ensuring their simultaneous movement.

Rheostat

It is an electrical device which is used to vary the resistance in a circuit. Rheostat work as current varying element in the electromagnet windings for the production of desired intensity of the magnetic field. The connector of the rheostat moves with the floating magnet and the variation in the current of the electric circuit occurs due to which amount of the current varies in the windings.

III. PROPOSED CIRCUIT AND ITS WORKING

There are two loops 1 and 2 as shown in fig.9. The current in the winding is drawn from the arm AB i.e. connector arm of the rheostat. The two windings are connected in series with the arm AB as in both cases, either the bump or depression; the poles generated are similar in nature and in the two zones also the intensity of the current is desired to change equally in both windings. This can be done by passing equal amount of the current in both the circuits hence; the two circuits are connected in series.

The direction of the current in two loops are opposite in nature. When the connector is in middle position the amount of resistances in two circuits are equal hence; the magnitude of the current is equal in two loops but the direction is opposite due to which in the arm AB there is no current.

When the connector moves right or left the value of the resistances in two loops also changes and hence the unequal amount of the current are generated in two loops therefore there is no current in the arm AB.

Figure 9
Proposed circuit for EMSS

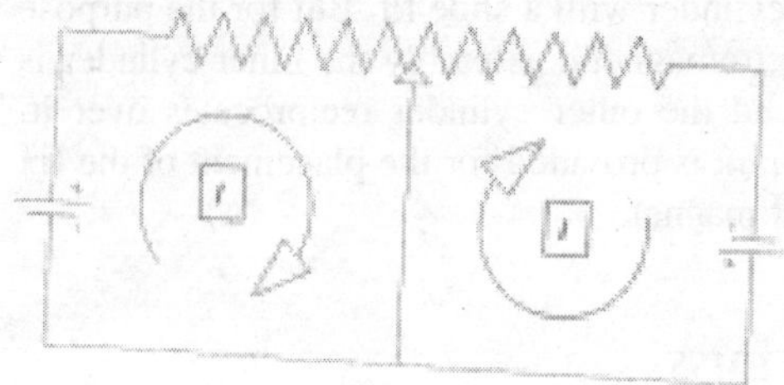

Case 1: When vehicle passes through the plane surface or kept in rest, then there is no current in the winding connected to the arm AB.

Case 2: When the vehicle passes through the bumps the wheel and hence floating magnet and rheostat connector is forced inwards. Let's say leftwards. Hence the resistance in loop 1 decreases and that in loop 2 increases. The change in the resistance of two loops is equal in amount, therefore the change in this current in the two loops are equal. Say the current in loop 1 increases with an amount i and that in loop 2 decreases with an amount i as shown in fig.10.

Hence, the overall current in the arm is 2i from A to B. The winding connected to the arm AB in series generates same nature of the poles. The winding is wound in such a way that the poles generated are south in both electromagnets. The value of current i depends upon the displacement of the connector which is further dependent upon the amount of shocks. It provides high current for the heavy shock and low current for the light shocks.

Figure 10

Circuit diagram for bump (leftward displacements)

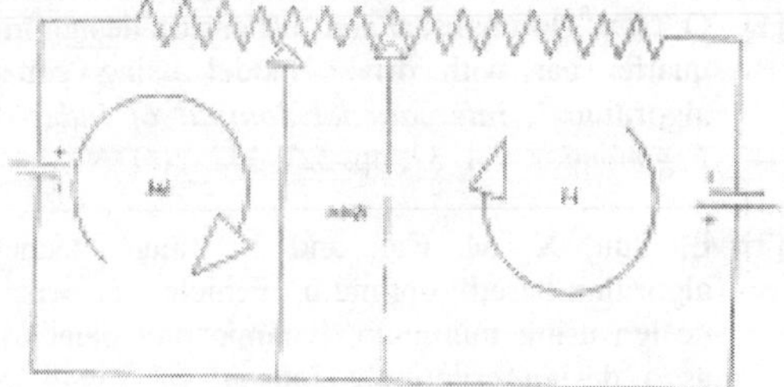

Case 3: When the vehicle passes through the sudden depression the connector moves right wards and hence resistance in the loop 2 decreases while in loop 1 increases due to which the current in loop 1 and 2 changes by equal amount I and it increases by the same amount in loop 2 as shown in fig.11.

The overall current in the arm AB changes by an amount 2i but the direction of the current in the arm AB is opposite to that the previous case. Hence the poles generated by the electromagnet are opposite in nature from the previous case.

Figure 11

Circuit diagram for bump (rightward displacement)

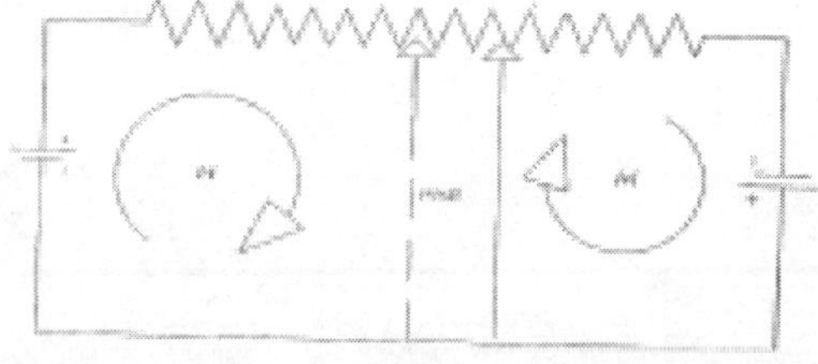

IV. MATHEMATICAL CALCULATION OF EMSS

Free body diagram of the floating magnet is shown in fig.12. It is clear from FBD of floating magnet that the forces acting on the floating magnet are two repulsive (F_R) and attractive (F_A) forces of the zone-1 and zone-2 and the reaction of the ground R. Hence the equation can be written as

$$F_R+F_A=R$$

We know that the magnetic force between the two magnets is given as:

$$F_R=\mu B_1 B_2/4\pi r^2$$
$$F_A= \mu B_1 B_2/4\pi r^2$$

The value of ground reaction is equal to the weight of the vehicle i.e. Mg

$$\mu B_1 B_2/4\pi r^2 + \mu B_1 B_2/4\pi r^2 = Mg$$

Hence it is clear from the above equation that if the value of the magnetic field intensity of the floating magnet is known then the intensity of the fix magnets can be calculated.

Figure 12

Free body diagram of floating magnet

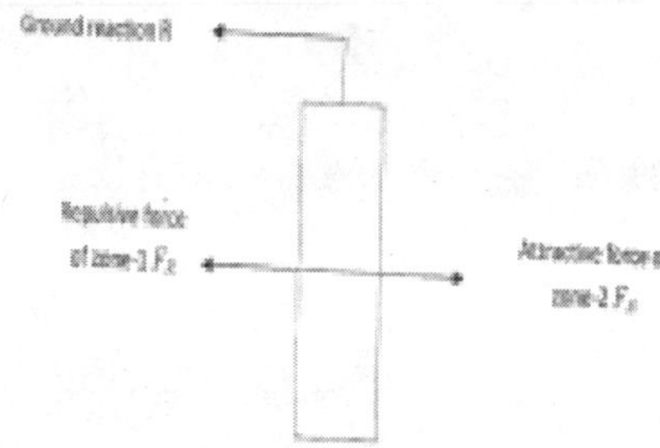

Calculation of Intensity of electromagnets

From the fig.13. it is clear that there are four forces on the floating electromagnet when it is displaced from the mid position i.e. repulsive force of EM-1(F_{EM1}), repulsive force of FPM-2 (F_{FPM1}), attractive force of EM-2 (F_{EM2}), attractive force of FPM-2 (F_{FPM2}). Consider the situation when the maximum possible shock is encountered by the vehicle, in this situation the floating magnet will be at extreme left position. If the clearance between the fix permanent magnet-1 (FPM-1) and the floating magnet is d then for the equilibrium of the floating magnet at this position can be given by the equation.

$$F_{EM1}+F_{FPM1}+F_{EM2}+F_{FPM2} = \text{weight of the vehicle} + \text{max shock intensity}$$

The value of the above forces are given by the equations

$$F_{FPM1}=\mu B_1 B_{EM1}/4\pi d^2$$

$$F_{EM1}=\mu B_1 B_{EM1}/4\pi d^2$$

$$F_{FPM2}=\mu B_1 B_{EM2}/4\pi (1-d)^2$$
$$F_{EM2}=\mu B_1 B_{EM2}/4\pi (1-d)^2$$

The value of the magnetic field intensity, generated by the two electromagnets is same as the current in the two windings are same; hence the above equation reduces to $2F_{EM}$

$$2F_{EM}+F_{EM1}+F_{FPM2} = \text{weight of the vehicle} + \text{max shock intensity}$$

The value of F_{EM1} and F_{FPM2} can be calculated by the above equation as the value of the d is known as per design consideration. The value of the F_{EM} can be calculated for the max value of the expected shock. Hence the optimum value of the current and the total number of the turn in each winding can be designed for the corresponding value of the desired output of the magnetic field of the electromagnets.

Figure 13

FBD of floating magnet under shock

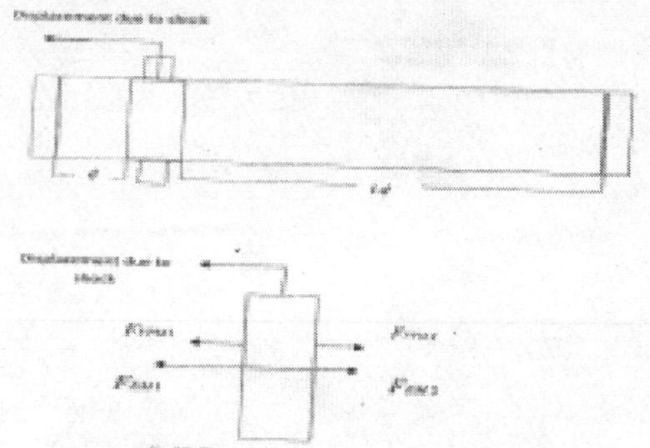

V. CONCLUSION

It can be concluded that electromagnetic suspension system can be designed and develop for providing better riding comforts in the existing automobiles.

VI. REFERENCES

[1] O. GD, "Optimal seat and suspension design for a quarter car with driver model using genetic algorithms", *International Journal of Industrial Ergonomics*, vol. 37, pp. 327-332, 2007.

[2]

[3] L. Sun, X. M. Cai, and J. Yang, "Genetic algorithm-based optimum vehicle suspension design using minimum dynamic pavement load as a design criterion", *Journal of Sound and Vibration*, vol. 301, pp.18-27, 2007.

[4] L. P. Chen, *Mechanical System Dynamic Analysis and ADAMS Application Tutorial*, Beijing: Tsinghua University Press, 2005, pp.80-120.

[5]   K.K.Jain, 2005 "Automobile Engineering" , Tata Mc Graw Hill and Co.

Solar Powered Air Conditioner- An Approach to Utilise the Solar Energy

[1]Naveen K.Gupta, [2]Manika Singh, [2]Monalisa Gloria James
[1]Assistant Professor, GLA University,Mathura
[2]B.Tech 3rd year,GLA University,Mathura

Abstract -**Electricity load increases in summer by a huge amount. Productivity suffers under such conditions, as most of the offices are fitted with air-conditioning systems that consumes lot of electric power resulting frequent breakdowns. Use of solar energy for air conditioning seems to a feasible option particularly in the country like India where solar energy is abundant in summers. Use of solar energy reduces operational cost of system and reduces dependency on fossil fuels and electricity thus helps in power saving. For this purpose a setup has to be put on a free area of the building where solar energy is ample. Solar air conditioning system works on the principle of vapour absorption cycle. This setup utilizes the solar energy which provides air conditioning effect to the building where it is installed. It also provides hot water and back up heating system to the building. These are mostly followed in European countries and still have to be followed in India.**

I. INTRODUCTION

Michael Faraday (1791-1867) was interested to know how gases behaved when they were forced into a liquid state and subsequently allowed to evaporate. Although, he experimented with a variety of liquids, one of his most relevant discoveries for our purposes was his work with ammonia (NH3).   In 1820, Faraday found that the he could compress ammonia gas into a liquefied state under significant pressure, but when he allowed it to evaporate it created a cooling. Thermal cooling was invented by Ferdinand Carré (1824-1894).   He used the basic principle developed by Faraday but instead of ammonia, he used a solution of water and sulfuric acid.
JOHN GORRIE discovered centralized ac system in 1842, he was a physician in Florida. . In 1902 the first modern electrical air conditioning was invented by Willis Carrie in Syracuse, New York.

In ancient times the Mediterranean homes were built to face the sun during the cold winter months to highly sophisticated thin-film photovoltaic cells, which could generate electricity from sunlight. In 1912, parabolic solar collectors i.e. solar cells were built on a small farming community on the Nile River 15 miles south of Cairo, Egypt, which was developed by a Philadelphia inventor, entrepreneur, and solar visionary named Frank Shuman. Each collector was 204 feet in length, 13 feet in width and was fitted with a mechanical tracker which kept it automatically tilted to appropriately absorb the sunlight. The heat collected by these reflectors was used to produce steam to run a series of large water pumps. Together they produced the equivalent of 55 horsepower and were capable of pumping 6000 gallons of water per minute, bringing irrigation water to vast areas of arid desert land.

Why solar energy?
Reduces operational costs of facilities.
Reduces dependency on fossil energy supply and/or electricity
Contributes to a healthy environment by cooling without the use of toxic gases
Use of renewable energy contributes to the reduction of CO2 emissions
System can be easily installed

II. ENVIRONMENTAL HAZARDS BY OTHER ELECTRICITY PRODUCTION RESOURCES

The conventional electricity resources like nuclear energy, hydro energy and energy generated through the burning of coal.

Nuclear power is hazardous to environment. The main problem faced by it is waste disposal which is causing problem to land and health of local people .Electricity through dams also create obstacle because the whole land and nearby villages have to be evacuated. This causes loss of fertile land. Electricity generation through coal and other such resources causes emission of harmful gases to the environment. Thus these resources though are more efficient than solar but cause harmful effect to environment. Electricity generation through solar power may not be able to fulfill the whole energy requirement but can be proved better with further advancement.

III. AIMS AND OBJECTIVE OF USING SOLAR AIR CONDITIONING SYSTEM

- Amortisation after only 5 years possible
- Conventional heating / cooling systems do not have an amortisation potential
- Low operational costs facilitate higher profits

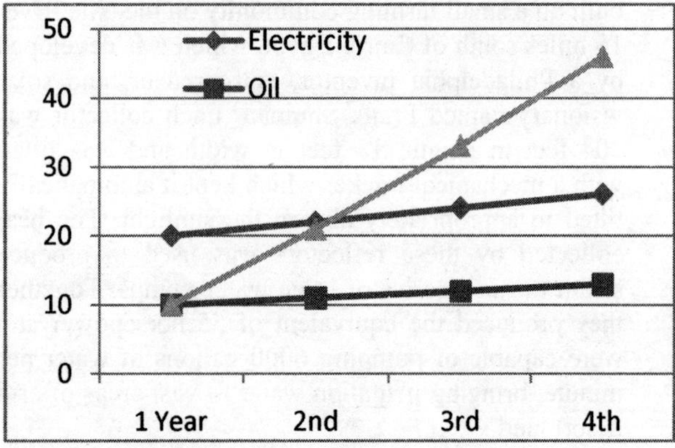

The graph shows the profits or savings by the use of solar air conditioning process, represented by green, compared to other resources like electricity and oil which is represented by red and yellow respectively.

Components used

- Solar collectors achieve yields up to 0.6 kW/m^2

- Modern heat exchangers have low heat losses

- Use of large buffer storage tanks to retain collected energy

- Absorption cooling requires heat > 80°C which is achievable through solar energy in summer

- Monitoring of the controls

Process involved in solar cooling

Vapour absorption cycle

It is one of the oldest ways of producing air conditioning effect. This system can be used in domestic as well as in industrial refrigeration plants. It utilizes heat energy to change the condition of refrigerant. The refrigerant used in this cycle is ammonia as it is readily soluble in cold water. In this system the compressor is replaced by an absorber, a pump, a generator and a pressure reducing valve. These components perform same function as by compressor in vapour compression cycle.

In this system the low pressure ammonia vapours leaving the evaporator enters the absorber where it is absorbed by cold water. The water has the capability to absorb large amount of ammonia, during this process pressure is reduced in the absorber which forces it to draw more ammonia from the evaporator. It raises the temperature of solution, so a cooling arrangement is also provided for the absorber as cold water can dissolve more ammonia. The strong solution thus formed is pumped to the generator by the liquid pump. Pump increases the pressure of solution up to 10 bar.

The strong solution of ammonia in the generator is heated by some external source like steam or gas, but in this set up we will use solar energy for heating. During heating process the ammonia solution is driven off at high pressure leaving behind weak ammonia solution which flows back to the absorber at low pressure after passing through pressure reducing valve. These high pressure ammonia vapours from the generator are condensed in the condenser to high pressure liquid ammonia. From there it is passed to the expansion valve through receiver. Here it is passed at a controlled rate reducing its pressure and temperature. Some of the liquid refrigerant evaporates as it passes through it, but mostly is vaporized at low temperature and pressure. After this it is then passed through evaporator, it consists of coils of pipe in which the liquid refrigerant at low temperature and pressure is evaporated. During this process it absorbs its latent heat of vaporization from the medium (here air) which is to be cooled and thus produce air conditioning effect.

Set up of solar cooling system

Solar plant

This is the whole set up which comprises of solar panel, solar tracker, control and sensing units etc. This is the basic unit which confines the rays of the sun and absorbs its heat which is further used in by the other units.

`HeatExchanger`

It is equipment built for efficient heat transfer from one medium to another. The media may be separated by a solid wall, so that they never mix, or they may be in direct contact. They are widely used in space heating refrigeration, air-conditioning, power plants, chemical plants, petrochemical plants, petrochemical refineries, and natural gas processing and sewage treatment.

Storage tank

It is a container used for holding liquid (sometimes for compressed gases (gas tank)). Storage tanks operate under no (or very little) pressure, distinguishing them from *pressure vessels*. Storage tanks are often cylindrical in shape. There are usually many environmental regulations applied to the design and operation of storage tanks, often depending on the nature of the fluid contained within.

Heating manifold

The primary function of the manifold is to evenly distribute the combustion mixture to each intake port in the cylinder head(s). Even distribution is important to optimize the efficiency and performance of the system.

Domestic Hot Water Circuit

These are the various Collector pumps through which heated water flows. Solar water heating systems are designed to deliver the optimum amount of hot water for most of the year. However, during winter there may not be sufficient solar heat gained to deliver sufficient hot water. In this case a gas or electric booster is normally used to heat the water. Hot water heated by the sun is used in many ways.

Space heating circuit

These are the other systems such as radiators, fan coil units, air handling units etc where the heat from the heating manifold could be used.

Back up heating system

This heating system is used only during emergency. Suppose there is a cloudy day so the heat is provided by this back up heating system, which could be in the form of boilers.

Absorption cooling machine

It is an absorption refrigerator which is a device that uses a heat source to provide the energy needed to drive the cooling system

Cooling tower

Cooling towers are used to remove heat from various sources such as machinery or heated process material. The primary use of large, industrial cooling towers is to remove the heat absorbed in the circulating cooling water systems used in power plants, petroleum refineries etc. The towers vary in size from small roof-top units to very large that can be up to 200 meters tall.

10. Cold water storage tank :This serves as a reservoir of Cold Water.

Back up chiller

This is a substitution for Absorption Refrigerator and is used in emergency.

Air handling unit

An Air handling unit is a device used to condition and circulate air as part of heating, ventilating, and air-conditioning system. An air handler is usually a large metal box containing a blower, heating or cooling elements and dampers.

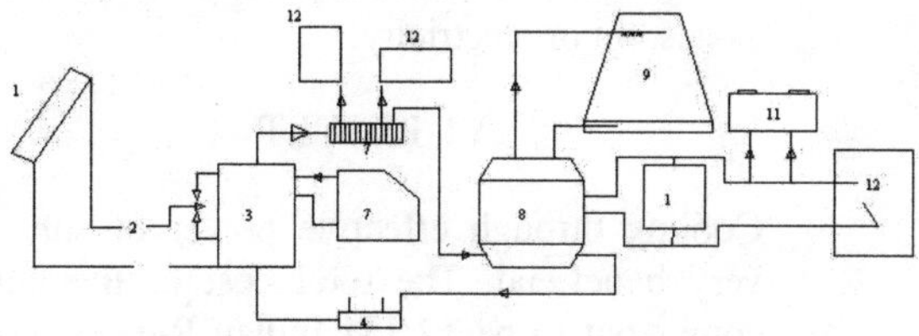

IV. OBSTACLES AND SOLUTION:

The main obstacle faced by this technology is the efficiency of solar power plant. One of the main problems with using solar panels is the small amount of electricity they generate compared to their size. A calculator might only require a single solar cell, but a solar-powered air conditioner would require several thousand. If the angle of the solar panels is changed even slightly, the efficiency can drop 50 percent. The panels are exposed to destructive weather elements, which can also seriously affect efficiency. The efficiency improvement is achieved by the use of an ultra-thin aluminum oxide layer at the front of the cell, and it brings a breakthrough in the use of solar energy a step closer.

Secondly, the power plant requires a large space where sunlight is ample. The large size is due to the use of vapour absorption cycle. But as this system is used to cool large systems thus its size factor can be ignored.

Thirdly, we cannot completely rely on this system thus we need another backup source to provide cooling in emergency.

Lots of research is going in this field and lots have to be done to make this system cheap and convenient to be used by common people.

V CONCLUSION

The use of solar cooling is very beneficial to environment with minimum energy loss. Its efficiency is little less but this is subdued by its beneficial effect over saving resources like electricity and fossil fuels.

Benefits resulted are

- The ammonia used in the closed cooling system is safe, odourless and non-toxic

- Ammonia carries no risk to the ozone layer

- The hot water heated by the sun used as the primary energy source carries no risk to the environment

- Should a support energizer be required, waste product combustion can be used as well as any conventional energy source such as gas, oil or electricity.

VI. RESULT

Cooling through effective power of sun is very beneficial. The total cost of the unit comes out to be 12,334 Indian Rupees. For large buildings like offices, hotels, apartments etc this will provide effective cooling system. A deep research should be taken regarding this to increase the efficiency of vapour absorption cycle. Major R&D efforts are required to exploit this technology.

Many researches are going in this field and is this technology is successfully followed in European countries.

VII. REFERENCES

[1] Cooling by Evaporation. Benjamin Franklin, London, June 17, 1758.
[2] History of Air Conditioning Source: Jones Jr., Malcolm. "Air Conditioning". Newsweek. Winter 1997
[3] The current status in Air Conditioning – papers & Presentation.
[4] Winnick, J. Chemical Engineering Thermodynamics.
[5] "NIST Guide to the SI". National Institute of Standards and Technology. Retrieved 2007 -05-18
[6] Shane Smith (2000).Greenhouse gardener's companion :growing food and flowers in your greenhouse or sunspace
[7] Dahlgren, Derek; Jewell, Amy; Li, Ruth et al. "History of Air Conditioning ". Bucknell University. Retrieved 2007-09-15.
[8] Air Conditioning Explained retrieved 19 May 2009
[9] EPA Rules & Regulations restricting refrigerant
[10] CFC worldwide ban
[11] Chemicals in the Environment : Freon 113

Study of Combustion of Soybean Oil in a Diesel Engine

Ankur Saxena

Department of Mechanical Engineering, Anand Engineering College, Agra

Abstract The use of vegetable oil methyl esters has been proposed as an alternative fuel for diesel engines. The purpose of this study is to investigate the combustion of soybean oil methyl ester in a direct injection diesel engine, and compare it to that of a conventional diesel fuel. Experimental measurements of performance, emissions, and rate of heat release were performed as a function of engine load for different fuel injection timings, and injector orifice diameters.

It was found that overall, the soybean oil methyl ester behaved comparably to diesel fuel in terms of performance and rate of heat release. The methyl ester fuel gave lower HC emissions and smoke number than diesel fuel at optimum operating conditions. The results for CO emissions were varied. NOx emissions were strongly related to the cylinder pressure development.

This paper presents a brief history of diesel engine technology and an overview of soybean oil, including performance characteristics, economics, and potential demand. The performance and economics of soybean oil are compared with those of petroleum diesel.

Keywords : - Alternative fuel, lower HC emissions, smoke number, injection timing.

I. INTRODUCTION

The idea of using vegetable oil for fuel has been around as long as the diesel engine. Rudolph Diesel, the inventor of the engine that bears his name, experimented with fuels ranging from powdered coal to peanut oil. In the early 20th century, however, diesel engines were adapted to burn petroleum distillate, which was cheap and plentiful. In the late 20th century, however, the cost of petroleum distillate rose, and by the late 1970s there was renewed interest in biodiesel. Commercial production of biodiesel in the United States began in the 1990s.

The most common sources of oil for biodiesel production in the United States are soybean oil and yellow grease (primarily, recycled cooking oil from restaurants). Blends of biodiesel and petroleum diesel are designated with the letter "B," followed by the volumetric percentage of biodiesel in the blend: B20, the blend most often evaluated, contains 20 percent biodiesel and 80 percent petroleum diesel; B100 is pure biodiesel. By several important measures biodiesel blends perform better than petroleum diesel, but its relatively high production costs and the limited availability of some of the raw materials used in its production continue to limit its commercial application.

Since the first oil crises of the 1970's alternative fuels have been investigated with the goal of replacing conventional petroleum supplies. The initial interest was mainly one of fuel supply but recently more attention has been focused on the use of renewable fuels in order to reduce the net production of CO_2 from combustion sources.

Methanol and ethanol are two accepted alternative fuels which possess the potential to be produced from biomass sources. Neither of these fuels is well suited for use in diesel engines, and the use of high compression ratios, ignition improvers and ignition assistance devices is common.

One type of fuel which is well suited for use in diesel engines is that of vegetable oil based fuels. Previous studies have shown that it is possible to use vegetable oils in combination with alcohols in diesel engines with acceptable performance. Initial studies by Barsic and Humke showed that untreated vegetable oils can be used as fuels in diesel engines, but that there are severe problems with injector fouling. Lubricating oil effects with vegetable oil based fuels have also been studied.

Experiments have shown that as far as combustion is concerned, one of the best ways to utilize alcohol in a diesel fuel is to combine it with a vegetable oil to make the corresponding ester of the vegetable oil. The most commonly used ester is the methyl ester, made by combining the vegetable oil with methanol. In fact, in some parts of Europe, rape seed methyl ester is already commercially available for use in vehicles.

II. HISTORY

The efficiency of the Carnot cycle increases with the increase in compression ratio—the ratio of gas volume at full expansion to its volume at full compression. This concept gave a motivation to Rudolph Diesel to build an engine with the highest possible compression ratio. He introduced fuel only when combustion was desired and allowed the fuel to ignite on its own in the hot compressed air. Diesel's engine achieved efficiency higher than that of the Otto engine and much higher than that of the steam engine. Diesel received a patent in 1893 and demonstrated a workable engine in 1897. Today, diesel engines are classified as "compression-ignition" engines.

Diesel's motivation was not only to improve efficiency but also to bring the benefits of powered machinery to smaller companies. Steam engines were so large that only the biggest firms could afford them, and Diesel wanted to enable smaller firms to compete against larger, steam-powered firms. He used peanut oil as the fuel for his demonstration engines at the 1900 World's Fair and thought that oils from locally grown crops would be used to power his engines.

The early 20th century saw the introduction of gasoline- powered automobiles. Oil companies were obliged to refine so much crude oil to supply gasoline that they were left with a surplus of distillate, which is an excellent fuel for diesel engines and much less expensive than vegetable oils. On the other hand, resource depletion has always been a concern with regard to petroleum, and farmers have always sought new markets for their products. Consequently, work has continued on the use of vegetable oils as fuel.

Early durability tests indicated that engines would fail prematurely when operating on fuel blends containing vegetable oil. Engines burning vegetable oil that had been transesterified with alcohols, however, exhibited no such problems and even performed better by some measures than engines using petroleum diesel. The formulation of what is now called biodiesel came out of those early experiments.

The term "biodiesel" means the monoalkyl esters of long chain fatty acids derived from plant or animal matter which meet (A) the registration requirements for fuels and fuel additives established by the Environmental Protection Agency under section 211 of the Clean Air Act (42 U.S.C. 7545), and (B) the requirements of the American Society of Testing and Materials D6751.

That definition of biodiesel is used here, although other processes also can be used to produce high-quality diesel fuel from vegetable oil or animal fat.

Bus running on soybean biodiesel. Photo: U.S. Department of Energy.

III. PERFORMANCE AND EMISSIONS CHARACTERISTICS

One of the most important characteristics of diesel fuel is its ability to autoignite, a characteristic that is quantified by a fuel's cetane number or cetane index, where a higher cetane number or index means that the fuel ignites more quickly.7 U.S. petroleum diesel typically has a cetane index in the low 40s, and European diesel typically has a cetane index in the low 50s.

Lubricity, another important characteristic of diesel fuel, is a measure of lubricating properties. Fuel injectors and some types of fuel pumps rely on fuel for lubrication. One study, published in 1998 and cited by the National Biodiesel Board, found that one-half of samples of petroleum diesel sold in the United States did not meet the recommended minimum standard for lubricity. Biodiesel has better lubricity than current low-sulfur petroleum diesel, which contains 500 parts per million (ppm) sulfur by weight. The petroleum diesel lubricity problem is expected to get worse when ultra-low-sulfur petroleum diesel (15 ppm sulfur by weight) is introduced in 2006. A 1- or 2-percent volumetric blend of biodiesel in low-sulfur petroleum diesel improves lubricity substantially. It should be noted, however, that the use of other lubricity additives may achieve the same effect at lower cost.

Biodiesel also has some performance disadvantages. The performance of biodiesel in cold conditions is markedly worse than that of petroleum diesel, and biodiesel made from yellow grease is worse than soybean biodiesel in this regard. At low temperatures, diesel fuel forms wax crystals, which can clog fuel lines and filters in a vehicle's fuel system. The "cloud point" is the temperature at which a sample of the fuel starts to appear cloudy, indicating that wax crystals have begun to form. At even lower temperatures, diesel fuel becomes a gel that cannot be pumped. The "pour point" is the temperature below which the fuel will not flow. The cloud and pour points for biodiesel are higher than those for petroleum diesel.

Another disadvantage of biodiesel is that it tends to reduce fuel economy. Energy efficiency is the percentage of the fuel's thermal energy that is delivered as engine output, and biodiesel has shown no significant effect on the energy efficiency of any test engine. Volumetric efficiency, a measure that is more familiar to most vehicle users, usually is expressed as miles traveled per gallon of fuel (or kilometers per liter of fuel). The energy content per gallon of biodiesel is approximately 11 percent lower than that of petroleum diesel.14 Vehicles running on B20 are therefore expected to achieve 2.2 percent (20 percent x 11 percent) fewer miles per gallon of fuel.

Since the possibility exists for the decomposition of the methyl ester during the ignition delay period, and the viscosity of the methyl is somewhat higher than that of typical diesel fuel, one might expect to observe some differences in the combustion behavior of methyl ester as compared to diesel fuel. From previous experiments, the differences are not expected to be large, but the present study was undertaken to examine the combustion process in more derail than had been done previously.

The goal of the study was to evaluate the effects of changes in the injection system on the combustion, performance and emissions of a direct injection diesel engine powered by soybean oil methyl ester.

IV. EXPERIMENTAL CONDITIONS

The engine used for the study was a 4-cylinder. 4- stroke, normally aspirated direct injection diesel engine. The engine and test condition specifications are given in Table 1.

Table 1. Specifications of the engine and test conditions

Bore	96.4 mm
Stroke	104.8 mm
Displacement	3.06 liter
Compression Ratio	16.5
Speed	1800 rpm
Intake Pressure	760 mm Hg
Exhaust Pressure	Atmospheric
Coolant Temperature	82 ^{0} C
Oil Temperature	71 ^{0} C

The injection system consisted of a standard rotary distributor pump with pencil type nozzles. The standard orifice diameter for the engine was 0.279 mm. Tests were also conducted with an orifice diameter of 0.229 mm. Injecting timing was changed by rotation of the pump with respect to the camshaft. The fuels used in the study were standard number 2 diesel fuel as the reference fuel, and soybean oil methyl ester. The characteristics of the fuels are shown in Table 2. The cetane numbers of the fuels tested were not measured for this study. Previous work has indicated that the cetane number of the diesel fuel is expected to be about 51 and that of the methyl ester is expected to be about 46.

Table 2 Specifications of the test fuels

Parameter	Diesel Fuel	Soybean Oil
High Heating Value KJ/Kg	45590	39750
Lower Heating Value KJ/Kg	42780	37260
Viscosity mm/s^2 @ 40^0 C	2.8	4.1
Specific Gravity	0.845	0.889
Carbon – mass %	86.5	78.0
Hydrogen – mass %	13.2	11.7
Oxygen – mass %	0	10.3
Sulphur – mass %	0.15	< 0.005
Nitrogen - ppm	-	29
Stoichiometric fuel air ratio	0.0687	0.0795

V. Testing Procedures

The engine was instrumented and engine exhaust emissions testing equipment were calibrated. Engine performance was checked at peak power and peak torque to ensure proper operation. The two-point check was performed throughout the project to maintain a record of engine performance.

The pre-mixed drum of blended fuel was connected to the fuel supply system of the engine. The engine was idled while the return fuel line from the engine drained. The fuel system was purged before the return fuel was redirected to the recirculation tank. Purged fuel was placed in a waste fuel tank.

The engine was operated at 1200 and 2100 rpm at 50% and 100% load for ten minutes. A two point power check was run. The engine was accelerated to rated power (2100 rpm full throttle) and held there until engine oil temperature stabilized (230 to 240oF). Exhaust backpressure was checked and then three logs of measured engine parameters were recorded. The engine was then decelerated to peak torque speed (1200 rpm full throttle). Engine exhaust backpressure and three measures of engine operating parameters were recorded.

VI. Engine Performance

The following observations were based on the data collected during this investigation:

- Peak torque was not affected by the addition of up to 40% biodiesel; however, a slight drop at rated speed was noticed at the 40% level.
- A steady drop in exhaust gas temperature was observed for both the rated and peak torque condition, indicating a shift in the peak pressure point toward top dead center, resulting from an increasingly shorter ignition delay. The lower exhaust gas temperature was a result of increased heat transfer into the coolant.
- The reduced ignition delay and increased peak pressure and temperature contributed to the increased oxides of nitrogen. This was most pronounced at peak torque, which generated the highest peak pressures and temperatures.
- Fuel consumption on a mass basis was very similar for all blends tested.

VII. EMISSIONS

Emissions results follow trends established by previous research. Increased levels of biodiesel increase NOx while reducing PM. Proportionally, PM reduction was slightly more than the increase in NOx, on a percentage basis. The decrease in PM was contributed to the oxygen in the fuel. The reduction in CO and THC was linear with the addition of biodiesel for the blends tested. These reductions indicate more complete combustion of the fuel. The presence of oxygen in the fuel was thought to promote complete combustion.

The BSFC and CO_2 did not change appreciably, and fuel consumption increased only minimally over the test cycle.

The 20% blend represented a good compromise between increased NOx and reduction of all other emissions. This biodiesel/diesel fuel blend was recommended.

VIII. RECOMMENDATIONS

The Environmental Protection agency has mandated that the level of oxides of nitrogen emissions be equal to or less than the emissions produced by the engine if fueled on reference diesel fuel. Economic conditions concerning the use of biodiesel as a transportation fuel dictate that the fuel should be used as a blend in a diesel engine. As such the following recommendations were made:

- Engine optimization strategies should be developed that fully takes advantage of the physical and chemical makeup of biodiesel.
- Additional tests should be conducted to evaluate other catalytic converters, and the durability of these devices should be documented over time.

IX. BIBLIOGRAPHY

[1] Feldman, M. E. and Peterson, C. L. (1992). Fuel injector timing and pressure optimization on a DI diesel engine for operation on biodiesel. Liquid Fuels from Renewable Resources- Proceedings of an Alternative Energy Conference. Nashville, TN.

[2] Marshall, W. F. (1993). Effects of methyl esters of tallow and grease on exhaust emissions and performance of a Cummins L10 engine. Itt Research Institute, National Institute for Petroleum and Energy Research, Bartlesville, OK. (Report prepared for Fats and Proteins Research Foundation, Inc, Ft. Myers, Beach, FL)

[3] Niehaus, R. A., Goering, C. E. Savage, L. D., and Sorenson, S. C. (1985). Cracked soybean oil as a fuel for a diesel engine. ASAE Paper No. 85-1560. ASAE, St. Joseph, MI.

[4] Reece, D. L. and Peterson, C. L. (1993). A report on the Idaho on-road vehicle test with RME and neat rapeseed oil as an alternative to diesel fuel. ASE Paper No. 93-5018. ASAE, St. Joseph, MI.

[5] Schumacher, L. G., Borgelt, S. C., Hires, W. G. and Humphrey, J. K. (1993). Biodiesel on the road- A report from Missouri. ASAE paper No. 93-5017. ASAE, St. Joseph, MI.

[6] Schumacher, L. G., Borgelt, S. C., and Hires, W. G. (1993). Soydiesel/Biodiesel Blend Research. ASAE paper No. 93-6523. ASAE, St. Joseph, MI.

[7] Srinivasa, R. P. and Gopalakrishnan, K. V. (1991). Vegetable oils and their methylesters as fuels for diesel engines. Indian Journal of Technology. 29: 292-297.

[8]Ziejewski, M., Kaufman, K. R., Schwab, A. W, and Pryde, E. H. (1984). Diesel engine evaluation of an nonionic sunflower oil-aqueous ethanol microemulsion. Journal of the American Oil Chemists Society. 61 (10): 1620-1626.